The Theory of
Error-Correcting Codes

North-Holland Mathematical Library

VOLUME 16

NORTH-HOLLAND
AMSTERDAM • LONDON • NEW YORK • TOKYO

The Theory of
Error-Correcting Codes

F. J. MacWilliams
N. J. A. Sloane

AT&T Shannon Labs
Florham Park
NJ 07932
U.S.A.

NORTH-HOLLAND
AMSTERDAM • LONDON • NEW YORK • TOKYO

ELSEVIER B.V.
Radarweg 29
P.O. Box 211, 1000 AE
Amsterdam, The Netherlands

ELSEVIER Inc.
525 B Street
Suite 1900, San Diego
CA 92101-4495, USA

ELSEVIER Ltd
The Boulevard
Langford Lane, Kidlington,
Oxford OX5 1GB, UK

ELSEVIER Ltd
84 Theobalds Road
London WC1X 8RR
UK

First edition: 1977
Second impression: 1978
Third impression: 1981
Fourth impression: 1983
Fifth impression: 1986
Sixth impression: 1988
Seventh impression: 1992
Eighth impression: 1993
Ninth impression: 1996
Tenth impression: 1998
Eleventh impression: 2003
Twelth impression: 2006

Library of Congress Cataloging in Publication Data

MacWilliams, Florence Jessiem 1917–1990
  The theory of error-correcting codes.
  North-Holland Mathematical Library; 16.
  1. Error-correcting codes (information theory)
  I. Sloane, Neil James Alexander, 1939– joint author.
  II. Title.
  QA268.M3   519.4   76-41296

ISBN 10 – 0 444 85193 3
ISBN 13 – 9 780444 851932

⊗ The paper used in this publication meets the requirements of ANSI/NISO Z39.48-1992 (Permanence of Paper).

Transferred to digital printing 2007

Preface

Coding theory began in the late 1940's with the work of Golay, Hamming and Shannon. Although it has its origins in an engineering problem, the subject has developed by using more and more sophisticated mathematical techniques. It is our goal to present the theory of error-correcting codes in a simple, easily understandable manner, and yet also to cover all the important aspects of the subject. Thus the reader will find both the simpler families of codes – for example, Hamming, BCH, cyclic and Reed–Muller codes – discussed in some detail, together with encoding and decoding methods, as well as more advanced topics such as quadratic residue, Golay, Goppa, alternant, Kerdock, Preparata, and self-dual codes and association schemes.

Our treatment of bounds on the size of a code is similarly thorough. We discuss both the simpler results – the sphere-packing, Plotkin, Elias and Garshamov bounds – as well as the very powerful linear programming method and the McEliece–Rodemich–Rumsey–Welch bound, the best asymptotic result known. An appendix gives tables of bounds and of the best codes presently known of length up to 512.

Having two authors has helped to keep things simple: by the time we both understand a chapter, it is usually transparent. Therefore this book can be used both by the beginner and by the expert, as an introductory textbook and as a reference book, and both by the engineer and the mathematician. Of course this has not resulted in a thin book, and so we suggest the following menus:

An elementary first course on coding theory for mathematicians: Ch. 1, Ch. 2 (§6 up to Theorem 22), Ch. 3, Ch. 4 (§§1–5), Ch. 5 (to Problem 5), Ch. 7 (not §§7, 8), Ch. 8 (§§1–3), Ch. 9 (§§1, 4), Ch. 12 (§8), Ch. 13 (§§1–3), Ch. 14 (§§1–3).

A second course for mathematicians: Ch. 2 (§§1–6, 8), Ch. 4 (§§6, 7 and part of 8), Ch. 5 (to Problem 6, and §§3, 4, 5, 7), Ch. 6 (§§1–3, 10, omitting the proof of Theorem 33), Ch. 8 (§§5, 6), Ch. 9 (§§2, 3, 5), Ch. 10 (§§1–5, 11), Ch. 11, Ch. 13 (§§4, 5, 9), Ch. 16 (§§1–6), Ch. 17 (§7, up to Theorem 35), Ch. 19 (§§1–3).

An elementary first course on coding theory for engineers: Ch. 1, Ch. 3, Ch. 4 (§§1–5), Ch. 5 (to Problem 5), Ch. 7 (not §7), Ch. 9 (§§1, 4, 6), Ch. 10 (§§1, 2, 5, 6, 7, 10), Ch. 13 (§§1–3, 6, 7), Ch. 14 (§§1, 2, 4).

A second course for engineers: Ch. 2 (§§1–6), Ch. 8 (§§1–3, 5, 6), Ch. 9 (§§2, 3, 5), Ch. 10 (§11), Ch. 12 (§§1–3, 8, 9), Ch. 16 (§§1, 2, 4, 6, 9), Ch. 17 (§7, up to Theorem 35).

There is then a lot of rich food left for an advanced course: the rest of Chapters 2, 6, 11 and 14, followed by Chapters 15, 18, 19, 20 and 21 – a feast!

The following are the principal codes discussed:

Alternant, Ch. 12;
BCH, Ch. 3, §§1, 3; Ch. 7, §6; Ch. 8, §5; Ch. 9; Ch. 21, §8;
Chien–Choy generalized BCH, Ch. 12, §7;
Concatenated, Ch. 10, §11; Ch. 18, §§5, 8;
Conference matrix, Ch. 2, §4;
Cyclic, Ch. 7, Ch. 8;
Delsarte–Goethals, Ch. 15, §5;
Difference-set cyclic, Ch. 13, §8;
Double circulant and quasi-cyclic, Ch. 16, §§6–8;
Euclidean and projective geometry, Ch. 13, §8;
Goethals generalized Preparata, Ch. 15, §7;
Golay (binary), Ch. 2, §6; Ch. 16, §2; Ch. 20;
Golay (ternary), Ch. 16, §2; Ch. 20;
Goppa, Ch. 12, §§3–5;
Hadamard, Ch. 2, §3;
Hamming, Ch. 1, §7, Ch. 7, §3 and Problem 8;
Irreducible or minimal cyclic, Ch. 8, §§3, 4;
Justesen, Ch. 10, §11;
Kerdock, Ch. 2, §8; Ch. 15, §5;
Maximal distance separable, Ch. 11;
Nordstrom–Robinson, Ch. 2, §8; Ch. 15, §§5, 6;
Pless symmetry, Ch. 16, §8;
Preparata, Ch. 2, §8; Ch. 15, §6; Ch. 18, §7.3;
Product, Ch. 18, §§2–6;
Quadratic residue, Ch. 16;
Redundant residue, Ch. 10, §9;
Reed–Muller, Ch. 1, §9; Chs. 13–15;
Reed–Solomon, Ch. 10;

Self-dual, Ch. 19;
Single-error-correcting nonlinear, Ch. 2, §7; Ch. 18, §7.3;
Srivastava, Ch. 12, §6.

Encoding methods are given for:

Linear codes, Ch. 1, §2;
Cyclic codes, Ch. 7, §8;
Reed–Solomon codes, Ch. 10, §7;
Reed–Muller codes, Ch. 13, §§6, 7; Ch. 14, §4.

Decoding methods are given for:

Linear codes, Ch. 1, §§3, 4;
Hamming codes, Ch. 1, §7;
BCH codes, Ch. 3, §3; Ch. 9, §6; Ch. 12, §9;
Reed–Solomon codes, Ch. 10, §10;
Alternant (including BCH, Goppa, Srivastava and Chien–Choy generalized
 BCH codes) Ch. 12, §9;
Quadratic residue codes, Ch. 16, §9;
Cyclic codes, Ch. 16, §9,

while other decoding methods are mentioned in the notes to Ch. 16.

When reading the book, keep in mind this piece of advice, which should be given in every preface: if you get stuck on a section, skip it, but keep reading! Don't hesitate to skip the proof of a theorem: we often do. Starred sections are difficult or dull, and can be omitted on the first (or even second) reading.

The book ends with an extensive bibliography. Because coding theory overlaps with so many other subjects (computers, digital systems, group theory, number theory, the design of experiments, etc.) relevant papers may be found almost anywhere in the scientific literature. Unfortunately this means that the usual indexing and reviewing journals are not always helpful. We have therefore felt an obligation to give a fairly comprehensive bibliography. The notes at the ends of the chapters give sources for the theorems, problems and tables, as well as small bibliographies for some of the topics covered (or not covered) in the chapter.

Only *block* codes for correcting *random* errors are discussed; we say little about codes for correcting other kinds of errors (bursts or transpositions) or about variable length codes, convolutional codes or source codes (see the

Notes to Ch. 1). Furthermore we have often considered only *binary* codes, which makes the theory a lot simpler. Most writers take the opposite point of view: they think in binary but publish their results over arbitrary fields.

There are a few topics which were included in the original plan for the book but have been reluctantly omitted for reasons of space:

(i) Gray codes and snake-in-the-box codes – see Adelson et al. [5, 6], Buchner [210], Cavior [253], Chien et al. [290], Cohn [299], Danzer and Klee [328], Davies [335], Douglas [382, 383], Even [413], Flores [432], Gardner [468], Gilbert [481], Guy [571], Harper [605], Klee [764–767], Mecklenberg et al. [951], Mills [956], Preparata and Nievergelt [1083], Singleton [1215], Tang and Liu [1307], Vasil'ev [1367], Wyner [1440] and Yuen [1448, 1449].

(ii) Comma-free codes – see Ball and Cummings [60, 61], Baumert and Cantor [85], Crick et al. [316], Eastman [399], Golomb [523, pp. 118–122], Golomb et al. [528], Hall [587, pp. 11–12], Jiggs [692], Miyakawa and Moriya [967], Niho [992] and Redinbo and Walcott [1102]. See also the remarks on codes for synchronizing in the Notes to Ch. 1.

(iii) Codes with unequal error protection – see Gore and Kilgus [549], Kilgus and Gore [761] and Mandelbaum [901].

(iv) Coding for channels with feedback – see Berlekamp [124], Horstein [664] and Schalkwijk et al. [1153–1155].

(v) Codes for the Gaussian channel – see Biglieri et al. [148–151], Blake [155, 156, 158], Blake and Mullin [162], Chadwick et al. [256, 257], Gallager [464], Ingemarsson [683], Landau [791], Ottoson [1017], Shannon [1191], Slepian [1221–1223] and Zetterberg [1456].

(vi) The complexity of decoding – see Bajoga and Walbesser [59], Chaitin [257a–258a], Gelfand et al. [471], Groth [564], Justesen [706], Kolmogorov [774a], Marguinaud [916], Martin-Löf [917a], Pinsker [1046a], Sarwate [1145] and Savage [1149–1152a].

(vii) The connections between coding theory and the packing of equal spheres in *n*-dimensional Euclidean space – see Leech [803–805], [807], Leech and Sloane [808–810] and Sloane [1226].

The following books and monographs on coding theory are our predecessors: Berlekamp [113, 116], Blake and Mullin [162], Cameron and Van Lint [234], Golomb [522], Lin [834], Van Lint [848], Massey [922a], Peterson [1036a], Peterson and Weldon [1040], Solomon [1251] and Sloane [1227a]; while the following collections contain some of the papers in the bibliography: Berlekamp [126], Blake [157], the special issues [377a, 678, 679], Hartnett [620], Mann [909] and Slepian [1224]. See also the bibliography [1022].

We owe a considerable debt to several friends who read the first draft very carefully, made numerous corrections and improvements, and frequently saved us from dreadful blunders. In particular we should like to thank I.F. Blake, P. Delsarte, J.-M. Goethals, R.L. Graham, J.H. van Lint, G. Longo, C.L. Mallows, J. McKay, V. Pless, H.O. Pollak, L.D. Rudolph, D.W. Sarwate, many other colleagues at Bell Labs, and especially A.M. Odlyzko for their help. Not all of their suggestions have been followed, however, and the authors are fully responsible for the remaining errors. (This conventional remark is to be taken seriously.) We should also like to thank all the typists at Bell Labs who have helped with the book at various times, our secretary Peggy van Ness who has helped in countless ways, and above all Marion Messersmith who has typed and retyped most of the chapters. Sam Lomonaco has very kindly helped us check the galley proofs.

Preface

Preface to the third printing

We should like to thank many friends who have pointed out errors and misprints. The corrections have either been made in the text or are listed below.

A Russian edition was published in 1979 by Svyaz (Moscow), and we are extremely grateful to L. A. Bassalygo, I. T. Grushko and V. A. Zinov'ev for producing a very careful translation. They supplied us with an extensive list of corrections. They also point out (in footnotes to the Russian edition) a number of places we did not cite the earliest source for a theorem. We have corrected the most glaring omissions, but future historians of coding theory should also consult the Russian edition.

Problem 17, page 75. Shmuel Schreiber has pointed out that not all ways of choosing the matrices A, B, C, D work. One choice which does work is

$$A = \begin{bmatrix} 0001 \\ 1000 \\ 0010 \\ 0100 \end{bmatrix}, \quad B = \begin{bmatrix} 0010 \\ 0100 \\ 0001 \\ 1000 \end{bmatrix}, \quad C = \begin{bmatrix} 0100 \\ 0010 \\ 1000 \\ 0001 \end{bmatrix}, \quad D = \begin{bmatrix} 1000 \\ 0001 \\ 0100 \\ 0010 \end{bmatrix}.$$

Page 36, Notes to §2. Add after Wu [1435, 1436]: K. Sh. Zigangirov, Some sequential decoding procedures, *Problems of Information Transmission*, **2** (4) (1966) 1–10; and K. Sh. Zigangirov, *Sequential Decoding Procedures* (Svyaz, Moscow, 1974).

Page 72, Research problem 2.4. It is now known that $A(10, 4) = 40$, $72 \leq A(11, 4) \leq 79$, and $144 \leq A(12, 4) \leq 158$. See M. R. Best, Binary codes with a minimum distance of four, *IEEE Trans. Info. Theory*, Vol. **IT-26** (6) (November 1980), 738–743.

Page 123, Research problem 4.1. Self-contained proofs have been given by O. Moreno, On primitive elements of Trace 1 in $GF(2^m)$, *Discrete Math.*, to appear, and L. R. Vermani, Primitive elements with nonzero trace, preprint.

Chapter 6, pp. 156 and 180. Theorem 33 was proved independently by Zinov'ev and Leont'ev [1472].

Page 166, Research problem 6.1 has been settled in the affirmative by I. I. Dumer, A remark about codes and tactical configurations, *Math. Notes*, to appear; and by C. Roos, Some results on *t*-constant codes, *IEEE Trans. Info. Theory*, to appear.

Page 175. Research problem 6.3 has also been solved by Dumer and Roos [op. cit.].

Page 178–179. Research problems 6.4 and 6.5 have been solved by Dumer [op. cit.]. The answer to 6.5 is No.

Page 267, Fig. 9.1. R. E. Kibler (private communication) has pointed out that the asterisks may be removed from the entries [127, 29, 43], [255, 45, 87] and [255, 37, 91], since there are minimum weight codewords which are low-degree multiples of the generator polynomial.

Page 280, Research problem 9.4. T. Helleseth (*IEEE Trans. Info. Theory*, Vol. **IT-25** (1979) 361–362) has shown that no other binary primitive BCH codes are quasi-perfect.

Page 299, Research problem 10.1. R. E. Kibler (private communication) has found a large number of such codes.

Page 323. The proof of Theorem 9 is valid only for $q = 2^m$. For odd q the code need not be cyclic. See G. Falkner, W. Heise, B. Kowol and E. Zehender, On the existence of cyclic optimal codes, *Atti Sem. Mat. Fis. Università di Modena*, to appear.

Page 394, line 9 from the bottom. As pointed out by Massey in [918, p. 100], the number of majority gates required in an *L*-step decoder need never exceed the dimension of the code.

Page 479. Research problem 15.2 is also solved in I. I. Dumer, Some new uniformly packed codes, in: *Proceedings MFTI*, Radiotechnology and Electronics Series (MFTI, Moscow, 1976) pp. 72–78.

Page 546. The answer to Research problem 17.6 is No, and in fact R. E. Kibler, Some new constant weight codes, *IEEE Trans. Info. Theory*, Vol. **IT-26** (May 1980) 364–365, shows that $27 \leq A(24, 10, 8) \leq 68$.

Appendix A, Figures 1 and 3. For later versions of the tables of $A(n, d)$ and $A(n, d, w)$ see R. L. Graham and N. J. A. Sloane, Lower bounds for constant weight codes, *IEEE Trans. Info. Theory*, Vol. **IT-26** (1980) 37–43; M. R. Best, Binary codes with a minimum distance of four, loc. cit., Vol. IT-26 (1980), 738–743; and other papers in this journal.

On page 682, line 6, the value of X corresponding to $F = 30$ should be changed from .039 to .093.

Contents

Chapter 3. An introduction to BCH codes and finite fields

Chapter 4. Finite fields

Chapter 5. Dual codes and their weight distribution

Chapter 6. Codes, designs and perfect codes

Chapter 7. Cyclic codes

Chapter 8. Cyclic codes (contd.): Idempotents and Mattson–Solomon polynomials

Chapter 9. BCH codes

Chapter 10. Reed–Solomon and Justesen codes

Chapter 11. MDS codes

Chapter 12. Alternant, Goppa and other generalized BCH codes

Chapter 13. Reed–Muller codes

Chapter 14. First-order Reed–Muller codes

Chapter 15. Second-order Reed–Muller, Kerdock and Preparata codes

Chapter 16. Quadratic-residue codes

Chapter 17. Bounds on the size of a code

Chapter 18. Methods for combining codes

Part I

Linear codes

§1. Linear codes

Codes were invented to correct errors on noisy communication channels. Suppose there is a telegraph wire from Boston to New York down which 0's and 1's can be sent. Usually when a 0 is sent it is received as a 0, but occasionally a 0 will be received as a 1, or a 1 as a 0. Let's say that on the average 1 out of every 100 symbols will be in error. I.e. for each symbol there is a probability $p = 1/100$ that the channel will make a mistake. This is called a *binary symmetric channel* (Fig. 1.1).

There are a lot of important messages to be sent down this wire, and they must be sent as quickly and reliably as possible. The messages are already written as a string of 0's and 1's – perhaps they are being produced by a computer.

We are going to *encode* these messages to give them some protection against errors on the channel. A block of k message symbols $\boldsymbol{u} = u_1 u_2 \ldots u_k$

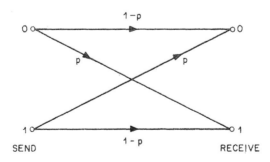

Fig. 1.1. The binary symmetric channel, with error probability p. In general $0 \le p \le \frac{1}{2}$.

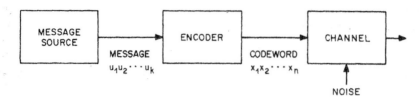

Fig. 1.2

($u_i = 0$ or 1) will be encoded into a *codeword* $x = x_1x_2 \ldots x_n$ ($x_i = 0$ or 1) where $n \geqslant k$ (Fig. 1.2); these codewords form a *code*.

The method of encoding we are about to describe produces what is called a *linear* code. The first part of the codeword consists of the message itself:

$$x_1 = u_1, \quad x_2 = u_2, \quad \ldots, \quad x_k = u_k,$$

followed by $n - k$ *check* symbols

$$x_{k+1}, \ldots, x_n.$$

The check symbols are chosen so that the codewords satisfy

$$H \begin{pmatrix} x_1 \\ x_2 \\ \vdots \\ x_n \end{pmatrix} = H x^{tr} = 0, \tag{1}$$

where the $(n - k) \times n$ matrix H is the *parity check matrix* of the code, given by

$$H = [A | I_{n-k}], \tag{2}$$

A is some fixed $(n - k) \times k$ matrix of 0's and 1's, and

$$I_{n-k} = \begin{pmatrix} 1 & & 0 \\ 0 & 1 & \\ & & \ddots & \\ & & & 1 \end{pmatrix}$$

is the $(n - k) \times (n - k)$ unit matrix. The arithmetic in Equation (1) is to be performed *modulo* 2, i.e. $0 + 1 = 1$, $1 + 1 = 0$, $-1 = +1$. We shall refer to this as *binary arithmetic*.

Example. *Code # 1.* The parity check matrix

$$H = \begin{bmatrix} 0 & 1 & 1 & 1 & 0 & 0 \\ 1 & 0 & 1 & 0 & 1 & 0 \\ 1 & 1 & 0 & 0 & 0 & 1 \end{bmatrix} \tag{3}$$

defines a code with $k = 3$ and $n = 6$. For this code

$$A = \begin{bmatrix} 0 & 1 & 1 \\ 1 & 0 & 1 \\ 1 & 1 & 0 \end{bmatrix}.$$

The message $u_1u_2u_3$ is encoded into the codeword $x = x_1x_2x_3x_4x_5x_6$, which begins with the message itself:

$$x_1 = u_1, \quad x_2 = u_2, \quad x_3 = u_3,$$

followed by three check symbols $x_4x_5x_6$ chosen so that $Hx^{tr} = 0$, i.e. so that

$$\begin{aligned} x_2 + x_3 + x_4 &= 0, \\ x_1 + x_3 + x_5 &= 0, \\ x_1 + x_2 + x_6 &= 0. \end{aligned} \qquad (4)$$

If the message is $u = 011$, then $x_1 = 0$, $x_2 = 1$, $x_3 = 1$, and the check symbols are

$$x_4 = -1 - 1 = 1 + 1 = 2 = 0,$$
$$x_5 = -1 = 1, \qquad x_6 = -1 = 1,$$

so the codeword is $x = 011011$.

The Equations (4) are called the *parity check equations*, or simply *parity checks*, of the code.

The first parity check equation says that the 2nd, 3rd and 4th symbols of every codeword must add to 0 modulo 2; i.e. their sum must have even parity (hence the name!).

Since each of the 3 message symbols $u_1u_2u_3$ is 0 or 1, there are altogether $2^3 = 8$ codewords in this code. They are:

000000	011011	110110
001110	100011	111000.
010101	101101	

In the general code there are 2^k codewords.

As we shall see, code # 1 is capable of correcting a single channel error (in any one of the six symbols), and using this code reduces the average probability of error per symbol from $p = .01$ to .00072 (see Problem 24). This is achieved at the cost of sending 6 symbols only 3 of which are message symbols.

We take (1) as our general definition:

Definition. Let H be any binary matrix. The *linear code* with *parity check matrix* H consists of all vectors x such that

$$Hx^{tr} = 0.$$

(where this equation is to be interpreted modulo 2).

It is convenient, but not essential, if H has the form shown in (2) and (3), in which case the first k symbols in each codeword are *message* or *information* symbols, and the last $n - k$ are *check* symbols.

Linear codes are the most important for practical applications and are the simplest to understand. Nonlinear codes will be introduced in Ch. 2.

Example. *Code # 2, a repetition code.* A code with $k = 1$, $n = 5$, and parity check matrix

$$H = \begin{bmatrix} 1 & 1 & & & \\ 1 & & 1 & & \\ 1 & & & 1 & \\ 1 & & & & 1 \end{bmatrix} \quad \text{(blanks denote zeros).}$$

Each codeword contains just one message symbol u. The parity check equations are

$$x_1 + x_2 = 0, \qquad x_1 + x_3 = 0, \qquad x_1 + x_4 = 0, \qquad x_1 + x_5 = 0,$$

i.e. $x_1 = x_2 = x_3 = x_4 = x_5 = u$. So there are only two codewords, 00000 and 11111. The message symbol is simply repeated 5 times: this is called a *repetition* code.

Example. *Code # 3, an even weight code.* A code with $k = 3$, $n = 4$ and parity check matrix $H = (1111)$. Each codeword contains 3 message symbols $x_1 x_2 x_3$ and one check symbol $x_4 = x_1 + x_2 + x_3$. The $2^3 = 8$ codewords are 0000, 0011, 0101, 1001, 0110, 1010, 1100, 1111, i.e. all vectors with an even number of 1's.

Problems. (1) *Code # 4* has parity check matrix

$$H = \begin{bmatrix} 1010 \\ 1101 \end{bmatrix}.$$

List all the codewords. Repeat for

$$H = \begin{bmatrix} 0111 \\ 1101 \end{bmatrix}.$$

How are these codes related?

(2) *Code # 5* has parity check matrix

$$H = \begin{bmatrix} 0111100 \\ 1011010 \\ 1101001 \end{bmatrix}.$$

List all the codewords.

(3) If $p > \frac{1}{2}$ in Fig. 1.1, show that interchanging the names of the received symbols changes this to a binary symmetric channel with $p < \frac{1}{2}$. If $p = \frac{1}{2}$ show that no communication is possible.

§2. Properties of a linear code

(i) The definition again: $x = x_1 \cdots x_n$ is a codeword if and only if

$$Hx^{\text{tr}} = 0. \tag{1}$$

(ii) Usually the parity check matrix H is an $(n-k) \times n$ matrix of the form

$$H = [A \mid I_{n-k}], \tag{2}$$

and as we have seen there are 2^k codewords satisfying (1). (This is still true even if H doesn't have this form, provided H has n columns and $n-k$ linearly independent rows.) When H has the form (2), the codewords look like this:

$$x = \underbrace{x_1 \cdots x_k}_{\substack{\text{message} \\ \text{symbols}}} \quad \underbrace{x_{k+1} \cdots x_n}_{\substack{\text{check} \\ \text{symbols}}} .$$

(iii) *The generator matrix.* If the message is $u = u_1 \cdots u_k$, what is the corresponding codeword $x = x_1 \cdots x_n$? First $x_1 = u_1, \ldots, x_k = u_k$, or

$$\begin{pmatrix} x_1 \\ \vdots \\ x_k \end{pmatrix} = I_k \begin{pmatrix} u_1 \\ \vdots \\ u_k \end{pmatrix}, \quad I_k = \text{unit matrix}. \tag{5}$$

Then from (1) and (2),

$$[A \mid I_{n-k}]\begin{pmatrix} x_1 \\ \vdots \\ x_n \end{pmatrix} = 0,$$

$$\begin{pmatrix} x_{k+1} \\ \vdots \\ x_n \end{pmatrix} = -A \begin{pmatrix} x_1 \\ \vdots \\ x_k \end{pmatrix}$$

$$= -A \begin{pmatrix} u_1 \\ \vdots \\ u_k \end{pmatrix} \quad \text{from (5).} \tag{6}$$

In the binary case $-A = A$, but later we shall treat cases where $-A \neq A$. Putting (5) on top of (6):

$$\begin{pmatrix} x_1 \\ \vdots \\ x_n \end{pmatrix} = \begin{bmatrix} I_k \\ -A \end{bmatrix}\begin{pmatrix} u_1 \\ \vdots \\ u_k \end{pmatrix}$$

and transposing, we get

$$x = uG \tag{7}$$

where

$$G = [I_k \mid -A^{\text{tr}}]. \tag{8}$$

If H is in standard form G is easily obtained from H – see (2). G is called a *generator matrix* of the code, for (7) just says that the codewords are all possible linear combinations of the rows of G. (We could have used this as the definition of the code.) (1) and (7) together imply that G and H are related by

$$GH^{tr} = 0 \quad \text{or} \quad HG^{tr} = 0. \tag{9}$$

Example. *Code #1* (cont.). A generator matrix is

$$G = [I_3 \mid -A^{tr}] = \begin{bmatrix} 100 & | & 011 \\ 010 & | & 101 \\ 001 & | & 110 \end{bmatrix} \begin{matrix} \text{row 1} \\ \text{row 2} \\ \text{row 3.} \end{matrix}$$

The 8 codewords are (from (7))

$$u_1 \cdot \text{row } 1 + u_2 \cdot \text{row } 2 + u_3 \cdot \text{row } 3 \quad (u_1, u_2, u_3 = 0 \text{ or } 1).$$

We see once again that the codeword corresponding to the message $u = 011$ is

$$\begin{aligned} x &= uG \\ &= \text{row } 2 + \text{row } 3 \\ &= 010101 + 001110 \\ &= 011011, \end{aligned}$$

the addition being done mod 2 as usual.

(iv) *The parameters of a linear code.* The codeword $x = x_1 \cdots x_n$ is said to have *length* n. This is measuring the length as a tailor would, not a mathematician. n is also called the block length of the code. If H has $n - k$ linearly independent rows, there are 2^k codewords. k is called the *dimension* of the code. We call the code an $[n, k]$ code.

This code uses n symbols to send k message symbols, so it is said to have *rate* or *efficiency* $R = k/n$.

(v) *Other generator and parity check matrices.* A code can have several different generator matrices. E.g.

$$\begin{bmatrix} 1110 \\ 0101 \end{bmatrix}, \quad \begin{bmatrix} 1011 \\ 0101 \end{bmatrix}$$

are both generator matrices for code #4. In fact any maximal set of linearly independent codewords taken from a given code can be used as the rows of a generator matrix for that code.

A *parity check* on a code $\mathscr{C}$ is any row vector h such that $hx^{tr} = 0$ for all codewords $x \in \mathscr{C}$. Then similarly any maximal set of linearly independent parity checks can be used as the rows of a parity check matrix H for $\mathscr{C}$. E.g.

$$\begin{bmatrix} 1010 \\ 1101 \end{bmatrix}, \quad \begin{bmatrix} 0111 \\ 1101 \end{bmatrix}$$

are both parity check matrices for code #4.

(vi) *Codes over other fields.* Instead of only using 0's and 1's, we could have allowed the symbols to be from any finite field (see Chapters 3 and 4). For example a *ternary* code has symbols 0, 1 and 2, and all calculations of parity checks etc. are done modulo 3 ($1+2=0$, $-1=2$, $2+2=1$, $2 \cdot 2 = 1$, etc.). If the symbols are from a finite field with q elements, the code is still defined by (1) and (2), or equivalently by (7) and (8), and an $[n, k]$ code contains q^k codewords.

Example. *Code # 6.* A $[4, 2]$ ternary code with parity check matrix

$$H = \begin{bmatrix} 1110 \\ 1201 \end{bmatrix} = [A \mid I_2].$$

The generator matrix is (from (8))

$$G = [I_2 \mid -A^{tr}] = \begin{bmatrix} 1022 \\ 0121 \end{bmatrix} \quad \begin{array}{l} \text{row 1} \\ \text{row 2.} \end{array}$$

There are 9 codewords $u_1 \cdot$ row 1 + $u_2 \cdot$ row 2 ($u_1, u_2 = 0$, 1 or 2), as follows:

message	codeword	message	codeword	message	codeword
u	x	u	x	u	x
00	0000	10	1022	20	2011
01	0121	11	1110	21	2102
02	0212	12	1201	22	2220

This code has rate $R = k/n = \frac{1}{2}$.

(vii) *Linearity.* If x and y are codewords of a given code, so is $x + y$, because $H(x + y)^{tr} = Hx^{tr} + Hy^{tr} = 0$. If c is any element of the field, then cx is also a codeword, because $H(cx)^{tr} = cHx^{tr} = 0$. E.g. in a ternary code if x is a codeword so is $2x = -x$. That is why these are called linear codes. Such a code is also an additive group, and a vector space over the field.

Problems. (4) *Code # 2* (cont.). Give parity check and generator matrices for the general $[n, 1]$ *repetition code*.

(5) *Code # 3* (cont.). Give parity check and generator matrices for the general $[n, n - 1]$ *even weight code*.

(6) If the code $\mathscr{C}$ has an invertible generator matrix, what is $\mathscr{C}$?

§3. At the receiving end

(We now return to binary codes.) Suppose the message $u = u_1 \cdots u_k$ is encoded into the codeword $x = x_1 \cdots x_n$, which is then sent through the channel. Because of channel noise, the received vector $y = y_1 \cdots y_n$ may be different from x. Let's define the *error vector*

$$e = y - x = e_1 \cdots e_n. \tag{10}$$

Then $e_i = 0$ with probability $1 - p$ (and the ith symbol is correct), and $e_i = 1$ with probability p (and the ith symbol is wrong). In the example of §1 p was equal to $1/100$, but in general p can be anywhere in the range $0 \leqslant p < \frac{1}{2}$. So we describe the action of the channel by saying it distorts the codeword x by adding the error vector e to it.

The decoder (Fig. 1.3) must decide from y which message u or (usually simpler) which codeword x was transmitted. Of course it's enough if the decoder finds e, for then $x = y - e$. Now the decoder can never be certain what e was. His strategy therefore will be to choose the *most likely* error vector e, given that y was received. Provided the codewords are all equally likely, this strategy is optimum in the sense that it minimizes the probability of the decoder making a mistake, and is called *maximum likelihood decoding*.

To describe how the decoder does this, we need two important definitions.

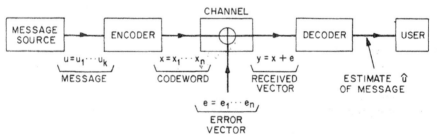

Fig. 1.3. The overall communication system.

Definition. The (*Hamming*) *distance* between two vectors $x = x_1 \cdots x_n$ and $y = y_1 \cdots y_n$ is the number of places where they differ, and is denoted by $\text{dist}(x, y)$. E.g.

$$\text{dist}(10111, 00101) = 2, \qquad \text{dist}(0122, 1220) = 3$$

(the same definition holds for nonbinary vectors).

Definition. The (*Hamming*) *weight* of a vector $x = x_1 \cdots x_n$ is the number of nonzero x_i, and is denoted by $\text{wt}(x)$. E.g.

$$\text{wt}(101110) = 4, \qquad \text{wt}(01212110) = 6.$$

Obviously

$$\text{dist}(x, y) = \text{wt}(x - y), \tag{11}$$

for both sides express the number of places where x and y differ.

Problem. (7) Define the *intersection* of binary vectors x and y to be the vector

$$x * y = (x_1 y_1, \ldots, x_n y_n),$$

which has 1's only where both x and y do. E.g. $11001 * 10111 = 10001$. Show that

$$\text{wt}(x + y) = \text{wt}(x) + \text{wt}(y) - 2\,\text{wt}(x * y). \tag{12}$$

Now back to decoding. Errors occur with probability p. For instance

$$\text{Prob}\{e = 00000\} = (1 - p)^5,$$
$$\text{Prob}\{e = 01000\} = p(1 - p)^4,$$
$$\text{Prob}\{e = 10010\} \doteq p^2(1 - p)^3.$$

In general if v is some fixed vector of weight a,

$$\text{Prob}\{e = v\} = p^a(1 - p)^{n-a}. \tag{13}$$

Since $p < \frac{1}{2}$, we have $1 - p > p$, and

$$(1 - p)^5 > p(1 - p)^4 > p^2(1 - p)^3 > \cdots.$$

Therefore a particular error vector of weight 1 is more likely than a particular error vector of weight 2, and so on. So the decoder's strategy is:

Decode y as the nearest codeword x (nearest in Hamming distance), i.e.:

Pick that error vector e which has least weight.

This is called *nearest neighbor decoding*.

A brute force decoding scheme then is simply to compare y with all 2^k codewords and pick the closest. This is fine for small codes. But if k is large this is impossible! One of the aims of coding theory is to find codes which can be decoded by a faster method than this.

The minimum distance of a code. The third important parameter of a code $\mathscr{C}$, besides the length and dimension, is the minimum Hamming distance between its codewords:

$$\mathbf{d} = \min \text{dist}(u, v)$$
$$= \min \text{wt}(u - v) \quad u \in \mathscr{C}, v \in \mathscr{C}, u \neq v. \tag{14}$$

d is called the *minimum distance* or simply the *distance* of the code. Any two codewords differ in at least d places.

A linear code of length n, dimension k, and minimum distance d will be called an $[n, k, d]$ code.

To find the minimum distance of a linear code it is not necessary to compare every pair of codewords. For if u and v belong to a linear code $\mathscr{C}$, $u - v = w$ is also a codeword, and (from (14))

$$d = \min_{w \in \mathscr{C}, w \neq 0} \text{wt}(w).$$

In other words:

Theorem 1. *The minimum distance of a linear code is the minimum weight of any nonzero codeword.*

Example. The minimum distances of codes #1, #2, #3, #6 are 3, 5, 2, 3 respectively.

How many errors can a code correct?

Theorem 2. *A code with minimum distance d can correct $[\frac{1}{2}(d-1)]$ errors.* If d is even, the code can simultaneously correct $\frac{1}{2}(d-2)$ errors and detect $d/2$ errors.*

Proof. Suppose $d = 3$ (Fig. 1.4). The *sphere* of radius r and center u consists of all vectors v such that dist $(u, v) \leqslant r$. If a sphere of radius 1 is drawn around each codeword, these spheres do not overlap. Then if codeword u is transmitted and one error occurs, so that the vector a is received, then a is inside the sphere around u, and is still closer to u than to any other codeword v. Thus nearest neighbor decoding will correct this error.

Similarly if $d = 2t + 1$, spheres of radius t around each codeword do not overlap, and the code can correct t errors.

Now suppose d is even (Fig. 1.5, where $d = 4$). Spheres of radius $\frac{1}{2}(d-2)$ around the codewords are disjoint and so the code can correct $\frac{1}{2}(d-2)$ errors. But if $d/2$ errors occur the received vector a may be midway between 2 codewords (Fig. 1.5). In this case the decoder can only detect that $d/2$ (or more) errors have occurred.   Q.E.D.

Thus code #1, which has minimum distance 3, is a single-error-correcting code.

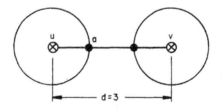

Fig. 1.4. A code with minimum distance 3 ($\otimes$=codeword).

*$[x]$ denotes the greatest integer less than or equal to x. E.g. $[3.5] = 3$, $[-1.5] = -2$.

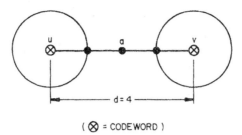

(⊗ = CODEWORD)

Fig. 1.5. A code with minimum distance 4.

On the other hand, if more than $d/2$ errors occur, the received vector may or may not be closer to some other codeword than to the correct one. If it is, the decoder will be fooled and will output the wrong codeword. This is called a *decoding error*. Of course with a good code this should rarely happen.

So far we have assumed that the decoder will *always* try to find the nearest codeword. This scheme, called *complete decoding*, is fine for messages which can't be retransmitted, such as a photograph from Mars, or an old magnetic tape. In such a case we want to extract as much as possible from the received vector.

But often we want to be more cautious, or cannot afford the most expensive decoding method. In such cases we might use an *incomplete* decoding strategy: if it appears that no more than l errors occurred, correct them, otherwise reject the message or ask for a retransmission.

Error detection is an extreme version of this, when the receiver makes no attempt to correct errors, but just tests the received vector to see if it is a codeword. If it is not, he *detects* that an error has occurred and asks for a retransmission of the message. This scheme has the advantages that the algorithm for detecting errors is very simple (see the next section) and the probability of an undetected error is very low (§5). The disadvantage is that if the channel is bad, too much time will be spent retransmitting, which is an inefficient use of the channel and produces unpleasant delays.

Nonbinary codes. Almost everything we have said applies equally well to codes over other fields. If the field F has q elements, then the message u, the codeword x, the received vector y, and the error vector

$$e = y - x = e_1 e_2 \cdots e_n$$

all have components from F.

We assume that e_i is 0 with probability $1 - p > \frac{1}{2}$, and e_i is any of the $q - 1$ nonzero elements of F with probability $p/(q - 1)$. In other words the channel is a *q-ary symmetric channel*, with q inputs, q outputs, a probability $1 - p > 1/q$ that no error occurs, and a probability $p < (q - 1)/q$ that an error does occur, each of the $q - 1$ possible errors being equally likely.

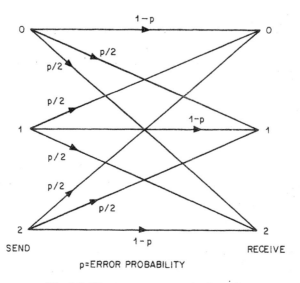

Fig. 1.6. The ternary symmetric channel.

Example. If $q = 3$, the ternary symmetric channel is shown in Fig. 1.6.

Problems. (8) Show that for binary vectors:

(a)
$$\text{wt} (x + y) \geq \text{wt} (x) - \text{wt} (y),$$
(15)

with equality iff $x_i = 1$ whenever $y_i = 1$.

(b)
$$\text{wt} (x + z) + \text{wt} (y + z) + \text{wt} (x + y + z)$$
$$\geq 2 \, \text{wt} (x + y + x * y) - \text{wt} (z),$$
(16)

with equality iff it never happens that x_i and y_i are 0 and z_i is 1.

(9) Define the *product* of vectors x and y from any field to be the vector

$$x * y = (x_1 y_1, \ldots, x_n y_n).$$

For binary vectors this is called the intersection – see Problem 7. Show that for ternary vectors,

$$\text{wt} (x + y) = \text{wt} (x) + \text{wt} (y) - f(x * y),$$
(17)

where, if $u = u_1 \cdots u_n$ is a ternary vector containing a 0's, b 1's and c 2's, then $f(u) = b + 2c$. Therefore

$$\text{wt} (x) + \text{wt} (y) - 2 \, \text{wt} (x * y) \leq \text{wt} (x + y) \leq \text{wt} (x) + \text{wt} (y) - \text{wt} (x * y). \quad (18)$$

(10) Show that in a linear binary code, either all the codewords have even weight, or exactly half have even weight and half have odd weight.

(11) Show that in a linear binary code, either all the codewords begin with 0, or exactly half begin with 0 and half with 1. Generalize!

Problems 10 and 11 both follow from:

(12) Suppose G is an abelian group which contains a subset A with three properties: (i) if $a_1, a_2 \in A$, then $a_1 - a_2 \in A$, (ii) if $b_1, b_2 \not\in A$, then $b_1 - b_2 \in A$, (iii) if $a \in A$, $b \not\in A$, then $a + b \not\in A$. Show that either $A = G$, or A is a subgroup of G and there is an element $b \not\in A$ such that

$$G = A \cup (b + A),$$

where $b + A$ is the set of all elements $b + a, a \in A$.

(13) Show that Hamming distance obeys the *triangle inequality*: for any vectors $x = x_1 \cdots x_n$, $y = y_1 \cdots y_n$, $z = z_1 \cdots z_n$,

$$\text{dist}(x, y) + \text{dist}(y, z) \geq \text{dist}(x, z). \tag{19}$$

Show that equality holds iff, for all i, either $x_i = y_i$ or $y_i = z_i$ (or both).

(14) Hence show that if a code has minimum distance d, the codeword x is transmitted, not more than $\frac{1}{2}(d - 1)$ errors occur, and y is received, then $\text{dist}(x, y) < \text{dist}(y, z)$ for all codewords $z \neq x$. (A more formal proof of Theorem 2.)

(15) Show that a code can simultaneously correct $\leq a$ errors and detect $a + 1, \ldots, b$ errors iff it has minimum distance at least $a + b + 1$.

(16) Show that if x and y are binary vectors, the Euclidean distance between the points x and y is

$$\left(\sum_{i=1}^{n} (x_i - y_i)^2 \right)^{1/2} = \sqrt{(\text{dist}(x, y))}.$$

(17) *Combining two codes (I)*. Let G_1, G_2 be generator matrices for $[n_1, k, d_1]$ and $[n_2, k, d_2]$ codes respectively. Show that the codes with generator matrices

$$\begin{pmatrix} G_1 & 0 \\ 0 & G_2 \end{pmatrix}$$

and $(G_1|G_2)$ are $[n_1 + n_2, 2k, \min\{d_1, d_2\}]$ and $[n_1 + n_2, k, d]$ codes, respectively, where $d \geq d_1 + d_2$.

(18) *Binomial coefficients*. The *binomial coefficient* $\binom{x}{m}$, pronounced "x choose m," is defined by

$$\binom{x}{m} = \begin{cases} \dfrac{x(x - 1) \cdots (x - m + 1)}{m!}, & \text{if } m \text{ is a positive integer}, \\ 1, & \text{if } m = 0, \\ 0, & \text{otherwise}, \end{cases}$$

where x is any real number, and $m! = 1 \cdot 2 \cdot 3 \cdot \ldots \cdot (m - 1)m, 0! = 1$. The reader should know the following properties:

(a) If $x = n$ is a nonnegative integer and $n \geq m \geq 0$,

$$\binom{n}{m} = \frac{n!}{m!(n-m)!},$$

which is the number of unordered selections of m objects from a set of n objects.

(b) There are $\binom{n}{m}$ binary vectors of length n and weight m. There are $(q-1)^m \binom{n}{m}$ vectors from a field of q elements which have length n and weight m. [For each of the m nonzero coordinates can be filled with any field element except zero.]

(c) The binomial series:

$$(a+b)^n = \sum_{m=0}^{n} \binom{n}{m} a^{n-m} b^m, \quad \text{if } n \text{ is a nonnegative integer;}$$

$$(1+b)^x = \sum_{m=0}^{\infty} \binom{x}{m} b^m, \quad \text{for } |b| < 1 \text{ and any real } x;$$

$$\frac{1}{(1-b)^{x+1}} = \sum_{r=0}^{\infty} \binom{x+r}{r} b^r, \quad \text{for } |b| < 1 \text{ and any real } x.$$

[Remember the student who, when asked to expand $(a+b)^n$ on an exam replied: ₍ₐ₊ᵦ₎ⁿ, ₍ₐ₊ᵦ₎ⁿ, $(a+b)^n$, $(a+b)^n$, $(a + b)^n$, ... ?]

(d) *Easy identities.* (Here n and m are nonnegative integers, x is any real number)

$$\binom{n}{m} = \binom{n}{n-m},$$

$$\binom{n}{m} = 0 \quad \text{for } m > n \geq 0,$$

$$\binom{n}{m} + \binom{n}{m-1} = \binom{n+1}{m},$$

$$(-1)^m \binom{-x}{m} = \binom{x+m-1}{m},$$

$$\sum_{m=0}^{n} \binom{n}{m} = 2^n,$$

$$\sum_{m \text{ even}} \binom{n}{m} = \sum_{m \text{ odd}} \binom{n}{m} = 2^{n-1} \quad \text{if } n \geq 1,$$

$$\sum_{m=0}^{n} (-1)^m \binom{n}{m} = 0 \quad \text{if } n \geq 1,$$

$$\sum_{m \text{ divisible by } 4} \binom{n}{m} = \tfrac{1}{4}[2^n + (1+i)^n + (1-i)^n], \, i = \sqrt{-1}. \tag{20}$$

(19) Suppose u and v are binary vectors with dist $(u, v) = d$. Show that the number of vectors w such that dist $(u, w) = r$ and dist $(v, w) = s$ is $\binom{d}{i}\binom{n-d}{r-i}$,

where $i = (d + r - s)/2$. If $d + r - s$ is odd this number is 0, while if $r + s = d$ it is $\binom{d}{r}$.

(20) Let u, v, w and x be four vectors which are pairwise distance d apart; d is necessarily even. Show that there is exactly one vector which is at a distance of $d/2$ from each of u, v and w. Show that there is at most one vector at distance $d/2$ from all of u, v, w and x.

§4. More about decoding a linear code

Definition of a coset. Let $\mathscr{C}$ be an $[n, k]$ linear code over a field with q elements. For any vector a, the set

$$a + \mathscr{C} = \{a + x : x \in \mathscr{C}\}$$

is called a *coset* (or *translate*) of $\mathscr{C}$. Every vector b is in some coset (in $b + \mathscr{C}$ for example). a and b are in the same coset iff $(a - b) \in \mathscr{C}$. Each coset contains q^k vectors.

Proposition 3. *Two cosets are either disjoint or coincide (partial overlap is impossible).*

Proof. If cosets $a + \mathscr{C}$ and $b + \mathscr{C}$ overlap, take $v \in (a + \mathscr{C}) \cap (b + \mathscr{C})$. Then $v = a + x = b + y$, where x and $y \in \mathscr{C}$. Therefore $b = a + x - y = a + x'$ $(x' \in \mathscr{C})$, and so $b + \mathscr{C} \subset a + \mathscr{C}$. Similarly $a + \mathscr{C} \subset b + \mathscr{C}$, and so $a + \mathscr{C} = b + \mathscr{C}$. Q.E.D.

Therefore the set F^n of all vectors can be partitioned into cosets of $\mathscr{C}$:

$$F^n = \mathscr{C} \cup (a_1 + \mathscr{C}) \cup (a_2 + \mathscr{C}) \cup \cdots \cup (a_t + \mathscr{C}) \tag{21}$$

where $t = q^{n-k} - 1$.

Suppose the decoder receives the vector y. y must belong to some coset in (21), say $y = a_i + x$ $(x \in \mathscr{C})$. What are the possible error vectors e which could have occurred? If the codeword x' was transmitted, the error vector is $e = y - x' = a_i + x - x' = a_i + x'' \in a_i + \mathscr{C}$. We deduce that:

> the possible error vectors are exactly the vectors in
> the coset containing y.

So the decoder's strategy is, given y, to choose a minimum weight vector $\hat{e}$ in the coset containing y, and to decode y as $\hat{x} = y - \hat{e}$. The minimum weight vector in a coset is called the *coset leader*. (If there is more than one vector with the minimum weight, choose one at random and call it the coset leader.)

We assume that the a_i's in (21) are the coset leaders.

The standard array. A useful way to describe what the decoder does is by a table, called a *standard array* for the code. The first row consists of the code itself, with the zero codeword on the left:

$$\boldsymbol{x}^{(1)} = \boldsymbol{0}, \boldsymbol{x}^{(2)}, \ldots, \boldsymbol{x}^{(s)}, \qquad (s = q^k);$$

and the other rows are the other cosets $\boldsymbol{a}_i + \mathscr{C}$, arranged in the same order and with the coset leader on the left:

$$\boldsymbol{a}_i + \boldsymbol{x}^{(1)}, \qquad \boldsymbol{a}_i + \boldsymbol{x}^{(2)}, \ldots, \boldsymbol{a}_i + \boldsymbol{x}^{(s)}.$$

Example. *Code #4* (cont.). The [4, 2] code with generator matrix

$$G = \begin{bmatrix} 1011 \\ 0101 \end{bmatrix}$$

has a standard array shown in Fig. 1.7 (ignore the last column for the moment).

message:	00	10	01	11	syndrome S
code:	0000	1011	0101	1110	$\binom{0}{0}$
coset:	1000	0011	1101	0110	$\binom{1}{1}$
coset:	0100	1111	0001	1010	$\binom{0}{1}$
coset:	0010	1001	0111	1100	$\binom{1}{0}$
	coset				
	leaders				

Fig. 1.7. *A standard array.*

Note that all 16 vectors of length 4 appear, divided into the 4 cosets forming the rows, and the coset leaders are on the left.

Here is how the decoder uses the standard array: When $\boldsymbol{y}$ is received (e.g. 1111) its position in the array is found. Then the decoder decides that the error vector $\hat{\boldsymbol{e}}$ is the coset leader found at the extreme left of $\boldsymbol{y}$ (0100), and $\boldsymbol{y}$ is decoded as the codeword $\hat{\boldsymbol{x}} = \boldsymbol{y} - \hat{\boldsymbol{e}} = 1011$ found at the top of the column containing $\boldsymbol{y}$. (The corresponding message is 10.)

Decoding using a standard array is maximum likelihood decoding.

The syndrome. There is an easy way to find which coset $\boldsymbol{y}$ is in: compute the vector

$$S = H\boldsymbol{y}^{tr},$$

which is called the *syndrome* of $\boldsymbol{y}$.

Properties of the syndrome. (1) S is a column vector of length $n - k$.

(2) The syndrome of $\boldsymbol{y}$, $S = H\boldsymbol{y}^{tr}$, is zero iff $\boldsymbol{y}$ is a codeword (by definition of the code). So if no errors occur, the syndrome of $\boldsymbol{y}$ is zero (but not

conversely). In general, if $y = x + e$ where $x \in \mathscr{C}$, then

$$S = Hy^{tr} = Hx^{tr} + He^{tr} = He^{tr}. \tag{22}$$

(3) For a binary code, if there are errors at locations $a, b, c, \ldots$, so that

$$e = 0 \cdots 0\underset{a}{1}0 \cdots \underset{b}{1} \cdots \underset{c}{1} \cdots 0,$$

then from Equation (22),

$$S = \sum_i e_i H_i \quad (H_i = i^{th} \text{ column of } H)$$

$$= H_a + H_b + H_c + \cdots .$$

In words:

Theorem 4. *For a binary code, the syndrome is equal to the sum of the columns of H where the errors occurred.* [Thus S is called the "syndrome" because it gives the symptoms of the errors.]

(4) Two vectors are in the same coset of $\mathscr{C}$ iff they have the same syndrome. For u and v are in the same coset iff $(u - v) \in \mathscr{C}$ iff $H(u - v)^{tr} = 0$ iff $Hu^{tr} = Hv^{tr}$. Therefore:

Theorem 5. *There is a 1-1-correspondence between syndromes and cosets.*

For example, the cosets in Fig. 1.7 are labeled with their syndromes.

Thus the syndrome contains all the information that the receiver has about the errors.

By Property (2), the pure *error detection* scheme mentioned in the last section just consists of testing if the syndrome is zero. To do this, we recompute the parity check symbols using the received information symbols, and see if they agree with the received parity check symbols. I.e., we re-encode the received information symbols. This only requires a copy of the encoding circuit, which is normally a very simple device compared to the decoder (see Fig. 7.8).

Problems. (21) Construct a standard array for code #1. Use it to decode the vectors 110100 and 111111.

(22) Show that if $\mathscr{C}$ is a binary linear code and $a \notin \mathscr{C}$, then $\mathscr{C} \cup (a + \mathscr{C})$ is also a linear code.

§5. Error probability

When decoding using the standard array, the error vector chosen by the decoder is always one of the coset leaders. The decoding is correct if and only if the true error vector is indeed a coset leader. If not, the decoder makes a decoding error and outputs the wrong codeword. (Some of the information symbols may still be correct, even so.)

Definition. The *probability of error*, or the *word error rate*, P_{err}, for a particular decoding scheme is the probability that the decoder output is the wrong codeword.

If there are M codewords $x^{(1)}, \ldots, x^{(M)}$, which we assume are used with equal probability, then

$$P_{err} = \frac{1}{M} \sum_{i=1}^{M} \text{Prob} \{\text{decoder output} \neq x^{(i)} \mid x^{(i)} \text{ was sent}\}. \qquad (23)$$

If the decoding is done using a standard array, a decoding error occurs iff e is not a coset leader, so

$$P_{err} = \text{Prob} \{e \neq \text{coset leader}\}.$$

Suppose there are α_i coset leaders of weight i. Then (using (13))

$$P_{err} = 1 - \sum_{i=0}^{n} \alpha_i p^i (1-p)^{n-i} \qquad (24)$$

(Since the standard array does maximum likelihood decoding, any other decoding scheme will have $P_{err} \geq (24)$.)

Examples. For code #4 (Fig. 1.7), $\alpha_0 = 1$, $\alpha_1 = 3$, so

$$P_{err} = 1 - (1-p)^4 - 3p(1-p)^3$$

$$= 0.0103 \ldots \quad \text{if } p = \frac{1}{100}.$$

For code #1, $\alpha_0 = 1$, $\alpha_1 = 6$, $\alpha_2 = 1$, so

$$P_{err} = 1 - (1-p)^6 - 6p(1-p)^5 - p^2(1-p)^4$$

$$= 0.00136 \ldots \quad \text{if } p = \frac{1}{100}.$$

If the code has minimum distance $d = 2t + 1$ or $2t + 2$, then (by Theorem 2) it can correct t errors. So every error vector of weight $\leq t$ is a coset leader.

I.e.

$$\alpha_i = \binom{n}{i} \quad \text{for } 0 \le i \le t. \tag{25}$$

But for $i > t$ the α_i are extremely difficult to calculate and are known for very few codes.

If the channel error probability p is small, $1 - p \doteq 1$ and $p^i(1-p)^{n-i} \gg p^{i+1}(1-p)^{n-i-1}$. In this case the terms in (24) with large i are negligible, and

$$P_{\text{err}} \doteq 1 - \sum_{i=0}^{t} \binom{n}{i} p^i (1-p)^{n-i} \tag{26}$$

or

$$P_{\text{err}} \doteq 1 - \sum_{i=0}^{t} \binom{n}{i} p^i (1-p)^{n-i} - \alpha_{t+1} p^{t+1} (1-p)^{n-t-1} \tag{27}$$

are useful approximations. In any event the RHS of (26) or (27) is an upper bound on P_{err}.

Perfect codes. Of course, if $\alpha_i = 0$ for $i > t = [(d-1)/2]$, then (26) is exact. Such a code is called *perfect*.

Thus a perfect t-error-correcting code can correct all errors of weight $\le t$, and none of weight greater than t. Equivalently, the spheres of radius t around the codewords are disjoint and together contain all vectors of length n.

We shall see much more about perfect codes in Chs. 6, 20.

Quasi-perfect codes. On the other hand, if $\alpha_i = 0$ for $i > t + 1$, then (27) is exact. Such a code is called *quasi-perfect*.

Thus a quasi-perfect t-error-correcting code can correct all errors of weight $\le t$, some of weight $t + 1$, and none of weight $> t + 1$. The spheres of radius $t + 1$ around the codewords may overlap, and together contain all vectors of length n.

We shall meet quasi-perfect codes again in Chapters 6, 9 and 15.

Sphere-packing or Hamming bound. Suppose $\mathscr{C}$ is a binary code of length n containing M codewords, which can correct t errors. The spheres of radius t around the codewords are disjoint. Each of these M spheres contain $1 + \binom{n}{1} + \cdots + \binom{n}{t}$ vectors (see Problem 18(b)). But the total number of vectors in the space is 2^n. Therefore we have established:

Theorem 6. (The sphere-packing or Hamming bound.)
A t-error-correcting binary code of length n containing M codewords must satisfy

$$M\left(1 + \binom{n}{1} + \binom{n}{2} + \cdots + \binom{n}{t}\right) \le 2^n. \tag{28}$$

Similarly, for a code over a field with q elements,

$$M\left(1+(q-1)\binom{n}{1}+\cdots+(q-1)^t\binom{n}{t}\right)\leqslant q^n. \tag{29}$$

For large n see Theorem 32 of Ch. 17. Since the proof of these two bounds did not assume linearity, they also hold for nonlinear codes (see Ch. 2).

By definition, a code is perfect iff equality holds in (28) or (29).

Symbol error rate. Since some of the message symbols may be correct even if the decoder outputs the wrong codeword, a more useful quantity is the symbol error rate, defined as follows.

Definition. Suppose the code contains M codewords $x^{(i)}=x_1^{(i)}\cdots x_n^{(i)}$, $i=1,\ldots,M$, and the first k symbols $x_1^{(i)}\cdots x_k^{(i)}$ in each codeword are information symbols. Let $\hat{x}=\hat{x}_1\cdots\hat{x}_n$ be the decoder output. Then the *symbol error rate* P_{symb} is the average probability that an information symbol is in error after decoding:

$$P_{\text{symb}}=\frac{1}{kM}\sum_{j=1}^{k}\sum_{i=1}^{M}\text{Prob}\,\{\hat{x}_j\neq x_j^{(i)}\mid x^{(i)}\text{ was sent}\} \tag{30}$$

Problem. (23) Show that if standard array decoding is used, and the messages are equally likely, then the number of information symbols in error after decoding does not depend on which codeword was sent. Indeed, if the codeword $x=x_1\cdots x_n$ is sent and is decoded as $\hat{x}=\hat{x}_1\cdots\hat{x}_n$, then

$$P_{\text{symb}}=\frac{1}{k}\sum_{j=1}^{k}\text{Prob}\,\{\hat{x}_j\neq x_j\}, \tag{31}$$

$$=\frac{1}{k}\sum_{e}f(e)\,\text{Prob}\,\{e\}, \tag{32}$$

where $f(e)$ is the number of incorrect information symbols after decoding, if the error vector is e, and so

$$P_{\text{symb}}=\frac{1}{k}\sum_{i=1}^{2^k}F_iP(c_i),$$

where F_i is the weight of the first k places of the codeword at the head of the ith column of the standard array, and $P(c_i)$ is the probability of all binary vectors in this column.

Example. The standard array of Fig. 1.7. Here $f(e)=0$ if e is in the first column of the standard array (a coset leader), $=1$ if e is in columns 2 or 3, and $=2$ if e is in the last column. From (32):

$$P_{\text{symb}}=\tfrac{1}{2}[1\cdot(p^4+3p^3q+3p^2q^2+pq^3)+2\cdot(p^3q+3p^2q^2)],\quad q=1-p,$$
$$=0.00530\ldots\quad\text{if }p=1/100.$$

Using this very simple code has lowered the average probability of error per symbol from 0.01 to 0.0053.

Problems. (24) Show that for code #1 with standard array decoding

$$P_{\text{symb}} = (22p^2q^4 + 36p^3q^3 + 24p^4q^2 + 12p^5q + 2p^6)/3$$
$$= 0.00072 \ldots \quad \text{if } p = 1/100.$$

As these examples suggest, P_{symb} is difficult to calculate and is not known for most codes.

(25) Show that for an $[n, k]$ code,

$$\frac{1}{k} P_{\text{err}} \leqslant P_{\text{symb}} \leqslant P_{\text{err}}$$

Incomplete decoding. An incomplete decoding scheme which corrects $\leqslant l$ errors can be described in terms of the standard array as follows. Arrange the cosets in order of increasing weight (i.e. decreasing probability) of the coset leader (Fig. 1.8).

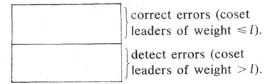

correct errors (coset leaders of weight $\leqslant l$).

detect errors (coset leaders of weight $> l$).

Fig. 1.8. Incomplete decoding using a standard array.

If the received vector y lies in the top part of the array, as before y is decoded as the codeword found at the top of its column. If y lies in the bottom half, the decoder just detects that more than l errors have occurred.

Error detection. When error detection is being used the decoder will make a mistake and accept a codeword which is not the one transmitted iff the error vector is a nonzero codeword. If the code contains A_i codewords of weight i, the error probability is

$$P_{\text{err}} = \sum_{i=1}^{n} A_i p^i (1-p)^{n-i}. \tag{33}$$

The probability that an error will be detected and the message retransmitted is

$$P_{\text{retrans.}} = 1 - (1-p)^n - P_{\text{err}}. \tag{34}$$

Example. For code #1, $A_0 = 1$, $A_3 = 4$, $A_4 = 3$, and

$$P_{\text{err}} = 4p^3(1-p)^3 + 3p^4(1-p)^2$$
$$= 0.00000391 \ldots \quad \text{if } p = \frac{1}{100}.$$

This is very much smaller than the error probability of 0.00136 obtained from standard array decoding of this code. The retransmission probability is $P_{\text{retrans.}} = 0.0585 \ldots$.

Research Problem 1.1. Find the distribution $\{\alpha_i\}$ of coset leaders for any of the common families of codes. This is still unsolved even for first-order Reed Muller codes – see Chapter 14, Berlekamp and Welch [133], Lechner [797–799], Sarwate [1144] and Sloane and Dick [1233].

§6. Shannon's Theorem on the existence of good codes

In the last section we saw that the weak code #4 reduced the average probability of error per symbol from 0.01 to 0.00530 (at the cost of using 4 symbols to send 2 message symbols), and that code #1, which also had rate $\frac{1}{2}$, further reduced it to 0.00072 (at the cost of using 6 symbols to send 3 message symbols).

In general we would like to know, for a given rate $R = k/n$, how small we can make P_{symb} with an $[n, k]$ code. The answer is given by a remarkable theorem of Shannon, which says that P_{err} (and hence P_{symb}, by Problem 25) can be made arbitrarily small, provided R is less than the capacity of the channel.

Definition. The *capacity* of a binary symmetric channel with error probability p (Fig. 1.1) is (see Fig. 1.9)

$$C(p) = 1 + p \log_2 p + (1 - p) \log_2 (1 - p). \tag{35}$$

Theorem 7. (Shannon's Theorem; proof omitted.) *For any $\epsilon > 0$, if $R < C(p)$ and n is sufficiently large, there is an $[n, k]$ binary code of rate $k/n \geq R$ with error probability $P_{\text{err}} < \epsilon$.*

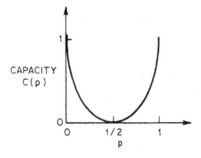

Fig. 1.9. Capacity of binary symmetric channel.

(A similar result holds for nonbinary codes, but with a different definition of capacity.)

Unfortunately this theorem has so far been proved only by probabilistic methods and does not tell how to construct these codes.

In practice (as we have seen) it is difficult to find P_{err} and P_{symb}, so the minimum distance d is used as a more convenient measure of how good the code is. For then the code can correct $[\frac{1}{2}(d-1)]$ errors (Theorem 2), and (26) is a good approximation to P_{err}, especially if p is small.

So one version of the *main problem of coding theory* is to find codes with large R (for efficiency) and large d (to correct many errors). Of course these are conflicting goals. Theorem 6 has already given an upper bound on the size of a code, and other upper bounds will be given in Ch. 17 (see especially Fig. 17.7). On the other hand Theorem 12 at the end of the chapter is a weaker version of Theorem 7 and implies that good linear codes exist. However, at the present time we do not know how to find such codes.

Of course for practical purposes one also wants a code which can be easily encoded and decoded.

§7. Hamming codes

The Hamming single-error-correcting codes are an important family of codes which are easy to encode and decode. In this section we discuss only binary Hamming codes.

According to Theorem 4, the syndrome of the receiver vector is equal to the sum of the columns of the parity check matrix H where the errors occurred. Therefore to design a single-error-correcting code we should make the columns of H nonzero (or else an error in that position would not affect the syndrome and would be undetectable) and distinct (for if two columns of H were equal, errors in those two positions would be indistinguishable).

If H is to have r rows (and the code to have r parity checks) there are only $2^r - 1$ columns available, namely the $2^r - 1$ nonzero binary vectors of length r. E.g. if $r = 3$, there are $2^3 - 1 = 7$ columns available:

$$
\begin{matrix}
0 & 0 & 0 & 1 & 1 & 1 & 1 \\
0 & 1 & 1 & 0 & 0 & 1 & 1 \\
1 & 0 & 1 & 0 & 1 & 0 & 1,
\end{matrix} \tag{36}
$$

the binary representations of the numbers 1 to 7. For a Hamming code we use them all, and get a code of length $n = 2^r - 1$.

Definition. A binary *Hamming code* $\mathcal{H}_r$ of length $n = 2^r - 1$ $(r \geq 2)$ has parity check matrix H whose columns consist of all nonzero binary vectors of length r, each used once. $\mathcal{H}_r$ is an $[n = 2^r - 1, k = 2^r - 1 - r, d = 3]$ code.

Example. Code #5, the [7, 4, 3] Hamming Code $\mathcal{H}_3$, with

$$H = \begin{bmatrix} 0 & 0 & 0 & 1 & 1 & 1 & 1 \\ 0 & 1 & 1 & 0 & 0 & 1 & 1 \\ 1 & 0 & 1 & 0 & 1 & 0 & 1 \end{bmatrix} \qquad (37)$$

Here we have taken the columns in the natural order (36) of increasing binary numbers. To get H in the standard form of (2), we take the columns in a different order:

$$H' = \begin{bmatrix} 0 & 1 & 1 & 1 & 1 & 0 & 0 \\ 1 & 0 & 1 & 1 & 0 & 1 & 0 \\ 1 & 1 & 0 & 1 & 0 & 0 & 1 \end{bmatrix}. \qquad (38)$$

In general,

$$H' = [A \mid I_r] \qquad (39)$$

where A contains all columns with at least two 1's.

Obviously changing the order of the columns doesn't affect the number of errors a code can correct or its error probability.

Definition. Two codes are called *equivalent* if they differ only in the order of the symbols. E.g.

0000	0000
0011,	0101
1100	1010
1111	1111

are equivalent [4, 2, 2] codes. So H and H' give equivalent codes.

Problems. (26) Show that the generator matrices

$$G = \begin{bmatrix} 1 & 1 & & \\ & 1 & 1 & \\ & & 1 & 1 \end{bmatrix}, \qquad G' = \begin{bmatrix} 1 & & & 1 \\ & 1 & & 1 \\ & & 1 & 1 \end{bmatrix}$$

generate equivalent codes.
   (27) Show that

$$G = \begin{bmatrix} 11 & & \\ & 11 & \\ & & 11 \end{bmatrix}, \qquad G' = \begin{bmatrix} 1 & 1 & 1 & 1 & 1 & 1 \\ & 1 & 1 & & 1 & 1 \\ & & 1 & & & 1 \end{bmatrix}$$

generate equivalent codes.
   (28) Show that the Hamming code $\mathcal{H}_r$ is unique, in the sense that any linear code with parameters $[2^r - 1, 2^r - 1 - r, 3]$ is equivalent to $\mathcal{H}_r$.

There may be very good engineering or aesthetic reasons for preferring one code to another which is equivalent to it. For example a third parity check matrix H for code #5 is

$$H'' = \begin{bmatrix} 1 & 1 & 1 & 0 & 1 & 0 & 0 \\ 0 & 1 & 1 & 1 & 0 & 1 & 0 \\ 0 & 0 & 1 & 1 & 1 & 0 & 1 \end{bmatrix} \tag{40}$$

– the same columns, but in yet another order. Now the code is *cyclic*, i.e. a cyclic end-around shift of a codeword is again a codeword. We shall see later (Ch. 7) that binary Hamming codes can always be made cyclic.

Problem. (29) *Code #7, the Hamming* [15, 11, 3] *code* $\mathcal{H}_4$. Give the three forms of H, corresponding to (37), (38) and (40). Using the (38) form, encode the message $u = 11111100000$, and decode the vector 111000111000111.

Decoding. Suppose we use H in the form (37), with the ith column $H_i =$ binary representation of i. If there is a single error in the l-th symbol, then from (22) the syndrome is $S = Hy^{\text{tr}} = He^{\text{tr}} = H_l =$ binary representation of l. So decoding is easy! (It will never be this easy again.)

Since the code can correct any single error, it has minimum distance $d \geqslant 3$ (by Theorem 2). In fact d is equal to 3, for it's easy to find codewords of weight 3 (e.g. $11100 \cdots 0$ if H is (37)). See also Theorem 10 below.

Theorem 8. *The Hamming codes are perfect single-error-correcting codes.*

Proof. Since the code can correct single errors, the spheres of radius 1 around the codewords are disjoint. There are $n + 1 = 2^r$ vectors in each sphere, and there are $2^k = 2^{2^r - 1 - r}$ spheres, giving a total of $2^{2^r - 1} = 2^n$ vectors. So every vector of length n is in one of the spheres and the code is perfect. Q.E.D.

Summary of Properties of Hamming Code $\mathcal{H}_r$

length $n = 2^r - 1$ (for $r = 2, 3, \ldots$),

dimension $k = 2^r - 1 - r$,

number of parity checks $= r$,

minimum distance $d = 3$.

$\mathcal{H}_r$ is a perfect single-error-correcting code, and is unique up to equivalence. The parity check matrix H is an $r \times n$ matrix whose columns are all nonzero r-tuples.

Fig. 1.10.

Problems. (30) Write down a generator matrix G for $\mathcal{H}_r$ and use it to show every nonzero codeword has weight ≥ 3. [Hint: if a codeword has weight ≤ 2 it must be the sum of ≤ 2 rows of G.]

(31) Show that the distribution of coset leaders for a Hamming code is $\alpha_0 = 1$, $\alpha_1 = n$. What is the error probability P_{err}?

§8. The Dual Code

If $u = u_1 \cdots u_n$, $v = v_1 \cdots v_n$ are vectors (with components from a field F), their *scalar product* is

$$u \cdot v = u_1 v_1 + \cdots + u_n v_n \tag{41}$$

(evaluated in F). For example the binary vectors $u = 1101$, $v = 1111$ have $u \cdot v = 1 + 1 + 0 + 1 = 1$.

If $u \cdot v = 0$, u and v are called *orthogonal*.

Problem. (32) For binary vectors $u \cdot v = 0$ iff wt$(u * v)$ is even, $= 1$ iff wt$(u * v)$ is odd. Also $u \cdot u = 0$ iff wt(u) is even.

As this problem shows, the scalar product in finite fields has rather different properties from the scalar product of real vectors used in physics.

Definition. If $\mathcal{C}$ is an $[n, k]$ linear code over F, its *dual* or *orthogonal code* $\mathcal{C}^\perp$ is the set of vectors which are orthogonal to all codewords of $\mathcal{C}$:

$$\mathcal{C}^\perp = \{u \mid u \cdot v = 0 \text{ for all } v \in \mathcal{C}\}. \tag{42}$$

Thus from §2, $\mathcal{C}^\perp$ is exactly the set of all parity checks on $\mathcal{C}$. If $\mathcal{C}$ has generator matrix G and parity check matrix H, then $\mathcal{C}^\perp$ has generator matrix $= H$, and parity check matrix $= G$. Thus $\mathcal{C}^\perp$ is an $[n, n - k]$ code. $\mathcal{C}^\perp$ is the orthogonal subspace to $\mathcal{C}$. (We shall discuss the minimum distance of $\mathcal{C}^\perp$ in Ch. 5.)

Problems. (33) (a) Show that $(\mathcal{C}^\perp)^\perp = \mathcal{C}$. (b) Let $\mathcal{C} + \mathcal{D} = \{u + v : u \in \mathcal{C}, v \in \mathcal{D}\}$. Show that $(\mathcal{C} + \mathcal{D})^\perp = \mathcal{C}^\perp \cap \mathcal{D}^\perp$.

(34) Show the dual of the $[n, 1, n]$ binary repetition code (# 2) is the $[n, n - 1, 2]$ even weight code (# 3).

If $\mathcal{C} \subset \mathcal{C}^\perp$, we call $\mathcal{C}$ *weakly self dual* (w.s.d.), while if $\mathcal{C} = \mathcal{C}^\perp$, $\mathcal{C}$ is called *(strictly) self dual*.

Thus $\mathcal{C}$ is w.s.d. if $u \cdot v = 0$ for every pair of (not necessarily distinct) codewords in $\mathcal{C}$. $\mathcal{C}$ is self-dual if it is w.s.d. and has dimension $k = \frac{1}{2}n$ (so n must be even).

For example the binary repetition code # 2 is w.s.d. iff n is even. When $n = 2$, the repetition code $\{00, 11\}$ is self-dual. So is the ternary code # 6.

Problems. (38) Construct binary self-dual codes of lengths 4 and 8.

(36) If n is odd, let $\mathscr{C}$ be an $[n, \frac{1}{2}(n-1)]$ w.s.d. binary code. Show $\mathscr{C}^\perp = \mathscr{C} \cup \{1 + \mathscr{C}\}$, where **1** is the vector of all 1's.

(37) Show that the code with parity check matrix $H = [A \mid I]$ over any field is strictly self-dual iff A is square matrix such that $AA^{tr} = -I$.

(38) If $\mathscr{C}$ is a binary w.s.d. code, show that every codeword has even weight. Furthermore, if each row of the generator matrix of $\mathscr{C}$ has weight divisible by 4, then so does every codeword.

(39) If $\mathscr{C}$ is a ternary w.s.d. code, show that every codeword has Hamming weight divisible by 3.

§9. Construction of new codes from old (II)

(I) *Adding an overall parity check.* Let $\mathscr{C}$ be an $[n, k, d]$ binary code in which some codewords have odd weight. We form a new code $\hat{\mathscr{C}}$ by adding a 0 at the end of every codeword of $\mathscr{C}$ with even weight, and a 1 at the end of every codeword with odd weight. $\hat{\mathscr{C}}$ has the property that every codeword has even weight, i.e. it satisfies the new parity check equation

$$x_1 + x_2 + \cdots + x_{n+1} = 0,$$

the "overall" parity check.

From (12), the distance between every pair of codewords is now even. If the minimum distance of $\mathscr{C}$ was odd, the new minimum distance is $d + 1$, and $\hat{\mathscr{C}}$ is an $[n + 1, k, d + 1]$ code. This technique, of adding more check symbols, is generally called *extending* a code.

If $\mathscr{C}$ has parity check matrix H, $\hat{\mathscr{C}}$ has parity check matrix

$$\hat{H} = \left[\begin{array}{cc} 1 \ 1 \cdots 1 & \\ & 0 \\ H & \vdots \\ & 0 \end{array} \right]$$

Example. *Code # 8.* Adding an overall parity check to code # 5 gives the $[8, 4, 4]$ *extended Hamming code* with

$$\hat{H} = \begin{bmatrix} 1 & 1 & 1 & 1 & 1 & 1 & 1 & 1 \\ 0 & 0 & 0 & 1 & 1 & 1 & 1 & 0 \\ 0 & 1 & 1 & 0 & 0 & 1 & 1 & 0 \\ 1 & 0 & 1 & 0 & 1 & 0 & 1 & 0 \end{bmatrix} \tag{43}$$

locations 1 2 3 4 5 6 7 0

Since this has $d = 4$, according to Theorem 2 it can correct any single error and detect any double error. Indeed, recalling that the syndrome S is the sum of the columns of $\hat{H}$ where the errors occurred, we have the following decoding scheme. If there are no errors, $S = 0$. If there is a single error, at location i, then

$$S = \begin{pmatrix} 1 \\ x \\ y \\ z \end{pmatrix},$$

where (xyz) is the binary representation of i (see (43)). Finally if there are two errors,

$$S = \begin{pmatrix} 0 \\ x \\ y \\ z \end{pmatrix},$$

and the decoder detects that two (or more) errors have occurred.

Problems. (40) Show that code # 8 is strictly self-dual.

(41) Show that the extended Hamming code is unique (in the same sense as Problem 28).

For nonbinary codes the same technique may or may not increase the minimum distance.

(42) If one adds an overall parity check (i.e. make the codewords satisfy $\Sigma_i x_i = 0 \bmod 3$) to the ternary codes with generator matrices

$$\begin{bmatrix} 11000 \\ 00111 \end{bmatrix} \quad \text{and} \quad \begin{bmatrix} 12000 \\ 00122 \end{bmatrix},$$

what happens to the minimum distance?

(II) *Puncturing a code by deleting coordinates.* The inverse process to extending a code $\mathscr{C}$ is called *puncturing*, and consists of deleting one or more coordinates from each codeword. E.g. puncturing the $[3, 2, 2]$ code # 9,

$$\text{code } \#9 \quad \begin{matrix} 0 \ 0 \ 0 \\ 0 \ 1 \ 1 \\ 1 \ 0 \ 1 \\ 1 \ 1 \ 0 \end{matrix}$$

by deleting the last coordinate gives the $[2, 2, 1]$ code

$$\begin{matrix} 0 \ 0 \\ 0 \ 1 \\ 1 \ 0 \\ 1 \ 1 \end{matrix}$$

The punctured code is usually denoted by $\mathscr{C}^*$.
In general each time a coordinate is deleted, the length n drops by 1, the number of codewords is unchanged, and (unless we are very lucky) the minimum distance d drops by 1.

(III) *Expurgating by throwing away codewords.* The commonest way to *expurgate* a code is the following. Suppose $\mathscr{C}$ is an $[n, k, d]$ binary code containing codewords of both odd and even weight. Then it follows from Problem 10 that half the codewords have even weight and half have odd weight. We expurgate $\mathscr{C}$ by throwing away the codewords of odd weight to get an $[n, k - 1, d']$ code. Often $d' > d$ (for instance if d is odd).

Example. Expurgating the $[7, 4, 3]$ code #5 gives a $[7, 3, 4]$ code.

(IV) *Augmenting by adding new codewords.* The commonest way to *augment* a code is by adding the all-ones vector **1**, provided it is not already in the code. This is the same as adding a row of 1's to the generator matrix. If $\mathscr{C}$ is an $[n, k, d]$ binary code which does not contain **1**, the augmented code is

$$\mathscr{C}^{(a)} = \mathscr{C} \cup \{\mathbf{1} + \mathscr{C}\}.$$

I.e. $\mathscr{C}^{(a)}$ consists of the codewords of $\mathscr{C}$ and their complements, and is an $[n, k + 1, d^{(a)}]$ code, where

$$d^{(a)} = \min\{d, n - d'\},$$

and d' is the largest weight of any codeword of $\mathscr{C}$.

Example. Augmenting code #9 gives the $[3, 3, 1]$ code consisting of all vectors of length 3.

(V) *Lengthening by adding message symbols.* The usual way to *lengthen* a code is to augment it by adding the codeword **1**, and then to extend it by adding an overall parity check. This has the effect of adding one more message symbol.

(VI) *Shortening a code by taking a cross-section.* An inverse operation to the lengthening process just described is to take the codewords which begin $x_1 = 0$ and delete the x_1 coordinate. This is called taking a cross-section of the code, and will be used in later chapters to shorten nonlinear codes.

Dual of Hamming code. We illustrate these six operations by performing them on the Hamming code (Fig. 1.11) and its dual (Fig. 1.13).

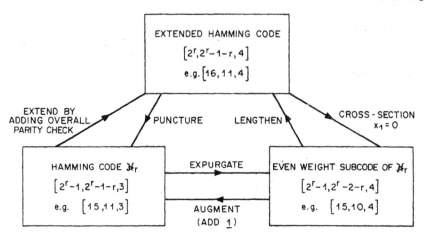

Fig. 1.11. Variations on a Hamming code.

Definition. The binary *simplex code* $\mathscr{S}_r$ is the dual of the Hamming code $\mathscr{H}_r$. By §8 we know $\mathscr{S}_r$ is a $[2^r - 1, r]$ code with generator matrix G_r which is the parity check matrix of $\mathscr{H}_r$. E.g. for $\mathscr{S}_2$,

$$G_2 = \begin{bmatrix} 011 \\ 101 \end{bmatrix}$$

and the codewords are

$$\mathscr{S}_2 = \begin{matrix} 000 \\ 011 \\ 101 \\ 110 \end{matrix}.$$

For $\mathscr{S}_3$,

$$G_3 = \begin{bmatrix} 000 & | & 1111 \\ \hline 011 & | & 0011 \\ 101 & | & 0101 \end{bmatrix},$$

$$= \begin{bmatrix} 000 & | & 1 & | & 111 \\ \hline G_2 & | & 0 & | & G_2 \end{bmatrix},$$

and the codewords are

$$\mathscr{S}_3 = \begin{array}{c|c|c}
\begin{matrix} 000 \\ 011 \\ 101 \\ 110 \\ \hline 000 \\ 011 \\ 101 \\ 110 \end{matrix} &
\begin{matrix} 0 \\ 0 \\ 0 \\ 0 \\ \hline 1 \\ 1 \\ 1 \\ 1 \end{matrix} &
\begin{matrix} 000 \\ 011 \\ 101 \\ 110 \\ \hline 111 \\ 100 \\ 010 \\ 001 \end{matrix}
\end{array}
\quad
\begin{matrix} & \\ & \mathscr{S}_2 \\ & \\ \hline & \\ & \mathscr{S}_2 \\ & \end{matrix}
\;=\;
\begin{array}{c|c}
\begin{matrix} \mathscr{S}_2 \\ \hline \mathscr{S}_2 \end{matrix} &
\begin{matrix} 0 \\ 0 \\ 0 \\ 0 \\ \hline 1 \\ 1 \\ 1 \\ 1 \end{matrix}
\end{array}
\begin{matrix} \mathscr{S}_2 \\ \hline \bar{\mathscr{S}}_2 \end{matrix}.$$

We see by induction that $\mathcal{S}_r$ consists of **0** and $2^r - 1$ codewords of weight 2^{r-1}. (The reader may recognize this inductive process as one of the standard ways of building Hadamard matrices – more about this in the next chapter.)

This is called a simplex code, because every pair of codewords is the same distance apart. So if the codewords were marked at the vertices of a unit cube in n dimensions, they would form a regular simplex. E.g. when $r = 2$, $\mathcal{S}_2 =$ code #9 forms a regular tetrahedron (the double lines in Fig. 1.12)

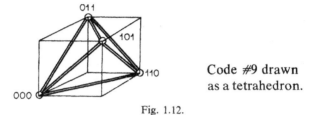

Code #9 drawn
as a tetrahedron.

Fig. 1.12.

The simplex code $\mathcal{S}_r$ will also reappear later under the name of a *maximal-length feedback shift register code* (see §4 of Ch. 3 and Ch. 14).

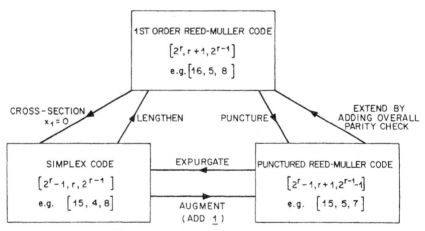

Fig. 1.13. Variations on the simplex code.

The dual of the extended Hamming code is also an important code, for it is a *first-order Reed–Muller* code (see Ch. 13). It is obtained by lengthening $\mathcal{S}_r$ as described in (V). For example lengthening $\mathcal{S}_3$ in this way we obtain the code in Fig. 1.14.

Problem. (43) A *signal set* is a collection of real vectors $x = (x_1 \cdots x_n)$. Define $x \cdot y = \sum_{i=1}^{n} x_i y_i$ (evaluated as a real number) and call $x \cdot x$ the *energy* of x. The unit vectors $s^{(1)}, \ldots, s^{(n)}$ (where $s^{(i)}$ has a 1 in the ith component and 0

```
0 0|0 0|0 0 0 0
0 1|0 1|0 1 0 1
0 0 1 1|0 0 1 1
0 1 1 0|0 1 1 0
0 0 0 0 1 1 1 1
0 1 0 1 1 0 1 0
0 0 1 1 1 1 0 0
0 1 1 0 1 0 0 1
1 1 1 1 1 1 1 1
1 0 1 0 1 0 1 0
1 1 0 0 1 1 0 0
1 0 0 1 1 0 0 1
1 1 1 1 0 0 0 0
1 0 1 0 0 1 0 1
1 1 0 0 0 0 1 1
1 0 0 1 0 1 1 0
```

Fig. 1.14. Code # 10, an [8, 4, 4] 1st order Reed–Muller Code.

elsewhere) form an *orthogonal* signal set, since $s^{(i)} \cdot s^{(j)} = \delta_{ij}$. Consider the translated signal set $\{t^{(i)} = s^{(i)} - a\}$. Show that the total energy $\sum_{i=1}^{n} t^{(i)} \cdot t^{(i)}$ is minimized by choosing

$$a = \left(\frac{1}{n}, \dots, \frac{1}{n} \right).$$

The resulting $\{t^{(i)}\}$ is called a *simplex* set (and is the continuous analog of the binary simplex code described above). The *biorthogonal* signal set $\{\pm s^{(i)}\}$ is the continuous analog of the first order Reed–Muller code.

§10. Some general properties of a linear code

To conclude this chapter we give several important properties of linear codes. The first three apply to linear codes over any field.

Theorem 9. *If H is the parity check matrix of a code of length n, then the code has dimension $n - r$ iff some r columns of H are linearly independent but no $r + 1$ columns are. (Thus r is the rank of H).*

This requires no proof.

Theorem 10. *If H is the parity check matrix of a code of length n, then the code has minimum distance d iff every d − 1 columns of H are linearly independent and some d columns are linearly dependent.*

Proof. There is a codeword x of weight w iff $Hx^{tr} = 0$ for some vector x of weight w iff some w columns of H are linearly dependent.   Q.E.D.

Theorem 10 gives another proof that Hamming codes have distance 3, for the columns of H are all distinct and therefore any 2 are independent, and there are three columns which are dependent.

Problem. (44) Let G be the generator matrix for an $[n, k, d]$ code $\mathscr{C}$. Show that any k linearly independent columns of G may be taken as information symbols – in other words, there is another generator matrix for $\mathscr{C}$ in which these k columns form a unit matrix.

Theorem 11. (The Singleton bound.) *If $\mathscr{C}$ is an $[n, k, d]$ code, then $n − k \geq d − 1$.*

1ˢᵗ Proof. $r = n − k$ is the rank of H and is the maximum number of linearly independent columns.

2ⁿᵈ Proof. A codeword with only one nonzero information symbol has weight at most $n − k + 1$. Therefore $d \leq n − k + 1$.   Q.E.D.

Codes with $r = d − 1$ are called *maximum distance separable* (abbreviated MDS), and are studied in Chapter 11.

Theorems 6 and 11 have provided upper bounds on the size of a code with given minimum distance. Our final theorem is a lower bound, which says that good linear codes do in fact exist.

Theorem 12. (The Gilbert–Varshamov bound.) *There exists a binary linear code of length n, with at most r parity checks and minimum distance at least d, provided*

$$1 + \binom{n-1}{1} + \cdots + \binom{n-1}{d-2} < 2^r. \tag{44}$$

Proof. We shall construct an $r \times n$ matrix H with the property that no $d − 1$ columns are linearly dependent. By Theorem 10, this will establish the theorem. The first column can be any nonzero r-tuple. Now suppose we have chosen i columns so that no $d-1$ are linearly dependent. There are at most

$$\binom{i}{1} + \cdots + \binom{i}{d-2}$$

distinct linear combinations of these i columns taken $d-2$ or fewer at a time. Provided this number is less than $2^r - 1$ we can add another column different from these linear combinations, and keep the property that any $d-1$ columns of the new $r \times (i+1)$ array are linearly independent. We keep doing this as long as

$$1 + \binom{i}{1} + \cdots + \binom{i}{d-2} < 2^r. \qquad\qquad \text{Q.E.D.}$$

For large n see Theorem 30 and Fig. 17.7 of Ch. 17.

Problem. (45) The Gilbert–Varshamov bound continued. Prove that there exists a linear code over a field of q elements, having length n, at most r parity checks, and minimum distance at least d, provided

$$\sum_{i=0}^{d-2} (q-1)^i \binom{n-1}{i} < q^r.$$

§11. Summary of Chapter 1

An $[n, k, d]$ binary linear code contains 2^k codewords $x = x_1 \cdots x_n$, $x_i = 0$ or 1, and any two codewords differ in at least d places. The code is defined either as those codewords x such that $Hx^{\text{tr}} = 0$, where H is a parity check matrix, or as all linear combinations of the rows of a generator matrix G. Such a code can correct $[\frac{1}{2}(d-1)]$ errors. Maximum likelihood decoding is done using a standard array. At rates below the capacity of the channel, the error probability can be made arbitrarily small by using sufficiently long codes.

A binary Hamming code $\mathcal{H}_r$ is a perfect single-error-correcting code with parameters $[n = 2^r - 1, k = 2^r - 1 - r, d = 3]$, $r \geq 2$, and is constructed from a parity check matrix H whose columns are all $2^r - 1$ distinct nonzero binary r-tuples.

Notes on Chapter 1

§1. The excellent books by Abramson [3], Gallager [464], Golomb [522] and Wozencraft and Jacobs [1433] show in more detail how codes fit into communication systems.

The following papers deal with the more practical aspects of coding theory and the use of codes on real channels. Asabe et al. [31], Baumert and McEliece [88, 89], Blythe and Edgcombe [167], Borel [171], Brayer et al. [191–194], Buchner [209], Burton [214], Burton and Sullivan [217], Chase [264, 265], Chien et al. [282, 287], Corr [309], Cowell [311], Dorsch [380], Elliott [408], Falconer [414], Forney [438], Franaszek [448], Franco and Saporta [449], Fredricksen [450], Freiman & Robinson [458], Frey and Kavanaugh [460], Goodman and Farrell [535], Hellman [639], Hsu and Kasami [671], the special issues [678, 679], Jacobs [686], Kasami et al. [735, 742], Klein and Wolf [769], Lerner [818], Murthy [977], Posner [1071], Potton [1073], Ralphs [1088], Rocher and Pickholtz [1120], Rogers [1121], Schmandt [1156], Tong [1333–1336], Townsend and Watts [1337], Tröndle [1340] and Wright [1434].

The following are survey articles on coding theory: Assmus and Mattson [47], Bose [176], Dobrushin [378], Goethals [494], Jacobs [686], Kasami [724], Kautz [749], Kautz and Levitt [753], Kiyasu [762], MacWilliams [876], Sloane [1225], Viterbi [1374], Wolf [1429], Wyner [1439] and Zadeh [1450].

We have described the channel as being a communication path, such as a telegraph line, but codes can be applied equally well in other situations, for example when data is stored in a computer and later retrieved. Various applications of codes to computers are described by Brown and Sellers [203], Chien [278, 281, 284], Davydov [337], Davydov and Tenengol'ts [338], Fischler [430], Hong and Patel [662], Hsiao et al. [668–670], Kasahara et al. [720], Lapin [794], Malhotra and Fisher [890], Oldham et al. [1011], Patel and Hong [1029] and Sloane [1230].

We usually consider only a binary symmetric channel, which has no memory from one symbol to the next, and doesn't lose (or add) symbols. Most real channels are not like this, but have bursts of noise, and lose synchronization. (For example, one channel has been described as "a very good channel, with errors predominantly due to a noisy Coke machine near the receiver" [438].) A lot of work has been done on using codes for synchronizing – see for example Bose and Caldwell [182], Brown [202], Calabi and Hartnett [226, 227], Eastman and Even [400], Freiman [456], Freiman and Wyner [459], Gilbert [482], Golomb et al. [526, 527], Hatcher [627], Hellman [640], Kautz [751], Levenshtein [820–822], Levy [829], Mandelbaum [900], McEliece [943], Pierce [1045], Rudner [1129], Scholtz et al. [1163–1166], Shimo et al. [1197], Shiva and Seguin [1203], Stanley and Yoder [1263], Stiffler [1278–1280], Tanaka and Kasai [1302], Tavares and Fukuda [1312, 1313], Tong [1333], Ullman [1358] and Varshamov [1362–1365]. See also the papers on comma-free codes mentioned in the introduction. Burst-correcting codes are discussed in Ch. 10.

§2. Other kinds of codes. Our codes are *block* codes: a block of message symbols becomes a codeword. Very powerful rival codes, not considered here, are convolutional and tree codes, which encode the message symbols continuously without breaking them up into blocks. These can be very efficiently decoded either by *sequential decoding* or by the *Viterbi algorithm*. See Forney [439, 441–444], Forney and Bower [445], Heller and Jacobs [638], Jelinek [691], Viterbi [1372, 1373] and Wu [1435, 1436].

On the other hand cryptographic codes have little in common with our codes: their goal is to conceal information, whereas our codes have the opposite aim. See Feistel et al. [422, 423], Gaines [463], Geffe [469, 470], Gilbert et al. [484], Hellman [640a], Kahn [707] and Shannon [1190].

Another class of codes are those for correcting errors in arithmetic operations (see for example Peterson and Weldon [1040] and Rao [1093]).

§3. Slepian [1224] calls Fig. 1.3 the information theorist's coat-of-arms. For binomial coefficients see Riordan [1114]. Problem 19 is from Shiva [1199].

§4. Linear codes. The basic papers are by Slepian [1217–1219]. To help with synchronizing, sometimes a coset of the code is used rather than the code itself – see Posner [1071, Fig. VII].

§5. Theorem 6 is due to Hamming [592]. For lots more about the error probability of codes see for example Batman and McEliece [80], Cain and Simpson [224], Crimmins et al. [317–319], Gallager [464], Jelinek [690], Leont'ev [817], Posner [1070], Redinbo and Wolf [1103], Slepian [1220], Sullivan [1292] and Wyner [1437, 1438]. Hobbs [656] gives an approximation to the distribution of coset leaders of any code.

§6. For a proof of Shannon's Theorem (Shannon [1188, 1189]) see Gallager [464, §6.2].

Remark. Incomplete decoding (see Fig. 1.8) which corrects all error patterns of weight $\leq l = [(d - 1)/2]$ and no others is called *bounded distance decoding*. It follows from the Elias or McEliece–Rodemich–Rumsey–Welch bounds (Theorem 34, 35, or 37 of Ch. 17) that bounded distance decoding does not achieve channel capacity – for details see Wyner [1437] or Forney [437]. On the other hand only a slightly more complicated decoding scheme *will* achieve capacity (e.g. Abramson [3, p. 167].

§7. Hamming codes were discovered by Golay [506, 514a] and Hamming [592]; see also Problem 8 of Ch. 7.

§8. In the language of vector spaces, a vector u is called *isotropic* if $u \cdot u = 0$, and a weakly self dual code is called *totally isotropic*. (See for example Lam [790]).

§9. Deza [373], Farber [416], Hall [580], Hall et al. [581], Landau and Slepian [792], Van Lint [852] and Tanner [1308, 1309] study simplex codes. For more about Problem 43 see Wozencraft and Jacobs [1433, p. 257].

§10. Although Theorem 11 is nowadays usually called the Singleton bound (referring to Singleton [1214]), Joshi [700] attributes this result to Komamiya [775]. It applies also to nonlinear codes (Problem 3 of Ch. 11 and Problem 17 of Ch. 17). For Theorem 12 see Gilbert [479], Varshamov [1362] and Sacks [1140].

2

Nonlinear codes, Hadamard matrices, designs and the Golay code

§1. Nonlinear codes

One basic purpose of codes is to correct errors on noisy communication channels, and for this purpose linear codes, introduced in Ch. 1, have many practical advantages. But if we want to obtain the largest possible number of codewords with a given minimum distance, we must sometimes use nonlinear codes.

For example, suppose we want a double-error-correcting code of length 11. The largest linear code has 16 codewords, whereas there is a nonlinear code, shown in Fig. 2.1, which contains 24 codewords, an increase of 50%. (This is a Hadamard code, see §3).

Our notation for nonlinear codes is given by the following.

Definition. An (n, M, d) code is a set of M vectors of length n (with components from some field F) such that any two vectors differ in at least d places, and d is the largest number with this property.

In this chapter all codes are binary. Note that square brackets denote a linear code, while round parentheses are used for a code which may or may not be linear. An $[n, k, d]$ binary linear code is an $(n, 2^k, d)$ code.

We usually assume that there is no coordinate place in which every

0	1	2	3	4	5	6	7	8	9	10
0	0	0	0	0	0	0	0	0	0	0
1	1		1	1	1				1	
	1	1		1	1	1				1
1		1	1		1	1	1			
	1		1	1		1	1	1		
		1		1	1		1	1	1	
			1		1	1		1	1	1
1				1		1	1		1	1
1	1				1		1	1		1
1	1	1				1		1	1	
	1	1	1				1		1	1
1		1	1	1				1		1
		1				1	1	1		1
1			1				1	1	1	
	1			1				1	1	1
1		1			1				1	1
1	1		1			1				1
1	1	1		1			1			
	1	1	1		1			1		
		1	1	1		1			1	
			1	1	1		1			1
1				1	1	1		1		
	1				1	1	1		1	
1	1	1	1	1	1	1	1	1	1	1

The first twelve rows form the (11, 12, 6) Hadamard code $\mathscr{A}_{12}$;
all 24 rows form the (11, 24, 5) Hadamard code $\mathscr{B}_{12}$

Fig. 2.1.

codeword is zero (otherwise it would be an $(n-1, M, d)$ code). Also, since the distances between codewords are unchanged if a constant vector is added to all the codewords, we may, if we wish, assume that the code contains the zero vector.

We say that two (n, M, d) binary codes $\mathscr{C}$ and $\mathscr{D}$ are *equivalent* if one can be obtained from the other by permuting the n symbols and adding a constant vector, or more formally if there is a permutation π and a vector a such that $\mathscr{D} = \{\pi(u) + a : u \in \mathscr{C}\}$. If $\mathscr{C}$ and $\mathscr{D}$ are linear this reduces to the definition of equivalence given in §7 of Ch. 1. As an example,

$$
\begin{array}{cc}
000 & 111 \\
011 & 100 \\
101 \quad \text{and} & 010 \\
110 & 001
\end{array}
$$

are equivalent codes (take $a = 111$).

[More generally, the equivalence of nonbinary codes is defined as follows. Let $\mathscr{C}$ and $\mathscr{D}$ be codes of length n over a field of q elements. Then $\mathscr{C}$ and $\mathscr{D}$ are *equivalent* if there exist n permutations $\pi_1, \ldots, \pi_n$ of the q elements and a permutation σ of the n coordinate positions such that

$$
\text{if } (u_1, \ldots, u_n) \in \mathscr{C} \quad \text{then} \quad \sigma(\pi_1(u_1), \ldots, \pi_n(u_n)) \in \mathscr{D}.
$$

In words, the field elements in each coordinate place are permuted and then the coordinate places themselves are permuted. If $\mathscr{C}$ and $\mathscr{D}$ are both linear only those π_i generated by scalar multiples and field automorphisms (§6 of Ch. 4) can be used.]

Definition. If $\mathscr{C}$ is an (n, M, d) code, let A_i be the number of codewords of weight i. The numbers $A_0, A_1, \ldots, A_n$ are called the *weight distribution* of $\mathscr{C}$. Of course $A_0 + A_1 + \cdots + A_n = M$. We have already used the weight distribution in calculating error probabilities in Ch. 1.

If $\mathbf{0}$ is a codeword, then an observer sitting on $\mathbf{0}$ would see A_i codewords at distance i from him. For a linear code the view would be the same from any codeword. (Why?) However, in a nonlinear code this need not be true, as shown by the code $\{00, 01, 11\}$. (If it *is* true, the code is said to be *distance invariant*. The Nordstrom–Robinson code (§8) is distance invariant, and other examples will be given in Ch. 6.) Therefore for a nonlinear code it is useful to consider the average number of codewords at distance i from a fixed codeword.

Definition. The *distance distribution* of $\mathscr{C}$ consists of the numbers $B_0, B_1, \ldots, B_n$, where

$$
B_i = \frac{1}{M} \cdot (\text{number of ordered pairs of codewords} \\
u, v \text{ such that dist}(u, v) = i).
$$

Note that $B_0 = 1$ and $B_0 + B_1 + \cdots + B_n = M$. For linear codes the weight and distance distributions coincide. Also a translated code $a + \mathscr{C}$ has the same distance distribution as $\mathscr{C}$.

It is helpful to think of these codes geometrically. A binary vector $(a_1, \ldots, a_n)$ of length n gives the coordinates of a vertex of a unit cube in n

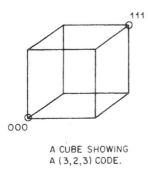

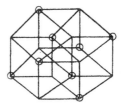

A CUBE SHOWING
A (3,2,3) CODE.

A 4-DIMENSIONAL CUBE
OR TESSERACT, SHOWING
A (4,8,2) CODE.

Fig. 2.2.

dimensions. Then an (n, M, d) code is just a subset of these vertices (Fig. 2.2).

In this geometrical language the coding theory problem is to choose as many vertices of the cube as possible while keeping them a certain distance apart.

This is a packing problem, for if the code has minimum distance d, the Euclidean distance between codewords is $\geq \sqrt{d}$. So finding an (n, M, d) code means finding M nonoverlapping spheres of radius $\frac{1}{2}\sqrt{d}$ with centers at the vertices of the cube.

Aside: **Research Problem 2.1.** The analogous problem of placing M points on the surface of a unit *sphere* in n dimensions is also unsolved. In other words, where should M misanthropes build houses on the surface of a planet, so as to maximize the smallest distance between any two houses? This problem is also important for communication theory – it is the problem of designing the best signals for transmitting over a band-limited channel.

§2. The Plotkin bound

Theorem 1. (The Plotkin bound.) *For any (n, M, d) code $\mathscr{C}$ for which $n < 2d$, we have*

$$M \leq 2\left[\frac{d}{2d - n}\right]. \tag{1}$$

Proof. We shall calculate the sum

$$\sum_{u \in \mathscr{C}} \sum_{v \in \mathscr{C}} \text{dist}(u, v)$$

in two ways. First, since $\text{dist}(u, v) \geq d$ if $u \neq v$, the sum is $\geq M(M - 1)d$. On the other hand, let A be the $M \times n$ matrix whose rows are the codewords. Suppose the ith column of A contains x_i 0's and $M - x_i$ 1's. Then this column contributes $2x_i(M - x_i)$ to the sum, so that the sum is equal to

$$\sum_{i=1}^{n} 2x_i(M - x_i).$$

If M is even this expression is maximized if all $x_i = \frac{1}{2}M$, and the sum is $\leq \frac{1}{2}nM^2$. Thus we have

$$M(M - 1)d \leq \frac{nM^2}{2},$$

or

$$M \leq \frac{2d}{2d - n}. \tag{2}$$

But M is even, so this implies

$$M \leq 2\left[\frac{d}{2d - n}\right].$$

On the other hand if M is odd, the sum is $\leq n(M^2 - 1)/2$, and instead of (2) we get

$$M \leq \frac{n}{2d - n} = \frac{2d}{2d - n} - 1$$

This implies

$$M \leq \left[\frac{2d}{2d - n}\right] - 1 \leq 2\left[\frac{d}{2d - n}\right],$$

using $[2x] \leq 2[x] + 1$.   Q.E.D.

Example. The top half of Fig. 2.1 forms an $(11, 12, 6)$ code ($\mathcal{A}_{12}$ in the notation of §3) for which equality holds in (1).

Problems. (1) Does there exist a $(16, 10, 9)$ code?

(2) Construct a $(16, 32, 8)$ code. [Hint: from the proof of Theorem 1, each column must contain half 0's and half 1's.]

Let $A(n, d)$ denote the largest number M of codewords in any (n, M, d) code. The next theorem shows it is enough to find $A(n, d)$ for even values of d:

Theorem 2.

$$A(n, 2r - 1) = A(n + 1, 2r)$$

Proof. Let $\mathscr{C}$ be an $(n, M, 2r - 1)$ code. By adding an overall parity check (Ch. 1, §9), we get an $(n + 1, M, 2r)$ code, thus

$$A(n, 2r - 1) \leqslant A(n + 1, 2r).$$

Conversely, given an $(n + 1, M, 2r)$ code, deleting one coordinate gives an $(n, M, d \geqslant 2r - 1)$ code, thus

$$A(n, 2r - 1) \geqslant A(n + 1, 2r). \qquad \text{Q.E.D.}$$

Theorem 3.

$$A(n, d) \leqslant 2A(n - 1, d).$$

Proof. Given an (n, M, d) code, divide the codewords into two classes, those beginning with 0 and those beginning with 1. One class must contain at least half of the codewords, thus

$$A(n - 1, d) \geqslant \tfrac{1}{2}A(n, d). \qquad \text{Q.E.D.}$$

Corollary 4. (The Plotkin bound.) *If d is even and $2d > n$, then*

$$A(n, d) \leqslant 2\left[\frac{d}{2d - n}\right], \tag{3}$$

$$A(2d, d) \leqslant 4d. \tag{4}$$

If d is odd and $2d + 1 > n$, then

$$A(n, d) \leqslant 2\left[\frac{d + 1}{2d + 1 - n}\right], \tag{5}$$

$$A(2d + 1, d) \leqslant 4d + 4. \tag{6}$$

Proof. To prove (4), we have from Theorem 3 and Equation (1)

$$A(4r, 2r) \leqslant 2A(4r - 1, 2r) \leqslant 8r.$$

If d is odd, then by Theorem 2

$$A(n, d) = A(n + 1, d + 1) \leqslant 2\left[\frac{d + 1}{2d + 1 - n}\right].$$

Equation (6) follows similarly. Q.E.D.

If Hadamard matrices exist of all possible orders (which has not yet been proved), then in fact equality holds in Equations (3)–(6). Thus the Plotkin bound is *tight*, in the sense that there exist codes which meet this bound. This is Levenshtein's theorem, which is proved in the next section.

§3. Hadamard matrices and Hadamard codes

Definition. A *Hadamard matrix* H of order n is an $n \times n$ matrix of $+1$'s and -1's such that

$$HH^T = nI. \tag{7}$$

In other words the real inner product of any two distinct rows of H is zero, i.e., distinct rows are orthogonal, and the real inner product of any row with itself is n. Since $H^{-1} = (1/n)H^T$, we also have $H^TH = nI$, thus the columns have the same properties.

It is easy to see that multiplying any row or column by -1 changes H into another Hadamard matrix. By this means we can change the first row and column of H into $+1$'s. Such a Hadamard matrix is called *normalized*.

Normalized Hadamard matrices of orders $1, 2, 4, 8$ are shown in Fig. 2.3 where we have written $-$ instead of -1, a convention we use throughout the book.

$$n = 1: \quad H_1 = (1) \qquad\qquad n = 2: \quad H_2 = \begin{pmatrix} 1 & 1 \\ 1 & - \end{pmatrix}$$

$$n = 4: \quad H_4 = \begin{pmatrix} 1 & 1 & 1 & 1 \\ 1 & - & 1 & - \\ 1 & 1 & - & - \\ 1 & - & - & 1 \end{pmatrix} \qquad n = 8: \quad H_8 = \begin{pmatrix} 1 & 1 & 1 & 1 & 1 & 1 & 1 & 1 \\ 1 & - & 1 & - & 1 & - & 1 & - \\ 1 & 1 & - & - & 1 & 1 & - & - \\ 1 & - & - & 1 & 1 & - & - & 1 \\ 1 & 1 & 1 & 1 & - & - & - & - \\ 1 & - & 1 & - & - & 1 & - & 1 \\ 1 & 1 & - & - & - & - & 1 & 1 \\ 1 & - & - & 1 & - & 1 & 1 & - \end{pmatrix}$$

Fig. 2.3. Sylvester-type Hadamard matrices of order 1, 2, 4 and 8.

Theorem 5. *If a Hadamard matrix H of order n exists, then n is $1, 2$ or a multiple of* 4.

Proof. Without loss of generality we may suppose H is normalized. Suppose $n \geq 3$ and the composition of the first three rows of H is as follows:

$$
\begin{array}{cccc}
11\cdots 1 & 11\cdots 1 & 11\cdots 1 & 11\cdots 1 \\
11\cdots 1 & 11\cdots 1 & --\cdots- & --\cdots- \\
\underline{11\cdots 1} & \underline{--\cdots-} & \underline{11\cdots 1} & \underline{--\cdots-} \\
i & j & k & l
\end{array}
$$

Since the rows are orthogonal, we have

$$i + j - k - l = 0,$$
$$i - j + k - l = 0,$$
$$i - j - k + l = 0,$$

which imply $i = j = k = l$, thus $n = 4i$ is a multiple of 4.   Q.E.D.

It is conjectured that Hadamard matrices exist *whenever* the order is a multiple of 4, although this has not yet been proved. A large number of constructions are known, and the smallest order for which a Hadamard matrix has not been constructed is (in 1977) 268. We give two constructions which are important for coding theory.

Construction I. If H_n is a Hadamard matrix of order n, then it is easy to verify that

$$H_{2n} = \begin{pmatrix} H_n & H_n \\ H_n & -H_n \end{pmatrix}$$

is a Hadamard matrix of order $2n$. Starting with $H_1 = (1)$, this gives H_2, H_4, H_8 (as shown in Fig. 2.3), ... and so Hadamard matrices of all orders which are powers of two. These are called *Sylvester* matrices.

For the second construction we need some facts about quadratic residues.

Quadratic residues

Definition. Let p be an odd prime. The nonzero squares modulo p, i.e., the numbers $1^2, 2^2, 3^2, \ldots$ reduced mod p, are called the *quadratic residues mod p*, or simply the *residues mod p*.

To find the residues mod p it is enough to consider the squares of the numbers from 1 to $p - 1$. In fact since $(p - a)^2 \equiv (-a)^2 \equiv a^2 \pmod{p}$, it is enough to consider the squares

$$1^2, 2^2, \ldots, \left(\frac{p-1}{2}\right)^2 \pmod{p}.$$

These are all distinct, for if $i^2 \equiv j^2 \pmod{p}$, with $1 \le i, j \le \frac{1}{2}(p - 1)$, then p divides $(i - j)(i + j)$, which is only possible if $i = j$.

Therefore there are $\frac{1}{2}(p - 1)$ quadratic residues mod p. The $\frac{1}{2}(p - 1)$ remaining numbers mod p are called *nonresidues*. Zero is neither a residue nor a nonresidue.

For example if $p = 11$, the quadratic residues mod 11 are

$$1^2 = 1, \quad 2^2 = 4, \quad 3^2 = 9, \quad 4^2 = 16 \equiv 5, \quad \text{and } 5^2 = 25 \equiv 3$$

mod 11, i.e. 1, 3, 4, 5, and 9. The remaining numbers 2, 6, 7, 8 and 10 are the nonresidues. [The reader is reminded that

$$A \equiv B \text{ (modulo } C),$$

pronounced "A is congruent to B mod C", means that

$$C \mid A - B$$

where the vertical slash means "divides," or equivalently

$$A - B \text{ is a multiple of } C.$$

Then A and B are in the same *residue class* mod C.]

We now state some properties of quadratic residues.

(Q1) The product of two quadratic residues or of two nonresidues is a quadratic residue, and the product of a quadratic residue and a nonresidue is a nonresidue. The proof is left as an exercise.

(Q2) If p is of the form $4k + 1$, -1 is a quadratic residue mod p. If p is of the form $4k + 3$, -1 is a nonresidue mod p.

The proof is postponed until Ch. 4.

(Q3) Let p be an odd prime. The function χ, called the *Legendre symbol*, is defined on the integers by

$$\chi(i) = 0 \quad \text{if } i \text{ is a multiple of } p,$$
$$\chi(i) = 1 \quad \text{if the remainder when } i \text{ is divided by } p \text{ is a quadratic residue mod } p, \text{ and}$$
$$\chi(i) = -1 \quad \text{if the remainder is a nonresidue.}$$

Theorem 6. *For any* $c \not\equiv 0 \pmod{p}$,

$$\sum_{b=0}^{p-1} \chi(b)\chi(b + c) = -1.$$

Proof. From (Q1) it follows that

$$\chi(xy) = \chi(x)\chi(y) \quad \text{for } 0 \leq x, y \leq p - 1.$$

The term $b = 0$ contributes zero to the sum. Now suppose $b \neq 0$ and let $z \equiv (b + c)/b \pmod{p}$. There is a unique integer z, $0 \leq z \leq p - 1$, for each b. As b runs from 1 to $p - 1$, z takes on the values $0, 2, \ldots, p - 1$ but not 1. Then

$$\sum_{b=0}^{p-1} \chi(b)\chi(b + c) = \sum_{b=1}^{p-1} \chi(b)\chi(bz)$$

$$= \sum_{b=1}^{p-1} \chi(b)^2\chi(z)$$

$$= \sum_{\substack{z=0 \\ z \neq 1}}^{p-1} \chi(z) = 0 - \chi(1) = -1. \qquad \text{Q.E.D.}$$

Remark. Very similar properties hold when p is replaced by any odd prime power p^m, the numbers $0, 1, \ldots, p-1$ are replaced by the elements of the finite field $GF(p^m)$ (see Ch. 3), and the quadratic residues are defined to be the nonzero squares in this field.

Construction II: *The Paley construction.* This construction gives a Hadamard matrix of any order $n = p + 1$ which is a multiple of 4 (or of order $n = p^m + 1$ if we use quadratic residues in $GF(p^m)$).

We first construct the *Jacobsthal matrix* $Q = (q_{ij})$. This is a $p \times p$ matrix whose rows and columns are labeled $0, 1, \ldots, p-1$, and $q_{ij} = \chi(j - i)$. See Fig. 2.4 for the case $p = 7$.

$$Q = \begin{array}{c}
\\
0 \\
1 \\
2 \\
3 \\
4 \\
5 \\
6
\end{array}
\begin{array}{c}
0\ 1\ 2\ 3\ 4\ 5\ 6 \\
\left[\begin{array}{ccccccc}
0 & 1 & 1 & - & 1 & - & - \\
- & 0 & 1 & 1 & - & 1 & - \\
- & - & 0 & 1 & 1 & - & 1 \\
1 & - & - & 0 & 1 & 1 & - \\
- & 1 & - & - & 0 & 1 & 1 \\
1 & - & 1 & - & - & 0 & 1 \\
1 & 1 & - & 1 & - & - & 0
\end{array}\right]
\end{array}$$

Fig. 2.4. A Jacobsthal matrix.

Note that $q_{ji} = \chi(i - j) = \chi(-1)\chi(j - i) = -q_{ij}$ since p is of the form $4k - 1$ (see property (Q2)), and so Q is skew-symmetric, i.e., $Q^T = -Q$.

Lemma 7. $QQ^T = pI - J$, and $QJ = JQ = 0$, where J is the matrix all of whose entries are 1.

Proof. Let $P = (p_{ij}) = QQ^T$. Then

$$p_{ii} = \sum_{k=0}^{p-1} q_{ik}^2 = p - 1,$$

$$p_{ij} = \sum_{k=0}^{p-1} q_{ik} q_{jk} = \sum_{k=0}^{p-1} \chi(k - i)\chi(k - j), \quad \text{for } i \neq j,$$

$$= \sum_{b=0}^{p-1} \chi(b)\chi(b + c) \quad \text{where } b = k - i, c = i - j,$$

$$= -1 \quad \text{by property (Q3)}.$$

Also $QJ = JQ = 0$ since each row and column of Q contains $\frac{1}{2}(p - 1)$ +1's and $\frac{1}{2}(p - 1)$ −1's.   Q.E.D.

Now let

$$H = \begin{pmatrix} 1 & \mathbf{1} \\ \mathbf{1}^T & Q - I \end{pmatrix}. \tag{8}$$

Then

$$HH^T = \begin{pmatrix} 1 & \mathbf{1} \\ \mathbf{1}^T & Q - I \end{pmatrix}\begin{pmatrix} 1 & \mathbf{1} \\ \mathbf{1}^T & Q^T - I \end{pmatrix} = \begin{pmatrix} p+1 & 0 \\ 0 & J + (Q-I)(Q^T-I) \end{pmatrix}$$

But from Lemma 7 $J + (Q-I)(Q^T-I) = J + pI - J - Q - Q^T + I = (p+1)I$, so $HH^T = (p+1)I_{p+1}$. Thus H is a normalized Hadamard matrix of order $p+1$, which is said to be of *Paley type*.

Example. Fig. 2.5 shows the Hadamard matrices of orders 8 and 12 obtained in this way.

$$H_8 = \begin{bmatrix} 1 & 1 & 1 & 1 & 1 & 1 & 1 & 1 \\ 1 & - & 1 & 1 & - & 1 & - & - \\ 1 & - & - & 1 & 1 & - & 1 & - \\ 1 & - & - & - & 1 & 1 & - & 1 \\ 1 & 1 & - & - & - & 1 & 1 & - \\ 1 & - & 1 & - & - & - & 1 & 1 \\ 1 & 1 & - & 1 & - & - & - & 1 \\ 1 & 1 & 1 & - & 1 & - & - & - \end{bmatrix}$$

$$H_{12} = \begin{bmatrix} 1 & 1 & 1 & 1 & 1 & 1 & 1 & 1 & 1 & 1 & 1 & 1 \\ 1 & - & 1 & - & 1 & 1 & 1 & - & - & - & 1 & - \\ 1 & - & - & 1 & - & 1 & 1 & 1 & - & - & - & 1 \\ 1 & 1 & - & - & 1 & - & 1 & 1 & 1 & - & - & - \\ 1 & - & 1 & - & - & 1 & - & 1 & 1 & 1 & - & - \\ 1 & - & - & 1 & - & - & 1 & - & 1 & 1 & 1 & - \\ 1 & - & - & - & 1 & - & - & 1 & - & 1 & 1 & 1 \\ 1 & 1 & - & - & - & 1 & - & - & 1 & - & 1 & 1 \\ 1 & 1 & 1 & - & - & - & 1 & - & - & 1 & - & 1 \\ 1 & 1 & 1 & 1 & - & - & - & 1 & - & - & 1 & - \\ 1 & - & 1 & 1 & 1 & - & - & - & 1 & - & - & 1 \\ 1 & 1 & - & 1 & 1 & 1 & - & - & - & 1 & - & - \end{bmatrix}$$

Fig. 2.5. Hadamard matrices of orders 8 and 12.

Constructions I and II together give Hadamard matrices of all orders $1, 2, 4, 8, 12, \ldots, 32$.

Let us call two Hadamard matrices equivalent if one can be obtained from the other by permuting rows and columns and multiplying rows and columns by -1. Then it is easy to see that there is only one equivalence class of Hadamard matrices of orders 1, 2 and 4. We shall see in Ch. 20 that there is only one class of order 8 (so the Hadamard matrices of order 8 shown in Figs. 2.3 and 2.5 are equivalent) and one class of order 12. Furthermore it is known that there are five classes of order 16, and 3 of order 20. The number of order 24 is unknown.

Problem. (3) If $n = 2^m$, let $u_1, \ldots, u_n$ denote the distinct binary m-tuples. Show that the matrix $H = (h_{ij})$ where $h_{ij} = (-1)^{u_i \cdot u_j}$ is a Hadamard matrix of order n which is equivalent to that obtained from Construction I.

Hadamard Codes. Let H_n be a normalized Hadamard matrix of order n. If $+1$'s are replaced by 0's and -1's by 1's, H_n is changed into the *binary Hadamard matrix* A_n. Since the rows of H_n are orthogonal, any two rows of A_n agree in $\frac{1}{2}n$ places and differ in $\frac{1}{2}n$ places, and so have Hamming distance $\frac{1}{2}n$ apart.

A_n gives rise to three Hadamard *codes*:

(i) an $(n-1, n, \frac{1}{2}n)$ code, $\mathcal{A}_n$, consisting of the rows of A_n with the first column deleted ($\mathcal{A}_8$ is shown in Fig. 2.7 and $\mathcal{A}_{12}$ is the top half of Fig. 2.1 if the rows are read backwards);

(ii) an $(n-1, 2n, \frac{1}{2}n - 1)$ code, $\mathcal{B}_n$, consisting of $\mathcal{A}_n$ together with the complements of all its còdewords ($\mathcal{B}_{12}$ is shown in Fig. 2.1); and

(iii) an $(n, 2n, \frac{1}{2}n)$ code, $\mathcal{C}_n$, consisting of the rows of A_n and their complements.

$\mathcal{A}_n$ is a simplex code, since the distance between two codewords is $\frac{1}{2}n$ (see Ch. 1, §9). In fact these three codes are nonlinear generalizations of the codes in Fig. 1.13. Furthermore if H_n, with $n = 2^r$, is obtained from Construction I, these are linear codes and are the same as the codes shown in Fig. 1.13.

On the other hand if H_n is obtained from Construction II, the resulting codes are nonlinear for $n > 8$. We get linear codes by taking the linear span of these codes; these are called *quadratic residue* codes, and will be studied in Ch. 16. But if H_n is not obtained from Constructions I or II, little appears to be known about the binary rank of A_n.

Problem. (4) Show that if n is a multiple of 8, $A_n A_n^T \equiv 0 \pmod{2}$, and hence the binary rank of A_n is $\leq \frac{1}{2}n$. [Hint: use the fact that if B and C are $n \times n$ matrices over any field then rank $(B) +$ rank $(C) \leq n +$ rank (BC) (see Theorem 3.11 of Marcus [914]).]

Levenshtein's theorem. In this section we prove Levenshtein's theorem, which says that codes exist which meet the Plotkin bound.

First a couple of simple constructions.

(i) The codewords in $\mathcal{A}_n$ which begin with 0 form an $(n-2, n/2, n/2)$ code $\mathcal{A}'_n$, if the initial zero is deleted ($\mathcal{A}'_{12}$ is shown in Fig. 2.7).

(ii) Suppose we have an (n_1, M_1, d_1) code $\mathcal{C}_1$ and an (n_2, M_2, d_2) code $\mathcal{C}_2$. We paste a copies of $\mathcal{C}_1$ side by side, followed by b copies of $\mathcal{C}_2$ (Fig. 2.6).

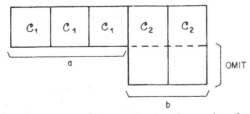

Fig. 2.6. Forming a new code by pasting together copies of two codes.

Now omit the last $M_2 - M_1$ rows of $\mathscr{C}_2$ (if $M_1 < M_2$), or omit the last $M_1 - M_2$ rows of $\mathscr{C}_1$ (if $M_1 > M_2$). The result is an (n, M, d) code with $n = an_1 + bn_2$, $M = \min\{M_1, M_2\}$, and $d \geq ad_1 + bd_2$, for any values of a and b. We denote this code by $a\mathscr{C}_1 \oplus b\mathscr{C}_2$.

Theorem 8. (Levenshtein.) *Provided enough Hadamard matrices exist, equality holds in the Plotkin bound* (3)–(6). *Thus if d is even,*

$$A(n, d) = 2\left[\frac{d}{2d - n}\right] \qquad \text{if } 2d > n \geq d, \tag{9}$$

$$A(2d, d) = 4d; \tag{10}$$

and if d is odd.

$$A(n, d) = 2\left[\frac{d + 1}{2d + 1 - n}\right] \quad \text{if } 2d + 1 > n \geq d, \tag{11}$$

$$A(2d + 1, d) = 4d + 4. \tag{12}$$

Proof. (11), (12) follow from (9), (10) using Theorem 2, so we may assume that d is even.

The Hadamard code $\mathscr{C}_{2d}$ above is a $(2d, 4d, d)$ code, which establishes (10).

To prove (9) we shall construct an (n, M, d) code with $M = 2[d/(2d - n)]$ for any n and even d satisfying $2d > n \geq d$. We need the following simple result, whose proof is left to the reader. If $2d > n \geq d$, define $k = [d/(2d - n)]$, and

$$a = d(2k + 1) - n(k + 1), \qquad b = kn - d(2k - 1). \tag{13}$$

Then a and b are nonnegative integers and

$$n = (2k - 1)a + (2k + 1)b$$
$$d = ka + (k + 1)b.$$

If n is even then so are a and b (from (13)). If n is odd and k even, then b is even. If n and k are odd, then a is even.

Now consider the code $\mathscr{C}$, where:

$$\text{if } n \text{ even,} \quad \mathscr{C} = \frac{a}{2}\mathscr{A}'_{4k} \oplus \frac{b}{2}\mathscr{A}'_{4k+4}$$

$$\text{if } n \text{ odd, } k \text{ even,} \quad \mathscr{C} = a\mathscr{A}_{2k} \oplus \frac{b}{2}\mathscr{A}'_{4k+4}$$

$$\text{if } n \text{ and } k \text{ odd,} \quad \mathscr{C} = \frac{a}{2}\mathscr{A}'_{4k} \oplus b\mathscr{A}_{2k+2}$$

Then in each case it is clear from remark (ii) above that $\mathscr{C}$ has length n, minimum distance d, and contains $2k = 2[d/(2d - n)]$ codewords. The existence of this code establishes (9). Q.E.D.

Note that the proof of (10) requires a Hadamard matrix of order $2d$ and that Hadamard matrices of orders $2k$ (if k even), $2k+2$ (if k odd), $4k$, $4k+4$ are sufficient for the proof of (9).

Example. We illustrate Levenshtein's method by constructing a $(27, 6, 16)$ code. Referring to the proof of the theorem, we find $k = 3$, $a = 4$, $b = 1$. This is the case n and k odd, so the code is obtained by pasting together two copies of $\mathscr{A}'_{12}$ and one copy of $\mathscr{A}_8$.

$\mathscr{A}_8$ and $\mathscr{A}'_{12}$ are obtained from H_8 and H_{12} in Fig. 2.4, and are shown in Fig. 2.7.

$$\mathscr{A}_8 = \begin{bmatrix} 0 & 0 & 0 & 0 & 0 & 0 & 0 \\ 1 & 0 & 0 & 1 & 0 & 1 & 1 \\ 1 & 1 & 0 & 0 & 1 & 0 & 1 \\ 1 & 1 & 1 & 0 & 0 & 1 & 0 \\ 0 & 1 & 1 & 1 & 0 & 0 & 1 \\ 1 & 0 & 1 & 1 & 1 & 0 & 0 \\ 0 & 1 & 0 & 1 & 1 & 1 & 0 \\ 0 & 0 & 1 & 0 & 1 & 1 & 1 \end{bmatrix} \qquad \mathscr{A}'_{12} = \begin{bmatrix} 0 & 0 & 0 & 0 & 0 & 0 & 0 & 0 & 0 & 0 \\ 1 & 1 & 0 & 1 & 0 & 0 & 0 & 1 & 1 & 1 \\ 1 & 1 & 1 & 0 & 1 & 1 & 0 & 1 & 0 & 0 \\ 0 & 1 & 1 & 1 & 0 & 1 & 1 & 0 & 1 & 0 \\ 0 & 0 & 1 & 1 & 1 & 0 & 1 & 1 & 0 & 1 \\ 1 & 0 & 0 & 0 & 1 & 1 & 1 & 0 & 1 & 1 \end{bmatrix}$$

Fig. 2.7. The $(7, 8, 4)$ code $\mathscr{A}_8$ and the $(10, 6, 6)$ code $\mathscr{A}'_{12}$.

The resulting $(27, 6, 16)$ code is shown in Fig. 2.8.

$$\begin{bmatrix} 0 & 0 & 0 & 0 & 0 & 0 & 0 & 0 & 0 & 0 & | & 0 & 0 & 0 & 0 & 0 & 0 & 0 & 0 & 0 & 0 & | & 0 & 0 & 0 & 0 & 0 & 0 & 0 \\ 1 & 1 & 0 & 1 & 0 & 0 & 0 & 1 & 1 & 1 & | & 1 & 1 & 0 & 1 & 0 & 0 & 0 & 1 & 1 & 1 & | & 1 & 0 & 0 & 1 & 0 & 1 & 1 \\ 1 & 1 & 1 & 0 & 1 & 1 & 0 & 1 & 0 & 0 & | & 1 & 1 & 1 & 0 & 1 & 1 & 0 & 1 & 0 & 0 & | & 1 & 1 & 0 & 0 & 1 & 0 & 1 \\ 0 & 1 & 1 & 1 & 0 & 1 & 1 & 0 & 1 & 0 & | & 0 & 1 & 1 & 1 & 0 & 1 & 1 & 0 & 1 & 0 & | & 1 & 1 & 1 & 0 & 0 & 1 & 0 \\ 0 & 0 & 1 & 1 & 1 & 0 & 1 & 1 & 0 & 1 & | & 0 & 0 & 1 & 1 & 1 & 0 & 1 & 1 & 0 & 1 & | & 0 & 1 & 1 & 1 & 0 & 0 & 1 \\ 1 & 0 & 0 & 0 & 1 & 1 & 1 & 0 & 1 & 1 & | & 1 & 0 & 0 & 0 & 1 & 1 & 1 & 0 & 1 & 1 & | & 1 & 0 & 1 & 1 & 1 & 0 & 0 \end{bmatrix}$$

Fig. 2.8. A $(27, 6, 16)$ code illustrating Levenshtein's construction.

Problem. (5) Construct $(28, 8, 16)$ and $(32, 8, 18)$ codes.

Other applications of Hadamard matrices.

(1) *Maximal determinants.* If H is a Hadamard matrix of order n, taking determinants of (7) we get

$$\det (HH^T) = \det H \cdot \det H^T = (\det H)^2 = n^n,$$

so H has determinant $\pm n^{n/2}$. An important theorem of Hadamard [572] states that if $A = (a_{ij})$ is any real $n \times n$ matrix with $-1 \leq a_{ij} \leq 1$, then

$$|\det A| \leq n^{n/2}.$$

Furthermore equality holds iff a Hadamard matrix of order n exists (and so n must be a multiple of 4). (See also Bellman [100, p. 127], Hardy et al. [601, p. 34], Ryser [1136, p. 105].) If n is not a multiple of 4, less is known about the largest possible determinant – see Problem 8.

(2) *Weighing designs.* By weighing several objects together instead of weighing them separately it is sometimes possible to determine the individual weights more accurately. Techniques for doing this are called *weighing designs*, and the best ones are based on Hadamard matrices.

These techniques are applicable to a great variety of problems of measurement, not only of weights, but of lengths, voltages, resistances, concentrations of chemicals, frequency spectra (Decker [341], Gibbs and Gebbie [478], Golay [507, 508], Harwit et al. [622], Phillips et al. [1042], Sloane et al. [1234–1236]), in fact to any experiment where the measure of several objects is the sum of the individual measurements. For simplicity, however, we shall just describe the weighing problem.

Suppose four light objects are to be weighed, using a balance with two pans which makes an error ϵ each time it is used. Assume that ϵ is a random variable with mean zero and variance σ^2, which is independent of the amount being weighed.

First, suppose the objects are weighed separately. If the unknown weights are a, b, c, d, the measurements are y_1, y_2, y_3, y_4, and the (unknown) errors made by the balance are $\epsilon_1, \epsilon_2, \epsilon_3, \epsilon_4$, then the four weighings give four equations:

$$a = y_1 + \epsilon_1, \qquad b = y_2 + \epsilon_2, \qquad c = y_3 + \epsilon_3, \qquad d = y_4 + \epsilon_4.$$

The estimates of the unknown weights $a, b, \ldots$ are

$$\hat{a} = y_1 = a - \epsilon_1, \qquad \hat{b} = y_2 = b - \epsilon_2, \ldots,$$

each with variance σ^2.

On the other hand, suppose the four weighings are made as follows:

$$\begin{aligned}
a + b + c + d &= y_1 + \epsilon_1, \\
a - b + c - d &= y_2 + \epsilon_2, \\
a + b - c - d &= y_3 + \epsilon_3, \\
a - b - c + d &= y_4 + \epsilon_4.
\end{aligned} \tag{14}$$

This means that in the first weighing all four objects are placed in the left hand pan of the balance, and in the other weighings two objects are in the left pan and two in the right. Since the coefficient matrix on the left is a Hadamard matrix, it is easy to solve for a, b, c, d. Thus the estimate for a is

$$\hat{a} = \frac{y_1 + y_2 + y_3 + y_4}{4}$$

$$= a - \frac{\epsilon_1 + \epsilon_2 + \epsilon_3 + \epsilon_4}{4}.$$

The variance of $c\epsilon$, where c is a constant, is c^2 times the variance of ϵ, and the variance of a sum of independent random variables is the sum of the individual variances. Therefore the variance of $\hat{a}$ is $4 \cdot (\sigma^2/16) = \sigma^2/4$, an improvement by a factor of 4. Similarly for the other unknowns.

In general if n objects are to be weighed in n weighings, and a Hadamard matrix of order n exists, this technique reduces the variance from σ^2 to σ^2/n. This is known to be the smallest that can be attained with any choice of signs on the left side of (14). For a proof of this, and for much more information about weighing designs, see Banerjee [65, 66], Geramita et al. [473–477], Hotelling [665], Mood [972], Payne [1032] and Sloane and Harwit [1236].

Now suppose the balance has only one scale pan, so only coefficients 0 and 1 can be used. In this case the variance of the estimates can also be reduced, though not by such a large factor. Again we illustrate the technique by an example. If seven objects $a, b, c, \ldots, g$ are to be weighed, use the following weighing design:

$$
\begin{aligned}
a \quad\;\; + c \quad\;\; + e \quad\;\; + g &= y_1 + \epsilon_1 \\
b + c \quad\quad\quad\;\; + f + g &= y_2 + \epsilon_2 \\
a + b \quad\quad\;\; + e + f \quad\;\; &= y_3 + \epsilon_3 \\
d + e + f + g &= y_4 + \epsilon_4 \\
a \quad\;\; + c + d \quad\;\; + f \quad\;\; &= y_5 + \epsilon_5 \\
b + c + d + e \quad\quad\;\; &= y_6 + \epsilon_6 \\
a + b \quad\;\; + d \quad\quad\;\; + g &= y_7 + \epsilon_7
\end{aligned}
\tag{15}
$$

The coefficients are determined in an obvious way from the Hadamard matrix H_8 of Fig. 2.3. Then the estimate for a is (cf. Problem 7)

$$
\begin{aligned}
\hat{a} &= \frac{y_1 - y_2 + y_3 - y_4 + y_5 - y_6 + y_7}{4} \\
&= a - \frac{\epsilon_1 - \epsilon_2 + \epsilon_3 - \epsilon_4 + \epsilon_5 - \epsilon_6 + \epsilon_7}{4},
\end{aligned}
$$

which has variance $7\sigma^2/16$, and similarly for the other weights. In general if there are n objects and a Hadamard matrix of order $n+1$ exists, this technique reduces the variance to $4n\sigma^2/(n+1)^2$, which is in some sense the best possible – see Mood [972], Raghavarao [1085, Ch. 17], and Sloane and Harwit [1236].

We can describe these techniques by saying that we have used a Hadamard code to encode, or transform, the data before measuring it.

Another type of weighing problem, also related to coding theory, will be discussed in Ch. 6.

(3) *The Hadamard Transform.* Let H be a Hadamard matrix of order n. If $X = (x_1, \ldots, x_n)$ is a real vector, its *Hadamard transform* (or *Walsh*, or *discrete Fourier transform*) is the vector

$$
\hat{X} = XH.
$$

Let F^m denote the set of all binary m-tuples. The entries $(-1)^{u \cdot v}$, for $u, v \in F^m$, form a Hadamard matrix of order $n = 2^m$ (Problem 3). If f is a mapping defined on F^m, its Hadamard transform $\hat{f}$ is given by

$$\hat{f}(u) = \sum_{v \in F^m} (-1)^{u \cdot v} f(v), \quad u \in F^m.$$

We shall use $\hat{f}$ in an important lemma (Lemma 2) in Ch. 5, and in studying Reed–Muller codes. Hadamard transforms are also widely used in communications and physics, and have an extensive and widely scattered literature – see for example Ahmed et al. [12–15], Harmuth [603–604, 680], Kennett [757], Pratt et al. [1079], Shanks [1187] and Wadbrook and Wollons [1375]. Analysis of the effect of errors on the transform involves detailed knowledge of the cosets of first order Reed–Muller codes – see Berlekamp [119].

Problems. (6) Let

$$H = \begin{pmatrix} 1 & 1 \\ 1^T & S \end{pmatrix}$$

be a normalized Hadamard matrix of order n. Show that $SS^T = nI - J$, $SJ = JS = -J$, $S^{-1} = (1/n)(S^T - J)$.

(7) Let $T = \frac{1}{2}(J - S)$ (so that T is the coefficient matrix in (15)). Show that $T^{-1} = -(2/n)S^T$.

(8) *Maximum determinant problems.* Suppose $A = (a_{ij})$ is a real matrix of order n. Let

$$f(n) = \max |\det(A)| \text{ subject to } a_{ij} = 0 \text{ or } 1,$$
$$g(n) = \max |\det(A)| \text{ subject to } a_{ij} = -1 \text{ or } 1,$$
$$h(n) = \max |\det(A)| \text{ subject to } a_{ij} = -1, 0, \text{ or } 1,$$
$$F(n) = \max |\det(A)| \text{ subject to } 0 \le a_{ij} \le 1,$$
$$G(n) = \max |\det(A)| \text{ subject to } -1 \le a_{ij} \le 1.$$

Show

(a) $f(n) = F(n)$,
(b) $g(n) = h(n) = G(n)$,
(c) $g(n) = 2^{n-1} f(n-1)$.

Thus all five problems are equivalent! Hadamard [572] proved that $f(n) \le 2^{-n}(n+1)^{(n+1)/2}$, with equality iff a Hadamard matrix of order $n + 1$ exists. The first few values of f are as follows:

n	1	2	3	4	5	6	7	8	9	10	11	12	13
$f(n)$	1	1	2	3	5	9	32	56	144	320	1458	3645	9477

See for example Brenner and Cummings [196], Cohn [298], Ehlich [403], Ehlich and Zeller [404] and Yang [1445].

§4. Conference Matrices

These are similar to Hadamard matrices but with a slightly different defining equation. They also give rise to good nonlinear codes. Our treatment will be brief.

Definition. A *conference matrix* C of order n is an $n \times n$ matrix with diagonal entries 0 and other entries $+1$ or -1, which satisfies

$$CC^T = (n-1)I. \tag{16}$$

(The name arises from the use of such matrices in the design of networks having the same attenuation between every pair of terminals.) These matrices are sometimes called C-matrices.

Properties. (1) As with Hadamard matrices we can *normalize* C by multiplying rows and columns by -1, so that C has the form

$$C = \begin{pmatrix} 0 & 1 \\ 1^T & S \end{pmatrix} \tag{17}$$

where S is a square matrix of order $n-1$ satisfying

$$SS^T = (n-1)I - J, \qquad SJ = JS = 0. \tag{18}$$

For example, normalized conference matrices of orders 2 and 4 are shown in Fig. 2.9

$$C_2 = \begin{pmatrix} 0 & 1 \\ 1 & 0 \end{pmatrix} \qquad C_4 = \begin{pmatrix} 0 & 1 & 1 & 1 \\ 1 & 0 & 1 & - \\ 1 & - & 0 & 1 \\ 1 & 1 & - & 0 \end{pmatrix}$$

Fig. 2.9. Conference matrices of orders 2 and 4.

(2) If C exists then n must be even. If $n \equiv 2 \pmod 4$ then C can be made symmetric by multiplying rows and columns by -1, while if $n \equiv 0 \pmod 4$ then C can be made skew-symmetric.

(3) Conversely if a symmetric C exists then $n \equiv 2 \pmod 4$ and $n-1$ can be written as

$$n - 1 = a^2 + b^2$$

where a and b are integers. If a skew-symmetric C exists then $n = 2$ or $n \equiv 0 \pmod 4$. (For the proof of these properties see Delsarte et al. [366] and Belevitch [97].)

Several constructions are known. The most useful for our purpose is the following

Construction. (Paley.) Let $n = p^m + 1 \equiv 2$ (mod 4), where p is an odd prime. As in §3 we define the $p^m \times p^m$ Jacobsthal matrix $Q = (q_{ij})$ where $q_{ij} = \chi(j - i)$. Now $p^m \equiv 1$ (mod 4), so Q is symmetric by property (Q2). Then

$$C = \begin{pmatrix} 0 & \mathbf{1} \\ \mathbf{1}^T & Q \end{pmatrix}$$

is a symmetric conference matrix of order n.

Examples. (i) $n = 6$, $p^m = 5$. The quadratic residues mod 5 are 1 and 4, and the construction gives the first matrix shown in Fig. 2.10.

(ii) $n = 10$, $p^m = 3^2$. Let the elements of the field GF(3^2) defined by $\alpha^2 + \alpha + 2 = 0$ be 0, 1, 2, α, $\alpha + 1$, $\alpha + 2$, 2α, $2\alpha + 1$, $2\alpha + 2$ (see Ch. 4). The quadratic residues in this field are 1, 2, $\alpha + 2$, $2\alpha + 1$. Then the construction gives the second matrix shown in Fig. 2.10.

$$
C_6 =
\begin{array}{cc|ccccc}
 & & 0 & 1 & 2 & 3 & 4 \\
\hline
 & 0 & 1 & 1 & 1 & 1 & 1 \\
0 & 1 & 0 & 1 & - & - & 1 \\
1 & 1 & 1 & 0 & 1 & - & - \\
2 & 1 & - & 1 & 0 & 1 & - \\
3 & 1 & - & - & 1 & 0 & 1 \\
4 & 1 & 1 & - & - & 1 & 0 \\
\end{array}
$$

$$
C_{10} =
\begin{array}{cc|ccc|ccc|ccc}
 & & 0 & 1 & 2 & \alpha & \alpha+1 & \alpha+2 & 2\alpha & 2\alpha+1 & 2\alpha+2 \\
\hline
 & 0 & 1 & 1 & 1 & 1 & 1 & 1 & 1 & 1 & 1 \\
0 & 1 & 0 & 1 & 1 & - & - & 1 & - & 1 & - \\
1 & 1 & 1 & 0 & 1 & 1 & - & - & - & - & 1 \\
2 & 1 & 1 & 1 & 0 & - & 1 & - & 1 & - & - \\
\alpha & 1 & - & 1 & - & 0 & 1 & 1 & - & - & 1 \\
\alpha+1 & 1 & - & - & 1 & 1 & 0 & 1 & 1 & - & - \\
\alpha+2 & 1 & 1 & - & - & 1 & 1 & 0 & - & 1 & - \\
2\alpha & 1 & - & - & 1 & - & 1 & - & 0 & 1 & 1 \\
2\alpha+1 & 1 & 1 & - & - & - & - & 1 & 1 & 0 & 1 \\
2\alpha+2 & 1 & - & 1 & - & 1 & - & - & 1 & 1 & 0 \\
\end{array}
$$

Fig. 2.10. Symmetric conference matrices of orders 6 and 10.

The above construction gives symmetric conference matrices of orders 6, 10, 14, 18, 26, 30, 38, 42, 50, ... Orders 22, 34, 58, ... do not exist from Property 3. A recent construction of Mathon [1477] gives matrices of order 46, 442, ...

Codes from conference matrices. Let C_n be a symmetric conference matrix of order n (so $n \equiv 2 \pmod 4$), and let S be as in Equation (17). Then the rows of $\frac{1}{2}(S + I + J)$, $\frac{1}{2}(-S + I + J)$, plus the zero and all-ones vectors form an

$$(n - 1, 2n, \tfrac{1}{2}(n - 2))$$

nonlinear *conference matrix code.* That the minimum distance is $\frac{1}{2}(n-2)$ follows easily from Equation (18).

Example. The conference matrix C_{10} in Fig. 2.10 gives the $(9, 20, 4)$ code $\mathscr{C}_9$ shown in Fig. 2.11.

```
000 000 000

111 001 010
111 100 001
111 010 100

010 111 001
001 111 100
100 111 010

001 010 111
100 001 111
010 100 111

100 110 101
010 011 110
001 101 011

101 100 110
110 010 011
011 001 101

110 101 100
011 110 010
101 011 001

111 111 111
```

Fig. 2.11. A $(9, 20, 4)$ conference matrix code $\mathscr{C}_9$.

(The zero codeword could be changed to a vector of weight 1 without decreasing the minimum distance. Therefore there are at least two inequivalent codes with parameters $(9, 20, 4)$.)

This important code was first found by Golay [509], using a different method. We shall give two other constructions for $(9, 20, 4)$ codes in §7. In §8 the code of Fig. 2.11 will be used to generate infinitely many nonlinear single-error-correcting codes.

The first few conference matrix codes are the following:

$$(5, 12, 2), \qquad (9, 20, 4), \qquad (13, 28, 6), \qquad (17, 36, 8),$$
$$(25, 52, 12), \qquad (29, 60, 14), \qquad (37, 76, 18), \qquad (41, 84, 20).$$

Research Problem 2.2. Since n is greater than $2d$ the Plotkin bound does not apply to these codes. However we will see in Ch. 17 that the $(9, 20, 4)$ code is optimal in the sense of having the largest number of codewords for this length and distance. (Thus $A(9, 4) = 20$.) We conjecture that all of these codes except the first and third are similarly optimal. (There exists a $(5, 16, 2)$ linear code and a $(13, 32, 6)$ nonlinear code – see §8.)

§5. t-designs

Definition. Let X be a v-set (i.e. a set with v elements), whose elements are called *points* or sometimes (for historical reasons) *varieties*. A *t-design* is a collection of distinct k-subsets (called *blocks*) of X with the property that any t-subset of X is contained in exactly λ blocks. In more picturesque language, it is a collection of committees chosen out of v people, each committee containing k persons, and such that *any t* persons serve together on exactly λ committees. We call this a t-(v, k, λ) design*. t-designs are also sometimes called *tactical configurations*.

Example. The seven points and seven lines (one of which is curved) of Fig. 2.12 form the *projective plane of order 2*. If we take the lines as blocks, this is a 2-$(7, 3, 1)$ design, since there is a unique line through any two of the seven points. The seven blocks are

$$013, \quad 124, \quad 235, \quad 346, \quad 450, \quad 561, \quad 602.$$

A 2-design is called a *balanced incomplete block design*, and the terminology of the subject comes from the original application of such designs in agricultural or biological experiments. For example suppose v varieties of fertilizer are to be compared in their effect on b different crops. Ideally each crop would be tested with each variety of fertilizer, giving b blocks of land (one for each crop) each of size v. This is a 2-(v, v, b) design known as a *complete* design. However for reasons of economy this is impossible, and we seek a design where each crop is tested with only k of the varieties of fertilizer (so each block now has size k), and where any two fertilizers are used together on the same crop a constant number λ of times. Thus the design is *balanced* so far as comparisons between pairs of fertilizers are concerned.

---

*Some authors put the letters in a different order.

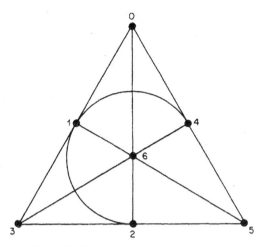

Fig. 2.12. The projective plane of order 2.

This is a 2-(v, k, λ) design with b blocks, and is *incomplete* if $k < v$.
Some other t-designs also have names of their own.

Definition. A *Steiner system* is a t-design with $\lambda = 1$, and a t-$(v, k, 1)$ design is usually called an $S(t, k, v)$. Thus the example of Fig. 2.12 is an $S(2, 3, 7)$.

Definition. A *projective plane of order n* is an $S(2, n + 1, n^2 + n + 1)$ with $n \geqslant 2$. That's why Fig. 2.12 is a projective plane of order 2. (For details see Appendix B.)

Definition. An *affine plane of order n* is an $S(2, n, n^2)$ with $n \geqslant 2$.

Theorem 9. *In a t-(v, k, λ) design, let $P_1, \ldots, P_t$ be any t distinct points, let λ_i be the number of blocks containing $P_1, \ldots, P_i$, for $1 \leqslant i \leqslant t$, and let $\lambda_0 = b$ be the total number of blocks. Then λ_i is independent of the choice of $P_1, \ldots, P_i$, and in fact*

$$\lambda_i = \frac{\lambda \binom{v - i}{t - i}}{\binom{k - i}{t - i}}, \quad \text{for } 0 \leqslant i \leqslant t,$$

$$= \lambda \frac{(v - i)(v - i - 1) \cdots (v - t + 1)}{(k - i)(k - i - 1) \cdots (k - t + 1)}. \tag{19}$$

This implies that a t-(v, k, λ) design is also an i-(v, k, λ_i) design for $1 \leqslant i \leqslant t$.

Proof. The result is true for $i = t$ by definition of a t-design, since any t points are contained in exactly λ blocks. We proceed by induction on i. Suppose we have already shown that λ_{i+1} is independent of the choice of $P_1, \ldots, P_{i+1}$. For each block B containing $P_1, \ldots, P_i$ and for each point Q distinct from $P_1, \ldots, P_i$ define $\chi(Q, B) = 1$ if $Q \in B$, $= 0$ if $Q \notin B$. Then from the induction hypothesis

$$\sum_Q \sum_B \chi(Q, B) = \lambda_{i+1}(v - i)$$

$$= \sum_B \sum_Q \chi(Q, B) = \lambda_i(k - i), \tag{20}$$

which proves that λ_i is independent of the choice of $P_1, \ldots, P_i$, and establishes (19).   Q.E.D.

Corollary 10. *In a t-(v, k, λ) design the total number of blocks is*

$$b = \frac{\lambda \binom{v}{t}}{\binom{k}{t}}, \tag{21}$$

and each point belongs to exactly r blocks, where

$$bk = vr. \tag{22}$$

Furthermore in a 2-design,

$$\lambda(v - 1) = r(k - 1). \tag{23}$$

Proof. $b = \lambda_0$ is given by (19). Since $r = \lambda_1$, (22) and (23) follow from (20).
                                                                      Q.E.D.

If $b = v$, and hence $r = k$, the design is called *symmetric*.

Corollary 11. *A necessary condition for a t-(v, k, λ) design to exist is that the numbers $\lambda \binom{v-i}{t-i} / \binom{k-i}{t-i}$ be integers for $0 \le i \le t$.*

In some cases, e.g., for Steiner systems $S(2, 3, v)$, $S(2, 4, v)$, $S(2, 5, v)$ and $S(3, 4, v)$, this condition is also sufficient, but not always. For example we shall see in Ch. 19 that an $S(2, 7, 43)$ or projective plane of order 6 does not exist, even though the above condition is satisfied.

Research Problem 2.3. It has been conjectured that for any given t, k and λ, if the conditions of Corollary 11 are satisfied and v is sufficiently large, then a t-(v, k, λ) design exists. But so far this has only been proved for $t = 2$, see

Wilson [1420–1422], and in fact no designs are presently known with $t > 5$, except trivial ones.

*Definition. In a t-(v, k, λ) design, let $P_1, \ldots, P_k$ be the points belonging to one of the blocks. Consider the blocks which contain $P_1, \ldots, P_j$ but do not contain $P_{j+1}, \ldots, P_i$, for $0 \leq j \leq i$. (For $j = 0$ we consider the blocks which do not contain $P_1, \ldots, P_i$, and for $j = i$ we consider the blocks which contain $P_1, \ldots, P_j$). If the number of such blocks is a constant, independent of the choice of $P_1, \ldots, P_i$, we denote it by λ_{ij}. The λ_{ij} are called *block intersection numbers*.

Theorem 12. *The λ_{ij} are well defined for $i \leq t$. In fact $\lambda_{ii} = \lambda_i$ for $i \leq t$, with $\lambda_{tt} = \lambda$, and are given by Theorem 9. Also the λ_{ij} satisfy the Pascal property*

$$\lambda_{ij} = \lambda_{i+1,j} + \lambda_{i+1,j+1}$$

whenever they are defined. Finally, if the design is a Steiner system so that $\lambda = 1$, then $\lambda_{tt} = \lambda_{t+1,t+1} = \cdots = \lambda_{kk} = 1$, and the λ_{ij} are therefore defined for all $0 \leq j \leq i \leq k$.

Proof. The Pascal property holds because λ_{ij} is equal to the number of blocks which contain $P_1, \ldots, P_j$ but do not contain $P_{j+2}, \ldots, P_{i+1}$. These blocks can be divided into those which contain P_{j+1}, i.e., $\lambda_{i+1,j+1}$, and those which do not, i.e., $\lambda_{i+1,j}$. That the λ_{ij} are defined for $i \leq t$ is an immediate consequence of Theorem 9. Indeed, both the total number of blocks, $\lambda_0 = \lambda_{00}$, and the number of blocks through P_1, $\lambda_1 = \lambda_{11}$, are independent of the choice of P_1, hence the number of blocks *not* containing P_1 is $\lambda_{10} = \lambda_0 - \lambda_1$, and is also constant, and so on. The last sentence of the theorem is obvious.   Q.E.D.

Thus for any t-(v, k, λ) design we may form the "*Pascal triangle*" of its block intersection numbers:

$$\lambda_{00} = \lambda_0$$
$$\lambda_{10} \quad \lambda_{11} = \lambda_1$$
$$\lambda_{20} \quad \lambda_{21} \quad \lambda_{22} = \lambda_2$$
$$\cdots$$

E.g. for the $S(2, 3, 7)$ of Fig. 2.12, we obtain the triangle

$$7$$
$$4 \ 3$$
$$2 \ 2 \ 1$$
$$0 \ 2 \ 0 \ 1$$

*The remainder of this section can be omitted on the first reading.

From a given t-(v, k, λ) design with block intersection numbers λ_{ij} we can obtain several other designs. Let $\mathscr{B}$ be the set of b blocks of the original design.

Suppose we omit one of the points, say P, from all the blocks of $\mathscr{B}$, and consider the two sets of blocks which remain: the first, $\mathscr{B}_1$ say, consists of the $\lambda_{1,0}$ blocks of k points that did not contain P in the first place, and the second, $\mathscr{B}_2$ say, consists of $\lambda_{1,1}$ blocks of $k-1$ points.

Theorem 13. *The blocks $\mathscr{B}_1$ form a $(t-1)$-$(v-1, k, \lambda_{t,t-1})$ design with block intersection numbers $\lambda_{ij}^{(1)} = \lambda_{i+1,j}$. The blocks $\mathscr{B}_2$ form a $(t-1)$-$(v-1, k-1, \lambda)$ design with block intersection numbers $\lambda_{ij}^{(2)} = \lambda_{i+1,j+1}$. These are called derived designs.*

The proof is left to the reader.

Corollary 14. *If a Steiner system $S(t, k, v)$ exists so does an $S(t-1, k-1, v-1)$.*

If $v - k \geqslant t$, taking the complements of all the blocks in $\mathscr{B}$ gives the *complementary* design, which is a t-$(v, v-k, \lambda_{t0})$ design with block intersection numbers $\lambda_{ij}^{(c)} = \lambda_{i,i-j}$.

Incidence Matrix. Given a t-(v, k, λ) design with v points $P_1, \ldots, P_v$ and b blocks $B_1, \ldots, B_b$, its $b \times v$ *incidence matrix* $A = (a_{ij})$ is defined by

$$a_{ij} = \begin{cases} 1 & \text{if } P_j \in B_i, \\ 0 & \text{if } P_j \notin B_i. \end{cases}$$

For example the incidence matrix of the design of Fig. 2.12 is

$$A = \begin{pmatrix} 1 & 1 & 0 & 1 & 0 & 0 & 0 \\ 0 & 1 & 1 & 0 & 1 & 0 & 0 \\ 0 & 0 & 1 & 1 & 0 & 1 & 0 \\ 0 & 0 & 0 & 1 & 1 & 0 & 1 \\ 1 & 0 & 0 & 0 & 1 & 1 & 0 \\ 0 & 1 & 0 & 0 & 0 & 1 & 1 \\ 1 & 0 & 1 & 0 & 0 & 0 & 1 \end{pmatrix}, \tag{24}$$

(which is a matrix the reader should recognize from §3).

Problem. (9) If $t \geqslant 2$, show that

(1) $A^T A = (r - \lambda_2)I + \lambda_2 J$,
(2) $\det(A^T A) = (r - \lambda_2)^{v-1}(v\lambda_2 - \lambda_2 + r)$,
(3) if $b > 1$, then $b \geqslant v$ (Fisher's inequality).

Codes and Designs. To every block in a Steiner system $S(t, k, v)$ corresponds a row of the incidence matrix A. If we think of these rows as codewords, the Steiner system forms a nonlinear code with parameters

$$\left(n = v, M = b = \binom{v}{t} \middle/ \binom{k}{t}, d \geq 2(k - t + 1) \right).$$

For two blocks cannot have more than $t - 1$ points in common (or else there would be t points contained in two blocks, a contradiction), and therefore the Hamming distance between two blocks is at least $2(k - t + 1)$. Note that every codeword has weight k: this is a *constant weight code.*

One can also consider the linear code generated by the blocks, but little seems to be known in general about such codes. (There are a few results for $t = 2$, concerning the codes generated by affine and projective planes, as we shall see in Ch. 13.)

Conversely, given an (n, M, d) code, one can sometimes obtain a design from the codewords of a fixed weight. For example:

Theorem 15. *Let $\mathcal{H}_m$ be an $[n = 2^m - 1, k = 2^m - 1 - m, d = 3]$ Hamming code (Ch. 1, §7), and let $\hat{\mathcal{H}}_m$ be obtained by adding an overall parity check to $\mathcal{H}_m$. Then the codewords of weight 3 in $\mathcal{H}_m$ form a Steiner system $S(2, 3, 2^m - 1)$, and the codewords of weight 4 in $\hat{\mathcal{H}}_m$ form an $S(3, 4, 2^m)$.*

Definition. A vector v *covers* a vector u if the ones in u are a subset of the ones in v; for example 1001 covers 1001, 1000, 0001, and 0000. Equivalent statements are

$$u * v = u,$$

$$\text{wt}(u + v) = \text{wt}(v) - \text{wt}(u).$$

Proof of Theorem 15. The first statement follows from the second using Corollary 14. To prove the second statement, let the coordinates of $\hat{\mathcal{H}}_m$ be labelled $P_0, P_1, \ldots, P_n$, where P_n is the overall parity check. Let u be an arbitrary vector of weight 3, with 1's at coordinates P_h, P_i, P_j say, with $h < i < j$. We must show that there is a unique codeword of weight 4 in $\hat{\mathcal{H}}_m$ which covers u. Certainly there cannot be two such codewords in $\hat{\mathcal{H}}_m$, or their distance apart would be 2, a contradiction. Case (i), $j < n$. Since $\mathcal{H}_m$ is a perfect single-error-correcting code, it contains a codeword c at distance at most 1 from u. Either $c = u$, in which case the extended codeword $\hat{c} = |c|1|$ has weight 4, is in $\hat{\mathcal{H}}_m$, and covers u, or else c has weight 4 and $\hat{c} = |c|0|$ works. Case (ii), $j = n$. The vector u' with 1's in coordinates P_h, P_i is covered by a unique codeword $c \in \mathcal{H}_m$ with weight 3, and $\hat{c} = |c|1|$ covers u. Q.E.D.

Corollary 16. *The number of codewords of weight 3 in $\mathcal{H}_m$ is*

$$\binom{2^m - 1}{2} \Big/ \binom{3}{2} = \frac{(2^m - 1)(2^{m-1} - 1)}{3},$$

and the number of weight 4 in $\hat{\mathcal{H}}_m$ is

$$\binom{2^m}{3} \Big/ \binom{4}{3} = \frac{2^{m-2}(2^m - 1)(2^{m-1} - 1)}{3}.$$

The codewords of weight 3 in $\mathcal{H}_m$ can be identified with the lines of the projective geometry $\mathrm{PG}(m - 1, 2)$, as we shall see in Ch. 13.

Problem. (10) Find the codewords of weight 3 in the Hamming code of length 7, and verify Corollary 16 for this case. Identify those codewords with the lines of Fig. 2.12.

Exactly the same proof as that of Theorem 15 establishes:

Theorem 17. *Let $\mathscr{C}$ be a perfect e-error-correcting code of length n, with e odd, and let $\hat{\mathscr{C}}$ be obtained by adding an overall parity check to $\mathscr{C}$. Then the codewords of weight $2e + 1$ in $\mathscr{C}$ form a Steiner system $S(e + 1, 2e + 1, n)$, and the codewords of weight $2e + 2$ in $\hat{\mathscr{C}}$ form an $S(e + 2, 2e + 2, n + 1)$.*

Generalizations of this Corollary and much more about codes and designs will be given in Ch. 6.

Problem. (11) If H_n is a Hadamard matrix of order $n \geqslant 8$, let S be defined as in Problem 6. Show that if -1's are replaced by 0's the rows of S form a symmetric

$$2 - \left(n - 1, \frac{n}{2} - 1, \frac{n}{4} - 1\right)$$

design, and conversely.

§6. An introduction to the binary Golay code

The Golay code is probably the most important of all codes, for both practical and theoretical reasons. In this section we give an elementary definition of the code, and establish a number of its properties. Further properties will be given in Chs. 16 and 20.

Definition. The extended Golay code $\mathcal{G}_{24}$ has the generator matrix shown in Fig. 2.13.

$G =$

	←					*l*					→	←					*r*					→	
	∞	0	1	2	3	4	5	6	7	8	9	10	∞	0	1	2	3	4	5	6	7	8	9	10	row
	1	1												1	1		1	1	1					1		0
	1		1												1	1		1	1	1					1	1
	1			1										1		1	1		1	1	1					2
	1				1										1		1	1		1	1	1				3
	1					1										1		1	1		1	1	1			4
	1						1										1		1	1		1	1	1		5
	1							1						1				1		1	1		1	1		6
	1								1					1	1				1		1	1		1		7
	1									1				1	1	1				1		1	1			8
	1										1				1	1	1				1		1	1		9
	1											1		1		1	1	1				1		1		10
													1	1	1	1	1	1	1	1	1	1	1	1	11

Fig. 2.13. Generator matrix for extended Golay code $\mathcal{G}_{24}$. The columns are labelled $l_\infty l_0 l_1 \cdots l_{10} r_\infty r_0 \cdots r_{10}$. The 11×11 matrix on the right is A_{11}.

The 11×11 binary matrix A_{11} on the right of the generator matrix is obtained from a Hadamard matrix of Paley type (cf. Figs. 2.1 and 2.5). This implies that the sum of any two rows of A_{11} has weight 6. Hence the sum of any two rows of G has weight 8.

Lemma 18. $\mathcal{G}_{24}$ *is self dual*: $\mathcal{G}_{24} = \mathcal{G}_{24}^{\perp}$.

Proof. If u and v are (not necessarily distinct) rows of G, then $\operatorname{wt}(u * v) \equiv 0$ (mod 2). Therefore every row of G is orthogonal to all the rows, and so $\mathcal{G}_{24} \subset \mathcal{G}_{24}^{\perp}$. But G has rank 12, so $\mathcal{G}_{24}$ has dimension 12, and therefore $\mathcal{G}_{24} = \mathcal{G}_{24}^{\perp}$.
Q.E.D.

Lemma 19. (i) *Every codeword of* $\mathcal{G}_{24}$ *has weight divisible by* 4.
(ii) $\mathcal{G}_{24}$ *contains the codeword* **1**.

Proof. (i) Problem 38 of Ch. 1. (ii) Add the rows of G.
Q.E.D.

Lemma 20. $\mathcal{G}_{24}$ *is invariant under the permutation of coordinates*

$$T = (l_\infty r_\infty)(l_0 r_0)(l_1 r_{10})(l_2 r_9) \cdots (l_{10} r_1),$$

which interchanges the two halves of a codeword. To be quite explicit:

If $\mathcal{G}_{24}$ contains the codeword $|L|R|$ with $L = a_\infty a_0 a_1 a_2, \ldots, a_{10}$, $R = b_\infty b_0 b_1 b_2, \ldots, b_{10}$, it also contains the codeword $|L'|R'|$ where

$$L' = b_\infty b_0 b_{10} b_9, \ldots, b_1, \qquad R' = a_\infty a_0 a_{10} a_9, \ldots, a_1.$$

Proof. T sends row 0 of G into

$$0, 1, 0100011101, 1, 1, 0000000000$$

which is easily verified to be the sum of rows $0, 2, 6, 7, 8, 10, 11$, and therefore is in the code. Similarly for rows 1 through 10. T sends the last row of G into $1^{12}0^{12}$, which is the complement of the last row, and is in the code by Lemma 19.   Q.E.D.

Remark. This lemma implies that whenever $\mathcal{G}_{24}$ contains a codeword $|L|R|$ with $\mathrm{wt}\,(L) = i$, $\mathrm{wt}\,(R) = j$, then it also contains a codeword $|L'|R'|$ with $\mathrm{wt}\,(L') = j$, $\mathrm{wt}\,(R') = i$.

The possible weights of codewords in $\mathcal{G}_{24}$ are, by Lemma 19,

$$0, 4, 8, 12, 16, 20, 24.$$

If u has weight 20, then $u + 1$ has weight 4. We show that there are no codewords of weight 4, hence none of weight 20.

Lemma 21. $\mathcal{G}_{24}$ *contains no codewords of weight* 4.

Proof. For any codeword $|L|R|$ of $\mathcal{G}_{24}$, $\mathrm{wt}\,(L) \equiv \mathrm{wt}\,(R) \equiv 0 \pmod 2$. By lemma 20 we may suppose that a codeword of weight 4 is of one of the types

     (1) $\mathrm{wt}\,(L) = 0, \mathrm{wt}\,(R) = 4$;   (2) $\mathrm{wt}\,(L) = 2, \mathrm{wt}\,(R) = 2$.

(1) is impossible, since if $\mathrm{wt}\,(L) = 0$, $\mathrm{wt}\,(R) = 0$ or 12. (2) is impossible, since if $\mathrm{wt}\,(L) = 2$, L is the sum of one or two rows of G, plus possibly the last row. In each case $\mathrm{wt}\,(R) = 6$ by the paragraph preceding Lemma 18.

     Thus the weights occurring in $\mathcal{G}_{24}$ are $0, 8, 12, 16, 24$. Let A_i be the number of words of weight i. Then $A_0 = A_{24} = 1$, $A_8 = A_{16}$. To each left side L there are two possible right sides, R and $\bar{R}$. If $\mathrm{wt}\,(L) = 0$, then $\mathrm{wt}\,(R) \neq 4$ (by Lemma 21) and $\mathrm{wt}\,(R) \neq 8$ (or else $\mathrm{wt}\,(\bar{R}) = 4$, again violating Lemma 21), so $\mathrm{wt}\,(R) = 0$ or 12. If $\mathrm{wt}\,(L) = 2$, then $\mathrm{wt}\,(R) = 6$ by a similar argument. Proceeding in this way we arrive at the following possibilities for codewords in $\mathcal{G}_{24}$:

Number	wt (L)	wt (R)	wt $(\bar{R})$	total	weight
1	0	0	12	0	12
$11 + \binom{11}{2}$	2	6	6	8	8
$\binom{11}{3} + \binom{11}{4}$	4	4	8	8	12
$\alpha = ?$	6	2	10	8	16
$\beta = ?$	6	6	6	12	12
$\binom{11}{7} + \binom{11}{8}$	8	4	8	12	16
$\binom{11}{9} + \binom{11}{10}$	10	6	6	16	16
1	12	0	12	12	24

But by Lemma 20, α is equal to the number of vectors of type $(2, 6)$, which is

$$2\left(11 + \binom{11}{2}\right).$$

Therefore

$$A_8 = 4\left(11 + \binom{11}{2}\right) + \binom{11}{3} + \binom{11}{4} = 759$$

and so $A_{12} = 2576$. Thus we have shown that the weight distribution of $\mathscr{G}_{24}$ is:

$$i: \quad 0 \quad 8 \quad 12 \quad 16 \quad 24$$
$$A_i: \quad 1 \quad 759 \quad 2576 \quad 759 \quad 1$$

Theorem 22. *Any binary vector of weight 5 and length 24 is covered by exactly one codeword of $\mathscr{G}_{24}$ of weight 8.*

Proof. If a vector of weight 5 were covered by two codewords u, v of weight 8, then dist $(u, v) \leq 6$, a contradiction. Each codeword of weight 8 covers $\binom{8}{5}$ vectors of weight 5, which are all distinct, and

$$759\binom{8}{5} = \binom{24}{5}. \qquad\qquad \text{Q.E.D.}$$

Corollary 23. *The codewords of weight 8 in the extended Golay code $\mathscr{G}_{24}$ form a Steiner system $S(5, 8, 24)$. Hence by Corollary 14 we get Steiner systems $S(4, 7, 23)$, $S(3, 6, 22)$, and $S(2, 5, 21)$.*

We shall refer to codewords in $\mathscr{G}_{24}$ of weights 8 and 12 as *octads* and *dodecads* respectively.

Theorem 24. *The block intersection numbers λ_{ij} for the Steiner system formed by the octads are shown in Fig. 2.14.*

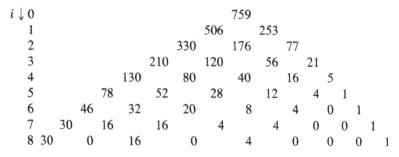

Fig. 2.14. Block intersection numbers λ_{ij} for the octads in the extended Golay code.

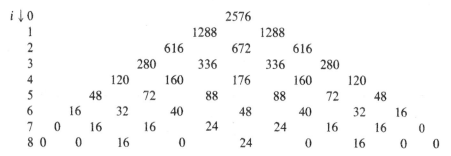

Fig. 2.15. Generalized block intersection members λ_{ij} for the dodecads in $\mathcal{G}_{24}$.

Corollary 25. *The codewords of weight 16 in $\mathcal{G}_{24}$ form a 5-(24, 16, 78) design, and the corresponding Pascal triangle is obtained by reflecting Fig. 2.14 about the middle and omitting the last three rows.*

Proof. This is the complementary design to that formed by the octads.

Q.E.D.

Theorem 26. *The dodecads in $\mathcal{G}_{24}$ form a 5-(24, 12, 48) design. Suppose $P_1, \ldots, P_8$ form an octad. Temporarily misusing the notation, let λ_{ij} be the number of dodecads containing $P_1, \ldots, P_j$ and not containing $P_{j+1}, \ldots, P_i$, for $0 \leq j \leq i \leq 8$. Then the λ_{ij} are as shown in Fig. 2.15.*

Proof. We first show that $\lambda_{66} = \lambda_{76} = \lambda_{86} = 16$. In fact there is an obvious 1-1-correspondence between dodecads containing $P_1, \ldots, P_6$ (and which cannot contain P_7 or P_8), and octads containing P_7 and P_8 but not containing $P_1, \ldots, P_6$. Therefore $\lambda_{66} = 16$ from Fig. 2.14. From this it follows that

$3 \times 16 = 48$ dodecads contain $P_1, \ldots, P_5$, thus $\lambda_{55} = 48$. Therefore the dodecads form a 5-design. The rest of the table can now be filled in from Theorem 9 and the Pascal property.   Q.E.D.

In Chapter 6 we shall give a sufficient condition for codewords of a fixed weight in any code to form a t-design, which includes these results as a special case.

Problems. (12) Show that the Golay code also gives rise to the following designs:

v	k	b	λ_1	λ_2	λ_3	λ_4
23	7	253	77	21	5	1
23	8	506	176	56	16	4
22	6	77	21	5	1	
22	7	176	56	16	4	
22	8	330	120	40	12	

(13) Show that the 4096 cosets of $\mathscr{G}_{24}$ have the following weight distributions:

Number	Weight 0	2	4	6	8	10	12	14	16	18	20	22	24
1	1				759		2576		759				1
276			1	77	352	946	1344	946	352	77		1	
1771				6	64	360	960	1316	960	360	64	6	

Number	Weight 1	3	5	7	9	11	13	15	17	19	21	23
24	1			253	506	1288	1288	506	253			1
2024		1	21	168	640	1218	1218	640	168	21	1	

Definition. The (unextended) *Golay code of length 23*, $\mathscr{G}_{23}$, is obtained by deleting the last coordinate from every codeword of $\mathscr{G}_{24}$. We shall see in Ch. 16 that deleting *any* coordinate from $\mathscr{G}_{24}$ gives an equivalent code.

Theorem 27. $\mathscr{G}_{23}$ *is a* [23, 12, 7] *code, with weight distribution*

i	0	7	8	11	12	15	16	23
A_i	1	253	506	1288	1288	506	253	1

Proof. This follows immediately from Fig. 2.14. For example the codewords of weight 7 in $\mathscr{G}_{23}$ are the octads in $\mathscr{G}_{24}$ with last coordinate equal to 1: the number of these is $\lambda_{11} = 253$.   Q.E.D.

Theorem 28. *The Golay code* $\mathscr{G}_{23}$ *is a perfect triple-error-correcting code.*

Proof. Since the code has minimum distance 7, the cosets of minimum weight $\leqslant 3$ are disjoint (by Theorem 2 of Ch. 1). The number of such cosets is

$$1 + \binom{23}{1} + \binom{23}{2} + \binom{23}{3} = 2^{11} = 2^{23-12},$$

so this includes *all* cosets of $\mathcal{G}_{23}$. Therefore $\mathcal{G}_{23}$ is perfect.   Q.E.D.

§7. The Steiner system S(5, 6, 12), and nonlinear single-error-correcting codes

Lemma 29. *The nonlinear code consisting of the rows of A_{11} (see Fig. 2.13), the sums of pairs of rows of A_{11}, and the complements of all these, is an* (11, 132, 3) *code.*

Proof. There are

$$2\left(11 + \binom{11}{2}\right) = 132$$

codewords. A typical codeword has the form $a + b$ or $a + b + 1$, where a, b are distinct rows of A_{11} (and b may be zero). To show that the distance between codewords is at least 3, we must check 4 cases. (i) dist $(a + b, a + c) =$ wt $(b + c) = 6$, by the paragraph proceeding Lemma 18. (ii) Similarly dist $(a + b, a + c + 1) = 5$. (iii) dist $(a + b, c + d) = $ wt $(a + b + c + d) \geqslant 4$. For if it were less than 4, the corresponding codeword of $\mathcal{G}_{24}$ would have weight less than 8, a contradiction. (iv) Similarly dist $(a + b, c + d + 1) \geqslant 3$.   Q.E.D.

Theorem 30. *The codewords of the extended* (12, 132, 4) *code form the blocks of a Steiner system* $S(5, 6, 12)$.

Proof. Any vector of weight 5 can be covered by at most one block, or else there would be two codewords at distance less than 4 apart, which is impossible. Therefore the 132 blocks cover $132.6 = \binom{12}{5}$ distinct vectors of weight 5. Thus every vector of weight 5 is covered by exactly one block.   Q.E.D.

From Corollary 14 we obtain Steiner systems $S(4, 5, 11)$, $S(3, 4, 10)$, and $S(2, 3, 9)$.

Nonlinear single-error-correcting codes. It will be convenient to rearrange the coordinates of the (12, 132, 4) code of Theorem 30 so that it contains the codeword (or block) 111 111 000 000. Let the coordinates be labelled $x_1, \ldots, x_{12}$. We may now include the 12 vectors of Fig. 2.16 in the code without decreasing the minimum distance, and obtain a $(12, 144, 4)$ code, which we denote by $\mathscr{C}_{12}$.

$$
\begin{array}{cccccccccccc}
x_1 & x_2 & x_3 & x_4 & x_5 & x_6 & x_7 & x_8 & x_9 & x_{10} & x_{11} & x_{12} \\
1 & 1 & 0 & 0 & 0 & 0 & 0 & 0 & 0 & 0 & 0 & 0 \\
0 & 0 & 1 & 1 & 0 & 0 & 0 & 0 & 0 & 0 & 0 & 0 \\
& & & & & \cdot & \cdot & & & & & \\
0 & 0 & 0 & 0 & 0 & 0 & 0 & 0 & 0 & 0 & 1 & 1 \\
0 & 0 & 1 & 1 & 1 & 1 & 1 & 1 & 1 & 1 & 1 & 1 \\
1 & 1 & 0 & 0 & 1 & 1 & 1 & 1 & 1 & 1 & 1 & 1 \\
& & & & & \cdot & \cdot & & & & & \\
1 & 1 & 1 & 1 & 1 & 1 & 1 & 1 & 1 & 1 & 0 & 0 \\
\end{array}
$$

Fig. 2.16. The 12 extra codewords.

Lemma 31. *Some codeword c in $\mathscr{C}_{12}$ is at distance 4 from 51 other codewords.*

Proof. We take $c = 111\,111\,000\,000$. The block intersection numbers λ_{ij} for $S(5, 6, 12)$ are as follows.

$$
\begin{array}{ccccccccccccc}
& & & & & & 132 & & & & & & \\
& & & & & 66 & & 66 & & & & & \\
& & & & 30 & & 36 & & 30 & & & & \\
& & & 12 & & 18 & & 18 & & 12 & & & \\
& & 4 & & 8 & & 10 & & 8 & & 4 & & \\
& 1 & & 3 & & 5 & & 5 & & 3 & & 1 & \\
1 & & 0 & & 3 & & 2 & & 3 & & 0 & & 1 \\
\end{array}
$$

The blocks at distance 4 from c must meet c in exactly 4 places. The number of such blocks is $\binom{6}{4}\lambda_{6,4} = 15.3 = 45$. Six of the codewords of Fig. 2.16 are also at distance 4 from c, for a total of 51.   Q.E.D.

(It is clear from this proof that no codeword is at distance 4 from *more* than 51 codewords.)

We shall construct 4 shorter codes from $\mathscr{C}_{12}$:

(i) Taking the codewords in $\mathscr{C}_{12}$ for which $x_{12} = 1$, we obtain an $(11, 72, 4)$ code $\mathscr{C}_{11}$. A similar argument to that of Lemma 31 shows that there is a codeword in $\mathscr{C}_{11}$ at distance 4 from 34 others.

(ii) Taking the codewords in $\mathscr{C}_{12}$ for which $x_{12} = 1$, $x_{10} = 0$, we obtain a $(10, 38, 4)$ code $\mathscr{C}_{10}^{(a)}$. There is a codeword in $\mathscr{C}_{10}^{(a)}$ at distance 4 from 22 others.

(iii) Taking the codewords in $\mathscr{C}_{12}$ for which $x_{12} = x_{11} = 1$, we obtain a $(10, 36, 4)$ code $\mathscr{C}_{10}^{(b)}$ which contains 0. There is a codeword at distance 4 from 30 others.

(iv) Taking the codewords in $\mathscr{C}_{12}$ for which $x_{12} = x_{11} = 1$, $x_{10} = 0$, we obtain a

$(9, 20, 4)$ code $\mathscr{C}_9$ containing 0. There is a codeword at distance 4 from 18 others. (Another $(9, 20, 4)$ code was given in §4.) This code becomes an $(8, 20, 3)$ code if one coordinate is deleted.

An alternative construction for an $(8, 20, 3)$ code is shown in Fig. 2.17. Note that this code is cyclic, in the sense that a cyclic shift of any codeword is again a codeword!

Since the number of codewords is not a power of 2, none of these codes are linear. As mentioned in §4 the $(9, 20, 4)$ code is optimal.

Research Problems (2.4). Are the $(10, 38, 4)$, $(11, 72, 4)$, $(12, 144, 4)$ codes optimal? At present the best bounds known are $A(10, 4) \leq 40$, $A(11, 4) \leq 80$ and $A(12, 4) \leq 160$ (see Ch. 17 and Appendix A).

(2.5) Generalize the $(12, 144, 4)$ code by finding other good nonlinear codes from the rows of a Hadamard matrix taken $1, 2, \ldots$ at a time.

(2.6) Generalize Fig. 2.17 by finding other good nonlinear cyclic (n, M, d) codes with n greater than $2d$.

```
0 0 0 0 0 0 0 0
1 1 0 1 0 0 0 0
0 1 1 0 1 0 0 0
0 0 1 1 0 1 0 0
0 0 0 1 1 0 1 0
0 0 0 0 1 1 0 1
1 0 0 0 0 1 1 0
0 1 0 0 0 0 1 1
1 0 1 0 0 0 0 1
1 1 1 0 0 1 0 0
0 1 1 1 0 0 1 0
0 0 1 1 1 0 0 1
1 0 0 1 1 1 0 0
0 1 0 0 1 1 1 0
0 0 1 0 0 1 1 1
1 0 0 1 0 0 1 1
1 1 0 0 1 0 0 1
1 0 1 0 1 0 1 0
0 1 0 1 0 1 0 1
1 1 1 1 1 1 1 1
```

Fig. 2.17. An $(8, 20, 3)$ code.

Because the 12 codewords of Fig. 2.16 can be permuted in many ways, there are several inequivalent versions of the codes $\mathscr{C}_9, \ldots, \mathscr{C}_{12}$.

§8. An introduction to the Nordstrom–Robinson code

The extended Golay code $\mathscr{G}_{24}$ may be used to construct some interesting nonlinear double-error-correcting codes.

The construction. It will be convenient to change the order of the columns of $\mathscr{G}_{24}$ so that $\mathscr{G}_{24}$ contains the codeword $1111\,1111\,00\ldots0 = 1^8 0^{16}$. Let G be a generator matrix for the new version of $\mathscr{G}_{24}$.

Now the first 7 columns of G are linearly independent, for otherwise $\mathscr{G}_{24}^{\perp}$ would contain a nonzero codeword of weight ≤ 7. But $\mathscr{G}_{24} = \mathscr{G}_{24}^{\perp}$, so this is impossible. Thus the first 7 coordinates may be taken as information symbols, and the 8th coordinate is the sum of the first 7.

We divide up the codewords according to their values on the first 7 coordinates: there are 2^7 possibilities, and for each of these there are $2^{12}/2^7 =$ 32 codewords. Thus there are $8 \times 32 = 256$ codewords which begin either with seven 0's (with 8th coordinate 0), or with six 0's and a 1 (with 8th coordinate 1).

Definition. The *Nordstrom–Robinson code* $\mathcal{N}_{16}$ is obtained by deleting the first 8 coordinates from these 256 vectors. (See Fig. 2.18, where $\mathcal{N}_{16}$ is enclosed within the double lines.)

Theorem 32. *The Nordstrom–Robinson code* $\mathcal{N}_{16}$ *is a* (16, 256, 6) *code.*

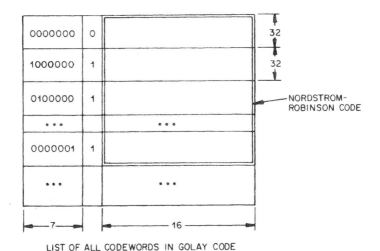

LIST OF ALL CODEWORDS IN GOLAY CODE

Fig. 2.18. Construction of the Nordstrom–Robinson code.

Proof. Let a, b be distinct codewords of $\mathcal{N}_{16}$, obtained by truncating the codewords a', b' of $\mathcal{G}_{24}$. Since dist $(a', b') \geq 8$, dist $(a, b) \geq 6$.   Q.E.D.

Observe that $\mathcal{N}_{16}$ is made up of a linear $[16, 5, 8]$ code $\mathcal{B}$, say (obtained from the codewords of $\mathcal{G}_{24}$ which are zero in the first 8 coordinates), plus 7 of its cosets in $\mathcal{G}_{24}$. In Ch. 15 we shall give several generalizations of the Nordstrom–Robinson code, namely the Kerdock and Preparata codes, which have this same kind of structure.

Problem. (14) Show that $\mathcal{N}_{16}$ is not linear.

Since $\mathcal{G}_{24}$ contains 1^{24}, the subcode $\mathcal{B}$ contains 1^{16}. All the other codewords of $\mathcal{B}$ must then have weight 8. Therefore the weight distribution of $\mathcal{B}$ is

i	0	8	16
A_i	1	30	1.

Consider one of the cosets of $\mathcal{B}$ in $\mathcal{N}_{16}$, say $\mathcal{B}_i = v_i + \mathcal{B}$. Clearly the only weights which can occur in $\mathcal{B}_i$ are

$$6 = 8 - 2, \qquad 10 = 12 - 2, \qquad 14 = 16 - 2$$

Since $1 \in \mathcal{B}$, if $\mathcal{B}_i$ contained a vector of weight 14, it would also contain one of weight 2, which is impossible. Also $A_6 = A_{10}$. Therefore the weight distribution of $\mathcal{B}_i$ is

i	6	10
A_i	16	16

(We shall see in Ch. 14 that $\mathcal{B}$ is a first-order Reed Muller code, and the vectors in $\mathcal{B}_i$ are bent functions.) Putting all this together we find that the weight distribution of $\mathcal{N}_{16}$ is as shown in Fig. 2.19.

i	0	6	8	10	16
A_i	1	112	30	112	1

Fig. 2.19. Weight distribution of Nordstrom–Robinson code $\mathcal{N}_{16}$.

Problems. (15) Show that if c is *any* codeword of $\mathcal{N}_{16}$, the number of codewords at distance i from c is also given by Fig. 2.19. Thus $\mathcal{N}_{16}$ is distance invariant. (Hint: look at the weight distribution of the translated code $c + \mathcal{N}_{16}$, and use Fig. 2.18).

Thus we know that the distance distribution of $\mathcal{N}_{16}$ is equal to its weight distribution and is given by Fig. 2.19.

By shortening $\mathcal{N}_{16}$ we obtain $(15, 128, 6)$, $(14, 64, 6)$ and $(13, 32, 6)$ codes. All four are optimal (Ch. 17). It is known that the $(16, 256, 6)$, $(15, 256, 5)$,

$(15, 128, 6)$, $(14, 128, 5)$, $(14, 64, 6)$, $(13, 64, 5)$ and $(13, 32, 6)$ codes are unique, but the $(12, 32, 5)$ shortened code is not.

The following problems give alternative constructions of a $(12, 32, 5)$ code and of $\mathcal{N}_{16}$.

(16) (Van Lint)

Let
$$I = \begin{pmatrix} 100 \\ 010 \\ 001 \end{pmatrix}, \qquad J = \begin{pmatrix} 111 \\ 111 \\ 111 \end{pmatrix}, \qquad P = \begin{pmatrix} 010 \\ 001 \\ 100 \end{pmatrix}, \qquad Q = P^2$$

$= P^T = J - I - P$. Show that $\mathbf{0}$ and the rows of the matrices $\alpha, \beta, \gamma, \delta$ form a $(12, 32, 5)$ code, where

$$\alpha = \begin{pmatrix} J-I & I & I & I \\ I & J-I & I & I \\ I & I & J-I & I \\ I & I & I & J-I \end{pmatrix}, \qquad \beta = \begin{pmatrix} J & P & I & Q \\ P & J & Q & I \\ I & Q & J & P \\ Q & I & P & J \end{pmatrix},$$

$$\gamma = (J - I, J - I, J - I, J - I),$$

$$\delta = \begin{pmatrix} 000 & 111 & 111 & 111 \\ 111 & 000 & 111 & 111 \\ 111 & 111 & 000 & 111 \\ 111 & 111 & 111 & 000 \end{pmatrix}.$$

(17) (Semakov and Zinov'ev.) Let A, B, C, D be 4×4 permutation matrices such that $A + B + C + D = J$, let $K = A + D$, $L = A + C$ and $M = A + B$. Show that the rows of $\alpha_1, \ldots, \alpha_8$ and their complements form a $(16, 256, 6)$ code, where

$$\alpha_1 = \begin{bmatrix} \bar{A} & A & A & A \\ A & \bar{A} & A & A \\ A & A & \bar{A} & A \\ A & A & A & \bar{A} \end{bmatrix}, \qquad \alpha_2 = \begin{bmatrix} \bar{A} & B & C & D \\ B & \bar{A} & D & C \\ C & D & \bar{A} & B \\ D & C & B & \bar{A} \end{bmatrix}, \qquad \alpha_3 = \begin{bmatrix} \bar{A} & C & D & B \\ C & \bar{A} & B & D \\ D & B & \bar{A} & C \\ B & D & C & \bar{A} \end{bmatrix},$$

$$\alpha_4 = \begin{bmatrix} \bar{A} & D & B & C \\ D & \bar{A} & C & B \\ B & C & \bar{A} & D \\ C & B & D & \bar{A} \end{bmatrix}, \qquad \alpha_5 = \begin{bmatrix} O & K & L & M \\ K & O & M & L \\ L & M & O & K \\ M & L & K & O \end{bmatrix}, \qquad \alpha_6 = \begin{bmatrix} O & L & M & K \\ L & O & K & M \\ M & K & O & L \\ K & M & L & O \end{bmatrix},$$

$$\alpha_7 = \begin{bmatrix} O & M & K & L \\ M & O & L & K \\ K & L & O & M \\ L & K & M & O \end{bmatrix}, \qquad \alpha_8 = \begin{bmatrix} P & P & P & P \\ P & \bar{P} & P & \bar{P} \\ P & P & \bar{P} & \bar{P} \\ P & \bar{P} & \bar{P} & P \end{bmatrix}, \qquad P = \begin{bmatrix} 0 & 0 & 0 & 0 \\ 0 & 1 & 0 & 1 \\ 0 & 0 & 1 & 1 \\ 0 & 1 & 1 & 0 \end{bmatrix},$$

and the bar denotes complementation.

Research Problem 2.7. Generalize the construction of Problems 16 and 17.

§9. Construction of new codes from old (III)

The direct sum construction. Given an (n_1, M_1, d_1) code $\mathscr{C}_1$ and an (n_2, M_2, d_2) code $\mathscr{C}_2$, their *direct sum* consists of all vectors* $|u|v|$, where $u \in \mathscr{C}_1$, $v \in \mathscr{C}_2$. This is clearly an $(n_1 + n_2, M_1 M_2, d = \min\{d_1, d_2\})$ code. Although simple, this construction is not very useful. More intelligent is:

The $|u|u+v|$ construction. Given an (n, M_1, d_1) code $\mathscr{C}_1$, and an (n, M_2, d_2) code $\mathscr{C}_2$, with the same lengths, we may form a new code $\mathscr{C}_3$ consisting of all vectors

$$|u|u+v|, \qquad u \in \mathscr{C}_1, \qquad v \in \mathscr{C}_2.$$

Theorem 33. $\mathscr{C}_3$ *is a* $(2n, M_1 M_2, d = \min\{2d_1, d_2\})$ *code.*

Proof. Let $a = |u|u+v|$, $b = |u'|u'+v'|$ be distinct codewords of $\mathscr{C}_3$, where $u, u' \in \mathscr{C}_1$, $v, v' \in \mathscr{C}_2$. If $v = v'$ then $\text{dist}(a, b) = 2 \, \text{dist}(u, u') \geq 2d_1$. Now suppose $v \neq v'$. Then

$$\text{dist}(a, b) = \text{wt}(u - u') + \text{wt}(u - u' + v - v')$$
$$\geq \text{wt}(u - u') + \text{wt}(v - v') - \text{wt}(u - u'),$$

by Problem 8 of Ch. 1

$$= \text{wt}(v - v') \geq d_2.$$

Q.E.D.

This construction builds up good codes very quickly:

Examples. Taking $\mathscr{C}_1 = [4, 3, 2]$ even weight code, $\mathscr{C}_2 = [4, 1, 4]$ repetition code, we get $\mathscr{C}_3 = [8, 4, 4]$ first order Reed–Muller code. Again with $\mathscr{C}_1 =$ this $[8, 4, 4]$ code, $\mathscr{C}_2 = [8, 1, 8]$ code, we get $\mathscr{C}_3 = [16, 5, 8]$ first order Reed–Muller code. In fact all Reed–Muller codes can be built up in this way, as we shall see in Ch. 13.

Remarks. (a) If $\mathscr{C}_1$ and $\mathscr{C}_2$ are of different lengths the construction still works if we add enough zeros at the end of the shorter code.

(b) If $\mathscr{C}_1$ and $\mathscr{C}_2$ are linear, say $\mathscr{C}_1 = [n_1, k_1, d_1]$, $\mathscr{C}_2 = [n_2, k_2, d_2]$, then $\mathscr{C}_3$ is an $[n_1 + \max\{n_1, n_2\}, k_1 + k_2, d = \min\{2d_1, d_2\}]$ linear code.

---

*If $u = u_1 \ldots u_m$, $v = v_1 \ldots v_n$ then $|u|v|$ denotes the vector $u_1 \ldots u_m v_1 \ldots v_n$ of length $m + n$.

An infinite family of nonlinear single-error-correcting codes. Starting with the $(8, 20, 3), \ldots, (11, 144, 3)$ codes given in §7, we can construct an infinite family:

$$
\begin{array}{ccc}
\mathscr{C}_1, & \mathscr{C}_2 \rightarrow & \mathscr{C}_3
\end{array}
$$

$$(8, 128, 2), (8, 20, 3) \rightarrow (16, \tfrac{20}{16} \cdot 2^{11}, 3)$$

$$(9, 256, 2), (9, 20, 4) \rightarrow (18, \tfrac{20}{16} \cdot 2^{12}, 4) \rightarrow (17, \tfrac{20}{16} \cdot 2^{12}, 3)$$

$$\cdots$$

$$(12, 2^{11}, 2), (12, 144, 4) \rightarrow (24, \tfrac{18}{16} \cdot 2^{18}, 4) \rightarrow (23, \tfrac{18}{16} \cdot 2^{18}, 3)$$

$$(16, 2^{15}, 2), (16, \tfrac{20}{16} \cdot 2^{11}, 3) \rightarrow (32, \tfrac{20}{16} \cdot 2^{26}, 3)$$

$$\cdots$$

Continuing in this way we obtain

Theorem 34. *For any block length n satisfying $2^m \leq n < 3.2^{m-1}$ there exists a nonlinear $(n, \lambda \cdot 2^{n-m-1}, 3)$ code, where $\lambda = \tfrac{20}{16}, \tfrac{19}{16}$ or $\tfrac{18}{16}$.*

Remarks. It follows from the sphere-packing bound (Theorem 6 of Ch. 1) that the largest single-error-correcting *linear* code of the same length, which is an $[n, n - m - 1, 3]$ shortened Hamming code, has only 2^{n-m-1} codewords.

Vasil'ev nonlinear single-error-correcting codes. The Hamming codes that were constructed in §7 of Ch. 1 are unique: any binary linear code with the parameters $[n = 2^m - 1, k = n - m, d = 3]$ is equivalent to a Hamming code. This is so because the parity-check matrix must contain all $2^m - 1$ nonzero binary m-tuples in some order.

But this is not true if the assumption of linearity is dropped.

Problems. (18) *Vasil'ev codes.* Let $\mathscr{C}$ be an $(n = 2^m - 1, M = 2^{n-m}, d = 3)$ perfect single-error-correcting binary code, not necessarily linear. Let λ be any mapping from $\mathscr{C}$ to GF(2), with $\lambda(0) = 0$, which is strictly nonlinear: $\lambda(u + v) \neq \lambda(u) + \lambda(v)$ for some $u, v \in \mathscr{C}$. Set $\pi(u) = 0$ or 1 depending on whether wt (u) is even or odd.

Show that the code

$$\mathscr{V} = \{|u|u + v|\pi(u) + \lambda(v)| : u \in F^n, v \in \mathscr{C}\}$$

is a $(2^{m+1} - 1, 2^{2n-m}, 3)$ perfect single-error-correcting code which is not equivalent to any linear code. Show that such codes exist for all $m \geq 3$.

(19) Let $\mathscr{C}_i$ $(i = 1, 2)$ be $[n, k_i]$ linear codes. Show that (i) if $\mathscr{C} = \{|u|u| : u \in \mathscr{C}_1\}$ then $\mathscr{C}^\perp = \{|a|a + v| : a \text{ arbitrary}, v \in \mathscr{C}_1^\perp\}$, (ii) if $\mathscr{C} = \{|u|u + v| : u \in \mathscr{C}_1, v \in \mathscr{C}_2\}$ then $\mathscr{C}^\perp = \{|a + b|b| ; a \in \mathscr{C}_1^\perp, b \in \mathscr{C}_2^\perp\}$, (iii) if $\mathscr{C} =$

$\{|a + x|b + x|a + b + x|: \ a, b \in \mathscr{C}_1, \ x \in \mathscr{C}_2\}$ then $\mathscr{C}^\perp = \{|u + w|v + w|u + v + w|:$ $u, v \in \mathscr{C}_1^\perp, \ w \in \mathscr{C}_2^\perp\}$.

Notes on Chapter 2

§1. The epigraph is from [876]. Plotkin [1064] was the first to seriously study nonlinear codes. That the largest double-error-correcting linear code of length 11 contains 16 codewords follows from Helgert and Stinaff's useful table [636]. For connections between codes and sphere-packing see Leech and Sloane [810].

§2. Theorem 1 and Corollary 4 are due to Plotkin [1064].

§3. References on Hadamard matrices are Baumert [81], Bussemaker and Seidel [221], Goethals and Seidel [501], Golomb [522], Hadamard [572], Hall [583, 586–588, 590], Van Lint [853], Paley [1019], Ryser [1136], Sylvester [1296], Thoene and Golomb [1320], Todd [1328], Turyn [1345–1347], Wallis et al. [1386] and Whiteman [1413].

For properties of quadratic residues see LeVeque [825, I, Ch. 5], Perron [1035], Ribenboim [1111], and Uspensky and Heaslet [1359, Ch. 10].

Plotkin [1064] and Bose and Shrikhande [187] construct binary codes from Hadamard matrices. Semakov et al. [1179–1182] have generalized these constructions to fields with q elements. Theorem 8 is from Levenshtein [819]. The largest possible $[n, k, d]$ *linear* code in the region $n \leqslant 2d$ is given in Theorem 27 of Ch. 17.

§4. Conference matrices were introduced by Belevitch [95–97]. Other valuable references are Delsarte et al. [366], Goethals and Seidel [500], Van Lint [853], Van Lint and Seidel [856], Paley [1019], Turyn [1346], and Wallis et al. [1386]. The codes C_n were given by Sloane and Seidel [1238]. Conference matrices have been used to construct Hadamard matrices (Paley [1019], Raghavarao [1085, §17.4]), in network theory (Belevitch [95–97]), in weighing designs (Raghavarao [1085, Ch. 17]), and in studying strongly regular graphs (Seidel [1175–1177]).

§5. References for t-designs are Alltop [21–24], Bose [173, 177], Carmichael [250], Collens [300], Dembowski [370], Doyen and Rosa [386], Hall [585, 587], Hanani [595–600], Hughes [673], Hughes and Piper [675], DiPaola et al. [1020], Raghavarao [1085], Rokowska [1122], Ryser [1136], Stanton and Collens [1266], Vajda [1360, 1361], Wilson [1420–1422] and Witt [1423, 1424].

§6. $\mathscr{G}_{23}$ was discovered by Golay in 1949 [506]. For the extensive literature about this code see Ch. 20.

§7. This construction of $S(5, 6, 12)$ is due to Leech [803]. This design is unique – see Problem 19 of Ch. 20. $\mathscr{C}_9 - \mathscr{C}_{12}$ were discovered by Golay [509] and Julin [701]. The constructions given here are from Leech and Sloane [810] and Sloane and Whitehead [1239].

§8. $\mathscr{N}_{16}$ was found by Nordstrom and Robinson [1002] (see also [1119, 1253]) and independently discovered by Semakov and Zinov'ev who gave the alternative construction of Problem 17 in [1180]. The construction from the Golay code is due to Goethals [493] and Semakov and Zinov'ev [1181]. The $(13, 64, 5)$ and $(12, 32, 5)$ codes were found by Stevens and Bouricius [1277] in 1959, and rediscovered by Nadler [982] and Green [556]. The $(15, 256, 5)$, $(14, 128, 5)$, $(13, 64, 5)$ and $(12, 32, 5)$ codes contain twice as many codewords as any linear code with the same length and minimum distance, since from Fontaine and Peterson [434] or Calabi and Myrvaagnes [228] there is no $[12, 5, 5]$ linear code. Problem 16 is due to Van Lint [851]. For the uniqueness of these codes see Goethals [497] and Snover [1247].

§9. The $|u|u + v|$ construction was apparently first given by Plotkin [1064], and rediscovered by Sloane and Whitehead [1239]. See also Liu et al. [858], where a generalization is used to construct the Nordstrom–Robinson code. Theorem 34 is from [1239], where other applications of this construction will be found. Problem 18 is from Vasil'ev [1366].

3

An introduction to BCH codes and finite fields

§1. Double-error-correcting BCH codes (I)

Hamming codes, we saw in Chapter 1, are single-error-correcting codes. The codes which in some sense generalize these to correct t errors are called Bose–Chaudhuri–Hocquenghem codes (or BCH codes for short), and we introduce them in this chapter. We shall also introduce one of the central themes in coding theory, namely the theory of finite fields.

We begin by attempting to find a generalization of the Hamming codes which will correct two errors.

The (binary) Hamming code of length $n = 2^m - 1$ needed m parity checks to correct one error. A good guess is that $2m$ parity checks will be needed to correct two errors. So let's try to construct the parity check matrix H' of the double-error-correcting code, by adding m more rows to the parity check matrix H of the Hamming code.

As an example take $m = 4$, $n = 15$. Then H has as columns all nonzero 4-tuples:

$$H = \begin{bmatrix} 0 & 0 & 0 & 0 & 0 & 0 & 0 & 1 & 1 & 1 & 1 & 1 & 1 & 1 & 1 \\ 0 & 0 & 0 & 1 & 1 & 1 & 1 & 0 & 0 & 0 & 0 & 1 & 1 & 1 & 1 \\ 0 & 1 & 1 & 0 & 0 & 1 & 1 & 0 & 0 & 1 & 1 & 0 & 0 & 1 & 1 \\ 1 & 0 & 1 & 0 & 1 & 0 & 1 & 0 & 1 & 0 & 1 & 0 & 1 & 0 & 1 \end{bmatrix},$$

which we abbreviate to

$$H = [1, 2, 3, \ldots, 14, 15], \tag{1}$$

where each entry i stands for the corresponding binary 4-tuple. We are going to add 4 more rows to H, say

$$H' = \begin{bmatrix} 1 & 2 & 3 & & 15 \\ f(1) & f(2) & f(3) & \cdots & f(15) \end{bmatrix}, \tag{2}$$

where each $f(i)$ is also a 4-tuple of 0's and 1's. The ith column of H' is

$$H_i = \binom{i}{f(i)},\tag{3}$$

a column vector of length 8.

How do we choose $f(i)$? Suppose 2 errors occurred, in positions i and j. The syndrome (from Theorem 4 of Ch. 1), is

$$S = H_i + H_j$$

$$= \binom{i+j}{f(i)+f(j)}$$

$$= \binom{z_1}{z_2} \quad \text{say.}$$

We must choose $f(i)$ so that the decoder can find i and j from S, i.e., can solve the simultaneous equations

$$i + j = z_1$$

$$f(i) + f(j) = z_2\tag{4}$$

for i and j, given z_1 and z_2. But all of $i, j, f(i), f(j), z_1, z_2$ are 4-tuples.

In order to solve these equations we would like to be able to add, subtract, multiply and divide 4-tuples. In other words we want to make 4-tuples into a *field*. We next describe the construction of this field and then return to the problem of finding double-error-correcting codes

Definition. A *field* is a set of elements in which it is possible to add, subtract, multiply and divide (except that division by 0 is not defined). Addition and multiplication must satisfy the commutative, associative, and distributive laws: for any α, β, γ in the field

$$\alpha + \beta = \beta + \alpha, \qquad \alpha\beta = \beta\alpha,$$

$$\alpha + (\beta + \gamma) = (\alpha + \beta) + \gamma, \qquad \alpha(\beta\gamma) = (\alpha\beta)\gamma,$$

$$\alpha(\beta + \gamma) = \alpha\beta + \alpha\gamma;$$

and furthermore elements $0, 1, -\alpha, \alpha^{-1}$ (for all α) must exist such that:

$$0 + \alpha = \alpha, \qquad (-\alpha) + \alpha = 0, \qquad 0\alpha = 0,$$

$$1\alpha = \alpha, \qquad \text{and if } \alpha \neq 0, \qquad (\alpha^{-1})\alpha = 1.$$

A *finite* field contains a finite number of elements, this number being called the *order* of the field. Finite fields are called Galois fields after their discoverer.

§2. Construction of the field GF(16)

The 4-tuples of 0's and 1's can clearly be added by vector addition, and in our case subtraction is the same as addition. Furthermore $a_0a_1a_2a_3 + a_0a_1a_2a_3 = 0$. We must however define a multiplication. To do this we associate with each 4-tuple a polynomial in α:

4-tuple	Polynomial
0000	0
1000	1
0100	α
1100	$1 + \alpha$
0010	α^2
1010	$1 + \alpha^2$
0001	α^3
. . .	. . .
1111	$1 + \alpha + \alpha^2 + \alpha^3$

Multiplying two of these polynomials will often give a polynomial of degree greater than 3, i.e., something which is not in our set of objects. E.g.

$$1101 \cdot 1001 \leftrightarrow (1 + \alpha + \alpha^3)(1 + \alpha^3) = 1 + \alpha + \alpha^4 + \alpha^6.$$

We want to reduce the answer to a polynomial of degree $\leqslant 3$. To do this we agree that α will satisfy a certain fixed equation of degree 4; a suitable equation is

$$\pi(\alpha) = 1 + \alpha + \alpha^4 = 0 \quad \text{or} \quad \alpha^4 = 1 + \alpha.$$

Then $\alpha^5 = \alpha + \alpha^2$, $\alpha^6 = \alpha^2 + \alpha^3$, so

$$1 + \alpha + \alpha^4 + \alpha^6 = 1 + \alpha + 1 + \alpha + \alpha^2 + \alpha^3 = \alpha^2 + \alpha^3.$$

This is equivalent to dividing by $\alpha^4 + \alpha + 1$ and keeping the remainder:

$$
\begin{array}{r}
\alpha^2 + 1 \\
\alpha^4 + \alpha + 1 \, \overline{)\alpha^6 + \alpha^4 + \alpha + 1} \\
\alpha^6 + \alpha^3 + \alpha^2 \\
\hline
\alpha^4 + \alpha^3 + \alpha^2 + \alpha + 1 \\
\alpha^4 + \alpha + 1 \\
\hline
\alpha^3 + \alpha^2 = \text{remainder}
\end{array}
$$

Thus the product 1101 times 1001 is

$$(1 + \alpha + \alpha^3)(1 + \alpha^3) = 1 + \alpha + \alpha^4 + \alpha^6 = (1 + \alpha^2)\pi(\alpha) + \alpha^2 + \alpha^3$$

$$= \alpha^2 + \alpha^3 \quad \text{since} \quad \pi(\alpha) = 0$$

$$\leftrightarrow 0011$$

Another way of describing this process is that we reduce a product of polynomials modulo $\pi(\alpha)$:

$$1 + \alpha + \alpha^4 + \alpha^6 = (1 + \alpha^2)\pi(\alpha) + \alpha^2 + \alpha^3$$
$$\equiv \alpha^2 + \alpha^3 \bmod \pi(\alpha).$$

Similarly $\alpha^4 \equiv 1 + \alpha \bmod \pi(\alpha)$, etc.

Now if this multiplication is to have an inverse, which it must if our system is to be a field, $\pi(x)$ must be irreducible over GF(2).

Definition. A polynomial is *irreducible* over a field if it is not the product of two polynomials of lower degree in the field. Loosely speaking an irreducible polynomial is like a prime number: it has no nontrivial factors. Any polynomial can be written uniquely (apart from a constant factor) as the product of irreducible polynomials (just as any number can be written uniquely as the product of prime numbers). We shall see in a moment that $x^4 + x + 1$ is irreducible over GF(2).

Theorem 1. *If $\pi(x)$ is irreducible, then every nonzero polynomial $B(\alpha)$ of degree $\leqslant 3$ has a unique inverse $B(\alpha)^{-1}$ such that*

$$B(\alpha) . B(\alpha)^{-1} \equiv 1 \bmod \pi(\alpha).$$

Proof. Look at the products $A(\alpha)B(\alpha)$ where $A(\alpha)$ runs through all the polynomials

$$1, \alpha, \alpha + 1, \alpha^2, \ldots, \alpha^3 + \alpha^2 + \alpha + 1 \tag{5}$$

of degree $\leqslant 3$. These products must all be distinct mod $\pi(\alpha)$, for if

$$A_1(\alpha)B(\alpha) \equiv A_2(\alpha)B(\alpha) \bmod \pi(\alpha)$$

then $\pi(\alpha) | (A_1(\alpha) - A_2(\alpha))B(\alpha)$ and (since $\pi(\alpha)$ is irreducible) either $\pi(\alpha) | A_1(\alpha) - A_2(\alpha)$ or $\pi(\alpha) | B(\alpha)$. Because the degrees of $A_1(\alpha), A_2(\alpha), B(\alpha)$ are less than the degree of $\pi(\alpha)$ this can only happen if $A_1(\alpha) = A_2(\alpha)$. Thus all the products $A(\alpha)B(\alpha)$ are distinct, and so they must also be equal to (5) in some order. In particular for just one $A(\alpha)$, $A(\alpha)B(\alpha) = 1$, and $A(\alpha) = B(\alpha)^{-1}$.   Q.E.D.

Example. (i) To find the inverse of α, note that $1 = \alpha + \alpha^4 = \alpha(1 + \alpha^3)$ so $\alpha^{-1} = 1 + \alpha^3$.

(ii) To find the inverse of $\alpha + \alpha^2$, suppose it is $a_0 + a_1\alpha + a_2\alpha^2 + a_3\alpha^3$. Then $(\alpha + \alpha^2)(a_0 + a_1\alpha + a_2\alpha^2 + a_3\alpha^3) = 1$, which implies

$$a_2 + a_3 = 1$$
$$a_0 + a_2 = 0$$
$$a_0 + a_1 + a_3 = 0$$
$$a_1 + a_2 = 0$$

whose solution is $a_0 = a_1 = a_2 = 1$, $a_3 = 0$. Therefore the inverse is $1 + \alpha + \alpha^2$.

Logarithm tables (e.g., Fig. 3.1) make finding an inverse much easier, as will be explained below.

Division. To find A/B, first find the inverse $B^{-1} = 1/B$ and then use the rule

$$\frac{A}{B} = A \cdot B^{-1}.$$

We must check that $\pi(x) = x^4 + x + 1$ is irreducible. It has degree 4 and so, if not irreducible, contains a factor of degree 1 or 2. The only polynomials of degree 1 are x and $x + 1$. Clearly $x \nmid \pi(x)$. If $x + 1 \mid \pi(x)$ then $\pi(-1) = 0$. But $\pi(-1) = 1 + 1 + 1 = 1 . \therefore x + 1 \nmid \pi(x)$. What about a factor of degree 2? We can rule out $x^2 + x$ and $x^2 + 1 = (x + 1)^2$. This only leaves $x^2 + x + 1$, which we test by a division using detached coefficients:

```
              11
     111)10011
          111
          ‾‾‾‾
          111
          111
          ‾‾‾‾
            1 = remainder
```

Therefore $x^4 + x + 1$ is irreducible.

Thus we have made the 16 4-tuples of 0's and 1's into a field. This is called the *Galois field* of order 16, abbreviated GF(2^4) or GF(16). The field elements can be written in several different ways, as shown in Fig. 3.1.

We note that the nonzero elements of the field form a cyclic group of order 15 with generator α, where $\alpha^{15} = 1$; and that we have been fortunate enough to choose an irreducible polynomial $\pi(x)$ which has a generator of this group as a zero (*cyclic groups* are defined on p. 96).

α (or any other generator of this cyclic group) is called a *primitive element* of GF(2^4). For example α, α^2, α^4 are primitive but α^3, α^5 are not. A polynomial having a primitive element as a zero is called a *primitive polynomial*. Not all irreducible polynomials are primitive, e.g. $x^4 + x^3 + x^2 + x + 1$ is irreducible, so could be used to generate the field, but is not a primitive polynomial.

Any nonzero element γ of GF(2^4) can be written uniquely as a power of α, say

$$\gamma = \alpha^i \quad \text{for } 0 \leqslant i \leqslant 14.$$

as a 4-tuple	as a polynomial	as a power of α	logarithm
0000	0	0	$-\infty$
1000	1	1	0
0100	α	α	1
0010	α^2	α^2	2
0001	α^3	α^3	3
1100	$1+\alpha$	α^4	4
0110	$\alpha+\alpha^2$	α^5	5
0011	$\alpha^2+\alpha^3$	α^6	6
1101	$1+\alpha+\alpha^3$	α^7	7
1010	$1+\alpha^2$	α^8	8
0101	$\alpha+\alpha^3$	α^9	9
1110	$1+\alpha+\alpha^2$	α^{10}	10
0111	$\alpha+\alpha^2+\alpha^3$	α^{11}	11
1111	$1+\alpha+\alpha^2+\alpha^3$	α^{12}	12
1011	$1+\alpha^2+\alpha^3$	α^{13}	13
1001	$1+\alpha^3$	α^{14}	14

Fig. 3.1. GF(2^4) generated by $\alpha^4+\alpha+1=0$.

(Of course $\alpha^0 = \alpha^{15} = 1$.) Then i is called the *logarithm* (or sometimes the *index*) of γ. It is convenient to say that $0 = \alpha^{-\infty}$.

It is helpful to think of the first representation (columns 1 and 2 of Fig. 3.1) as resembling the representation of a complex number z in rectangular coordinates:

$$z = x + iy,$$

and the second (columns 3 and 4) as the representation in polar coordinates:

$$z = re^{i\theta}.$$

The rectangular representation is best for addition, while the polar representation is best for multiplication.

Indeed, to multiply two field elements, take logarithms and add, remembering that $\alpha^{15} = 1$. (So that the logarithms are manipulated mod 15.)

Example: To multiply 0111 and 1111:

element	log
0111	11
1111	+ 12
	23

But $\alpha^{15} = 1$, so the answer is $0111.1111 = \alpha^{23} = \alpha^{23-15} = \alpha^8 = 1010$.

To find a reciprocal:

$$(1010)^{-1} = (\alpha^8)^{-1} = \alpha^{-8} = \alpha^{15-8} = \alpha^7 = 1101.$$

To find a square root:

$$(0110)^{1/2} = (\alpha^5)^{1/2} = (\alpha^{20})^{1/2} = \alpha^{10} = 1110.$$

We shall see in Chapter 4 that any finite field can be constructed in exactly the same way, and has the property that the multiplicative group of nonzero elements is cyclic, with a primitive element as generator. We shall also see that the number of elements in a finite field is a prime power, and that there is essentially only one field with a given number of elements.

In particular, if we take $\pi(x)$ to be a primitive irreducible polynomial over GF(2) of degree m, we get the field GF(2^m) of all 2^m binary m-tuples.

§3. Double-error-correcting BCH codes (II)

Now that our new field GF(16) makes it possible to do arithmetic with 4-tuples, let's return to the problem of designing the double-error-correcting BCH code of length 15. How should we choose $f(i)$ in Equation (4) so that these equations can be solved (in GF(16))?

A *bad choice* would be $f(i) = ci$, where c is a constant. For then (4) becomes

$$i + j = z_1$$
$$c(i + j) = z_2$$

which are redundant and can't be solved. Another bad choice is $f(i) = i^2$, for $i^2 + j^2 = (i + j)^2 \pmod 2$, and (4) becomes

$$i + j = z_1$$
$$(i + j)^2 = z_2$$

which are also redundant.

A *good choice* is $f(i) = i^3$, for then (4) becomes

$$i + j = z_1 \neq 0$$
$$i^3 + j^3 = z_2 \tag{6}$$

which we can solve. We have

$$z_2 = i^3 + j^3 = (i + j)(i^2 + ij + j^2) = z_1(z_1^2 + ij)$$

$$\therefore \quad ij = \frac{z_2}{z_1} + z_1^2. \tag{7}$$

From (6), (7), i and j are the roots of

$$x^2 + z_1 x + \left(\frac{z_2}{z_1} + z_1^2\right) = 0 \qquad (z_1 \neq 0). \tag{8}$$

Note that if there are no errors, $z_1 = z_2 = 0$; while if there is a single error at location i, $z_2 = i^3 = z_1^3$. Thus we have:

Decoding scheme for double-error-correcting BCH code. Receive y, calculate the syndrome $S = Hy^{tr} = \begin{pmatrix} z_1 \\ z_2 \end{pmatrix}$ say. Then

(i) If $z_1 = z_2 = 0$, decide that no errors occurred.

(ii) If $z_1 \neq 0$, $z_2 = z_1^3$, correct a single error at location $i = z_1$.

(iii) If $z_1 \neq 0$, $z_2 \neq z_1^3$, form the quadratic (8). If this has 2 distinct roots i and j, correct errors at these locations.

(iv) If (8) has no roots, or if $z_1 = 0$, $z_2 \neq 0$, detect that at least 3 errors occurred.

Let us repeat that i, j, z_1, z_2 are all elements of GF(16), and that (8) is to be solved in this field. (Unfortunately the usual formula for solving a quadratic equation doesn't work in GF(2^4). One way of finding the roots is by trying each element of the field in turn, and another method will be described in §7 of Ch. 9).

The parity check matrix. Now let us rearrange the matrix H so that the first row is in the order 1, α, α^2, α^3 ... ; i.e., our matrix is

$$H = \begin{pmatrix} 1 & \alpha & \alpha^2 & \alpha^3 & \alpha^4 & \alpha^5 & \alpha^6 & \alpha^7 & \alpha^8 & \alpha^9 & \alpha^{10} & \alpha^{11} & \alpha^{12} & \alpha^{13} & \alpha^{14} \\ 1 & \alpha^3 & \alpha^6 & \alpha^9 & \alpha^{12} & 1 & \alpha^3 & \alpha^6 & \alpha^9 & \alpha^{12} & 1 & \alpha^3 & \alpha^6 & \alpha^9 & \alpha^{12} \end{pmatrix} \quad (9)$$

This has the important advantage, as we shall see in Ch. 7, of making the code cyclic.

Notice that not all powers of α appear in the second row: this is because α^3 is not a primitive element (since $(\alpha^3)^5 = 1$).

Expanding this in binary we obtain

$$H = \begin{pmatrix} 1 & 0 & 0 & 0 & 1 & 0 & 0 & 1 & 1 & 0 & 1 & 0 & 1 & 1 & 1 \\ 0 & 1 & 0 & 0 & 1 & 1 & 0 & 1 & 0 & 1 & 1 & 1 & 1 & 0 & 0 \\ 0 & 0 & 1 & 0 & 0 & 1 & 1 & 0 & 1 & 0 & 1 & 1 & 1 & 1 & 0 \\ 0 & 0 & 0 & 1 & 0 & 0 & 1 & 1 & 0 & 1 & 0 & 1 & 1 & 1 & 1 \\ 1 & 0 & 0 & 0 & 1 & 1 & 0 & 0 & 0 & 1 & 1 & 0 & 0 & 0 & 1 \\ 0 & 0 & 0 & 1 & 1 & 0 & 0 & 0 & 1 & 1 & 0 & 0 & 0 & 1 & 1 \\ 0 & 0 & 1 & 0 & 1 & 0 & 0 & 1 & 0 & 1 & 0 & 0 & 1 & 0 & 1 \\ 0 & 1 & 1 & 1 & 1 & 0 & 1 & 1 & 1 & 1 & 0 & 1 & 1 & 1 & 1 \end{pmatrix} \quad (10)$$

Example of decoding procedure. First let's suppose two errors occurred, say in places 6, 8 (i.e. in the columns (α^6, α^3), (α^8, α^9).) Then $z_1 = 1001 = \alpha^{14}$, $z_2 = 0100 = \alpha$, and $(z_2/z_1) + z_1^2 = \alpha^2 + \alpha^{13} = 1001 = \alpha^{14}$. Thus the equation (8) for i, j is

$$x^2 + \alpha^{14} x + \alpha^{14} = (x + \alpha^6)(x + \alpha^8)$$

and indeed the roots give the locations of the errors.

On the other hand, suppose three errors occurred, in places 0, 1, 3 say. Then $z_1 = 1101 = \alpha^7$, $z_2 = 1100 = \alpha^4$, and Equation (8) is

$$x^2 + \alpha^7 x + \alpha^5.$$

By trying each element in turn the decoder finds that this equation has no zeros in the field, and so decides that at least three errors occurred.

Nothing in our construction depends on the length being 15, and plainly we can use any field GF(2^m) to get a double-error-correcting BCH code of length $2^m - 1$. The parity check matrix is

$$H = \begin{pmatrix} 1 & \alpha & \alpha^2 & \cdots & \alpha^{2m-2} \\ 1 & \alpha^3 & \alpha^6 & \cdots & \alpha^{3(2m-2)} \end{pmatrix}, \tag{11}$$

where each entry is to be replaced by the corresponding binary m-tuple.

The decoding scheme given above shows that this code does indeed correct double errors. We return to these codes, and construct t-error-correcting BCH codes, in Chapters 7 and 9.

Problems. (1) Find the locations of the errors if the syndrome is $S = (1001\ 0110)^{tr}$ or $(0101\ 1111)^{tr}$.

(2) If the received vector is $11000 \cdots 0$, what was transmitted?

§4. Computing in a finite field

Since elements of GF(2^4) are represented by 4-tuples of 0's and 1's, they are easy to manipulate using digital circuits or in a binary computer. In this section we give a brief description of some circuits for carrying out computations in GF(2^4). (A similar description could be given for any field GF(2^m).) For further information see Bartee and Schneider [72], Peterson and Weldon [1040, Ch. 7], and Berlekamp [113, Chs. 1–5]. Good references for shift registers are Gill [485], Golomb [523], and Kautz [750].

The basic building blocks are:

Storage element (or flip-flop), contains a 0 or a 1.   Binary adder* (output is 1 iff an odd number of inputs are 1). Also called an EXCLUSIVE-OR gate.

Binary multiplier (output is 1 iff all inputs are 1). Also called an AND gate.

*Strictly speaking this should be called a half-adder, since there is no carry.

Thus an element of $GF(2^4)$ generated by $\alpha^4 + \alpha + 1 = 0$ (see Fig. 3.1) is represented by 4 0's and 1's, which can be stored in a row of 4 storage elements (called a *register*):

$\boxed{0}\,\boxed{0}\,\boxed{1}\,\boxed{1}$ contains the element $0011 \leftrightarrow \alpha^2 + \alpha^3$.

To multiply by α. The circuit shown in Fig. 3.2, called a *linear feedback shift register*, multiplies the contents of the register by α in $GF(2^4)$.

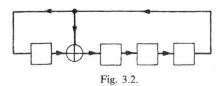

Fig. 3.2.

For if initially it contains

$\boxed{a_0}\,\boxed{a_1}\,\boxed{a_2}\,\boxed{a_3} \leftrightarrow a_0 + a_1\alpha + a_2\alpha^2 + a_3\alpha^3,$

then one time instant later it contains

$$\underset{a_3}{\square}\ \underset{a_0+a_3}{\square}\ \underset{a_1}{\square}\ \underset{a_2}{\square} \leftrightarrow \alpha(a_0 + a_1\alpha + a_2\alpha^2 + a_3\alpha^3)$$
$$= a_3 + (a_0 + a_3)\alpha + a_1\alpha^2 + a_2\alpha^3.$$

If initially this register contains $1000 \leftrightarrow 1$, then at successive time instants it contains $1, \alpha, \alpha^2, \ldots, \alpha^{14}, \alpha^{15} = 1, \alpha, \ldots$, since α is primitive. So the output of the circuit in Fig. 3.3 is periodic with period 15. This is the maximum possible period with 4 storage elements (since there are just $2^4 - 1 = 15$ nonzero states). Segments of length 15 of the output sequence are codewords in a *maximal-length feedback shift register code*, which is a simplex code (see §9 of Ch. 1 and Ch. 14).

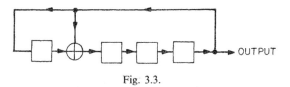

Fig. 3.3.

To multiply by a fixed element. E.g. to multiply an arbitrary element $a_0 + \cdots + a_3\alpha^3$ by $1 + \alpha^2$:

$$(a_0 + a_1\alpha + a_2\alpha^2 + a_3\alpha^3)(1 + \alpha^2) = a_0 + a_1\alpha + (a_0 + a_2)\alpha^2$$
$$+ (a_1 + a_3)\alpha^3 + a_2\alpha^4 + a_3\alpha^5$$
$$= (a_0 + a_2) + (a_1 + a_2 + a_3)\alpha + (a_0 + a_2 + a_3)\alpha^2 + (a_1 + a_3)\alpha^3;$$

which is accomplished by the circuit in Fig. 3.4.

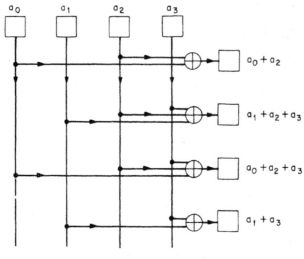

Fig. 3.4.

Similarly to divide by a fixed element, multiply by the inverse.

To multiply two arbitrary field elements. Unfortunately (because this is an essential step, for example, in decoding BCH and Goppa codes, see Chapters 9 and 12), this is considerably more difficult.

Suppose we want to form the product

$$c = a . b = (a_0 + a_1\alpha + a_2\alpha^2 + a_3\alpha^3)(b_0 + b_1\alpha + b_2\alpha^2 + b_3\alpha^3)$$

given a and b.

Methods. (1) (Brute force.) If $c = c_0 + c_1\alpha + c_2\alpha^2 + c_3\alpha^3$ then

$$c_0 = a_0b_0 + a_1b_3 + a_2b_2 + a_3b_1$$
$$c_1 = a_0b_1 + a_1(b_0 + b_3) + a_2(b_2 + b_3) + a_3(b_1 + b_2), \text{etc.,}$$

and a large and complicated circuit is needed to form the c_i's from the a_i's and b_i's.

(2) Mimic long multiplication as done by hand. Thus we write

$$c = b_0a + b_1(\alpha a) + b_2(\alpha^2 a) + b_3(\alpha^3 a),$$

and use the circuit in Fig. 3.5. At each step, add $\alpha^i a$ to c iff $b_i = 1$, and then multiply $\alpha^i a$ by α.

Laws and Rushforth [796] have recently described a cellular array circuit which also multiplies in this way, but is iterative in space rather than in time, and so is faster than the above circuit.

(3) Use log and antilog tables. To multiply two elements of $GF(2^4)$, take

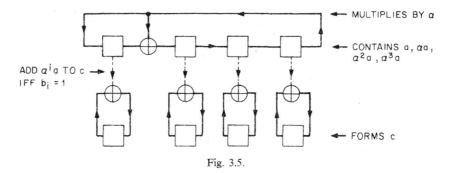

Fig. 3.5.

their logarithms to base α (as in §2), where α is a primitive element of the field, add the logs as integers modulo 15, and take the antilog of the answer. Figure 3.6 shows this schematically.

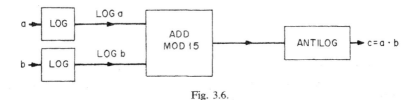

Fig. 3.6.

Unlike the logarithm of a real number, the logarithm in a finite field is an extremely irregular function. No good shortcut is known for finding $\log a$, $a \in \mathrm{GF}(2^4)$. Either one calculates it directly, by computing successive powers of α until a is reached (which is slow), or, better, a log table is used, as in Fig. 3.1 above, or Figs. 4.1, 4.2. This method is fine for $\mathrm{GF}(2^4)$ but is not practicable for $\mathrm{GF}(2^m)$ if m is large, especially as an antilog table of the same length is needed.

(4) Zech's logarithms (Conway [301]). In this scheme only the polar representation (i.e. as a power of a primitive element α) of the field elements is used. Multiplication is now easy, but what about addition? This is carried out by using Zech's logarithms. The Zech's logarithm of n is defined by the equation

$$1 + \alpha^n = \alpha^{Z(n)}$$

(see Fig. 3.7). Then to add α^m, α^n:

$$\alpha^m + \alpha^n = \alpha^m(1 + \alpha^{n-m}) = \alpha^{m+Z(n-m)}.$$

Thus the antilog table has been eliminated. For example,

$$\alpha^3 + \alpha^5 = \alpha^3(1 + \alpha^2) = \alpha^3 \alpha^{Z(2)} = \alpha^{11}.$$

n	$Z(n)$	n	$Z(n)$
$-\infty$	0	7	9
0	$-\infty$	8	2
1	4	9	7
2	8	10	5
3	14	11	12
4	1	12	11
5	10	13	6
6	13	14	3

Gives $Z(n)$ where
$1 + \alpha^n = \alpha^{Z(n)}$

Fig. 3.7. Zech's logarithms in GF(2^4).

Notes on Chapter 3

§2. The theorem that any polynomial over a field can be written uniquely as the product of irreducible polynomials may be found in any textbook on algebra – see for example Albert [19, p. 49] or Van der Waerden [1376, Vol. 1, p. 60].

§3. *The decoding scheme.* Note that not all quadratic equations can be solved in GF(2^4) – see Berlekamp [113, p. 243]. This decoding scheme is *incomplete*, for it doesn't correct those triple errors that the code is capable of correcting – see Ch. 9. Other references on computations in Galois fields are Beard [92], Levitt and Kautz [827] and Tanaka et al. [1300]. See also the Notes to Ch. 4.

4

Finite fields

§1. Introduction

Finite fields are used in most of the known construction of codes, and for decoding. They are also important in many branches of mathematics, e.g. in constructing block designs, finite geometries (see Appendix B), etc. This chapter gives a description of these fields.

The field $GF(2^4)$ was defined in Ch. 3 to consist of all polynomials in x with binary coefficients and degree at most 3, with calculations performed modulo the irreducible polynomial $\pi(x) = x^4 + x + 1$. This chapter will show that all finite fields can be obtained in this way.

The fields $GF(p)$, $p = prime$. The simplest fields are the following. Let p be a prime number. Then the integers modulo p form a field of order p, denoted by $GF(p)$ or Z_p. The elements of $GF(p)$ are $\{0, 1, 2, \ldots, p-1\}$, and $+, -, \times, \div$ are carried out mod p.

E.g. $GF(2)$ is the binary field $\{0, 1\}$. $GF(3)$ is the ternary field $\{0, 1, 2\}$, with $1 + 2 = 3 = 0 \pmod 3$, $2 \cdot 2 = 4 = 1 \pmod 3$, $1 - 2 = -1 = 2 \pmod 3$, etc.

Problems. (1) Check that $GF(p)$ is a field.
(2) Write out the addition and multiplications tables for $GF(5)$ and $GF(7)$.

If a field E contains a field F we say that E is an *extension* of F. E.g. the field of real numbers is an extension of the field of rational numbers.

Construction of $GF(p^m)$. The same construction that was used in Ch. 3 to construct $GF(2^4)$ from $GF(2)$ will work in general, provided we know a polynomial $\pi(x)$ which (i) has coefficients from $GF(p)$, and (ii) is irreducible over

GF(p); i.e., is not the product of two polynomials of lower degree with coefficients from GF(p). (Such a polynomial will be described briefly as being *irreducible over* GF(p). Corollary 16 will show that such polynomials always exist.)

E.g. $x^2 + x + 1 = (x + 2)^2$ is not irreducible over GF(3), but $x^2 + x + 2$ and $x^2 + 1$ are.

If $\pi(x)$ has degree m, we proceed as in Ch. 3 and obtain:

Theorem 1. *Suppose $\pi(x)$ is irreducible over* GF(p) *and has degree m. Then the set of all polynomials in x of degree $\leq m - 1$ and coefficients from* GF(p), *with calculations performed modulo $\pi(x)$, forms a field of order p^m.*

We shall see later (Theorem 6) that there is essentially only one field of order p^m. This is called a Galois field and is denoted by GF(p^m).

Any member of GF(p^m) can also be written as an m-tuple of elements from GF(p), just as we saw in Ch. 3.

The members of GF(p^m) may also be described as residue classes (cf. §3 of Ch. 2) of the polynomials in x with coefficients from GF(p), reduced modulo $\pi(x)$. Let α denote the residue class of x itself, i.e. the element $0100 \cdots 0$ of GF(p^m). Then from the construction of GF(p^m), $\pi(\alpha) = 0$. Thus the equation $\pi(x) = 0$ has a root α in GF(p^m). We say that GF(p^m) was obtained from GF(p) by *adjoining to* GF(p) *a zero of* $\pi(x)$.

Then GF(p^m) consists of all polynomials in α of degree $\leq m - 1$, with coefficients from GF(p). An element of GF(p^m) is:

$$a_0 a_1 \cdots a_{m-1} \leftrightarrow a_0 + a_1 \alpha + \cdots + a_{m-1} \alpha^{m-1},$$

where $a_i \in$ GF(p) and $\pi(\alpha) = 0$.

This construction also works for infinite fields. For example, let Q denote the rational numbers. If we adjoin to Q a zero of the polynomial $x^2 - 2$ we obtain the field $Q(\sqrt{2})$ consisting of all numbers $a + b\sqrt{2}$, with addition and multiplication just as one would expect. Finite fields are a good deal simpler than infinite fields.

Example. Using the irreducible polynomial $\pi(x) = x^2 + x + 2$ over GF(3), we obtain the representation of GF(3^2) shown in Fig. 4.1.

Problems. (3) Find binary irreducible polynomials of degrees 2, 3 and 5. List the elements of GF(8) and carry out some calculations in this field.

(4) Find a polynomial of degree 3 which is irreducible over GF(3).

Outline of chapter. §2 and §3 give the basic theory of finite fields. Any finite field contains p^m elements, for some prime p and integer $m \geq 1$, and there is

$$00 = 0 \qquad = \alpha^{\times}$$
$$10 = 1 \qquad = \alpha^{0}$$
$$01 = \qquad \alpha = \alpha^{1}$$
$$12 = 1 + 2\alpha = \alpha^{2}$$
$$22 = 2 + 2\alpha = \alpha^{3}$$
$$20 = 2 \qquad = \alpha^{4}$$
$$02 = \qquad 2\alpha = \alpha^{5}$$
$$21 = 2 + \quad \alpha = \alpha^{6}$$
$$11 = 1 + \quad \alpha = \alpha^{7}.$$

Fig. 4.1. GF(3^2) with $\pi(\alpha) = \alpha^2 + \alpha + 2 = 0$.

essentially only one such field, which is denoted by GF(p^m) (Theorem 6). Furthermore such a field exists for every p and m (Theorem 7). Theorem 4 shows that every finite field contains a primitive element, enabling us to take logarithms.

Associated with each element of the field is an irreducible polynomial called its minimal polynomial. §3 studies these polynomials, which are important for cyclic codes. In §4 we show how to find irreducible polynomials, using the important formula (8).

§5 contains tables of small fields and primitive polynomials. In §6 it is shown that the automorphism group of GF(p^m) is a cyclic group of order m (Theorem 12). §7 contains a formula (Theorem 15) for the number of irreducible polynomials. The last two §'s discuss different kinds of bases for GF(p^m) considered as a vector space over GF(p). In particular §9 gives a proof of the important (but difficult) normal basis theorem (Theorem 25).

§2. Finite fields: the basic theory

> In Galois fields, full of flowers,
> Primitive elements dance for hours ...
> S.B. Weinstein

The characteristic of a field. Let F be an arbitrary finite field, of order q say. F contains the unit element 1, and since F is finite the elements 1, $1 + 1 = 2$, $1 + 1 + 1 = 3, \ldots$ cannot all be distinct. Therefore there is a smallest number p such that $p = 1 + 1 + \cdots + 1$ (p times) $= 0$. This p must be a prime number (for if $rs = 0$ then $r = 0$ or $s = 0$) and is called the *characteristic* of the field. The field GF(2^4) constructed in Ch. 3 has characteristic 2. If F has characteristic p, then $p\beta = 0$ for any element $\beta \in F$.

Thus F contains the p elements

$$0, 1, 1 + 1 = 2, \qquad 1 + 1 + 1 = 3, \ldots, \qquad 1 + \cdots + 1 = p - 1.$$

The sum and product of such elements have the same form, so these p

elements form a subfield GF(p) of F. If $q = p$, there are no other elements in F, and $F = GF(p)$.

Other elements of F. Supposing $q > p$, we choose a maximal set of elements of F which are linearly independent over GF(p), say $\beta_0 = 1, \beta_1, \ldots, \beta_{m-1}$. Then F contains all the elements

$$a_0\beta_0 + a_1\beta_1 + \cdots + a_{m-1}\beta_{m-1}, \quad a_i \in GF(p),$$

and no others. Thus F is a vector space of dimension m over GF(p), and contains $q = p^m$ elements for some prime p and some integer $m \geq 1$ (or, F has order p^m). Let F^* stand for the set of $q - 1$ nonzero elements of F.

The important special property of finite, as distinct from infinite, fields is:

Theorem 2. F^* *is a cyclic multiplicative group of order* $p^m - 1$. [*A finite multiplicative group is cyclic if it consists of the elements* $1, a, a^2, \ldots, a^{r-1}$, *with* $a^r = 1$. *Then* a *is called a generator of the group.*]

Proof. That F^* is a multiplicative group follows from the definition of a field. Let $\alpha \in F^*$. Since F^* has size $p^m - 1$, α^i has at most $p^m - 1$ distinct values. Therefore there are integers r and i, with $1 \leq r \leq p^m - 1$, such that $\alpha^{r+i} = \alpha^i$, or $\alpha^r = 1$. The smallest such r is called the *order* of α.

Now choose α so that r is as large as possible. We shall show that the order l of any element $\beta \in F^*$ divides r. For any prime π, suppose $r = \pi^a r'$, $l = \pi^b l'$, where r' and l' are not divisible by π. Then α^{π^a} has order r', $\beta^{l'}$ has order π^b, and $\alpha^{\pi^a}\beta^{l'}$ has order $\pi^b r'$ (by Problem 7). Hence $b \leq a$ or else r would not be maximal. Thus every prime power that is a divisor of l is also a divisor of r, and so l divides r. Hence every β in F^* satisfies the equation $x^r - 1 = 0$. This means that $x^r - 1$ is divisible by $\prod_{\beta \in F^*}(x - \beta)$. Since there are $p^m - 1$ elements in F^*, $r \geq p^m - 1$. But $r \leq p^m - 1$, hence $r = p^m - 1$. Thus $\prod_{\beta \in F^*}(x - \beta) = x^{p^m - 1} - 1$, and the nonzero elements of F form the cyclic group $\alpha, \alpha^2, \ldots, \alpha^{p^m - 2}, \alpha^{p^m - 1} = 1$.   Q.E.D.

Corollary 3. (Fermat's theorem.) *Every element* β *of a field* F *of order* p^m *satisfies the identity*

$$\beta^{p^m} = \beta,$$

or equivalently is a root of the equation

$$x^{p^m} = x.$$

Thus

$$x^{p^m} - x = \prod_{\beta \in F}(x - \beta). \tag{1}$$

If F is a field of order p^m, an element α of F is called *primitive* if it has order $p^m - 1$ (cf. §2 of Ch. 3). Then it follows that any nonzero element of F is a power of α. A second corollary to Theorem 2 is:

Theorem 4. *Any finite field F contains a primitive element.*

Proof. Take α to be a generator of the cyclic group F^*.   Q.E.D.

The following lemma is very useful.

Lemma 5. *In any field of characteristic p,*

$$(x + y)^p = x^p + y^p.$$

Proof. From problem 18 of Ch. 1,

$$(x + y)^p = \sum_{k=0}^{p} \binom{p}{k} x^{p-k} y^k.$$

where

$$\binom{p}{0} = \binom{p}{p} = 1.$$

Also if $1 \leqslant k \leqslant p - 1$,

$$\binom{p}{k} = \frac{p(p-1)\cdots(p-k+1)}{1 \cdot 2 \cdots k}$$

$$\equiv 0 \; (\bmod \, p)$$

since the numerator contains a factor of p but the denominator does not.
          Q.E.D.

E.g. in a field of characteristic 2, we have

$$(x + y)^2 = x^2 + y^2, \qquad (x + y)^4 = x^4 + y^4, \qquad (x + y)^8 = x^8 + y^8, \ldots.$$

We end this section with an application of Corollary 3. Using it in the field $GF(p)$ we get the important result that

$$b^{p-1} \equiv 1 \; (\bmod \, p) \tag{2}$$

for any integer b which is not a multiple of p. This is the usual version of Fermat's theorem, which implies for example, that

$$2^6 \equiv 1 \; (\bmod \, 7), \qquad 2^{10} \equiv 1 \; (\bmod \, 11),$$

i.e.,

$$7 \mid 2^6 - 1 = 63, \qquad 11 \mid 2^{10} - 1 = 1023.$$

Problems. (5) Derivative of a polynomial over a finite field. If

$$f(x) = \sum_{i=0}^{n} a_i x^i, \qquad a_i \in GF(p^m),$$

define the derivative

$$\frac{df(x)}{dx} = f(x)' = \sum_{i=1}^{n} i a_i x^{i-1}.$$

Show that (i) $(f(x) + g(x))' = f(x)' + g(x)'$. (ii) $(f(x)g(x))' = f(x)'g(x) + f(x)g(x)'$. (iii) If $(x - \alpha)^k$ divides $f(x)$, show that $(x - \alpha)^{k-1}$ divides $f(x)'$. (iv) If $f(x)$ has no multiple zeros in any extension field, then $f(x)$ and $f(x)'$ are relatively prime. (v) If $p = 2$, $f(x)'$ contains only even powers and is a perfect square. Also $f(x)'' = 0$.

(6) Suppose that F is a finite extension field of $GF(p)$ which contains all the zeros of $x^{p^m} - x$. (i) Show that $x^{p^m} - x$ has distinct zeros in F. [Hint: show that this polynomial and its derivative are relatively prime.] (ii) Prove directly that these zeros form a field.

(7) Let G be a commutative group, containing elements g and h of orders r and s respectively. (i) Show that if $g^n = 1$ then $r \mid n$. (ii) Show that if r and s are relatively prime then gh has order rs. (iii) Show that if $r = r_1 r_2$ then g^{r_1} has order r_2.

(8) The Euler φ-function (or totient function) $\varphi(m)$ is the number of positive integers $\leqslant m$ that are relatively prime to m, for any positive integer m.

(i) Show

$$\varphi(m) = m \prod_{p \mid m} \left(1 - \frac{1}{p}\right),$$

where p runs through the primes dividing m.

(ii) Show

$$\sum_{d \mid m} \varphi(d) = m,$$

where the sum is over all divisors of m, including 1 and m.

(iii) Prove the Fermat–Euler theorem: If a and m are relatively prime, then $a^{\varphi(m)} \equiv 1 \pmod{m}$.

(9) Let A_m be the number of primitive elements in $GF(p^m)$. Show that $A_m = \varphi(p^m - 1)$, where φ is the Euler function defined in Problem 8. Hence show $\lim \inf_{m \to \infty} A_m / p^m = 0$.

§3. Minimal polynomials

Fermat's theorem (Corollary 3) implies that every element β of GF(q), $q = p^m$, satisfies the equation

$$x^q - x = 0. \tag{3}$$

This polynomial has all its coefficients from the prime field GF(p), and is *monic* (has leading coefficient 1). But β may satisfy a lower degree equation than (3).

Definition. The *minimal polynomial over* GF(p) of β is the lowest degree monic polynomial $M(x)$ say with coefficients from GF(p) such that

$$M(\beta) = 0.$$

Example. In GF(2^4), where the minimal polynomials have coefficients equal to 0 or 1, we have:

Element	Minimal Polynomial
0	x
1	$x + 1$
α	$x^4 + x + 1$
α^1	$x^4 + x^3 + 1$
α^3	$x^4 + x^3 + x^2 + x + 1$
α^5	$x^2 + x + 1$

We shall see how to find minimal polynomials in §4.

Properties of minimal polynomials. Suppose $M(x)$ is the minimal polynomial of $\beta \in$ GF(p^m).

Property (M1). $M(x)$ *is irreducible.*

Proof. If $M(x) = M_1(x)M_2(x)$ with the degrees of $M_1(x)$ and $M_2(x)$ both > 0, then $M(\beta) = M_1(\beta)M_2(\beta) = 0$ and so either $M_1(\beta)$ or $M_2(\beta) = 0$, contradicting the fact that $M(x)$ is the *lowest* degree polynomial with β as a root.

Q.E.D.

Property (M2). *If $f(x)$ is any polynomial (with coefficients in* GF(p)*) such that $f(\beta) = 0$, then $M(x) \mid f(x)$.*

Proof. By dividing $M(x)$ into $f(x)$, write

$$f(x) = M(x)a(x) + r(x),$$

where the degree of the remainder $r(x)$ is less than that of $M(x)$. Put $x = \beta$:

$$0 = 0 + r(\beta),$$

and so $r(x)$ is a polynomial of lower degree than $M(x)$ having β as a root. This is a contradiction unless $r(x) = 0$, and then $f(x)$ is divisible by $M(x)$.

Q.E.D.

Property (M3).

$$M(x) \mid x^{p^m} - x$$

Proof. From (M2) and Corollary 3.   Q.E.D.

Property (M4). deg $M(x) \leqslant m$.

Proof. $GF(p^m)$ is a vector space of dimension m over $GF(p)$. Therefore any $m + 1$ elements, such as $1, \beta, \ldots, \beta^m$, are linearly dependent, i.e., there exist coefficients $a_i \in GF(p)$, not all zero, such that

$$\sum_{i=0}^{m} a_i \beta^i = 0.$$

Thus

$$\sum_{i=0}^{m} a_i x^i$$

is a polynomial of degree $\leqslant m$ having β as a root. Therefore deg $M(x) \leqslant m$.

Q.E.D.

Property (M5). *The minimal polynomial of a primitive element of* $GF(p^m)$ *has degree m. Such a polynomial is called a primitive polynomial.*

Proof. Let β be a primitive element of $GF(p^m)$, with minimal polynomial $M(x)$ of degree d. As in Theorem 1 we may use $M(x)$ to generate a field F of order p^d. But F contains β and hence all of $GF(p^m)$, so $d \geqslant m$. By (M4) $d = m$.

Q.E.D.

Note. If an irreducible polynomial $\pi(x)$ is used to construct $GF(p^m)$ and $\alpha \in GF(p^m)$ is a root of $\pi(x)$, then obviously $\pi(x)$ is the minimal polynomial of α.

We can now prove the uniqueness of $GF(p^m)$.

Theorem 6. *All finite fields of order p^m are isomorphic. [Two fields F, G are said to be isomorphic if there is a one-to-one mapping from F onto G which preserves addition and multiplication.]*

Proof. Let F and G be fields of order p^m, and suppose α is a primitive element of F with minimal polynomial $M(x)$. By (M3), $M(x) \mid x^{p^m} - x$. Therefore from Corollary 3 there is an element β (say) of G which has minimal polynomial $M(x)$. Now F can be considered to consist of all polynomials in α of degree $\leq m - 1$, i.e. F consists of polynomials modulo $M(x)$. Furthermore G contains (and therefore consists of) all polynomials in β of degree $\leq m - 1$. Therefore the mapping $\alpha \leftrightarrow \beta$ is an isomorphism $F \leftrightarrow G$.   Q.E.D.

For example, consider the two versions of GF(2^3) shown in Fig. 4.2, one defined by $\pi(x) = x^3 + x + 1$, the other by $\pi(x) = x^3 + x^2 + 1$.

defined by $x^3 + x + 1$	defined by $x^3 + x^2 + 1$
$000 = 0$	$000 = 0$
$100 = 1$	$100 = 1$
$010 = \alpha$	$010 = \gamma$
$001 = \alpha^2$	$001 = \gamma^2$
$110 = \alpha^3$	$101 = \gamma^3$
$011 = \alpha^4$	$111 = \gamma^4$
$111 = \alpha^5$	$110 = \gamma^5$
$101 = \alpha^6$	$011 = \gamma^6$
$(\alpha^7 = 1)$	$(\gamma^7 = 1)$

Fig. 4.2. Two versions of GF(2^3)

Then α and $\beta = \gamma^3$ both have minimal polynomial $x^3 + x + 1$, and $\alpha \leftrightarrow \gamma^3$ is an isomorphism between the two versions. For example $1 + \alpha^2 = \alpha^6$ in the first version becomes $1 + (\gamma^3)^2 = (\gamma^3)^6$ in the second version. We take the first version as our standard version (see Fig. 4.5).

Finite fields can also be represented as irreducible cyclic codes – see Ch. 8. Conversely, GF(p^m) always exists:

Theorem 7. *For any prime p and integer $m \geq 1$, there is a field of order p^m, which is denoted by* GF(p^m). *(By the previous theorem, this field is essentially unique.)*

Proof. For $m = 1$, GF(p) = Z_p was defined in §1. Suppose then that $m > 1$, and set $F_1 = $ GF(p). The idea of the proof is to construct a sequence of fields until we reach one, F_r say, which contains all the zeros of $x^{p^m} - x$. Then from Problem 6 the zeros of $x^{p^m} \to x$ in F_r form the desired field of order p^m. Let

$f_1(x)$ be an irreducible factor of degree ≥ 2 of $x^{p^m} - x$ over F_1 (if there is one), and use the construction of Theorem 1 with $\pi(x) = f_1(x)$ to obtain a new field F_2. Let $f_2(x)$ be an irreducible factor of degree ≥ 2 of $x^{p^m} - x$ over F_2 (if there is one), and again use the construction of Theorem 1 with $\pi(x) = f_2(x)$ to obtain a new field F_3. After a finite number of steps we arrive at a field F_r which contains all the zeros $x^{p^m} - x$. This completes the proof of the theorem.

Alternative proof. From Corollary 16 below there exists an irreducible polynomial of degree m over GF(p). The theorem then follows from the construction of Theorem 1.   Q.E.D.

Problem. (10) Show that the mapping $\gamma \rightarrow \gamma^p$ is an isomorphism from GF(p^m) onto itself.

Subfields of GF(p^m). We could iterate the construction of Theorem 1 (as we did in the proof of Theorem 7). First, obtain the field GF(p^m) from GF(p) by adjoining a zero of a polynomial $\pi(x)$ of degree m which is irreducible over GF(p). Now let $f(x)$ be a polynomial of degree n which is irreducible over GF(p^m). Form a new field from GF(p^m) by adjoining a zero of $f(x)$. The argument used in Ch. 3 now shows that this new field has p^{mn} elements. Hence by Theorem 6 it is GF(p^{mn}). So iterating the construction doesn't give any new fields.

But going from GF(p) to GF(p^{mn}) in two steps does allow us to deduce the following useful theorem..

Theorem 8. (i) GF(p^r) *contains a subfield (isomorphic to)* GF(p^s) *iff s divides r.* (ii) *If $\beta \in$ GF(p^r) then β is in* GF(p^s) *iff $\beta^{p^s} = \beta$. In any field if $\beta^2 = \beta$ then β is 0 or 1.*

The proof requires a lemma.

Lemma 9. *If n, r, s are integers with $n \geq 2$, $r \geq 1$, $s \geq 1$, then*

$$n^s - 1 \mid n^r - 1 \quad \textit{iff} \quad s \mid r.$$

(Recall that the vertical slash means "divides".)

Proof. Write $r = Qs + R$, where $0 \leq R < s$. Then

$$\frac{n^r - 1}{n^s - 1} = n^R \cdot \frac{n^{Qs} - 1}{n^s - 1} + \frac{n^R - 1}{n^s - 1}.$$

Now $n^{Qs} - 1$ is always divisible by $n^s - 1$. The last term is less than 1 and so is an integer iff $R = 0$.   Q.E.D.

Problem. (11) (i) Show that in any field

$$x^s - 1 \mid x^r - 1 \quad \text{iff} \quad s \mid r.$$

(ii) Show that g.c.d. $\{x^r - 1, x^s - 1\} = x^d - 1$, where

$$d = \text{g.c.d.} \{r, s\}.$$

Proof of Theorem 8. (i) If $s \mid r$ then from Problem 6 $GF(p^r)$ contains a subfield isomorphic to $GF(p^s)$. For the converse, let β be a primitive element of $GF(p^s)$. Then

$$\beta^{p^s-1} = 1, \ \beta^{p^r-1} = 1.$$

So $p^s - 1 \mid p^r - 1$, and $s \mid r$ by the lemma.

(ii) The first statement is immediate from Corollary 3, and the second statement is obvious.   Q.E.D.

For example, the subfields of $GF(2^{12})$ are shown in Fig. 4.3.

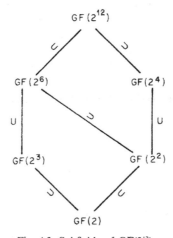

Fig. 4.3. Subfields of $GF(2^{12})$.

Conjugates and cyclotomic cosets.

Property (M6). *β and β^p have the same minimal polynomial. In particular, in $GF(2^m)$, β and β^2 have the same minimal polynomial.*

Proof by example. Suppose $\beta \in GF(2^4)$ has minimal polynomial $x^4 + x^3 + 1$. Then

$$(\beta^2)^4 + (\beta^2)^3 + 1 = (\beta^4 + \beta^3 + 1)^2 \quad \text{by Lemma 5}$$
$$= 0.$$

So by (M2) the minimal polynomial of β^2 divides $x^4 + x^3 + 1$. But $(\beta^2)^8 = \beta$, so we can use the same argument to show that the minimal polynomial of β divides that of β^2. Therefore they are equal. Q.E.D.

Elements of the field with the same minimal polynomial are called *conjugates*. (This is the reason i and $-i$ are called conjugate *complex* numbers – both have minimal polynomial $x^2 + 1$ over the reals.)

Let's look at what happens in GF(2^4). By (M6), the following elements all have the same minimal polynomial: α, α^2, $(\alpha^2)^2 = \alpha^4$, $(\alpha^4)^2 = \alpha^8$ (and $(\alpha^8)^2 = \alpha$ again). Likewise α^3, α^6, α^{12}, $\alpha^{24} = \alpha^9$ (and $\alpha^{18} = \alpha^3$ again) all have the same minimal polynomial, and so on. We see that the powers of α fall into disjoint sets, which we shall call cyclotomic cosets. All α^j where j runs through a cyclotomic coset have the same minimal polynomial.

Definition. The operation of multiplying by p divides the integers mod $p^m - 1$ into sets called the *cyclotomic cosets* mod $p^m - 1$.

The cyclotomic coset containing s consists of

$$\{s, ps, p^2 s, p^3 s, \ldots, p^{m_s - 1} s\}$$

where m_s is the smallest positive integer such that $p^{m_s} \cdot s \equiv s \pmod{p^m - 1}$.

E.g. the cyclotomic cosets mod 15 (with $p = 2$) are:

$$C_0 = \{0\},$$
$$C_1 = \{1, 2, 4, 8\},$$
$$C_3 = \{3, 6, 12, 9\},$$
$$C_5 = \{5, 10\},$$
$$C_7 = \{7, 14, 13, 11\}.$$

Our notation is that if s is the smallest number in the coset, the coset is called C_s. The subscripts s are called the *coset representatives* mod $p^m - 1$.

Problems (12) Verify the cyclotomic cosets for $p = 2$ shown in Fig. 4.4.

(13) Find the cyclotomic cosets mod 8 and 26 ($p = 3$).

Definition of $M^{(i)}(x)$. Let $M^{(i)}(x)$ be the minimal polynomial of $\alpha^i \in$ GF(p^m). Of course by (M6)

$$M^{(pi)}(x) = M^{(i)}(x). \tag{4}$$

From the preceding discussion it follows that if i is in the cyclotomic coset C_s, then in GF(p^m)

$$\prod_{j \in C_s} (x - \alpha^j) \text{ divides } M^{(i)}(x). \tag{5}$$

mod 7	mod 31	mod 63
$C_0 = \{0\}$	$C_0 = \{0\}$	$C_0 = \{0\}$
$C_1 = \{1, 2, 4\}$	$C_1 = \{1, 2, 4, 8, 16\}$	$C_1 = \{1, 2, 4, 8, 16, 32\}$
$C_3 = \{3, 6, 5\}$	$C_3 = \{3, 6, 12, 24, 17\}$	$C_3 = \{3, 6, 12, 24, 48, 33\}$
	$C_5 = \{5, 10, 20, 9, 18\}$	$C_5 = \{5, 10, 20, 40, 17, 34\}$
	$C_7 = \{7, 14, 28, 25, 19\}$	$C_7 = \{7, 14, 28, 56, 49, 35\}$
	$C_{11} = \{11, 22, 13, 26, 21\}$	$C_9 = \{9, 18, 36\}$
	$C_{15} = \{15, 30, 29, 27, 23\}$	$C_{11} = \{11, 22, 44, 25, 50, 37\}$
		$C_{13} = \{13, 26, 52, 41, 19, 38\}$
		$C_{15} = \{15, 30, 60, 57, 51, 39\}$
		$C_{21} = \{21, 42\}$
		$C_{23} = \{23, 46, 29, 58, 53, 43\}$
		$C_{27} = \{27, 54, 45\}$
		$C_{31} = \{31, 62, 61, 59, 55, 47\}$

mod 127

$C_0 = \{$	0	$\}$				
$C_1 = \{$	1	2	4	8	16	32	64$\}$
$C_3 = \{$	3	6	12	24	48	96	65$\}$
$C_5 = \{$	5	10	20	40	80	33	66$\}$
$C_7 = \{$	7	14	28	56	112	97	67$\}$
$C_9 = \{$	9	18	36	72	17	34	68$\}$
$C_{11} = \{$	11	22	44	88	49	98	69$\}$
$C_{13} = \{$	13	26	52	104	81	35	70$\}$
$C_{15} = \{$	15	30	60	120	113	99	71$\}$
$C_{19} = \{$	19	38	76	25	50	100	73$\}$

mod 127 (*cont.*)

$C_{21} = \{$	21	42	84	41	82	37	74$\}$
$C_{23} = \{$	23	46	92	57	114	101	75$\}$
$C_{27} = \{$	27	54	108	89	51	102	77$\}$
$C_{29} = \{$	29	58	116	105	83	39	78$\}$
$C_{31} = \{$	31	62	124	121	115	103	79$\}$
$C_{43} = \{$	43	86	45	90	53	106	85$\}$
$C_{47} = \{$	47	94	61	122	117	107	87$\}$
$C_{55} = \{$	55	110	93	59	118	109	91$\}$
$C_{63} = \{$	63	126	125	123	119	111	95$\}$

Fig. 4.4. Cyclotomic cosets mod 7, 31, 63 and 127.

Problem (14) Show that the coefficients of the LHS of (5), which are the elementary symmetric functions of the α^i's, are in $GF(p)$. [Hint: use Theorem 8(ii).] Conclude that the LHS of (5) is equal to $M^{(i)}(x)$.

Property (M7). *If i is in C_s then*

$$M^{(i)}(x) = \prod_{i \in C_s} (x - \alpha^i) .$$ (6)

Furthermore from Equation (1)

$$x^{p^m-1} - 1 = \prod_s M^{(s)}(x),$$ (7)

where s runs through the coset representatives mod $p^m - 1$.

Problems. (15) When $p = 2$ and $n = 2^m - 1$, show that $C_1 \neq C_3$ and $|C_1| = |C_3| = m$, provided $m \geqslant 3$. Hence deg $M^{(1)}(x) =$ deg $M^{(3)}(x) = m$. What about C_5?

(16) If $\pi(x)$ is an irreducible polynomial over GF(q), say that it *belongs to exponent* e if all its roots have order e. Show that this implies $\pi(x) | x^e - 1$ but $\pi(x) \nmid x^n - 1$ for $n < e$.

(17) Let $\beta \in$ GF(q) have minimal polynomial $M(x)$. Show that deg $M(x) = d$ iff d is the smallest positive integer such that $\beta^{q^d} = \beta$.

(18) If

$$f(x) = \prod_{i=0}^{m-1} (x - \alpha^{sq^i})$$

show that

$$f(x) = [M^{(s)}(x)]^d$$

where $d = m/m_s$.

Representation of the Field by Matrices. The *companion matrix* of the polynomial $a(x) = a_0 + a_1 x + \cdots + a_{r-1}x^{r-1} + x^r$ is defined to be the $r \times r$ matrix

$$M = \begin{bmatrix} 0 & 1 & 0 & \cdots & 0 \\ 0 & 0 & 1 & \cdots & 0 \\ 0 & 0 & 0 & \cdots & 1 \\ -a_0 & -a_1 & -a_2 & \cdots & -a_{r-1} \end{bmatrix}$$

Problems. (19) Show that the characteristic polynomial of M, i.e., det $(M - \lambda I)$, is equal to $a(\lambda)$. Hence deduce that $a(M) = 0$.

(20) Let M be the companion matrix of a primitive polynomial $\pi(x)$ of degree m over GF(q). Show that $M^i = I$ if $i = q^m - 1$, $M^i \neq I$ for $1 \leqslant i < q^m - 1$. Deduce that the powers of M are the nonzero elements of GF(q^m).

For example, take $\pi(x) = x^3 + x + 1$ over GF(2). The elements of GF(2^3) can then be represented as:

$$0 \quad M \quad M^2 \quad M^3 \quad M^4 \quad M^5 \quad M^6 \quad M^7$$

$$\begin{pmatrix} 000 \\ 000 \\ 000 \end{pmatrix} \begin{pmatrix} 010 \\ 001 \\ 110 \end{pmatrix} \begin{pmatrix} 001 \\ 110 \\ 011 \end{pmatrix} \begin{pmatrix} 110 \\ 011 \\ 111 \end{pmatrix} \begin{pmatrix} 011 \\ 111 \\ 101 \end{pmatrix} \begin{pmatrix} 111 \\ 101 \\ 100 \end{pmatrix} \begin{pmatrix} 101 \\ 100 \\ 010 \end{pmatrix} \begin{pmatrix} 100 \\ 010 \\ 001 \end{pmatrix},$$

where addition and multiplication in the field correspond to addition and multiplication of these matrices.

This is a very laborious way of describing the field. It is less trouble to write the first row of each matrix as a polynomial in α, i.e., $M \leftrightarrow \alpha$, $M^2 \leftrightarrow \alpha^2$, $M^3 \leftrightarrow 1+\alpha$ etc., and perform multiplication modulo $\alpha^3 + \alpha + 1$. Of course this gives the same representation of the field that we had in Theorem 1.

§4. How to find irreducible polynomials

The first two theorems present the key formulas.

Theorem 10.

$$x^{p^m} - x = product\ of\ all\ monic\ polynomials,$$
$$irreducible\ over\ GF(p),\ whose$$
$$degree\ divides\ m. \tag{8}$$

Proof. (i) Let $\pi(x)$ be an irreducible polynomial over $GF(p)$ of degree d, where $d \mid m$. The case $\pi(x) = x$ is trivial, so assume $\pi(x) \neq x$. If we use $\pi(x)$ to construct a field, then $\pi(x)$ is the minimal polynomial of one of the elements and $\pi(x) \mid x^{p^{d-1}} - 1$ from (M3). From Lemma 9 and Problem 11, $p^d - 1 \mid p^m - 1$, and $x^{p^{d-1}} - 1 \mid x^{p^{m-1}} - 1$. Therefore $\pi(x) \mid x^{p^m} - x$.

(ii) Conversely let $\pi(x)$ be a divisor of $x^{p^m} - x$, irreducible and of degree d. We must show $d \mid m$. Again we can assume $\pi(x) \neq x$, so that $\pi(x) \mid x^{p^{m-1}} - 1$. As in part (i) we use $\pi(x)$ to construct a field F of order p^d. Let $\alpha \in F$ be a root of $\pi(x)$ and let β be a primitive element of F, say

$$\beta = a_0 + a_1\alpha + \cdots + a_{d-1}\alpha^{d-1}. \tag{9}$$

Now $\pi(\alpha) = 0$, so $\alpha^{p^m} = \alpha$ and from (9) and Lemma 5 $\beta^{p^m} = \beta$. Thus $\beta^{p^{m-1}} = 1$, so the order of β, $p^d - 1$, must divide $p^m - 1$. Therefore $d \mid m$ from Lemma 9.
$$\text{Q.E.D.}$$

A similar argument shows:

Theorem 11. *For any field* $GF(q)$, $q = prime\ power$,

$$x^{q^m} - x = product\ of\ all\ monic\ polynomials,$$
$$irreducible\ over\ GF(q),\ whose$$
$$degree\ divides\ m. \tag{10}$$

We use Equation (7) and Theorem 10 to find irreducible polynomials and minimal polynomials. For example when $q = 2$ we proceed as follows.

$m = 1$: Theorem 10 says

$$x^2 + x = x(x + 1).$$

There are two irreducible polynomials of degree 1, namely x and $x + 1$. The minimal polynomials of 0 and 1 in $GF(2)$ are respectively x and $x + 1$.

$m = 2$:

$$x^{2^2} + x = x^4 + x = x(x + 1)(x^2 + x + 1).$$

There is one irreducible polynomial of degree 2, $x^2 + x + 1$. The minimal polynomials of the elements of $GF(2^2)$ are:

element	minimal polynomial
0	x
1	$M^{(0)}(x) = x + 1$
α, α^2	$M^{(1)}(x) = M^{(2)}(x) = x^2 + \lambda + 1$

$m = 3$:

$$x^{2^3} + x = x^8 + x = x(x + 1)(x^6 + x^5 + x^4 + x^3 + x^2 + x + 1)$$
$$= x(x + 1)(x^3 + x + 1)(x^3 + x^2 + 1). \tag{11}$$

There are two irreducible polynomials of degree 3, namely

$$x^3 + x + 1 \quad \text{and} \quad x^3 + x^2 + 1.$$

In $GF(2^3)$ defined by $\alpha^3 + \alpha + 1 = 0$ we have:

element	minimal polynomial
0	x
1	$M^{(0)}(x) = x + 1$
$\alpha, \alpha^2, \alpha^4$	$M^{(1)}(x) = M^{(2)}(x) = M^{(4)}(x) = x^3 + x + 1$
$\alpha^3, \alpha^6, \alpha^5$	$M^{(3)}(x) = M^{(6)}(x) = M^{(5)}(x) = M^{(-1)}(x) = x^3 + x^2 + 1$

Observe that (11) does agree with Theorem 10: $m = 3$ is divisible by 1 and 3, and $x^{2^3} + x$ is the product of the two irreducible polynomials of degree 1 and the two of degree 3.

$x^3 + x + 1$ and $x^3 + x^2 + 1$ are called *reciprocal* polynomials. In general the reciprocal polynomial of

$$a_n x^n + a_{n-1} x^{n-1} + \cdots + a_1 x + a_0$$

is

$$a_0 x^n + a_1 x^{n-1} + \cdots + a_{n-1} x + a_n$$

obtained by reversing the order of the coefficients. Another way of saying this is that the reciprocal polynomial of $f(x)$ is

$$x^{\deg f(x)} f(x^{-1}).$$

The roots of the reciprocal polynomial are the reciprocals of roots of the original polynomial. The reciprocal of an irreducible polynomial is also irreducible.

So if α has minimal polynomial $M^{(1)}(x)$, we know immediately that α^{-1} has minimal polynomial $M^{(-1)}(x) = $ reciprocal polynomial of $M^{(1)}(x)$.

$m = 4$: We know these factors of $x^{2^4} + x$: $x, x + 1, x^2 + x + 1$ (irreducible of degree 1 and 2); $x^4 + x + 1$ and its reciprocal $x^4 + x^3 + 1$ (irreducible of degree 4). By division we find the remaining irreducible polynomial of degree 4:

$x^4 + x^3 + x^2 + x + 1$, and so

$$x^{24} + x = x(x+1)(x^2+x+1)(x^4+x+1)(x^4+x^3+1)(x^4+x^3+x^2+x+1).$$

In $GF(2^4)$ defined by $\alpha^4 + \alpha + 1 = 0$ we have:

element	minimal polynomial
0	x
1	$M^{(0)}(x) = x + 1$
$\alpha, \alpha^2, \alpha^4, \alpha^8$	$M^{(1)} = M^{(2)} = M^{(4)} = M^{(8)} = x^4 + x + 1$
$\alpha^3, \alpha^6, \alpha^{12}, \alpha^9$	$M^{(3)} = M^{(6)} = M^{(12)} = M^{(9)} = x^4 + x^3 + x^2 + x + 1$
α^5, α^{10}	$M^{(5)} = M^{(10)} = x^2 + x + 1$
$\alpha^7, \alpha^{14}, \alpha^{13}, \alpha^{11}$	$M^{(7)} = M^{(14)} = M^{(13)} = M^{(11)} = M^{(-1)} = x^4 + x^3 + 1$

Theorem 4 implies that the polynomial $\pi(x)$ used to generate the field may always be chosen to be a primitive polynomial. (Simply take $\pi(x)$ to be the minimal polynomial of a primitive element.) As we saw in Fig. 3.1, this has the advantage that an element of the field can be written either as a polynomial in α of degree $\leq m - 1$, or as a power of α, where α is a zero of $\pi(x)$.

But it is a lot harder to find which of the irreducible polynomials are primitive: see §3 of Ch. 8.

Problem. (21) Show that

$$x^{2^5} + x = x(x+1)(x^5+x^2+1)(x^5+x^3+1)(x^5+x^4+x^3+x^2+1)$$

$$.(x^5+x^4+x^3+x+1)(x^5+x^4+x^2+x+1)(x^5+x^3+x^2+1)$$

and that in $GF(2^5)$ defined by $\alpha^5 + \alpha^2 + 1 = 0$:

element	minimal polynomial
0	x
1	$M^{(0)}(x) = x + 1$
$\alpha, \alpha^2, \alpha^4, \alpha^8, \alpha^{16}$	$M^{(1)}(x) = x^5 + x^2 + 1$
$\alpha^3, \alpha^6, \alpha^{12}, \alpha^{24}, \alpha^{17}$	$M^{(3)}(x) = x^5 + x^4 + x^3 + x^2 + 1$
$\alpha^5, \alpha^{10}, \alpha^{20}, \alpha^9, \alpha^{18}$	$M^{(5)}(x) = x^5 + x^4 + x^2 + x + 1$
$\alpha^7, \alpha^{14}, \alpha^{28}, \alpha^{25}, \alpha^{19}$	$M^{(7)}(x) = x^5 + x^3 + x^2 + x + 1$
$\alpha^{11}, \alpha^{22}, \alpha^{13}, \alpha^{26}, \alpha^{21}$	$M^{(11)}(x) = x^5 + x^4 + x^3 + x + 1$
$\alpha^{15}, \alpha^{30} = \alpha^{-1}, \alpha^{29}, \alpha^{27}, \alpha^{23}$	$M^{(15)}(x) = M^{(-1)}(x) = x^5 + x^3 + 1$

§5. Tables of small fields

Figure 4.5 gives the fields $GF(2)$, $GF(2^2)$, $GF(2^3)$, $GF(2^5)$, $GF(2^6)$, $GF(3)$, and $GF(3^3)$. The first column gives the element γ as an m-tuple, while the second gives the logarithm i, where $\gamma = \alpha^i$. ($GF(2^4)$ and $GF(3^2)$ will be found in Figs. 3.1 and 4.1.)

GF(2)	
element	log
0	$-\infty$
1	0

GF(2^2)

$\pi(\alpha) = \alpha^2 + \alpha + 1 = 0$

element	log
00	$-\infty$
10	0
01	1
11	2

GF(2^3)

$\pi(\alpha) = \alpha^3 + \alpha + 1 = 0$

element	log
000	$-\infty$
100	0
010	1
001	2
110	3
011	4
111	5
101	6

GF(2^5) with $\pi(\alpha) = \alpha^5 + \alpha^2 + 1 = 0$

element	log	element	log
00000	$-\infty$	11111	15
10000	0	11011	16
01000	1	11001	17
00100	2	11000	18
00010	3	01100	19
00001	4	00110	20
10100	5	00011	21
01010	6	10101	22
00101	7	11110	23
10110	8	01111	24
01011	9	10011	25
10001	10	11101	26
11100	11	11010	27
01110	12	01101	28
00111	13	10010	29
10111	14	01001	30

GF(2^6) with $\pi(\alpha) = \alpha^6 + \alpha + 1 = 0$

element	log	element	log	element	log
000000	$-\infty$	110111	21	010111	42
100000	0	101011	22	111011	43
010000	1	100101	23	101101	44
001000	2	100010	24	100110	45

Fig. 4.5. Some Galois fields.

$$\text{GF}(2^6) \text{ with } \pi(\alpha) = \alpha^6 + \alpha + 1 = 0$$

element	log	element	log	element	log
000100	3	010001	25	010011	46
000010	4	111000	26	111001	47
000001	5	011100	27	101100	48
110000	6	001110	28	010110	49
011000	7	000111	29	001011	50
001100	8	110011	30	110101	51
000110	9	101001	31	101010	52
000011	10	100100	32	010101	53
110001	11	010010	33	111010	54
101000	12	001001	34	011101	55
010100	13	110100	35	111110	56
001010	14	011010	36	011111	57
000101	15	001101	37	111111	58
110010	16	110110	38	101111	59
011001	17	011011	39	100111	60
111100	18	111101	40	100011	61
011110	19	101110	41	100001	62
001111	20				

GF(3)

element	log
0	$-\infty$
1	0
2	1

$$\text{GF}(3^3)$$

$$\pi(\alpha) = \alpha^3 + 2\alpha + 1 = 0$$

element	log	element	log
000	$-\infty$		
100	0	200	13
010	1	020	14
001	2	002	15
210	3	120	16
021	4	012	17
212	5	121	18
111	6	222	19
221	7	112	20
202	8	101	21
110	9	220	22
011	10	022	23
211	11	122	24
201	12	102	25

Fig. 4.5. (*cont.*)

Problem. (22) Of course GF(2^6) contains GF(2^3), from Theorem 8. Let ξ be a primitive element of GF(2^6). Show that GF(2^3) = $\{0, 1, \xi^9, \xi^{18}, \xi^{27}, \xi^{36}, \xi^{45}, \xi^{54}\}$. Show that if $\xi^6 + \xi + 1 = 0$ then to get the first version of GF(2^3) (shown in Fig. 4.2) we must take the primitive element of GF(2^3) to be ξ^{27}, ξ^{45} or ξ^{54}.

Figure 4.6 gives a short list of primitive polynomials, chosen so that the number of terms in the polynomial is a minimum. For binary polynomials only the exponents are given. E.g. the fifth line 520 means that $x^5 + x^2 + 1$ is primitive.

Over GF(2)					Over GF(3)
exponents of terms					$x + 1$
1	0				$x^2 + x + 2$
2	1	0			$x^3 + 2x + 1$
3	1	0			$x^4 + x + 2$
4	1	0			$x^5 + 2x + 1$
5	2	0			$x^6 + x + 2$
6	1	0			
7	1	0			Over GF(5)
8	6	5	4	0	
9	4	0			$x + 1$
10	3	0			$x^2 + x + 2$
11	2	0			$x^3 + 3x + 2$
12	7	4	3	0	$x^4 + x^2 + 2x + 2$
13	4	3	1	0	
14	12	11	1	0	Over GF(7)
15	1	0			
16	5	3	2	0	$x + 1$
17	3	0			$x^2 + x + 3$
18	7	0			$x^3 + 3x + 2$
19	6	5	1	0	
20	3	0			

Fig. 4.6. Some primitive polynomials over GF(p).

§6. The automorphism group of GF(p^m)

Associated with the field GF(p^m) is the set of mappings, called automorphisms, of the field onto itself which fix every element of the subfield GF(p), and preserve addition and multiplication. Such a mapping will be denoted by

$$\sigma: \beta \to \beta^\sigma.$$

Definition. An *automorphism* of GF(p^m) over GF(p) is a mapping σ which fixes the elements of GF(p) and has the properties

(i) $(\alpha + \beta)^\sigma = \alpha^\sigma + \beta^\sigma$
(ii) $(\alpha\beta)^\sigma = \alpha^\sigma\beta^\sigma$.

The set of all automorphisms of GF(p^m) forms a group if we define the product of σ and τ by

$$\alpha^{\sigma \cdot \tau} = (\alpha^\sigma)^\tau.$$

This group is called the *automorphism group* or *Galois group* of GF(p^m).

Example. The field GF(4) = $\{0, 1, \alpha, \alpha^2\}$, with $\alpha^2 = \alpha + 1$, $\alpha^3 = 1$. The automorphism group of GF(4) consists of the identity mapping 1 and the mapping

$$\sigma: 1 \to 1, \qquad \alpha \to \alpha^2, \qquad \alpha^2 \to \alpha.$$

Clearly $\sigma^2 = 1$.

Theorem 12. *The automorphism group of* GF(p^m) *is the cyclic group of order m consisting of the mapping $\sigma_p: \beta \to \beta^p$ and its powers.*

Proof. Using Lemma 5 it is clear that σ_p and its powers are automorphisms of GF(p^m). Let α be a primitive element and σ an automorphism of GF(p^m). From the definition of an automorphism, α and α^σ have the same minimal polynomial. By Problem 14, α^σ is one of $\alpha, \alpha^p, \alpha^{p^2}, \ldots, \alpha^{p^{m-1}}$. But if $\alpha^\sigma = \alpha^{p^i}$, then $\sigma = (\sigma_p)^i$.   Q.E.D.

This theorem shows that in a finite field of characteristic p every element has a unique p^{th} root. We shall occasionally use the fact that every element of GF(2^m) has a unique square root, given by $\sqrt{}(\beta) = \beta^{2^{m-1}}$.

If $p \neq 2$ exactly half the nonzero field elements have square roots. These elements are the quadratic residues, mentioned in Ch. 2. If α is a primitive field element the quadratic residues are α^{2i}, the even powers of α. It is now obvious that they form a group and that property (Q1) of Ch. 2 holds, namely:

$$\text{residue} \cdot \text{residue} \quad = \text{residue},$$
$$\text{nonresidue} \cdot \text{nonresidue} = \text{residue},$$
$$\text{residue} \cdot \text{nonresidue} = \text{nonresidue}.$$

We can also prove property (Q2), namely:

Theorem 13. *If $p^m = 4k + 1$, then -1 is a quadratic residue; if $p^m = 4k - 1$, -1 is a nonresidue.*

Proof. $\alpha^{\frac{1}{2}(p^m-1)} = -1$. If $p^m = 4k+1$ then $\frac{1}{2}(p^m-1) = 2k$, and -1 is an even power of α; while if $p^m = 4k-1$ then $\frac{1}{2}(p^m-1) = 2k-1$, and -1 is an odd power of α.   Q.E.D.

Problems. (23) Prove Dedekind's theorem: if $\varphi_1, \ldots, \varphi_n$ are distinct automorphisms of a field F, then it is impossible to find elements $a_1, \ldots, a_n$, not all 0, in F, such that $a_1\beta^{\varphi_1} + \cdots + a_n\beta^{\varphi_n} = 0$ for all $\beta \in F$.

(24) Regard σ_p as a linear transformation and let T be the matrix of this transformation. Show that $f(x) = x^m - 1$ is the least degree polynomial such that $f(T) = 0$.

§7. The number of irreducible polynomials

Let $I_q(m)$ be the number of monic polynomials of degree m which are irreducible over $\mathrm{GF}(q)$, where q is any prime power. This number can be expressed in terms of the Möbius function, which is defined by:

$$\mu(n) = \begin{cases} 1 & \text{if } n = 1 \\ (-1)^r & \text{if } n \text{ is the product of } r \text{ distinct primes} \\ 0 & \text{otherwise} \end{cases}$$

The fundamental properties of the Möbius function are given by

Theorem 14.

(i)
$$\sum_{d|n} \mu(d) = \begin{cases} 1 & \text{if } n = 1 \\ 0 & \text{if } n > 1 \end{cases}$$

(ii) *The Möbius inversion formula:*

$$\text{if} \quad f(n) = \sum_{d|n} g(d) \quad \text{for all positive integers } n,$$

$$\text{then} \quad g(n) = \sum_{d|n} \mu(d)f(n/d).$$

(iii)
$$\varphi(n) = \sum_{d|n} d\mu(n/d) \quad \text{for } n \geq 1.$$

Problem. (25) Prove the above theorem.

Then we have:

Theorem 15.

$$I_q(m) = \frac{1}{m} \sum_{d|m} \mu(d) q^{m/d}.$$

Proof. From Theorem 11,

$$q^m = \sum_{d|m} d I_q(d).$$

The result then follows by Möbius inversion.   Q.E.D.

E.g. when $q = 2$, Theorem 15 gives $I_2(1) = 2$, $I_2(2) = 1$, $I_2(3) = 2$, $I_2(4) = 3$, in agreement with the calculations above.

Corollary 16. $I_q(m) \geq 1$ *for all* q, m.

Proof. To get a lower bound from Theorem 15, replace $\mu(d)$ by -1 for $d > 1$. Then $I_q(m) > (q^m - q^{m-1} - \cdots - 1)/m > 0$ by a simple calculation.   Q.E.D.

Thus there is a polynomial of degree m, irreducible over GF(p), for all p and m, giving the second proof of Theorem 7.

§8. Bases of GF(p^m) over GF(p).

GF(p^m) is a vector space of dimension m over GF(p). Any set of m linearly independent elements can be used as a basis for this vector space.

In constructing GF(p^m) from a primitive irreducible polynomial $\pi(x)$, we used the basis 1, α, $\alpha^2, \ldots, \alpha^{m-1}$, where α is a zero of $\pi(x)$. However there are other possibilities.

Trace

Definition. The sum

$$T_m(\beta) = \beta + \beta^p + \beta^{p^2} + \cdots + \beta^{p^{m-1}} = \sum_{j=0}^{m-1} \beta^{p^j}$$

is called the *trace* of $\beta \in GF(p^m)$. Since

$$T_m(\beta)^p = \sum_{j=0}^{m-1} \beta^{p^{j+1}} = T_m(\beta),$$

from Theorem 8 $T_m(\beta)$ is one of $0, 1, 2, \ldots, p-1$ (i.e., an element of $GF(p)$). Thus the trace of an element of $GF(2^m)$ is 0 or 1.

Problem. (26) Show that the trace has the following properties.

(i) $T_m(\beta + \gamma) = T_m(\beta) + T_m(\gamma)$, $\beta, \gamma \in GF(p^m)$.

(ii) $T_m(\beta)$ takes on each value in $GF(p)$ equally often, i.e., p^{m-1} times. [Hint: from Problem 23, T_m is not identically zero.]

(iii) $T_m(\beta^p) = T_m(\beta)^p = T_m(\beta)$.

(iv) $T_m(1) \equiv m \pmod{p}$.

(v) Let $T_m(x)$ be the polynomial

$$\sum_{j=0}^{m-1} x^{p^j}.$$

For $s \in GF(p)$, show

$$T_m(x) - s = \prod_{T_m(\beta)=s} (x - \beta),$$

$$x^{p^m} - x = \prod_{s \in GF(p)} (T_m(x) - s).$$

(vi) If $M(z) = z^r + M_{r-1}z^{r-1} + \cdots$ is the minimal polynomial of $\beta \in GF(p^m)$, show that $T_m(\beta) = -mM_{r-1}/r$.

Lemma 17. (The Vandermonde matrix.) The matrix

$$\begin{pmatrix} 1 & a_1 & a_1^2 & a_1^3 & \cdots & a_1^{n-1} \\ 1 & a_2 & a_2^2 & a_2^3 & \cdots & a_2^{n-1} \\ \multicolumn{6}{c}{\cdots\cdots\cdots\cdots\cdots} \\ 1 & a_n & a_n^2 & a_n^3 & \cdots & a_n^{n-1} \end{pmatrix}$$

(*where the a_i's are from any finite or infinite field*), is called a Vandermonde matrix. Its determinant is

$$\prod_{j=1}^{n-1} \prod_{i=j+1}^{n} (a_i - a_j),$$

which is nonzero if the a_i's are distinct.

Proof. It is easy to show that the determinant is equal to

$$\det \begin{pmatrix} (a_2-a_1) & (a_2-a_1)a_2 & (a_2-a_1)a_2^2 & \cdots & (a_2-a_1)a_2^{n-2} \\ (a_3-a_1) & (a_3-a_1)a_3 & (a_3-a_1)a_3^2 & \cdots & (a_3-a_1)a_3^{n-2} \\ \cdots & \cdots & \cdots & \cdots & \cdots \\ (a_n-a_1) & (a_n-a_1)a_n & (a_n-a_1)a_n^2 & \cdots & (a_n-a_1)a_n^{n-2} \end{pmatrix}.$$

The result then follows by induction.   Q.E.D.

The complementary basis. In the remainder of this section we restrict ourselves to fields of characteristic $p = 2$.

Two bases $\beta_1, \beta_2, \ldots, \beta_m$ and $\lambda_1, \lambda_2, \ldots, \lambda_m$ are said to be *complementary* if

$$T_m(\beta_i \lambda_j) = 0 \quad \text{for } i \neq j,$$

$$T_m(\beta_i \lambda_i) = 1.$$

These are also sometimes called *dual* bases.

Lemma 18. *Let* $\beta_1, \beta_2, \ldots, \beta_m$ *be a basis. The matrix*

$$B = \begin{pmatrix} \beta_1 & \beta_1^2 & \beta_1^{2^2} & \cdots & \beta_1^{2^{m-1}} \\ \beta_2 & \beta_2^2 & \beta_2^{2^2} & \cdots & \beta_2^{2^{m-1}} \\ \cdots & \cdots & \cdots & \cdots & \cdots \\ \beta_m & \beta_m^2 & \beta_m^{2^2} & \cdots & \beta_m^{2^{m-1}} \end{pmatrix} \tag{12}$$

is invertible, and $\det B = 1$.

Proof. Expressing the basis $1, \alpha, \ldots, \alpha^{m-1}$ in terms of the basis $\beta_1, \beta_2, \ldots, \beta_m$, we have

$$\begin{pmatrix} 1 \\ \alpha \\ \alpha^2 \\ \cdot \\ \cdot \\ \alpha^{m-1} \end{pmatrix} = C \begin{pmatrix} \beta_1 \\ \beta_2 \\ \cdot \\ \cdot \\ \beta_m \end{pmatrix}$$

for some binary matrix C. Then

$$CB = \begin{pmatrix} 1 & 1 & \cdots & 1 \\ \alpha & \alpha^2 & \cdots & \alpha^{2^{m-1}} \\ \cdots & \cdots & \cdots & \cdots \\ \alpha^{m-1} & (\alpha^2)^{m-1} & \cdots & (\alpha^{2^{m-1}})^{m-1} \end{pmatrix}$$

*Starred sections can be omitted on first reading.

By Lemma 17 this matrix is invertible, hence B is invertible. Now

$$BB^T = \begin{pmatrix} T_m(\beta_1) & T_m(\beta_1\beta_2) & \cdots & T_m(\beta_1\beta_m) \\ T_m(\beta_1\beta_2) & T_m(\beta_2) & \cdots & T_m(\beta_2\beta_m) \\ \cdots\cdots\cdots\cdots\cdots\cdots\cdots\cdots\cdots \\ T_m(\beta_1\beta_m) & T_m(\beta_2\beta_m) & \cdots & T_m(\beta_m) \end{pmatrix}$$

is a binary matrix, so has determinant 0 or 1. Thus $\det BB^T = (\det B)^2 = 1$, which implies $\det B = 1$.   Q.E.D.

Problem. (27) If $\beta_1, \beta_2, \ldots, \beta_m$ is a basis then $T_m(\beta_i) = 1$ for at least one β_i.

Theorem 19. *Every basis has a complementary basis.*

Proof. Recall that the inverse of a matrix $[a_{ij}]$ is obtained by replacing a_{ij} by the cofactor of a_{ji}, divided by $\det [a_{ij}]$. It is readily seen that B^{-1} is of the form

$$\begin{pmatrix} \lambda_1 & \lambda_2 & \cdots & \lambda_m \\ \lambda_1^2 & \lambda_2^2 & \cdots & \lambda_m^2 \\ \cdots\cdots\cdots\cdots\cdots\cdots \\ \lambda_1^{2^{m-1}} & \lambda_2^{2^{m-1}} & \cdots & \lambda_m^{2^{m-1}} \end{pmatrix}$$

where $\lambda_1, \lambda_2, \ldots, \lambda_m$ are linearly independent over GF(2).
    Then the equation $BB^{-1} = I$ shows that $\beta_1, \beta_2, \ldots, \beta_m$ and $\lambda_1, \lambda_2, \ldots, \lambda_m$ are complementary bases.   Q.E.D.

Problem. (28) If $\gamma \in \mathrm{GF}(2^m)$ then

$$\gamma = \sum_{i=1}^{m} T_m(\gamma\lambda_i)\beta_i.$$

*§9. Linearized polynomials and normal bases

    Let $q = p^a$ be a power of the prime p.

Definition. *A linearized polynomial over* $\mathrm{GF}(q^s)$ *is a polynomial of the form*

$$L(z) = \sum_{i=0}^{h} l_i z^{q^i},$$

where $l_i \in \mathrm{GF}(q^s)$, $l_h = 1$. E.g. $z + z^2 + z^4$ is a linearized polynomial over $\mathrm{GF}(2^s)$ for any s.

Let the zeros of $L(z)$ lie in the extension field $GF(q^m)$, $m \geqslant s$.

Lemma 20. *The zeros of $L(z)$ form a subspace of $GF(q^m)$, where $GF(q^m)$ is regarded as a vector space over $GF(q)$.*

Proof. If β_1, β_2 are zeros of $L(z)$ then so is $\lambda_1\beta_1 + \lambda_2\beta_2$, for $\lambda_i \in GF(q)$, by Corollary 3 and Lemma 5.   Q.E.D.

Problem. (29) If r divides m, show that the zeros of $L(z)$ which lie in $GF(q^r)$ form a subspace of $GF(q^m)$.

Since $L'(z) = l_0$, if $l_0 \neq 0$ the q^h zeros of $L(z)$ are distinct and form an h-dimensional subspace of $GF(q^m)$.

Lemma 21. *Conversely, let U be an h-dimensional subspace of $GF(q^m)$. Then*

$$L(z) = \prod_{\beta \in U} (z - \beta)$$

is a linearized polynomial over $GF(q^m)$, i.e.

$$L(z) = z^{q^h} + l_{h-1}z^{q^{h-1}} + \cdots + l_0z. \tag{13}$$

Proof. Let $\beta_0, \ldots, \beta_{h-1} \in GF(q^m)$ be a basis for U over $GF(q)$. By Lemma 18 the matrix

$$\begin{bmatrix} \beta_0 & \beta_0^q & \beta_0^{q^2} & \cdots & \beta_0^{q^{h-1}} \\ \beta_1 & \beta_1^q & \beta_1^{q^2} & \cdots & \beta_1^{q^{h-1}} \\ \cdots\cdots\cdots\cdots\cdots\cdots\cdots \\ \beta_{h-1} & \beta_{h-1}^q & & \cdots & \beta_{h-1}^{q^{h-1}} \end{bmatrix}$$

is invertible. Thus there is a solution $l_0, \ldots, l_{h-1}$ in $GF(q^m)$ to the equations

$$\beta_i^{q^h} + \sum_{j=0}^{h-1} l_j\beta_i^{q^j} = 0, \quad i = 0, \ldots, h - 1.$$

Therefore $\beta_0, \ldots, \beta_{h-1}$ are the zeros of the polynomial

$$L(z) = z^{q^h} + \sum_{j=0}^{h-1} l_j z^{q^j}.$$

Any linear combination of the β_i, i.e. any element of U, is also a zero of $L(z)$. Hence

$$L(z) = \prod_{\beta \in U} (z - \beta).$$ Q.E.D.

Example. Let U be the subspace of $GF(2^3)$ consisting of the points 000, 100, 010, 110. Then (from Fig. 4.5)

$$\prod_{\beta \in U} (z - \beta) = z(z - 1)(z - \alpha)(z - \alpha^3) = z^4 + \alpha^5 z^2 + \alpha^4 z$$

is a linearized polynomial over $GF(2^3)$.

Normal bases.

Definition. A *normal basis* of $GF(p^m)$ over $GF(p)$ is a basis of the form γ, γ^p, $\gamma^{p^2}, \ldots, \gamma^{p^{m-1}}$.

For example $GF(2^2)$ clearly has the normal basis α, α^2 over $GF(2)$. We shall occasionally use a normal basis in future chapters, hence this section is devoted to proving that there always is such a basis.

A linearized polynomial

$$L(z) = \sum_{i=0}^{h} l_i z^{p^i}$$

will be called a *p-polynomial* if the coefficients l_i are restricted to $GF(p)$.

By Lemmas 20 and 21 the zeros of a p-polynomial form a subspace M of $GF(p^m)$ over $GF(p)$ for some choice of m. They have the additional property that if μ belongs to M so does μ^p (since $L(\mu)^p = L(\mu^p)$). A subspace with this property will be called a *modulus*. If $l_0 \neq 0$, $L(z)$ has only simple zeros, and we suppose this is always the case.

On the other hand, if M is a modulus, then by Lemma 21 $\prod_{\beta \in M} (z - \beta)$ is a linearized polynomial $L(z)$. Now M consists of a union of sets of the form $\{\beta, \beta^p, \beta^{p^2}, \ldots\}$, and so by Problem 14 the coefficients of $L(z)$ are in $GF(p)$. Thus we have:

Lemma 22. *Suppose $L(z)$ is a linearized polynomial over $GF(p^m)$. The zeros of $L(z)$ form a modulus iff $L(z)$ is a p-polynomial.*

Example. From the table of $GF(2^4)$ in Fig. 3.1 it is easy to see that the elements 0, 1, α, α^2, α^4, α^8, α^5, α^{10} of $GF(2^4)$ form a modulus. The corresponding p-polynomial is $z^8 + z^4 + z^2 + z$.

The ordinary product of two p-polynomials is not a p-polynomial. We define the *symbolic product* of two p-polynomials $F(z)$ and $G(z)$ to be

$$F(z) \otimes G(z) = F(G(z)).$$

Problems. (30) Show that this multiplication is commutative, i.e. $F(G(z)) = G(F(z))$.

To the p-polynomial

$$F(z) = \sum_{i=0}^{h} l_i z^{p^i}$$

we associate the ordinary polynomial

$$f(z) = \sum_{i=0}^{h} l_i z^i.$$

(31) Show that the ordinary polynomial associated with $F(z) \otimes G(z)$ is $f(z) \cdot g(z)$.

Lemma 23. *If a p-polynomial $H(z)$ is a symbolic product $F(z) \otimes G(z)$ of p-polynomials then $H(z)$ is divisible in the ordinary sense by $F(z)$ and $G(z)$. Conversely, if $H(z)$ is divisible by $G(z)$, then $H(z) = F(z) \otimes G(z)$ for some $F(z)$.*

Proof. First let $\bar{f}(z) = (1/z)F(z)$. Then $H(z) = \bar{f}(G(z)) \cdot G(z)$, so is divisible by $G(z)$, and similarly by $F(z)$. Conversely, suppose $H(z) = A(z) \cdot G(z)$. Then dividing $H(z)$ symbolically by $G(z)$ we obtain

$$H(z) = F(z) \otimes G(z) + R(z) = \bar{f}(G(z)) \cdot G(z) + R(z)$$

where $\deg R(z) < \deg G(z)$. But since $H(z) = A(z) \cdot G(z)$ we must have $R(z) = 0$.   Q.E.D.

Example. Consider the p-polynomial of the last example. We have

$$z^8 + z^4 + z^2 + z = (z^2 + z) \otimes (z^4 + z) = (z^4 + z + 1)(z^4 + z)$$
$$= (z^6 + z^5 + z^4 + z^3 + 1)(z^2 + z).$$

If $L(z)$ cannot be written in the form $F(z) \otimes G(z)$ it is said to be *symbolically irreducible*. A zero μ of $L(z)$ is called a *primitive* zero if it is not a zero of a p-polynomial of lower degree.

Theorem 24. *A p-polynomial $L(z)$ for which $l_0 \neq 0$ always has a primitive zero.*

Proof. Let

$$L(z) = F_1(z)^{e_1} \otimes F_2(z)^{e_2} \otimes \cdots \otimes F_r(z)^{e_r} \tag{14}$$

be the decomposition of $L(z)$ into symbolically irreducible factors, where the $F_i(z)$ are distinct and $F_i(z)$ has degree p^{m_i}. Then $L(z)$ has degree p^m, where $m = \sum e_i m_i$. Since $l_0 \neq 0$, $L(z)$ has only simple zeros.

If μ is an element of $GF(p^s)$ there is a unique p-polynomial of smallest degree, $U(z)$ say, which has μ as a zero; i.e. $U(z)$ is the polynomial with zeros μ, μ^p, μ^{p^2}, ... and all linear combinations of these. Furthermore if μ is a zero of $L(z)$ then $L(z)$ is divisible by $U(z)$. Hence a zero of $L(z)$ is a primitive zero if it is not a zero of any of the symbolic divisors of $L(z)$. Using the principle of inclusion and exclusion the number of such zeros is

$$p^m - \sum_i p^{m-m_i} + \sum_{i,j} p^{m-m_i-m_j} - \cdots$$

$$= p^m \left(1 - \frac{1}{p^{m_1}}\right)\left(1 - \frac{1}{p^{m_2}}\right) \cdots \left(1 - \frac{1}{p^{m_r}}\right)$$

$$> 0. \tag{15}$$

Thus $L(z)$ has a primitive zero.   Q.E.D.

Example. The p-polynomial

$$x^8 + x^4 + x^2 + x = (x^2 + x) \otimes (x^2 + x) \otimes (x^2 + x),$$

and $x^2 + x$ is symbolically irreducible. Therefore the number of primitive zeros is $2^3(1 - \frac{1}{2}) = 4$. Indeed, the primitive zeros are α, α^2, α^4, α^8, where $\alpha^4 + \alpha + 1 = 0$.

Now let

$$L(z) = z^{p^m} + l_{m-1} z^{p^{m-1}} + \cdots + l_1 z^p + l_0 z, \qquad l_i \in \mathrm{GF}(p), \quad l_0 \neq 0,$$

be any p-polynomial. By Theorem 24, $L(z)$ has a primitive zero μ. $L(z)$ is the unique p-polynomial of lowest degree having μ as a zero. The zeros of $L(z)$ are the p^m elements

$$\epsilon_0\mu + \epsilon_1\mu^p + \epsilon_2\mu^{p^2} + \cdots + \epsilon_{m-1}\mu^{p^{m-1}},$$

where $\epsilon_i \in \mathrm{GF}(p)$, and they form a modulus M by Lemma 22. Clearly this modulus has the normal basis

$$\mu, \mu^p, \mu^{p^2}, \ldots, \mu^{p^{m-1}}.$$

We can now prove the main theorem of this section.

Theorem 25. (The normal basis theorem.) *A normal basis exists in any field* $\mathrm{GF}(p^m)$.

Proof. Choose $L(z) = z^{p^m} - z$. We know that the zeros of $L(z)$ are all the elements of $\mathrm{GF}(p^m)$. The preceding remarks show that $\mathrm{GF}(p^m)$ has a normal basis over $\mathrm{GF}(p)$.   Q.E.D.

Problem. (32) For characteristic 2, show that the complementary basis of a normal basis is also a normal basis.

Now suppose p is 2 and m is odd, so that $x^m + 1$ has no multiple zeros. Let

$$x^m + 1 = \prod_{i=1}^{r} f_i(x), \quad \deg f_i = m_i,$$

be the factorization of $x^m + 1$ into distinct irreducible factors. By Problem 31,

$$z^{2^m} + z = F_1(z) \otimes \cdots \otimes F_r(z).$$

From (15), the number of primitive zeros of this polynomial, and hence the number of elements μ which generate normal bases of GF(2^m), is

$$N(2, m) = 2^{\Sigma m_i} \prod_{i=1}^{r} \left(1 - \frac{1}{2^{m_i}}\right)$$

$$= \prod_{i=1}^{r} (2^{m_i} - 1)$$

which is odd. The number of normal bases is $N(2, m)/m$, which is also odd. Thus we have proved:

Theorem 26. *If m is odd, GF(2^m) has a self-complementary normal basis.*

Example. We have

$$x^3 + 1 = (x + 1)(x^2 + x + 1),$$

$$z^{2^3} + z = (z^2 + z) \otimes (z^2 + z^2 + z).$$

Hence the number of elements which generate normal bases for GF(2^3) over GF(2) is $2^2 - 1 = 3$, and there is one such normal basis, i.e. α^3, α^6, α^5 using the table of Fig. 4.5. This basis is self-complementary.

We state without proof the following refinement of the normal basis theorem.

Theorem 27. (Davenport.) *Any finite field GF(p^m) contains a primitive element γ such that γ, γ^p, ..., $\gamma^{p^{m-1}}$ is a normal basis.*

Corollary 28. GF(2^m) *contains a primitive element of trace* 1.

Research Problem. (4.1) Find a simple direct proof of Corollary 28.

Problem. (33) Show that the only symbolically irreducible p-polynomials over GF(2) of degree < 16 are: z, z^2, $z^2 + z$, $z^4 + z^2 + z$, $z^8 + z^2 + z$ and $z^8 + z^4 + z$.

Notes on Chapter 4

Good references on finite fields are Albert [19], Artin [30], Berlekamp [113, Ch. 4], Birkhoff and Bartee [153], Conway [301], Jacobson [687, Vol. 3], Peterson and Weldon [1040, Ch. 6], Van der Waerden [1376, Vol. 1], and Zariski and Samuel [1454, Vol. 1]. Further properties will be found in Berlekamp et al. [131], Carlitz [242–246], Cazacu and Simovici [254], Davenport [329], Daykin [340], Roth [1126], Zierler [1466, 1467] and Zierler and Brillhart [1469].

§2. The epigraph is from Weinstein [1395]. We are taking for granted two theorems. Let $f(x)$ be a polynomial of degree n with coefficients in a field F. (1) If $f(\beta) = 0$ for $\beta \in F$ then $f(x) = (x - \beta)g(x)$ for some $g(x)$ with coefficients in F. (See for example p. 40 of [19] or p. 31, Vol. I of [1454].) (2) Hence $f(x)$ has at most n zeros in F. For Problem 8 see Hardy and Wright [602].

§4. Berlekamp [113, p. 112] describes more sophisticated methods for finding minimal polynomials.

§5. For more extensive tables of small fields, see Alanan and Knuth [17], Bussey [222, 223] and Conway [301]. Extensive tables of irreducible polynomials can be found in Alanen and Knuth [17], Green and Taylor [554], Marsh [917], Mossige [973], Peterson and Weldon [1040] and Stahnke [1260]. See also Beard and West [93] and Golomb [524a]. Factoring polynomials over finite fields is discussed in Ch. 9.

§6. For Problem 23 see e.g. [30, p. 25] or [687, Vol. 3, p. 25].

§7. For the Möbius function see Hardy and Wright [602]. Knuth [772, p. 36] gives the inverse of a Vandermonde matrix. See also Althaus and Leake [27].

§9. The usual proof of the normal basis theorem (see for example Jacobson [687, Vol. 3, p. 61]) requires much more background. The proof given here is due to Ore [1015]. For another proof see Berger and Reiner [107]. Linearized polynomials have been studied by Ore [1015, 1016], Pele [1034], and Berlekamp [113, Ch. 11]. Theorem 27 is from Davenport [330] (see also Carlitz [243]). Lempel [814] shows that every field GF(2^m) has a self-complementary basis. See also Mann [910].

Dual codes and their weight distribution

§1. Introduction

The main result of the chapter, Theorem 1, is the surprising fact that the weight enumerator of the dual $\mathscr{C}^{\perp}$ of a binary linear code $\mathscr{C}$ is uniquely determined by the weight enumerator of $\mathscr{C}$ itself. In fact it is given by a linear transformation of the weight enumerator of $\mathscr{C}$. This theorem is proved in §2. If the same transformation is applied to the distance distribution of a nonlinear code, we obtain a set of nonnegative numbers with useful properties (§5). In order to study nonlinear codes we need to develop some algebraic machinery, namely Krawtchouk polynomials (end of §2), the group algebra (§3) and characters (§4). In §6 we return to linear codes and consider several different types of weight enumerators of nonbinary codes. For each of these there is a result analogous to Theorem 1, namely that the enumerator of the dual code $\mathscr{C}^{\perp}$ is given by a linear transformation of the enumerator of $\mathscr{C}$. Theorem 14 is a very general result of this type. However, the results for the complete weight enumerator (Theorem 10) and the Hamming weight enumerator (Theorem 13) are the most useful. The last section (§7) gives further properties of Krawtchouk polynomials.

§2. Weight distribution of the dual of a binary linear code

Recall (from §8 of Ch. 1) that if $\mathscr{C}$ is a linear code over a finite field, the dual code $\mathscr{C}^{\perp}$ consists of all vectors u having inner product zero with every codeword of $\mathscr{C}$. If $\mathscr{C}$ is an $[n, k]$ code, $\mathscr{C}^{\perp}$ is an $[n, n-k]$ code.

Weight enumerators. As in Ch. 2, A_i will denote the number of codewords of weight i in $\mathscr{C}$. We will call the polynomial

$$\sum_{i=0}^{n} A_i x^{n-i} y^i$$

the *weight enumerator* of $\mathscr{C}$, and denote it by $W_{\mathscr{C}}(x, y)$.

Observe that there are two ways of writing $W_{\mathscr{C}}(x, y)$:

$$W_{\mathscr{C}}(x, y) = \sum_{i=0}^{n} A_i x^{n-i} y^i$$

$$= \sum_{u \in \mathscr{C}} x^{n - \text{wt}(u)} y^{\text{wt}(u)}. \tag{1}$$

Here x and y are indeterminates, and $W_{\mathscr{C}}(x, y)$ is a homogeneous polynomial of degree n in x and y. It is frequently useful for $W_{\mathscr{C}}(x, y)$ to be a homogeneous polynomial. But we can always get rid of x by setting $x = 1$, and still have a perfectly good weight enumerator

$$W_{\mathscr{C}}(1, y) = W_{\mathscr{C}}(y) = \sum_{i=0}^{n} A_i y^i. \tag{2}$$

Likewise A_i' will denote the number of codewords of weight i in $\mathscr{C}^{\perp}$. The weight enumerator of $\mathscr{C}^{\perp}$ is

$$W_{\mathscr{C}^{\perp}}(x, y) = \sum_{i=0}^{n} A_i' x^{n-i} y^i$$

$$= \sum_{u \in \mathscr{C}^{\perp}} x^{n - \text{wt}(u)} y^{\text{wt}(u)} \tag{3}$$

Examples. (i) Consider the even weight code $\{000, 011, 101, 110\}$, denoted by $\mathscr{C}_3$. The dual $\mathscr{C}_3^{\perp}$ is $\{000, 111\}$ (Problem 34 of Ch. 1), and the weight enumerators are:

$$W_{\mathscr{C}_3}(x, y) = x^3 + 3xy^2,$$
$$W_{\mathscr{C}_3^{\perp}}(x, y) = x^3 + y^3.$$

(ii) The code $\{00, 11\}$, denoted by $\mathscr{C}_2$, is self-dual: $\mathscr{C}_2^{\perp} = \mathscr{C}_2$, and

$$W_{\mathscr{C}_2}(x, y) = x^2 + y^2.$$

(iii) The $[7, 4, 3]$ Hamming code $\mathscr{H}_3$. From §9 of Ch. 1,

$$W_{\mathscr{H}_3}(x, y) = x^7 + 7x^4 y^3 + 7x^3 y^4 + y^7,$$
$$W_{\mathscr{H}_3^{\perp}}(x, y) = x^7 + 7x^3 y^4.$$

MacWilliams theorem for binary linear codes. The main result of this chapter is that $W_{\mathscr{C}^{\perp}}(x, y)$ is given by a linear transformation of $W_{\mathscr{C}}(x, y)$. We give first the version for binary codes (Theorem 1).

Let F^n be the set of all binary vectors of length n. This is a vector space of dimension n over the binary field $F = \{0, 1\}$.

Theorem 1. (MacWilliams theorem for binary linear codes.) *If $\mathscr{C}$ is an $[n, k]$ binary linear code with dual code $\mathscr{C}^\perp$ then*

$$W_{\mathscr{C}^\perp}(x, y) = \frac{1}{|\mathscr{C}|} W_{\mathscr{C}}(x + y, x - y), \tag{4}$$

where $|\mathscr{C}| = 2^k$ is the number of codewords in $\mathscr{C}$. Equivalently,

$$\sum_{k=0}^{n} A'_k x^{n-k} y^k = \frac{1}{|\mathscr{C}|} \sum_{i=0}^{n} A_i (x + y)^{n-i} (x - y)^i, \tag{5}$$

or

$$\sum_{u \in \mathscr{C}^\perp} x^{n - \mathrm{wt}(u)} y^{\mathrm{wt}(u)} = \frac{1}{|\mathscr{C}|} \sum_{u \in \mathscr{C}} (x + y)^{n - \mathrm{wt}(u)} (x - y)^{\mathrm{wt}(u)}. \tag{6}$$

Equations (4), (5), and (6) are sometimes called the MacWilliams identities.

The proof depends on an important lemma. Let f be any mapping defined on F^n. We must be able to add and subtract the values $f(u)$, but otherwise f can be arbitrary. The *Hadamard transform $\hat{f}$* of f (see §3 of Ch. 2) is defined by

$$\hat{f}(u) = \sum_{v \in F^n} (-1)^{u \cdot v} f(v), \quad u \in F^n. \tag{7}$$

Lemma 2. *If $\mathscr{C}$ is an $[n, k]$ binary linear code (i.e., a k-dimensional subspace of F^n),*

$$\sum_{u \in \mathscr{C}^\perp} f(u) = \frac{1}{|\mathscr{C}|} \sum_{u \in \mathscr{C}} \hat{f}(u). \tag{8}$$

Proof.

$$\sum_{u \in \mathscr{C}} \hat{f}(u) = \sum_{u \in \mathscr{C}} \sum_{v \in F^n} (-1)^{u \cdot v} f(v)$$

$$= \sum_{v \in F^n} f(v) \sum_{u \in \mathscr{C}} (-1)^{u \cdot v}.$$

Now if $v \in \mathscr{C}^\perp$, $u \cdot v$ is always zero, and the inner sum is $|\mathscr{C}|$. But if $v \notin \mathscr{C}^\perp$ then by Problem 12 of Ch. 1, $u \cdot v$ takes the values 0 and 1 equally often, and the inner sum is 0. Therefore

$$\sum_{u \in \mathscr{C}} \hat{f}(u) = |\mathscr{C}| \sum_{v \in \mathscr{C}^\perp} f(v). \qquad \text{Q.E.D.}$$

Proof of Theorem 1. We apply the lemma with

$$f(u) = x^{n - \mathrm{wt}(u)} y^{\mathrm{wt}(u)}.$$

Then we have

$$\hat{f}(u) = \sum_{v \in F^n} (-1)^{u \cdot v} x^{n - \mathrm{wt}(v)} y^{\mathrm{wt}(v)}. \tag{9}$$

Let $u = (u_1 \cdots u_n)$, $v = (v_1 \cdots v_n)$. Then

$$\hat{f}(u) = \sum_{v \in F^n} (-1)^{u_1 v_1 + \cdots + u_n v_n} \prod_{i=1}^{n} x^{1-v_i} y^{v_i}$$

$$= \sum_{v_1=0}^{1} \sum_{v_2=0}^{1} \cdots \sum_{v_n=0}^{1} \prod_{i=1}^{n} (-1)^{u_i v_i} x^{1-v_i} y^{v_i}. \tag{10}$$

Just as

$$a_0 b_0 c_0 + a_0 b_0 c_1 + a_0 b_1 c_0 + a_0 b_1 c_1 + a_1 b_0 c_0 + a_1 b_0 c_1 + a_1 b_1 c_0 + a_1 b_1 c_1$$
$$= (a_0 + a_1)(b_0 + b_1)(c_0 + c_1),$$

so (10) is equal to

$$\prod_{i=1}^{n} \sum_{w=0}^{1} (-1)^{u_i w} x^{1-w} y^{w}.$$

If $u_i = 0$, the inner sum is $x + y$. If $u_i = 1$, it is $x - y$. Thus

$$\hat{f}(u) = (x + y)^{n - \text{wt}(u)} (x - y)^{\text{wt}(u)}. \tag{11}$$

Then Equation (8) reads

$$\sum_{u \in \mathscr{C}^\perp} x^{n - \text{wt}(u)} y^{\text{wt}(u)} = \frac{1}{|\mathscr{C}|} \sum_{u \in \mathscr{C}} (x + y)^{n - \text{wt}(u)} (x - y)^{\text{wt}(u)}. \qquad \text{Q.E.D.}$$

Examples. We apply Theorem 1 to the examples at the beginning of this section.

(i) $W_{\mathscr{C}_3}(x, y) = x^3 + 3xy^2$,

$$\frac{1}{4} W_{\mathscr{C}_3}(x + y, x - y) = \frac{1}{4}[(x + y)^3 + 3(x + y)(x - y)^2]$$
$$= x^3 + y^3,$$

which is indeed $W_{\mathscr{C}_3^\perp}(x, y)$. Again,

$$\frac{1}{2} W_{\mathscr{C}_3^\perp}(x + y, x - y) = \frac{1}{2}[(x + y)^3 + (x - y)^3]$$
$$= x^3 + 3xy^2 = W_{\mathscr{C}_3}(x, y),$$

illustrating that the theorem is symmetric with respect to the rôles of $\mathscr{C}$ and $\mathscr{C}^\perp$.

(ii) $W_{\mathscr{C}_2}(x, y) = x^2 + y^2$. So

$$\frac{1}{2} W_{\mathscr{C}_2}(x + y, x - y) = \frac{1}{2}[(x + y)^2 + (x - y)^2]$$
$$= x^2 + y^2 = W_{\mathscr{C}_2}(x, y),$$

which is correct since $\mathscr{C}_2$ is self-dual.

Problems. (1) Verify that

$$\frac{1}{16} W_{\mathscr{H}_3}(x + y, x - y) = x^7 + 7x^3y^4 = W_{\mathscr{H}_3^\perp}(x, y).$$

(2) Show that Theorem 1 is symmetric with respect to the rôles of $\mathscr{C}$ and $\mathscr{C}^\perp$, i.e.,

$$W_{\mathscr{C}}(x, y) = \frac{1}{|\mathscr{C}^\perp|} W_{\mathscr{C}^\perp}(x + y, x - y).$$

[Hint: Put $u = x + y$, $v = x - y$ in Equation (5) and remember that W is homogeneous.]

(3) Show that the weight enumerator of the $[n = 2^m - 1, n - m, 3]$ Hamming code $\mathscr{H}_m$ is

$$\frac{1}{n + 1} [(x + y)^n + n(x + y)^{(n-1)/2}(x - y)^{(n+1)/2}].$$

(4) (a) If $\mathscr{C}$ is a code with minimum distance $\geqslant 3$, show that there are iA_i vectors of weight $i - 1$ at distance 1 from $\mathscr{C}$, and $(n - i)A_i$ vectors of weight $i + 1$ at distance 1 from $\mathscr{C}$.

(b) Hence show that for a Hamming code,

$$\sum_{i=1}^{n} iA_iy^{i-1} + \sum_{i=0}^{n} A_iy^i + \sum_{i=0}^{n-1} (n - i)A_iy^{i+1} = (1 + y)^n,$$

i.e. the weight distribution satisfies the recurrence $A_0 = 1$, $A_1 = 0$,

$$(i + 1)A_{i+1} + A_i + (n - i + 1)A_{i-1} = \binom{n}{i}.$$

(5) The extended Golay code (§6 of Ch. 2) is self-dual. Verify this by working out the RHS of Equation (4) for this code.

Relationship between the A_i's and the A_i's. What is the relationship between the weight distributions $\{A_i'\}$ and $\{A_i\}$ obtained from (5)? If we write

$$(x + y)^{n-i}(x - y)^i = \sum_{k=0}^{n} P_k(i)x^{n-k}y^k \tag{12}$$

then from (5)

$$A_k' = \frac{1}{|\mathscr{C}|} \sum_{i=0}^{n} A_iP_k(i). \tag{13}$$

$P_k(i)$ is a Krawtchouk polynomial. The formal definition of these polynomials follows.

Krawtchouk polynomials.

Definition. For any positive integer n, the *Krawtchouk polynomial* $P_k(x; n) = P_k(x)$ is defined by

$$P_k(x; n) = \sum_{j=0}^{k} (-1)^j \binom{x}{j} \binom{n-x}{k-j}, \quad k = 0, 1, 2, \ldots \tag{14}$$

where x is an indeterminate, and the binomial coefficients are defined in Problem 18 of Ch. 1. Thus $P_k(x; n)$ is a polynomial in x of degree k. If there is no danger of confusion we omit the n.

From the binomial series (Problem 18 again) $P_k(x; n)$ has the generating function

$$(1+z)^{n-x}(1-z)^x = \sum_{k=0}^{\infty} P_k(x)z^k. \tag{15}$$

If i, n are integers with $0 \le i \le n$ this becomes

$$(1+z)^{n-i}(1-z)^i = \sum_{k=0}^{n} P_k(i)z^k. \tag{16}$$

The first few values are

$$P_0(x) = 1,$$
$$P_1(x) = n - 2x,$$
$$P_2(x) = \binom{n}{2} - 2nx + 2x^2,$$
$$P_3(x) = \binom{n}{3} - \left(n^2 - n + \frac{2}{3}\right)x + 2nx^2 - \frac{4x^3}{3}.$$

For further properties see §7.

The A'_k's given by (13) are interesting even if $\mathscr{C}$ is a nonlinear code (cf. Theorems 6 and 8). But to deal with nonlinear codes we need some new machinery.

*Moments of the weight distribution. Before proceeding with nonlinear codes we digress to give some other identities relating the weight distribution $\{A_i\}$ of an $[n, k]$ code, and the weight distribution $\{A'_i\}$ of the dual code.

Our starting point is equation (5) with $x = 1$, rewritten as

$$\sum_{i=0}^{n} A_i y^i = \frac{1}{2^{n-k}} \sum_{i=0}^{n} A'_i (1+y)^{n-i}(1-y)^i. \tag{17}$$

Setting $y = 1$ gives (since $A'_0 = 1$)

$$\sum_{i=0}^{n} \frac{A_i}{2^k} = 1$$

as it should.

Differentiating (17) with respect to y and setting $y = 1$ gives the first moment:

$$\sum_{i=1}^{n} \frac{iA_i}{2^k} = \frac{1}{2}(n - A_1') = \frac{1}{2}n \quad \text{if } A_1' = 0.$$

So if $A_1' = 0$, the mean weight is $\frac{1}{2}n$. Continuing in this way we obtain:

$$\sum_{i=\nu}^{n} \binom{i}{\nu} \frac{A_i}{2^k} = \frac{1}{2^\nu} \sum_{i=0}^{\nu} (-1)^i \binom{n-i}{\nu-i} A_i' \tag{18}$$

for $\nu = 0, 1, \ldots, n$. The LHS of (18) is called a *binomial moment* of the A_i's. Now suppose the dual code has minimum distance d', so that

$$A_1' = \cdots = A_{d'-1}' = 0.$$

If $\nu < d'$, the RHS of (18) no longer depends on the code: thus

$$\sum_{i=\nu}^{n} \binom{i}{\nu} \frac{A_i}{2^k} = \frac{1}{2^\nu} \binom{n}{\nu}. \tag{19}$$

This will be used in the next chapter. Equation (19) shows that the νth binomial moment, for $\nu = 0, 1, \ldots, d'-1$, is independent of the code, and is equal to that of the $[n, n, 1]$ code F^n.

Problems. (6) Prove the following alternative version of (18):

$$\sum_{i=0}^{n-\nu} \binom{n-i}{\nu} A_i = 2^{k-\nu} \sum_{i=0}^{\nu} \binom{n-i}{n-\nu} A_i'$$

for $\nu = 0, 1, \ldots, n$.

(7) If instead of differentiating (17) with respect to y we apply the operator $y(d/dy)$ ν times and set $y = 1$, we obtain the *power moments* of the A_i's. For $\nu < d'$ these are particularly simple. Show that

$$\sum_{i=0}^{n} i^2 \frac{A_i}{2^k} = \frac{1}{4} n(n+1), \quad \text{if } 2 < d', \tag{20}$$

and

$$\sum_{i=0}^{n} i^\nu \frac{A_i}{2^k} = \frac{1}{2^n} \sum_{j=0}^{n} j^\nu \binom{n}{j}, \quad \text{if } 0 \le \nu < d'. \tag{21}$$

In general, show that for $\nu = 0, 1, \ldots$

$$\sum_{i=0}^{n} i^\nu A_i = \sum_{j=0}^{n} (-1)^j A_j' \left\{ \sum_{l=0}^{\nu} l! S(\nu, l) 2^{k-l} \binom{n-j}{n-l} \right\}, \tag{22}$$

where

$$S(\nu, l) = \frac{1}{l!} \sum_{r=0}^{l} (-1)^{l-r} \binom{l}{r} r^\nu$$

is a Stirling number of the second kind (and the conventions for binomial coefficients given in Ch. 1 apply). The equations (22) are frequently called the *Pless identities*.

(8) Show that if there are only r unknown A_i's, and $A'_1, \ldots, A'_{r-1}$ are known, then all the A_i's can be determined. [Hint: the coefficients of the LHS of (22) form a Vandermonde matrix – see Lemma 17 of Ch. 4.]

(9) *Moments about the mean.* (a) Show that (17) implies

$$\sum_{j=0}^{n} A_j e^{(n-2j)x} = 2^k \sum_{j=0}^{n} A'_j \cosh^{n-j} x \, \sinh^j x. \tag{23}$$

(b) Hence show that for $r = 0, 1, \ldots$

$$\frac{1}{2^k} \sum_{i=0}^{n} \left(\frac{n}{2} - i \right)^r A_i = \frac{1}{2^r} \sum_{j=0}^{n} F_r^{(j)}(n) A'_j, \tag{24}$$

where

$$F_r^{(j)}(n) = \left[\frac{d^r}{dx^r} \cosh^{n-j} x \, \sinh^j x \right]_{x=0}.$$

(c) Show that $F_r^{(j)}(n) \geq 0$ and

$$F_r^{(0)}(n) = \frac{1}{2^{n-r}} \sum_{i=0}^{n} \left(\frac{n}{2} - i \right)^r \binom{n}{i}.$$

(d) Hence, for $r = 0, 1, \ldots$,

$$\frac{1}{2^k} \sum_{i=0}^{n} \left(\frac{n}{2} - i \right)^r A_i \geq \frac{1}{2^n} \sum_{i=0}^{n} \left(\frac{n}{2} - i \right)^r \binom{n}{i}. \tag{25}$$

Note that the RHS of (25) is the rth moment about the mean of the weight distribution of the $[n, n, 1]$ code F^n.

(e) Show that the equality holds in (25) if $r \leq d' - 1$.

Research Problem (5.1). Let $\mathscr{C}$ be a linear code. As in §5 of Ch. 1 let α_i be the number of coset leaders of $\mathscr{C}$ of weight i, and α'_i the corresponding number for $\mathscr{C}^\perp$. To what extent do the numbers $\{\alpha_i\}$ determine $\{\alpha'_i\}$?

§3. The group algebra

We are going to describe binary vectors of length n by polynomials in $z_1, \ldots, z_n$. For example, $100 \cdots 0$ will be represented by z_1, $1010 \cdots 0$ by $z_1 z_3$, and so on. In general $v = v_1 v_2 \cdots v_n$ is represented by $z_1^{v_1} z_2^{v_2} \cdots z_n^{v_n}$, which we abbreviate z^v. Clearly if we know z^v we can recover v. Thus $\sum_{v \in \mathscr{C}} z^v$ is just a very fancy weight enumerator for $\mathscr{C}$. (We shall use this type of weight enumerator again in §6 for codes over $GF(q)$.) We make the convention that $z_i^2 = 1$ for all i. This makes the set of all z^v into a multiplicative group denoted by G. Thus F^n and G are isomorphic groups, with addition in F^n

$$v + w = (v_1, \ldots, v_n) + (w_1, \ldots, w_n) = (v_1 + w_1, \ldots, v_n + w_n)$$

corresponding to multiplication in G:

$$z^v z^w = z_1^{v_1} \cdots z_n^{v_n} \cdot z_1^{w_1} \cdots z_n^{w_n} = z_1^{v_1 + w_1} \cdots z_n^{v_n + w_n} = z^{v+w}.$$

Definition. The *group algebra* QG of G over the rational numbers Q consists of all formal sums

$$\sum_{v \in F^n} a_v z^v, \quad a_v \in Q, \; z^v \in G.$$

Addition and multiplication of elements of QG are defined in the natural way by

$$\sum_{v \in F^n} a_v z^v + \sum_{v \in F^n} b_v z^v = \sum_{v \in F^n} (a_v + b_v) z^v,$$

$$r \sum_{v \in F^n} a_v z^v = \sum_{v \in F^n} r a_v z^v, \quad r \in Q$$

and

$$\sum_{v \in F^n} a_v z^v \cdot \sum_{w \in F^n} b_w z^w = \sum_{v, w \in F^n} a_v b_w z^{v+w}.$$

QG gives us an algebraic notation for subsets of F^n, or codes: corresponding to the code $\mathcal{C} \subset F^n$ we have the element

$$C = \sum_{u \in \mathcal{C}} z^u$$

of QG. For example, corresponding to the code $\{000, 011, 101, 110\}$ we have

$$C = 1 + z_2 z_3 + z_1 z_3 + z_1 z_2.$$

In general it may be helpful to think of the elements of QG as "generalized codes"; the coefficient a_v being the number of times v occurs in the "code."

Problem (10). For $n = 3$ show that

$$(z_1 + z_2 + z_3)^2 = 3 + 2(z_1 z_2 + z_2 z_3 + z_1 z_3).$$

One of the nice things about the group algebra notation is that it provides a concise way of stating certain properties of codes. Let

$$Y_i = \sum_{\text{wt}(u) = i} z^u$$

represent the vectors of weight i. For example,

$$Y_0 = 1$$
$$Y_1 = z_1 + z_2 + \cdots + z_n,$$
$$Y_2 = z_1 z_2 + z_1 z_3 + \cdots + z_{n-1} z_n.$$

The sphere of radius e around v is then described by

$$z^v(Y_0 + Y_1 + \cdots + Y_e).$$

If $\mathscr{C}$ is a perfect $(n, M, 2e + 1)$ code, this fact is expressed by the identity

$$C \cdot (Y_0 + Y_1 + \cdots + Y_e) = \sum_{u \in F^n} z^u, \tag{26}$$

where

$$.C = \sum_{v \in \mathscr{C}} z^v.$$

Example. For the perfect single-error-correcting code $\{000, 111\}$, $C = 1 + z_1 z_2 z_3$, $Y_0 + Y_1 = 1 + z_1 + z_2 + z_3$, and indeed

$$C \cdot (Y_0 + Y_1) = 1 + z_1 + z_2 + z_3 + z_1 z_2 + z_1 z_3 + z_2 z_3 + z_1 z_2 z_3.$$

Problem. (11) Describe the direct sum construction and the $|u|u + v|$ construction (§9 of Ch. 2) using the group algebra.

§4. Characters

To each $u \in F^n$ we associate the mapping χ_u from G to the rational numbers given by

$$\chi_u(z^v) = (-1)^{u \cdot v}, \tag{27}$$

where $u \cdot v$ is the scalar product of the vectors u, v over Q. χ_u is called a *character* of G. χ_u is extended to act on QG by linearity:

$$\chi_u \left(\sum_{v \in F^n} a_v z^v \right) = \sum_{v \in F^n} a_v \chi_u(z^v) = \sum_{v \in F^n} (-1)^{u \cdot v} a_v. \tag{28}$$

Note that

$$\chi_u(z^v) = \begin{cases} 1 & \text{if } u, v \text{ are orthogonal,} \\ -1 & \text{if not.} \end{cases}$$

Problems. (12) (i) $\chi_u(z^v) = \chi_v(z^u)$.
   (ii) $\chi_u(z^v) \chi_u(z^w) = \chi_u(z^{v+w})$.
   (iii) If C_1, C_2 are arbitrary elements of QG,

$$\chi_u(C_1) \chi_u(C_2) = \chi_u(C_1 \cdot C_2).$$

   (iv) $\chi_u(z^w) \chi_v(z^w) = \chi_{u+v}(z^w)$.
   (13) If $\mathscr{C}$ is a linear code and $C = \sum_{u \in \mathscr{C}} z^u$, then

$$\chi_v(C) = \begin{cases} |\mathscr{C}| & \text{if } v \in \mathscr{C}^\perp \\ 0 & \text{if } v \notin \mathscr{C}^\perp \end{cases}$$

(14) Show that if $u \in F^n$ has weight i, then

$$\sum_{\text{wt}(v)=k} (-1)^{u \cdot v} = \chi_u(Y_k) = P_k(i).$$

(15) *More about characters.* More precisely, a character χ of an abelian group G is any homomorphism from G to the multiplicative group of the complex numbers of absolute value 1. For G as defined at the beginning of Section 3, the characters take the values ± 1. Show that the characters of G form a group X which is isomorphic to G (and to F^n). Then (27) is just one example of an isomorphism between F^n and X.

(16) (An inversion formula) Let

$$C = \sum_{v \in F^n} c_v z^v$$

and suppose the numbers

$$\chi_u(C) = \sum_{v \in F^n} c_v (-1)^{u \cdot v}$$

are known. Show that

$$c_v = \frac{1}{2^n} \sum_{u \in F^n} (-1)^{u \cdot v} \chi_u(C),$$

and so C is determined by the $\chi_u(C)$.

§5. MacWilliams theorem for nonlinear codes

Weight enumerator of an element of the group algebra. Let

$$C = \sum_{v \in F^n} c_v z^v$$

be an arbitrary element of the group algebra QG, with the property that

$$M = \sum_{v \in F^n} c_v \neq 0.$$

We call the $(n+1)$-tuple $\{A_0, \ldots, A_n\}$, where

$$A_i = \sum_{\text{wt}(v)=i} c_v,$$

the *weight distribution* of C. This is the natural generalization of the weight distribution of a code. Of course $\sum A_i = M$. As in §2 we also define the *weight enumerator* of C to be

$$W_C(x, y) = \sum_{v \in F^n} c_v x^{n - \text{wt}(v)} y^{\text{wt}(v)}$$

$$= \sum_{i=0}^{n} A_i x^{n-i} y^i.$$

Definition. The *transform* of C is the element C' of QG given by

$$C' = \frac{1}{M} \sum_{u \in F^n} \chi_u(C) z^u, \tag{29}$$

where χ was defined in §4.

Suppose

$$C' = \sum_{u \in F^n} c'_u z^u,$$

so that

$$c'_u = \frac{1}{M} \chi_u(C)$$

$$= \frac{1}{M} \sum_{v \in F^n} (-1)^{u \cdot v} c_v, \quad u \in F^n. \tag{30}$$

(The c'_u's are proportional to the Hadamard transform of the c_v's – see §3 of Ch. 2.) Then the weight distribution of C' is $\{A'_0, \ldots, A'_n\}$, where

$$A'_i = \sum_{\text{wt}(u)=i} c'_u = \frac{1}{M} \sum_{\text{wt}(u)=i} \chi_u(C), \tag{31}$$

and the weight enumerator of C' is

$$W_{C'}(x, y) = \sum_{i=0}^{n} A'_i x^{n-i} y^i.$$

As in Theorem 1, $W_{C'}$ is given by a linear transformation of W_C.

Theorem 3.

$$W_{C'}(x, y) = \frac{1}{M} W_C(x + y, x - y), \tag{32}$$

or equivalently, using (15),

$$A'_k = \frac{1}{M} \sum_{i=0}^{n} A_i P_k(i), \quad k = 0, \ldots, n. \tag{33}$$

This theorem and Theorem 5 can be thought of as MacWilliams theorems for nonlinear codes.

Proof. (32) is also equivalent to

$$\sum_{v \in F^n} c'_v x^{n-\text{wt}(v)} y^{\text{wt}(v)} = \frac{1}{M} \sum_{u \in F^n} c_u (x + y)^{n-\text{wt}(u)} (x - y)^{\text{wt}(u)}.$$

From (30), the LHS is equal to

$$\frac{1}{M} \sum_{u \in F^n} c_u \sum_{v \in F^n} (-1)^{u \cdot v} x^{n - \mathrm{wt}(v)} y^{\mathrm{wt}(v)}$$

$$= \frac{1}{M} \sum_{u \in F^n} c_u (x + y)^{n - \mathrm{wt}(u)} (x - y)^{\mathrm{wt}(u)},$$

by (9) and (11) of §2.   Q.E.D.

For any $(n + 1)$-tuple $\{A_0, \ldots, A_n\}$ with

$$M = \sum_{i=0}^{n} A_i \neq 0$$

we call $\{A_0', \ldots, A_n'\}$ the *MacWilliams transform* of the A_i's. By Theorem 3 this can be obtained either from (31) or from (33).

Problems. (17) Show that $(C')' = C/c_0$, provided $c_0 \neq 0$.

(18) Show that $A_0' = 1$.

(19) Show that

$$A_0 = \frac{M}{2^n} \sum_{i=0}^{n} A_i'.$$

(20) Let $\mathscr{C}$ be an $[n, k]$ linear code and suppose the coset $v + \mathscr{C}$ contains E_i vectors of weight i. Use (31) to show that the transform E_i' is $\alpha_i - \beta_i$, where α_i is the number of codewords of weight i in $\mathscr{C}^\perp$ which are orthogonal to v, and β_i is the number which are not. Hence show

$$\sum_{i=0}^{n} E_i x^{n-i} y^i = \frac{1}{2^{n-k}} \sum_{i=0}^{n} (\alpha_i - \beta_i)(x + y)^{n-i}(x - y)^i.$$

Distance distribution of a nonlinear code. Now let $\mathscr{C}$ be a linear or nonlinear (n, M, d) code. $\mathscr{C}$ is described by the element

$$C = \sum_{v \in \mathscr{C}} z^v \tag{34}$$

of QG.

Problem. (21) Show that if $\mathscr{C}$ is linear then

$$C' = \sum_{v \in \mathscr{C}^\perp} z^v.$$

Thus Theorem 1 is a special case of Theorem 3.

Now let $\mathscr{C}$ be linear or nonlinear. Let $D = (1/M)C^2$. If we expand

$$D = \sum_{w \in F^n} d_w z^w,$$

the weight distribution of D is $\{B_0, \ldots, B_n\}$, where

$$B_i = \sum_{\text{wt}(w)=i} d_w.$$

Lemma 4. $\{B_0, \ldots, B_n\}$ *is the distance distribution of* $\mathscr{C}$.

Proof.

$$D = \frac{1}{M} C^2 = \frac{1}{M} \sum_{u \in \mathscr{C}} z^u \sum_{v \in \mathscr{C}} z^v$$

$$= \frac{1}{M} \sum_{u \in \mathscr{C}} \sum_{v \in \mathscr{C}} z^{u+v}$$

$$= \sum_{i=0}^{n} \frac{1}{M} \sum_{\substack{u,v \in \mathscr{C} \\ \text{dist}(u,v)=i}} z^{u+v}$$

$$= \sum_{w \in F^n} d_w z^w.$$

Therefore

$$B_i = \sum_{\text{wt}(w)=i} d_w$$

$$= \frac{1}{M} \sum_{\substack{u,v \in \mathscr{C} \\ \text{dist}(u,v)=i}} 1. \qquad\qquad \text{Q.E.D.}$$

By applying Theorem 3 to the element $D = (1/M)C^2$ of QG we obtain:

Theorem 5. *The transform of the distance distribution is*

$$B'_s = \frac{1}{M} \sum_{\text{wt}(u)=s} \chi_u(D),$$

$$= \frac{1}{M} \sum_{i=0}^{n} B_i P_s(i), \quad s = 0, \ldots, n. \qquad (35)$$

Proof. Use Theorem 3 and note that

$$\sum_{w \in F^n} d_w = \frac{1}{M} \left(\sum_{v \in F^n} c_v \right)^2 = M. \qquad\qquad \text{Q.E.D.}$$

Properties of the B_i's.

Theorem 6. $B'_i \geq 0$, $i = 0, \ldots, n$.

Proof.

$$B'_i = \frac{1}{M} \sum_{\text{wt}(u)=i} \chi_u(D) = \frac{1}{M^2} \sum_{\text{wt}(u)=i} \chi_u(C^2)$$

$$= \frac{1}{M^2} \sum_{\text{wt}(u)=i} \chi_u(C)^2, \quad \text{by Problem 12}, \tag{36}$$

$$\geq 0. \qquad\qquad\qquad\qquad \text{Q.E.D.}$$

.This innocent-looking result turns out to be quite useful – see the linear programming bound (§4 of Ch. 17). Another proof is given in Theorem 12 of Ch. 21.

Theorem 7. (a) If $0 \in \mathscr{C}$, $B_i = 0 \Rightarrow A_i = 0$.
  (b) $B'_i = 0 \Rightarrow \chi_u(C) = 0$ for all u of weight $i \Rightarrow A'_i = 0$.
  (c) $B'_i \neq 0 \Rightarrow \chi_u(C) \neq 0$ for some u of weight i.

Proof. (a) is obvious. (b) and (c) follow from (36) and (31). Q.E.D.

Dual distance and orthogonal arrays.

Definition. The *dual distance* d' of a code $\mathscr{C}$ is defined by $B'_i = 0$ for $1 \leq i \leq d' - 1$, $B'_{d'} \neq 0$. If $\mathscr{C}$ is linear, d' is the minimum distance of $\mathscr{C}^\perp$.

Note that with this definition of d', Equations (18) to (25) and Problems (6) to (9) hold for the distance distributions of nonlinear codes.
  Let $[\mathscr{C}]$ be the $M \times n$ array of all code words of $\mathscr{C}$. If $\mathscr{C}$ is linear, any set of $r \leq d' - 1$ columns of $[\mathscr{C}]$ must be linearly independent, otherwise $\mathscr{C}^\perp$ would contain a vector of weight $r < d'$. (This is the dual statement to Theorem 10 of Ch. 1.) This statement is also true for nonlinear codes. We rephrase it slightly.

Theorem 8. *Any set of $r \leq d' - 1$ columns of $[\mathscr{C}]$ contains each r-tuple exactly $M/2^r$ times, and d' is the largest number with this property.*

Remark. An array with this property is called an *orthogonal array* of n constraints, 2 levels, strength $d' - 1$ and index $M/2^{d'-1}$ (see §8 of Ch. 11).

Proof. To prove the first half, from the proof of Theorem 7 we have $\chi_u(C) = 0$ for all u of weight $1, 2, \ldots, d' - 1$. $\chi_u(C) = 0$ for all u of weight 1 implies that each column of $[\mathscr{C}]$ has $M/2$ zeros and $M/2$ ones. Then $\chi_u(C) = 0$ for all u of weight 2 implies that each pair of columns contains $M/4$ occurrences of each of 00, 10, 01, 11. Clearly one can go on like this up to $d' - 1$. Conversely, since $B'_{d'} \neq 0$, there is a u of weight d' such that $\chi_u(C) \neq 0$.   Q.E.D.

Examples. (i) The (11, 12, 6) Hadamard code $\mathscr{A}_{12}$ (§3 of Ch. 2). Since this is a simplex code and contains 0, the weight and distance distributions are equal:

$$i: \quad 0 \quad 6$$
$$A_i = B_i: \quad 1 \quad 11 \quad, \qquad W_{\mathscr{A}_{12}}(x, y) = x^{11} + 11x^5 y^6.$$

The transformed distribution is obtained from

$$\frac{1}{12} W_{\mathscr{A}_{12}}(x + y, x - y) = \frac{1}{12}((x + y)^{11} + 11(x + y)^5(x - y)^6)$$

$$= \sum_{i=0}^{11} B'_i x^{11-i} y^i,$$

and is equal to

$$i: \quad 0 \quad 3 \quad 4 \quad 5 \quad 6 \quad 7 \quad 8 \quad 11$$
$$B'_i: \quad 1 \quad 18\tfrac{1}{3} \quad 36\tfrac{2}{3} \quad 29\tfrac{1}{3} \quad 29\tfrac{1}{3} \quad 36\tfrac{2}{3} \quad 18\tfrac{1}{3} \quad 1$$

Note that $\Sigma B'_i = 170\tfrac{2}{3} = 2^{11}/12$. Also $B'_i \geq 0$, as required by Theorem 6. This table shows that $d' = 3$.

The list of codewords $[\mathscr{A}_{12}]$ is found in the top half of Fig. 2.1. To illustrate Theorem 8 we observe that in any two columns of $[\mathscr{A}_{12}]$ the vectors 00, 01, 10, 11 each occur 3 times.

(ii) The (8, 16, 2) code shown in Fig. 5.1.

```
0 0 0 0 0 0 0 0    1 1 1 1 1 1 1 1
1 1 0 0 0 0 0 0    0 0 1 1 1 1 1 1
1 0 1 0 0 0 0 0    0 1 0 1 1 1 1 1
1 0 0 1 0 0 0 0    0 1 1 0 1 1 1 1
1 0 0 0 1 0 0 0    0 1 1 1 0 1 1 1
1 0 0 0 0 1 0 0    0 1 1 1 1 0 1 1
1 0 0 0 0 0 1 0    0 1 1 1 1 1 0 1
1 0 0 0 0 0 0 1    0 1 1 1 1 1 1 0
```

Fig. 5.1. An (8, 16, 2) nonlinear code.

See Problem 22 of Ch. 15. Again the weight and distance distributions coincide:

$$i: \quad 0 \; 2 \; 6 \; 8$$
$$A_i = B_i: \quad 1 \; 7 \; 7 \; 1 \qquad W(x, y) = x^{\circ} + 7x^6 y^2 + 7x^2 y^6 + y^8.$$

The transformed distribution is obtained from $\frac{1}{16} W(x + y, x - y)$. But, after simplification, this turns out to be *equal* to $W(x, y)$! So that for this code $B'_i = B_i$. Also $d' = 2$.

This code has another unusual property, namely $D' = D$. The proof of this is a good exercise in the use of the group algebra:

Problem. (22) Show that

$$C = 1 + z_1 z_2 + \cdots + z_1 z_8 + z_2 \cdots z_8 \left(\frac{1}{z_2} + \cdots + \frac{1}{z_8} \right) + z_1 z_2 \cdots z_8,$$

and that

$$D = \frac{1}{16} C^2$$

$$= 1 + \frac{1}{4} \sum_{i<j} z_i z_j + \frac{1}{4} z_1 \cdots z_8 \sum_{i<j} \frac{1}{z_i z_j} + z_1 \cdots z_8,$$

$$D' = D.$$

Note that if we only know D, in general C cannot be recovered.

(iii) The $(16, 256, 6)$ Nordstrom–Robinson code $\mathcal{N}_{16}$ (§8 of Ch. 2). Once again $A_i = B_i$ – see Fig. 2.19. This code also has the property that $B'_i = B_i$, although it is laborious to show this directly. It will follow from the results of Ch. 15.

Problem. (23) Let $\mathcal{C}$ be a code and C be given by (34). (i) Show $D = C$ iff $\mathcal{C}$ is linear.

(ii) Show $C' = C$ iff $\mathcal{C}$ is linear and self-dual.

*§6. Generalised MacWilliams theorems for linear codes

In this section we describe several weight enumerators of linear codes over an arbitrary field $F = \mathrm{GF}(q) = \mathrm{GF}(p^m)$, where p is a prime. Let the elements of $\mathrm{GF}(q)$ be denoted by $\omega_0 = 0, \omega_1, \ldots, \omega_{q-1}$, in some fixed order.

Complete weight enumerator. The first weight enumerator to be considered classifies codewords u in F^n according to the number of times each field element ω_i appears in u.

Definition. The *composition* of $u = (u_1, \ldots, u_n)$, denoted by comp (u), is $(s_0, s_1, \ldots, s_{q-1})$ where $s_i = s_i(u)$ is the number of components u_j equal to ω_i. Clearly

$$\sum_{i=0}^{q-1} s_i = n.$$

Let $\mathscr{C}$ be a linear $[n, k]$ code over $GF(q)$ and let $A(t)$ be the number of codewords $u \in \mathscr{C}$ with comp $(u) = t = (t_0, \ldots, t_{q-1})$. Then the *complete weight enumerator* of $\mathscr{C}$ is

$$\mathscr{W}_{\mathscr{C}}(z_0, \ldots, z_{q-1}) = \sum A(t) z_0^{t_0} \cdots z_{q-1}^{t_{q-1}}$$

$$= \sum_{u \in \mathscr{C}} z_0^{s_0} \cdots z_{q-1}^{s_{q-1}}. \tag{37}$$

For example, let $\mathscr{C}_4$ be the $[4, 2, 3]$ ternary code #6 of Ch. 1. The complete weight enumerator is

$$\mathscr{W}_{\mathscr{C}_4}(z_0, z_1, z_2) = z_0^4 + z_0 z_1^3 + 3 z_0 z_1^2 z_2 + 3 z_0 z_1 z_2^2 + z_0 z_2^3$$

$$= z_0 (z_0^3 + (z_1 + z_2)^3). \tag{38}$$

Characters of $GF(q)$. In order to state the theorem we need to define the characters of $GF(q)$. Recall from Ch. 4 that any element β of $GF(q)$ can be written in the form

$$\beta = \beta_0 + \beta_1 \alpha + \beta_2 \alpha^2 + \cdots + \beta_{m-1} \alpha^{m-1},$$

or equivalently as an m-tuple

$$\beta = (\beta_0, \beta_1, \ldots, \beta_{m-1}),$$

where α is a primitive element of $GF(q)$, and $0 \le \beta_i \le p - 1$. Let ξ be the complex number $e^{2\pi i/p}$. This is a primitive p^{th} root of unity, i.e., $\xi^p = e^{2\pi i} = 1$, while $\xi^l \ne 1$ for $0 < l < p$.

Definition. For each $\beta = (\beta_0, \ldots, \beta_{m-1}) \in GF(q)$ define χ_β to be the complex-valued mapping defined on $GF(q)$ by

$$\chi_\beta(\gamma) = \xi^{\beta_0 \gamma_0 + \cdots + \beta_{m-1} \gamma_{m-1}} \tag{39}$$

for $\gamma = (\gamma_0, \ldots, \gamma_{m-1}) \in GF(q)$. χ_β is called a *character* of $GF(q)$.

Problems. (24) Show $\chi_\beta(\gamma) = \chi_\gamma(\beta)$, for all $\beta, \gamma \in GF(q)$.
  (24) Show that

$$\chi_\beta(\gamma + \gamma') = \chi_\beta(\gamma) \chi_\beta(\gamma') \tag{40}$$

for all $\beta, \gamma, \gamma' \in GF(q)$. Thus χ_β is a homomorphism from the additive group of $GF(q)$ into the multiplicative group of complex numbers of magnitude 1.
  (26) Show that

$$\chi_{\beta+\beta'}(\gamma) = \chi_\beta(\gamma)\chi_{\beta'}(\gamma)$$

for all $\beta, \beta', \gamma \in GF(q)$. Thus the set of all q characters χ_β forms a group which is isomorphic to the additive group of $GF(q)$.

Example. $q = p = 3$,   with $GF(3) = \{0, 1, 2\}$ and $\xi = \omega = e^{2\pi i/3} = \cos 120° + i \sin 120°$. There are three characters:

$\chi_0(0) = 1,$ $\chi_0(1) = 1,$ $\chi_0(2) = 1$ (the trivial character);
$\chi_1(0) = 1,$ $\chi_1(1) = \omega,$ $\chi_1(2) = \omega^2$;
$\chi_2(0) = 1,$ $\chi_2(1) = \omega^2,$ $\chi_2(2) = \omega.$

Lemma 9. *For any nonzero* $\beta \in GF(q)$,

$$\sum_{\gamma \in GF(q)} \chi_\beta(\gamma) = 0.$$

Proof. The sum is equal to

$$\sum_{\gamma \in GF(q)} \xi^{\beta_0\gamma_0 + \cdots + \beta_{m-1}\gamma_{m-1}} = \prod_{j=0}^{m-1} \left(\sum_{\gamma_j = 0}^{p-1} \xi^{\beta_j\gamma_j} \right).$$

Since $\beta \neq 0$ there is a nonzero β_j, say $\beta_r \neq 0$. Then the rth factor in the above product is

$$\sum_{\gamma_r=0}^{p-1} \xi^{\beta_r\gamma_r} = \sum_{k=0}^{p-1} \xi^k = \frac{1 - \xi^p}{1 - \xi} = 0. \qquad\qquad \text{Q.E.D.}$$

Example. In $GF(3)$ with $\beta = 1$ the lemma says

$$\sum_{\gamma=0}^{2} \omega^\gamma = 1 + \omega + \omega^2 = 0.$$

  To give the next theorem we must choose any one of the characters χ_β with $\beta \neq 0$. For concreteness we choose $\beta = 1$, i.e., the character χ_1 defined by

$$\chi_1(\gamma) = \xi^{\gamma_0}, \quad \text{for } \gamma = (\gamma_0, \ldots, \gamma_{m-1}) \in GF(q). \qquad (41)$$

If q is a prime p, this is simply $\chi_1(\gamma) = \xi^\gamma$, $\gamma \in GF(p)$, where $\xi = e^{2\pi i/p}$.

MacWilliams theorem for complete weight enumerators.

Theorem 10. *If $\mathscr{C}$ is a linear $[n, k]$ code over $GF(q)$ with complete weight* enumerator $\mathcal{W}_{\mathscr{C}}$, the complete weight enumerator of the dual code $\mathscr{C}^{\perp}$ is

$$\mathcal{W}_{\mathscr{C}^{\perp}}(z_0, \ldots, z_r, \ldots, z_{q-1}) = \frac{1}{|\mathscr{C}|} \mathcal{W}_{\mathscr{C}}\left(\sum_{s=0}^{q-1} \chi_1(\omega_0\omega_s)z_s, \ldots, \sum_{s=0}^{q-1} \chi_1(\omega_r\omega_s)z_s, \ldots\right)$$

(42)

Example. For a code over GF(3) the theorem states that

$$\mathcal{W}_{\mathscr{C}^{\perp}}(z_0, z_1, z_2) = \frac{1}{|\mathscr{C}|} \mathcal{W}_{\mathscr{C}}(z_0 + z_1 + z_2, z_0 + \omega z_1 + \omega^2 z_2, z_0 + \omega^2 z_1 + \omega z_2) \quad (43)$$

where $\omega = e^{2\pi i/3}$. I.e., $\mathcal{W}_{\mathscr{C}^{\perp}}$ is obtained by applying the linear transformation

$$\begin{pmatrix} 1 & 1 & 1 \\ 1 & \omega & \omega^2 \\ 1 & \omega^2 & \omega \end{pmatrix}$$

(44)

to $\mathcal{W}_{\mathscr{C}}$, and dividing the result by $|\mathscr{C}|$. E.g., for the code $\mathscr{C}_4$ given above,

$$\mathcal{W}_{\mathscr{C}_4^{\perp}}(z_0, z_1, z_2) = \tfrac{1}{9}(z_0 + z_1 + z_2)[(z_0 + z_1 + z_2)^3 + (2z_0 - z_1 - z_2)^3].$$

Problem. (27) Show that $\mathscr{C}_4$ is self dual. Check this by showing

$$\mathcal{W}_{\mathscr{C}_4^{\perp}} = \mathcal{W}_{\mathscr{C}_4}.$$

The theorem depends on the following lemma.

Lemma 11. *For $u, v \in F^n$ let $\chi_u(v) = \chi_1(u \cdot v)$. As in Lemma 2, the Hadamard transform $\hat{f}$ of a mapping f defined on F^n is given by*

$$\hat{f}(u) = \sum_{v \in F^n} \chi_u(v)f(v).$$

Then if $\mathscr{C}$ is any $[n, k]$ code over GF(q),

$$\sum_{u \in \mathscr{C}^{\perp}} f(u) = \frac{1}{|\mathscr{C}|} \sum_{u \in \mathscr{C}} \hat{f}(u).$$

(45)

The proof is essentially the same as that of Lemma 2, but requires the use of Lemma 9.

Proof of Theorem 10. We apply Lemma 11 with

$$f(u) = z_0^{s_0(u)} \cdots z_{q-1}^{s_{q-1}(u)},$$

$$\hat{f}(u) = \sum_{v \in F^n} \chi_u(v)z_0^{s_0(v)} \cdots z_{q-1}^{s_{q-1}(v)},$$

$$= \prod_{r=0}^{q-1} \left(\sum_{s=0}^{q-1} \chi_1(\omega_r\omega_s)z_s\right)^{s_r(u)},$$

in the same way that (11) is obtained from (9). Then Equation (45) gives the theorem.   Q.E.D.

Lee weight enumerator. By setting certain variables equal to each other in the complete weight enumerator we obtain the Lee and Hamming weight enumerators, which give progressively less and less information about the code, but become easier to handle.

Definition. Suppose now that $q = 2\delta + 1$ is an odd prime power, and let the elements of GF(q) be labeled $\omega_0 = 0, \omega_1, \ldots, \omega_\delta, \omega_{\delta+1}, \ldots, \omega_{q-1}$, where $\omega_{q-i} = -\omega_i$ for $1 \le i \le \delta$. E.g., we take

$$GF(5) = \{\omega_0 = 0, \omega_1 = 1, \omega_2 = 2, \omega_3 = -2 = 3, \omega_4 = -1 = 4\}.$$

The *Lee composition* of a vector $u \in F^n$, denoted by Lee (u), is $(l_0, l_1, \ldots, l_\delta)$ where $l_0 = s_0(u)$, $l_i = s_i(u) + s_{q-i}(u)$ for $1 \le i \le \delta$.

E.g., for $q = 5$, the Lee composition classifies codewords according to the number of components which are 0, the number which are ± 1, and the number which are ± 2.

The *Lee enumerator* of code $\mathscr{C}$ is

$$\mathscr{L}_\mathscr{C}(z_0, \ldots, z_\delta) = \sum_{u \in \mathscr{C}} z_0^{l_0} z_1^{l_1} \cdots z_\delta^{l_\delta}.$$

For example, the self-dual code $\mathscr{A}$ over GF(5) of length 2 consisting of the codewords

$$00, \ 12, \ 2-1, \ -21, \ -1-2,$$

has Lee enumerator $z_0^2 + 4z_1z_2$.

MacWilliams theorem for Lee enumerators.

Theorem 12. *The Lee enumerator for the dual code $\mathscr{C}^\perp$ is obtained from the Lee enumerator of $\mathscr{C}$ by replacing each z_i by*

$$z_0 + \sum_{s=1}^\delta \{\chi_1(\omega_i\omega_s) + \chi_1(-\omega_i\omega_s)\}z_s.$$

and dividing the result by $|\mathscr{C}|$.

Proof. Set $z_{q-i} = z_i$ for $1 \le i \le \delta$ in Theorem 10.   Q.E.D.

For the code $\mathscr{A}$ in the preceding example, we have $\chi_1(\omega_j\omega_s) = \alpha^{js}$ where $\alpha = e^{2\pi i/5} = \cos 72° + i \sin 72°$, and the transformation of the theorem replaces

$$
\begin{pmatrix} z_0 \\ z_1 \\ z_2 \end{pmatrix} \quad \text{by} \quad \begin{pmatrix} 1 & 2 & 2 \\ 1 & \alpha + \alpha^4 & \alpha^2 + \alpha^3 \\ 1 & \alpha^2 + \alpha^3 & \alpha + \dot{\alpha}^4 \end{pmatrix} \begin{pmatrix} z_0 \\ z_1 \\ z_2 \end{pmatrix} \tag{46}
$$

Since $\mathcal{A}$ is self-dual the theorem asserts correctly that

$$
z_0 + 4z_1z_2 = \tfrac{1}{5}[(z_0 + 2z_1 + 2z_2)^2 + 4\{z_0 + (\alpha + \alpha^4)z_1 + (\alpha^2 + \alpha^3)z_2\}\{z_0 + (\alpha^2 + \alpha^3)z_1
$$
$$
+ (\alpha + \alpha^4)z_2\}].
$$

Problem. (28) Verify this identity.

Hamming weight enumerator. Now let q be any prime power. As in §3 of Ch. 1 the *Hamming weight*, or simply the *weight*, of a vector $u = (u_1, \dots, u_n) \in F^n$ is the number of nonzero components u_i, and is denoted by $\mathrm{wt}(u)$. We shall use the notation of §2 and let A_i be the number of codewords of weight i, and

$$
W_{\mathscr{C}}(x, y) = \sum_{i=0}^{n} A_i x^{n-i} y^i
$$
$$
= \sum_{u \in \mathscr{C}} x^{n - \mathrm{wt}(u)} y^{\mathrm{wt}(u)}
$$

be the *Hamming weight enumerator* of a code $\mathscr{C}$.

MacWilliams theorem for Hamming weight enumerators.

Theorem 13.

$$
W_{\mathscr{C}^\perp}(x, y) = \frac{1}{|\mathscr{C}|} W_{\mathscr{C}}(x + (q-1)y, x - y). \tag{47}
$$

Proof. In Theorem 10 put $z_0 = x$, $z_1 = z_2 = \cdots = z_{q-1} = y$, and use Lemma 9.
     Q.E.D.

Example. For the code $\mathcal{A}$ of the preceding example, $W_{\mathcal{A}} = x^2 + 4y^2$, and the theorem asserts correctly that

$$
x^2 + 4y^2 = \frac{1}{5}[(x + 4y)^2 + 4(x - y)^2].
$$

When $q = 2$, Theorem 13 reduces to Theorem 1.

A weight enumerator which completely specifies the code. By introducing enough variables it is possible to specify the code *completely*. An example will make this clear. Suppose $F = \mathrm{GF}(5)$. We describe the vector $u = (2, 0, 4) \in F^3$ by the polynomial $z_{12} z_{20} z_{34}$. In general, the variables z_{ij} means that the ith place in the vector u is the jth element ω_j of F. The vector $u = (\omega_{a_1}, \omega_{a_2}, \ldots, \omega_{a_n})$ is described by the polynomial

$$f(u) = z_{1a_1} z_{2a_2} \cdots z_{na_n}. \tag{48}$$

Thus u is uniquely determined by $f(u)$. This requires the use of nq variables z_{ij} $(1 \leq i \leq n, 0 \leq j \leq q - 1)$. (This is similar to what we did in §3 for binary codes).

What we shall call the *exact enumerator* of a code $\mathscr{C}$ is then defined as

$$\mathscr{E}_{\mathscr{C}} = \sum_{u \in \mathscr{C}} f(u).$$

MacWilliams theorem for exact enumerators.

Theorem 14. $\mathscr{E}_{\mathscr{C}^\perp}$ *is obtained from* $\mathscr{E}_{\mathscr{C}}$ *by replacing each z_{ir} by*

$$\sum_{s=0}^{q-1} \chi_1(\omega_r \omega_s) z_{is} \tag{49}$$

and dividing the result by $|\mathscr{C}|$.

Proof. Use Lemma 11 with $f(u)$ defined by (48).   Q.E.D.

Theorem 14 is a very general version of the MacWilliams theorem and of course all the earlier theorems follow from it. In the remainder of this section we give a few more corollaries which give information about the contents of certain coordinate positions of the codewords.

Joint weight enumerators. The joint weight enumerator of two codes $\mathscr{A}$ and $\mathscr{B}$ measures the overlap between the zeros in a typical codeword of $\mathscr{A}$ and a typical codeword of $\mathscr{B}$. This generalizes the Hamming weight enumerator just as a joint probability-density function generalizes a single density function. For simplicity we only consider the binary case, with $F = GF(2)$.

For $u = (u_1, \ldots, u_n), v = (v_1, \ldots, v_n) \in F^n$ let

$$i(u, v) = \text{number of } r \text{ such that } u_r = 0, \quad v_r = 0,$$
$$j(u, v) = \text{number of } r \text{ such that } u_r = 0, \quad v_r = 1,$$
$$k(u, v) = \text{number of } r \text{ such that } u_r = 1, \quad v_r = 0,$$
$$l(u, v) = \text{number of } r \text{ such that } u_r = 1, \quad v_r = 1.$$

Of course

$$i(u, v) + j(u, v) + k(u, v) + l(u, v) = n$$
$$j(u, v) + l(u, v) = \text{wt}(v),$$
$$k(u, v) + l(u, v) = \text{wt}(u).$$

The *joint weight enumerator* of $\mathscr{A}$ and $\mathscr{B}$ is

$$\mathscr{J}_{\mathscr{A},\mathscr{B}}(a, b, c, d) = \sum_{u \in \mathscr{A}} \sum_{v \in \mathscr{B}} a^{i(u,\,v)} b^{j(u,\,v)} c^{k(u,\,v)} d^{l(u,\,v)} \qquad (50)$$

The joint weight enumerator of a code $\mathscr{A}$ with itself is called the *biweight enumerator* of $\mathscr{A}$.

Problems. Establish the following properties of the joint weight enumerator.

(29)
$$\mathscr{J}_{\mathscr{A},\mathscr{B}}(1, 1, 1, 1) = |\mathscr{A}\|\mathscr{B}|,$$
$$\mathscr{J}_{\mathscr{B},\mathscr{A}}(a, b, c, d) = \mathscr{J}_{\mathscr{A},\mathscr{B}}(a, c, b, d).$$

The single weight enumerators are given by

$$W_{\mathscr{A}}(x, y) = \frac{1}{|\mathscr{B}|} \mathscr{J}_{\mathscr{A},\mathscr{B}}(x, x, y, y),$$

$$W_{\mathscr{B}}(x, y) = \frac{1}{|\mathscr{A}|} \mathscr{J}_{\mathscr{A},\mathscr{B}}(x, y, x, y),$$

$$W_{\mathscr{A}}(x, y) = \mathscr{J}_{\mathscr{A},\mathscr{B}}(x, 0, y, 0), \quad \text{provided } \mathbf{0} \in \mathscr{B}$$
$$W_{\mathscr{B}}(x, y) = \mathscr{J}_{\mathscr{A},\mathscr{B}}(x, y, 0, 0). \quad \text{provided } \mathbf{0} \in \mathscr{A}$$

Also

$$W_{\mathscr{A}}(x, y)W_{\mathscr{B}}(z, t) = \mathscr{J}_{\mathscr{A},\mathscr{B}}(xz, xt, yz, yt).$$

(30) *If* $\mathscr{A} = \{0, 1\}$ = repetition code of length n,

$$\mathscr{J}_{\mathscr{A},\mathscr{A}}(a, b, c, d) = a^n + b^n + c^n + d^n.$$

If $\mathscr{A} = \{0\}$, $\mathscr{B}$ arbitrary,

$$\mathscr{J}_{\mathscr{A},\mathscr{B}}(a, b, c, d) = W_{\mathscr{B}}(a, b).$$

If $\mathscr{A}$ arbitrary, $\mathscr{B} = F^n$ = all codewords of length n,

$$\mathscr{J}_{\mathscr{A},\mathscr{B}}(a, b, c, d) = W_{\mathscr{A}}(a + b, c + d)$$

If $\mathscr{A}$ arbitrary, $\mathscr{B} = \{$all even weight vectors$\}$,

$$\mathscr{J}_{\mathscr{A},\mathscr{B}}(a, b, c, d) = \frac{1}{2} W_{\mathscr{A}}(a + b, c + d) + \frac{1}{2} W_{\mathscr{A}}(a - b, c - d).$$

(31)
$$\mathscr{J}_{\mathscr{A}^{\perp},\mathscr{B}}(a, b, c, d) = \frac{1}{|\mathscr{A}|} \mathscr{J}_{\mathscr{A},\mathscr{B}}(a + c, b + d, a - c, b - d).$$

$$\mathscr{J}_{\mathscr{A},\mathscr{B}^{\perp}}(a, b, c, d) = \frac{1}{|\mathscr{B}|} \mathscr{J}_{\mathscr{A},\mathscr{B}}(a + b, a - b, c + d, c - d).$$

(32)
$$\mathscr{J}_{\mathscr{A}^{\perp},\mathscr{B}^{\perp}}(a, b, c, d) = \frac{1}{|\mathscr{A}\|\mathscr{B}|} \mathscr{J}_{\mathscr{A},\mathscr{B}}(a + b + c + d, a - b + c - d,$$
$$a + b - c - d, a - b - c + d). \qquad (51)$$

(33) For the $[n = 2^m - 1, m, 2^{m-1}]$ simplex code = dual of Hamming code, for $m \geq 2$,

$$\mathscr{J}_{\mathscr{A}\mathscr{A}}(a, b, c, d) = a^n + na^{(n-1)/2}(b^{(n+1)/2} + c^{(n+1)/2} + d^{(n+1)/2})$$

$$+ \frac{n(n-1)}{a}(abcd)^{(n+1)/4}$$

(34) For the $[n = 2^m - 1, 2^m - m - 1, 3]$ Hamming code, for $m \geq 2$,

$$\mathscr{J}_{\mathscr{A}\mathscr{A}}(a, b, c, d) = \frac{1}{(n+1)^2}\left[\sigma_1^n + \frac{n(n-1)}{\sigma_1}(\sigma_4 - 2\sigma_{22} + 8\sigma_{1111})^{(n+1)/4}\right.$$

$$+ n\sigma_1^{(n-1)/2}\{(a - b + c - d)^{(n+1)/2}$$

$$\left. + (a + b - c - d)^{(n+1)/2} + (a - b - c + d)^{(n+1)/2}\}\right],$$

where σ denotes a symmetric function of a, b, c, d:

$$\sigma_i = a^i + b^i + c^i + d^i$$

$$\sigma_{ij} = a^i(b^j + c^j + d^j) + b^i(a^j + c^j + d^j) + \cdots, \quad i \neq j,$$

$$\sigma_{ii} = a^i b^i + a^i c^i + a^i d^i + b^i c^i + b^i d^i + c^i d^i$$

$$\cdots\cdots\cdots\cdots$$

$$\sigma_{iiii} = a^i b^i c^i d^i.$$

(35) For the $[n = 2^m, m + 1, 2^{m-1}]$ first-order Reed–Muller code, for $m \geq 2$,

$$\mathscr{J}_{\mathscr{A}\mathscr{A}}(a, b, c, d) = \sigma_n + 2(n-1)\sigma_{n/2,n/2} + 4(n-1)(n-2)\sigma_{1111}^{n/4}.$$

(36) To emphasize that the biweight enumerator gives more information about a code than does the weight enumerator, show that the codes generated by {110000, 001100, 000011} and {110000, 011000, 001111} have the same weight enumerator but different biweight enumerators. Another such pair consists of the [32, 16, 8] second-order Reed–Muller code and the quadratic residue code with the same parameters (see Ch. 19).

Split weight enumerator. In many codes the vectors are divided naturally into a left half and a right half. For example, codes formed from the $|u \mid u + v|$ construction (§9 of Ch. 2), Reed–Muller codes and codes obtained from them (see Chs. 13 and 15). For such codes it is useful to keep track of the weights of the two halves separately.

The *left* and *right weight* of a vector $v = (v_1, \ldots, v_m, v_{m+1}, \ldots, v_{2m})$ are respectively $w_L = \text{wt}(v_1, \ldots, v_m)$, $w_R = \text{wt}(v_{m+1}, \ldots, v_{2m})$. The *split weight enumerator* of a $[2m, k]$ code $\mathscr{C}$ is

$$\mathscr{S}_{\mathscr{C}}(x, y, X, Y) = \sum_{v \in \mathscr{C}} x^{m - w_L(v)} y^{w_L(v)} X^{m - w_R(v)} Y^{w_R(v)}.$$

Problems. (37) Let $\mathscr{C}$ and $\mathscr{D}$ be codes of length n with weight enumerators $W_1(x, y)$, $W_2(x, y)$ and split weight enumerators $\mathscr{S}_1(x, y, X, Y)$, $\mathscr{S}_2(x, y, X, Y)$.

Show that the direct sum

$$\mathscr{C} \oplus \mathscr{D} = \{|u \mid v|: \quad u \in \mathscr{C}, v \in \mathscr{D}\}$$

has weight enumerator $W_1(x, y)W_2(x, y)$ and split weight enumerator $W_1(x, y)W_2(X, Y)$. On the other hand the code

$$\mathscr{C} \| \mathscr{D} = \{|u' \mid v' \mid u'' \mid v''|: \quad u = |u' \mid u''| \in \mathscr{C}, v = |v' \mid v''| \in \mathscr{D}\},$$

(where u and v have each been broken into two equal halves) has the same weight enumerator but its split weight enumerator is $\mathscr{S}_1(x, y, X, Y)\mathscr{S}_2(x, y, X, Y)$. Notice that $\mathscr{C} \| \mathscr{D}$ can be obtained from $\mathscr{C} \oplus \mathscr{D}$ by permuting the coordinates of the codewords. As an example take $\mathscr{C} = \mathscr{D} = \{00, 11\}$, with

$$W_1(x, y) = x^2 + y^2, \quad \mathscr{S}_1(x, y, X, Y) = xX + yY.$$

Then the split weight enumerators of

$$\mathscr{C} \oplus \mathscr{D} = \{0000, 0011, 1100, 1111\}$$

and

$$\mathscr{C} \| \mathscr{D} = \{0000, 0101, 1010, 1111\}$$

are respectively $W_1(x, y)W_1(X, Y) = (x^2 + y^2)(X^2 + Y^2)$ and $\mathscr{S}_1(x, y, X, Y)^2 = (xX + yY)^2$.

(38) Prove

$$\mathscr{S}_{\mathscr{C}^\perp}(x, y, X, Y) = \frac{1}{|\mathscr{C}|} \mathscr{S}_{\mathscr{C}}(x + y, x - y, X + Y, X - Y). \tag{52}$$

(39) Give the split weight enumerator of a first-order Reed–Muller code.

(40) Let $\mathscr{C}$ be a linear code of length $2m$, and suppose the first m symbols of each codeword are sent over a binary symmetric channel with error probability p, and the last m symbols over a channel with error probability P. Let

$$d_1 = \min\{\alpha_1 w_L(u) + \alpha_2 w_R(u)\},$$

where the minimum is taken over all nonzero u in $\mathscr{C}$, and $\alpha_1 = \log(1 - p)/p$, $\alpha_2 = \log(1 - P)/P$. Show that $\mathscr{C}$ can correct all error vectors e satisfying

$$\alpha_1 w_L(e) + \alpha_2 w_R(e) < \frac{1}{2} d_1.$$

(If the split weight enumerator of $\mathscr{C}$ is known, d_1 can be easily obtained.)

§7. Properties of Krawtchouk polynomials

The results of §3–§5 dealing with weight enumerators of nonlinear codes can also be generalized to $GF(q)$. To do so requires a slightly more general version of the Krawtchouk polynomials which we give in this section.

Definition. For any prime power q and positive integer n, define the Krawtchouk polynomial

$$P_k(x; n) = P_k(x) = \sum_{j=0}^{k} (-1)^j \gamma^{k-j} \binom{x}{j} \binom{n-x}{k-j}, \quad k = 0, 1, \ldots, n, \quad (53)$$

where $\gamma = q - 1$, and the binomial coefficients are defined in Problem 18 of Ch. 1. These polynomials have the generating function

$$(1 + \gamma z)^{n-x} (1 - z)^x = \sum_{k=0}^{\infty} P_k(x) z^k. \quad (54)$$

If x is an integer with $0 \leqslant x \leqslant n$ the upper limit of summation can be replaced by n.

Theorem 15. (Alternative expressions.)

(i)
$$P_k(x) = \sum_{j=0}^{k} (-q)^j \gamma^{k-j} \binom{n-j}{k-j} \binom{x}{j}. \quad (55)$$

(ii)
$$P_k(x) = \sum_{j=0}^{k} (-1)^j q^{k-j} \binom{n-k+j}{j} \binom{n-x}{k-j}. \quad (56)$$

Proof. (i)

$$(1 + \gamma z)^{n-x} (1 - z)^x = (1 + \gamma z)^n \left(1 - \frac{qz}{1 + \gamma z}\right)^x$$

$$= \sum_{j=0}^{\infty} (-qz)^j (1 + \gamma z)^{n-j} \binom{x}{j}.$$

The coefficient of z^k in this is

$$P_k(x) = \sum_{j=0}^{k} (-q)^j \gamma^{k-j} \binom{n-j}{k-j} \binom{x}{j}.$$

(ii) The proof is similar, starting from

$$(1 + \gamma z)^{n-x} (1 - z)^x = (1 - z)^n \left(1 + \frac{qz}{1 - z}\right)^{n-x}. \quad \text{Q.E.D.}$$

Thus $P_k(x)$ is a polynomial of degree k in x, with leading coefficient $(-q)^k/k!$ and constant term

$$P_k(0) = \binom{n}{k} \gamma^k. \quad (57)$$

Theorem 16. (Orthogonality relations.) *For nonnegative integers* r, s,

$$\sum_{i=0}^{n} \binom{n}{i} \gamma^i P_r(i) P_s(i) = q^n \gamma^r \binom{n}{r} \delta_{r,s}, \quad (58)$$

where $\delta_{r,s} = 1$ *if* $r = s$, $\delta_{r,s} = 0$ *if* $r \neq s$ *is the Kronecker symbol.*

Proof. The LHS is the coefficient of $y^r z^s$ in

$$\sum_{i=0}^{n} \binom{n}{i} \gamma^i (1+\gamma y)^{n-i}(1-y)^i(1+\gamma z)^{n-i}(1-z)^i$$
$$= [(1+\gamma y)(1+\gamma z)+\gamma(1-y)(1-z)]^n$$
$$= q^n(1+\gamma yz)^n. \qquad \text{Q.E.D.}$$

Theorem 17. *For nonnegative integers* i, s,

$$\gamma^i \binom{n}{i} P_s(i) = \gamma^s \binom{n}{s} P_i(s). \tag{59}$$

Proof. This follows at once from (53) by rearranging the binomial coefficients.

Corollary 18.

$$\sum_{i=0}^{n} P_r(i) P_i(s) = q^n \delta_{r,s}.$$

Proof. This is immediate from Theorems 16 and 17. \qquad Q.E.D.

Theorem 19. (Recurrence.) *The Krawtchouk polynomials satisfy a three-term recurrence*:

$$(k+1)P_{k+1}(x) = [(n-k)\gamma + k - qx]P_k(x) - \gamma(n-k+1)P_{k-1}(x), \tag{60}$$

for $k = 1, 2, \ldots$, with initial values $P_0(x) = 1$, $P_1(x) = \gamma n - qx$.

Proof. Differentiate (54) with respect to z, multiply by $(1+\gamma z)(1-z)$, and equate coefficients of z^k. \qquad Q.E.D.

Theorem 20. *If the Krawtchouk expansion of a polynomial* $\alpha(x)$ *of degree t is*

$$\alpha(x) = \sum_{k=0}^{t} \alpha_k P_k(x), \tag{61}$$

then the coefficients are given by

$$\alpha_k = q^{-n} \sum_{i=0}^{n} \alpha(i) P_i(k).$$

Proof. Multiply (61) by $P_i(l)$, set $x = i$, sum on i, and use Corollary 18. Q.E.D.

Problems. (41) Show

$$\sum_{k=0}^{n} \binom{n-k}{n-j} P_k(x) = q^i \binom{n-x}{j}.$$

(42) Show

$$P_0(x; n) + P_1(x; n) + \cdots + P_e(x; n) = P_e(x-1; n-1).$$

In the remaining problems $q = 2$.

(43) Show that $P_k(i) = (-1)^i P_{n-k}(i)$, $0 \le i \le n$.

(44) Show $P_n(k) = (-1)^k$, $P_k(1) = (1 - (2k/n))\binom{n}{k}$.

(45) Show that

$$\sum_{i=0}^{n} P_i(k) = 2^n \delta_{k,0},$$

$$\sum_{i=0}^{n} i P_i(k) = 2^{n-1}(n\delta_{k,0} - \delta_{k,1}),$$

$$\sum_{i=0}^{n} i^2 P_i(k) = 2^{n-2}\{n(n+1)\delta_{k,0} - 2n\delta_{k,1} + 2\delta_{k,2}\}.$$

(46) For nonnegative integers i and k show that

$$(n-k)P_i(k+1) = (n-2i)P_i(k) - kP_i(k-1).$$

[Hint: Theorems 17 and 19.]

(47) Because the Krawtchouk polynomials are orthogonal (Theorem 16), many of the results of Szegö's book [1297] apply to them. For example, prove the Christoffel–Darboux formula (cf. Theorem 3.2.2 of [1297]):

$$\binom{n}{t} \sum_{i=0}^{t} P_i(x)P_i(y) \bigg/ \binom{n}{i} = \frac{1}{2}(t+1)\frac{P_{t+1}(x)P_t(y) - P_t(x)P_{t+1}(y)}{y-x}.$$

[Hint: Use Theorem 19 to show that

$$(t+1)\{P_{t+1}(x)P_t(y) - P_t(x)P_{t+1}(y)\} \bigg/ \binom{n}{t}$$

$$= t\{P_t(x)P_{t-1}(y) - P_{t-1}(x)P_t(y)\} \bigg/ \binom{n}{t-1}$$

$$- 2(x-y)P_t(x)P_t(y) \bigg/ \binom{n}{t},$$

then sum on t.]

Notes on Chapter 5

§2. Theorems 1, 10, and 13 are due to MacWilliams [871, 872]. Lemma 2 is from Van Lint [848, p. 120]. Lemmas 2 and 11 are versions of the Poisson summation formula – see Dym and McKean [396, p. 220]. Theorem 1 has also been proved by the methods of combinatorial geometry – see Greene [560].

Krawtchouk polynomials were defined by Krawtchouk [782]; see also Dunkl [393], Dunkl and Ramirez [394], Eagleson [398], Krawtchouk [783], Vere–Jones [1370], and especially Szegö [1297]. They were first explicitly used in coding theory by Delsarte [350–352]. We shall meet these polynomials again in Ch. 21 as the eigenvalues of a certain association scheme.

The power moment identities were given by Pless [1051]. See also Stanley [1261] and Zierler [1464]. Problem 8 is also from Pless [1051]. For more about Stirling numbers see Riordan [1113, p. 33]. Berlekamp [113, §16.2] gives a more general family of moments.

Let the burst-length of a vector be the distance between the first 1 and the last 1. Korzhik [777] considers the distribution of codewords according to their burst length. Problem 4 was suggested by our colleague John I. Smith.

§§3,4. The group algebra (also called the group ring) and characters can be found in many books on algebra, e.g. Mann [907, p. 73]. They can be generalized to codes over GF(q), along with the results of §5.

§5. The nonlinear MacWilliams theorem can be found in MacWilliams et al. [886] and in Zierler [1468]. McEliece [942] has given a version of this theorem in which the underlying alphabet need not be a field.

Theorems 6, 7, and 8 are due to Delsarte [350]. Orthogonal arrays were defined by Bose and Bush [181]; see also Bush [220], Hall [587] and Raghavarao [1085]. Problem 20 is due to Assmus and Mattson [47].

§6. For the Lee enumerator and codes for the Lee metric see Lee [801], Berlekamp [115], Golomb et al. [524, 525, 529, 531, 532], Mazur [934] and the Notes to Ch. 6. Theorem 12 is from MacWilliams et al. [883]. The remaining weight enumerators in this section were defined in [883] and by Mallows and Sloane [895]. Katayama [747] has given a version of the MacWilliams theorem which applies to r-tuples of codewords (generalizing (50) and (51), which are the case $r = 2$), and has applied it to the Hamming code and its dual.

§7. Szegö [1297] and Erdelyi et al. [410, Ch. 10] are good references for orthogonal polynomials.

6

Codes, designs and perfect codes

§1. Introduction

This chapter continues the study of set-theoretic properties of codes begun in Chapter 5.

We begin in §2 by defining four fundamental parameters d, d', s, s' associated with any linear or nonlinear code $\mathscr{C}$. d is the minimum distance between codewords of $\mathscr{C}$, and s is the *number* of different nonzero distances between codewords. For a linear code d' and s' are the analogous quantities for the dual code $\mathscr{C}^{\perp}$. For a nonlinear code d', still called the dual distance, was defined in §5 of Ch. 5; and s' is the number of subscripts $i \neq 0$ such that $B_i' \neq 0$, where B_i' is as usual the transform of the distance distribution. s' is called the *external distance* of the code, because any vector is at distance $\leqslant s'$ from at least one codeword of $\mathscr{C}$. (Theorem 21 of §6. However, s' need not be the smallest integer with this property: see Problem 11.)

A number of interesting codes are such that either $s \leqslant d'$ or $s' \leqslant d$. When this happens the minimum distance (or the dual distance) is at least as large as the number of unknowns in the distance distribution, and the MacWilliams identities can be solved. From this it follows that such codes have three remarkable properties, which we establish in §§3, 4, 7:

(a) The number of codewords at distance i from a codeword v is independent of the choice of v. Speaking loosely, the view of the code from any codeword is the same as the view from any other codeword (Theorem 3 and Corollary 5 of §3). Such a code is called *distance invariant*. If the code contains the codeword 0, then the weight and distance distributions coincide. (Of course property (a) always holds for linear codes.)

(b) If $0, \tau_1, \ldots, \tau_s$ are the weights occurring in the code, then the number of codewords of weight τ_i is an explicit function of n, M, and the numbers τ_j (Theorems 2 and 4 of §3 and Theorem 7 of §4).

(c) The codewords of each weight form a t-design, where t is at least $d' - s$ or $d - s'$. (Actually we prove a slightly stronger result than this. See Theorem 9 of §4, Corollary 14 of §5, and Theorem 24 of §7.)

In §5 we show that if the codewords of each weight in any binary linear code $\mathscr{C}$ form t-designs then so do the codewords of each weight in $\mathscr{C}^\perp$ (Theorem 13).

§6 studies the weight distribution of the translates of any linear or non-linear code.

§8 establishes two properties of perfect codes: the fact that a code is perfect iff $s' = e$ (Theorem 27), and Lloyd's necessary condition for a code to be perfect (Theorem 28).

Up to this point in the chapter all the codes considered have been binary. Then in §9 we discuss how the results may be generalized to codes over $GF(q)$. We only consider constructing designs from linear codes over $GF(q)$ (Theorem 29). When restricted to binary codes Theorem 29 gives a result which is apparently stronger than Corollary 14 (see Corollary 31).

Finally in §10 we prove the Tietäväinen–Van Lint theorem (Theorem 33) that the only nontrivial perfect codes over any field are those with the parameters n, M, and d of the Hamming or Golay codes.

§2. Four fundamental parameters of a code

Let $\mathscr{C}$ be an (n, M, d) binary code, not necessarily linear, which contains $\mathbf{0}$. Suppose $\{B_i\}$ is the distance distribution of $\mathscr{C}$, i.e., B_i is the number of ordered pairs of codewords at a distance i apart, divided by M. Thus $B_0 = 1$. (See §1 of Ch. 2.)

Let $0, \tau_1, \tau_2, \ldots, \tau_s$ be the subscripts i for which $B_i \neq 0$, where

$$0 < \tau_1 < \cdots < \tau_s \leq n.$$

Then $d = \tau_1$ is the *minimum distance* between codewords of $\mathscr{C}$, and s is the *number* of distinct (nonzero) distances between codewords.

Let $\{B_i'\}$ be the MacWilliams transform of $\{B_i\}$, given by (see §5 of Ch. 5)

$$1 + \sum_{j=1}^{s} B_{\tau_j} y^{\tau_j} = M 2^{-n} \sum_{i=0}^{n} B_i'(1 + y)^{n-i}(1 - y)^i, \tag{1}$$

or equivalently by Equations (35) or (36) of Ch. 5. Suppose $0, \sigma_1, \ldots, \sigma_{s'}$, are the subscripts i for which $B_i' \neq 0$, where

$$0 < \sigma_1 < \cdots < \sigma_{s'} \leq n.$$

We call s' the *external distance* of $\mathscr{C}$, and $d' = \sigma_1$ is the *dual distance*. We saw in Theorem 8 of Ch. 5 that d' is the largest number such that each $(d'-1)$-subset of the coordinates of $\mathscr{C}$ contains all $(d'-1)$-tuples an equal number of times.

The weight distribution of $\mathscr{C}$ is $\{A_i\}$, where A_i is the number of codewords of weight i, and $\{A_i'\}$ is the MacWilliams transform of $\{A_i\}$.

Lemma 1. *The number of nonzero A_i is at most s, the number of nonzero A_i' is at most s', and*

$$A_i = 0 \quad for \quad 0 < i < d,$$
$$A_i' = 0 \quad for \quad 0 < i < d'.$$

Proof. The statements about A_i follow from the definitions of d and s. The statements about A_i' follow from Theorem 7 of Ch. 5. Q.E.D.

Suppose now we change the origin, i.e., replace $\mathscr{C}$ by $\mathscr{C}^* = \mathscr{C} + v$, where v is a codeword. The new code $\mathscr{C}^*$ is still an (n, M, d) code with the distance distribution $\{B_i\}$ and a possibly different weight distribution $\{A_i^*\}$ say. However Lemma 1 is still true for $\{A_i^*\}$.

Note: If $\mathscr{C}$ is linear, then $A_i = B_i =$ weight distribution of $\mathscr{C}$, $A_i' = B_i' =$ weight distribution of $\mathscr{C}^\perp$, and d' is the minimum distance of $\mathscr{C}^\perp$.

Our first goal in this chapter is to study the interesting properties of codes for which either $s \leq d'$ or $s' \leq d$. Some examples of such codes are:

(E1) The $[n, n-1, 2]$ even weight code, whose dual is the $[n, 1, n]$ repetition code. Here $d = 2$, $s = [\frac{1}{2}n]$, $d' = n$, $s' = 1$.

(E2) The $[n = 2^m - 1, m, 2^{m-1}]$ simplex code, whose dual is the $[2^m - 1, 2^m - 1 - m, 3]$ Hamming code. Here $d = 2^{m-1}$, $s = 1$, $d' = 3$, $s' = n - 4$.

Problem. (1) More generally, for the $(n, n+1, \frac{1}{2}(n+1))$ Hadamard code $\mathscr{A}_{n+1}$ of §3 of Ch. 2, (where $n \equiv 3 \pmod 4$), show that $d = \frac{1}{2}(n+1)$, $s = 1$, $d' = 3$, $s' = n - 4$.

(E3) The $[n = 2^m, m+1, 2^{m-1}]$ first-order Reed–Muller code (§9 of Ch. 1), whose dual is the $[2^m, 2^m - m - 1, 4]$ extended Hamming code. Here $d = 2^{m-1}$, $s = 2$, $d' = 4$, $s' = \frac{1}{2}n - 2$.

Problem. (2) Show that the Hadamard code $\mathscr{C}_n$ also has these parameters.

(E4) The $[24, 12, 8]$ extended Golay code (§6 of Ch. 2), which is self-dual, has $d = d' = 8$, $s = s' = 4$.

(E5) The $(16, 256, 6)$ Nordstrom–Robinson code $\mathscr{N}_{16}$ (§8 of Ch. 2), has $d = d' = 6$, $s = s' = 4$ (see the end of §5 of Ch. 5).

Other examples (which include some quadratic residue codes, Reed–Muller, Kerdock and Preparata codes) will be found in later chapters.

Problem. (3) Show that for the code $\{00000, 00011, 00101, 11011\}$, $A_i = B_i$ for all i, yet the code is not distance invariant.

§3. An explicit formula for the weight and distance distribution

Theorem 2. *If $s \leqslant d'$, then an explicit formula for the distance distribution $\{B_i\}$ in terms of n, M, and the τ_j's is*

$$B_{\tau_i} = -\prod_{\substack{j=1 \\ j \neq i}}^{s} \frac{\tau_j}{\tau_j - \tau_i} + \frac{1}{N} \sum_{t=0}^{n} \binom{n}{t} \prod_{\substack{j=1 \\ j \neq i}}^{s} \frac{\tau_j - t}{\tau_j - \tau_i}, \quad 1 \leqslant i \leqslant s,$$

where $N = 2^n/M$. (N.B. An empty product is equal to 1, by convention.)

Proof. As in §2 of Ch. 5, we differentiate (1) j times, divide by $j!$, and set $y = 1$. This gives s equations

$$\sum_{i=1}^{s} B_{\tau_i} = \frac{2^n}{N} - 1,$$

$$\sum_{i=1}^{s} \binom{\tau_i}{j} B_{\tau_i} = \frac{2^{n-j}}{N} \binom{n}{j}, \quad 1 \leqslant j \leqslant s - 1. \tag{2}$$

To solve these we set $B_{\tau_i} = a_{\tau_i} + b_{\tau_i}$, where

$$\sum_{i=1}^{s} \binom{\tau_i}{j} a_{\tau_i} = \frac{2^{n-j}}{N} \binom{n}{j}, \quad 0 \leqslant j \leqslant s - 1,$$

$$\sum_{i=1}^{s} \binom{\tau_i}{j} b_{\tau_i} = -\delta_{0j}, \quad 0 \leqslant j \leqslant s - 1. \tag{3}$$

The matrix of coefficients for both systems is

$$T = \begin{bmatrix} 1 & 1 & \cdots & 1 \\ \tau_1 & \tau_2 & \cdots & \tau_s \\ \binom{\tau_1}{2} & \binom{\tau_2}{2} & \cdots & \binom{\tau_s}{2} \\ \cdots\cdots\cdots\cdots\cdots\cdots \\ \binom{\tau_1}{s-1} & \binom{\tau_2}{s-1} & \cdots & \binom{\tau_s}{s-1} \end{bmatrix}$$

The inverse of T is obtained as follows. Let

$$f_{\tau_i}(x) = \prod_{\substack{j=1 \\ j \neq i}}^{s} \frac{\tau_j - x}{\tau_j - \tau_i}.$$

Expand $f_{\tau_i}(x)$ in binomial coefficients, viz.

$$f_{\tau_i}(x) = \sum_{j=0}^{s-1} f_{ij} \binom{x}{j},$$

where

$$\binom{x}{j} = x(x-1)\cdots(x-j+1)/j!.$$

Clearly

$$\sum_{j=0}^{s-1} f_{ij} \binom{\tau_k}{j} = f_{\tau_i}(\tau_k) = \delta_{ik},$$

and so

$$T^{-1} = [f_{ij}].$$

Now multiply the column vectors on the right hand side of Equation (3) by T^{-1}. We obtain

$$a_{\tau_i} = \frac{1}{N} \sum_{j=0}^{s-1} f_{ij} 2^{n-j} \binom{n}{j}.$$

Now

$$2^{n-j}\binom{n}{j} = \sum_{k=0}^{n-j}\binom{n-j}{k}\binom{n}{j} = \sum_{t=j}^{n}\binom{n}{t}\binom{t}{j}.$$

Thus

$$a_{\tau_i} = \frac{1}{N}\sum_{t=0}^{n}\binom{n}{t}\sum_{j=0}^{s-1} f_{ij}\binom{t}{j} = \frac{1}{N}\sum_{t=0}^{n}\binom{n}{t}\prod_{\substack{j=1\\j\neq i}}^{s}\frac{\tau_j - t}{\tau_j - \tau_i},$$

$$b_{\tau_i} = -f_{i0} = -\prod_{\substack{j=1\\j\neq i}}^{s}\frac{\tau_j}{\tau_j - \tau_i}. \qquad\qquad \text{Q.E.D.}$$

Theorem 3. *If $s \leqslant d'$, $A_i = B_i$ for all i.*

Proof. $A_i \neq 0$ implies $B_i \neq 0$, so the only possible nonzero A_i's are $A_0, A_{\tau_1}, \ldots,$ A_{τ_s}. From Lemma 1, $A_i' = 0$ for $1 \leqslant i \leqslant d' - 1$. Thus the A_{τ_i}'s satisfy (2), and the proof of Theorem 2 applies. $\qquad\qquad$ Q.E.D.

Problem. (4) Given that the only distances occurring in the Nordstrom–Robinson code $\mathcal{N}_{16}$ are 0, 6, 8, 10, 16, and that $d' = 6$, use Theorem 2 to obtain the distance distribution.

Theorem 4. *Suppose* $s' \le d$. *Then* $A_i' = B_i'$ *for all* i, *and*

$$A_{\sigma_i}' = B_{\sigma_i}' = -\prod_{\substack{j=1 \\ j \ne i}}^{s'} \frac{\sigma_j}{\sigma_j - \sigma_i} + \frac{1}{M} \sum_{t=0}^{n} \binom{n}{t} \prod_{\substack{j=1 \\ j \ne i}}^{s'} \frac{\sigma_j - t}{\sigma_j - \sigma_i}, \quad 1 \le i \le s'.$$

Proof. Exactly as for Theorems 2 and 3.   Q.E.D.

Corollary 5. *If* $s' \le d$, $A_i = B_i$ *for all* i.

Proof. $\{A_i\}$, $\{B_i\}$ are the MacWilliams transforms of $\{A_i'\}$, $\{B_i'\}$.   Q.E.D.

Theorem 6. *If* $s \le d'$ *or* $s' \le d$ *the code is distance invariant.*

Proof. Theorem 3 and Corollary 5 hold for any translate of $\mathcal{C}$ by a codeword.
                                                            Q.E.D.

Remark. Although the conditions $s \le d'$ or $s' \le d$ are sufficient for $A_i = B_i$, they are not necessary, as shown by the code of Fig. 5.1 or by problem 3 of this chapter.

§4. Designs from codes when $s \le d'$

In this section we show that if $s \le d'$ then the codewords of each weight form a t-design (see the definition in §5 of Ch. 2). To begin with we restate slightly the formula for A_{τ_i}.

From Theorem 3, if $s \le d'$, then $A_n = B_n = 0$ or 1. We now assume that A_n is known (it usually is), so that the unknown A_i's are $A_{\tau_1}, \dots, A_{\tau_s}$, where

$$\bar{s} = \begin{cases} s & \text{if } A_n = 0 \\ s - 1 & \text{if } A_n = 1. \end{cases}$$

Theorem 7. *If* $\bar{s} \le d'$, *then for* $1 \le i \le \bar{s}$,

$$A_{\tau_i} = -\prod_{\substack{j=1 \\ j \ne i}}^{\bar{s}} \frac{\tau_j}{\tau_j - \tau_i} + \frac{1}{N} \sum_{r=0}^{n} \binom{n}{r} \prod_{\substack{j=1 \\ j \ne i}}^{\bar{s}} \frac{\tau_j - r}{\tau_j - \tau_i} - A_n \prod_{\substack{j=1 \\ j \ne i}}^{\bar{s}} \frac{\tau_j - n}{\tau_j - \tau_i}. \tag{4}$$

Proof. As for Theorem 2, except that Equation (2) is now written as

$$\sum_{i=1}^{\bar{s}} \binom{\tau_i}{j} A_{\tau_i} = \left(\frac{2^{n-j}}{N} - \delta_{0j} - A_n\right)\binom{n}{j}. \tag{5}$$

Q.E.D.

If $A_n = 1$, then $\tau_{\bar{s}+1-i} = n - \tau_i$. Thus the 1st and 3rd terms in (4) can be combined into

$$\left(-1 + \frac{(-1)^{\bar{s}}\tau_i A_n}{n - \tau_i}\right) \cdot \prod_{\substack{j=1 \\ j \neq i}}^{\bar{s}} \frac{\tau_j}{\tau_j - \tau_i}.$$

Let $S(x) = \prod_{j=1}^{\bar{s}} (\tau_j - x)$. Then (4) can be written:

$$A_{\tau_i} \cdot \prod_{\substack{j=1 \\ j \neq i}}^{\bar{s}} (\tau_j - \tau_i) = \frac{S(0)}{\tau_i}\left(-1 + \frac{(-1)^{\bar{s}}\tau_i A_n}{n - \tau_i}\right) + \frac{1}{N}\sum_{r=0}^{n}\binom{n}{r}\frac{S(r)}{\tau_i - r}. \tag{6}$$

The codewords of weight τ_i form a t-design. Let u be a fixed vector of F^n of weight t, where $0 < t < d'$. For $\tau_i \geq t$, let $\lambda_{\tau_i}(u)$ be the number of codewords of $\mathscr{C}$ of weight τ_i which cover u.

Theorem 8. *The numbers $\lambda_{\tau_i}(u)$ satisfy the $d' - t$ equations*

$$\sum_{i=1}^{\bar{s}}\binom{\tau_i - t}{j}\lambda_{\tau_i}(u) = \frac{M}{2^{t+j}}\binom{n-t}{j} = \frac{2^{n-t-j}}{N}\binom{n-t}{j}, \quad 0 \leq j \leq d' - 1 - t. \tag{7}$$

Proof. Use Theorem 8 of Ch. 5 to count in two ways the vectors of weight $t + j$ which cover u and are covered by a codeword of $\mathscr{C}$.   Q.E.D.

Since the vector **1** covers everything, we may write (7) as

$$\sum_{i=1}^{\bar{s}}\binom{\tau_i - t}{j}\lambda_{\tau_i}(u) = \left(\frac{2^{n-t-j}}{N} - A_n\right)\binom{n-t}{j}. \tag{8}$$

If we can choose t so that $d' - t = \bar{s}$, we have $\bar{s}$ linearly independent equations for the $\bar{s}$ unknowns $\lambda_{\tau_i}(u)$. Then the λ_{τ_i} do not depend on the particular choice of u. Thus the codewords of each weight τ_i form a t-design, where $t = d' - \bar{s}$. The parameters of this design are obtained as follows.

Equations (8) are of the same form as (5), and the solution is:

$$\lambda_{\tau_i}g_{\tau_i - t}(\tau_i - t) = \frac{1}{N}\sum_{r=0}^{n-t}\binom{n-t}{r}g_{\tau_i - t}(r) - A_n g_{\tau_i - t}(n - t),$$

where

$$g_{\tau_i - t}(x) = \prod_{\substack{j=1 \\ j \neq i}}^{\bar{s}} (\tau_j - t - x).$$

Clearly $g_{\tau_i-t}(x-t) = g_{\tau_i}(x)$, so the solution is

$$g_{\tau_i}(\tau_i)\lambda_{\tau_i} = \frac{1}{N} \sum_{r=t}^{n} \binom{n-t}{r-t} g_{\tau_i}(r) - A_n g_{\tau_i}(n),$$

or

$$\lambda_{\tau_i} \cdot \prod_{\substack{j=1 \\ j \neq i}}^{\bar{s}} (\tau_j - \tau_i) = \frac{A_n S(n)}{n - \tau_i} + \frac{1}{N} \sum_{r=t}^{n} \binom{n-t}{r-t} \frac{S(r)}{\tau_i - r}. \tag{9}$$

Thus we have proved:

Theorem 9. *If $\bar{s} < d'$, then the codewords of weight τ_i in $\mathscr{C}$ form a $(d' - \bar{s}) -$ $(n, \tau_i, \lambda_{\tau_i})$ design, with λ_{τ_i} given by (9), provided that $\tau_i \geq d' - \bar{s}$.*

Examples. (cont.) (E1) Supposing n to be even, the even weight code contains the all-ones vector, so $\bar{s} = s - 1 = \frac{1}{2}n - 1$. From the theorem, the codewords of any weight $2r \geq \frac{1}{2}n + 1$ form an $(\frac{1}{2}n + 1)$-design. However, since this code contains *all* vectors of weight $2r$, this is in fact a $2r$-design. Thus the conclusion of the theorem is not always the strongest possible result.

(E2) From the theorem, the codewords of weight $\frac{1}{2}(n + 1)$ in the Hadamard code $\mathscr{A}_{n+1}$ form a 2-design, (in agreement both with problem 11 of Ch. 2 and with example (i) at the end of §5 of Ch. 5).

(E3) Similarly we get 3-designs from the codewords of the first-order Reed–Muller code (and the Hadamard code $\mathscr{C}_n$).

If the extended Hamming code is considered to be the main code, the primes are interchanged, and $d' - \bar{s} = 2^{m-1} - (\frac{1}{2}n - 3) = 3$. Thus the codewords of each weight in the extended Hamming code form a 3-design, agreeing with Theorem 15 of Ch. 2.

(E4) Similarly the codewords of the extended Golay code form 5-designs (cf. Corollaries 23, 25 and Theorem 26 of Ch. 2).

(E5) Similarly the codewords of the Nordstrom–Robinson code give 3- $(16, 6, 4)$, 3-$(16, 8, 3)$, and 3-$(16, 10, 24)$ designs.

Problem. (5) Check this.

Identities satisfied by the weights. Since the codewords of weight τ_i form a $(d' - \bar{s})$-design, then they certainly form a ρ-design for all $\rho < d' - \bar{s}$. Suppose this is a $\rho - (n, \tau_i, \lambda_\rho^{(\tau_i)})$-design.

From (9) for $\rho = 1$ we obtain

$$\lambda_1^{(\tau_i)} \prod_{\substack{j=1 \\ j \neq i}}^{\bar{s}} (\tau_j - \tau_i) = \frac{A_n S(n)}{n - \tau_i} + \frac{1}{N} \sum_{r=1}^{n} \binom{n-1}{r-1} \frac{S(r)}{\tau_i - r}.$$

Since
$$\tau_i A_{\tau_i} = n\lambda_1^{(\tau_i)},$$

and A_{τ_i} is given by (6), we obtain the following identity for the weights of the code:

$$- S(0) + A_n \left[(-1)^s \frac{S(0)\tau_i}{n - \tau_i} - \frac{nS(n)}{n - \tau_i} \right] + \frac{1}{N} \sum_{r=0}^{n} \binom{n}{r} S(r) = 0. \tag{10}$$

Also, if $A_n = 1$,
$$S(n) = (-1)^{\bar{s}} \prod_{i=1}^{\bar{s}} (n - \tau_i) = (-1)^{\bar{s}} S(0).$$

We now consider two cases. If $A_n = 0$, then (10) becomes

$$S(0) = \frac{1}{N} \sum_{r=0}^{n} \binom{n}{r} S(r). \tag{11}$$

Similarly, if $A_n = 1$, (10) becomes

$$S(0)(1 + (-1)^{\bar{s}}) = \frac{1}{N} \sum_{r=0}^{n} \binom{n}{r} S(r). \tag{12}$$

Problem. (6) Prove (12) directly from the expressions for A_{τ_i} given by Theorem 2 and Equation (6). Show that (12) is trivial for $\bar{s}$ odd.

More generally, since the codewords form a ρ-design for $2 \le \rho \le d' - \bar{s}$, we have

$$(n - \rho + 1)\lambda_\rho^{(\tau_i)} = (\tau_i - \rho + 1)\lambda_{\rho-1}^{(\tau_i)}. \tag{13}$$

Using the expressions for $\lambda_\rho^{(\tau_i)}$ and $\lambda_{\rho-1}^{(\tau_i)}$ from (9), we obtain

$$A_n S(n) = \frac{1}{N} \sum_{r=\rho-1}^{n} \binom{n - \rho + 1}{r - \rho + 1} S(r). \tag{14}$$

Thus we have proved:

Theorem 10. *Suppose $\bar{s} \le d'$. Then the weights of the codewords satisfy the following numerical conditions (where*

$$S(x) = \prod_{j=1}^{\bar{s}} (\tau_j - x)):$$

(i) *If $A_n = 0$, then*

$$S(0) = \frac{1}{N} \sum_{r=0}^{n} \binom{n}{r} S(r),$$

$$0 = \frac{1}{N} \sum_{r=\rho-1}^{n} \binom{n - \rho + 1}{r - \rho + 1} S(r),$$

for $2 \le \rho \le d' - \bar{s}$.

(ii) *If $A_n = 1$, then*

$$S(0)(1 + (-1)^s) = \frac{1}{N} \sum_{r=0}^{n} \binom{n}{r} S(r),$$

$$(-1)^s S(0) = S(n) = \frac{1}{N} \sum_{r=\rho-1}^{n} \binom{n-\rho+1}{r-\rho+1} S(r),$$

for $2 \le \rho \le d' - \bar{s}$.

Problem. (7) If $A_n = 0$, $d' \ge 3$ and $s = \bar{s} = 1$ (all nonzero codewords have the same weight τ) show that the codewords can form at most a 2-design; and if they do then $\tau = \frac{1}{2}(n + 1)$, $M = n + 1$, and M is divisible by 4.

§5. The dual code also gives designs

In this section we let $\mathscr{C}$ be any $[n, k, d]$ binary linear code with the property that for some $t < d$ the codewords of each weight $w > 0$ form a t-design.

Let T be a set of t coordinate places. The code obtained by deleting these places from $\mathscr{C}$ will be denoted by $\mathscr{C}^T$. Thus $\mathscr{C}^T$ is an $[n - t, k, d - t]$ code. Let $\{A_j^T\}$ be the weight distribution of $\mathscr{C}^T$.

Lemma 11. *A_j^T is independent of the choice of T.*

Proof. Let λ_i^v be the number of codewords of $\mathscr{C}$ of weight v which contain exactly i coordinates of T. By §5 of Ch. 2, λ_i^v does not depend on the choice of T. Then neither does

$$A_j^T = \lambda_0^j + t\lambda_1^{j+1} + \binom{t}{2}\lambda_2^{j+2} + \cdots + \binom{t}{t}\lambda_t^{j+t}. \qquad \text{Q.E.D.}$$

Let $\mathscr{C}^{0@T}$ be the shortened code obtained by taking those codewords of $\mathscr{C}$ which are zero on T and deleting the coordinates of T. Thus $\mathscr{C}^{0@T}$ is an $[n - t, k - t, d]$ code. Also $(\mathscr{C}^\perp)^{0@T}$ is the dual of $\mathscr{C}^T$ (cf. Figs. 1.11, 1.13).

Corollary 12. *The weight distribution of $(\mathscr{C}^\perp)^{0@T}$ is independent of the choice of T.*

Proof. By Theorem 1 of Ch. 5. Q.E.D.

Theorem 13. *Let $\mathscr{C}$ be an $[n, k, d]$ binary linear code with $k > 1$, such that for each weight $w > 0$ the codewords of weight w form a t-design, where $t < d$. Then the codewords of each weight in $\mathscr{C}^{\perp}$ also form a t-design.*

Remark. If $k = 1$, in order to have a t-design $\mathscr{C}$ must be the repetition code. Then $\mathscr{C}^{\perp}$ consists of all even weight vectors and gives trivial designs – see example (E1) above.

Proof. Since $k > 1$, $\mathscr{C}$ contains codewords of some weight v in the range $0 < v < n$, and by hypothesis these codewords form a t-design.

Pick w such that $\mathscr{C}^{\perp}$ contains a codeword b (say) of weight w. If $w = n$ there is nothing to prove, so we may suppose $w < n$.

The first step is the show that $w < n - t$. (a) Suppose $w = n - t$. If $v - t$ is odd, pick a codeword a in $\mathscr{C}$ of weight v, which is 1 where b is 0.

$$
\begin{array}{cccc}
\overleftarrow{\quad\quad w \quad\quad\longrightarrow} & \overleftarrow{\quad t \quad\longrightarrow} & \\
b = 1\ 1\ 1 & 1\ 1\ 1 & 0\ 0\ 0\ 0\ 0\ 0 \\
a = 0\ 0\ 0 & 1\ 1\ 1 & 1\ 1\ 1\ 1\ 1\ 1 \\
& \overleftarrow{\ }\overrightarrow{\ } & \\
& v - t &
\end{array}
$$

Then $a \cdot b \equiv v - t \equiv 1 \pmod 2$, which is impossible. On the other hand if $v - t$ is even, pick a to be 1 on all but one of the zeros of b.

$$
\begin{array}{cccc}
\overleftarrow{\quad\quad w \quad\quad\longrightarrow} & \overleftarrow{\quad t \quad\longrightarrow} & \\
b = 1\ 1\ 1 & 1\ 1\ 1 & 0\ 0\ 0\ 0\ 0\ 0 \\
a = 0\ 0\ 0 & 1\ 1\ 1 & 1\ 1\ 1\ 1\ 1\ 0 \\
& \overleftarrow{\ }\overrightarrow{\ } & \\
& v - t + 1 &
\end{array}
$$

That this can be done follows from the fact that, in the notation of §5 of Ch. 2, $\lambda_{t-1} > \lambda_t$. Now $a \cdot b \equiv v - t + 1 \equiv 1 \pmod 2$, again a contradiction. This proves $w \neq n - t$.

(b) Suppose $w = n - t + i = n - (t - i)$, where $i > 0$. But this is impossible by (a), since a t-design is automatically a $(t - i)$-design. This proves that $w < n - t$.

We shall now show that if $w < n - t$, then the set of codewords of weight w in $\mathscr{C}^{\perp}$ forms a t-design. Let $c_1, \ldots, c_s$ be the *complements* of these codewords, where $s = A'_w$. Let T be any set of t coordinates.

The number of c_i's which are 1 on T is exactly the number of codewords of weight w in $(\mathscr{C}^{\perp})^{0@T}$. Thus by Corollary 12 this number is independent of the choice of T, and so the c_i's form a t-design. Hence the codewords of weight w form the complementary design (§5 of Ch. 2).   Q.E.D.

Example. (E2) (cont.) From Theorem 9 the codewords in the simplex code form a 2-design. From Theorem 13, the codewords of each weight in the Hamming code also form a 2-design.

Combining Theorems 9 and 13 we have:

Corollary 14. *Let $\mathscr{C}$ be a linear code with parameters d, s, d', s'. Let $\bar{s}$ be as above, and*

$$\bar{s}' = \begin{cases} s' & \text{if } A_n' = 0, \\ s' - 1 & \text{if } A_n' = 1. \end{cases}$$

If either $\bar{s} < d'$ or $\bar{s}' < d$, then the codewords of weight w in $\mathscr{C}$ form a t-design, where

$$t = \max\{d' - \bar{s}, d - \bar{s}'\},$$

provided that $t < a$.

Research Problem (6.1). In all nontrivial examples known to us, $d' - \bar{s}$ is less than d and the final proviso of the theorem is unnecessary. Is this always so?

§6. Weight distribution of translates of a code

In this section we study the weight distribution of translates of any (n, M, d) code $\mathscr{C}$. If f is any vector of F^n, let $A_i(f)$ be the number of vectors of weight i in the translate $\mathscr{C} + f$. Of course if $\mathscr{C}$ is linear, $\mathscr{C} + f$ is the coset of $\mathscr{C}$ containing f. The MacWilliams transform of the $\{A_i(f)\}$ is, from (31) of Ch. 5,

$$A_i'(f) = \frac{1}{M} \sum_{\text{wt}(u)=i} \chi_u(C)(-1)^{u \cdot f},$$

where

$$C = \sum_{v \in \mathscr{C}} z^v$$

represents the code in the group algebra notation of Ch. 5.

When $f = 0$, $A_i(0) = A_i$ is the weight distribution of $\mathscr{C}$, and $A_i'(0) = A_i'$. Note that (cf. Problems 18, 19 of Ch. 5)

$$A_0'(f) = 1,$$

$$\sum_{i=0}^{n} A_i'(f) = \frac{1}{M} A_0(f) 2^n,$$

which is zero if $f \notin \mathscr{C}$.

Theorem 15. *The $A_i'(f)$ are orthogonal:*

$$\sum_{f \in F^n} A_i'(f) A_j'(f) = \begin{cases} 0 & \text{if } i \ne j, \\ 2^n B_i' & \text{if } i = j. \end{cases}$$

Proof.

$$M^2 \sum_{f \in F^n} A_i'(f) A_j'(f) = \sum_{\text{wt}(u)=i} \chi_u(C) \sum_{\text{wt}(v)=j} \chi_v(C) \sum_{f \in F^n} (-1)^{f \cdot (u+v)}.$$

Now

$$\sum_{f \in F^n} (-1)^{f \cdot (u+v)} = \begin{cases} 0 & \text{if } u + v \ne 0, \\ 2^n & \text{if } u + v = 0. \end{cases}$$

If $i \ne j$, then $u + v \ne 0$, which proves the first part of the theorem. If $i = j$, by (36) of Ch. 5,

$$M^2 \sum_{f \in F^n} A_i'(f)^2 = 2^n \left(\sum_{\text{wt}(u)=i} \chi_u(C) \right)^2 = 2^n M^2 B_i'. \tag{15}$$

Q.E.D.

Corollary 16. $B_i' = 0$ *iff* $A_i'(f) = 0$ *for all* $f \in F^n$.

Proof. Theorem 7 of Ch. 5, and Equation (15).   Q.E.D.

As in §3 of Ch. 5 let

$$Y_i = \sum_{\text{wt}(v)=i} z^v.$$

Lemma 17. *If* $u \in F^n$ *has weight* s,

$$\chi_u(Y_i) = \sum_{r=0}^{i} (-1)^r \binom{n-s}{i-r} \binom{s}{r} = P_i(s),$$

where $P_i(x)$ *is a Krawtchouk polynomial.*

Proof.

$$\chi_u(Y_i) = \sum_{\text{wt}(v)=i} (-1)^{u \cdot v}.$$

There are $\binom{n-s}{i-r}\binom{s}{r}$ vectors v of weight i which have $i-r$ 1's in the $n-s$ coordinates where u is 0, and r 1's in the s coordinates where u is 1. Each of these vectors v contributes $(-1)^r$ to the sum.   Q.E.D.

The annihilator polynomial of $\mathscr{C}$ *is defined to be*

$$\alpha(x) = \frac{2^n}{M} \prod_{j=1}^{s'} \left(1 - \frac{x}{\sigma_j}\right), \tag{16}$$

where $0, \sigma_1, \sigma_2, \ldots, \sigma_{s'}$ are the subscripts i for which $B_i' \neq 0$. Note that for $0 < i \leqslant n$ either $\alpha(i) = 0$ or $B_i' = 0$. Hence by Corollary 16 $A_i'(f) \neq 0$ for some f implies $\alpha(i) = 0$.

The expansion of $\alpha(x)$ in terms of Krawtchouk polynomials,

$$\alpha(x) = \sum_{i=0}^{s'} \alpha_i P_i(x),$$

is called the *Krawtchouk expansion* of $\alpha(x)$, and the α_i are called the *Krawtchouk coefficients*. The expansion stops at $P_{s'}(x)$ since $\alpha(x)$ is of degree s'. Also $\alpha_{s'} \neq 0$ and (from Theorem 20 of Ch. 5)

$$\alpha_i = \frac{1}{2^n} \sum_{k=0}^{n} \alpha(k) P_k(i).$$

Lemma 18.

$$C \sum_{i=0}^{s'} \alpha_i Y_i = \sum_{u \in F^n} z^u.$$

Proof. To prove that these two elements of the group algebra are equal it is enough (by Problem 16 of Ch. 5) to show that $\chi_v(\text{LHS}) = \chi_v(\text{RHS})$ for all $v \in F^n$. Now

$$\chi_v(\text{RHS}) = \sum_{u \in F^n} (-1)^{u \cdot v} = 2^n \delta_{0,v},$$

$$\chi_v(\text{LHS}) = \chi_v(C) \chi_v \left(\sum_{i=0}^{s'} \alpha_i Y_i\right),$$

by Problem 12 of Ch. 5. First, if v has weight $w > 0$, by Lemma 17

$$\chi_v \left(\sum_{i=0}^{s'} \alpha_i Y_i\right) = \sum_{i=0}^{s'} \alpha_i P_i(w) = \alpha(w).$$

If w is one of the σ_j's, $\alpha(w) = 0$ by definition, but if not, $\chi_v(C) = 0$ by Theorem 7 of Ch. 5. Thus in either case

$$\chi_v(\text{LHS}) = \chi_v \left(C \sum_{i=0}^{s'} \alpha_i Y_i\right) = 0, \quad v \neq 0. \tag{17}$$

Second, if $v = 0$, $\chi_0(\text{LHS}) = \chi_0(C)\alpha(0) = M \cdot 2^n/M = 2^n$.   Q.E.D.

Property (17) is the reason $\alpha(x)$ is called the annihilator polynomial of $\mathscr{C}$.

Lemma 19. *If*

$$\sum_{i=0}^{t} \beta_i Y_i$$

has the property that

$$\chi_u \left(C \sum_{i=0}^{t} \beta_i Y_i \right) = 0 \quad \text{for all } u \in F^n, \, u \neq 0,$$

then the annihilator polynomial $\alpha(x)$ of $\mathscr{C}$ divides

$$\beta(x) = \sum_{i=0}^{t} \beta_i P_i(x).$$

Problem. (9) Prove Lemma 19.

Theorem 20. *For each $f \in F^n$ the numbers $A_i(f)$ are uniquely determined by $A_0(f), \ldots, A_{s'-1}(f)$.*

Proof. (i) We first show that $A_{s'}(f)$ is given in terms of $A_0(f), \ldots, A_{s'-1}(f)$ by

$$\alpha_{s'} A_{s'}(f) = 1 - \sum_{i=0}^{s'-1} \alpha_i A_i(f). \tag{18}$$

(Recall $\alpha_{s'} \neq 0$.) To prove this we calculate

$$CY_i = \sum_{v \in \mathscr{C}} z^v \cdot \sum_{\text{wt}(w)=i} z^w = \sum_{v \in \mathscr{C}} \sum_{\text{wt}(w)=i} z^{v+w}$$

$$= \sum_{v \in \mathscr{C}} \sum_{\substack{f \in F^n \text{ with} \\ \text{wt}(f+v)=i}} z^f, \quad \text{where } f = v + w,$$

$$= \sum_{f \in F^n} A_i(f) z^f,$$

since $A_i(f)$ is the number of codewords v at distance i from f. Then

$$\sum_{f \in F^n} \sum_{i=0}^{s'} \alpha_i A_i(f) z^f = C \sum_{i=0}^{s'} \alpha_i Y_i = \sum_{f \in F^n} z^f, \quad \text{(by Lemma 18)}$$

so that

$$\sum_{i=0}^{s'} \alpha_i A_i(f) = 1, \tag{19}$$

which proves (18).

(ii) If we expand

$$x\alpha(x) = \sum_{i=0}^{s'+1} \beta_i P_i(x),$$

then the proof of Lemma 18 shows

$$C \sum_{i=0}^{s'+1} \beta_i Y_i = 0,$$

hence

$$\sum_{i=0}^{s'+1} \beta_i A_i(f) = 0$$

giving $A_{s'+1}(f)$. To obtain $A_{s'+2}(f)$ we expand $x^2 \alpha(x)$, and so on. Q.E.D.

Remark. The recurrence formula for Krawtchouk polynomials (Theorem 19 of Ch. 5) is helpful in obtaining the expansion of $x\alpha(x)$ from that of $\alpha(x)$.

Example (1) For the Hamming code $\mathcal{H}$ of length n, since the code is perfect there are just two types of cosets, namely the code itself, and n cosets with coset leader of weight 1. If $f \in \mathcal{H}$, $A_i(f) = A_i$. If $f \notin \mathcal{H}$, $A_i + nA_i(f) = \binom{n}{i}$, so $A_i(f) = (1/n)(\binom{n}{i} - A_i)$. For $n = 7$, see Fig. 6.1

number	0	1	2	3	4	5	6	7
1	1			7	7			1
7		1	3	4	4	3	1	

Fig. 6.1. Weight distribution of cosets of Hamming code of length 7.

We illustrate Theorem 20 by verifying the second line of this figure.

For this code $s' = 1$, $\sigma_1 = 4$, so $\alpha(x) = 8(1 - \tfrac{1}{4}x) = 8 - 2x = P_1(x) + P_0(x)$. Hence $\alpha_0 = \alpha_1 = 1$, and $A_1(f) = 1 - A_0(f)$. Since $f \notin \mathcal{H}$, $A_0(f) = 0$, $A_1(f) = 1$.

The recurrence (Theorem 19 of Ch. 5) gives

$$xP_k(x) = -\tfrac{1}{2}(k+1)P_{k+1}(x) + \tfrac{7}{2}P_k(x) - (4 - \tfrac{1}{2}k)P_{k-1}(x),$$

i.e.,

$$xP_0(x) = -\tfrac{1}{2}P_1(x) + \tfrac{7}{2}P_0(x),$$

$$xP_1(x) = -P_2(x) + \tfrac{7}{2}P_1(x) - \tfrac{7}{2}P_0(x),$$

$$xP_2(x) = -\tfrac{3}{2}P_3(x) + \tfrac{7}{2}P_2(x) - 3P_1(x).$$

Next, $x\alpha(x) = xP_1(x) + xP_0(x) = -P_2(x) + 3P_1(x)$. Therefore $3A_1(f) - A_2(f) = 0$, so $A_2(f) = 3$. Again,

$$x^2\alpha(x) = -xP_2(x) + 3xP_1(x)$$

$$= \tfrac{3}{2}P_3(x) - \tfrac{13}{2}P_2(x) + \tfrac{27}{2}P_1(x) - \tfrac{21}{2}P_0(x),$$

$$\tfrac{3}{2}A_3(f) - \tfrac{13}{2}A_2(f) + \tfrac{27}{2}A_1(f) - \tfrac{21}{2}A_0(f) = 0,$$

which gives $A_3(f) = 4$, verifying the second line of Fig. 6.1.

Example (2) The Nordstrom–Robinson code $\mathcal{N}_{16}$ (§8 of Ch. 2) has distance distribution $B_0 = B_{16} = 1$, $B_6 = B_{10} = 112$, $B_8 = 30$, and the transformed distribution is the same, $B_i' = B_i$. Thus the annihilator polynomial is

$$\alpha(x) = 2^8 \left(1 - \frac{x}{6}\right)\left(1 - \frac{x}{8}\right)\left(1 - \frac{x}{10}\right)\left(1 - \frac{x}{16}\right)$$

$$= \tfrac{1}{20}P_4(x) + \tfrac{1}{5}P_3(x) + \tfrac{3}{10}P_2(x) + P_1(x) + P_0(x),$$

$$x\alpha(x) = -\tfrac{5}{40}P_5(x) + \tfrac{33}{40}P_3(x) + \tfrac{21}{4}P_1(x),$$

. . . .

Continuing in this way we find the weight distributions of the translates of $\mathcal{N}_{16}$ shown in Fig. 6.2.

Fig. 6.2. Weight distribution of translates of $\mathcal{N}_{16}$.

Number\Weight	0	2	4	6	8	10	$\cdots$
1		1		112	30	112	$\cdots$
120			1	14	63	100	63 $\cdots$
7				20	48	120	48 $\cdots$

Number\Weight	1	3	5	7	9	$\cdots$
16	1		42	85	85	$\cdots$
112		5	33	90	90	$\cdots$

Example (3) *Cosets of double-error-correcting BCH codes.* Let $\mathscr{C}$ be a double-error-correcting BCH code with parameters $[2^m - 1, 2^m - 1 - 2m, 5]$ for odd $m \geqslant 3$. It will be shown in §4 of Ch. 15 that $\mathscr{C}^\perp$ has just 3 nonzero weights, namely $2^{m-1} \pm 2^{(m-1)/2}$ and 2^{m-1}. Therefore the annihilator polynomial (16) of $\mathscr{C}$ is

$$\alpha(x) = 2^{2m}\left(1 - \frac{x}{2^{m-1} - 2^{(m-1)/2}}\right)\left(1 - \frac{x}{2^{m-1} + 2^{(m-1)/2}}\right)\left(1 - \frac{x}{2^{m-1}}\right),$$

$$= \sum_{i=0}^{3} \alpha_i P_i(x).$$

For the present purpose we need only to find α_3. The coefficient of x^3 in $\alpha(x)$ is

$$-\frac{2^{2m}}{(2^{2m-2} - 2^{m-1})2^{m-1}}$$

and the coefficient of x^3 in

$$P_3(x) = \binom{n-x}{3} - \binom{n-x}{2}x + (n-x)\binom{x}{2} - \binom{x}{3}$$

is $-\frac{4}{3}$. Therefore

$$\alpha_3 = \frac{3}{4} \cdot \frac{2^{2m}}{(2^{2m-2} - 2^{m-1})2^{m-1}} = \frac{3}{2^{m-1} - 1}.$$

Thus in a coset of $\mathscr{C}$ with minimum weight 3, from (18)

$$A_3(f) = \frac{2^{m-1} - 1}{3}.$$

Therefore all cosets with minimum weight 3 have the same weight distribution. Equation (18) also shows that all cosets must have minimum weight 0, 1, 2 or 3. Since the code is double-error-correcting, there is one coset of weight 0, n of weight 1, and $\binom{n}{2}$ of weight 2. The rest have weight 3. Thus $\mathscr{C}$ is a quasi-perfect code (see §5 of Ch. 1).

For even values of m the double-error-correcting BCH code of length $2^m - 1$ is still quasi-perfect, as we shall see in §8 of Ch. 9, although the dual code now contains more than 3 nonzero weights. The following theorem explains why s' is called the external distance of the code.

Theorem 21. *For any vector $f \in F^n$ there is a codeword at distance $\leq s'$ from f.*

Proof. From (18), not all of the numbers $A_i(f)$ for $i \leq s'$ can be zero.

Q.E.D.

Example. For the Hamming code $s' = 1$, and indeed since the code is perfect there is a codeword at distance ≤ 1 from every vector.

Remark. We may define the *covering radius* of $\mathscr{C}$ to be

$$t = \max_{f \in F^n} \min_{u \in \mathscr{C}} \text{dist } (u, f).$$

Thus t is the true external distance of the code, i.e., t is the maximum of the smallest weight in any translate of $\mathscr{C}$. Theorem 21 says that $t \leq s'$. But t may be less than s', as we see in Problem 11.

Two other metric properties of a code are its *diameter*,

$$\delta = \max_{u,v \in \mathscr{C}} \text{dist } (u, v),$$

and *radius*,

$$\rho = \min_{u \in F^n} \max_{v \in \mathscr{C}} \text{dist } (u, v).$$

ρ can be obtained if we know the weight distributions of the translates of $\mathscr{C}$, for ρ is the minimum of the largest weight in any translate.

Problems. (10) Show that

$$\tfrac{1}{2}\delta \leqslant \rho \leqslant \delta. \tag{20}$$

(11) (i) Show that $t = 2^{m-1} - 1$ for the $[n = 2^m - 1, k = m, d = 2^{m-1}]$ simplex code. [Hint: use the Plotkin bound.] (ii) Show $s' = n - 4$, and thus $t < s'$. (iii) Show $\rho = \delta = 2^{m-1}$. (iv) Hence show there are infinitely many codes for which equality holds on the RHS of (20), and similarly for the LHS.

Research Problem (6.2). Find bounds on t for any code, and a good method for calculating it.

Theorem 22. *Let $\mathscr{C}$ be an $[n, k, d]$ binary linear code, and $C = \sum_{v \in \mathscr{C}} z^v$. For $t \leqslant [\tfrac{1}{2}(d - 1)]$ let $D_t = CY_t$, representing the vectors which are at distance t from some codeword of $\mathscr{C}$, i.e., the union of the cosets with coset leaders of weight t. Let f_{st} be the number of vectors of weight s in D_t. Then*

$$\sum_{s=0}^{n} f_{st} x^{n-s} y^s = \frac{1}{2^{n-k}} \sum_{j=0}^{n} A'_j P_t(j)(x + y)^{n-j}(x - y)^j. \tag{21}$$

Proof. Write $D_t = \sum_{v \in F^n} d_v z^v$. Then by Problems 16 and 14 of Ch. 5,

$$d_v = \frac{1}{2^n} \sum_{u \in F^n} (-1)^{u \cdot v} \chi_u(C) P_t(\mathrm{wt}\,(u))$$

$$= \frac{1}{2^{n-k}} \sum_{u \in \mathscr{C}^\perp} (-1)^{u \cdot v} P_t(\mathrm{wt}\,(u))$$

by Problem 13 of Ch. 5. Therefore the LHS of (21) is

$$\sum_{v \in F^n} d_v x^{n-\mathrm{wt}(v)} y^{\mathrm{wt}(v)} = \frac{1}{2^{n-k}} \sum_{u \in \mathscr{C}^\perp} \sum_{v \in F^n} (-1)^{u \cdot v} P_t(\mathrm{wt}\,(u)) x^{n-\mathrm{wt}(v)} y^{\mathrm{wt}(v)}. \tag{22}$$

On the other hand the RHS is

$$\frac{1}{2^{n-k}} \sum_{u \in \mathscr{C}^\perp} P_t(\mathrm{wt}\,(u))(x + y)^{n-\mathrm{wt}(u)}(x - y)^{\mathrm{wt}(u)}$$

which is equal to (22) by Equations (9) and (11) of Ch. 5.   Q.E.D.

Problem. (12) Suppose a code has minimum distance $d = 2e + 1$, and consider the incomplete decoding algorithm (§5 of Ch. 1) which corrects all error patterns of weight $\leqslant t$ and no more, for some fixed $t \leqslant e$. Show that the probability of correct decoding is

$$\sum_{i=0}^{t} \binom{n}{i} p^i (1-p)^{n-i},$$

and that the probability of the decoder making an incorrect decision is

$$\sum_{i=0}^{t} \sum_{s=i+1}^{n} f_{si} p^s (1-p)^{n-s}.$$

§7. Designs from nonlinear codes when $s' < d$

If the code is nonlinear, the case $s' < d$ is not covered by Corollary 14. The following weaker result applies to this case.

Theorem 23. *Let $\mathscr{C}$ be an (n, M, d) code. If $d - s' \leqslant s' < d$, the codewords of weight d in $\mathscr{C}$ form a*

$$(d - s') - \left(n, d, \frac{1 - \alpha_{d-s'}}{\alpha_{s'}}\right) \tag{23}$$

design.

Proof. Let f be a vector of weight $d - s'$. If $d - s' < s'$, in the translate $\mathscr{C} + f$ we have

$$A_0(f) = \cdots = A_{d-s'-1}(f) = A_{d-s'+1}(f) = \cdots = A_{s'-1}(f) = 0,$$

$$A_{d-s'}(f) = 1,$$

and $A_{s'}(f)$ is the number of codewords of weight d which cover f. By (18),

$$A_{s'}(f) = \frac{1 - \alpha_{d-s'}}{\alpha_{s'}},$$

which is independent of the choice of f. If $s' = d - s'$ (i.e., $s' = \frac{1}{2}d$), then

$$A_0(f) = \cdots = A_{s'-1}(f) = 0, \quad \alpha_{s'} A_{s'}(f) = 1.$$

In this case $A_{s'}(f) - 1$ is the number of codewords of weight d which cover f, and

$$A_{s'}(f) - 1 = \frac{1 - \alpha_{s'}}{\alpha_{s'}},$$

which is also independent of f. In either case then, the number of codewords of weight d which cover f is the same for all f of weight $d - s'$, so these codewords form a

$$(d - s') - (n, d, (1 - \alpha_{d-s'})/\alpha_{s'}) \text{ design.} \qquad \text{Q.E.D.}$$

Theorem 24. *If* $d - s' \leqslant s' < d$, *the codewords of any fixed weight w in $\mathscr{C}$ form a* $(d - s')$-*design.*

Proof. Let f have weight $d - s'$. The theorem is true for codewords of weight d by Theorem 23. The number of codewords of weight $d + 1$ which cover f is $A_{s'+1}(f)$. Since $A_{s'+1}(f)$ is independent of the choice of f by Theorem 20, the theorem is true for $w = d + 1$.

We may write $A_{s'+2}(f) = T_1 + T_2$, where T_1 is the number of codewords of weight $d + 2$ which cover f, and T_2 is the number of codewords of weight d which cover exactly $d - s' - 1$ ones of f. T_2 is determined by Theorem 23, and is independent of the choice of f. Therefore T_1 is also independent of the choice of f, and so the theorem is true for $w = d + 2$. Clearly we may continue in this way. Q.E.D.

Theorem 24 *is* weaker than Corollary 14. For the codewords of weight 8 in the extended Golay code form a 5-design (by Corollary 14 or even by Theorem 9), but Theorem 24 only says they form a 4-design.

Research Problem (6.3). Strengthen Theorem 24 for nonlinear codes.

§8. Perfect codes

In this section we shall prove that an $(n, M, d = 2e + 1)$ code is perfect if and only if $s' = e$. We conclude the section with an important necessary condition (Lloyd's theorem) for a code to be perfect.

Theorem 25. *For any code*, $s' \geqslant \frac{1}{2}(d - 1)$.

Proof. Assume that $\mathscr{C}$ contains the codeword 0. First suppose d is odd and let v be a vector of weight $\frac{1}{2}(d - 1)$. If $s' < \frac{1}{2}(d - 1)$ then by Theorem 21 there is a codeword c at distance $< \frac{1}{2}(d - 1)$ from v. But then wt $(c) < d$, a contradiction. Similarly if d is even. Q.E.D.

Lemma 26. *If* $s' = \frac{1}{2}(d - 1)$ $(d$ *is odd) then* $\alpha(x) = P_0(x) + P_1(x) + \cdots + P_{\frac{1}{2}(d-1)}(x)$.

Proof. Let f be a vector of weight $i < \frac{1}{2}(d - 1)$. In $\mathscr{C} + f$ we have

$$A_0(f) = \cdots = A_{i-1}(f) = 0, \quad A_i(f) = 1,$$

$$A_{i+1}(f) = \cdots = A_s(f) = 0.$$

Then by (18), $\alpha_i = 1$ for $i = 0, 1, \ldots, s' - 1$. If f has weight $s' = \frac{1}{2}(d - 1)$, then

$$A_0(f) = \cdots = A_{s'-1}(f) = 0, \quad A_{s'}(f) = 1,$$

so again $\alpha_{s'} = 1$ from (18).   Q.E.D.

Theorem 27. *An $(n, M, 2e + 1)$ code $\mathscr{C}$ is perfect iff $s' = e$.*

Proof. Suppose $s' = e$. Then by Theorem 21 every vector is at distance $\leq e$ from some codeword, which says that the spheres of radius e about the codewords fill F^n, and $\mathscr{C}$ is perfect. Conversely, suppose $\mathscr{C}$ is perfect, so that (26) of Ch. 5 holds.

We apply Lemma 19 with $\beta(x) = P_0(x) + \cdots + P_e(x)$ and deduce that the annihilator polynomial $\alpha(x)$ of $\mathscr{C}$ must divide $P_0(x) + \cdots + P_e(x)$. Hence

$$e = \deg\{P_0(x) + \cdots + P_e(x)\} \geq \deg \alpha(x) = s'.$$

But $s' \geq e$ by Theorem 25.   Q.E.D.

Thus the annihilator polynomial of a perfect code is

$$\alpha(x) = P_0(x) + \cdots + P_e(x).$$

This is called Lloyd's polynomial and is denoted by $L_e(x)$. By Problem 42 of Ch. 5, $L_e(x) = P_e(x - 1; n - 1)$. It follows that:

Theorem 28. (Lloyd.) *If there exists a binary $(n, M, 2e + 1)$ perfect code then $L_e(x)$ has e integer zeros $\sigma_1, \ldots, \sigma_e$ satisfying*

$$0 < \sigma_1 < \cdots < \sigma_e < n.$$

Example. For the $[7, 4, 3]$ Hamming code, $L_e(x) = 8 - 2x$.

§9. Codes over GF(q)

Most of this chapter can be generalized to codes over GF(q). To begin with, the parameters d, s, d', s' are defined as in §2 with weight and distance replaced respectively by Hamming weight and Hamming distance.

Equation (1) now reads

$$1 + \sum_{j=1}^{s} B_{\tau_j} y^{\tau_j} = Mq^{-n} \sum_{i=0}^{n} B_i'(1 + \gamma y)^{n-i}(1 - y)^i,$$

where $\gamma = q - 1$, and Theorem 2 becomes

Theorem 2.* *If* $s \leqslant d'$, *then*

$$B_{\tau_i} = -\prod_{\substack{j=1 \\ j \neq i}}^{s} \frac{\tau_j}{\tau_j - \tau_i} + \frac{1}{N} \sum_{t=0}^{n} \gamma' \binom{n}{t} \prod_{\substack{j=1 \\ j \neq i}}^{s} \frac{\tau_j - t}{\tau_j - \tau_i},$$

for $1 \leqslant i \leqslant s$, *where* $N = q^n / M$.

Theorem 4 is changed similarly, while Lemma 1 and Theorems 3, 5, 6 are unchanged.

Since there is no longer a unique vector of weight n, we cannot introduce $\bar{s}$, and there is no analog of Theorem 7.

t-Designs from nonbinary codes. If $\mathscr{C}$ is a linear code of length n over GF(q), we shall try to get a t-design from the codewords of weight w, $w > 0$, in the following way. The set of w coordinates where a codeword is nonzero is called the *support* of the codeword. Of course (since the code is linear) the $q - 1$ scalar multiples of this codeword have the same support. Then this support (counted once, not $q - 1$ times) will form a block of the t-design.

The t-design may still contain repeated blocks. However this will not be so for binary codes, nor for codes over GF(q) if w is equal to the minimum weight d of the code. For suppose c and c' are codewords of weight d with the same support which are not scalar multiples of each other. Then there is a scalar multiple of c, tc say, such that $0 < \text{dist}\,(tc, c') < d$, a contradiction.

For example, consider code #6 of Ch. 1. The codewords 0121 and 0212 both have support 0111. Thus the 8 codewords of weight 3 give the four blocks

$$0111, \quad\quad 1011, \quad\quad 1101, \quad\quad 1110,$$

forming a trivial 3-(4, 3, 1) design.

A sufficient condition for this construction to produce a t-design is given by:

Theorem 29. (Assmus and Mattson.) *Let* $\mathscr{C}$ *be an* $[n, k, d]$ *linear code over* GF(q). *Suppose we can find an integer* t, *with* $0 < t < d$, *such that there are at most* $d - t$ σ_i's *in the range* $1 \leqslant \sigma_i \leqslant n - t$. *(Here the* σ_i's *are the weights occurring in* $\mathscr{C}^\perp$, *as in* §2.) *Then the supports of the codewords of weight* d *in* $\mathscr{C}$ *form a* t-design.

Proof. Let T be any set of t coordinate places, and (as in §5) let $\mathscr{C}^T$ be formed by deleting those coordinates from $\mathscr{C}$. Thus $\mathscr{C}^T$ is an $[n - t, k, d - t]$ code. The dual code is $(\mathscr{C}^T)^\perp = (\mathscr{C}^\perp)^{0@^T}$. By hypothesis $(\mathscr{C}^T)^\perp$ has $\leqslant d - t$ nonzero weights and its dual has minimum distance $d - t$. So we may apply Theorem 2* to obtain the weight distribution of $(\mathscr{C}^T)^\perp$, and hence by Theorem 13 of Ch. 5, the weight distribution of $\mathscr{C}^T$. In particular the number of codewords of weight $d - t$ in $\mathscr{C}^T$, α say, is independent of the choice of T.

Each of these α codewords comes from a codeword in $\mathscr{C}$ with weight d, whose support contains T. Now (as we showed above) all codewords with weight d with the same support are scalar multiples of each other. Therefore the number of blocks which contain T is $\alpha/(q-1)$, and is independent of the choice of T.   Q.E.D.

Example. The self-dual code #6 of Ch. 1 contains codewords of weight 0 and 3 only. Thus $n = 4$, $d = \sigma_1 = 3$. Therefore we may take $t = 2$, and deduce from Theorem 29 that the supports of the codewords of weight 3 form a 2-design. In fact, as we saw earlier, they form a 3-design, so the conclusion of the theorem is not always the strongest possible result.

Corollary 30. *Assume the same hypotheses as Theorem 29, and $q > 2$. Then for each weight τ_i in the range $d \leq \tau_i \leq v_0$, the supports of the codewords of weight τ_i in $\mathscr{C}$ form a t-design, where v_0 is the largest integer satisfying*

$$v_0 - \left[\frac{v_0 + q - 2}{q - 1} \right] < d.$$

Proof. The definition of v_0 implies that if two codewords of weight $\tau_i \leq v_0$ have the same support, then they must be scalar multiples of each other. The proof then proceeds as in Theorem 29.   Q.E.D.

For example, the $[12, 6, 6]$ ternary Golay code $\mathscr{G}_{12}$ (see Ch. 20) is a self-dual code with weights 0, 6, 9, 12. Then $v_0 = 11$, so the supports of the codewords of weights 6 and 9 form 5-designs.

In the binary case we can never have repeated blocks, therefore:

Corollary 31. *Let $\mathscr{C}$ be an $[n, k, d]$ binary linear code. If the hypothesis of Theorem 29 is satisfied then the codewords of any weight τ_i form a t-design.*

Example. For the $[24, 12, 8]$ extended Golay code we may take $t = 5$, for there are $d - t = 3$ σ_i's in the range $[1, 19]$, namely 8, 12, 16. Thus the codewords of each weight form a 5-design, as we have seen before.

We have not been able to generalize Theorem 13, and state this as:

Research Problem (6.4). Generalize Theorem 13 to $GF(q)$.

Problem. (13) Show that Corollary 31 implies Corollary 14.

Research Problem (6.5). Is Corollary 31 truly stronger than Corollary 14?

Weight distribution of translates, and perfect codes. All the results of §§6, 8 hold for codes over GF(q). The proofs for the binary case used the group algebra introduced in Ch. 5. The proofs for codes over GF(q) are the same, but require a more elaborate notation for the group algebra, which we do not propose to give.

For example, Lloyd's theorem becomes:

Theorem 32. (Lloyd.) *If there exists an* $(n, M, 2e + 1)$ *perfect code over* GF(q), *then the Lloyd polynomial*

$$L_e(x) = P_0(x; n) + \cdots + P_e(x; n)$$

$$= P_e(x - 1; n - 1)$$

$$= \sum_{j=0}^{e} (-1)^j (q-1)^{e-j} \binom{x-1}{j} \binom{n-x}{e-j} \cdot \tag{24}$$

has e integer zeros $\sigma_1, \ldots, \sigma_e$ *satisfying*

$$0 < \sigma_1 < \cdots < \sigma_e < n.$$

Now $P_k(x; n)$ *is given by Equation* (53) *of Ch. 5.*

Problem. (14) Let $\mathscr{C}$ be an $[n, k, 2e + 1]$ linear code over GF(q). Show that $\mathscr{C}$ is perfect iff the codewords of weight $2e + 1$ form an $(e + 1) - (n, 2e + 1, (q - 1)^e)$ design. If $\mathscr{C}$ is perfect, show that the number of codewords of weight $2e + 1$ in $\mathscr{C}$ is

$$A_{2e+1} = \frac{(q-1)^{e+1}\binom{n}{e+1}}{\binom{2e+1}{e+1}}.$$

*§10. There are no more perfect codes

Three types of perfect codes were discovered in the late 1940's:

(i) The linear single-error-correcting Hamming codes. We have seen the binary version in §7 of Ch. 1. In Ch. 7 we shall construct the Hamming codes over GF(q). These have the parameters

$$\left[n = \frac{q^m - 1}{q - 1}, k = n - m, d = 3 \right].$$

(ii) The binary [23, 12, 7] Golay code $\mathcal{G}_{23}$ (§6 of Ch. 2, and Ch. 20).
(iii) The ternary [11, 6, 5] Golay code $\mathcal{G}_{11}$ (Ch. 20).
As we shall see in Ch. 20, the parameters of the two Golay codes determine them uniquely: any code with these parameters must be equivalent to one of the Golay codes. But for single-error-correcting perfect codes the situation is different. In 1962 Vasil'ev constructed a family of nonlinear binary single-error-correcting codes with the same parameters as Hamming codes (see §9 of Ch. 2). Later Schönheim and Lindström gave nonlinear codes with the same parameters as Hamming codes over GF(q) for all q. This question is still not completely settled:

Research Problem (6.6). Find all perfect nonlinear single-error-correcting codes over GF(q).

Finally there are the *trivial* perfect codes: a code containing just one codeword, or the whole space, or a binary repetition code of odd length. Subsequently many people attempted to discover other perfect codes, or when this failed, to prove no others existed. Van Lint made considerable progress on this problem, which was finally Finn-ished by Tietäväinen in 1973. The final result is:

Theorem 33. (Tietäväinen and Van Lint.) *A nontrivial perfect code over any field* GF(q) *must have the same parameters n, M and d as one of the Hamming or Golay codes.*

We break this up into five parts, Theorems 37–41. The essence of the proof is to show that the only cases in which Lloyd's polynomial has distinct integer zeros in the range $[1, n-1]$ are those given above.
Throughout §10 let $\mathcal{C}$ be an $(n, M, 2e+1)$ perfect code over GF(q), where $q = p^r$, $p = $ prime, and let $\sigma_1 < \sigma_2 < \cdots < \sigma_e$ be the integer zeros of $L_e(x)$ (Theorem 32).
First some lemmas.

Lemma 34. (The sphere packing condition.) *The number of codewords M is a power of q, and*

$$\sum_{i=0}^{e} (q-1)^i \binom{n}{i} = q^l \tag{25}$$

for some integer l.

Proof. Since the code is perfect (cf. Theorem 6 of Ch. 1),

$$M \sum_{i=0}^{e} \binom{n}{i} (q-1)^i = q^n = p^{nr}.$$

Therefore $M = p^j$ and

$$\sum_{i=0}^{e} \binom{n}{i} (q-1)^i = p^{nr-j}.$$

Thus $q - 1 = p^r - 1$ divides $p^{nr-j} - 1$. By Lemma 9 of Ch. 4, r divides j and M is a power of q.   Q.E.D.

Lemma 35. $L_e(0) = q^l$; *and if $\mathscr{C}$ is nontrivial, $L_e(1)$ and $L_e(2)$ are nonzero.*

Proof. From (24), using $\binom{-1}{i} = (-1)^i$ (see Problem 18 of Ch. 1),

$$L_e(0) = \sum_{j=0}^{e} (q-1)^{e-j} \binom{n}{e-j}$$

$$= q^l \quad \text{(from Lemma 34)}.$$

Also from (24),

$$L_e(1) = (q-1)^e \binom{n-1}{e}$$

which is nonzero since $e \le n - 1$; and

$$L_e(2) = \frac{(q-1)^{e-1}}{e} \binom{n-2}{e-1} \{q(n-e-1) - n + 1\}, \tag{26}$$

which is zero only if $q = 1 + e/(n - e - 1)$. But $q \ge 2$, so this implies $n \le 2e + 1$, and $\mathscr{C}$ is trivial.   Q.E.D.

Lemma 36.

$$\sigma_1 + \cdots + \sigma_e = \frac{e(n-e)(q-1)}{q} + \frac{e(e+1)}{2},$$

$$\sigma_1 \sigma_2 \cdots \sigma_e = e!\, q^{l-e}.$$

Proof. From Theorem 15 of Ch. 5, $L_e(x)$ is a polynomial in x of degree e. The coefficient of x^e is $(-q)^e/e!$, and the coefficient of x^{e-1} is

$$-\frac{(-q)^e}{e!} \left\{ \frac{e(e+1)}{2} + \frac{q-1}{q} e(n-e) \right\}. \qquad \text{Q.E.D.}$$

Theorem 37. *A perfect single-error-correcting code over* GF(q) *has the same parameters n, M and d as the Hamming code.*

Proof. The sphere packing condition (Lemma 34) is

$$1 + (q - 1)n = q^l,$$

and Lloyd's theorem (Theorem 32) says that

$$L_1(x) = P_0(x) + P_1(x) = 1 + \{(q - 1)n - qx\}$$

has an integer root $x = \sigma_1$. Thus

$$1 + (q - 1)n - q\sigma_1 = 0.$$

Hence $\sigma_1 = q^{l-1}$ and $n = (q^l - 1)/(q - 1)$. The size of this code is q^{n-l} from the sphere packing bound. Q.E.D.

Furthermore this perfect code has the same weight and distance distribution as the Hamming code. This follows from the GF(q) version of Theorem 4, since $d = 3$ and $s' = e = 1$.

Theorem 38. *There is no nontrivial binary perfect double-error-correcting code.*

Proof. The sphere packing condition says

$$1 + n + \binom{n}{2} = 2^l$$

or

$$(2n + 1)^2 = 2^{l+3} - 7. \tag{27}$$

Lloyd's polynomial is $2L_2(x) = y^2 - 2(n + 1)y + 2^{l+1}$, where $y = 2x$. Therefore by Lemma 35 this polynomial must have distinct roots $y_1 = 2\sigma_1$ and $y_2 = 2\sigma_2$ which are even integers greater than 4. Since $y_1 y_2 = 2^{l+1}$, $y_1 = 2^a$ and $y_2 = 2^b$ for $3 \leqslant a < b$. Then $y_1 + y_2 = 2^a + 2^b = 2(n + 1)$, so (27) becomes

$$(2^a + 2^b - 1)^2 = 2^{l+3} - 7 = 2^{a+b+2} - 7.$$

Reducing both sides modulo 16 gives $1 \equiv -7$, a contradiction. Q.E.D.

Theorem 39. *The only possible parameters for a nontrivial perfect double-error-correcting code over a field* GF(q), $q = p^r$, *are the parameters of the ternary Golay code.*

Proof. The sphere packing condition is now

$$1 + (q-1)n + (q-1)^2 \binom{n}{2} = q^l, \tag{28}$$

which implies

$$2(q-1)n = q - 3 + \sqrt{(q^2 - 6q + 1 + 8q^l)}.$$

From Lemma 36,

$$\sigma_1 \sigma_2 = 2q^{l-2}$$

$$\sigma_1 + \sigma_2 = \frac{2(n-2)(q-1)}{q} + 3.$$

Eliminating n gives

$$q(\sigma_1 + \sigma_2) = 1 + \sqrt{(q^2 - 6q + 1 + 8q^l)}. \tag{29}$$

Since σ_1, $\sigma_2 > 2$ by Lemma 35, we have $\sigma_1 = p^\lambda$, $\sigma_2 = 2p^\mu$ for λ, $\mu \geq 1$ and $\lambda + \mu = r(l-2)$. Substituting in (29), squaring and dividing through by q we obtain

$$-2(p^\lambda + 2p^\mu) + q(p^\lambda + 2p^\mu)^2 = q - 6 + 8q^{l-1}. \tag{30}$$

All the terms except -6 contain a factor of p. Therefore p is 2 or 3. If p is 2, then $q - 6$ must be divisible by 4 and so $q = 2$, which is impossible by Theorem 38.

Now suppose $p = 3$. If $q = 3$, (30) becomes

$$-2(3^\lambda + 2 \cdot 3^\mu) + 3(3^\lambda + 2 \cdot 3^\mu)^2 = -3 + 8 \cdot 3^{l-1}.$$

If $l \leq 3$ this has no solutions. For $l \geq 4$ after dividing by 3 we have

$$3^{2\lambda} + 4 \cdot 3^{\lambda+\mu} + 4 \cdot 3^{2\mu} - 2 \cdot 3^{\lambda-1} - 4 \cdot 3^{\mu-1} = -1 + 8 \cdot 3^{l-2}.$$

Therefore $\mu = 1$, and so $\lambda = 2$. This implies $l = 5$ and (from (28)) $n = 11$, which are the parameters of the ternary Golay code.

Finally suppose $q = 3^r$ with $r > 1$. Equation (30) modulo 9 is

$$7(3^\lambda + 2 \cdot 3^\mu) \equiv 3 - q^{l-1} \pmod 9$$

which shows $l > 1$, so

$$7(3^\lambda + 2 \cdot 3^\mu) \equiv 3 \pmod 9,$$

i.e., (multiplying by 4),

$$3^\lambda + 2 \cdot 3^\mu \equiv 3 \pmod 9.$$

Therefore $\lambda = 1$ and $\mu > 1$. $1 + \mu = r(l-2)$ implies $3^{1+\mu} = q^{l-2}$. Then (30) becomes

$$-2(3 + 2 \cdot 3^\mu) + 3^r(3 + 2 \cdot 3^\mu)^2 = 3^r - 6 + 8 \cdot 3^{r+\mu+1},$$

$$-4 \cdot 3^\mu + 8 \cdot 3^r = 4 \cdot 3^{r+\mu+1} - 4 \cdot 3^{2\mu+r}.$$

But now the RHS is divisible by a higher power of 3 than the LHS.

Q.E.D.

Remark. A computer search has shown that the only solutions to Equation (25) in the range $n \leqslant 1000$, $e \leqslant 1000$, $q \leqslant 100$ are the trivial solutions, the parameters of the Hamming and Golay codes, and one more solution:

$$n = 90, \qquad k = 78, \qquad e = 2 \quad \text{when } q = 2.$$

But Theorem 38 shows there is no code with the latter parameters.

Research Problem (6.7). How close can one get: is there a [90, 77, 5] binary code?

Theorem 40. *There is no nontrivial perfect e-error-correcting code over* GF(q), $q = p^r$, *if* $q > 2$ *and* $e > 2$.

Proof. (i) We first show $2\sigma_1 \leqslant \sigma_e$. If s is an integer, $s = p^a t$, where $(p, t) = 1$, we define $a_p(s) = t$. Clearly $a_p(s_1 s_2) = a_p(s_1) a_p(s_2)$, and $a_p(s) \leqslant s$. From Lemma 36, $\sigma_1 \sigma_2 \cdots \sigma_e = e! \, q^{1-e}$, so

$$a_p(\sigma_1) a_p(\sigma_2) \cdots a_p(\sigma_e) = a_p(e!) \leqslant e!.$$

So either two of the $a_p(\sigma_i)$'s are equal, say $a_p(\sigma_i) = a_p(\sigma_j)$, or else the numbers $a_p(\sigma_1), \ldots, a_p(\sigma_e)$ are equal to $1, \ldots, e$ in some order and $p > e \geqslant 3$. In the first case $\sigma_i = p^a t$, $\sigma_j = p^\beta t$, and if $i < j$, $2\sigma_i \leqslant \sigma_j$. In the second case for some i and j, $a_p(\sigma_i) = 1$, $a_p(\sigma_j) = 2$, $\sigma_i = p^a$, $\sigma_j = 2p^\beta$, and (interchanging i and j if necessary) $2\sigma_i \leqslant \sigma_j$. Therefore $2\sigma_1 \leqslant 2\sigma_i \leqslant \sigma_j \leqslant \sigma_e$.

(ii) Next we show

$$\sigma_1 \sigma_e \leqslant \frac{8}{9} \left(\frac{\sigma_1 + \sigma_e}{2} \right)^2. \tag{31}$$

Writing $x = \sigma_e / \sigma_1$ this becomes

$$x \leqslant \tfrac{2}{9}(1 + x)^2 \quad \text{for } x \geqslant 2$$

which is immediate.

(iii)

$$\sigma_1 \sigma_2 \cdots \sigma_e = \frac{(-1)^e L_e(0)}{\text{coeff. of } x^e \text{ in } L_e(x)}$$

$$= q^{-e} e! \sum_{j=0}^{e} (q-1)^{e-j} \binom{n}{e-j} \tag{32}$$

$$> q^{-e}(q-1)^e n(n-1) \cdots (n-e+1),$$

taking only the term $j = 0$ in the sum,

$$> q^{-e}(q-1)^e n^e \left(1 - \frac{e(e-1)}{2n}\right). \tag{33}$$

(iv) We shall now use a weighted version of the arithmetic-mean geometric-mean inequality, which states that for real numbers $x_i \geq 0$, and weights $\alpha_i > 0$ satisfying

$$\sum_{i=1}^{N} \alpha_i = 1,$$

we have

$$\prod_{i=1}^{N} x_i^{\alpha_i} \leq \sum_{i=1}^{N} \alpha_i x_i. \tag{34}$$

Thus

$$\prod_{i=1}^{e} \sigma_i = (\sigma_1 \sigma_e) \cdot (\sigma_2 \cdots \sigma_{e-1})$$

$$\leq \frac{8}{9} \left(\frac{\sigma_1 + \sigma_e}{2}\right)^2 \left(\frac{\sigma_2 + \cdots + \sigma_{e-1}}{e-2}\right)^{e-2}$$

by (ii) and (34),

$$\leq \frac{8}{9} \left(\frac{\sigma_1 + \cdots + \sigma_e}{e}\right)^e$$

using (34) again.

Thus we have an improved version of the arithmetic-mean geometric-mean inequality which applies to the σ_i's. From Lemma 36, the last expression is

$$= \frac{8}{9} \left\{\frac{(n-e)(q-1)}{q} + \frac{e+1}{2}\right\}^e \tag{35}$$

$$\leq \tfrac{8}{9} q^{-e}(q-1)^e n^e \tag{36}$$

after a bit of algebra using $q > 2$. Combining (33) and (36) gives

$$1 - \frac{e(e-1)}{2n} < \frac{8}{9},$$

$$n < \tfrac{9}{2} e(e-1). \tag{37}$$

(v) Since

$$\prod_{i=1}^{e} (\sigma_i - 1) = \frac{e!}{q^e} L_e(1) = \frac{(q-1)^e (n-1) \cdots (n-e)}{q^e} \tag{38}$$

must be an integer, we have

$$p^{re} | (n-1)(n-2) \cdots (n-e).$$

Suppose α is the largest power of p dividing any of $n-1, \ldots, n-e$. What can we say about the power of p dividing the product $(n-1)\cdot\ldots\cdot(n-e)$? This is at most

$$A = p^{\alpha + [e/p] + [e/p^2] + \cdots}$$

and we must have $A \geqslant p^{re}$. Therefore

$$\alpha \geqslant re - \left[\frac{e}{p}\right] - \left[\frac{e}{p^2}\right] - \cdots \geqslant e\left(r - \frac{1}{p-1}\right)$$
$$\geqslant \frac{er}{2}.$$

Thus $p^{r[e/2]} = q^{[e/2]}$ divides $n - i$ for some i, and so

$$n > q^{[e/2]}. \tag{39}$$

(vi) From (37), (39)

$$3^{[e/2]} \leqslant q^{[e/2]} < n < \tfrac{9}{2}e(e-1)$$

which implies that $e \leqslant 11$, $q \leqslant 8$, and $n < 495$. By the computer search mentioned above, there are no perfect codes in this range.   Q.E.D.

Theorem 41. *The only possible parameters for a nontrivial binary perfect e-error-correcting code with $e > 2$ are those of the Golay code.*

Proof. From (26),

$$\prod_{i=1}^{e}(\sigma_i - 2) = e!\,L_e(2)/2^e$$
$$= (n-2)(n-3)\cdots(n-e)(n-2e-1)/2^e. \tag{40}$$

Since $(\sigma_i - 1)(\sigma_i - 2)$ is even, (38) and (40) imply $2^{3e} | (n-1)(n-2)^2 \cdots (n-e)^2(n-2e-1)$, and hence $n > \tfrac{2}{e} \cdot 2^{e/2}$. The rest of the proof follows that of Theorem 40, using (32) and (35) to get an upper bound on n analogous to (37).   Q.E.D.

A code with the same parameters as the [23, 12, 7] binary Golay code has the same weight and distance distributions as this code. This follows from Theorem 4, since $d = 7$ and $s' = e = 3$. Similarly for the ternary Golay code.

Notes on Chapter 6

Paige [1018] and Bose [174] were probably the first to obtain designs from codes, while Assmus and Mattson [37, 41, 47] were the first to give results similar to Theorems 9 and 23. Our treatment follows Delsarte [350–352]. Semakov and Zinov'ev [1180, 1181] study the connection between constant weight codes and t-designs.

§3. Theorems 2 and 4 appear to be new.

§5. This section is based on Shaughnessy [1195].

§6. Most of this section follows Delsarte [351], although some of the proofs are different. An alternative proof of Theorem 21 can be found in Assmus and Mattson [41].

In Euclidean space of dimension n, Equation (20) can be improved to

$$\tfrac{1}{2}\delta \leqslant \rho \leqslant \delta \sqrt{\left(\frac{n}{2(n+1)}\right)}$$

– see Jung [702]. Problem 11 shows that no such improvement is possible in Hamming space. Theorem 22 is from MacWilliams [872].

§8. Theorems 25, 27 are from MacWilliams [871, 872]. Problem 13 is due to Assmus and Mattson [37] (see also [47]).

Theorem 28 is due to Lloyd [859]. For Theorem 32 and other generalizations see Assmus and Mattson [41], Bassalygo [76a, 77], Biggs [144], Delsarte [380], Lenstra [815] and Roos [1123].

A code is said to be uniformly packed if $s' = e + 1$ [505].

§9. Theorem 29 and Corollary 30 are due to Assmus and Mattson [41, 47]. Delsarte [351] has given a different generalization of Theorems 9 and 24 to codes over $GF(q)$.

§10. Van Lint [855] is an excellent survey article on perfect codes, and includes references to the early work (which we have omitted). Tietäväinen's work on perfect codes will be found in [1321–1324], and Van Lint's in [845–850, 854, 855]. For some recent Russian work see [1471, 1472].

Schönheim [1159] (see also [1160]) and Lindström [842] show that for any $q \geqslant 3$ and $m \geqslant 2$ there is a genuinely nonlinear perfect single-error-correcting code of length $n = (q^{m+1} - 1)/(q - 1)$ over $GF(q)$.

The proof of Theorem 33 given here follows Van Lint [848] and [854] and Tietäväinen [1323], with contributions by H.W. Lenstra and D.H. Smith (unpublished) in the proof of Theorem 41. The computer search used in proving Theorems 40 and 41 was made by Van Lint [845]; Lenstra and A.M. Odlyzko (unpublished) have shown that it can be avoided by tightening the inequalities. For the arithmetic-mean geometric-mean inequality see for example Beckenbach and Bellman [94, p. 13].

Generalizations of perfect codes. For perfect codes in the Lee metric, or over mixed alphabets, or in graphs, see Astola [54], Bassalygo [77], Baumert et al. [84], Biggs [144–147], Cameron et al. [236], Golomb et al. [524, 525, 529, 531, 532], Hammond [593, 594], Heden [628–630], Herzog and Schönheim [643], Johnson [698], Lindström [843], Post [1072], Racsmány [1084], Schönheim [1159, 1160, 1162] and Thas [1319].

7

Cyclic codes

§1. Introduction

Cyclic codes are the most studied of all codes, since they are easy to encode, and include the important family of BCH codes. Furthermore they are building blocks for many other codes, such as the Kerdock, Preparata, and Justesen codes (see later chapters).

In this chapter we begin by defining a cyclic code to be an ideal in the ring of polynomials modulo $x^n - 1$ (§2). A cyclic code of length n over GF(q) consists of all multiples of a generator polynomial $g(x)$, which is the monic polynomial of least degree in the code, and is a divisor of $x^n - 1$ (§3). The polynomial $h(x) = (x^n - 1)/g(x)$ is called the check polynomial of the code (§4).

In §5 we study the factors of $x^n - 1$. We always assume that n and q are relatively prime. Then the zeros of $x^n - 1$ lie in the field GF(q^m), where m is the least positive integer such that n divides $q^m - 1$.

At the end of §3 it is shown that Hamming and double-error-correcting BCH codes are cyclic. Then in §6 we give the general definition of t-error-correcting BCH codes over GF(q). In §7 we look more generally at how a matrix over GF(q^m) defines a code over GF(q). The last section describes techniques for encoding cyclic codes.

Further properties of cyclic codes are dealt with in the next chapter.

§2. Definition of a cyclic code

A code $\mathscr{C}$ is *cyclic* if it is linear and if any cyclic shift of a codeword is also a codeword, i.e., whenever $(c_0, c_1, \ldots, c_{n-1})$ is in $\mathscr{C}$ then so is

$(c_{n-1}, c_0, \ldots, c_{n-2})$. For example, the Hamming code (40) of Ch. 1 is cyclic. So is the code $\mathscr{C}_3 = \{000, 110, 101, 011\}$.

To get an algebraic description, we associate with the vector $c = (c_0, c_1, \ldots, c_{n-1})$ in F^n (where F is any finite field $GF(q)$) the polynomial $c(x) = c_0 + c_1 x + \cdots + c_{n-1} x^{n-1}$. E.g. $\mathscr{C}_3$ corresponds to the polynomials 0, $1 + x$, $1 + x^2$, $x + x^2$.

We shall use the following notation. If F is a field, $F[x]$ denotes the set of polynomials in x with coefficients from F.

In fact, $F[x]$ is a *ring*.

Definition. A *ring* R (loosely speaking) is a set where addition, subtraction, and multiplication are possible. Formally, R is an additive abelian group, together with a multiplication satisfying $ab = ba$, $a(b + c) = ab + ac$, $(ab)c = a(bc)$, and which contains an identity element 1 such that $1 \cdot a = a$. [Thus our ring is sometimes called a commutative ring with identity.]

The ring $R_n = F[x]/(x^n - 1)$. For our purposes another ring is more important than $F[x]$. This is the ring $R_n = F[x]/(x^n - 1)$, consisting of the residue classes of $F[x]$ modulo $x^n - 1$. Each polynomial of degree $\leq n - 1$ belongs to a different residue class, and we take this polynomial as representing its residue class. Thus we can say that $c(x)$ belongs to R_n. R_n is a vector space of dimension n over F.

Multiplying by x corresponds to a cyclic shift. If we multiply $c(x)$ by x in R_n we get

$$xc(x) = c_0 x + c_1 x^2 + \cdots + c_{n-2} x^{n-1} + c_{n-1} x^n$$
$$= c_{n-1} + c_0 x + \cdots + c_{n-2} x^{n-1},$$

since $x^n = 1$ in R_n. But this is the polynomial associated with the vector $(c_{n-1}, c_0, \ldots, c_{n-2})$. Thus multiplying by x in R_n corresponds to a cyclic shift!

Ideals.

Definition. An *ideal* $\mathscr{C}$ of R_n is a linear subspace of R_n such that:

(i) if $c(x) \in \mathscr{C}$ then so is $r(x)c(x)$ for all $r(x) \in R_n$. Clearly (i) can be replaced by:

(ii) if $c(x) \in \mathscr{C}$ then so is $xc(x)$.

Our initial definition can now be simply written as:

Definition. A *cyclic code* of length n is an ideal of R_n.

Example. $\mathcal{C}_3 = \{0, 1 + x, 1 + x^2, x + x^2\}$ is an ideal in R_3. For $\mathcal{C}_3$ is closed under addition (hence linear), and any multiple of $c(x) \in \mathcal{C}_3$ is again in $\mathcal{C}_3$ (e.g., $x^2(1 + x^2) = x^2 + x^4 = x^2 + x$, since $x = x^4$ in R_3).

The group algebra FG. A second description of R_n is often helpful. Let $G = \{1, x, x^2, \dots, x^{n-1}\}$, $x^n = 1$, be a cyclic group of order n. The group algebra FG of G over F, (cf. §3 of Ch. 5), consists of all formal sums

$$c(x) = \sum_{i=0}^{n-1} c_i x^i, \quad c_i \in F.$$

Addition in FG is by coordinates, and multiplication is modulo $x^n - 1$. Clearly FG coincides with R_n.

Problems. (1) What is the ideal describing the cyclic code $\{0000, 0101, 1010, 1111\}$?

(2) Describe the smallest cyclic code containing the vector 0011010.

(3) Show that R_n is not a field (Hint: $x - 1$ has no multiplicative inverse).

(4) Show that a polynomial $f(x)$ has a multiplicative inverse in R_n. if and only if $f(x)$ is relatively prime to $x^n - 1$ in $F[x]$.

(5) Show that in an $[n, k]$ cyclic code any k consecutive symbols may be taken as information symbols.

§3. Generator polynomial ·

A particularly simple kind of ideal is a *principal ideal*, which consists of all multiples of a fixed polynomial $g(x)$ by elements of R_n. It will be denoted by

$$\langle g(x) \rangle.$$

$g(x)$ is called a *generator polynomial* of the ideal.

In fact every ideal in R_n is a principal ideal; every cyclic code has a generator polynomial. The next theorem proves this and other basic properties of cyclic codes.

Theorem 1. *Let $\mathcal{C}$ be a nonzero ideal in R_n, i.e., a cyclic code of length n.*

(a) *There is a unique monic polynomial $g(x)$ of minimal degree in $\mathcal{C}$.*

(b) *$\mathcal{C} = \langle g(x) \rangle$, i.e., $g(x)$ is a generator polynomial of $\mathcal{C}$.*

(c) *$g(x)$ is a factor of $x^n - 1$.*

(d) *Any $c(x) \in \mathcal{C}$ can be written uniquely as $c(x) = f(x)g(x)$ in $F[x]$, where $f(x) \in F[x]$ has degree $< n - r$, $r = \deg g(x)$. The dimension of $\mathcal{C}$ is $n - r$. Thus the message $f(x)$ becomes the codeword $f(x)g(x)$.*

(e) *If $g(x) = g_0 + g_1 x + \cdots + g_r x^r$, then $\mathcal{C}$ is generated (as a subspace of F^n)*

by the rows of the generator matrix

$$
G = \begin{bmatrix}
g_0 & g_1 & g_2 & \cdots & g_r & & 0 \\
 & g_0 & g_1 & \cdots & g_{r-1} & g_r \\
 & & \cdots & \cdots & \cdots \\
0 & & g_0 & \cdots & \cdots & & g_r
\end{bmatrix},
$$

$$
= \begin{bmatrix}
g(x) \\
 & xg(x) \\
 & & \cdots \\
 & & & x^{n-r-1}g(x)
\end{bmatrix}
\tag{1}
$$

using an obvious notation.

Proof. (a) Suppose $f(x), g(x) \in \mathscr{C}$ both are monic and have the minimal degree r. But then $f(x) - g(x) \in \mathscr{C}$ has lower degree, a contradiction unless $f(x) = g(x)$. (b) Suppose $c(x) \in \mathscr{C}$. Write $c(x) = q(x)g(x) + r(x)$ in R_n, where $\deg r(x) < r$. But $r(x) = c(x) - q(x)g(x) \in \mathscr{C}$ since the code is linear, so $r(x) = 0$. Therefore $c(x) \in \langle g(x) \rangle$. (c) Write $x^n - 1 = h(x)g(x) + r(x)$ in $F[x]$, where $\deg r(x) < r$. In R_n this says $r(x) = -h(x)g(x) \in \mathscr{C}$, a contradiction unless $r(x) = 0$. (d), (e): From (b), any $(c)x \in \mathscr{C}$, $\deg c(x) < n$, is equal to $q(x)g(x)$ in R_n. Thus

$$
\begin{aligned}
c(x) &= q(x)g(x) + e(x)(x^n - 1) \text{ in } F[x], \\
&= (q(x) + e(x)h(x))g(x) \text{ in } F[x], \\
&= f(x)g(x) \text{ in } F[x],
\end{aligned}
\tag{2}
$$

where $\deg f(x) \leq n - r - 1$. Thus the code consists of multiples of $g(x)$ by polynomials of degree $\leq n - r - 1$, evaluated in $F[x]$ (not in R_n). There are $n - r$ linearly independent multiples of $g(x)$, namely $g(x), xg(x), \ldots, x^{n-r-1}g(x)$. The corresponding vectors are the rows of G. Thus the code has dimension $n - r$.   Q.E.D.

We next give some examples of cyclic codes.

Binary Hamming codes. Recall from §7 of Ch. 1 that the parity check matrix of a binary Hamming code of length $n = 2^m - 1$ has as columns all $2^m - 1$ distinct nonzero m-tuples. Now if α is a primitive element of GF(2^m) (§2 of Ch. 3, §2 of Ch. 4) then $1, \alpha, \alpha^2, \ldots, \alpha^{2^m-2}$ are distinct and can be represented by distinct nonzero binary m-tuples.

So the binary Hamming code $\mathscr{H}_m$ with parameters

$$
[n = 2^m - 1, k = n - m, d = 3]
$$

has a parity check matrix which can be taken to be

$$H = (1, \alpha, \alpha^2, \ldots, \alpha^{2^m-2}),$$ (3)

where each entry is to be replaced by the corresponding column vector of m 0's and 1's.

E.g. for $\mathcal{H}_3$,

$$H = (1, \alpha, \alpha^2, \alpha^3, \alpha^4, \alpha^5, \alpha^6)$$

$$= \begin{pmatrix} 0010111 \\ 0101110 \\ 1001011 \end{pmatrix},$$ (4)

where $\alpha \in GF(2^3)$ satisfies $\alpha^3 + \alpha + 1 = 0$.

A vector $c = (c_0, c_1, \ldots, c_{n-1})$ belongs to $\mathcal{H}_m$

$$\text{iff } Hc^T = 0$$

$$\text{iff } \sum_{i=0}^{n-1} c_i\alpha^i = 0$$

$$\text{iff } c(\alpha) = 0$$

where $c(x) = c_0 + c_1x + \cdots + c_{n-1}x^{n-1}$. From property (M2) of Ch. 4, $c \in \mathcal{H}_m$ iff the minimal polynomial $M^{(1)}(x)$ divides $c(x)$. Thus $\mathcal{H}_m$ consists of all multiples of $M^{(1)}(x)$, or in other words:

Theorem 2. *The Hamming code $\mathcal{H}_m$ as defined above is a cyclic code with generator polynomial $g(x) = M^{(1)}(x)$.*

From Theorem 1 a generator matrix for $\mathcal{H}_m$ is

$$G = \begin{bmatrix} M^{(1)}(x) \\ \quad xM^{(1)}(x) \\ \qquad x^2M^{(1)}(x) \\ \cdots\cdots\cdots\cdots\cdots\cdots\cdots\cdots \\ \qquad\qquad x^{n-m-1}M^{(1)}(x) \end{bmatrix}$$ (5)

E.g. for $\mathcal{H}_3$,

$$G = \begin{bmatrix} 1\ 1\ 0\ 1 \\ \quad 1\ 1\ 0\ 1 \\ \qquad 1\ 1\ 0\ 1 \\ \qquad\quad 1\ 1\ 0\ 1 \end{bmatrix}$$ (6)

Problem. (6) Verify that the rows of (4) are orthogonal to those of (6).

Double-error-correcting BCH codes. In Equation (11) of Ch. 3 a double-error-correcting code $\mathcal{C}$ of length $n = 2^m - 1$ was defined to have the parity check matrix

$$H = \begin{pmatrix} 1 & \alpha & \alpha^2 & \cdots \alpha^{2^m-2} \\ 1 & \alpha^3 & \alpha^6 & \cdots \alpha^{3(2^m-2)} \end{pmatrix},$$ (7)

where again each entry is to be replaced by the corresponding binary m-tuple. Now

$$c \in \mathscr{C} \text{ iff } Hc^T = 0$$

$$\text{iff } \sum_{i=0}^{n-1} c_i\alpha^i = 0 \quad \text{and} \quad \sum_{i=0}^{n-1} c_i\alpha^{3i} = 0$$

$$\text{iff } c(\alpha) = 0 \quad \text{and} \quad c(\alpha^3) = 0$$

$$\text{iff } M^{(1)}(x) \,|\, c(x) \quad \text{and} \quad M^{(3)}(x) \,|\, c(x), \text{ by (M2),}$$

where $M^{(3)}(x)$ is the minimal polynomial of α^3,

$$\text{iff l.c.m. } \{M^{(1)}(x), M^{(3)}(x)\} \,|\, c(x).$$

But $M^{(1)}(x)$ and $M^{(3)}(x)$ are irreducible (by (M1)) and distinct (by Problem 15 of Ch. 4), so finally we have

$$c \in \mathscr{C} \text{ iff } M^{(1)}(x)M^{(3)}(x) \,|\, c(x).$$

Thus we have proved:

Theorem 3. *The double-error-correcting BCH code $\mathscr{C}$ has parameters*

$$[n = 2^m - 1, k = n - 2m, d \geqslant 5], \quad m \geqslant 3,$$

and is a cyclic code with generator polynomial

$$g(x) = M^{(1)}(x)M^{(3)}(x).$$

Proof. $\deg {}_g(x) = 2m$ follows from Problem 15 of Chapter 4. The minimum distance was established in Ch. 3 (another proof is given by Theorem 8 below).   Q.E.D.

Problem. (7) Show that the double-error-correcting BCH code of length 15 given in §3 of Ch. 3 has generator polynomial $g(x) = (x^4 + x + 1)(x^4 + x^3 + x^2 + x + 1)$ (use §4 of Ch. 4). Give a generator matrix.

Remark. So far nothing has been said about the minimum distance d of a cyclic code. This is because in general it is very difficult to find d. The BCH bound (Theorem 8 below) will give a lower bound to d if the zeros of $g(x)$ are known. In Ch. 8 we will see that in some cases the Mattson–Solomon polynomial enables one to find d.

Problems. (8) *Nonbinary Hamming codes.* (a) The Hamming code $\mathscr{H}_m(q)$ over $GF(q)$ has an $m \times (q^m - 1)/(q - 1)$ parity check matrix whose columns are all nonzero m-tuples from $GF(q)$ with first nonzero entry equal to 1. Code #6 of Ch. 1 shows $\mathcal{H}_2(3)$. Prove that $\mathcal{H}_m(q)$ is a perfect $[n = (q^m - 1)/(q - 1), k = n - m, d = 3]$ code.

(b) If m and $q - 1$ are relatively prime prove that $\mathcal{H}_m(q)$ is equivalent to the cyclic code with zeros $\alpha, \alpha^q, \alpha^{q^2}, \ldots$ where $\alpha \in GF(q^m)$ is a primitive n^{th} root of unity.

(9) *Shortened cyclic codes.* Engineering constraints sometimes call for a code of a length for which there is no good cyclic code. In this case a *shortened cyclic code* $\mathcal{C}^*$ can be used, obtained by taking those codewords of a cyclic code $\mathcal{C}$ which begin with i consecutive zeros, and deleting these zeros. The resulting code is of course not cyclic. However, show that there is a polynomial $f(x)$ such that $\mathcal{C}^*$ is an ideal in the ring of polynomials mod $f(x)$, and conversely, any ideal in such a ring is a shortened cyclic code.

§4. The check polynomial

Let $\mathcal{C}$ be a cyclic code with generator polynomial $g(x)$. From Theorem 1, $g(x)$ divides $x^n - 1$. Then

$$h(x) = (x^n - 1)/g(x)$$

$$= \sum_{i=0}^{k} h_i x^i \text{ (say)}, \quad (h_k \neq 0),$$

is called the *check polynomial* of $\mathcal{C}$. The reason for this name is as follows. If

$$c(x) = \sum_{i=0}^{n-1} c_i x^i = f(x)g(x)$$

is any codeword of $\mathcal{C}$, then

$$c(x)h(x) = \sum_{i=0}^{n-1} c_i x^i \cdot \sum_{j=0}^{k} h_j x^j$$

$$= f(x)g(x)h(x)$$

$$= 0 \text{ in } R_n.$$

The coefficient of x^j in this product is

$$\sum_{i=0}^{n-1} c_i h_{j-i} = 0, \quad j = 0, 1, \ldots, n - 1, \tag{8}$$

where the subscripts are taken modulo n. Thus the Equations (8) are parity check equations satisfied by the code. Let

$$H = \begin{bmatrix} & & & h_k & \cdots & h_2 & h_1 & h_0 \\ & & h_k & \cdots & h_2 & h_1 & h_0 \\ & \multicolumn{7}{c}{\cdots\cdots\cdots\cdots\cdots\cdots\cdots\cdots} \\ h_k & \cdots & & h_2 & h_1 & h_0 & & \end{bmatrix} \tag{9}$$

$$= \begin{bmatrix} & & & \overleftarrow{h(x)} \\ & & x\overleftarrow{h(x)} & \\ & \cdot\cdot & & \\ x^{n-k-1}\overleftarrow{h(x)} & & & \end{bmatrix}$$

using an obvious notation. Then (8) says that if $c \in \mathscr{C}$ then $Hc^T = 0$. Since

$$k = \deg h(x) = n - \deg g(x) = \text{dimension } \mathscr{C},$$

and the rows of H are obviously linearly independent, the condition $Hc^T = 0$ is also sufficient for c to be in the code. Thus H is a parity check matrix for $\mathscr{C}$.

Example. For the Hamming code $\mathscr{H}_3$, $h(x) = (x^7 + 1)/(x^3 + x + 1) = (x + 1)(x^3 + x^2 + 1) = x^4 + x^2 + x + 1$. Thus

$$H = \begin{bmatrix} 1 & 0 & 1 & 1 & 1 \\ & 1 & 0 & 1 & 1 & 1 \\ & & 1 & 0 & 1 & 1 & 1 \end{bmatrix}. \tag{10}$$

This is the same as (4).

Problem. (10) For the code of Problem 7, give $h(x)$, H, and verify that this matrix defines the same code as Equation (10) of Ch. 3.

Note that Equation (8) says that the codeword c must satisfy the parity check equations

$$\begin{aligned} c_{n-k-1}h_k + c_{n-k}h_{k-1} + \cdots + c_{n-1}h_0 &= 0, \\ c_{n-k-2}h_k + c_{n-k-1}h_{k-1} + \cdots + c_{n-2}h_0 &= 0, \\ \multicolumn{2}{c}{\cdots\cdots\cdots\cdots\cdots\cdots\cdots\cdots\cdots\cdots\cdots\cdots} \\ c_0 h_k + c_1 h_{k-1} + \cdots + c_k h_0 &= 0. \end{aligned} \tag{11}$$

I.e., c satisfies the linear recurrence

$$c_t h_k + c_{t+1}h_{k-1} + \cdots + c_{t+k}h_0 = 0 \tag{12}$$

for $0 \le t \le n - k - 1$. Thus if $c_{n-1}, \ldots, c_{n-k}$ are taken as message symbols, Equations (12) successively define the check symbols $c_{n-k-1}, \ldots, c_0$ (since $h_k \ne 0$).

The dual code. Let $\mathscr{C}$ be a cyclic code with generator polynomial $g(x)$ and check polynomial $h(x) = (x^n - 1)/g(x)$.

Theorem 4. *The dual code $\mathscr{C}^{\perp}$ is cyclic and has generator polynomial*

$$g^{\perp}(x) = x^{\deg h(x)} h(x^{-1})$$

Proof. From Equation (9).   Q.E.D.

By this theorem the code with generator polynomial $h(x)$ is equivalent to $\mathscr{C}^{\perp}$. In fact it consists of the codewords of $\mathscr{C}^{\perp}$ written backwards.

Problem. (11) Show that the $[7, 4, 3]$ code with $g(x) = x^3 + x + 1$ and the $[7, 3, 4]$ code with $g(x) = x^4 + x^3 + x^2 + 1$ are duals.

§5. Factors of $x^n - 1$

Since the generator polynomial of a cyclic code of length n over $\mathrm{GF}(q)$ must be a factor of $x^n - 1$, in this section we study these factors, and introduce the splitting field of $x^n - 1$.

We assume always that n and q are relatively prime. (Thus n is odd in the binary case.) By Problem 8 of Ch. 4 there is a smallest integer m such that n divides $q^m - 1$. This m is called the *multiplicative order of q modulo n*. By Problem 11 of Ch. 4, $x^n - 1$ divides $x^{q^{m-1}} - 1$ but does not divide $x^{q^{s-1}} - 1$ for $0 < s < m$.

Thus the zeros of $x^n - 1$, which are called n^{th} *roots of unity*, lie in the extension field $\mathrm{GF}(q^m)$ and in no smaller field.

The derivative of $x^n - 1$ is nx^{n-1}, which is relatively prime to $x^n - 1$, since n and q are relatively prime. Therefore $x^n - 1$ has n distinct zeros.

Factoring $x^n - 1$ over $\mathrm{GF}(q^m)$. Thus there are n distinct elements $\alpha_0, \alpha_1, \ldots, \alpha_{n-1}$ in $\mathrm{GF}(q^m)$ (the nth roots of unity) such that

$$x^n - 1 = \prod_{i=0}^{n-1} (x - \alpha_i).$$

$\mathrm{GF}(q^m)$ is therefore called the *splitting field* of $x^n - 1$.

Problems. (12) Show that the zeros of $x^n - 1$ form a cyclic subgroup of $\mathrm{GF}(q^m)^*$. I.e., there is an element α in $\mathrm{GF}(q^m)$, called a *primitive n^{th} root of unity*, such that

$$x^n - 1 = \prod_{i=0}^{n-1} (x - \alpha^i). \tag{13}$$

Throughout this chapter α will denote a primitive n^{th} root of unity.

(13) Show that the zeros of $x^n - 1$ form the multiplicative group of a field iff $n = q^m - 1$.

Factoring $x^n - 1$ *over* $GF(q)$. Cyclotomic cosets mod $p^m - 1$ were defined in §3 of Ch. 4. More generally the *cyclotomic coset mod n over* $GF(q)$ which contains s is

$$C_s = \{s, sq, sq^2, \ldots, sq^{m_s - 1}\},$$

where $sq^{m_s} \equiv s \bmod n$. (It is convenient but not essential to choose s to be the smallest integer in C_s.) Then the integers mod n are partitioned into cyclotomic cosets:

$$\{0, 1, \ldots, n - 1\} = \bigcup_s C_s,$$

where s runs through a set of *coset representatives mod n*. Note that $m = m_1$ is the number of elements in C_1.

E.g. for $n = 9$, $q = 2$,

$$C_0 = \{0\},$$
$$C_1 = \{1, 2, 4, 8, 7, 5\},$$
$$C_3 = \{3, 6\}.$$

Thus $m = 6$, and $x^9 - 1$ splits into linear factors over $GF(2^6)$.

Then as in Ch. 4 (see especially Problem (14)) the *minimal polynomial* of α^s is

$$M^{(s)}(x) = \prod_{i \in C_s} (x - \alpha^i).$$

This is a monic polynomial with coefficients from $GF(q)$, and is the lowest degree such polynomial having α^s as a root.

Also

$$x^n - 1 = \prod_s M^{(s)}(x) \tag{14}$$

where s runs through a set of coset representatives mod n. This is the factorization of $x^n - 1$ into irreducible polynomials over $GF(q)$.

E.g. $n = 9$, $q = 2$:

$$x^9 + 1 = M^{(0)}(x) M^{(1)}(x) M^{(3)}(x),$$

where

$$M^{(0)}(x) = x + 1,$$
$$M^{(1)}(x) = x^6 + x^3 + 1,$$
$$M^{(3)}(x) = x^2 + x + 1.$$

Figure 7.1 gives the factors of $x^n + 1$ over GF(2) for $n \leqslant 63$ and $n = 127$. Of course $x^{2m} + 1 = (x^m + 1)^2$ (Lemma 5 of Ch. 4), so only odd values of n are given. Also for $n = 3, 5, 11, 13, 19, 29, 37, 53, 59, 61, \ldots$ the factorization is $x^n + 1 = (x + 1)(x^{n-1} + \cdots + x + 1)$, since for these primes there are only two cyclotomic cosets, C_0 and C_1. The factors are given in octal, with the lowest degree terms on the left. Thus the first line of the table means that

$$1 + x^7 = (1 + x)(1 + x^2 + x^3)(1 + x + x^3).$$

Problems. (14) Show that if s is relatively prime to n, then C_s contains m elements.

(15) Let

$$f(x) = \prod_{i \in K} (x - \alpha^i),$$

where K is a subset of $\{0, 1, \ldots, n - 1\}$. Show that $f(x)$ has coefficients in GF(q) iff $k \in K \Rightarrow qk \pmod{n} \in K$.

n	Factors (in octal, lowest degree on left)
7	6.54.64.
9	6.7.444.
15	6.7.46.62.76.
17	6.471.727.
21	6.7.54.64.534.724.
23	6.5343.6165.
25	6.76.4102041.
27	6.7.444.4004004.
31	6.45.51.57.67.73.75.
33	6.7.4522.6106.7776.
35	6.54.64.76.57134.72364.
39	6.7.57074.74364.77774.
41	6.5747175.6647133.
43	6.47771.52225.64213.
45	6.7.46.62.76.444.40044.44004.
47	6.43073357.75667061.
49	6.54.64.40001004.40200004.
51	6.7.433.471.637.661.727.763.
55	6.76.7776.5551347.7164555.
57	6.7.5604164.7565674.7777774.
63	6.7.54.64.414.444.534.554.604.634.664.714.724.
127	6.406.422.436.442.472.516.526.562.576.602.626.646.652.712.736.742.756.772.

Fig. 7.1. Factors of $1 + x^n$ over GF(2).

(16) Show that if

$$h(x) = \prod_i (x - \alpha_i)$$

is a divisor of $x^n - 1$ over $GF(q)$, then

$$x^{\deg h(x)} h(x^{-1}) = \text{constant} \cdot \prod_i (x - \alpha_i^{-1}).$$

(17) Show that properties (M1)–(M6) given in Ch. 4 hold for this more general definition of minimal polynomials.

The Zeros of a code. Let $\mathscr{C}$ be a cyclic code with generator polynomial $g(x)$. Since $g(x)$ is a divisor of $x^n - 1$ over $GF(q)$ we have

$$g(x) = \prod_{i \in K} (x - \alpha^i), \tag{15}$$

where (by Problem 15) $i \in K \Rightarrow qi \pmod{n} \in K$. K is a union of cyclotomic cosets. The nth roots of unity $\{\alpha^i : i \in K\}$ are called the *zeros of the code*. (Naturally the other n^{th} roots of unity are called the *nonzeros* – these are the zeros of $h(x) = (x^n - 1)/g(x)$.)

Clearly if $c(x) \in R_n$ then $c(x)$ belongs to $\mathscr{C}$ iff $c(\alpha^i) = 0$ for all $i \in K$. Thus a cyclic code is defined in terms of the zeros of $c(x)$. By Problem 16, the zeros of the dual code are the inverses of the nonzeros of the original code. I.e. if $\mathscr{C}$ has zeros α^i where i runs through $C_{u_1}, C_{u_2}, \ldots$, then $\mathscr{C}^\perp$ has nonzeros α^j where j runs through $C_{-u_1}, C_{-u_2}, \ldots$.

Up to now we have taken the generator polynomial of a code to be $g(x) = $ the lowest degree monic polynomial in the code. But other generators are possible.

Lemma 5. *If $p(x) \in R_n$ does not introduce any new zeros, i.e., if $p(\alpha^i) \neq 0$ for all $i \notin K$, then $g(x)$ and $p(x)g(x)$ generate the same code. (E.g. $g(x)^2$ generates the same code as $g(x)$.)*

Proof. Clearly $\langle g(x) \rangle \supseteq \langle p(x)g(x) \rangle$. By hypothesis $p(x)$ and $h(x)$ are relatively prime, so by Corollary 15 of Ch. 12 there exist polynomials $a(x)$, $b(x)$ such that

$$1 = a(x)p(x) + b(x)h(x) \qquad \text{in } F[x].$$

$$\therefore \quad g(x) = a(x)p(x)g(x) + b(x)g(x)h(x) \quad \text{in } F[x],$$

$$= a(x)p(x)g(x) \qquad\qquad\quad \text{in } R_n.$$

$$\therefore \quad \langle g(x) \rangle \subseteq \langle p(x)g(x) \rangle. \qquad\qquad\qquad\qquad\qquad \text{Q.E.D.}$$

Problem. (18) If $\mathscr{C} = \langle f(x) \rangle$ let $K = \{i, 0 \leqslant i \leqslant n - 1 : f(\alpha^i) = 0\}$. Show that $\mathscr{C} = \langle g(x) \rangle$ where

$$g(x) = \text{l.c.m.} \{M^{(i)}(x) : \quad i \in K\}.$$

Lemma 6. *Let $\xi \in GF(q^m)$ be any zero of $x^n - 1$. Then*

$$\sum_{i=0}^{n-1} \xi^i = \begin{cases} 0 & \text{if } \xi \neq 1 \\ n & \text{if } \xi = 1. \end{cases}$$

Proof. If $\xi = 1$, the sum is equal to n. Suppose $\xi \neq 1$. Then

$$\sum_{i=0}^{n-1} \xi^i = (1 - \xi^n)/(1 - \xi) = 0. \qquad\qquad \text{Q.E.D.}$$

Lemma 7. *An inversion formula. The vector $c = (c_0, c_1, \ldots, c_{n-1})$ may be recovered from $c(x) = c_0 + c_1 x + \cdots + c_{n-1} x^{n-1}$ by*

$$c_i = \frac{1}{n} \sum_{j=0}^{n-1} c(\alpha^i) \alpha^{-ij}. \tag{16}$$

Proof.

$$\sum_{j=0}^{n-1} c(\alpha^i) \alpha^{-ij} = \sum_{j=0}^{n-1} \sum_{k=0}^{n-1} c_k \alpha^{j(k-i)}$$

$$= \sum_{k=0}^{n-1} c_k \sum_{j=0}^{n-1} \alpha^{j(k-i)} = n c_i \quad \text{by Lemma 6.} \qquad \text{Q.E.D.}$$

Problems. (19) Let A be the following Vandermonde matrix over $GF(2^m)$ (cf. Lemma 17 of Ch. 4):

$$A = \begin{pmatrix} 1 & 1 & & 1 \\ \alpha_0 & \alpha_1 & \cdots & \alpha_{n-1} \\ \alpha_0^2 & \alpha_1^2 & \cdots & \alpha_{n-1}^2 \\ \cdots\cdots\cdots\cdots\cdots \\ \alpha_0^{n-1} & \alpha_1^{n-1} & \cdots & \alpha_{n-1}^{n-1} \end{pmatrix},$$

where the α_i are the nth roots of unity. Show that $\det (A) = 1$ and

$$A^{-1} = \begin{pmatrix} 1 & \alpha_0^{-1} & \cdots & \alpha_0^{-(n-1)} \\ 1 & \alpha_1^{-1} & \cdots & \alpha_1^{-(n-1)} \\ \cdots\cdots\cdots\cdots\cdots \\ 1 & \alpha_{n-1}^{-1} & \cdots & \alpha_{n-1}^{-(n-1)} \end{pmatrix}.$$

(20) Let $c(x) = c_0 + c_1 x + \cdots + c_{n-1} x^{n-1} \in R_n$ and

$$M = \begin{pmatrix} c_0 & c_1 & \cdots & c_{n-1} \\ c_{n-1} & c_0 & \cdots & c_{n-2} \\ \cdots\cdots\cdots\cdots\cdots\cdots\cdots \\ c_1 & c_2 & \cdots & c_0 \end{pmatrix}.$$

Show that $A^{-1}MA = \text{diag}\,[c(\alpha_0), c(\alpha_1), \dots, c(\alpha_{n-1})]$. (See also Problem 7 of Ch. 16.) Hence give another proof that $\dim \mathscr{C} = n - \deg g(x)$.

(21) Show that **1** is in a cyclic code iff $g(1) \neq 0$. Show that a binary cyclic code contains a codeword of odd weight iff it contains **1**.

§6. t-Error-correcting BCH codes

Theorem 8. (The BCH bound.) *Let $\mathscr{C}$ be a cyclic code with generator polynomial $g(x)$ such that for some integers $b \geq 0$, $\delta \geq 1$,*

$$g(\alpha^b) = g(\alpha^{b+1}) = \cdots = g(\alpha^{b+\delta-2}) = 0.$$

I.e. the code has a string of $\delta - 1$ consecutive powers of α as zeros. Then the minimum distance of the code is at least δ.

Proof. If $c = (c_0, c_1, \dots, c_{n-1})$ is in $\mathscr{C}$ then

$$c(\alpha^b) = c(\alpha^{b+1}) = \cdots = c(\alpha^{b+\delta-2}) = 0,$$

so that $H'c^T = 0$ where

$$H' = \begin{pmatrix} 1 & \alpha^b & \alpha^{2b} & \cdots & \alpha^{(n-1)b} \\ 1 & \alpha^{b+1} & \alpha^{2(b+1)} & \cdots & \alpha^{(n-1)(b+1)} \\ \cdots\cdots\cdots\cdots\cdots\cdots\cdots\cdots\cdots\cdots \\ 1 & \alpha^{b+\delta-2} & & \cdots & \alpha^{(n-1)(b+\delta-2)} \end{pmatrix}.$$

(Note that H' need not be the full parity check matrix of $\mathscr{C}$.) The idea of the proof is to show that any $\delta - 1$ or fewer columns of H' are linearly independent over $GF(q^m)$. Suppose c has weight $w \leq \delta - 1$, i.e., $c_i \neq 0$ iff $i \in \{a_1, a_2, \dots, a_w\}$. Then $H'c^T = 0$ implies

$$\begin{pmatrix} \alpha^{a_1 b} & \cdots & \alpha^{a_w b} \\ \alpha^{a_1(b+1)} & \cdots & \alpha^{a_w(b+1)} \\ \cdots\cdots\cdots\cdots\cdots\cdots \\ \alpha^{a_1(b+w-1)} & \cdots & \alpha^{a_w(b+w-1)} \end{pmatrix} \begin{pmatrix} c_{a_1} \\ \vdots \\ c_{a_w} \end{pmatrix} = 0.$$

Hence the determinant of the matrix on the left is zero. But this determinant is equal to $\alpha^{(a_1+\cdots+a_w)b}$ times

$$\det \begin{pmatrix} 1 & \cdots & 1 \\ \alpha^{a_1} & \cdots & \alpha^{a_w} \\ \cdots\cdots\cdots\cdots\cdots\cdots \\ \alpha^{a_1(w-1)} & \cdots & \alpha^{a_w(w-1)} \end{pmatrix},$$

which is a Vandermonde determinant and is nonzero by Lemma 17 of Ch. 4, a contradiction.   Q.E.D.

Examples. The binary Hamming code $\mathcal{H}_m$ has generator polynomial $M^{(1)}(x)$ (see §3). Now $M^{(1)}(\alpha) = M^{(1)}(\alpha^2) = 0$ (by (M6) of Ch. 4), so the code has a pair of consecutive zeros, α and α^2. Hence by Theorem 8 the minimum distance is ≥ 3.

The binary double-error-correcting BCH code has $g(x) = M^{(1)}(x)M^{(3)}(x)$. Now $M^{(1)}(\alpha) = M^{(1)}(\alpha^2) = M^{(1)}(\alpha^4) = 0$ and $M^{(3)}(\alpha^3) = 0$. Therefore there are 4 consecutive zeros: $\alpha, \alpha^2, \alpha^3, \alpha^4$; and thus $d \geq 5$, in agreement with Ch. 3.

The minimum distance of $\mathscr{C}$ may in fact be greater than δ, for more than $\delta - 1$ columns of H' may be linearly independent when the entries are replaced by m-tuples from $\mathrm{GF}(q)$.

Corollary 9. *A cyclic code of length n with zeros $\alpha^b, \alpha^{b+r}, a^{b+2r}, \ldots, \alpha^{b+(\delta-2)r}$, where r and n are relatively prime, has minimum distance at least δ.*

Proof. Let $\beta = \alpha^r$. Since r and n are relatively prime, β is also a primitive n^{th} root of unity. Therefore $\alpha^b = \beta^t$ for some t, and the codes has zeros $\beta^t, \beta^{t+1}, \ldots, \beta^{t+\delta-2}$. The result follows from the proof of Theorem 8 with β replaced by α.

BCH codes.

Definition. A cyclic code of length n over $\mathrm{GF}(q)$ is a *BCH code of designed distance δ* if, for some integer $b \geq 0$,

$$g(x) = \text{l.c.m.} \{M^{(b)}(x), M^{(b+1)}(x), \ldots, M^{(b+\delta-2)}(x)\}. \tag{17}$$

I.e. $g(x)$ is the lowest degree monic polynomial over $\mathrm{GF}(q)$ having $\alpha^b, \alpha^{b+1}, \ldots, \alpha^{b+\delta-2}$ as zeros. Therefore

$$c \text{ is in the code iff } c(\alpha^b) = c(\alpha^{b+1}) = \cdots = c(\alpha^{b+\delta-2}) = 0. \tag{18}$$

Thus the code has a string of $\delta - 1$ consecutive powers of α as zeros. From Theorem 8 we deduce that the minimum distance is greater than or equal to the designed distance δ.

Equation (18) also shows that a parity check matrix for the code is

$$H = \begin{pmatrix} 1 & \alpha^b & \alpha^{2b} & \cdots & \alpha^{(n-1)b} \\ 1 & \alpha^{b+1} & \alpha^{2(b+1)} & \cdots & \alpha^{(n-1)(b+1)} \\ \cdots\cdots\cdots\cdots\cdots\cdots\cdots\cdots\cdots\cdots \\ 1 & \alpha^{b+\delta-2} & \cdots & & \alpha^{(n-1)(b+\delta-2)} \end{pmatrix}, \tag{19}$$

where each entry is replaced by the corresponding column of m elements from GF(q).

After this replacement the rows of the resulting matrix over GF(q) are the parity checks satisfied by the code. There are $m(\delta - 1)$ of these, but they need not all be linearly independent. Thus the dimension of the code is at least $n - m(\delta - 1)$. For a second proof, from (M4) of Ch. 4 deg $M^{(i)}(x) \le m$, hence deg $g(x) = n -$ dimension of code $\le m(\delta - 1)$. Thus we have proved:

Theorem 10. *A BCH code over* GF(q) *of length n and designed distance δ has minimum distance $d \ge \delta$, and dimension $\ge n - m(\delta - 1)$. (m was defined in §5.)*

The dimension will be greater than this if some of the rows of the GF(q) version of H are linearly dependent, or (equivalently) if the degree of the RHS of (17) is less than $m(\delta - 1)$. Examples of this are given below.

A generator matrix and an alternative form for the parity check matrix are given by Equations (1) and (9) respectively.

Remarks. (1) If $b = 1$ these are sometimes called *narrow sense* BCH codes. If $n = q^m - 1$ they are called *primitive*, for then α is a primitive element of the field GF(q^m) (and not merely a primitive n^{th} root of unity).

If some α^i is a zero of the code then so are all α^l, for l in the cyclotomic coset C_i. Since the cyclotomic cosets are smallest if $n = q^m - 1$, this is the most important case.

(2) If b is fixed, BCH codes are *nested*. I.e., the code of designed distance δ_1 contains the code of designed distance δ_2 iff $\delta_1 \le \delta_2$.

(3) In general the dual of a BCH code is not a BCH code.

Binary BCH codes. When $q = 2$, by property (M6) of Ch. 4,

$$M^{(2i)}(x) = M^{(i)}(x),$$

and so the degree of $g(x)$ can be reduced. For example if $b = 1$ we may always assume that the designed distance δ is odd. For the codes with designed distance $2t$ and $2t + 1$ coincide – both have

$$g(x) = \text{l.c.m.} \{M^{(1)}(x), M^{(3)}(x), \ldots, M^{(2t-1)}(x)\} \tag{20}$$

Thus deg $g(x) \le mt$, and the dimension of the code is $\ge n - mt$. The parity check matrix is

$$H = \begin{pmatrix} 1 & \alpha & \alpha^2 & \cdots & & \alpha^{n-1} \\ 1 & \alpha^3 & \alpha^6 & \cdots & & \alpha^{3(n-1)} \\ \cdots\cdots\cdots\cdots\cdots\cdots\cdots\cdots\cdots\cdots \\ 1 & \alpha^{2t-1} & & & & \alpha^{(2t-1)(n-1)} \end{pmatrix}, \qquad (21)$$

where each entry is replaced by the corresponding binary m-tuple. Of course the second column of H need only contain $\alpha, \alpha^{i_1}, \alpha^{i_2}, \ldots$ where $1, i_1, i_2, \ldots$ are in different cyclotomic cosets.

Examples. We begin by listing in Figs. 7.2, 7.3 all the (narrow-sense, primitive) binary BCH codes of length 15 and 31. Fortunately the minimal polynomials of the elements of GF(16) and GF(32) were given in §4 of Ch. 4.

designed distance δ	generator polynomial $g(x)$	exponents of roots of $g(x)$	dimension $= n - \deg g(x)$	actual distance d
1	1	–	15	1
3	$M^{(1)}(x)$	1, 2, 4, 8	11	3
5	$M^{(1)}(x)M^{(3)}(x)$	1–4, 6, 8, 9, 12	7	5
7	$M^{(1)}(x)M^{(3)}(x)M^{(5)}(x)$	1–6, 8–10, 12	5	7
9, 11, 13 or 15	$M^{(1)}M^{(3)}M^{(5)}M^{(7)}$ $= (x^{15}+1)/(x+1)$	1–14	1	15

Fig. 7.2 BCH codes of length 15.

designed distance δ	generator polynomial $g(x)$	dimension $= n - \deg g(x)$	actual distance d
1	1	31	1
3	$M^{(1)}$	26	3
5	$M^{(1)}M^{(3)}$	21	5
7	$M^{(1)}M^{(3)}M^{(5)}$	16	7
9 or 11	$M^{(1)}M^{(3)}M^{(5)}M^{(7)}$	11	11
13 or 15	$M^{(1)}M^{(3)}M^{(5)}M^{(7)}M^{(11)}$	6	15
17, 19, ..., 31	$M^{(1)}M^{(3)}M^{(5)}M^{(7)}M^{(11)}M^{(15)}$	1	31

Fig. 7.3. BCH codes of length 31.

Note that the codes of designed distance 9 and 11 coincide. This is because the latter code has generator polynomial

$$g(x) = \text{l.c.m.} \{M^{(1)}(x), M^{(3)}(x), M^{(5)}(x), M^{(7)}(x), M^{(9)}(x)\}.$$

But $9 \in C_5$, so the minimal polynomials of α^9 and α^5 are the same: $M^{(9)}(x) = M^{(5)}(x)$. Therefore

$$g(x) = M^{(1)}(x)M^{(3)}(x)M^{(5)}(x)M^{(7)}(x)$$

which is also the generator polynomial of the code of designed distance 9.

This example shows that a BCH code of designed distance δ may coincide with a BCH code of designed distance δ', where $\delta' > \delta$. The largest such δ' is called the *Bose distance* of the code. From the BCH bound the true minimum distance is at least equal to δ', but may be greater, as the next examples show.

Finally Fig. 7.4 gives the binary (nonprimitive) BCH codes of length $n = 23$. The cyclotomic cosets are

$$C_0 = \{0\},$$
$$C_1 = \{1, 2, 4, 8, 16, 9, 18, 13, 3, 6, 12\},$$
$$C_5 = \{5, 10, 20, 17, 11, 22, 21, 19, 15, 7, 14\}.$$

Since $|C_1| = 11$, the multiplicative order of 23 mod 2 is 11 (see §5). Thus $x^{23} + 1$ splits into linear factors over $GF(2^{11})$, and α is a primitive 23rd root of unity in $GF(2^{11})$.

Over $GF(2)$, $x^{23} + 1$ factors into

$$x^{23} + 1 = (x + 1)M^{(1)}(x)M^{(5)}(x),$$

see Fig. 7.1, where

$$M^{(1)}(x) = x^{11} + x^9 + x^7 + x^6 + x^5 + x + 1,$$
$$M^{(5)}(x) = x^{11} + x^{10} + x^6 + x^5 + x^4 + x^2 + 1.$$

The BCH code with designed distance $\delta = 5$ (and $b = 1$) has generator polynomial

$$g(x) = \text{l.c.m.}\{M^{(1)}(x), M^{(3)}(x)\}.$$

But $M^{(1)}(x) = M^{(3)}(x)$. Therefore $g(x) = M^{(1)}(x)$, and the parity check matrix is

$$H = (1, \alpha, \alpha^2, \ldots, \alpha^{22}),$$

where each entry is replaced by a binary column of length 11. The dimension of the code is $23 - \deg g(x) = 12$.

Figure 7.4 shows that the Bose distance of this code is also 5. However, as we shall see in Ch. 20, this BCH code is equivalent to the Golay code $\mathcal{G}_{23}$, and has minimum distance 7. Thus here also the minimum distance is greater than the designed distance, illustrating the fact that the BCH bound is not tight.

designed distance δ	generator polynomial $g(x)$	dimension $= n - \deg g(x)$	actual distance d
1	1	23	1
3 or 5	$M^{(1)}$	12	7
$7, 9, \ldots, 23$	$M^{(1)}M^{(5)}$	1	23

Fig. 7.4. BCH codes of length 23.

We return to BCH codes in Ch. 9.

Problems. (22) Using a table of cyclotomic cosets for $n = 63$, find the dimensions of the BCH codes for $\delta = 3, 5, 7, 9, 11$. Do the same for $n = 51$.

(23) Use the BCH bound to show that the $[2^m - 1, m]$ simplex code with $h(x) = M^{(1)}(x)$ has $d = 2^{m-1}$.

(24) (Hard) *Hartmann and Tzeng's generalization of the BCH bound.* Suppose that, for some integers c_1 and c_2 relatively prime to n, a cyclic code has zeros $\alpha^{b+i_1 c_1 + i_2 c_2}$ for all $i_1 = 0, 1, \ldots, \delta - 2$ and $i_2 = 0, 1, \ldots, s$. Show that the minimum distance is at least $\delta + s$. (Thus if $c_1 = 1$ the code has $s + 1$ strings of $\delta - 1$ consecutive zeros. The BCH bound is the special case $s = 0$.)

Reversible codes.

Definition. A code $\mathscr{C}$ is *reversible* if $(c_0, c_1, \ldots, c_{n-2}, c_{n-1}) \in \mathscr{C}$ implies $(c_{n-1}, c_{n-2}, \ldots, c_1, c_0) \in \mathscr{C}$. For example $\{000, 110, 101, 011\}$ is a reversible code. So is the $[15, 6, 6]$ binary BCH code of length 15 with $g(x) = M^{(-1)}(x)M^{(0)}(x)M^{(1)}(x)$.

Problems. (25) Show that the BCH code with $b = -t$ and designed distance $\delta = 2t + 2$ is reversible.

(26) Show that a cyclic code is reversible iff the reciprocal of every zero of $g(x)$ is also a zero of $g(x)$.

(27) Show that if -1 is a power of $q \bmod n$ then every cyclic code over $GF(q)$ of length n is reversible.

(28) Melas's double-error-correcting codes. Show, using Problem 24, that if m is odd the $[n = 2^m - 1, k = n - 2m]$ reversible binary code with $g(x) = M^{(-1)}(x)M^{(1)}(x)$ has $d \geq 5$.

(29) Zetterberg's double-error-correcting codes. Let $n = 2^{2i} + 1$ $(i > 1)$ and let $\alpha \in GF(2^{4i})$ be a primitive n^{th} root of unity. Again using Problem 24, show that the code with $g(x) = M^{(1)}(x)$ is a reversible code with $d \geq 5$.

(30) (Korzhik) Show that if $h(x)$ is divisible by $x^T - 1$, then the minimum distance is at most n/T.

*§7. Using a matrix over GF(q^m) to define a code over GF(q)

This section studies more carefully how a matrix over the big field GF(q^m) can be used to define a code over the little field GF(q).

First suppose the code is to be defined by a parity check matrix H over GF(q^m). More precisely, let $H = (H_{ij})$, where $H_{ij} \in$ GF(q^m) for $1 \le i \le r$, $1 \le j \le n$, be an $r \times n$ matrix of rank r over GF(q^m). Then let $\mathscr{C}_H$ be the code over GF(q) consisting of all vectors $a = (a_1, \ldots, a_n)$, $a_i \in$ GF(q), such that $Ha^T = 0$.

Another way of getting $\mathscr{C}_H$ is as follows. Pick a basis $\alpha_1, \ldots, \alpha_m$ for GF(q^m) over GF(q), and write

$$H_{ij} = \sum_{l=1}^{m} H_{ijl}\alpha_l, \ H_{ijl} \in \text{GF}(q).$$

Define $\bar{H}$ to be the $rm \times n$ matrix obtained from H by replacing each entry H_{ij} by the corresponding column vector $(H_{ij1}, \ldots, H_{ijm})^T$ from GF(q). Thus

$$\bar{H} = \begin{pmatrix} H_{111} & H_{121} & \cdots & H_{1n1} \\ H_{112} & H_{122} & \cdots & H_{1n2} \\ \cdots\cdots\cdots\cdots\cdots\cdots\cdots \\ H_{11m} & H_{12m} & \cdots & H_{1nm} \\ \cdots & & \cdots & \cdots \\ H_{r1m} & H_{r2m} & \cdots & H_{rnm} \end{pmatrix}.$$

Then

$$a \in \mathscr{C}_H \text{ iff } \sum_{j=1}^{n} H_{ij}a_j = 0 \quad \text{for } i = 1, \ldots, r$$

$$\text{iff } \sum_{j=1}^{n} H_{ijl}a_j = 0 \quad \text{for } i = 1, \ldots, r; \ l = 1, \ldots, m$$

$$\text{iff } \bar{H}a^T = 0.$$

Thus either H or $\bar{H}$ can be used to define $\mathscr{C}_H$. The rank of $\bar{H}$ over GF(q) is at most rm, so $\mathscr{C}_H$ is an $[n, k \ge n - rm]$ code, assuming $rm \le n$.

Of course we could also consider the code $\mathscr{C}_H^\#$ over GF(q^m) consisting of all vectors $b = (b_1, \ldots, b_n)$, $b_i \in$ GF(q^m), such that $Hb^T = 0$. Then $\mathscr{C}_H^\#$ is an $[n, n - r]$ code over GF(q^m). Since GF(q) $\subset$ GF(q^m), every codeword in $\mathscr{C}_H$ is in $\mathscr{C}_H^\#$. In fact $\mathscr{C}_H$ consists of exactly those codewords of $\mathscr{C}_H^\#$ which have components from GF(q). We will denote this by writing

$$\mathscr{C}_H = \mathscr{C}_H^\# \,|\, \text{GF}(q), \tag{22}$$

and call $\mathscr{C}_H$ a *subfield subcode* of $\mathscr{C}_H^\#$.

In general, if $\mathscr{C}^*$ is any $[n, k^*, d^*]$ code over $GF(q^m)$, the *subfield subcode* $\mathscr{C}^*|GF(q)$ consists of the codewords of $\mathscr{C}^*$ which have components from $GF(q)$. Then $\mathscr{C}^*|GF(q)$ is an $[n, k, d]$ code with $n - m(n - k^*) \leqslant k \leqslant k^*$ and $d \geqslant d^*$.

For example, let $\mathscr{C}^*$ be the $[7, 6, 2]$ BCH code over $GF(2^3)$ with generator polynomial $x + \alpha$, where $\alpha \in GF(2^3)$ satisfies $\alpha^3 + \alpha + 1 = 0$. Let $\mathscr{C}$ be the subfield subcode $\mathscr{C}^*|GF(2)$. The codeword $a(x) = (x + \alpha)(x + \alpha^2)(x + \alpha^4) = x^3 + x + 1$ is in $\mathscr{C}^*$ and hence in $\mathscr{C}$. Thus $\mathscr{C}$ contains the $[7, 4, 3]$ code $\mathscr{H}_3$. In fact $\mathscr{C} = \mathscr{H}_3$, since $\mathscr{C}$ has minimum distance at least 2.

The trace mapping T_m from $GF(q^m)$ to $GF(q)$ can be used to express the dual of $\mathscr{C}^*|GF(q)$ in terms of the dual of $\mathscr{C}^*$. This mapping is defined by

$$T_m(x) = x + x^q + x^{q^2} + \cdots + x^{q^{m-1}}, \quad x \in GF(q^m),$$

– see §8 of Ch. 4. Let $T_m(\mathscr{C}^*)$ be the code over $GF(q)$ consisting of all distinct vectors

$$T_m(b) = (T_m(b_1), \ldots, T_m(b_n)), \quad \text{for } b \in \mathscr{C}^*. \tag{23}$$

Then $T_m(\mathscr{C}^*)$ is an $[n, k, d]$ code over $GF(q)$ with $k^* \leqslant k \leqslant mk^*$ and $d \leqslant d^*$.

Theorem 11. (Delsarte.) *The dual of a subfield subcode is the trace of the dual of the original code, or*

$$(\mathscr{C}^* \,|\, GF(q))^{\perp} = T_m((\mathscr{C}^*)^{\perp}). \tag{24}$$

Proof. (i) $T_m((\mathscr{C}^*)^{\perp}) \subset (\mathscr{C}^* \,|\, GF(q))^{\perp}$. To prove this, if $a \in \text{LHS}$ we must show that $a \cdot c = 0$ for all $c \in \mathscr{C}^* \,|\, GF(q)$. In fact $a = (T_m(b_1), \ldots, T_m(b_n))$ for $b \in (\mathscr{C}^*)^{\perp}$. Therefore

$$a \cdot c = \sum_{i=1}^{n} T_m(b_i)c_i = T_m\left(\sum_{i=1}^{n} b_i c_i\right) = T_m(0) = 0.$$

(ii) $(\mathscr{C}^*|GF(q))^{\perp} \subset T_m((\mathscr{C}^*)^{\perp})$, or equivalently

$$(T_m(\mathscr{C}^*)^{\perp})^{\perp} \subset \mathscr{C}^* \,|\, GF(q) \tag{25}$$

To prove (25), if $a \in \text{LHS}$ we must show that $a \in \mathscr{C}^*$. By definition $a \cdot c = 0$ for all $c = (T_m(b_1), \ldots, T_m(b_n))$ where $b \in (\mathscr{C}^*)^{\perp}$. If $b \in (\mathscr{C}^*)^{\perp}$ so is λb for all $\lambda \in GF(q^m)$. Therefore

$$\sum_{i=1}^{n} a_i T_m(\lambda b_i) = T_m\left(\lambda \sum_{i=1}^{n} a_i b_i\right) = 0 \quad \text{for all } \lambda \in GF(q^m),$$

so

$$\sum_{i=1}^{n} a_i b_i = 0 \quad \text{and} \quad a \in \mathscr{C}^*. \qquad \text{Q.E.D.}$$

For example the dual of the above $[7, 6, 2]$ code has generator polynomial $(x^7 + 1)/(x + \alpha^6) = x^6 + \alpha^6 x^5 + \alpha^5 x^4 + \alpha^4 x^3 + \alpha^3 x^2 + \alpha^2 x + \alpha$. The trace of this is $x^6 + x^5 + x^4 + x^2$, and in this way we obtain the $[7, 3, 4]$ code $\mathscr{H}_3^{\perp}$.

Problems. (31) Let C be any invertible $r \times r$ matrix over $GF(q^m)$ and set $H_1 = CH$. Show that $\mathscr{C}_{H_1} = \mathscr{C}_H$; i.e. CH and H define the same code.

(32) As a converse to the preceding, let $\mathscr{C}$ be any $[n, k]$ code over $GF(q)$. Show that if $rm \geq n - k$ there is an $r \times n$ matrix H over $GF(q^m)$ such that $\mathscr{C} = \mathscr{C}_H$.

(33) (a) Let $\mathscr{C}^*$ be a cyclic code over $GF(q^m)$ with zeros α^i for $i \in K$, where $\alpha \in GF(q^s)$. Show that $\mathscr{C}^* | GF(q)$ is a cyclic code with zeros $\alpha^i, \alpha^{iq}, \alpha^{iq^2}, \ldots, \alpha^{iq^{s-1}}$ for $i \in K$.

(b) As an illustration of (a), take $\mathscr{C}^*$ to be the $[7, 5]$ cyclic code over $GF(2^3)$ with generator polynomial $(x + \alpha)(x + \alpha^2)$, where α is a primitive element of $GF(2^3)$. Show that $\mathscr{C}^* | GF(2) = \mathscr{H}_3$.

§8. Encoding cyclic codes

In this section two encoding circuits are described which can be used for any cyclic code. We illustrate the technique by two examples.

Examples. (E1) The $[15, 11, 3]$ Hamming code $\mathscr{H}_4$, a cyclic code with generator polynomial $g(x) = x^4 + x + 1$.

(E2) The $[15, 7, 5]$ double-error-correcting BCH code with

$$g(x) = (x^4 + x + 1)(x^4 + x^3 + x^2 + x + 1)$$
$$= x^8 + x^7 + x^6 + x^4 + 1.$$

Suppose the message $u = u_0, u_1, \ldots, u_{10}$ is to be encoded by code (E1), and the corresponding codeword is

$$c = \underbrace{c_0, \ldots, c_3}_{\substack{\text{check} \\ \text{symbols}}}, \underbrace{c_4, \ldots, c_{14}}_{\substack{\text{message} \\ \text{symbols}}}$$

See Fig. 7.5.

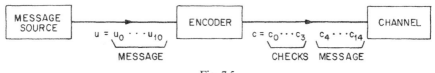

Fig. 7.5.

Encoder #1. The first encoder requires $\deg g(x)$ delay elements. c is in the code iff the polynomial $c(x)$ is divisible by $g(x) = x^4 + x + 1$. So we must choose $c_0, \ldots, c_3$ to make this happen. One way to do this is to divide

$$c'(x) = c_{14} x^{14} + c_{13} x^{13} + \cdots + c_4 x^4$$

by $x^4 + x + 1$, giving a remainder $r(x) = r_3x^3 + \cdots + r_0$. Then set $c_i = r_i$ ($i = 0, \ldots, 3$), for $c(x) = c'(x) + r(x)$ *is* divisible by $x^4 + x + 1$.

To implement this, a circuit is needed which divides by $x^4 + x + 1$. A simple example will show how to construct such a circuit. Suppose we divide

$$x^9 + x^8 + x^5 + x^4 \quad \text{by} \quad x^4 + x + 1,$$

using detached coefficients. (I.e. we write 10011 instead of $x^4 + x + 1$, etc.)

$$
\begin{array}{r}
110110 = \text{quotient} \\
\overline{10011)1100110000} = \text{dividend} \\
10011 \\
\overline{10101} \\
10011 \\
\overline{01100} \\
00000 \\
\overline{11000} \\
10011 \\
\overline{10110} \\
10011 \\
\overline{01010} \\
00000 \\
\overline{1010} = \text{remainder}
\end{array}
$$

The quotient is $x^5 + x^4 + x^2 + x$ and the remainder $r(x)$ is $x^3 + x$.

The key point to observe is that each time there is a 1 in the quotient, the dividend is changed 3 and 4 places back.

Therefore the circuit shown in Fig. 7.6 performs the same calculation.

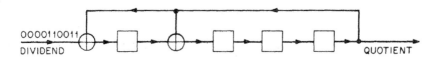

Fig. 7.6. A circuit to divide by $x^4 + x + 1$.

The remainder ($0101 = x + x^3$) is what is left in the register when the dividend has been completely fed in.

So our first attempt at encoding is: Feed in the dividend (message symbols followed by zeros)

$$0 \ 0 \ 0 \ 0 \ c_4, c_5, \ldots, c_{14}.$$

The remainder when all 15 have been fed in is

$$c_0 c_1 c_2 c_3.$$

A circuit to do this is shown in Fig. 7.7.

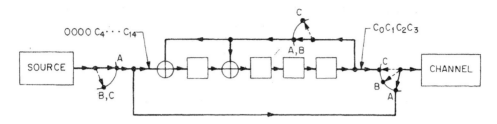

Fig. 7.7. Preliminary version of encoder #1.

The switches have three positions: at A for 11 clock cycles, during which time the message is fed into the channel and into the register; at B for 4 cycles, while 4 zeros enter the register; and at C for 4 cycles, while the remainder enters the channel.

The disadvantage of this scheme is obvious: the channel is idle while the switches are at B.

To overcome this difficulty, we feed the message into the right-hand end of the shift register. This has the effect of premultiplying the symbols by x^4 as they come in. So instead of the divisor circuit of Fig. 7.6 we use that of Fig. 7.8.

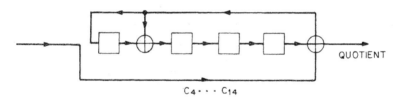

Fig. 7.8. Division circuit, with premultiplication by x^4.

The remainder is now available in the register as soon as c_4 has been fed in. The final encoder is shown in Fig. 7.9. The switches are at A for 11 cycles, and B for 4 cycles.

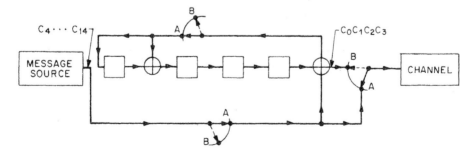

Fig. 7.9. Final version of encoder #1.

It is clear that a similar encoder will work for any cyclic code, and requires deg $g(x)$ delay elements in the shift register. Figure 7.10 shows the encoder for Example (E2).

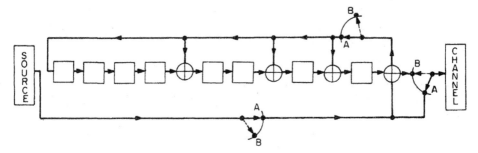

Fig. 7.10. Encoder #1 for a BCH code.

Encoder #2. The second encoder requires deg $h(x)$ delay elements. We saw above that the check symbols are defined by Equations (11).

For Example (E2),

$$h(x) = (x^{15} + 1)/g(x)$$
$$= (x + 1)(x^2 + x + 1)(x^4 + x^3 + 1)$$
$$= x^7 + x^6 + x^4 + 1.$$

So the codeword satisfies

$$c_7 + c_8 + c_{10} + c_{14} = 0,$$
$$c_6 + c_7 + c_9 + c_{13} = 0,$$
$$\dots\dots\dots\dots\dots\dots$$

If $c_{14}, \dots, c_8$ are the message symbols, this defines the check symbols $c_7, c_6, \dots, c_0$. Figure 7.11 shows the encoder to do this. The switch is

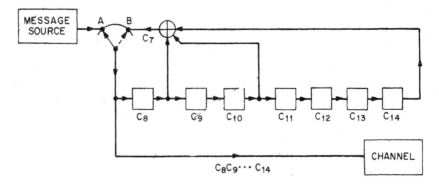

Fig. 7.11. Encoder #2 for a BCH code.

at A for 7 cycles, at B for 8 cycles. The circuit is shown immediately after the last message symbol, c_8, has been fed in, and the first check symbol

$$c_7 = c_8 + c_{10} + c_{14}$$

is being calculated.

Clearly Encoder #2 will work for any cyclic code, and requires deg $h(x) = n - \deg g(x)$ delay elements. Often one chooses that encoder having the smaller number of delay elements.

Problems. (34) Design Encoder #2 for Example E1.

(35) (a) Show that Encoder #1 corresponds to the following generator matrix

$$G_1 = \begin{bmatrix} a_{00} & a_{01} & \cdots & a_{0r-1} & 1 & & \\ a_{10} & a_{11} & \cdots & a_{1r-1} & & 1 & 0 \\ \multicolumn{5}{c}{\dots\dots\dots\dots\dots\dots} & \\ a_{k-10} & a_{k-11} & \cdots & a_{k-1r-1} & 0 & & 1 \end{bmatrix}, \quad (26)$$

in the sense that if the message $u = u_0 u_1 \cdots u_{k-1} = c_{n-k} c_{n-k-1} \cdots c_{n-1}$ is input to Encoder #1, the codeword $c = u G_1$ is obtained. Here $a_{i0} + a_{i1} x + \cdots + a_{ir-1} x^{r-1}$ is the remainder when x^{r+i} is divided by $g(x)$, for $i = 0, \dots, k-1$. Thus (26) is obtained from (1) by diagonalizing the last k columns. If we write (26) as $G_1 = [A_1 | I]$, the corresponding parity check matrix is $H_1 = [I | A_1^T]$.

(b) Similarly show that Encoder #2 corresponds to the following parity check matrix

$$H_2 = \begin{bmatrix} 1 & & 0 & b_{r-1k-1} & b_{r-1k-2} & \cdots & b_{r-10} \\ & 1. & & b_{r-2k-1} & b_{r-2k-2} & \cdots & b_{r-20} \\ & & \ddots & \multicolumn{4}{c}{\dots\dots\dots\dots\dots\dots} \\ 0 & & \cdot 1 & b_{0k-1} & b_{0k-2} & \cdots & b_{00} \end{bmatrix} \quad (27)$$

$$= [I | B_2] \text{ (say)},$$

and to the generator matrix $G_2 = [B_2^T | I]$, where $b_{ik-1} x^{k-1} + b_{ik-2} x^{k-2} + \cdots + b_{i0}$ is the remainder when x^{k+i} is divided by $h(x)$, for $i = 0, \dots, r-1$. Thus (27) is obtained from (9) by diagonalizing the first r columns.

Calculating the syndrome. Techniques for decoding cyclic codes will be described in §6 of Ch. 9 (BCH codes), §9 of Ch. 12 (alternant codes), §7 of Ch. 13 (RM codes), and §6 of Ch. 16 (cyclic codes in general, and especially quadratic-residue codes). The first decoding step is always to calculate the syndrome which essentially means re-encoding the received vector (see §4 of Ch. 1). If the code is being used for error *detection* (not error correction) this is all the receiver has to do.

Problem. (36) (a) Suppose the received vector $y = y_0 y_1 \cdots y_{n-1}$ is input to Encoder #1. Let $s_0 s_1 \cdots s_{r-1}$ be the contents of the shift register when all n

digits of y have been fed in. Show that $S = (s_0 s_1 \cdots s_{r-1})^T$ satisfies $S = H_1 y^T$ and hence is the syndrome of y. [Hint: write $y(x) = (y_0 + \cdots + y_{k-1} x^{k-1}) + (y_k x^k + \cdots + y_{n-1} x^{n-1}) = y_h(x) + y_t(x)$, and $y_t(x) = q(x) g(x) + f(x)$, then $s(x) = \Sigma s_i x^i = f(x) + y_h(x)$, etc.]

(b) Show that after one additional clock cycle the shift register contains the syndrome corresponding to $xy(x)$, a cyclic shift of y.

(c) If Encoder #2 is used to calculate the syndrome the circuit (e.g. Fig. 7.11) must be modified slightly. Show that Fig. 7.12 will indeed form the syndrome $S = H_2 y^T$. **The switches are in position A for $k = 7$ clock cycles, during which time the (possibly distorted) information symbols are fed into the shift register. Then the switches are in position B for $r = 8$ cycles, and the output gives S.**

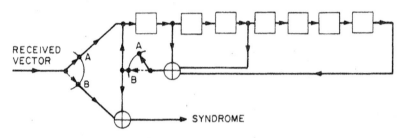

Fig. 7.12. Encoder #2 modified to calculate syndrome.

Notes on Chapter 7

§2. Prange [1074–1077] seems to have been the first to study cyclic codes. See also Abramson [2].

§3. Much more about binary and nonbinary Hamming and related codes can be found in Abramson [1], Azumi and Kasami [55], Bose and Burton [180], Cocke [296], Farrell and Al-Bandar [419], Golay [512], Hsiao [667], Van Lint [848, 855], Lytle [867], Marcovitz [913], Peterson and Weldon [1040], Sankar and Krishnamurthy [1143] and Stirzaker and Yuen [1281].

Two combinatorial problems which are related to perfect codes and especially to nonbinary Hamming codes are the *coin-weighing* and the *football pool* problems – for details see Bellman [101], Golay [510], Kamps and Van Lint [712, 713], Katona [748], Van Lint [855], Stanton [1265], Stanton et al. [1267] and Zaremba [1452, 1453]. See also the covering problem mentioned in §3 of Appendix A.

For shortened cyclic codes see Peterson and Weldon [1040, §8.10] and Tokura and Kasami [1332]. Kasami [728] has shown that these codes meet the Gilbert–Varshamov bound.

§5. The *real* factors of $x^n - 1$ are called *cyclotomic* polynomials – see Berlekamp [113], Kurshan [788], Lehmer [811]. See also [1200]. Problem 20 holds for circulant matrices over any field. See Muir [974, pp. 444–445]. Lemma 7 is a special case of the inversion formula for characters of an Abelian group (cf. Mann [907, p. 75]).

§6. Binary BCH codes were discovered simultaneously by Bose and Ray-Chaudhuri [184, 185] and Hocquenghem [658]. The nonbinary codes are due to Gorenstein and Zierler [552]. Levinson [826] gives a brief description of BCH codes. See also Lum [863]. Problem 24 is from Hartmann and Tzeng [615]; see also Hartmann [608, 610]. The references for reversible codes are Hartmann and Tzeng [615], Hong and Bossen [661], Massey [919], Melas [953], Tzeng and Hartmann [1350] and Zetterberg [1455]. Problem 30 is from Korzhik [778].

BCH codes are not always the best cyclic codes. For example Chen [266] found [63, 45, 8], [63, 24, 16] and [63, 28, 15] cyclic codes which are better than BCH codes (see Fig. 9.1). See also Berlekamp and Justesen [127] and Appendix A.

§7. Theorem 11 is from Delsarte [359].

§8. The encoding techniques follow Peterson and Weldon [1040], who give an excellent treatment of digital circuitry.

8

Cyclic codes (cont.): Idempotents and Mattson–Solomon polynomials

§1. Introduction

This chapter continues the study of cyclic codes begun in Chapter 7. In §2 of that chapter we saw that a cyclic code consists of all multiples of its generator polynomial $g(x)$.

Another useful polynomial in a cyclic code is its idempotent $E(x)$, defined by the property $E(x)^2 = E(x)$ (§2). (It is sometimes easier to find the idempotent than the generator polynomial.) The smallest cyclic codes are the minimal ideals (§3). These are important for several reasons: (i) any cyclic code is a direct sum of minimal ideals (Theorem 7); (ii) a minimal ideal is isomorphic to a field (Theorem 9); (iii) minimal ideals include the important family of simplex codes.

The automorphism group of a code (i.e. the set of all permutations of coordinates which fix the code) is discussed in §5. This group is important for understanding the structure of a code and also for decoding.

A useful device for getting the weight distribution of a code is its Mattson–Solomon (MS) polynomial (§6). The last section uses the MS polynomial to calculate the weight distribution of some cyclic codes.

Throughout this chapter $\mathscr{C}$ is a cyclic code of length n over $F = GF(q)$, where n and q are relatively prime. As in §5 of Ch. 7, m is the multiplicative order of q modulo n, and $\alpha \in GF(q^m)$ is a primitive n^{th} root of unity. Also R_n is the ring $GF(q)[x]/(x^n - 1)$ consisting of all polynomials of degree $\le n - 1$ with coefficients from $GF(q)$.

§2. Idempotents

In Sections 2–4 we restrict ourselves to binary codes of length n, where n is odd.

Definition. A polynomial $E(x)$ of R_n is an *idempotent* if

$$E(x) = E(x)^2 = E(x^2).$$

For example $x + x^2 + x^4$ is an idempotent in R_7 since $(x + x^2 + x^4)^2 = x + x^2 + x^4$. So are 1 and $x^3 + x^6 + x^5$. In general

$$E(x) = \sum_{i=0}^{n-1} \epsilon_i x^i$$

is an idempotent iff $\epsilon_i = \epsilon_{2i}$ (subscripts mod n). Thus the exponents of the nonzero terms are a union of cyclotomic cosets.

Plainly if $E(x)$ is an idempotent so is $1 + E(x)$.

Theorem 1. (i) *A cyclic code or ideal* $\mathscr{C} = \langle g(x) \rangle$ *contains a unique idempotent* $E(x)$ *such that* $\mathscr{C} = \langle E(x) \rangle$. *Also* $E(x) = p(x)g(x)$ *for some polynomial* $p(x)$, *and*

$$E(\alpha^i) = 0 \text{ iff } g(\alpha^i) = 0.$$

(ii) $c(x) \in \mathscr{C}$ *if and only if* $c(x)E(x) = c(x)$.

($\mathscr{C}$ may contain several idempotents, but only one of them generates $\mathscr{C}$.)

Proof. Let $x^n + 1 = g(x)h(x)$, where $g(x)$, $h(x)$ are relatively prime. An easy consequence of the Euclidean algorithm (Corollary 15 of Ch. 12) is that there exist polynomials $p(x), q(x)$ such that

$$p(x)g(x) + q(x)h(x) = 1, \quad \text{in } F[x]. \tag{1}$$

Set $E(x) = p(x)g(x)$. Then from (1)

$$p(x)g(x)[p(x)g(x) + q(x)h(x)] = p(x)g(x),$$

i.e.,

$$E(x)^2 + 0 = E(x) \text{ in } R_n,$$

so $E(x)$ is an idempotent. An nth root of unity is a zero of either g or h but not both. From (1), p and h are relatively prime. So if there is an nth root of unity which is a zero of p, it must also be a zero of g. Since $p(x)$ doesn't introduce any new zeros, by Lemma 5 of Ch. 7 $E(x)$ and $g(x)$ generate the same code. To prove (ii), if $c(x) = c(x)E(x)$ then clearly $c(x) \in \mathscr{C}$. Conversely, if $c(x) \in \mathscr{C}$ then $c(x) = b(x)E(x)$, and $c(x)E(x) = b(x)E(x)^2 = b(x)E(x) =$

$c(x)$. Finally to show that $E(x)$ is the unique idempotent which generates $\mathscr{C}$, suppose $F(x)$ is another such. Then from (ii), $F(x)E(x) = E(x) = F(x)$.

                                                         Q.E.D.

For example, the $[7, 4, 3]$ Hamming code $\mathscr{H}_3$ has $g(x) = x^3 + x + 1$, $h(x) = (x + 1)(x^3 + x^2 + 1) = x^4 + x^2 + x + 1$, and

$$xg(x) + h(x) = 1.$$

Thus the idempotent of this code is $E(x) = xg(x) = x^4 + x^2 + x$.

Problem. (1) Convince yourself that $E(x)$ does generate this code.

Lemma 2. $E(x)$ *is an idempotent iff*

$$E(\alpha^i) = 0 \text{ or } 1 \quad \text{for } i = 0, 1, \ldots, n - 1.$$

Proof. Suppose $E(x)$ is an idempotent. Then $E(\alpha^i)^2 = E(\alpha^i)$, so is equal to 0 or 1 by Theorem 8 of Ch. 4. For the converse, suppose

$$E(x) = \sum_{i=0}^{n-1} \epsilon_i x^i.$$

Since $E(\alpha^i)$ is 0 or 1, $E(\alpha^{2i}) = E(\alpha^i)^2 = E(\alpha^i)$. By the inversion formula of Lemma 7 of Ch. 7,

$$\epsilon_i = \sum_{j=0}^{n-1} E(\alpha^j)\alpha^{-ij} = \sum_{s} \sum_{j \in C_s} \alpha^{-ij},$$

where s runs through a subset of the cyclotomic cosets. Thus $\epsilon_i = \epsilon_{2i}$, and $E(x)$ is an idempotent.   Q.E.D.

Corollary 3. *Dimension of $\mathscr{C}$*

$$= \textit{number of } \alpha^i \textit{ for which } g(\alpha^i) \neq 0,$$
$$= \textit{number of } \alpha^i \textit{ for which } E(\alpha^i) = 1.$$

Problem. (2) Show that

$$g(x) = (E(x), x^n + 1),$$

where (a, b) denotes the greatest common divisor of a and b.

If $a(x) = a_0 + a_1 x + \cdots + a_{n-1} x^{n-1}$ we define

$$a^*(x) = a_0 + a_{n-1} x + \cdots + a_1 x^{n-1}$$

(so that the constant term is unchanged while the other coefficients are reversed). Plainly $a(\alpha^{-1}) = a^*(\alpha)$.

Lemma 4. *If $E(x)$ is an idempotent so is $E^*(x)$.*

Proof. If

$$E(x) = \sum_s \sum_{j \in C_s} x^j$$

then

$$E^*(x) = \sum_s \sum_{j \in C_{-s}} x^j.$$ Q.E.D.

Theorem 5. *Let $\mathscr{C}$ be a code with idempotent $E(x)$. Then $\mathscr{C}^\perp$ has idempotent $(1 + E(x))^*$.*

Proof. Let $\alpha_1, \ldots, \alpha_n$ be the nth roots of unity; suppose that $\alpha_1, \ldots, \alpha_t$ are the zeros of $\mathscr{C}$; i.e., $E(\alpha_i) = 0$ for $1 \leq i \leq t$, $E(\alpha_i) = 1$ for $t + 1 \leq i \leq n$. Then $1 + E(x)$ has zeros $\alpha_{t+1}, \ldots, \alpha_n$, and $(1 + E(x))^*$ has zeros $\alpha_{t+1}^{-1}, \ldots, \alpha_n^{-1}$. These are the zeros of $\mathscr{C}^\perp$ by Theorem 4 of Ch. 7.   Q.E.D.

Problems. (3) Show that $1 + E(x)$ is the idempotent of the code with the generator polynomial $h(x)$.

(4) Find the generator polynomial and idempotent for each cyclic code of length 15, and identify the duals.

§3. Minimal ideals, irreducible codes, and primitive idempotents

A *minimal* ideal is one which does not contain any smaller nonzero ideal. The corresponding cyclic code is called a *minimal* or *irreducible* code, and the idempotent of the ideal is called a *primitive* idempotent. We shall see that every idempotent is a sum of primitive idempotents, and that any vector in R_n can be written uniquely as a sum of vectors from minimal ideals.

The nonzeros of a minimal ideal must be $\{\alpha^i : i \in C_s\}$ for some cyclotomic coset C_s. We denote this minimal ideal by $\mathscr{M}_s$, and the corresponding primitive idempotent by θ_s, and often write $\mathscr{M}_s = \langle \theta_s \rangle$. Thus

$$\theta_s(\alpha^j) = \begin{cases} 1 & \text{if } j \in C_s, \\ 0 & \text{otherwise.} \end{cases} \tag{2}$$

In particular, $\theta_0(x)$ has the single nonzero $x = 1$ and is given by

$$\theta_0(x) = (x^n + 1)/(x + 1) = \sum_{i=0}^{n-1} x^i.$$

We give an explicit construction for $\theta_s(x)$, using the inversion formula of Lemma 7 of Ch. 7.

Theorem 6.

$$\theta_s(x) = \sum_{i=0}^{n-1} \epsilon_i x^i$$

where

$$\epsilon_i = \sum_{j \in C_s} \alpha^{-ij} \quad \text{for } i \geq 0.$$

Proof. From Lemma 7 of Ch. 7,

$$\epsilon_i = \sum_{j=0}^{n-1} \theta_s(\alpha^j) \alpha^{-ij} = \sum_{j \in C_s} \alpha^{-ij}. \qquad \text{Q.E.D.}$$

For example, if $n = 7$ the coefficients of θ_1, θ_3 are

$$\theta_1: \quad \epsilon_i = \alpha^{-i} + \alpha^{-2i} + \alpha^{-4i},$$
$$\theta_3: \quad \epsilon_i = \alpha^{-3i} + \alpha^{-6i} + \alpha^{-5i}.$$

Since $7 = 2^3 - 1$, α is in this case a primitive element of $GF(2^3)$.

Problems. (5) Use the table of $GF(2^3)$ in Fig. 4.5 to show that $\theta_0 = 1 + x + \cdots + x^6$, $\theta_1 = 1 + x + x^2 + x^4$, $\theta_3 = 1 + x^3 + x^5 + x^6$.

(6) If instead $GF(2^3)$ is defined by $\alpha^3 + \alpha^2 + 1 = 0$ show that θ_1 and θ_3 are interchanged.

Thus using a different polynomial to define the field has the effect of relabelling the θ_s's.

Theorem 7. *The primitive idempotents satisfy:*

(i) $\Sigma_s \, \theta_s = 1$.

(ii) $\theta_i \theta_j = 0 \quad$ *if* $i \neq j$.

(iii) R_n *is the direct sum of the minimal ideals generated by the* θ_s. *Thus any vector* $a(x) \in R_n$ *can be written uniquely in the form*

$$a(x) = \sum_s a_s(x),$$

where $a_s(x)$ *is in the ideal generated by* θ_s.

(iv) *If* $E(x)$ *is any idempotent, then for some* $a_s \in GF(2)$, $E(x)$ *can be written as*

$$\sum_s a_s \theta_s.$$

Conversely, any such expression is an idempotent.

Proof. (i)

$$\sum_s \theta_s = \sum_s \sum_{i=0}^{n-1} \left(\sum_{j \in C_s} \alpha^{-ij} \right) x^i$$

$$= \sum_{i=0}^{n-1} x^i \sum_{j=0}^{n-1} \alpha^{-ij}$$

$$= 1 \quad \text{by Lemma 6 of Ch. 7.}$$

(ii) Let $\mathcal{M}_i$, $\mathcal{M}_j$ be the minimal ideals with idempotents θ_i, θ_j. $\mathcal{M}_i \cap \mathcal{M}_j$ is a proper sub-ideal of $\mathcal{M}_i$, hence 0. $\theta_i \theta_j \in \mathcal{M}_i \cap \mathcal{M}_j$, thus $\theta_i \theta_j = 0$.

(iii) From (i),

$$a(x) \cdot 1 = a(x) \sum_s \theta_s = \sum_s a_s(x),$$

where $a_s(x)$ is in the ideal generated by θ_s.

(iv) The nonzeros of $E(x)$ are a union of sets of nonzeros of minimal idempotents. The result follows from Lemma 2 and (2). The converse then follows from (ii).   Q.E.D.

The polynomial $\theta_s^*(x)$ is also a primitive idempotent, and so there is a unique smallest s' such that

$$\theta_s^*(x) = \theta_{s'}(x).$$

Thus $s' \in C_{-s}$. The nonzeros of $\theta_s^*(x)$ are $\{\alpha^i : i \in C_{-s}\}$.

If $n = 2^m - 1$, $\theta_1(x)$ generates the $[2^m - 1, m, 2^{m-1}]$ simplex code $\mathcal{S}_m = \mathcal{H}_m^\perp$, and has weight 2^{m-1}. The nonzero codewords are the $2^m - 1$ cyclic shifts of $\theta_1(x)$. If $n = 2^m - 1$ and s is prime to n, α^s is also a primitive element and so $\theta_s(x)$ generates a code equivalent to $\mathcal{S}_m$. Stated another way, if $h(x)$ is any primitive polynomial, the code with check polynomial $h(x)$ is equivalent to $\mathcal{S}_m$.

Problems. (7) Using the table of GF(2^4) given in Fig. 3.1 of §3.2, show that the primitive idempotents for $n = 15$ are:

$$\theta_0 = \theta_0^* = x^{14} + x^{13} + \cdots + x^2 + x + 1,$$

$$\theta_1 = \theta_7^* = x^{12} + x^9 + x^8 + x^6 + x^4 + x^3 + x^2 + x,$$

$$\theta_7 = \theta_1^* = x^{14} + x^{13} + x^{12} + x^{11} + x^9 + x^7 + x^6 + x^3,$$

$$\theta_3 = \theta_3^* = (x^{14} + x^{13} + x^{12} + x^{11}) + (x^9 + x^8 + x^7 + x^6) + (x^4 + x^3 + x^2 + x),$$

$$\theta_5 = \theta_5^* = (x^{14} + x^{13}) + (x^{11} + x^{10}) + (x^8 + x^7) + (x^5 + x^4) + (x^2 + x).$$

(8) Similarly show that the primitive idempotents for $n = 31$ are (only the

exponents are given):

$$\theta_0 = \theta_0^* = (0, 1, 2, \ldots, 30),$$

$$\theta_1 = \theta_{15}^* = (0, 5, 7, 9, 10, 11, 13, 14, 18, 19, 20, 21, 22, 25, 26, 28),$$

$$\theta_3 = \theta_7^* = (0, 3, 6, 7, 12, 14, 15, 17, 19, 23, 24, 25, 27, 28, 29, 30),$$

$$\theta_5 = \theta_{11}^* = (0, 1, 2, 4, 5, 8, 9, 10, 15, 16, 18, 20, 23, 27, 29, 30),$$

$$\theta_7 = \theta_3^* = (0, 1, 2, 3, 4, 6, 7, 8, 12, 14, 16, 17, 19, 24, 25, 28),$$

$$\theta_{11} = \theta_5^* = (0, 1, 2, 4, 8, 11, 13, 15, 16, 21, 22, 23, 26, 27, 29, 30),$$

$$\theta_{15} = \theta_1^* = (0, 3, 5, 6, 9, 10, 11, 12, 13, 17, 18, 20, 21, 22, 24, 26).$$

The primitive idempotents for $n = 63$ are shown in Fig. 8.1. Here the idempotents are written in octal with lowest degree terms on the left, e.g., $\theta_1 = z + z^2 + z^4 + z^8 + z^{13} + \cdots$.

$$\theta_1 = 3\ 2\ 1\ 0\ 2\ 6\ 2\ 5\ 1\ 1\ 7\ 0\ 1\ 5\ 6\ 3\ 0\ 7\ 2\ 7\ 7$$
$$\theta_3 = 0\ 1\ 2\ 2\ 3\ 1\ 3\ 0\ 1\ 2\ 2\ 3\ 1\ 3\ 0\ 1\ 2\ 2\ 3\ 1\ 3$$
$$\theta_5 = 0\ 4\ 4\ 1\ 6\ 0\ 2\ 7\ 7\ 1\ 2\ 4\ 3\ 1\ 7\ 3\ 5\ 3\ 2\ 3\ 3$$
$$\theta_7 = 0\ 4\ 4\ 0\ 4\ 4\ 0\ 4\ 4\ 0\ 4\ 4\ 0\ 4\ 4\ 0\ 4\ 4\ 0\ 4\ 4$$
$$\theta_9 = 7\ 2\ 3\ 5\ 1\ 6\ 4\ 7\ 2\ 3\ 5\ 1\ 6\ 4\ 7\ 2\ 3\ 5\ 1\ 6\ 4$$
$$\theta_{11} = 0\ 1\ 0\ 3\ 0\ 5\ 1\ 7\ 2\ 1\ 6\ 2\ 2\ 6\ 7\ 3\ 1\ 5\ 2\ 7\ 7$$
$$\theta_{13} = 3\ 7\ 5\ 2\ 6\ 3\ 3\ 5\ 5\ 1\ 1\ 6\ 1\ 3\ 6\ 2\ 4\ 3\ 0\ 2\ 0$$
$$\theta_{15} = 3\ 2\ 3\ 1\ 1\ 2\ 0\ 3\ 2\ 3\ 1\ 1\ 2\ 0\ 3\ 2\ 3\ 1\ 1\ 2\ 0$$
$$\theta_{21} = 3\ 3\ 3\ 3\ 3\ 3\ 3\ 3\ 3\ 3\ 3\ 3\ 3\ 3\ 3\ 3\ 3\ 3\ 3\ 3\ 3$$
$$\theta_{23} = 3\ 3\ 1\ 3\ 2\ 7\ 3\ 6\ 3\ 0\ 5\ 2\ 3\ 7\ 5\ 0\ 1\ 6\ 0\ 4\ 4$$
$$\theta_{27} = 4\ 5\ 6\ 2\ 7\ 1\ 3\ 4\ 5\ 6\ 2\ 7\ 1\ 3\ 4\ 5\ 6\ 2\ 7\ 1\ 3$$
$$\theta_{31} = 3\ 7\ 5\ 3\ 4\ 3\ 1\ 6\ 6\ 0\ 3\ 6\ 2\ 2\ 5\ 1\ 5\ 0\ 2\ 1\ 3$$

Fig. 8.1. Primitive idempotents for length $n = 63$. (Here α is a root of $1 + x + x^2 + x^5 + x^6 = 0$.)

(9) Find the primitive idempotents in R_n for $n = 5, 7, 9, 11$.

(10) If $\mathcal{A}_1$ has idempotent E_1 and $\mathcal{A}_2$ has idempotent E_2, then (a) $\mathcal{A}_1 \cap \mathcal{A}_2$ has idempotent $E_1 E_2$,

(b) $\mathcal{A}_1 \cup \mathcal{A}_2$ (the smallest ideal containing both $\mathcal{A}_1$ and $\mathcal{A}_2$) has idempotent $E_1 + E_2 + E_1 E_2$.

(11) If the idempotent of $\mathscr{C}$ is

$$\theta_i + \theta_j + \cdots + \theta_k,$$

the nonzeros of the code are α^ν for $\nu \in C_i \cup \cdots \cup C_k$. If the idempotent is

$$1 + \theta_i + \theta_j + \cdots + \theta_k,$$

the nonzeros of $\mathscr{C}$ are α^ν for $\nu \notin C_i \cup \cdots \cup C_k$.

(12) Show that the Hamming code $\mathcal{H}_m$ has idempotent $1 + \theta_1(x)$, and the $[2^m - 1, 2^m - 1 - 2m, 5]$ double-error-correcting BCH code has idempotent $1 + \theta_1(x) + \theta_3(x)$.

(13) Study the $[9, 6, 2]$ irreducible code with idempotent $\theta_1(x) = x^3 + x^6$. Find all the codewords and show that the weight distribution is $A_0 = 1$, $A_2 = 9$, $A_4 = A_6 = 27$.

(14) Call two codes $\mathcal{C}_1$, $\mathcal{C}_2$ of length n *disjoint* if $\mathcal{C}_1 \cap \mathcal{C}_2 = 0$. Let $\mathcal{C}_1$, $\mathcal{C}_2$ be cyclic codes with idempotents E_1, E_2 and generator polynomials g_1, g_2. (i) Show that $\mathcal{C}_1$, $\mathcal{C}_2$ are disjoint iff, when E_1, E_2 are written as sums of primitive idempotents θ_i (cf. Theorem 7), no θ_i occurs in both. (ii) Show $\mathcal{C}_1$, $\mathcal{C}_2$ are disjoint if and only if $x^n + 1 \mid g_1 g_2$. (iii) Show that $\mathrm{dist}(\mathcal{C}_1, \mathcal{C}_2) = \min\{\mathrm{dist}(u, v) : u \in \mathcal{C}_1, v \in \mathcal{C}_2, \text{ not both zero}\} = \text{min.dist. of the code with}$ generator polynomial g.c.d. $\{g_1, g_2\}$ and idempotent $E_1 + E_2 + E_1 E_2$.

(15) (Lempel) Let

$$\eta_s(x) = \sum_{i \in C_s} x^i,$$

Show that in $GF(2)[x]$ — i.e. not mod $x^n + 1$ —

$$\eta_s(x)(1 + \eta_s(x)) = \eta_s^{(0)}(x)(1 + x^n),$$

where $\eta_s^{(0)}(x)$ is the sum of the odd powers of x occurring in $\eta_s(x)$.

(16) Show that there are $\varphi(n)/m$ minimal ideals equivalent to $\mathscr{S}_m$ (and the same number of cyclic codes equivalent to $\mathcal{H}_m$), where $\varphi(n)$ is the Euler function of Problem 8, Ch. 4.

(17) (Hard) Let $h(x)$ be a polynomial which divides $x^n + 1$. Let $E(x)$ be the idempotent of the ideal with generator polynomial $g(x) = (x^n + 1)/h(x)$. Show that

$$E(x) = (x^n + 1)\left(\frac{xh'(x)}{h(x)} + \delta\right),$$

where $h'(x)$ is the derivative of $h(x)$, and $\delta = 0$ if the degree of $h(x)$ is even, and 1 if the degree of $h(x)$ is odd. Hence show that

$$E(x) = x^{\deg h(x)-1} g(x) r'\left(\frac{1}{x}\right),$$

where $r(x) = x^{\deg h(x)} h(1/x)$. [Hint: Show $\deg E(x) \le n - 1$. Then show $E(\alpha_i) = 1$ if $h(\alpha_i) = 0$ and $E(\alpha_i) = 0$ if $h(\alpha_i) \ne 0$.]

Degenerate cyclic codes. A cyclic code which consists of several repetitions of a code of smaller block length is said to be *degenerate*. For instance $\langle\theta_0\rangle = \{0, 1\}$ is degenerate since it consists of several repetitions of the code $\{0, 1\}$.

Problem. (18) For $n = 15$, verify that θ_3 and θ_5 are idempotents of degenerate ideals. Find the dimensions of these ideals.

Lemma 8. $\langle g(x)\rangle$ *is degenerate if and only if the check polynomial $h(x)$ divides* $x^r + 1$ *for some* $r < n$.

Proof. If $\langle g(x)\rangle$ is degenerate, every codeword, including $g(x)$, is of the form $s(x)(1 + x^r + x^{2r} + \cdots + x^{n-r})$ for some r. Thus r divides n (Problem 11 of Ch. 4) and

$$g(x) = s(x)(x^n + 1)/(x^r + 1), \qquad h(x) = (x^r + 1)/s(x).$$

Conversely, let $r < n$ be the smallest integer such that $h(x)$ divides $x^r + 1$. Then r divides n (for if not, $h(x)$ divides $x^{r'} + 1$ where $r' = (r, n)$). Thus

$$g(x) = \frac{x^n + 1}{h(x)} = \frac{x^n + 1}{x^r + 1}\frac{x^r + 1}{h(x)} = s(x)(1 + x^r + \cdots + x^{n-r}).$$

Every codeword is of the form

$$a(x)s(x)(1 + x^r + x^{2r} + \cdots + x^{n-r}),$$

where by Theorem 1 of Ch. 7, $\deg a(x)s(x) < r$.   Q.E.D.

Problem. (19) Show that for $n = 15$, if $s = 1, 5$ or 7, $\langle\theta_s\rangle$ consists of **0** and all cyclic shifts of θ_s. On the other hand $\langle\theta_3\rangle$ consists of the vectors $|u|u|u|$ where u is in the $[5, 4, 2]$ even weight code. Algebraically $\langle\theta_3\rangle$ consists of all elements $a(x)(x + x^2 + x^3 + x^4)(1 + x^5 + x^{10})$, where $\deg a(x) \leqslant 3$.

Remark. We now have a method of telling whether an irreducible polynomial $h(x)$ of degree m is primitive. (See §2 of Ch. 3, §4 of Ch. 4.) We form the idempotent $E(x) = (x^n + 1) \cdot (xh'(x)/h(x) + \delta)$ where $n = 2^m - 1$ (Problem 17). If this idempotent has $2^m - 1$ distinct cyclic permutations the polynomial is primitive; if not, the code generated by $E(x)$ is degenerate.

Problems. (20) Do this for the polynomial $h(x) = x^4 + x^3 + x^2 + x + 1$.

(21) Show that:

(a) If $(s, n) = 1$ then $|C_s| = m$, $\deg M^{(s)}(x) = m$, and the ideal generated by θ_s has dimension m and is nondegenerate.

(b) If $(s, n) > 1$ the ideal is degenerate, and its dimension divides m and may equal m.

(c) If s is relatively prime to $n = 2^m - 1$, the ideal consists of the codewords **0**, $x^i\theta_s$ for $i = 0, 1, \ldots, 2^m - 2$.

(22) Let $\mathscr{C}$ be a linear code with no coordinate always zero, in which there

is only one nonzero weight. Show that there is a minimal ideal $\mathcal{M}_1$ such that $\mathscr{C}$ is equivalent to the code $\{|u|u| \cdots |u|: u \in \mathcal{M}_1\}$.

A minimal ideal is isomorphic to a field.

Theorem 9. *The minimal ideal $\mathcal{M}_s = \langle \theta_s \rangle$ of dimension m_s is isomorphic to the field* $GF(2^{m_s})$.

Proof. This will follow if we show that if $a_1(x)$, $a_2(x)$ are in $\mathcal{M}_s$ and $a_1(x)a_2(x) = 0$, then either $a_1(x)$ or $a_2(x)$ is zero.
  Suppose $a_1(x) \neq 0$. Let

$$\mathcal{N} = \{b(x) \in \mathcal{M}_s : b(x)a_1(x) = 0\}.$$

$\mathcal{N}$ is an ideal. Since $\mathcal{M}_s$ is minimal and $\mathcal{N} \subset \mathcal{M}_s$, either $\mathcal{N} = \mathcal{M}_s$ or $\mathcal{N} = 0$. By Theorem 1 $\theta_s a_1(x) = a_1(x)$, so $\mathcal{N} \neq \mathcal{M}_s$. Therefore $a_2(x) = 0$.   Q.E.D.

  For example, the ideal $\langle \theta_0 \rangle$ is a highly redundant picture of the ground field $GF(2)$.

Lemma 10. *An isomorphism φ between $\mathcal{M}_s$ and $GF(2^{m_s})$ is given by*:

$$if \ a(x) \in \mathcal{M}_s, \quad a(x)^{\varphi} = a(\beta),$$

where $\beta \in GF(2^{m_s})$ is a primitive n^{th} root of unity which is a nonzero of $\mathcal{M}_s$. (Of course different choices for β give different mappings.)
  For example, consider the $[7, 3, 4]$ code with idempotent $\theta_1(x) = 1 + x + x^2 + x^4$. Using Fig. 4.5 we may take $\beta = \alpha$, α^2 or α^4, giving the mappings from the code to $GF(2^3)$ shown in Fig. 8.2.

Codeword $a(x)$	Elements of $GF(2^3)$		
0 1 2 3 4 5 6	$a(\alpha)$	$a(\alpha^2)$	$a(\alpha^4)$
0 0 0 0 0 0 0	0	0	0
1 1 1 0 1 0 0	1	1	1
0 1 1 1 0 1 0	α	α^2	α^4
0 0 1 1 1 0 1	α^2	α^4	α
1 0 0 1 1 1 0	α^3	α^6	α^5
0 1 0 0 1 1 1	α^4	α	α^2
1 0 1 0 0 1 1	α^5	α^3	α^6
1 1 0 1 0 0 1	α^6	α^5	α^3

Fig. 8.2. Three mappings from $\mathscr{S}_3$ to $GF(2^3)$.

Proof. Suppose for simplicity that $\beta = \alpha^s$. Let ξ be a primitive element of $GF(2^{m_s})$, and consider the mapping ψ from $GF(2^{m_s})$ to $\mathcal{M}$, defined by

$$(\xi^i)^\psi = \sum_{j=0}^{n-1} T_{m_s}(\xi^i \beta^{-j}) x^j. \tag{3}$$

We shall show that ψ is the inverse mapping to φ. Let the RHS of (3) be denoted by $a^{(i)}(x)$. We first show that $a^{(i)}(x) \in \mathcal{M}_s$.

$$
\begin{aligned}
a^{(i)}(\alpha^k) &= \sum_{j=0}^{n-1} T_{m_s}(\xi^i \alpha^{-js}) \alpha^{jk} \\
&= \sum_{l=0}^{m_s-1} \sum_{j=0}^{n-1} \xi^{i2^l} \alpha^{(k-s2^l)j} \\
&= \begin{cases} \xi^{i2^l} & \text{if } k = s2^l \text{ for some } l, \\ 0 & \text{otherwise.} \end{cases}
\end{aligned}
$$

Thus $a^{(i)}(\alpha^k) = 0$ unless $k \in C_s$; hence $a^{(i)}(x) \in \mathcal{M}_s$. Then it is immediate that $a^{(i)}(x)^\varphi = a^{(i)}(\beta) = \xi^i$.

From Theorem 9, φ must be the inverse of ψ and both maps are 1-1 and onto. Also φ clearly preserves addition and multiplication, hence is an isomorphism.   Q.E.D.

The idempotent of $\mathcal{M}_s$ maps onto the unit of $GF(2^{m_s})$. If $c_1, c_2 \in \mathcal{M}_s$ and $c_1(x)c_2(x) = \theta_s(x)$ then c_1 and c_2 are inverses in $\mathcal{M}_s$. (Of course c_1, c_2 have no inverses in R_n.) Note also that $(xa(x))^\varphi = \beta a(\beta)$.

Idempotents of cyclic codes over $GF(q)$. Let $\mathcal{C}$ be a cyclic code of length n over $GF(q)$, where n and q are relatively prime. An element $E(x)$ of R_n, the ring $GF(q)[x]/(x^n - 1)$, is an idempotent if $E(x) = E(x)^2$.

Problems. (23) Show that there is a unique polynomial in $\mathcal{C}$ which is both an idempotent and a generator.

(24) If $\mathcal{C}$ has idempotent $E(x)$, show that $\mathcal{C}^\perp$ has idempotent $(1 - E(x))^*$.

(25) Show that there is a set of primitive idempotents $\theta_0, \theta_1, \ldots, \theta_t$ such that

$$\theta_i^2 = \theta_i, \quad \theta_i \theta_j = 0 \quad \text{if } i \neq j, \quad \sum_{i=0}^t \theta_i = 1.$$

Also R_n is the direct sum of the minimal ideals generated by the θ_i.

(26) Show that the minimal ideal of dimension m_s generated by θ_s is isomorphic to the field $GF(q^{m_s})$.

§4. Weight distribution of minimal codes

Let $\mathscr{C}$ be an $[n, k]$ nondegenerate minimal cyclic code, with nonzeros $\beta, \beta^2, \ldots, \beta^{2^{k-1}}$. Then $ns = 2^k - 1$, and k is the smallest integer for which such an equation is possible. (Thus $k = m$.) We suppose $s > 1$.

Let ξ be a primitive element of $GF(2^k)$, with $\beta = \xi^s$. Let $\lambda(x) \in \mathscr{C}$ be the codeword such that $\lambda(x)^\circ = \xi$. Then $(x\lambda(x))^\circ = \xi^{s+1}$; in fact, the n cyclic shifts of $\lambda(x)$ correspond to $\xi, \xi^{s+1}, \xi^{2s+1}, \ldots, \xi^{(n-1)s+1}$. Thus the codewords of $\mathscr{C}$ which correspond to $1, \xi, \xi^2, \ldots, \xi^{s-1}$ are a complete set of cycle representatives for $\mathscr{C}$, and determine the weight distribution of $\mathscr{C}$.

A way to find the cycle representatives. Consider the $[2^k - 1, k]$ simplex code $\mathscr{S}_k$ with nonzeros $\xi, \xi^2, \ldots, \xi^{2^{k-1}}$. Codewords of $\mathscr{S}_k$ are written as polynomials in y, where $y^{2^k-1} = 1$. The idempotent of $\mathscr{S}_k$ is

$$E(y) = \sum_{i=0}^{2^k-2} e_i y^i, \quad e_i = T_k(\xi^{-i}).$$

The coefficients of $E(y)$ may be arranged in an $s \times n$ array

$$
\begin{array}{llll}
e_0 & e_s & \cdots & e_{(n-1)s} \\
e_1 & e_{s+1} & \cdots & e_{(n-1)s+1} \\
\cdots\cdots\cdots\cdots\cdots\cdots\cdots\cdots \\
e_{s-1} & e_{2s-1} & \cdots & e_{ns-1}
\end{array}
\qquad (4)
$$

Let

$$c_j(x) = \sum_{i=0}^{n-1} e_{j+is} x^i, \quad \text{for } 0 \leqslant j \leqslant s - 1.$$

Theorem 11. $c_j(x) \in \mathscr{C}$, and $c_j(x)^\circ = \xi^{-j}$. *Thus the $c_j(x)$ are a set of cycle representatives for $\mathscr{C}$.*

Proof.

$$c_0(x) = \sum_{i=0}^{n-1} T_k(\xi^{-si}) x^i = \sum_{i=0}^{n-1} T_k(\beta^{-i}) x^i,$$

which is the idempotent of $\mathscr{C}$. Then

$$c_j(x) = \sum_{i=0}^{n-1} T_k(\xi^{-j}\beta^{-i}) x^i$$

and hence $c_j(x)^\circ = \xi^{-j}$.   Q.E.D.

Clearly any cyclic shift of the idempotent will also give a set of cycle representatives. A cyclic shift of the idempotent of $\mathcal{S}_k$ is easily found as the output of the associated shift register (see Figs. 14.2, 14.3).

For example, the $[15, 4, 8]$ code $\mathcal{S}_4$ has idempotent 000 100 110 101 111 from Equation (1) of Ch. 14, or Problem 7 of Ch. 8. Arranged as in (4) with $s = 3$, $n = 5$ this becomes

$$0 \ 1 \ 1 \ 1 \ 1$$
$$0 \ 0 \ 1 \ 0 \ 1,$$
$$0 \ 0 \ 0 \ 1 \ 1$$

giving a set of cycle representatives for the $[5, 4, 2]$ code.

Problem. (27) Show that the weight distribution of the $[21, 6]$ code is

$$i: \quad 0 \ \ 8 \ \ 12$$
$$A_i: \quad 1 \ \ 21 \ \ 42$$

A special case. Suppose the cycle representatives are $\theta(x), \lambda_1(x), \ldots, \lambda_{s-1}(x)$, where $\theta(x)^\varphi = 1$, $\lambda_i(x)^\varphi = \xi^i$. Clearly $\lambda_2(x) = \lambda_1(x)^2 = \lambda_1(x^2)$, so $\lambda_2(x)$ has the same weight as $\lambda_1(x)$. A particularly simple case is where all the $\lambda_i(x)$ have the same weight as $\lambda_1(x)$. For example, if $n = 51 = (2^8 - 1)/5$, $\lambda(x)^\varphi = \xi$, $\lambda(x^2)^\varphi = \xi^2$, $\lambda(x^4)^\varphi = \xi^4$, $\lambda(x^8)^\varphi = \xi^8 = \xi^{3+5}$. Thus $\lambda(x^8)$ is a cyclic shift of the codeword corresponding to ξ^3. Hence the code has only two nonzero weights, namely wt $(\theta(x))$ and wt $(\lambda(x))$.

In general, if s is any prime for which 2 is a *primitive root* [i.e., there are only two cyclotomic cosets mod s, C_0 and C_1], the code $\mathcal{M}_1$ has at most two weights, τ_1 and τ_2 say. $\mathcal{M}_1$ contains n codewords of weight τ_1 which are the cyclic shifts of θ_1, and $n(s - 1)$ of weight τ_2.

The weight distribution when there are two nonzero weights. Suppose $\mathcal{M}_1$ has only two nonzero weights τ_1, τ_2, where the idempotent θ_1 has weight τ_1. The dual code has minimum distance 3 by the BCH bound. From Theorem 1 of Ch. 5,

$$1 + ny^{\tau_1} + n(s - 1)y^{\tau_2} = 2^{m-n}\{(1 + y)^n + A_3'(1 + y)^{n-3}(1 - y)^3 + \cdots\}.$$

Differentiating twice and setting $y = 1$ we obtain

$$\tau_1 + (s - 1)\tau_2 = 2^{m-1},$$
$$\tau_1(\tau_1 - 1) + (s - 1)\tau_2(\tau_2 - 1) = (n - 1)2^{m-2}.$$

Solving for τ_2 we find

$$\tau_2 = 2^{m/2-1}\left(\frac{2^{m/2} \pm 1}{s}\right).$$

Thus m must be even, and

$$\tau_1 = \begin{cases} 2^{m/2-1}\left(\dfrac{2^{m/2}+1}{s} - 1\right) & \text{if } s \text{ divides } 2^{m/2}+1 \\[2ex] 2^{m/2-1}\left(\dfrac{2^{m/2}-1}{s} + 1\right) & \text{if } s \text{ divides } 2^{m/2}-1. \end{cases}$$

Figure 8.3 gives some examples of two weight codes.

n	m	s	τ_1	τ_2
21	6	3	8	12
51	8	5	32	24
85	8	3	48	40
93	10	11	32	48
315	12	13	128	160
341	10	3	160	176
819	12	5	384	416
1365	12	3	704	672
13797	18	19	26.2^8	27.2^8

Fig. 8.3.

Remark. Even if 2 is not a primitive root of s there are many other cases where the code has only 2 nonzero weights – see Problem 5 of Ch. 15.

§5. The automorphism group of a code

Let $\mathscr{C}$ be a binary code of length n. Any permutation of the n coordinate places changes $\mathscr{C}$ into an *equivalent* code having many of the same properties as $\mathscr{C}$ (same minimum weight, weight distribution, etc.).

Definition. The permutations of coordinate places which send $\mathscr{C}$ into itself-codewords go into (possibly different) codewords – form the *automorphism group* of $\mathscr{C}$, denoted by Aut $(\mathscr{C})$.

Problem. (28) Show that Aut $(\mathscr{C})$ is indeed a group.

A typical permutation π of the symbols $\{1, 2, \ldots, n\}$ sends each i into $\pi(i)$ (or, in a more convenient notation, into $i\pi$). The vector $c = (c_1, \ldots, c_n)$ goes into $c\pi = (c_{\pi(1)}, \ldots, c_{\pi(n)})$.* If ρ is another permutation, the product $\pi\rho$ means

*An equally good, although different, definition is $c\pi = (c_{\pi^{-1}(1)}, \ldots, c_{\pi^{-1}(n)})$.

"first apply π, then ρ". Thus $c(\pi\rho) = (c\pi)\rho$. E.g. If $\pi = (123)$, $\rho = (14)$,

$$(c_1, c_2, c_3, c_4)\pi = (c_2, c_3, c_1, c_4),$$

$$(c_1, c_2, c_3, c_4)\pi\rho = (c_2, c_3, c_4, c_1).$$

Thus Aut $(\mathscr{C})$ is a subgroup of the *symmetric group* $\mathscr{S}_n$ consisting of all $n!$ permutations of n symbols.

Examples. (1) The automorphism groups of the repetition and even weight codes are both equal to $\mathscr{S}_n$.

(2) The group of the code

$$\begin{array}{cccc} 1 & 2 & 3 & 4 \end{array}$$
$$\begin{pmatrix} 0 & 0 & 0 & 0 \\ 0 & 0 & 1 & 1 \\ 1 & 1 & 0 & 0 \\ 1 & 1 & 1 & 1 \end{pmatrix}$$

consists of these 8 permutations:

$$(1), (12), (34), (12)(34), (13)(24), (14)(23), (1324), (1423).$$

Problems. (29) If $\mathscr{C}$ is linear, Aut $\mathscr{C}$ = Aut $\mathscr{C}^\perp$.

(30) If $\mathscr{C}$ is linear and $\mathscr{C}_1$ is obtained from $\mathscr{C}$, (a) by adding a parity check, or (b) by adding the vector $\mathbf{1}$, then Aut $\mathscr{C}_1 \supset$ Aut $\mathscr{C}$. In case (b), if $\mathscr{C}$ has odd length and only even weights, then Aut $\mathscr{C}_1$ = Aut $\mathscr{C}$.

(31) Show that $\mathscr{C}_1 \subset \mathscr{C}_2$ does not imply that Aut $\mathscr{C}_1 \supset$ Aut $\mathscr{C}_2$.

(32) Let $\mathscr{C}$ be the [12, 6, 3] code with generator matrix specified by Fig. 8.4. Thus row 1 has ones in coordinates {1236}, row 2 in {345}, ... Show that Aut $(\mathscr{C})$ contains only the identity permutation.

It is in general difficult to determine the complete automorphism group of a linear code, and even more difficult if the code is not linear. We shall see in Ch. 16 how the automorphism group may be used for decoding.

Let $\mathscr{C}$ be an $[n, k]$ linear code with generator matrix M; M contains k linearly independent rows.

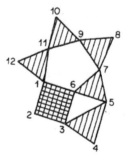

Fig. 8.4. A [12, 6, 3] code with trivial automorphism group.

Lemma 12. *The permutation of coordinate places represented by the $n \times n$ matrix A is in* Aut $\mathscr{C}$ *if and only if*

$$KM = MA$$

for some invertible $k \times k$ matrix K.

Proof. MA is a generator matrix for $\mathscr{C}$ if and only if the corresponding permutation is in Aut $(\mathscr{C})$. MA can be obtained from M by the linear transformation K.   Q.E.D.

Example. The $[7, 3, 4]$ simplex code $\mathscr{S}_3$ has

$$M = \begin{matrix} 0\ 1\ 2\ 3\ 4\ 5\ 6 \\ \begin{pmatrix} 1\ 1\ 1\ 0\ 1\ 0\ 0 \\ 0\ 1\ 1\ 1\ 0\ 1\ 0 \\ 0\ 0\ 1\ 1\ 1\ 0\ 1 \end{pmatrix} \end{matrix}$$

The permutation $\sigma_2 = (0)(124)(365)$ sends codewords into codewords. In fact

$$M\sigma_2 = \begin{matrix} 0\ 1\ 2\ 3\ 4\ 5\ 6 \\ \begin{pmatrix} 1\ 1\ 1\ 0\ 1\ 0\ 0 \\ 0\ 1\ 0\ 0\ 1\ 1\ 1 \\ 0\ 1\ 1\ 1\ 0\ 1\ 0 \end{pmatrix} \end{matrix} = \begin{pmatrix} 1\ 0\ 0 \\ 0\ 1\ 1 \\ 0\ 1\ 0 \end{pmatrix} M.$$

Definition. The set of all invertible $k \times k$ matrices over a field F is called the *general linear group* and is denoted by $GL(k, F)$. If F is a finite field $GF(q)$ we write this as $GL(k, q)$.

Theorem 13. *The general linear group* $GL(k, q)$ *has order*

$$(q^k - 1)(q^k - q)(q^k - q^2) \cdots (q^k - q^{k-1}).$$

Proof. Let K be a matrix in $GL(k, q)$. The first column of K can be any nonzero vector over $GF(q)$, hence can be chosen in $q^k - 1$ ways. The second column must not be a multiple of the first, hence can be chosen in $q^k - q$ ways. And so on.   Q.E.D.

By Lemma 12, if the columns of the generator matrix are distinct the automorphism group of a binary linear code of dimension k is isomorphic to a subgroup of $GL(k, 2)$.

Problems. (33) (Hard.) (a) Suppose no coordinate of $\mathscr{C}$ is always zero. Then for any $K \in GL(k, 2)$, KM is another generator matrix for the same code $\mathscr{C}$. Show that K corresponds to an element π of Aut ($\mathscr{C}$) (i.e., the $n \times n$ matrix A corresponding to π satisfies $KM = MA$) iff K preserves the weight of every codeword. (b) Hence show that the automorphism group of the simplex code $\mathscr{S}_m$ is isomorphic to $GL(m, 2)$.

(34) **Other groups associated with a code.** Suppose $\mathscr{C}$ is an $[n, k]$ code over $GF(q)$, and let

$$\mathscr{G}_1 = \{A \in GL(n, q): uA = u \text{ for all } u \in \mathscr{C}\},$$
$$\mathscr{G}_2 = \{A \in GL(n, q): uA \in \mathscr{C} \text{ for all } u \in \mathscr{C}\}.$$

Show that

$$|\mathscr{G}_1| = q^{k(n-k)} \prod_{i=1}^{n-k-1} (q^{n-k} - q^{n-k-i}),$$

$$|\mathscr{G}_2| = q^{\binom{n}{2}} \prod_{i=1}^{k} (q^i - 1) \cdot \prod_{i=1}^{n-k} (q^i - 1).$$

Next we give a useful property of weight distributions.

Theorem 14. *Suppose $\mathscr{C}$ is a code of length N in which all weights are even and with the property that no matter which coordinate is deleted, the resulting punctured code (cf. §9 of Ch. 1) has the same weight distribution. If $\{A_i\}$ is the weight distribution of $\mathscr{C}$, and $\{a_i\}$ is that of any of the punctured codes, then*

$$a_{2j-1} = \frac{2jA_{2j}}{N},$$

$$a_{2j} = \frac{N - 2j}{N} A_{2j}.$$

Furthermore the punctured codes have odd minimum distance.

Proof. Consider the array $\mathscr{L}$ whose rows are the A_{2j} codewords of weight $2j$ in $\mathscr{C}$. This array contains $2jA_{2j}$ 1's. By hypothesis the number of 1's in each column of $\mathscr{L}$ is the same, and is equal to a_{2j-1}. Thus

$$a_{2j-1} = \frac{2jA_{2j}}{N}. \tag{5}$$

The second formula follows because $A_{2j} = a_{2j} + a_{2j-1}$. \hfill Q.E.D.

A group G of permutations of the symbols $\{1, \ldots, n\}$ is *transitive* if for any symbols i, j there is a permutation $\pi \in G$ such that $i\pi = j$. More generally G is *t-fold transitive* if, given t distinct symbols $i_1, i_2, \ldots, i_t$, and t distinct symbols $j_1, j_2, \ldots, j_t$, there is a $\pi \in G$ such that $i_1\pi = j_1, \ldots, i_t\pi = j_t$.

Corollary 15. *Suppose* $\mathscr{C}$ *is fixed by a transitive permutation group. Then* (i) *deleting any coordinate place gives an equivalent code* $\mathscr{C}^*$, *and* (ii) *if all weights in* $\mathscr{C}$ *are even then* $\mathscr{C}^*$ *has odd minimum weight.*

We shall see later that many extended cyclic codes do have transitive automorphism groups. But beware! Theorem 14 does not apply to all cyclic codes, as Problem (35) shows.

Problems. (35) Let $\mathscr{C}$ be the degenerate [9, 3, 3] cyclic code generated by $x^6 + x^3 + 1$. Show that the extended code does not satisfy the hypothesis of Theorem 14.

(36) Show that if $\mathscr{C}$ has length N and is fixed by a transitive permutation group then $N \mid iA_i$ where A_i is the number of codewords of weight i.

(37) Let $\mathscr{C}$ be a code of length N in which all weights are even, which is invariant under a transitive permutation group, and has weight enumerator $W(x, y)$. Let $\mathscr{C}_1$ with weight enumerator $W_1(x, y)$ be obtained from $\mathscr{C}$ by puncturing any one coordinate, and let $\mathscr{C}_2$ with weight enumerator $W_2(x, y)$ be the even weight subcode of $\mathscr{C}_1$. Show that

$$ W_2(x, y) = \frac{1}{N} \frac{\partial}{\partial x} W(x, y) \quad \text{and} \quad W_1(x, y) = \frac{1}{N} \left(\frac{\partial}{\partial x} + \frac{\partial}{\partial y} \right) W(x, y). $$

(38) (Camion.) Suppose Aut $(\mathscr{C})$ is t-fold transitive, where $\mathscr{C}$ is an $[n, k, d]$ cyclic code over $GF(q)$. Show that $k \geq (n - t + 1)/(d - t + 1)$.

The automorphism group of a cyclic code. By definition the automorphism group of a cyclic code contains all the cyclic permutations, i.e., the cyclic permutation $(0, 1, 2, \ldots, n - 1)$ and all its powers.

Because n is odd, the map $\sigma_2 : x \to x^2$ is a permutation of R_n (for σ_2 permutes the basis $1, x, x^2, \ldots, x^{n-1}$). Now $a(x)\sigma_2 = a(x^2) = a(x)^2$ is in the same code as $a(x)$. Therefore the automorphism group of a cyclic code contains σ_2 as well as all the cyclic permutations. σ_2 is a permutation of order m, where $m = |C_1|$.

Problems. (39) Find an automorphism group of the code of Fig. 5.1.

(40) Show that σ_2 and $T = (0, 1, \ldots, n - 1)$ together generate a group of order mn consisting of the permutations $\{\sigma_2^i T^j$, for $0 \leq i < m$ and $0 \leq j < n\}$.

Example. The effect of σ_2 on the codewords of the simplex code $\mathscr{S}_3$ is shown in Fig. 8.5.

1	x	x^2	x^3	x^4	x^5	x^6	
1	1	1	0	1	0	0	$E(x) = E(x)^2$
0	1	1	1	0	1	0	$a(x) = a(x)^8$
0	0	1	1	1	0	1	$a(x)^2$
0	1	0	0	1	1	1	$a(x)^4$
1	0	0	1	1	1	0	$b(x) = b(x)^8$
1	1	0	1	0	0	1	$b(x)^2$
1	0	1	0	0	1	1	$b(x)^4$

Fig. 8.5.

Equivalence of cyclic codes. We now consider R_n to be the group algebra of a cyclic group G of order n (rather than as a polynomial ring). The mappings $\sigma_\mu : x^i \to x^{i\mu}$, where μ is an integer prime to n, form a group $\mathcal{G}$ of automorphisms of G. [An *automorphism* of a group G is a mapping σ from G onto itself which preserves multiplication: $\sigma(ab) = \sigma(a)\sigma(b)$.] Thus $\mathcal{G}$ permutes the coordinate places of R_n, and sends cyclic codes into cyclic codes.

For example, if $n = 7$, $\mu = 3$, σ_3 is the permutation $(1)(x, x^3, x^2, x^6, x^4, x^5)$ of G (or of the coordinate places). Thus σ_3 interchanges the idempotents $\theta_1 = 1 + x + x^2 + x^4$ and $\theta_3 = 1 + x^3 + x^6 + x^5$, and so also interchanges the cyclic codes generated by these idempotents.

$\mathcal{G}$ is a multiplicative abelian group, isomorphic to the multiplicative group of integers less than and prime to n, and has order $\varphi(n)$ (see Problem 8 of Ch. 4).

The mapping $i \to i\mu$, where μ is prime to n, permutes the cyclotomic cosets. For example if $n = 31$ the mapping $i \to 3i$ has the following effect:

$$C_0 \to C_0, \qquad C_1 \to C_3 \to C_5 \to C_{15} \to C_7 \to C_{11} \to C_1,$$

(cf. Fig. 4.4).

On the other hand if μ is a power of 2, the mapping $\sigma_\mu : i \to i\mu$ fixes the cyclotomic cosets. Similarly the mapping σ_μ fixes every cyclic code. Hence to find the permutations which actually change cyclic codes we must factor out the subgroup $\{\sigma_i : i \in C_1\}$ from $\mathcal{G}$. The quotient group consists of one σ_μ from each cyclotomic coset containing numbers prime to n, and has order $\varphi(n)/m$.

For example, when $n = 63$ the cyclotomic cosets containing numbers prime to n are

$$C_5 = \{\, 5 \ \ 10 \ \ 20 \ \ 40 \ \ 17 \ \ 34\},$$
$$C_{11} = \{11 \ \ 22 \ \ 44 \ \ 25 \ \ 50 \ \ 37\},$$
$$C_{31} = \{31 \ \ 62 \ \ 61 \ \ 59 \ \ 55 \ \ 47\},$$
$$C_{23} = \{23 \ \ 46 \ \ 29 \ \ 58 \ \ 53 \ \ 43\},$$
$$C_{13} = \{13 \ \ 26 \ \ 52 \ \ 41 \ \ 19 \ \ 38\},$$
$$C_1 = \{\, 1 \ \ \ 2 \ \ \ 4 \ \ \ 8 \ \ 16 \ \ 32\}.$$

The boldface numbers are the powers of 5 mod 63; therefore in this case the quotient group is a cyclic group of order 6.

The effect of σ_5 on the primitive idempotents (or on the cyclotomic cosets) is

$$\theta_1 \to \theta_5 \to \theta_{11} \to \theta_{31} \to \theta_{23} \to \theta_{13} \to \theta_1$$
$$\theta_{21} \to \theta_{21}$$
$$\theta_3 \to \theta_{15} \to \theta_3$$
$$\theta_7 \to \theta_7$$
$$\theta_9 \to \theta_{27} \to \theta_9$$

We observe that σ_5 has sorted the primitive idempotents into classes. Those in the first class are the nondegenerate cyclic codes of length 63. Those in the second are repetitions of a cyclic code of length 3, those in the third of length 21, in the fourth, 9, and in the fifth, 7. A moment's reflection shows that this must always happen: any nondegenerate idempotent of block length n may be obtained by applying a suitable σ_μ to θ_1.

Thus all nondegenerate minimal ideals of the same block length are equivalent.

Of course using σ_5 we can see that many other codes are equivalent. E.g. the codes of length 63 with idempotents

$$\theta_1 + \theta_5, \ \theta_5 + \theta_{11}, \ \theta_{11} + \theta_{31}, \ \theta_{31} + \theta_{23}, \ \theta_{23} + \theta_{13}, \ \theta_{13} + \theta_1$$

are equivalent.

Problem. (41) (i) For $n = 7$ show that $i \to 3i$ maps

$$C_1 \to C_3 \to C_1.$$

(ii) For $n = 15$, $i \to 7i$ maps

$$C_1 \to C_7 \to C_1, \qquad C_3 \to C_3, \qquad C_5 \to C_5.$$

(iii) For $n = 127$, $i \to 3i$ maps

$$C_1 \to C_3 \to C_9 \to C_{27} \to C_{13} \to C_{29} \to C_{47} \to$$
$$\to C_7 \to C_{21} \to C_{63} \to C_{31} \to C_{55} \to C_{19} \to$$
$$\to C_{23} \to C_{11} \to C_5 \to C_{15} \to C_{43} \to C_1.$$

The automorphism group of BCH codes. Let $\mathcal{C}$ be a code of length $n = 2^m - 1$. We label the coordinate places by nonzero elements of $GF(2^m)$, i.e., $1, \alpha, \alpha^2, \ldots, \alpha^{2^m - 2}$. We add an overall parity check place labeled by ∞, corresponding to the zero of $GF(2^m)$:

coordinate:	0	1	2	$\cdots$	$n-1$	∞
corresponding element of $GF(2^m)$:	1	α	α^2	$\cdots$	α^{n-1}	0

Definition. *The affine group.* Let $P_{u,v}$ permute the elements of GF(2^m) by:

$$P_{u,v}: \alpha^i \to u\alpha^i + v, \qquad 0 \to v$$

where $u, v \in$ GF(2^m), $u \neq 0$. $P_{\alpha^k,0}$ is a cyclic shift of k places, fixing 0, i.e. fixing the coordinate ∞.

E.g. for length $n = 7$, with GF(2^3) as in Fig. 4.5, the permutation $P_{\alpha,\alpha}$ sends

$$1 \to 0 \to \alpha \to \alpha^4 \to \alpha^6 \to \alpha^3 \to \alpha^2 \to 1, \qquad \alpha^5 \to \alpha^5.$$

Thus

$$1 \quad \alpha \quad \alpha^2 \quad \alpha^3 \quad \alpha^4 \quad \alpha^5 \quad \alpha^6 \quad 0$$

$$\text{codeword:} \quad c_0 \ \ c_1 \ \ c_2 \ \ c_3 \ \ c_4 \ \ c_5 \ \ c_6 \ \ c_\infty$$

$$\text{permuted codeword:} \quad c_\infty \ \ c_4 \ \ c_0 \ \ c_2 \ \ c_6 \ \ c_5 \ \ c_3 \ \ c_1$$

The set of all $P_{u,v}$ forms a group called the *affine group* on GF(2^m). This group has order $2^m(2^m - 1)$ and is doubly transitive. (We shall use a k-dimensional generalization of this group in Ch. 13).

Problem. (42) Show that this group *is* doubly transitive: if $i_1 \neq i_2$, $j_1 \neq j_2$, the equations $u\alpha^{i_1} + v = \alpha^{j_1}$, $u\alpha^{i_2} + v = \alpha^{j_2}$, or $u\alpha^{i_1} + v = \alpha^{j_1}$, $u\alpha^{i_2} + v = 0$ have a solution for $u, v \in$ GF(2^m).

Theorem 16. *Let $\mathscr{C}$ be a primitive BCH code of length $n = 2^m - 1$ and designed distance δ, and let $\hat{\mathscr{C}}$ be the extended code. Then the automorphism group of $\hat{\mathscr{C}}$ contains the affine group on GF(2^m).*

Proof. Let $P_{u,v}$ be any permutation in the affine group. Let $c = (c_0, c_1, \ldots, c_{n-1}, c_\infty)$ be a codeword of weight w in $\hat{\mathscr{C}}$, with 1's in coordinates corresponding to the elements $X_1, \ldots, X_w$ in GF(2^m). Thus $X_i = \alpha^{a_i}$ if $c_{a_i} = 1$, $X_w = 0$ if $c_\infty = 1$.

Let $S_k = \sum_{i=1}^{w} X_i^k$, $k = 0, 1, 2, \ldots$, where $0^0 = 1$. Then $S_0 = \sum_{i=1}^{w} 1 = w = 0$ since $c \in \hat{\mathscr{C}}$. Also $S_k = c(\alpha^k) = 0$ for $k = 1, 2, \ldots, \delta - 1$ since $\mathscr{C}$ is a BCH code.

Let $X_i' = uX_i + v$ be the locations of the 1's in the permuted codeword. Then

$$S_k' = \sum_i (X_i')^k = \sum_i (uX_i + v)^k$$

$$= \sum_{l=0}^{k} \binom{k}{l} u^l v^{k-l} \sum_i X_i^l = \sum_{l=0}^{k} \binom{k}{l} u^l v^{k-l} S_l$$

$$= 0 \quad \text{for } 0 \leqslant k \leqslant \delta - 1.$$

Therefore the permuted codeword is also in $\hat{\mathscr{C}}$.   Q.E.D.

Corollary 17. *Let $\mathscr{C}$ be cyclic and $\hat{\mathscr{C}}$ the extended code. If* Aut $(\hat{\mathscr{C}})$ *is transitive, then $\mathscr{C}$ has an odd minimum distance d, and also contains codewords of (even) weight $d + 1$. In particular this applies if $\mathscr{C}$ is a primitive binary* BCH *code.*

Proof. From Theorem 14, if a_i is the number of codewords in $\mathscr{C}$ of weight i,

$$(N - 2j)a_{2j-1} = 2ja_{2j}. \qquad \text{Q.E.D.}$$

Problem. (43) Modify the proof of Theorem 16 to obtain the following generalization due to Kasami, Lin and Peterson. For a positive integer i let $J(i)$ denote the set of positive integers whose binary expansion is "covered" by that of i. Thus if $i = \Sigma\, \delta_r 2^r$, where $\delta_r = 0$ or 1, then

$$J(i) = \{j: j = \Sigma\, \epsilon_r 2^r, \quad \epsilon_r = 0 \text{ or } 1, \quad \epsilon_r \le \delta_r\}.$$

Let $\mathscr{C}$ be a cyclic code of length $2^m - 1$ for which 1 is a nonzero, and let $\hat{\mathscr{C}}$ be the extended code. Then $\hat{\mathscr{C}}$ is invariant under the affine group iff whenever α^i is a zero of $\mathscr{C}$, so is α^j for all $j \in J(i)$.

Theorem 18. *If $\mathscr{C}$ is a binary code which is invariant under a t-fold transitive group G, then the codewords of each fixed weight i in $\mathscr{C}$ form a $t\text{-}(n, i, \lambda)$ design, where $\lambda = A_i\binom{i}{t}/\binom{n}{t}$.*

Proof. Let $S_i(P_1, \ldots, P_t)$ be the set of codewords of weight i which contain 1's in coordinates $P_1, \ldots, P_t$. Since G is t-fold transitive, $|S_i(P_1, \ldots, P_t)|$ is independent of the particular choice of $P_1, \ldots, P_t$. Therefore the codewords of weight i form a $t\text{-}(n, i, \lambda)$ design. From Theorem 9 of Ch. 2, $A_i = \lambda\binom{n}{t}/\binom{i}{t}$.
$$\text{Q.E.D.}$$

Corollary 19. *The codewords of each weight in an extended primitive binary* BCH *code form a 2-design. (But compare Theorem 15 of Ch. 2.)*

Problems. (44) Show that the minimum distance of the double-error-correcting BCH code of length $2^m - 1$ is exactly 5 if $m \ge 4$.

(45) Let $a + \mathscr{C}$ be a coset of $\mathscr{C}$ and $\pi \in$ Aut $(\mathscr{C})$. Show that $\pi(a + \mathscr{C})$ is also a coset of $\mathscr{C}$.

(46) (Camion.) Corresponding to the rth matrix (a_{ij}) in GL(m, q) define the column vector $x_r = (a_{11}, a_{12}, \ldots, a_{mm})^T$ of length m^2, for $r = 1, \ldots, n = |$GL$(m, q)|$. Show that the code over GF(q) with $m^2 \times n$ generator matrix

$G = (x_1, x_2, \ldots, x_n)$ has parameters

$$n = \prod_{i=0}^{m-1} (q^m - q^i), \quad k = m^2, \qquad d = n(q^m - q - 1)(q^m - q^{m-1})/(q^m - q)(q^m - 1).$$

The automorphism group of a nonbinary code.

Definition. A *monomial* matrix is a matrix with exactly one nonzero entry in each row and column. Thus a monomial matrix over GF(2) is a permutation matrix, and a monomial matrix over an arbitrary field is a permutation matrix times an invertible diagonal matrix.

Let $\mathscr{C}$ be a code of length n over GF(q). We first suppose that $q = p = a$ prime.

Definition. The *automorphism group* Aut ($\mathscr{C}$) of a linear code $\mathscr{C}$ consists of all $n \times n$ monomial matrices A over GF(p) such that $cA \in \mathscr{C}$ for all $c \in \mathscr{C}$.

For example the $[3, 1, 3]$ code $\mathscr{C} = \{000, 111, 222\}$ over GF(3) has automorphism group of order $2.3! = 12$. For the code is fixed by any permutation of the coordinates, and by multiplying each codeword by 2, i.e. by the monomial matrix

$$\begin{pmatrix} 200 \\ 020 \\ 002 \end{pmatrix}.$$

Thus Aut ($\mathscr{C}$) consists of the monomials

$$\begin{pmatrix} 100 \\ 010 \\ 001 \end{pmatrix} \begin{pmatrix} 100 \\ 001 \\ 010 \end{pmatrix} \begin{pmatrix} 001 \\ 010 \\ 100 \end{pmatrix} \begin{pmatrix} 010 \\ 100 \\ 001 \end{pmatrix} \begin{pmatrix} 010 \\ 001 \\ 100 \end{pmatrix} \begin{pmatrix} 001 \\ 100 \\ 010 \end{pmatrix}$$

$$\begin{pmatrix} 200 \\ 020 \\ 002 \end{pmatrix} \begin{pmatrix} 200 \\ 002 \\ 020 \end{pmatrix} \begin{pmatrix} 002 \\ 020 \\ 200 \end{pmatrix} \begin{pmatrix} 020 \\ 200 \\ 002 \end{pmatrix} \begin{pmatrix} 020 \\ 002 \\ 200 \end{pmatrix} \begin{pmatrix} 002 \\ 200 \\ 020 \end{pmatrix}.$$

Problem. (47) Show the automorphism group of code #6 of Ch. 1 has order 48.

If q is a prime power, the automorphism group of $\mathscr{C}$ also contains any field automorphisms of GF(q) which preserve $\mathscr{C}$. An example will be given in Ch. 16.

§6. The Mattson–Solomon polynomial

In Ch. 7 the vector $a = (a_0, a_1, \ldots, a_{n-1})$ was represented by the polynomial $a(x) = a_0 + a_1 x + \cdots + a_{n-1} x^{n-1}$. We now introduce another polynomial associated with a, the Mattson–Solomon polynomial $A(z)$. Let $F = \mathrm{GF}(q)$, $\mathscr{F} = \mathrm{GF}(q^m)$, and $\alpha \in \mathscr{F}$ be a primitive n^{th} root of unity.

Definition. *The Mattson–Solomon* (MS) *polynomial* associated with a vector $a = (a_0, a_1, \ldots, a_{n-1})$, $a_i \in \mathscr{F}$, is the following polynomial in $\mathscr{F}[z]$:

$$A(z) = \sum_{j=1}^{n} A_j z^{n-j}, \tag{6}$$

where

$$A_j = a(\alpha^j) = \sum_{i=0}^{n-1} a_i \alpha^{ij}, \quad j = 0, \pm 1, \pm 2, \ldots. \tag{7}$$

(N.B. $A(z)$ is not to be taken mod $z^n - 1$.) Alternative forms for $A(z)$ are

$$A(z) = \sum_{j=0}^{n-1} A_{-j} z^j$$

$$= \sum_{i=0}^{n-1} a_i \sum_{j=0}^{n-1} (\alpha^{-i} z)^j. \tag{8}$$

For example the MS polynomials of the codewords $1 + x + x^2 + x^4$ and $x(1 + x + x^2 + x^4)$ in the $[7, 3, 4]$ simplex code $\mathscr{S}_3$ are $z^3 + z^5 + z^6$ and $\alpha^4 z^3 + \alpha^2 z^5 + \alpha z^6$ respectively, where $\alpha \in \mathrm{GF}(2^3)$ (using Fig. 4.5).

Remarks. (1) The coefficients A_j are given by

$$
\begin{bmatrix} A_0 \\ A_1 \\ A_2 \\ \vdots \\ A_{n-1} \end{bmatrix}
=
\begin{bmatrix}
1 & 1 & 1 & \cdots & 1 \\
1 & \alpha & \alpha^2 & \cdots & \alpha^{n-1} \\
1 & \alpha^2 & \alpha^4 & \cdots & \alpha^{2(n-1)} \\
\multicolumn{5}{c}{\dotfill} \\
1 & \alpha^{n-1} & & \cdots & \alpha^{(n-1)^2}
\end{bmatrix}
\begin{bmatrix} a_0 \\ a_1 \\ a_2 \\ \vdots \\ a_{n-1} \end{bmatrix}.
$$

For this reason $A(z)$ is sometimes called a *discrete Fourier transform* of a; however, we shall always refer to it as the Mattson–Solomon polynomial.

(2) If the $a_i \in F$, then $(A_j)^q = A_{jq}$ for all j (subscripts mod n).

(3) A narrow-sense BCH code of designed distance δ can now be defined as all vectors a for which $A_1 = A_2 = \cdots = A_{\delta-1} = 0$.

(4) We apologize for using A_i both for the number of codewords of weight

i and for the coefficients of the MS polynomial. We hope the meaning will be clear from the context.

Theorem 20. (Inversion formula.) *The vector a is recovered from $A(z)$ by*

$$a_i = \frac{1}{n} A(\alpha^i), \quad i = 0, 1, \ldots, n-1,$$

$$a = \frac{1}{n} (A(1), A(\alpha), \ldots, A(\alpha^{n-1})), \qquad (9)$$

$$a(x) = \frac{1}{n} \sum_{i=0}^{n-1} A(\alpha^i) x^i.$$

Proof. Use (16) of Ch. 7 and (6).   Q.E.D.

Notation. If $f(y)$ is any polynomial, the remainder when $f(y)$ is divided by $y^n - 1$ will be denoted by $[f(y)]_n$. The *componentwise product* of two polynomials

$$f(y) = \sum_{i=0}^{n-1} f_i y^i \quad \text{and} \quad g(y) = \sum_{i=0}^{n-1} g_i y^i$$

is defined to be

$$f(y) * g(y) = \sum_{i=0}^{n-1} f_i g_i y^i.$$

Lemma 21. *If a is a binary vector then $A(z)$ is an idempotent in the ring of polynomials over $GF(2^m)$ taken modulo $z^n - 1$, i.e.*

$$[A(z)^2]_n = A(z).$$

Proof. $A(\alpha^i) = 0$ or 1 from Theorem 20. The result then follows from Lemma 2.   Q.E.D.

Theorem 22. (Other properties.)
   (i) *If $c(x) = a(x) + b(x)$, then*

$$C(z) = A(z) + B(z).$$

   (ii)

$$c(x) = [a(x)b(x)]_n$$

iff

$$C(z) = A(z) * B(z).$$

(iii) *Similarly*

$$c(x) = a(x) * b(x)$$

iff

$$C(z) = \frac{1}{n} [A(z)B(z)]_n.$$

(iv) *The MS polynomial of a cyclic shift of a*, $(a_1, a_2, \ldots, a_{n-1}, a_0)$, *is* $A(\alpha z)$.

(v) *The MS polynomials of* **0** *and* **1** *are* 0 *and* n *respectively.*

(vi) *An overall parity check on a is given by*

$$\sum_{i=0}^{n-1} a_i = A(0) = A_0. \tag{10}$$

Proof of (ii). If $c(x) = [a(x)b(x)]_n$, then $C_j = c(\alpha^j) = a(\alpha^j)b(\alpha^j) = A_j B_j$, therefore $C(z) = A(z) * B(z)$. Conversely, suppose $C_j = A_j B_j$ for all j. Then we must show

$$\sum_{k=0}^{n-1} c_k x^k = \left[\sum_{i=0}^{n-1} a_i x^i \cdot \sum_{l=0}^{n-1} b_l x^l \right]_n. \tag{11}$$

From (16) of Ch. 7, the RHS of (11) is

$$\left(\sum_{i=0}^{n-1} \frac{1}{n} \sum_{j=0}^{n-1} a(\alpha^j) \alpha^{-ij} x^i \right) \left(\sum_{l=0}^{n-1} \frac{1}{n} \sum_{J=0}^{n-1} b(\alpha^J) \alpha^{-lJ} x^l \right)$$

reduced modulo $x^n - 1$. The coefficient of x^k in this product is

$$\sum_{l=0}^{n-1} \left(\frac{1}{n} \sum_{j=0}^{n-1} a(\alpha^j) \alpha^{-lj} \right) \left(\frac{1}{n} \sum_{J=0}^{n-1} b(\alpha^J) \alpha^{-(n+k-l)J} \right)$$

$$= \frac{1}{n^2} \sum_{j=0}^{n-1} a(\alpha^j) \sum_{J=0}^{n-1} b(\alpha^J) \alpha^{-(n+k)J} \sum_{l=0}^{n-1} \alpha^{l(J-j)}.$$

By Lemma 6 of Ch. 7 the inner sum is zero unless $j = J$, so this expression becomes

$$= \frac{1}{n} \sum_{j=0}^{n-1} a(\alpha^j) b(\alpha^j) \alpha^{-jk}$$

$$= \frac{1}{n} \sum_{j=0}^{n-1} A_j B_j \alpha^{-jk}$$

$$= \frac{1}{n} \sum_{j=0}^{n-1} C_j \alpha^{-jk}$$

which by Theorem 20 is equal to c_k, the coefficient of x^k on the LHS of (11).

The proof of the rest of the theorem is straightforward and is omitted.

Q.E.D.

We now examine more carefully the mapping between $a(x)$ and its Mattson–Solomon polynomial $A(z)$. (See Fig. 8.6.)

Let $T(x)$ be the set of all polynomials in x with coefficients in $GF(q^m)$, of degree $< n$. $T(x)$ can be made into a ring (see §2 of Ch. 7) in two ways. Addition is as usual in both rings. In $T(x)_\odot$ the multiplication of two polynomials is performed modulo $x^n - 1$, i.e. $a(x) \cdot b(x) = [a(x)b(x)]_n$. In $T(x)_*$ the product of $a(x)$ and $b(x)$ is the componentwise product $a(x) * b(x)$, as defined above.

Then the mapping which sends $a(x)$ into its Mattson–Solomon polynomial $A(z)$ is, from Theorem 22, a ring isomorphism from $T(x)_\odot$ onto $T(z)_*$, and also from $T(x)_*$ onto $T(z)_\odot$. The inverse mapping is given by Equation (9). [A *ring homomorphism* is a mapping φ from a ring A into a ring B which preserves addition and multiplication, i.e. $\varphi(a_1 + a_2) = \varphi(a_1) + \varphi(a_2)$ and $\varphi(a_1 a_2) = \varphi(a_1)\varphi(a_2)$. If φ is 1-to-1 and onto it is called a ring *isomorphism*.]

In the binary case we can say a bit more. Set $q^m = 2^m$, and let $S(x)$ be the subset of $T(x)$ consisting of polynomials with coefficients restricted to $GF(2)$. Define $S(x)_\odot$ and $S(x)_*$ as before. Any binary polynomial $a(x) \in S(x)_*$ is an idempotent, i.e. satisfies $a(x) * a(x) = a(x)$. From Lemma 21, $a(x)$ is mapped into an idempotent $A(z)$ in $T(z)_\odot$. Note that the idempotents of $T(z)_\odot$ form a ring $E(z)_\odot$ say, for in characteristic 2 the sum of two idempotents is again an idempotent. In fact $E(z)_\odot$ is the image of $S(x)_*$ under the MS mapping. For the inverse image of an idempotent of $T(z)_\odot$ is in $S(x)_*$ by Theorem 20 and Lemma 2. Thus $S(x)_* \xrightarrow{\text{MS}} E(z)_\odot$ is a ring isomorphism.

Let $E(z)_*$ be the ring consisting of the polynomials in $E(z)_\odot$ but with componentwise multiplication. (Note that the elements of $E(z)_*$ are not idempotents under componentwise multiplication.) Then $S_\odot \xrightarrow{\text{MS}} E_*$ is also a ring isomorphism – see Fig. 8.6.

$T(x) =$ polynomials in x, coefficients from $GF(q^m)$, degree $< n$.

$$T(x)_* = (T(x), +, *) \xrightarrow[1-1]{\text{MS}} T(z)_\odot = (T(z), +, \cdot)$$

$$T(x)_\odot = (T(x), +, \cdot) \xrightarrow[1-1]{\text{MS}} T(z)_* = (T(z), +, *)$$

If $q = 2$, and $S(x) =$ subset of $T(x)$ with coefficients from $GF(2)$:

$$S(x)_* \xrightarrow[1-1]{\text{MS}} E(z)_\odot = \text{idempotents} \subset T(z)_\odot$$

$$S(x)_\odot \xrightarrow[1-1]{\text{MS}} E(z)_* \subset T(z)_*.$$

Fig. 8.6. The Mattson–Solomon mapping.

Binary linear codes are the linear subspaces of $S(x)_\odot$ (or $S(x)_*$), hence of $E(z)_*$ (or $E(z)_\odot$). Binary cyclic codes are the ideals in $S(x)_\odot$, and become ideals in $E(z)_*$. An ideal in $S(x)_\odot$ consists of all multiples of a fixed polynomial, as we saw in Theorem 1. Ideals in $E(z)_*$ also have a simple structure, as follows.

Lemma 23. *An ideal in $E(z)_*$ consists of the set of all polynomials $A(z) = \Sigma\, A_{-i} z^i$ such that $A_{i_1} = \cdots = A_{i_l} = 0$ for some fixed subscripts $i_1, \ldots, i_l$.*

Proof. Let $\mathscr{C}$ be the image of this ideal in $S(x)_\odot$. Since $\mathscr{C}$ is also an ideal, it has a certain set of zeros $\alpha^{i_1}, \ldots, \alpha^{i_l}$ in $GF(2^m)$ – i.e. the zeros of the generator polynomial of $\mathscr{C}$ – such that

$$a(x) \in \mathscr{C} \text{ iff } a(\alpha^{i_1}) = \cdots = a(\alpha^{i_l}) = 0$$

$$\text{iff } A_{i_1} = \cdots = A_{i_l} = 0. \qquad \text{Q.E.D.}$$

For example a narrow-sense BCH code of designed distance δ is the ideal of $E(z)_*$ with $A_1 = A_2 = \cdots = A_{\delta-1} = 0$.

In Ch. 12 it will be shown that Goppa codes can be described in terms of multiples of a fixed polynomial in $E(z)_\odot$. But ideals in $E(z)_\odot$ are not particularly interesting – see Problem 50.

Problems. (48) For $n = 3$ show that $E(z)_\odot$ consists of 0, 1, $z + z^2$, $1 + z + z^2$, $\alpha z + \alpha^2 z^2$, $\alpha^2 z + \alpha z^2$, $1 + \alpha z + \alpha^2 z^2$, $1 + \alpha^2 z + \alpha z^2$.

(49) Let $\mathscr{A}$ be a "cyclic code" in $E(z)_\odot$, i.e. a subspace of $E(z)_\odot$ such that $A(z) \in \mathscr{A} \Rightarrow [zA(z)]_n \in \mathscr{A}$. Show that $\mathscr{A} = \{0\}$ or $\mathscr{A} = \{0, 1 + z + z^2 + \cdots + z^{n-1}\}$.

(50) Let $\mathscr{A}$ be an ideal in $E(z)_\odot$. Show that the corresponding code in $S(x)_*$ consists of all vectors which are zero in certain specified coordinates. Thus these codes are not interesting. [Hint: work in $S(x)_*$.]

Locator polynomial. Suppose the vector $a = (a_0, a_1, \ldots, a_{n-1})$, $a_i \in F$, has nonzero components

$$a_{i_1}, a_{i_2}, \ldots, a_{i_w}$$

and no others, where $w = \text{wt}(a)$. We associate with a the following elements of $\mathscr{F}$:

$$X_1 = \alpha^{i_1}, \ldots, X_w = \alpha^{i_w},$$

called the *locators* of a, and the following elements of F,

$$Y_1 = a_{i_1}, \ldots, Y_w = a_{i_w},$$

giving the values of the nonzero components. Thus a is completely specified by the list $(X_1, Y_1), \ldots, (X_w, Y_w)$. Of course if a is a binary vector the Y_i's are 1.

Note that

$$a(\alpha^j) = A_j = \sum_{i=1}^{w} Y_i X_i^j.$$

Definition. The *locator polynomial* of the vector a is

$$\sigma(z) = \prod_{i=1}^{w} (1 - X_i z)$$

$$= \sum_{i=0}^{w} \sigma_i z^i, \quad \sigma_0 = 1.$$

(The roots of $\sigma(z)$ are the reciprocals of the locators.) Thus the coefficients σ_i are the elementary symmetric functions of the X_i:

$$\sigma_1 = -(X_1 + \cdots + X_w),$$
$$\sigma_2 = X_1 X_2 + X_1 X_3 + \cdots + X_{w-1} X_w,$$
$$\cdots\cdots\cdots\cdots\cdots\cdots\cdots$$
$$\sigma_w = (-1)^w X_1 \cdots X_w.$$

Generalized Newton identities. The A_i's and the σ_i's are related by a set of simultaneous linear equations.

Theorem 24. *For all j, the A_i's satisfy the recurrence*

$$A_{j+w} + \sigma_1 A_{j+w-1} + \cdots + \sigma_w A_j = 0. \tag{12}$$

In particular, taking $j = 1, 2, \ldots, w$,

$$\begin{pmatrix} A_w & A_{w-1} & \cdots & A_1 \\ A_{w+1} & A_w & \cdots & A_2 \\ \cdots\cdots\cdots\cdots\cdots \\ A_{2w-1} & A_{2w-2} & \cdots & A_w \end{pmatrix} \begin{pmatrix} \sigma_1 \\ \sigma_2 \\ \vdots \\ \sigma_w \end{pmatrix} = - \begin{pmatrix} A_{w+1} \\ A_{w+2} \\ \vdots \\ A_{2w} \end{pmatrix}. \tag{13}$$

Proof. In the equation

$$\prod_{i=1}^{w} (1 - X_i z) = 1 + \sigma_1 z + \cdots + \sigma_w z^w$$

put $z = 1/X_i$ and multiply by $Y_i X_i^{j+w}$:

$$Y_i X_i^{j+w} + \sigma_1 Y_i X_i^{j+w-1} + \cdots + \sigma_w Y_i X_i^j = 0.$$

Summing on $i = 1, \ldots, w$ gives (12).   Q.E.D.

Problems. (51) Let a be a vector of weight w. Show that

$$\begin{vmatrix} A_\nu & A_{\nu-1} & \cdots & A_1 \\ A_{\nu+1} & A_\nu & \cdots & A_2 \\ \multicolumn{4}{c}{\cdots\cdots\cdots\cdots\cdots} \\ A_{2\nu-1} & A_{2\nu-2} & \cdots & A_\nu \end{vmatrix}$$

is nonsingular if $\nu = w$, but is singular if $\nu > w$.

(52) *The usual form of Newton's identities.* Let $X_1, \ldots, X_w$ be indeterminates, and

$$\sigma(z) = \prod_{i=1}^{w} (1 - X_i z) = \sum_{i=0}^{w} \sigma_i z^i,$$

where σ_i is an elementary symmetric function of the X_i, $\sigma_0 = 1$, and $\sigma_i = 0$ for $i > w$. Define the power sums

$$P_i = \sum_{r=1}^{w} X_r^i, \quad \text{for all } i.$$

(a) If $P(z) = \sum_{i=1}^{\infty} P_i z^i$, show that

$$\sigma(z)P(z) + z\sigma'(z) = 0.$$

(b) By equating coefficients show that

$$\begin{aligned} & P_1 + \sigma_1 = 0 \\ & P_2 + \sigma_1 P_1 + 2\sigma_2 = 0 \\ & \cdots\cdots\cdots\cdots\cdots\cdots \\ & P_w + \sigma_1 P_{w-1} + \cdots + \sigma_{w-1} P_1 + w\sigma_w = 0 \end{aligned} \tag{14}$$

and, for $i > w$,

$$P_i + \sigma_1 P_{i-1} + \cdots + \sigma_w P_{i-w} = 0. \tag{15}$$

Observe that (15) agrees with (12) in the case that all Y_i are 1 (e.g. if a is a binary vector).

(53) Suppose the X_i belong to a field of characteristic p. For s fixed, show that $P_i = 0$ for $1 \leq l \leq s$ if $\sigma_i = 0$ for all l in the range $1 \leq l \leq s$ which are not divisible by p.

(54) In the binary case show that Equations (14), (15) imply

$$\begin{pmatrix} 1 & 0 & 0 & 0 & 0 & \cdots & 0 \\ A_2 & A_1 & 1 & 0 & 0 & \cdots & 0 \\ A_4 & A_3 & A_2 & A_1 & 1 & \cdots & 0 \\ \multicolumn{7}{c}{\cdots\cdots\cdots\cdots\cdots\cdots\cdots\cdots} \\ A_{2w-2} & A_{2w-3} & & \cdots & & \cdots & A_{w-1} \end{pmatrix} \begin{pmatrix} \sigma_1 \\ \sigma_2 \\ \sigma_3 \\ \vdots \\ \sigma_w \end{pmatrix} = \begin{pmatrix} A_1 \\ A_3 \\ A_5 \\ \vdots \\ A_{2w-1} \end{pmatrix} \tag{16}$$

(55) (Chien.) Use Equation (12) to give another proof of the BCH bound.

(56) (Chien and Choy.) In the binary case $(q = 2)$ show that the MS polynomial of $a(x)$ can be obtained from the locator polynomial $\sigma(z)$ of $a(x)$ by the formula

$$A(z) = \frac{z(z^n + 1)\sigma_R'(z)}{\sigma_R(z)} + (z^n + 1)w,$$

where $\sigma_R(z) = z^w \sigma(1/z)$ and $w = a(1) = \mathrm{wt}(a)$. Hence $\sigma_R(z)A(z) \equiv 0 \bmod z^n + 1$.

Application to decoding BCH codes. Equations (12)–(16) are important for decoding a BCH code of designed distance δ, as we shall see in the next chapter. In this application a is the error vector and the decoder can easily find $A_1, \ldots, A_{\delta-1}$. Then Equations (12)–(16) are used to determine $\sigma(z)$. To obtain the Y_i's, define the *evaluator polynomial*

$$\omega(z) = \sigma(z) + \sum_{i=1}^{w} zX_iY_i \prod_{\substack{j=1 \\ j \neq i}}^{w} (1 - X_jz). \tag{17}$$

Once $\omega(z)$ is known, Y_i is given by

$$Y_i = \omega(X_i^{-1}) \Big/ \prod_{j \neq i} (1 - X_jX_i^{-1})$$

$$= - X_i\omega(X_i^{-1})/\sigma'(X_i^{-1}). \tag{18}$$

Theorem 25.

$$\omega(z) = (1 + S(z))\sigma(z), \tag{19}$$

where

$$S(z) = \sum_{i=1}^{\infty} A_iz^i. \tag{20}$$

Note that since $\deg \omega(z) \leqslant \deg \sigma(z) = w$, only $A_1, \ldots, A_w$ are needed to determine $\omega(z)$ from (19).

Proof.

$$\frac{\omega(z)}{\sigma(z)} = 1 + \sum_{i=1}^{w} \frac{zX_iY_i}{1 - zX_i}$$

$$= 1 + \sum_{i=1}^{w} Y_i \sum_{j=1}^{\infty} (zX_i)^j$$

$$= 1 + S(z). \qquad \text{Q.E.D.}$$

The weight of a vector.

Theorem 26. *The weight of a is $n - r$, where r is the number of nth roots of unity which are zeros of the Mattson–Solomon polynomial $A(z)$.*

Proof. From (9).

Corollary 27. *If a has* MS *polynomial* $A(z)$, *then* $\mathrm{wt}\,(a) \geq n - \deg A(z)$.

We use Theorem 26 to give another proof of the BCH bound.

Theorem 28. (The BCH bound.) *Let $\mathscr{C}$ be a cyclic code with generator polynomial $g(x) = \Pi_{i \in K}(x - \alpha^i)$, where K contains a string of $d - 1$ consecutive integers $b, b + 1, \ldots, b + d - 2$, for some b. Then the minimum weight of any nonzero codeword $a \in \mathscr{C}$ is at least d.*

Proof. Since $a(x)$ is a multiple of $g(x)$, by hypothesis $a(\alpha^j) = 0$, for $b \leq j \leq b + d - 2$. Therefore

$$A(z) = a(\alpha)z^{n-1} + \cdots + a(\alpha^{b-1})z^{n-b+1} + a(\alpha^{b+d-1})z^{n-b-d+1} + \cdots + a(\alpha^n).$$

Let

$$\hat{A}(z) = z^{b-1}A(z) - (a(\alpha)z^{b-2} + \cdots + a(\alpha^{b-1}))(z^n - 1)$$
$$= a(\alpha^{b+d-1})z^{n-d} + \cdots + a(\alpha^n)z^{b-1} + a(\alpha)z^{b-2} + \cdots + a(\alpha^{b-1}).$$

Clearly the number of nth roots of unity which are zeros of $A(z)$ is the same as the number which are zeros of $\hat{A}(z)$. This number is $\leq \deg \hat{A}(z) \leq n - d$. Thus the weight of a is at least d by Theorem 26.   Q.E.D.

Mattson-Solomon polynomials of minimal ideals. In the remainder of this section we restrict ourselves to binary codes.

The MS polynomials of θ_s, θ_s^*, $x^i\theta_s^*$ are respectively

$$\sum_{j \in C_s} z^{n-j}, \qquad \sum_{j \in C_s} z^j, \qquad \sum_{j \in C_s} \alpha^{-ij}z^j.$$

We see that it is easier to work with θ_s^* than with θ_s. Provided we are careful, the notation can be simplified by using the trace function. Let

$$T_l(z) = z + z^2 + z^{2^2} + z^{2^3} + \cdots + z^{2^{l-1}},$$

where exponents greater than n are immediately reduced modulo n. For example, if $n = 15$, $T_4(z^3) = z^3 + z^6 + z^{12} + z^9$, and *not* $z^3 + z^6 + z^{12} + z^{24}$. Failure to observe this convention leads to errors.

Problem. (57) For $n = 31$, show that $T_5(z^3) = \omega^2$ when $z = \omega$, ω being a primitive cube root of unity.

Thus the MS polynomials of θ_s^*, $x^i\theta_s^*$ and $x^i\theta_s$ are respectively $T_{m_s}(z^s)$, $T_{m_s}(\alpha^{-is}z^s)$ and $T_{m_s}(\alpha^{+is}z^{n-s})$.

We can also handle minimal ideals which are not just cyclic shifts of the idempotent. Let $\mathcal{M}_s = \langle \theta_s^* \rangle$ be any minimal ideal. The elements $a(\alpha^{-s})$ where $a(x) \in \mathcal{M}_s$ form a subfield $\mathcal{F}_s$ of $GF(2^m)$ which is isomorphic to $GF(2^{m_s})$, as in Theorem 9. Therefore the MS polynomial of $a(x)$ is

$$A(z) = \sum_{j \in C_s} a(\alpha^{-j})z^j = T_{m_s}(a(\alpha^{-s})z^s)$$

$$= T_{m_s}(\beta z^s),$$

where $\beta = a(\alpha^{-s})$ is an element of $\mathcal{F}_s$. Exponents of β in the latter expression are reduced modulo $2^{m_s} - 1$, and the exponents of z are reduced modulo n.

Mattson–Solomon polynomials of codewords. Any element $a(x) \in R_n$ can be written, by Theorem 7 (iii), as

$$a(x) = \sum_{s \geqslant 0} a_s(x)\theta_s^*(x). \tag{21}$$

Its MS polynomial is

$$A(z) = \sum_{s \geqslant 0} T_{m_s}(\beta_s z^s), \quad \beta_s \in \mathcal{F}_s, \tag{22}$$

where s runs through a complete set of coset representatives modulo n. $a_s(x)$ may be zero, in which case β_s is also zero.

Note (21) may be written as

$$a(x) = \sum_{s \in S} x^{i_s} \theta_s^*(x), \tag{23}$$

where S is a set of coset representatives with repetitions allowed. The MS polynomial of $a(x)$ is then

$$A(z) = \sum_{s \in S} \sum_{j \in C_s} (\alpha^{-i_s} z)^j. \tag{24}$$

Let $\mathscr{C}$ be a cyclic code whose nonzeros are in $C_{-u_1}, \ldots, C_{-u_l}$. A codeword $a(x)$ in $\mathscr{C}$ can be written as

$$a(x) = \sum_{i=1}^{l} a_{u_i}(x)\theta_{u_i}^*(x),$$

and has MS polynomial

$$A(z) = \sum_{i=1}^{l} T_{n_i}(\beta_i z^{u_i}),$$

where $n_i = |C_{-u_i}|$ and $\beta_i \in \mathcal{F}_{n_i}$. A codeword in $\mathscr{C}^\perp$ has the form

$$b(x) = \sum_{u \neq u_i} b_u(x)\theta_u(x),$$

with MS polynomial

$$B(z) = \sum_{u \neq u_i} T_{|C_u|}(\beta_u z^{n-u}).$$

Example. *Codes of length 15* (cf. Problem 7). The $[15, 4, 8]$ simplex code $\langle \theta_1^* \rangle = \langle \theta_7 \rangle$ consists of the vectors with MS polynomials

$$A(z) = T_4(\beta z), \quad \beta \in \mathrm{GF}(2^4).$$

In general the $[n = 2^m - 1, m, 2^{m-1}]$ simplex code $\mathscr{S}_m$ is given by

$$A(z) = T_m(\beta z), \quad \beta \in \mathrm{GF}(2^m). \tag{25}$$

The $[15, 2, 10]$ simplex code $\langle \theta_5 \rangle = \langle \theta_5^* \rangle$ consists of $0, \theta_5, x\theta_5, x^2\theta_5$. The MS polynomials are

$$T_2(\beta z^5), \quad \beta \in \{0, 1, \alpha^5, \alpha^{10}\} \cong \mathrm{GF}(2^2).$$

I.e.

$$0, \; z^5 + z^{10}, \; \alpha^5 z^5 + \alpha^{10} z^{10}, \; \alpha^{10} z^5 + \alpha^5 z^{10}.$$

The $[15, 4, 6]$ code $\langle \theta_3 \rangle = \langle \theta_3^* \rangle$ consists of the MS polynomials

$$T_4(\beta z^3), \quad \beta \in \mathrm{GF}(2^4).$$

The $[15, 11, 3]$ Hamming code consists of all vectors of the form

$$a_0 \theta_0(x) + a_1 x^{i_1} \theta_1(x) + f_3(x)\theta_3(x) + a_5 x^{i_5}\theta_5(x),$$

where $a_0, a_1, a_5 \in \mathrm{GF}(2)$; $i_1, i_5 \in \{0, 1, \dots, 14\}$; $\deg f_3(x) \leqslant 3$. The MS polynomials are

$$\sum_{s \neq -1} T_{m_s}(\beta_s z^s), \quad \beta_s \in \mathscr{F}_s.$$

A similar expression holds for any Hamming code.

Problem. (58) Show that the $[23, 12, 7]$ Golay code consists of the vectors with MS polynomial

$$a + T_{11}(\beta z^5), \quad a \in \mathrm{GF}(2), \beta \in \mathrm{GF}(2^{11}).$$

Another formula for the weight of a vector. Theorem 31 gives a formula which will sometimes enable us to find the weight distribution of a code explicitly. The proof depends on an identity (Theorem 29) satisfied by the MS polynomial.

Let $A'(z)$ denote the derivative of $A(z)$. We note that in characteristic 2, $zA'(z)$ consists of those terms γz^j in $A(z)$ for which j is odd.

Theorem 29. *The MS polynomial $A(z)$ of any vector $\mathbf{a} \in R_n$ satisfies*

$$zA'(z) = \frac{A(z)(A(z) + 1)}{z^n + 1}.$$

Proof. By (24) it suffices to prove the theorem for $A(z) = \sum_{j \in C_s} (\beta z)^j$, where $\beta^n = 1$. Let $[r]_s$ denote the remainder when r is divided by s. Then

$$A(z) = \sum_{i=0}^{m_s-1} (\beta z)^{[s2^i]_n},$$

$$A(z)^2 = \sum_{i=0}^{m_s-1} (\beta z)^{2[s2^i]_n}.$$

Note that the exponents in $A(z)^2$ may be greater than n. If $[s2^i]_n$ is an even number, then

$$[s2^i]_n = 2[s2^{i-1}]_n$$

and in $A(z)^2 + A(z)$ the term with exponent $[s2^i]_n$ in $A(z)$ cancels the term with exponent $2[s2^{i-1}]_n$ in $A(z)^2$. If the exponent $[s2^i]_n$ is odd, then

$$[s2^i]_n = 2[s2^{i-1}]_n - n.$$

Combining the term with exponent $2[s2^i]_n - n$ in $A(z)$ and the term with exponent $2[s2^i]_n$ in $A(z)^2$ we obtain

$$(\beta z)^{[s2^i]_n}(1 + z^n) \text{ (recall that } \beta^n = 1).$$

Thus we obtain $(z^n + 1)$ multiplied by the terms with odd exponents in $A(z)$, i.e. $zA'(z)(z^n + 1)$.   Q.E.D.

Problem. (59) Check the theorem for $n = 9$, $C_1 = \{1, 2, 4, 8, 7, 5\}$.

Corollary 30. *If ζ is a zero of $A'(z)$ or $z^{n+1} + z$, then $A(\zeta)$ is 0 or 1.*

Define

$$f_1(z) = (A(z), z^n + 1),$$
$$f_2(z) = (A(z), zA'(z)),$$

where $(a(z), b(z))$ denotes the monic greatest common divisor of $a(z)$ and $b(z)$. Since $A(z)$, $A(z) + 1$ are relatively prime, Theorem 29 implies

$$A(z) = \beta f_1(z) f_2(z), \quad \text{for some } \beta \in GF(2^m), \ \beta \neq 0.$$

We have therefore proved:

Theorem 31. *The weight of the vector with MS polynomial $A(z)$ is*

$$n - \deg f_1(z) = n - \deg A(z) + \deg f_2(z).$$

Problem. (60) Let the binary expansion of the weight w of vector

$a = (a_0 a_1 \cdots a_{n-1})$ be

$$w = \Gamma_1(a) + 2\Gamma_2(a) + 4\Gamma_4(a) + \cdots$$

where $\Gamma_i(a) = 0$ or 1. Show:

(i) $\Gamma_2(a) \equiv \binom{w}{2} \bmod 2$ [Hint: Lucas' Theorem 28 of Ch. 13];

(ii) $\binom{w}{2} = \sum_{i<j} a_i a_j$;

(iii) Hence if w is even,

$$\Gamma_2(a) = \sum_{i=1}^{(n-1)/2} A_i A_{n-i}.$$

§7. Some weight distributions

Theorem 31 is useful when it is easy to find the degree of $f_2(z)$.

Examples. (1) The simplex code of length $n = 2^m - 1$. From (25) any codeword has MS polynomial

$$A(z) = \sum_{i=0}^{m-1} (\beta z)^{2^i}.$$

Hence $zA'(z) = \beta z$ and $f_2(z) = \gcd(A(z), zA'(z)) = z$. The degree of $f_2(z)$ is 1, and the weight of the codeword is

$$2^m - 1 - 2^{m-1} + 1 = 2^{m-1}.$$

(2) The degenerate simplex code $\langle \theta_s^* \rangle$ of length $n = 2^m - 1$, where $m = 2u \geq 4$ is even, and $s = 2^u + 1$.

The cyclotomic coset C_s is

$$s = 2^u + 1, \qquad 2^{u+1} + 2, \ldots, 2^{m-1} + 2^{u-1}$$

of length u. (Check for $n = 15$, $s = 5$, and $n = 63$, $s = 9$.) The ideal with idempotent θ_s^* has dimension u, and its nonzero codewords are $x^j \theta_s^*$, $0 \leq j \leq 2^u - 2$. The MS polynomial of a typical codeword is

$$A(\beta, z) = \sum_{j \in C_s} (\beta z)^j,$$

of degree $2^{m-1} + 2^{u-1}$. Thus $zA'(\beta, z) = \beta^s z^s$, and $f_2(z) = z^s$ has degree $2^u + 1$, so the weight of each codeword is

$$2^{2u-1} + 2^{u-1}.$$

(3) The code with idempotent $\theta_1^* + \theta_s^*$, with n, m, s as in Example 2. This code has dimension $3u$, and a typical codeword is

$$ax^i \theta_1^* + bx^j \theta_s^*,$$

where $a, b \in GF(2)$ and $0 \le i \le 2^{2u} - 2$, $0 \le j \le 2^u - 2$. If $a = 1$, $b = 0$ this codeword has weight 2^{2u-1}. If $a = 0$, $b = 1$ the weight is $2^{2u-1} + 2^{u-1}$. If $a = b = 1$ we have a cyclic permutation of a codeword of the form

$$x^i \theta_1^* + \theta_s^*.$$

The MS polynomial is

$$A(\beta, z) = \sum_{j \in C_1} (\beta z)^j + \sum_{j \in C_s} z^j,$$

$$A'(\beta, z) = \beta + z^r, \quad \text{where } r = 2^u.$$

We shall find $f_2(z)$. Now $A'(\beta, z) = 0$ for $\beta = z^r$, or

$$z = z^{2^m} = (z^r)^r = \beta^r.$$

Then

$$A(\beta, \beta^r) = T_m(\beta^{r+1}) + \sum_{j \in C_s} \beta^{rj}.$$

Now β^{r+1} is in $GF(2^u)$ by Theorem 8 of Ch. 4. Thus

$$T_m(\beta^{r+1}) = 2T_u(\beta^{r+1}) = 0,$$

$$\sum_{j \in C_s} \beta^{rj} = \beta^{r+1} + \beta^{2r+2} + \cdots + \beta^{rs/2},$$

$$= T_u(\beta^{r+1}).$$

This is either 0 or 1, depending on β. If $A(\beta, \beta^r) = 0$, the degree of $f_2(z)$ is $r + 1$ and the codeword has weight $2^{2u-1} + 2^{u-1}$. If $A(\beta, \beta^r) = 1$, the degree of $f_2(z)$ is 1, and the codeword has weight $2^{2u-1} - 2^{u-1}$. Thus we have:

Theorem 32. *For $m = 2u \ge 4$, and $s = 2^u + 1$, the code with idempotent $\theta_1^* + \theta_s^*$ has three nonzero weights, namely $\tau_1 = 2^{2u-1} - 2^{u-1}$, $\tau_2 = 2^{2u-1}$, $\tau_3 = 2^{2u-1} + 2^{u-1}$.*

Theorem 33. *The weight distribution of this code is:*

Weight	Number of Codewords
0	1
$2^{2u-1} - 2^{u-1}$	$2^{u-1}(2^{2u} - 1)$
2^{2u-1}	$2^{2u} - 1$
$2^{2u-1} + 2^{u-1}$	$2^{3u-1} - 2^{2u} + 2^{u-1}$.

Proof. Since the code contains $\mathcal{H}_m^\perp$, its dual is contained in $\mathcal{H}_m$. Thus $d' \ge 3$ and we may apply Theorem 2 of Ch. 6.   Q.E.D.

Problems. (61) Show that the dual code has minimum distance 3.

(62) Let $\mathscr{C}$ be the code with idempotent $\theta_0 + \theta_1^* + \theta_s^*$, formed by adding the

codeword 1. Show that $\mathscr{C}$ has dimension $3u + 1$, and $\mathscr{C}^\perp$ has minimum distance 4. Let $\hat{\mathscr{C}}$ be the code obtained from $\mathscr{C}$ by adding an overall parity check. Show that the weight distribution of $\hat{\mathscr{C}}$ is:

Weight	Number of Codewords
0	1
$\tau_1 = 2^{2u-1} - 2^{u-1}$	$A_{\tau_1} = 2^{3u} - 2^{2u}$
$\tau_2 = 2^{2u-1}$	$A_{\tau_2} = 2(2^{2u} - 1)$
$\tau_3 = 2^{2u-1} + 2^{u-1}$	$A_{\tau_3} = 2^{3u} - 2^{2u}$
2^{2u}	1.

Some Examples. The $[16, 7, 6]$ code obtained by extending the cyclic code with idempotent $\theta_0 + \theta_1^* + \theta_5^*$:

$$i:\quad 0\quad 6\quad 8\quad 10\quad 16$$
$$A_i:\quad 1\quad 48\quad 30\quad 48\quad 1$$

The $[64, 10, 28]$ code obtained by extending the cyclic code with idempotent $\theta_0 + \theta_1^* + \theta_9^*$:

$$i:\quad 0\quad 28\quad 32\quad 36\quad 64$$
$$A_i:\quad 1\quad 448\quad 126\quad 448\quad 1$$

Example. (4) The code of length $n = 2^m - 1$, where $m = 2u + 1$ is odd, which has idempotent $\theta_1^* + \theta_l^*$, $l = 2^u + 1$. Now l and n are relatively prime and

$$C_l = \{2^u + 1, 2^{u+1} + 2, \ldots, 2^{2u} + 2^u, 2^{u+1} + 1, \ldots, 2^{2u} + 2^{u-1}\}$$

of length m. (Check for $n = 31$, $l = 5$, and $n = 127$, $l = 9$.) The code has dimension $2m$, and consists of the codewords

$$ax^i\theta_1^* + bx^i\theta_l^*.$$

If either a or b is zero, the codeword has weight 2^{2u}. Every codeword with $a = b = 1$ is a cyclic permutation of a codeword of the form

$$x^i\theta_1^* + \theta_l^*.$$

The MS polynomial, of degree $2^{2u} + 2^u$, is

$$A(\beta, z) = \sum_{j \in C_1} (\beta z)^j + \sum_{j \in C_l} z^j, \quad \beta \in \mathrm{GF}(2^m).$$

Let $r = 2^u$, then $A'(\beta, z) = \beta + z^r + z^{2r} = (\beta^{2r} + z + z^2)^r$. Then $\tau(\beta, z) = \beta^{2r} + z + z^2$ may have 0, 1 or 2 zeros in common with $A(z)$, and the degree of $f_2(z)$ is correspondingly 1, $r + 1$, or $2r + 1$.

Thus the only possible weights are

$$2^m - 1 - (2^{2u} + 2^u - 1) = 2^{2u} - 2^u,$$
$$2^m - 1 - (2^{2u} + 2^u - 1 - 2^u) = 2^{2u},$$
$$2^m - 1 - (2^{2u} + 2^u - 1 - 2^{u+1}) = 2^{2u} + 2^u.$$

This argument does not prove that both extremal weights occur. However, since $\mathscr{C}^{\perp}$ has minimum distance ≥ 3, we may use Theorem 2 of Ch. 6 to calculate the number of codewords of each weight:

Theorem 34. *The $[n = 2^m - 1, 2m, 2^{2u} - 2^u]$ cyclic code, where $m = 2u + 1$, with idempotent $\theta_1^* + \theta_l^*$, $l = 2^u + 1$, has the following weight distribution.*

Weight	Number of Codewords
0	1
$\tau_1 = 2^{2u} - 2^u$	$A_{\tau_1} = n(2^{2u-1} + 2^{u-1}),$
$\tau_2 = 2^{2u}$	$A_{\tau_2} = n(2^{2u} + 1),$
$\tau_3 = 2^{2u} + 2^u$	$A_{\tau_3} = n(2^{2u-1} - 2^{u-1}).$

Some examples of Theorem 34 are shown in Fig. 8.7.

Problem. (63) Prove the following for $n = 31$, $m = 5$.
   (1) If $T(\alpha) = 0$ then $\alpha = \lambda + \lambda^2 = (\lambda + 1) + (\lambda + 1)^2$ for some λ in GF(2^5). Thus $\tau(\beta, z)$, $\alpha = \beta^{2r}$, has two zeros in GF(2^5).
   (2) In this case exactly one of these is a zero of $A(\beta, z)$.
   (3) If $T(\beta) = 1$, $\tau(\beta, z)$ is irreducible over GF(2^5).

Corollary 35. $\mathscr{C}^{\perp}$ *has minimum distance 5. Thus the codewords of each weight form a 2-design with parameters*

$$\lambda_{\tau_1} = A_{\tau_1} \frac{\tau_1(\tau_1 - 1)}{(2^m - 1)(2^m - 2)} = 2^{2u-2}(2^{2u} - 2^u - 1),$$

$$\lambda_{\tau_2} = A_{\tau_2} \frac{\tau_2(\tau_2 - 1)}{(2^m - 1)(2^m - 2)} = 2^{2u-1}(2^{2u} + 1),$$

$$\lambda_{\tau_3} = A_{\tau_3} \frac{\tau_3(\tau_3 - 1)}{(2^m - 1)(2^m - 2)} = 2^{2u-2}(2^{2u} + 2^u - 1).$$

Problem. (64) Form the $[2^m, 2m + 1]$ code $\hat{\mathscr{C}}$ by adding a parity check to the code with idempotent $\theta_0 + \theta_1^* + \theta_l^*$. Show that the vectors of each weight form a 3-design and find the parameters.

[31, 10, 12]	i:	0	12	16	20
	A_i:	1	10.31	17.31	6.31
[127, 14, 56]	i:	0	56	64	72
	A_i:	1	36.127	65.127	28.127
[511, 18, 240]	i:	0	240	256	272
	A_i:	1	136.511	257.511	120.511

Fig. 8.7. Examples of Theorem 34.

Notes on Chapter 8

§2. Some of §§2, 3 follow immediately from ring theory. (See Burrow [212], Curtis and Reiner [321]). The group $G = \{1, x, \ldots, x^{n-1}\}$ (of §2 of Ch. 7) has order n which is relatively prime to q, so any representation of G over $GF(q)$ is completely reducible (Maschke's theorem). Therefore the group algebra FG (and so R_n) is semisimple. By Wedderburn's first theorem, R_n is a direct sum of minimal ideals, and each minimal ideal contains a unique idempotent generator. For properties of idempotents see Cohen et al. [297], Goppa [540] and MacWilliams [874, 882].

§3. Theorem 6 is not recommended for machine computation of the primitive idempotents. Prange's algorithm ([1077], see also MacWilliams [874]) is superior. Problems 14, 15, 17, 22 are due to Goodman [534], Lempel [813], Goppa [540] and Weiss [1397].

§4. The following papers study the weight distribution of cyclic and related codes: Baumert [83], Baumert and McEliece [86], Berlekamp [113, 114, 118] Buchner [208], Delsarte and Goethals [362] for other two-weight codes, Goethals [488], Hartmann et al. [613], Kasami [727, 729], Kerdock et al. [759], MacWilliams [874], McEliece [940, 941, 945], McEliece and Rumsey [949], Oganesyan et al. [1003–1006], Robillard [1117], Seguin [1174], Solomon [1249], Stein et al. [1275] and Willett [1418].

§5. Problem 33, Theorem 14, Theorem 16, Problem 43, and Problems 38, 46 are due to MacWilliams [870, 871], Prange [1074–1077], Peterson [1038], Kasami et al. [738] (see also Delsarte [344]), and Camion [239] respectively. Lin [832] proved Theorem 18 in the case $t = 2$.

§6. MS polynomials were introduced by Mattson and Solomon [928]; see also

Chien and Choy [286], Kerdock et al. [759]; and Dym and McKean [396, Ch. 4] for the discrete Fourier transform. For Newton's identities and other properties of symmetric functions see David and Barton [331, Ch. 17], David, Kendall and Barton [332], Foulkes [447], Kendall and Stuart [756, Ch. 3], Turnbull [1342, §32] and Van der Waerden [1376, vol. 1, p. 81]. Problem 55 is from Chien [283] (see also Delsarte [346]), Problem 56 from Chien and Choy [286], and Theorem 25 is due to Forney [435].

Problem 60 is from Solomon and McEliece [1256]. Other properties of the $\Gamma_i(a)$ will be found there and in McEliece [939, 940]. These properties are useful in finding weight distributions.

§7. For odd m the weight distributions of all codes of length $2^m - 1$ with idempotents of the form $\theta_1^* + \theta_s^*$, $s = 2^i + 1$, have been given in [759]. Berle-kamp [118] and Kasami [729] have given very general results on the weight distribution of subcodes of second-order Reed–Muller codes, i.e. of codes with idempotents $\Sigma_j\, \theta_j^*$ where j runs through a set of numbers of the form $2^i + 1$ (see Ch. 15).

9

BCH codes

§1. Introduction

BCH codes are of great practical importance for error correction, particularly if the expected number of errors is small compared with the length. These codes were first introduced in Ch. 3, where double-error-correcting BCH codes were constructed as a generalization of Hamming codes. But BCH codes are best considered as cyclic codes, and so it was not until §6 of Ch. 7 that the general definition of a t-error-correcting BCH code was given. Naturally most of the theory of cyclic codes given in Chs. 7 and 8 applies to BCH codes.

For example, BCH codes are easily encoded by either of the methods of §8 of Ch. 7. Decoding will be dealt with in §6 below.

We recall from Ch. 7 that a BCH code over $GF(q)$ of length n and designed distance δ is the largest possible cyclic code having zeros

$$\alpha^b, \alpha^{b+1}, \ldots, \alpha^{b+\delta-2},$$

where $\alpha \in GF(q^m)$ is a primitive n^{th} root of unity, b is a nonnegative integer, and m is the multiplicative order of q modulo n (see §5 of Ch. 7).

Important special cases are $b = 1$ (called *narrow-sense* BCH codes), or $n = q^m - 1$ (called *primitive* BCH codes). A BCH code is assumed to be narrow-sense and primitive unless stated otherwise. BCH codes with $n = q - 1$ (i.e. $m = 1$, $\alpha \in GF(q)$) are another important subclass. These are known as *Reed–Solomon* codes, and the next chapter is devoted to their special properties.

The following bounds on the dimension k and minimum distance d of any BCH code were obtained in §6 of Ch. 7.

Theorem 1. (a) *The* BCH *code over* $\mathrm{GF}(q)$ *of length* n *and designed distance* δ *has dimension* $k \geq n - m(\delta - 1)$ *and minimum distance* $d \geq \delta$.

(b) *The binary* BCH *code of length* n *and designed distance* $\delta = 2t + 1$ *has dimension* $k \geq n - mt$ *and minimum distance* $d \geq \delta$.

Naturally one wishes to know how close these bounds are to the true values of k and d, and we shall study this question in §§2 and 3 of this chapter. The answer is roughly that they are quite close. However the precise determination of d is still unsolved in general.

§4 contains a table of binary BCH codes of length ≤ 255. In this range (and in fact for lengths up to several thousand) BCH codes are among the best codes we know (compare Appendix A). But as shown in §5 their performance deteriorates if the rate $R = k/n$ is held fixed and the length approaches infinity.

There is a large literature on decoding BCH codes, and very efficient algorithms exist. §6 contains an introduction to the decoding, giving a simple description of the main steps.

The final step is to find the zeros of the error locator polynomial. If this should be a quadratic equation (over a field of characteristic 2) the zeros can be easily found, as we see in §7. The results of §7 are also used in §8 to show that binary double-error-correcting BCH codes are quasi-perfect.

In §9 a deep result from number theory is used to show that if $\mathscr{C}$ is a BCH code of designed distance not exceeding about $n^{\frac{1}{2}}$, then all the weights of $\mathscr{C}^{\perp}$ are close to $\frac{1}{2}n$.

The weight distributions of many codes can be observed to have the approximate shape of a normal probability density function. §10 gives a partial explanation of this phenomenon.

In the final section of the chapter we describe an interesting family of non-BCH triple-error-correcting codes.

Notation. $\mathscr{C}$ will usually denote an $[n, k, d]$ (narrow-sense, primitive) BCH code over $\mathrm{GF}(q)$ of designed distance δ. As usual $\alpha \in \mathrm{GF}(q^m)$ is a primitive n^{th} root of unity, where m is the multiplicative order of $q \bmod n$ (see §5 of Ch. 7).

Let $a = (a_0, \ldots, a_{n-1})$ be a vector over $\mathrm{GF}(q)$ of weight w, with

$$a(x) = \sum_{i=0}^{n-1} a_i x^i.$$

If $a_{i_1}, \ldots, a_{i_w}$ are the non-zero components of a, we define the *locators* of a to be $X_r = \alpha^{i_r}$, $r = 1, \ldots, w$, the *locator polynomial* to be

$$\sigma(z) = \prod_{r=1}^{w} (1 - X_r z) = \sum_{r=0}^{w} \sigma_r z^r, \tag{1}$$

and let $Y_r = a_{i_r}$, $r = 1, \ldots, w$. Also the power sums are defined by

$$A_l = a(\alpha^l) = \sum_{r=1}^{w} Y_r X_r^l, \quad l = 0, 1, \ldots. \tag{2}$$

(See §6 of Ch. 8.) In particular, if a belongs to $\mathscr{C}$, then

$$A_l = 0, \quad \text{for } 1 \leq l \leq \delta - 1.$$

§2. The true minimum distance of a BCH code

The first three theorems say that under certain conditions the true minimum distance of a BCH code is equal to the designed distance. The last theorem gives an upper bound on the true minimum distance for any BCH code. The first two results are easy:

Theorem 2. (Farr.) *The binary BCH code of length* $n = 2^m - 1$ *and designed distance* $\delta = 2t + 1$ *has minimum distance* $2t + 1$ *if* (a)

$$\sum_{i=0}^{t+1} \binom{2^m - 1}{i} > 2^{mt}, \tag{3}$$

or if (b)

$$m > \log_2(t + 1)! + 1 \sim t \log_2(t/e) \text{ as } t \to \infty.$$

Proof. (a) By Corollary 17 of Ch. 8 the minimum distance d is odd. Suppose $d \geq 2t + 3$. The dimension of the code is $\geq n - mt$, so from the sphere-packing bound

$$2^{n-mt} \sum_{i=0}^{t+1} \binom{n}{i} \leq 2^n.$$

But this contradicts (3). That (b) implies (a) is a routine calculation which we omit.   Q.E.D.

Example. Let $m = 5$, $t = 1, 2, 3$. Then it is readily checked that

$$\sum_{i=0}^{t+1} \binom{31}{i} > 2^{5t}.$$

Thus as shown in Fig. 9.1, the codes of length 31 with $\delta = 3, 5, 7$ actually have $d = 3, 5, 7$.

Theorem 3. (Peterson.) *If* $n = ab$ *then the* (*not necessarily primitive*) *binary*

BCH *code* $\mathscr{C}$ *of length n and designed distance a has minimum distance exactly a.*

Proof. Let α be a primitive n^{th} root of unity, so that $\alpha^{ib} \neq 1$ for $i < a$. Since $x^n - 1 = (x^b - 1)(1 + x^b + x^{2b} + \cdots + x^{(a-1)b})$, the elements $\alpha, \alpha^2, \ldots, \alpha^{a-1}$ are not zeros of $x^b - 1$ and are zeros of $1 + x^b + \cdots + x^{(a-1)b}$. Therefore the latter is a codeword of weight a in $\mathscr{C}$.   Q.E.D.

Example. The code of length 255 and designed distance 51 has minimum distance exactly 51.

The following Lemma characterizes locator polynomials of vectors of 0's and 1's (over any finite field).

Lemma 4. *Let*

$$\sigma(z) = \sum_{i=0}^{w} \sigma_i z^i$$

be any polynomial over $\text{GF}(q^m)$. *Then* $\sigma(z)$ *is the locator polynomial of a codeword a of 0's and 1's in a BCH code over* $\text{GF}(q)$ *of designed distance* δ *iff conditions* (i) *and* (ii) *hold.*
    (i) *The zeros of* $\sigma(z)$ *are distinct* n^{th} *roots of unity.*
    (ii) σ_i *is zero for all i in the range* $1 \leq i \leq \delta - 1$ *which are not divisible by p, where q is a power of the prime p.*

Proof. If $\sigma(z)$ is the locator polynomial of such a vector a, then $A_l = 0$ for $1 \leq l \leq \delta - 1$. The result then follows by Problem (53) of Ch. 8. On the other hand suppose (i), (ii) hold, and let $X_1, \ldots, X_w$ be the reciprocals of the zeros of $\sigma(z)$. Then the vector a with 1's in these locations satisfies $A_l = 0$ for $1 \leq l \leq \delta - 1$ (again by Problem (53)) and so belongs to the BCH code.
                                                                                    Q.E.D.

We use this lemma to prove

Theorem 5. *A BCH code of length* $n = q^m - 1$ *and designed distance* $\delta = q^h - 1$ *over* $\text{GF}(q)$ *has true minimum distance* δ.

Proof. We shall find a codeword of weight δ. Let U be an h-dimensional subspace of GF(q^m). By Lemma 21 of Ch. 4

$$L(z) = \prod_{\beta \in U} (z - \beta)$$

$$= z^{q^h} + l_{h-1} z^{q^{h-1}} + \cdots + l_0 z \tag{4}$$

is a linearized polynomial over GF(q^m). Consider the vector a which has as locators all nonzero β in U, and has 1's in these locations. The locator polynomial of a is therefore

$$\sigma(z) = \prod_{\substack{\beta \neq 0 \\ \beta \in U}} (1 - \beta z)$$

$$= z^{q^h} L(z^{-1})$$

$$= \sum_{i=0}^{q^h-1} \sigma_i z^i,$$

where $\sigma_i = 0$ if i is not equal to $q^h - q^j$, for $j = 0, 1, \ldots, h$. By Lemma 4 a is a vector of weight $q^h - 1$ in the BCH code of designed distance $q^h - 1$.

Q.E.D.

Example. Let $q = 2$, $m = 3$, $h = 2$ – i.e. consider the BCH code of length 7 and designed distance 3. This is the [7, 4, 3] Hamming code. The proof of the theorem shows that the code contains the incidence vectors of all lines in the projective plane over GF(2) (see Fig. 2.12 and Equation (24) of Ch. 2), and hence the true minimum distance is 3.

Theorem 6. *The true minimum distance d of a primitive BCH code $\mathscr{C}$ of designed distance δ over GF(q) is at most $q\delta - 1$.*

Proof. Define d_0 by $q^{h-1} \leq \delta < d_0 = q^h - 1$. Since BCH codes are nested, $\mathscr{C}$ contains the code of designed distance d_0, which by the previous theorem has true minimum distance d_0. Therefore

$$d \leq d_0 \leq q\delta - 1. \qquad \text{Q.E.D.}$$

Although a number of results similar to Theorems 2, 3, 5 are known (see notes at the end of the chapter), the following problem remains unsolved:

Research Problem (9.1). Find necessary and sufficient conditions on n and the designed distance δ for the minimum distance d to equal δ.

The exact determination of d, in the case $d > \delta$, will be even harder.

§3. The number of information symbols in BCH codes

Suppose $\mathscr{C}$ is a BCH code over GF(q) of length $n = q^m - 1$ and designed distance δ. In Theorem 1 we saw that the number of information symbols, which we denote by $I(n, \delta)$, is at least $n - m(\delta - 1)$. But the precise number depends on the degree of the generator polynomial $g(x)$, for

$$I(n, \delta) = n - \deg g(x).$$

The degree of $g(x)$. By definition $g(x)$ is the lowest degree polynomial having $\alpha, \alpha^2, \ldots, \alpha^{\delta-1}$ as zeros, where α is a primitive element of GF(q^m). If α^i is a zero of $g(x)$ so are all the conjugates of α^i, namely $\alpha^{iq}, \alpha^{iq^2}, \ldots$. The number of different conjugates of α^i is the smallest integer m_i such that $iq^{m_i} \equiv i$ (mod $q^m - 1$). In the notation of §3 of Ch. 4, m_i is the size of the cyclotomic coset containing i.

There is an easy way to find m_i. Write i to the base q as

$$i = i_{m-1}q^{m-1} + \cdots + i_1 q + i_0,$$

where $0 \leqslant i_\nu \leqslant q - 1$. Then

$$qi \equiv i_{m-2}q^{m-1} + \cdots + i_1 q^2 + i_0 q + i_{m-1} \,(\text{mod } q^m - 1).$$

I.e. the effect of multiplying by q is a cyclic shift of the m-tuple $(i_{m-1}, \ldots, i_1, i_0)$.

For example, suppose $q = 2$, $m = 4$, $q^m - 1 = 15$. Then $i = 1$ gives the 4-tuple (0001), which has period 4 under cyclic shifts: (0001), (0010), (0100), (1000), (0001),...; hence $m_1 = 4$. Equivalently the cyclotomic coset $C_1 = \{1, 2, 4, 8\}$ has size 4. On the other hand $i = 5 = (0101)$ has period 2: (0101), (1010), (0101),...; hence $m_5 = 2$. Thus $C_5 = \{5, 10\}$.

If i is sufficiently small compared to n, then $m_i = m$. More precisely, we have

Theorem 7. *For* $q = 2$ *and* $n = 2^m - 1$, *the cyclotomic cosets* $C_1, C_3, C_5, \ldots, C_i$ *are distinct and each contain m elements, provided*

$$i < 2^{\lceil m/2 \rceil} + 1 \tag{5}$$

(where $\lceil x \rceil$ denotes the smallest integer $\geqslant x$). Thus $m_i = m$ if i satisfies (5).

Proof. (i) We first prove that these cosets are distinct. Consider the coset C_i. The binary expansion of i written as an m-tuple is

$$\underbrace{00 \cdots 0}_{\lceil m/2 \rceil} XX \cdots X1, \quad X = 0 \text{ or } 1. \tag{6}$$

Could there be a cyclic shift of (6) which also begins with $[m/2]$ zeros and ends with a 1? Obviously not. Therefore there cannot be an odd $j < 2^{\lceil m/2 \rceil} + 1$ with $C_i = C_j$.

(ii) The number of elements in C_i is the number of distinct cyclic shifts of the binary expansion of i written as an m-tuple. It is easy to see that all m cyclic shifts are distinct if $i < 2^{\lceil m/2 \rceil} + 1$. Q.E.D.

Example. $n = 63$, $m = 6$. The first few cyclotomic cosets are:

	binary 6-tuple
$C_1 = \{1, 2, 4, 8, 16, 32\},$	000001
$C_3 = \{3, 6, 12, 24, 48, 33\},$	000011
$C_5 = \{5, 10, 20, 40, 17, 34\},$	000101
$C_7 = \{7, 14, 28, 56, 49, 35\},$	000111
$C_9 = \{9, 18, 36\}.$	001001

Indeed, $C_1, \ldots, C_7$ are distinct and have size 6, as predicted by the theorem.

This theorem has an immediate corollary.

Corollary 8. *Let $\mathscr{C}$ be a binary BCH code of length $n = 2^m - 1$ and designed distance $\delta = 2t + 1$, where*

$$2t - 1 < 2^{\lceil m/2 \rceil} + 1.$$

Then dim $\mathscr{C} = 2^m - 1 - mt.$

This corollary applies when the designed distance δ is small. For arbitrary δ we have

Theorem 9. *The degree of $g(x)$ is the number of integers i in the range $1 \leq i \leq q^m - 1$ such that some cyclic shift of the q-ary expansion of i (written as an m-tuple) is $\leq \delta - 1$.*

Proof. The condition simply states that α^i is a root of $g(x)$ iff α^i has conjugate α^{iq^r} with $1 \leq iq^r \leq \delta - 1$. Q.E.D.

For example, consider the binary BCH code of length 15 and designed distance $\delta = 5$. $g(x)$ has roots $\alpha, \alpha^2, \alpha^4, \alpha^8$, and $\alpha^3, \alpha^6, \alpha^{12}, \alpha^9$. The exponents i are

$$0001, \ 0010, \ 0100, \ 1000,$$
$$0011, \ 0110, \ 1100, \ 1001,$$

and deg $g(x) = 8$, verifying the theorem. The number of information symbols $I(15, 5) = 15 - 8 = 7$.

But if $\delta = 7$, $g(x)$ has the additional zeros α^5, α^{10}, with $i = (0101)$, (1010), and deg $g(x) = 10$, $I(15, 7) = 15 - 10 = 5$.

Problems. (1) How many information symbols are there in the binary BCH codes of length 63 and with $\delta = 5, 7, 11, 21$, and 27?

(2) Repeat for the BCH codes over GF(3) of length 80 and $\delta = 4, 7$, and 11.

Designed distance $\delta = q^\lambda$. In the rest of this section we shall find an upper bound to $I(n, \delta)$ in the special case where $n = q^m - 1$ and $\delta = q^\lambda$. Note that

$$\frac{\delta}{n} = \frac{q^\lambda}{q^m - 1} \sim q^{-r}, \quad \text{as } \lambda \to \infty,$$

where $r = m - \lambda$.

In the m-tuple $i_0, i_1, \dots, i_{m-1}$ a maximal string of consecutive zeros is called a *run*. We distinguish between *straight* runs and *circular* runs. For example 0100100 contains two straight runs of length 2 and one of length 1, and one circular run of length 3 and one of length 2.

Theorem 9 now implies:

Lemma 10. *The degree of $g(x)$ is the number of integers i in the range $1 \leq i \leq q^m - 1$ such that the q-ary expansion of i (written as an m-tuple) contains a circular run of length at least r.*

Proof. $i = i_{m-1}q^{m-1} + \cdots + i_1 q + i_0$ is $\leq q^\lambda - 1$ iff $i_\lambda = i_{\lambda+1} = \cdots = i_{m-1} = 0$, i.e. iff $(i_{m-1} \cdots i_1 i_0)$ starts with a run of length at least $m - \lambda = r$. Q.E.D.

From now on let r be fixed. Let $s(m)$ denote the number of m-tuples of elements from $\{0, 1, \dots, q - 1\}$ which contain a straight run of length at least r; and let $c(m)$ be the number of m-tuples which contain a circular run of length at least r but no such straight run. Then from Lemma 10,

$$\deg g(x) = s(m) + c(m) - 1 \tag{7}$$

(the last term because $i = 0$ is excluded), and

$$I(q^m - 1, q^\lambda) = q^m - s(m) - c(m). \tag{8}$$

Thus the problem is reduced to the purely combinatorial one of determining $s(m)$ and $c(m)$. For our present purpose we can drop $c(m)$:

$$I(q^m - 1, q^\lambda) \leq q^m - s(m) \tag{9}$$

(but see Problem 3 at the end of this section). It is easy to find a recurrence for $s(m)$:

Lemma 11.

$$s(k) = 0 \quad \text{for} \quad 0 \le k < r, \quad s(r) = 1, \tag{10}$$

$$s(m) = qs(m-1) + (q-1)\{q^{m-r-1} - s(m-r-1)\}, \quad m > r. \tag{11}$$

Proof. (10) is immediate. To prove (11), observe that an m-tuple counted in $s(m)$ either contains a run of length r in the last $m-1$ places (giving the first term), or else must have the form

$$0 \cdots 0 i_{m-r-1} i_{m-r-2} \cdots i_0$$

where $i_{m-r-1} \ne 0$ and there is no run of length r in the last $m-r-1$ places.
Q.E.D.

If we let $\varphi(m) = q^m - s(m)$ then

$$I(q^m - 1, q^x) \le \varphi(m) \tag{12}$$

and, from Lemma 11,

$$\varphi(k) = q^k \quad \text{for} \quad 0 \le k < r, \quad \varphi(r) = q^r - 1, \tag{13}$$

$$\varphi(m) = q\varphi(m-1) - (q-1)\varphi(m-r-1), \quad m > r. \tag{14}$$

The most general solution to the recurrence (14) is

$$\varphi(m) = \sum_{i=0}^{r} a_i \rho_i^m, \tag{15}$$

where $a_0, \dots, a_r$ are determined by the initial conditions (13) and $\rho_0, \dots, \rho_r$ are the real or complex roots of

$$\rho^{r+1} = q\rho^r - (q-1). \tag{16}$$

Now all the roots of (16) satisfy $|\rho_i| < q$. This is clearly true for real roots. For a complex root $\rho = be^{i\theta}$, $\theta \ne 0$, it is even true that $b < 1$. For suppose $b \ge 1$. From (16), using the triangle inequality,

$$b^{r+1} + q - 1 > qb^r. \tag{17}$$

On the other hand (16) also implies

$$\rho^r = (q-1)(\rho^{r-1} + \cdots + 1),$$

$$b^r \le (q-1)(b^{r-1} + \cdots + 1),$$

$$b^{r+1} \le qb^r - (q-1),$$

which contradicts (17).

We have therefore proved:

Theorem 12. (Mann.) *The number of information symbols in the narrow-sense* BCH *code over* GF(q) *of length* $q^m - 1$ *and designed distance* q^λ *is*

$$I(q^m - 1, q^\lambda) \leqslant \sum_{i=0}^{r} a_i \rho_i^m$$

where $r = m - \lambda$, *and* $a_0, \ldots, a_r, \rho_0, \ldots, \rho_r$ *depend on* r *(but not on* m*), and satisfy* $|\rho_i| < q$.

We shall use this result in §5.

Problem. (3) (Mann.) Show that $I(q^m - 1, q^\lambda)$ can be found exactly, as follows:
  (i) If $m \geqslant 2r$ then

$$c(m) = (q - 1)^2 \sum_{k=0}^{r-2} (r - k - 1)\{q^{m-r-k-2} - s(m - r - k - 2)\}.$$

  (ii) Hence (12) may be replaced by the equality

$$I(q^m - 1, q^\lambda) = \varphi(m) - (q - 1)^2 \sum_{k=0}^{r-2} (r - k - 1)\varphi(m - r - k - 2),$$

for $m \geqslant 2r$.
  (iii) The roots of (16) are $\rho_0 = 1$, ρ_1 real with $qr/(r + 1) < \rho_1 < q$, and $|\rho_i| < 1$ for $2 \leqslant i \leqslant r$.
  (iv) In (15),

$$a_i = \frac{\rho_i(\rho_i^r - 1)}{\rho_i^{r+1} - (q - 1)r}.$$

  (v) Finally, after some manipulation,

$$I(q^m - 1, q^\lambda) = \sum_{i=0}^{r} \rho_i^m,$$

$$= \text{nearest integer to } 1 + \rho_1^m, \text{ for large } m.$$

§4. A table of BCH codes

Even though the performance of BCH codes deteriorates for large n if the rate is held fixed (as we shall see in §5), for modest lengths, up to a few thousand, they are among the best codes known. Figure 9.1 gives a table of primitive binary BCH codes of length $\leqslant 255$. For each code we give the length

n	k	d	n	k	d
7	4	3	127	29	43*
				22	47
15	11	3		15	55
	7	5		8	63
	5	7			
			255	247	3
31	26	3		239	5
	21	5		231	7
	16	7		223	9
	11	11		215	11
	6	15		207	13
				199	15
63	57	3		191	17
	51	5		187	19
	45	7		179	21*
	39	9		171	23
	36	11		163	25*
	30	13		155	27
	24	15		147	29*
	18	21		139	31
	16	23		131	37*
	10	27		123	39*
	7	31		115	43*
				107	45*
127	120	3		99	47
	113	5		91	51
	106	7		87	53*
	99	9		79	55
	92	11		71	59*
	85	13		63	61*
	78	15		55	63
				47	85
	71	19		45	87*
	64	21		37	91*
	57	23		29	95
	50	27		21	111
	43	31 #		13	119
	36	31		9	127

$\# d = \delta + 2$

*lower bound on d

Fig. 9.1. A table of BCH codes.

n, dimension k, and minimum distance d. If d is marked with an asterisk, it is only a lower bound on the true minimum distance.

For only one code in this Fig., the $[127, 43, 31]$ code of designed distance $\delta = 29$, is it known that $d > \delta$. However (see Notes), there are infinitely many such codes for larger values of m. It is easy to find examples of nonprimitive BCH codes with $d > \delta$, as we saw in Ch. 7.

Interesting nonprimitive BCH codes. Sometimes nonprimitive BCH codes (when shortened) contain more information symbols for the same number of check symbols than do primitive BCH codes.

For example, let $n = 33$. The first two cyclotomic cosets are

$$C_1 = \{1, 2, 4, 8, 16, -1, -2, -4, -8, -16\},$$

$$C_3 = \{3, 6, 12, 24, 15, -3, -6, -12, -24, -15\}.$$

In general, if $n = 2^m + 1$, then $|C_1| = 2m$, so $x^n + 1$ has zeros in $\mathrm{GF}(2^{2m})$. If β is a zero of a minimal polynomial, so is β^{-1}. Hence any cyclic code of length $2^m + 1$ is reversible (§6 of Ch. 7).

Let $\mathscr{C}$ be the (wide-sense, nonprimitive!) BCH code with generator polynomial $M^{(0)}(x) M^{(1)}(x) \cdots M^{(u)}(x)$. Then $g(x)$ has zeros α^i for $i = 0, \pm 1, \ldots, \pm(u + 1)$, and so by the BCH bound has minimum distance $\geq 2u + 4$.

For $u = 1$, $\mathscr{C}$ is a $[2^m + 1, 2^m - 2m, 6]$ code, which has one more information symbol and the same number of check symbols as the $[2^m, 2^m - 1 - 2m, 6]$ extended primitive BCH code. (See §7.3 of Ch. 18.)

Problem. (4) Let T be the permutation $(0, 1, \ldots, n - 1)$, n odd, and let $R = (0, n - 1)(1, n - 2), \ldots, ((n - 3)/2, (n + 1)/2), ((n - 1)/2)$ be the reversing permutation. Of course T generates a cyclic group $G_1 = \{T^i : 0 \leq i < n\}$. (i) Show that T and R generate a group $G_2 = \{T^i, RT^i : 0 \leq i < n\}$ of order $2n$. This is called the *dihedral* group, and is the group of rotations and reflections of a regular polygon with n sides. Show that

$$T^n = R^2 = (TR)^2 = I. \tag{18}$$

(ii) Let G be a group containing elements T and R such that (18) holds, and $T^i \neq I$ for $1 \leq i < n$, $R \neq I$, $TR \neq I$. Show that the subgroup of G generated by T and R is isomorphic to the dihedral group of order $2n$. (iii) A reversible cyclic code $\mathscr{C}$ of length n is invariant under the group G_3 generated by $\sigma_2 : i \rightarrow 2i \pmod{n}$, T and R. Show that (a) if m, the multiplicative order of $2 \bmod n$, is even and $n \mid 2^{m/2} + 1$, then $R = \sigma_2^{m/2} T^{-1}$, and G_3 has order mn and is the group given in Problem 40 of Ch. 8; but (b) otherwise $G_3 = \{\sigma_2^i T^j, \sigma_2^i R T^j : 0 \leq i < m, 0 \leq j < n\}$ has order $2mn$.

§5. Long BCH codes are bad

The Gilbert–Varshamov bound (Theorem 12 of Ch. 1, Theorem 30 of Ch. 17) states that if R is fixed, $0 \leq R \leq 1$, then there exist binary $[n, k, d]$ codes with $k/n \geq R$ and $d/n \geq H_2^{-1}(1 - R)$, where $H_2^{-1}(x)$ is the inverse of the entropy function (see §11 of Ch. 10). However except in the case $R = 0$ or $R = 1$, no family of codes yet constructed meets this bound.

Definition. A family of codes over GF(q), for q fixed, is said to be *good* if it contains an infinite sequence of codes $\mathscr{C}_1, \mathscr{C}_2, \ldots$, where $\mathscr{C}_i$ is an $[n_i, k_i, d_i]$ code, such that both the rate $R_i = k_i/n_i$ and d_i/n_i approach a nonzero limit as $i \to \infty$.

Unfortunately, as we shall now see, primitive BCH codes do not have this property – asymptotically they are *bad*!

Theorem 13. *There does not exist an infinite sequence of primitive* BCH *codes over* GF(q) *with both d/n and k/n bounded away from zero.*

Proof. Suppose $\mathscr{C}_1, \mathscr{C}_2, \ldots$ is a sequence of such codes, where $\mathscr{C}_i$ is an $[n_i = q^{m_i} - 1, k_i, d_i]$ code, and $k_i/n_i \geq B_1 > 0$, $d_i/n_i \geq B_2 > 0$. Let $\mathscr{C}'_i$ be the BCH code of length n_i and designed distance $\delta_i = [(d_i + 1)/q]$. By Theorem 6, the true minimum distance of $\mathscr{C}'_i$ is $\leq d_i$, hence $\mathscr{C}'_i \supseteq \mathscr{C}_i$.

Choose λ_i so that $q^{\lambda_i + 1} > \delta_i \geq q^{\lambda_i}$, and let $\mathscr{C}''_i$ be the BCH code of length n_i and designed distance q^{λ_i}. Clearly $\mathscr{C}''_i \supseteq \mathscr{C}'_i \supseteq \mathscr{C}_i$, so

$$I(q^{m_i} - 1, q^{\lambda_i}) \geq k_i. \tag{19}$$

Now

$$\frac{q^{\lambda_i}}{n_i} > \frac{\delta_i}{q n_i} > \frac{d_i}{q^2 n_i} - \frac{1}{q n_i} \geq \frac{B_2}{q^2} - \frac{1}{q n_i} > 0.$$

Let $r_i = m_i - \lambda_i$. Since

$$\frac{q^{\lambda_i}}{n_i} = \frac{q^{\lambda_i}}{q^{m_i} - 1} \approx q^{-r_i},$$

r_i cannot increase indefinitely, say $r_i \leq A_1$. From Theorem 12,

$$I(q^{m_i} - 1, q^{\lambda_i}) \leq \sum_{j=0}^{r} a_j^{(i)} (\rho_j^{(i)})^{m_i}.$$

Since r_i is an integer between 0 and A_1, there are only finitely many different $a_j^{(i)}$'s. Let $A_2 = \max |a_j^{(i)}|$. Similarly let $A_3 = \max |\rho_j^{(i)}| < q$. Then

$$I(q^{m_i} - 1, q^{\lambda_i}) \leq (A_1 + 1) A_2 A_3^{m_i},$$

therefore

$$\frac{I(q^{m_i} - 1, q^{\lambda_i})}{q^{m_i} - 1} \to 0 \quad \text{as} \quad i \to \infty.$$

But from (19),

$$\frac{I(q^{m_i} - 1, q^{\lambda_i})}{q^{m_i} - 1} \geq \frac{k_i}{n_i} \geq B_1 > 0,$$

a contradiction.   Q.E.D.

Even though long BCH codes are bad, what about other cyclic codes? The answer to this important question remains unknown.

Research Problem (9.2). Are cyclic codes over $GF(q)$ good, for q fixed?

§6. Decoding BCH codes

A good deal of work has been done on decoding BCH codes, and efficient algorithms now exist. This section contains a simple description of the main steps. Much more information will be found in the extensive list of references given at the end of the chapter.

To begin with let $\mathscr{C}$ be an $[n, k, d]$ binary BCH code, of odd designed distance δ. Suppose the codeword $c = c_0 c_1 \cdots c_{n-1}$ is transmitted and the distorted vector $y = c + e$ is received, where $e = e_0 e_1 \cdots e_{n-1}$ is the error vector (Fig. 9.2). The decoding can be divided into 3 stages. (I) Calculation of the syndrome. (II) Finding the error locator polynomial $\sigma(z)$. (III) Finding the roots of $\sigma(z)$. We shall describe each stage in turn.

Stage. (I) *Calculation of syndrome.* The parity check matrix can be taken to be

$$H = \begin{pmatrix} 1 & \alpha & \alpha^2 & \cdots & \alpha^{n-1} \\ 1 & \alpha^3 & \alpha^6 & \cdots & \alpha^{3(n-1)} \\ \cdots\cdots\cdots\cdots\cdots\cdots\cdots\cdots \\ 1 & \alpha^{\delta-2} & \alpha^{2(\delta-2)} & \cdots & \alpha^{(\delta-2)(n-1)} \end{pmatrix}.$$

Fig. 9.2.

As usual let $c(x) = \Sigma\, c_i x^i$, $e(x) = \Sigma\, e_i x^i$, $y(x) = \Sigma\, y_i x^i$. The syndrome of y is (§4 of Ch. 1)

$$
S = Hy^T = \begin{pmatrix} 1 & \alpha & \alpha^2 & \cdots & \alpha^{n-1} \\ 1 & \alpha^3 & \alpha^6 & \cdots & \alpha^{3(n-1)} \\ \cdots\cdots\cdots\cdots\cdots\cdots\cdots \\ 1 & \alpha^{\delta-2} & & \cdots & \alpha^{(\delta-2)(n-1)} \end{pmatrix} \begin{pmatrix} y_0 \\ y_1 \\ \vdots \\ y_{n-1} \end{pmatrix}
$$

$$
= \begin{pmatrix} \Sigma\, y_i \alpha^i \\ \Sigma\, y_i \alpha^{3i} \\ \cdots\cdots\cdots \\ \Sigma\, y_i \alpha^{(\delta-2)i} \end{pmatrix} = \begin{pmatrix} y(\alpha) \\ y(\alpha^3) \\ \vdots \\ y(\alpha^{\delta-2}) \end{pmatrix} = \begin{pmatrix} A_1 \\ A_3 \\ \vdots \\ A_{\delta-2} \end{pmatrix}, \tag{20}
$$

where $A_i = y(\alpha^i)$. (Note that $A_{2r} = y(\alpha^{2r}) = y(\alpha^r)^2 = A_r^2$.) The decoder can easily calculate the A_i from $y(x)$, as follows. Divide $y(x)$ by the minimal polynomial $M^{(i)}(x)$ of α^i, say

$$
y(x) = Q(x)M^{(i)}(x) + R(x), \quad \deg R(x) < \deg M^{(i)}(x).
$$

Then $A_i = y(\alpha^i)$ is equal to $R(x)$ evaluated at $x = \alpha^i$. The circuitry for this was given in §4 of Ch. 3 and §8 of Ch. 7.

For example, consider the $[15,7,5]$ BCH code of Ch. 3, with $\delta = 5$. Suppose $y(x) = y_0 + y_1 x + \cdots + y_{14}x^{14}$ is received. To calculate A_1, divide $y(x)$ by $M^{(1)}(x) = x^4 + x + 1$ (see §4 of Ch. 4) and set $x = \alpha$ in the remainder, giving $A_1 = R(\alpha) = u_0 + u_1\alpha + u_2\alpha^2 + u_3\alpha^3$. This is carried out by the top half of Fig. 9.3. To calculate A_3, divide $y(x)$ by $M^{(3)}(x) = x^4 + x^3 + x^2 + x + 1$ and obtain the remainder $R(x) = R_0 + R_1 x + \cdots + R_3 x^3$. Then

$$
\begin{aligned}
A_3 = R(\alpha^3) &= R_0 + R_1\alpha^3 + R_2\alpha^6 + R_3\alpha^9 \\
&= R_0 + R_1\alpha^3 + R_2(\alpha^2 + \alpha^3) + R_3(\alpha + \alpha^3) \\
&= R_0 + R_3\alpha + R_2\alpha^2 + (R_1 + R_2 + R_3)\alpha^3 \\
&= v_0 + v_1\alpha + v_2\alpha^2 + v_3\alpha^3.
\end{aligned}
$$

This calculation is carried out by the bottom half of Fig. 9.3.

Thus by the end of Stage (I), the decoder has found $A_1, A_3, \ldots, A_{\delta-2}$. Also $A_2 = A_1^2$, $A_4 = A_2^2, \ldots, A_{\delta-1} = A_{(\delta-1)/2}^2$ are easily found if needed.

Stage. (II) *Finding the error locator polynomial $\sigma(z)$.* Suppose e has weight w and contains nonzero components $e_{i_1}, \ldots, e_{i_w}$. Then $i_1, \ldots, i_w$ are the coordinates of y which are in error. As at the end of §1 define the locators $X_r = \alpha^{i_r}$, $r = 1, \ldots, w$ and the error locator polynomial

$$
\sigma(z) = \prod_{i=1}^{w} (1 - X_i z) = \sum_{i=0}^{w} \sigma_i z^i. \tag{21}
$$

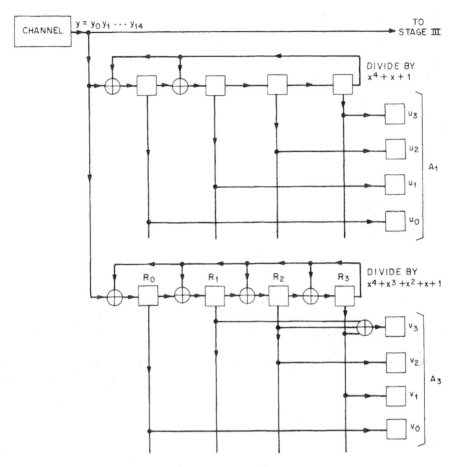

Fig. 9.3. Stage I of decoder: Finding the syndrome.

Now we have

$$A_l = y(\alpha^l) = c(\alpha^l) + e(\alpha^l)$$
$$= e(\alpha^l), \quad \text{for } 1 \le l \le \delta - 1,$$

since $c(\alpha^l) = 0$ follows from the definition of a BCH code of designed distance δ, thus

$$A_l = \sum_{i=1}^{w} X_i^l.$$

In Stage (I) the decoder has found the power sums $A_1, A_2, \ldots, A_{\delta-1}$. Stage (II), which is much harder than the other stages, is to determine $\sigma(z)$ from A_1, $A_2, \ldots, A_{\delta-1}$. The σ_i's and A_i's are related by Newton's identities (in two forms, Equation (12) and Problems 52, 54 of Ch. 8). However, as we saw in

§3 of Ch. 1, the syndrome, i.e. the A_i's, do not determine e or $\sigma(z)$ uniquely. The decoder must find the vector e of lowest weight w, or the lowest degree $\sigma(z)$, that will satisfy Newton's identities. (It is this uncertainty in w which makes Stage (II) so difficult.)

Several techniques are known for finding $\sigma(z)$.

To begin with a simple example, consider the double-error-correcting BCH code of §3 of Ch. 3. The decoding algorithm given there can be restated as follows.

Stages. (I) Compute the syndrome

$$S = \binom{A_1}{A_3}.$$

(II)

(i) If $A_1 = A_3 = 0$, set $\sigma(z) = 0$.

(ii) If $A_1 \neq 0$, $A_3 = A_1^3$, set $\sigma(z) = 1 + A_1 z$.

(iii) If $A_1 \neq 0$, $A_3 \neq A_1^3$, set

$$\sigma(z) = 1 + A_1 z + \left(\frac{A_3}{A_1} + A_1^2\right) z^2. \tag{22}$$

(iv) If $A_1 = 0$, $A_3 \neq 0$, detect that at least three errors occurred.

Once $\sigma(z)$ is found (in cases (i), (ii), (iii)) the decoder proceeds to Stage (III) to find the reciprocals of the roots of $\sigma(z)$, the X_i's, which say which locations are in error.

For a general t-error-correcting BCH code, there are two approaches to finding $\sigma(z)$, based on the two versions of Newton's identities.

Method. (1) *Using Newton's identities in the form of Equations* (14), (16) *of Chapter* 8. We begin with this method because of its simplicity, even though it is not practical unless t is small. From Problem 54 of Ch. 8, if w errors occurred, the σ_i's and A_i's are related by

$$\begin{bmatrix} 1 & 0 & 0 & 0 & 0 & \cdots & 0 \\ A_2 & A_1 & 1 & 0 & 0 & \cdots & 0 \\ A_4 & A_3 & A_2 & A_1 & 1 & \cdots & 0 \\ \cdots\cdots\cdots\cdots\cdots\cdots\cdots\cdots\cdots \\ A_{2w-4} & A_{2w-5} & \cdots & & \cdots & A_{w-3} \\ A_{2w-2} & A_{2w-3} & \cdots & & \cdots & A_{w-1} \end{bmatrix} \begin{bmatrix} \sigma_1 \\ \sigma_2 \\ \sigma_3 \\ \cdots \\ \sigma_{w-1} \\ \sigma_w \end{bmatrix} = \begin{bmatrix} A_1 \\ A_3 \\ A_5 \\ \cdots\cdots \\ A_{2w-3} \\ A_{2w-1} \end{bmatrix} \tag{23}$$

For example, for a double-error-correcting code this equation is (if $w = 2$)

$$\begin{bmatrix} 1 & 0 \\ A_2 & A_1 \end{bmatrix}\begin{bmatrix} \sigma_1 \\ \sigma_2 \end{bmatrix} = \begin{bmatrix} A_1 \\ A_3 \end{bmatrix}.$$

This solution is

$$\sigma_1 = A_1, \qquad \sigma_2 = \frac{A_3}{A_1} + A_2 = \frac{A_3}{A_1} + A_1^2,$$

and so the error locator polynomial is

$$\sigma(z) = 1 + \sigma_1 z + \sigma_2 z^2$$

$$= 1 + A_1 z + \left(\frac{A_3}{A_1} + A_1^2\right) z^2,$$

in agreement with (22).

In general, Equations (23) can be solved iteratively using the following theorem.

Theorem 14. (Peterson.) *Let*

$$A_l = \sum_{i=1}^{w} X_i^l.$$

The $\nu \times \nu$ matrix

$$M_\nu = \begin{bmatrix} 1 & 0 & 0 & 0 & 0 & \cdots & 0 \\ A_2 & A_1 & 1 & 0 & 0 & \cdots & 0 \\ A_4 & A_3 & A_2 & A_1 & 1 & \cdots & 0 \\ \multicolumn{7}{c}{\cdots\cdots\cdots\cdots\cdots\cdots\cdots} \\ A_{2\nu-4} & A_{2\nu-5} & \cdots & \cdots & \cdots & & A_{\nu-3} \\ A_{2\nu-2} & A_{2\nu-3} & \cdots & \cdots & \cdots & & A_{\nu-1} \end{bmatrix}$$

is nonsingular if $w = \nu$ or $w = \nu - 1$, and is singular if $w < \nu - 1$.

Proof. (i) Suppose $w < \nu - 1$. Then

$$M_\nu \begin{bmatrix} 0 \\ 1 \\ \sigma_1 \\ \vdots \\ \sigma_{\nu-2} \end{bmatrix} = 0$$

from (23), and so M_ν is singular.

(ii) Suppose $w = \nu$. Then

$$\det M_\nu = \prod_{i<j} (X_i + X_j).$$

For if we put $X_i = X_j$, $\det M_\nu = 0$ by (i). Therefore $\det M_\nu = \text{const.} \ \Pi_{i<j} (X_i + X_j)$, and the constant is easily found to be 1.

(iii) Finally, if $w = \nu - 1$, M_ν is nonsingular from (ii).   Q.E.D.

Using this theorem we have the following iterative algorithm for finding $\sigma(z)$, for a BCH code of designed distance $\delta = 2t + 1$, assuming w errors occur where $w \leqslant t$.

Assume t errors occurred, and try to solve Equations (23) with w replaced by t. By Theorem 14, if t or $t-1$ errors have occurred, a solution exists and we go to Stage (III). But if fewer than $t-1$ errors occurred, the equations have no solution. In this case assume $t-2$ errors occurred, and again try to solve (23) with w now replaced by $t-2$. Repeat until a solution is found.

The difficulty with this method is that it requires repeated evaluation of a large determinant over GF(2^m). For this reason if t is large (e.g. bigger than 3 or 4) the next method is to be preferred.

Method. (2) *Using the generalized Newton's identities – the Berlekamp algorithm.* Assuming that w errors occurred, the σ_i's and A_i's are related by Equations (12), (13) of Ch. 8. Equation (12) can be interpreted as saying that the A_i's are the output from a linear feedback shift register of w stages, with initial contents $A_1, A_2, \ldots, A_w$ – see Fig. 9.4. The register is shown at the

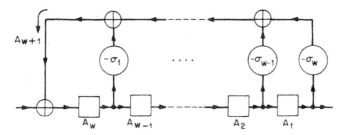

Fig. 9.4. The A_i's are produced by a shift register.

instant when it contains $A_1, A_2, \ldots, A_w$ and

$$A_{w+1} = -\sigma_1 A_w - \sigma_2 A_{w-1} - \cdots - \sigma_{w-1} A_2 - \sigma_w A_1$$

is being formed.

The decoder's problem is: given the sequence $A_1, A_2, \ldots, A_{\delta-1}$, find that linear feedback shift register of shortest length w which produces $A_1, \ldots, A_{\delta-1}$ as output when initially loaded with $A_1, \ldots, A_w$.

There is an efficient algorithm for finding such a shift register (and hence the error locator polynomial $\sigma(z)$) which is due to Berlekamp. In Chapter 12 we shall describe a version of this algorithm which applies to a wide class of codes, including Goppa codes and other generalization of BCH codes.

Whichever method is used, at the end of Stage (II) the decoder knows the error locator polynomial $\sigma(z)$.

Stage. (III) *Finding the roots of* $\sigma(z)$. Since $\sigma(z) = \Pi_{i=1}^{w}(1 - X_i z)$, the reciprocals of the zeros of $\sigma(z)$ are $X_1 = \alpha^{i_1}, \ldots, X_w = \alpha^{i_w}$, and errors have occurred at coordinates $i_1, \ldots, i_w$.

If $\sigma(z)$ has degree 1 or 2, the zeros can be found directly (see §7). But in

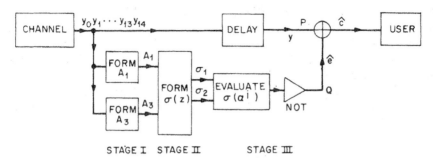

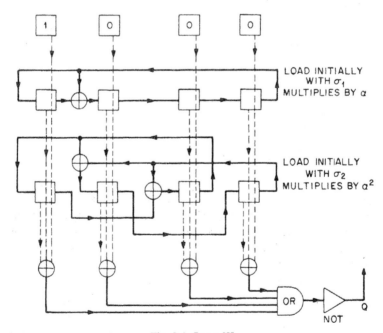

STAGE I   STAGE II   STAGE III

Fig. 9.5. The complete decoder.

general the simplest technique is just to test each power of α in turn to see if it is a zero of $\sigma(z)$. This part of the decoding is often called the *Chien search*. There is an error in coordinate i iff $\sigma(\alpha^{-i}) = 0$.

Figure 9.5 shows all three stages of the decoder. To illustrate the circuitry for Stage (III) we again consider the $[15, 7, 5]$ code. See Fig. 9.6. The first digit of y to reach point P is y_{14}, and y_{14} is in error iff $\sigma(\alpha^{-14}) = \sigma(\alpha) = 1 + \sigma_1\alpha + \sigma_2\alpha^2 = 0$. The next digit to arrive is y_{13}, which is in error iff $\sigma(\alpha^{-13}) = \sigma(\alpha^2) = 1 + \sigma_1\alpha^2 + \sigma_2\alpha^4 = 0$, and so on.

The circuit shown in Fig. 9.6 does exactly what is required. Initially the three registers are loaded with 1, σ_1 and σ_2 (obtained from Stage II). The

Fig. 9.6. Stage III.

second register is wired to multiply by α, the third register by α^2. After one clock cycle the registers contain 1, $\sigma_1\alpha$, $\sigma_2\alpha^2$, and the output from the OR gate at the bottom is 0 iff $1 + \sigma_1\alpha + \sigma_2\alpha^2 = \sigma(\alpha) = 0$. Otherwise the output is 1. The NOT gate complements this output, and so a 1 arrives at Q iff y_{14} is in error. Then the adder at P corrects any error in y_{14}.

One clock cycle later, a 1 reaches Q iff y_{13} is in error, and so on.

Thus Stage III finds the zeros of $\sigma(z)$, and corrects the corresponding errors in y.

Decoding nonbinary BCH codes. Stage I is much the same as in the binary case – the decoder finds $A_1, \ldots, A_{\delta-1}$. In Stage II, Equation (13) of Ch. 8 must be used instead of Newton's identities (16). After finding $\sigma(z)$, Equation (19) of Ch. 8 may be used to find the error evaluator polynomial $\omega(z)$. Alternatively, Berlekamp's algorithm gives an efficient way of finding $\sigma(z)$ and $\omega(z)$ simultaneously. In Stage (III), when a zero of $\sigma(z)$ is found, indicating the presence of an error, Equation (18) of Ch. 8 is used to find the value of the error.

Correcting more than t errors. The decoding algorithms we have described only correct t or fewer errors in a BCH code of designed distance $2t + 1$. *Complete* decoding algorithms (in the sense of §3 of Ch. 1) are known for all double- and some triple-error-correcting codes (see Notes), but the following problem remains unsolved.

Research Problem (9.3). Find a complete decoding algorithm for all BCH codes.

Problem. (5) What is the expression for $\sigma(z)$ analogous to Equation (22), if three errors occur in a triple-error-correcting BCH code?

§7. Quadratic equations over GF(2^m)

The last step in decoding a BCH code is to find the roots of the error locator polynomial $\sigma(z)$. Over GF(2^m) quadratic equations can be solved almost as easily as linear equations, as we see in this section. The results obtained here will also be used in §5 of Ch. 12 when studying double-error-correcting Goppa codes.

In §9 of Ch. 4 it was shown that GF(2^m) has a basis of the form γ, γ^2, $\gamma^4, \ldots, \gamma^{2^{m-1}}$, called a *normal basis*. A typical element β of GF(2^m) can be written

$$\beta = b_0\gamma + b_1\gamma^2 + b_2\gamma^4 + \cdots + b_{m-1}\gamma^{2^{m-1}}, \tag{24}$$

$b_i = 0$ or 1, and the trace of β is (since $T_m(\gamma^2) = T_m(\gamma) = 1$)

$$T_m(\beta) = b_0 + b_1 + \cdots + b_{m-1}. \tag{25}$$

Theorem 15. *The quadratic equation $x^2 + x + \beta = 0$, $\beta \in \mathrm{GF}(2^m)$, has two roots in $\mathrm{GF}(2^m)$ if $T_m(\beta) = 0$, and has no roots in $\mathrm{GF}(2^m)$ if $T_m(\beta) = 1$. Thus*

$$x^2 + x + \beta = (x + \eta)(x + \xi), \quad for \;\; \eta, \, \xi \in \mathrm{GF}(2^m),$$

if $T_m(\beta) = 0$, but $x^2 + x + \beta$ is irreducible over $\mathrm{GF}(2^m)$ if $T_m(\beta) = 1$.

Proof. Express x in terms of the normal basis as

$$x = x_0\gamma + x_1\gamma^2 + \cdots + x_{m-1}\gamma^{2^{m-1}},$$
$$x^2 = x_{m-1}\gamma + x_0\gamma^2 + \cdots + x_{m-2}\gamma^{2^{m-1}}.$$

If x is a solution to $x^2 + x + \beta = 0$ then (equating coefficients of γ^{2^i}),

$$x_0 + x_{m-1} = b_0, \quad x_1 + x_0 = b_1, \ldots, x_{m-1} + x_{m-2} = b_{m-1}.$$

Adding these equations gives

$$0 = \sum_{i=0}^{m-1} b_i = T_m(\beta), \quad \text{by (25)}.$$

Thus $T_m(\beta) = 0$ is a necessary condition for $x^2 + x + \beta = 0$ to have a solution. It is also sufficient, for if $T_m(\beta) = 0$, then there are two solutions given by $x_0 = \delta$, $x_1 = \delta + b_1$, $x_2 = \delta + b_1 + b_2, \ldots, x_{m-1} = \delta + b_1 + \cdots + b_{m-1}$, where $\delta = 0$ or 1.   Q.E.D.

Examples. In $\mathrm{GF}(2)$, $T_1(1) = 1$ and $x^2 + x + 1$ is irreducible. In $\mathrm{GF}(4)$, $T_2(\alpha) = \alpha + \alpha^2 = 1$, and $x^2 + x + \alpha$ is irreducible.

Theorem 16. *Let β be a fixed element of $\mathrm{GF}(2^m)$ with trace 1. Any irreducible quadratic over $\mathrm{GF}(2^m)$ can be transformed into $\xi(x^2 + x + \beta)$, for some $\xi \in \mathrm{GF}(2^m)$, by an appropriate change of variable.*

Proof. Suppose $ax^2 + bx + c$ is irreducible, with $a \neq 0$, $c \neq 0$. Furthermore $b \neq 0$ or this is a perfect square. Replacing x by bx/a changes this to

$$\frac{b^2}{a}\left(x^2 + x + \frac{ac}{b^2}\right) = \xi(x^2 + x + d), \quad d = \frac{ac}{b^2}.$$

Now $T_m(d) = 1$ by Theorem 15. Therefore $T_m(\beta + d) = T_m(\beta) + T_m(d) = 0$. Again by Theorem 15 there exists an ϵ in $\mathrm{GF}(2^m)$ such that $\epsilon^2 + \epsilon + (\beta + d) = 0$. Replacing x by $x + \epsilon$ changes the quadratic to $\xi(x^2 + x + \beta)$.   Q.E.D.

Problem. (6) Let β be a fixed element of $GF(2^m)$ with trace 0. Show that any reducible quadratic over $GF(2^m)$ with distinct roots can be transformed into $\xi(x^2 + x + \beta)$, for some $\xi \in GF(2^m)$, by an appropriate change of variable.

§8. Double-error-correcting BCH codes are quasi-perfect

Theorem 17. (Gorenstein, Peterson, and Zierler.) *Let $\mathscr{C}$ be a double-error-correcting binary* BCH *code of length $n = 2^m - 1$. Then $\mathscr{C}$ is quasi-perfect (see §5 of Ch. 1), i.e. any coset of $\mathscr{C}$ contains a vector of weight $\leqslant 3$.*

Proof. If m is odd this result was established in §6 of Ch. 6.

Suppose m is even. ($\mathscr{C}^{\perp}$ now has five weights – see Fig. 15.4 – so the argument used in Ch. 6 fails.) Let the parity check matrix of $\mathscr{C}$ be

$$H = \begin{pmatrix} 1 & \alpha & \alpha^2 & \cdots & \alpha^{n-1} \\ 1 & \alpha^3 & \alpha^6 & \cdots & \alpha^{3(n-1)} \end{pmatrix}.$$

If u is any vector of length n, with locators $X_1, \ldots, X_t$, $t = \text{wt}(u)$, the syndrome of u is

$$Hu^T = \begin{pmatrix} A_1 \\ A_3 \end{pmatrix}, \quad A_1 = \sum_{i=1}^{t} X_i, \quad A_3 = \sum_{i=1}^{t} X_i^3.$$

By Theorem 5 of Ch. 1 there is a one–one correspondence between cosets and syndromes. So we must show that for any syndrome A_1, A_3 there is a corresponding vector u of weight $\leqslant 3$. In other words, given any $A_1, A_3 \in GF(2^m)$, we must find X_1, X_2, X_3 in $GF(2^m)$ such that

$$X_1 + X_2 + X_3 = A_1,$$
$$X_1^3 + X_2^3 + X_3^3 = A_3. \qquad (26)$$

Put $y_i = X_i + A_1$, $i = 1, 2, 3$. Then X_1, X_2, X_3 exist satisfying (26) iff y_1, y_2, y_3 exist satisfying

$$y_1 + y_2 + y_3 = 0$$
$$y_1^3 + y_2^3 + y_3^3 = s,$$

where $s = A_1^3 + A_3$. Substituting $y_3 = y_1 + y_2$ into the second equation gives

$$y_1 y_2 (y_1 + y_2) = s.$$

We will find a solution with $y_2 \neq 0$. Setting $y = y_1/y_2$ gives

$$y^2 + y + \frac{s}{y_2^3} = 0. \qquad (27)$$

By Theorem 15 we must find a y_2 such that $T_m(s/y_2^3) = 0$. If $s = 0$, (27) has the solutions $y = 0$ and 1. Suppose $s = \alpha^{3k+\nu}$, where $\nu = 0$, 1, or 2. If $\nu = 0$, take $y_2 = \alpha^k$ and $T_m(s/y_2^3) = T_m(1) = 0$ as required. There are $n/3$ elements with $\nu = 0$. Since $n/3 < n/2$, there is a $\theta \in GF(2^m)$ with $T_m(\theta) = 0$ which is not a cube, say $\theta = \alpha^{3l+1}$ (the case $\theta = \alpha^{3l+2}$ is similar). Then $\theta^2 = \alpha^{3l'+2}$ also has trace 0. Now if for example $s = \alpha^{3k+1}$, the choice $y_2 = \alpha^{k-l}$ gives $T_m(s/y_2^3) = T_m(\theta) = 0$ as required. Q.E.D.

Problem. (7) (Gorenstein, Peterson and Zierler.) (i) Let $\mathscr{C}_1$ be a BCH code over $GF(q)$ of Bose distance $d_1 = 2t_1 + 1$ (see §6 of Ch. 7). Suppose the BCH code $\mathscr{C}_2$ of designed distance $2t_1 + 2$ has Bose distance d_2. Show that $\mathscr{C}_2$ has a coset of weight $\geq d_1$.

(ii) Hence show that a primitive triple-error-correcting binary BCH code of length at least 15 has a coset of weight ≥ 5, and so is not quasi-perfect.

Research Problem (9.4). Show that no other BCH codes are quasi-perfect.

§9. The Carlitz–Uchiyama bound

Theorem 18. *Suppose $\mathscr{C}$ is a binary BCH code of length $n = 2^m - 1$ with designed distance $\delta = 2t + 1$, where*

$$2t - 1 < 2^{\lceil m/2 \rceil} + 1. \tag{28}$$

Then the weight w of any nonzero codeword in $\mathscr{C}^\perp$ lies in the range

$$2^{m-1} - (t-1)2^{m/2} \leq w \leq 2^{m-1} + (t-1)2^{m/2}. \tag{29}$$

Note that w must be even.

Proof. The idea of the proof is this. The number of 0's in a codeword $a = (a_0, \ldots, a_{n-1}) \in \mathscr{C}^\perp$, minus the number of 1's, is equal to $\Sigma(-1)^{a_i}$. Using the Mattson–Solomon polynomial this is written as an exponential sum (30) involving the trace function. Then a deep theorem of Carlitz and Uchiyama is invoked to show this sum is small. Therefore the number of 0's is approximately equal to the number of 1's, and the weight of a is roughly $\frac{1}{2}n$.

The zeros of $\mathscr{C}$ are in the cyclotomic cosets $C_1, C_3, \ldots, C_{2t-1}$. From (28) and Corollary 8 these cosets are all distinct and have size m. By §5 of Ch. 7, the nonzeros of $\mathscr{C}^\perp$ are $C_{-1}, C_{-3}, \ldots, C_{-2t+1}$. By §6 of Ch. 8 the MS polynomial of $a \in \mathscr{C}^\perp$ is

$$A(z) = \sum_{\substack{i=1 \\ i \text{ odd}}}^{2t-1} T_m(\beta_i z^i), \quad \beta_i \in GF(2^m),$$

$$= T_m\left(\sum_{\substack{i=1 \\ i \text{ odd}}}^{2t-1} \beta_i z^i \right) = T_m f(z),$$

where $f(z)$ has degree $2t - 1$ and $f(0) = 0$. Now

$$\sum_{i=0}^{n-1} (-1)^{a_i} = \text{number of 0's in } a - \text{number of 1's in } a$$

$$= n - 2\,\text{wt}\,(a)$$

$$= \sum_{i=0}^{n-1} (-1)^{A(\alpha^i)}, \quad \text{by Theorem 20 of Ch. 8,}$$

$$= \sum_{i=0}^{n-1} (-1)^{T_m(f(\alpha^i))}. \tag{30}$$

We quote without proof the following:

Theorem 19. (Carlitz and Uchiyama.) *If $f(z)$ is a polynomial over $GF(2^m)$ of degree r such that $f(z) \neq g(z)^2 + g(z) + b$ for all polynomials $g(z)$ over $GF(2^m)$ and constants $b \in GF(2^m)$, then*

$$\left| \sum_{\beta \in GF(2^m)} (-1)^{T_m(f(\beta))} \right| \leq (r - 1)2^{m/2}.$$

Certainly our $f(z)$ satisfies the hypothesis since it has odd degree. Therefore

$$\left| (-1)^{T_m(f(0))} + \sum_{i=0}^{n-1} (-1)^{T_m(f(\alpha^i))} \right| \leq (2t - 2)2^{m/2},$$

$$|1 + n - 2\,\text{wt}\,(a)| \leq (2t - 2)2^{m/2},$$

which implies (29).   Q.E.D.

Corollary 20. *Let $\mathscr{C}$ be any binary BCH code of length $n = 2^m - 1$ and designed distance $\delta = 2t + 1$. Then the minimum distance of $\mathscr{C}^\perp$ is at least*

$$2^{m-1} - (t - 1)2^{m/2}. \tag{31}$$

Proof. If $2t - 1 < 2^{\lceil m/2 \rceil} + 1$ the result follows from Theorem 18. But if $2t - 1 \geq 2^{\lceil m/2 \rceil} + 1$, the expression (31) is negative.   Q.E.D.

Examples. When $t = 1$, $\mathscr{C}^\perp$ is a simplex code with all nonzero weights equal to 2^{m-1}, and in fact (29) states $w = 2^{m-1}$. When $t = 2$, (29) gives $2^{m-1} - 2^{m/2} \leq w \leq 2^{m-1} + 2^{m/2}$, which is consistent with Figs. 15.3 and 15.4.

As a final example let $\mathscr{C}$ be the BCH code of length 127 and designed distance 11. From Fig. 4.4 of Ch. 4, $\mathscr{C}^\perp$ has 15 consecutive zeros $\alpha^{113}, \ldots, \alpha^{126}$, 1, and so by the BCH bound (Theorem 8 of Ch. 7), $d \geq 16$. However Theorem 18 gives $d \geq 19$, and since d must be even, $d \geq 20$.

In fact the data suggest that a slightly stronger result should hold.

Research Problem (9.5). Can the conclusion of Theorem 18 be strengthened to

$$2^{m-1} - (t-1)2^{[m/2]} \leqslant w \leqslant 2^{m-1} + (t-1)2^{[m/2]}?$$

A simpler bound on the minimum distance of the dual of a BCH code is given by the following

Theorem 21. (Sikel'nikov.) *Let $\mathscr{C}$ be a binary* BCH *code of length $2^m - 1$ and designed distance $2t + 1$. Then $\mathscr{C}^{\perp}$ has minimum distance*

$$d' \geqslant 2^{m-1-\lceil \log_2 (2t-1) \rceil}$$

Proof. The MS polynomial of $a \in \mathscr{C}^{\perp}$ has the form

$$A(z) = \sum_i \sum_{j \in C_i} \beta_j z^j, \quad \beta_j \in \mathrm{GF}(2^m),$$

where i runs through the distinct coset representatives among $C_1, C_3, \dots,$ C_{2t-1}. The binary m-tuple corresponding to any such i has the form

$$(0, \dots, 0, l_{r-1}, \dots, l_1, l_0), \quad l_i = 0 \text{ or } 1,$$

where $r \leqslant 1 + [\log_2 (2t-1)]$. This contains a run of at least $m - 1 - [\log_2 (2t-1)]$ zeros. The binary m-tuple corresponding to $j \in C_i$ is some cyclic shift of the m-tuple corresponding to i, hence

$$j \leqslant 2^m - 2^{m-1-\lceil \log_2 (2t-1) \rceil}$$

By Theorem 26 of Ch. 8, since 0 is a zero of $A(z)$,

$$\begin{aligned}
\mathrm{wt}\,(a) &\geqslant 2^m - 1 - (\deg A(z) - 1) \\
&\geqslant 2^{m-1} - [\log_2 (2t-1)].
\end{aligned} \qquad \text{Q.E.D.}$$

*§10. Some weight distributions are asymptotically normal

Let $\mathscr{C}$ be an $[n, k, d]$ binary code, with weight distribution $(A_0, A_1, \dots, A_n)$, where A_i is the number of codewords of weight i. Then

$$\boldsymbol{a} = (a_0, a_1, \dots, a_n), \quad \text{where } a_j = A_j/2^k, \qquad (32)$$

is a vector with $a_0 = 2^{-k}$, $a_1 = \dots = a_{d-1} = 0$, $a_i \geqslant 0$, and $\Sigma a_i = 1$.

Example. Let $\mathscr{C}$ be the $[31, 21, 5]$ double-error-correcting BCH code. From Fig. 15.3, $\mathscr{C}^{\perp}$ has weight distribution

i:	0	12	16	20
A'_i:	1	310	527	186

By Theorem 1 of Ch. 5, the weight enumerator of $\mathscr{C}$ is

$$\frac{1}{2^{10}}[(x+y)^{31}+310(x+y)^{19}(x-y)^{12}+527(x+y)^{15}(x-y)^{16}+186(x+y)^{11}(x-y)^{20}].$$

Therefore the weight distribution of $\mathscr{C}$ is

i:	0	5	6	7	8	9	10	11	12
$A_i/31$:	$\frac{1}{31}$	6	26	85	255	610	1342	2760	4600

13	14	15	16	17	$\cdots$
6300	8100	9741	9741	8100	$\cdots$

The numbers $a_0, \ldots, a_{31}$ are plotted in Fig. 9.7. The reader will immediately be struck by how smooth and regular this figure is. The same phenomenon can be observed in many codes. A rule-of-thumb which works for many codes is that a_j is well approximated by the binomial distribution, i.e. by

$$a_j \approx b_j = \frac{1}{2^n}\binom{n}{j} \quad \text{for } j \ge d. \tag{33}$$

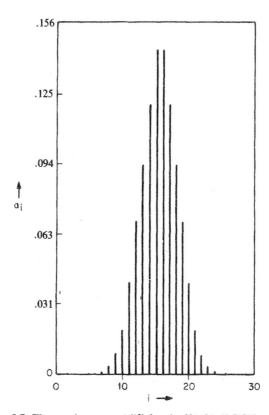

Fig. 9.7. The numbers $a_i = A_i/2^{21}$ for the [31, 21, 5] BCH code.

Note that equality holds in (33) if $\mathscr{C}$ is the $[n, n, 1]$ code F^n. Also it is not difficult to show that (33) holds for an $[n, k]$ code chosen at random (see Problem 8).

In this section we give, in Theorem 23, some justification for (33) in the case that the dual code $\mathscr{C}^{\perp}$ has large minimum distance. Some knowledge of probability theory is assumed.

So far we haven't said what we mean by the symbol $\approx$ in (33). To get a precise statement we shall define a cumulative distribution function $A(z)$ associated with the code.

Let $\boldsymbol{a} = (a_0, \ldots, a_n)$ be any real vector with $a_j \geq 0$ and $\Sigma a_j = 1$. The *mean* and *variance* of $\boldsymbol{a}$ are defined by

$$\mu = \mu(\boldsymbol{a}) = \sum_{j=0}^{n} j a_j,$$

$$\sigma^2 = \sigma^2(\boldsymbol{a}) = \sum_{j=0}^{n} (\mu - j)^2 a_j.$$

The rth *central moment* is

$$\mu_r(\boldsymbol{a}) = \sum_{j=0}^{n} \left(\frac{\mu - j}{\sigma}\right)^r a_j, \quad r = 0, 1, \ldots.$$

Thus $\mu_0(\boldsymbol{a}) = 1$, $\mu_1(\boldsymbol{a}) = 0$, $\mu_2(\boldsymbol{a}) = 1$.

Of course the binomial vector $\boldsymbol{b} = (b_0, \ldots, b_n)$, where $b_j = 2^{-n}\binom{n}{j}$, plays a special rôle. For this vector

$$\mu(\boldsymbol{b}) = \sum_{j=1}^{n} \frac{j}{2^n} \binom{n}{j} = \tfrac{1}{2}n,$$

$$\sigma^2(\boldsymbol{b}) = \sum_{j=1}^{n} \frac{j^2}{2^n} \binom{n}{j} - \frac{n^2}{4} = \frac{n}{4},$$

$$\mu_r(\boldsymbol{b}) = \sum_{j=0}^{n} \frac{1}{2^n} \binom{n}{j} \left(\frac{n/2 - j}{\sqrt{n}/2}\right)^r. \tag{34}$$

Lemma 22. Let $\mathscr{C}$ be a code and $\boldsymbol{a}$ be as in Equation (32). If $\mathscr{C}^{\perp}$ has minimum distance d', then

$$\mu_r(\boldsymbol{a}) = \mu_r(\boldsymbol{b}) \quad \text{for } r = 0, 1, \ldots, d' - 1.$$

Proof. Follows from Problem 9e of Ch. 5.   Q.E.D.

Definition. The *cumulative distribution function* (c.d.f.) $A(z)$ associated with $\boldsymbol{a}$ is given by

$$A(z) = \sum_{j \geq \mu - \sigma z}^{n} a_j,$$

for any real number z.

The classical central limit theorem then states that the cumulative distribution function $B(z)$ of b approaches the normal or Gaussian c.d.f. defined by

$$\Phi(z) = \frac{1}{\sqrt{(2\pi)}} \int_{-\infty}^{z} e^{-t^2/2} \, dt,$$

as $n \to \infty$. This is a special case of the main theorem of this section:

Theorem 23 (Sidel'nikov.) *Let $\mathscr{C}$ be an $[n, k, d]$ binary code, and let $d' \geqslant 3$ be the minimum distance of the dual code $\mathscr{C}^\perp$. Then*

$$|A(z) - \Phi(z)| \leqslant \frac{20}{\sqrt{(d')}}.$$

Proof. (Some details are omitted.) Let a be any codeword of $\mathscr{C}$. By (20) of Ch. 5, $\sigma^2(a) = n/4$. Define the characteristic function of $\alpha(t)$ of a by

$$\alpha(t) = \sum_{r=0}^{\infty} \mu_r(a) \frac{(it)^r}{r!} \tag{35}$$

$$= \sum_{r=0}^{\infty} \sum_{j=0}^{n} a_j \frac{1}{r!} \left\{ \frac{(\mu - j)it}{\sigma} \right\}^r$$

$$= \sum_{j=0}^{n} a_j \, e^{it(\mu-j)/\sigma}. \tag{36}$$

Equation (35) is the Taylor series of $\alpha(t)$, and we shall need a bound on the tail of this series. One first shows (e.g. by induction on r) that for $r = 1, 2, \ldots$ and $t > 0$,

$$\left| e^{it} - 1 - \frac{it}{1!} - \cdots - \frac{(it)^{r-1}}{(r-1)!} \right| \leqslant \frac{t^r}{r!}.$$

Using (36) this implies that, for r even and any t,

$$\left| \alpha(\xi + t) - \alpha(\xi) - \frac{t}{1!} \alpha'(\xi) - \cdots - \frac{t^{r-1}}{(r-1)!} \alpha^{(r-1)}(\xi) \right| \leqslant \mu_r(a) \frac{|t|^r}{r!},$$

and so, putting $\xi = 0$,

$$\left| \alpha(t) - \sum_{s=0}^{r-1} \mu_s(a) \frac{(it)^s}{s!} \right| \leqslant \mu_r(a) \frac{|t|^r}{r!}. \tag{37}$$

Now set $r = 2[(d' - 1)/2]$, the largest even integer less than d'. Then from Lemma 22, (37) can be replaced by

$$\left| \alpha(t) - \sum_{s=0}^{r-1} \mu_s(a) \frac{(it)^s}{s!} \right| \leqslant \mu_r(b) \frac{|t|^r}{r!}. \tag{38}$$

As starting point for the proof we use a classical formula from probability theory (see for example Feller [427, Equation (3.13) p. 512 and Equation (5.3) p. 516]):

$$|A(z) - \Phi(z)| \leq \frac{1}{\pi} \int_{-T}^{T} \left| \frac{\alpha(t) - e^{-t^2/2}}{t} \right| dt + \frac{24}{T\pi \sqrt{(2\pi)}}, \tag{39}$$

for all $T > 0$. Using (38) the integral in (39) is

$$\leq \frac{1}{\pi} \int_{-T}^{T} \frac{1}{|t|} \left| \sum_{s=0}^{\infty} \mu_s(b) \frac{(it)^s}{s!} - e^{-t^2/2} \right| dt$$

$$+ \frac{2}{\pi} \int_{-T}^{T} \mu_r(b) \frac{|t|^{r-1}}{r!} dt. \tag{40}$$

The first of these two integrals is equal to

$$\frac{1}{\pi} \int_{-T}^{T} \frac{1}{|t|} \left| \cos^n tn^{-\frac{1}{2}} - e^{-t^2/2} \right| dt$$

$$= \frac{2}{\pi} \int_{0}^{T} \frac{e^{-t^2/2}}{t} \left| e^{n\{\frac{1}{2}(t/\sqrt{(n)})^2 + \log\cos(t/\sqrt{(n)})\}} - 1 \right| dt.$$

Let $\tau = tn^{-\frac{1}{2}}$. We estimate $\frac{1}{2}\tau^2 + \log\cos\tau$ by expanding $\log(1 - (1 - \cos\tau))$ in powers of $1 - \cos\tau$. This gives

$$\left| \frac{\tau^2}{2} + \log\cos\tau \right| \leq \left| \cos\tau - 1 + \frac{\tau^2}{2} \right|$$

$$+ \left| \sum_{k=2}^{\infty} \frac{1}{k} (1 - \cos\tau)^k \right| \leq \frac{\tau^4}{4!} + \frac{\tau^4}{4} = \frac{7\tau^4}{24}, \quad \text{for } |\tau| < 1,$$

using

$$1 - \frac{\tau^2}{2!} \leq \cos\tau \leq 1 - \frac{\tau^2}{2!} + \frac{\tau^4}{4!}.$$

Thus the integrand is

$$\leq \frac{e^{-t^2/2}}{\tau} (e^{7t^4/24n} - 1) \leq \frac{7t^3}{24n} e^{7t^4/24n - t^2/2},$$

using $e^x - 1 \leq x e^x$ for $x \geq 0$,

$$\leq \frac{7t^3}{24n} e^{-5t^2/24} \quad \text{provided } T < n^{\frac{1}{2}}.$$

Therefore the first integral in (40) is

$$\leq \frac{7}{12\pi n} \int_{0}^{\infty} t^3 e^{-5t^2/24} dt$$

$$= \frac{168}{25\pi n}.$$

The second integral in (40) is

$$\frac{4\mu_r(b)T^r}{\pi r \cdot r!}.$$

By Problem 9, this is

$$\leq \frac{4T^r}{\pi r \cdot r!} \cdot \left(\frac{r}{e}\right)^{r/2} r^{\frac{1}{2}} e^{1/6}.$$

Collecting these results gives, for all $T < n^{\frac{1}{2}}$,

$$|A(z) - \Phi(z)| \leq \frac{168}{25\pi n} + \frac{4T^r e^{1/6}}{\pi r! \sqrt{(r)}} \left(\frac{r}{e}\right)^{r/2} + \frac{24}{T\pi \sqrt{(2\pi)}}.$$

The choice of

$$T = \left(\frac{r}{e}\right)^{\frac{1}{2}} \left(\frac{6}{e^{1/6}}\right)^{1/r}$$

now leads to the desired result.   Q.E.D.

From the Carlitz–Uchiyama bound (Theorem 18), and Theorem 21, this theorem applies to BCH codes of small designed distance. Other applications will be given in the Notes and in the quadratic residue code chapter.

Corollary 24. *A version of the classical central limit theorem. The cumulative distribution function $B(z)$ of the binomial distribution b satisfies*

$$|B(z) - \Phi(z)| \leq \frac{20}{\sqrt{(n)}}.$$

Proof. Let $\mathscr{C}$ be the $[n, n, 1]$ code F^n. Then $d' = n$.   Q.E.D.

Of course the constant, 20, in Theorem 23 is very poor, for d' must exceed **400 before Theorem 23 says anything. On the other hand Fig. 9.7 suggests that** $\Phi(z)$ is a good approximation to $A(z)$ even for short codes.

Research Problem (9.6). Strengthen Theorem 23.

Problems. (8) (See also Problem 31 of Ch. 17.) Consider an $[n, k]$ binary code $\mathscr{C}$ with generator matrix $G = [I \mid A]$, where A is a $k \times (n - k)$ matrix of 0's and 1's (cf. Equation (8) of Ch. 1). Suppose each entry of A is chosen at random, to be 0 or 1 with probability $\frac{1}{2}$, and then choose one of the codewords u of $\mathscr{C}$

at random. Show that

$$\text{Prob}\{\text{wt}(u) = 0\} = 2^{-k},$$

$$\text{Prob}\{\text{wt}(u) = j\} = 2^{-n}\left\{\binom{n}{j} - \binom{n-k}{j}\right\}, \quad j > 0.$$

Thus we can say that (33) holds for an $[n, k]$ code chosen at random.

(9) Show that $\mu_r(b)$ given by Equation (34) satisfies

$$\mu_r(b) \leq \left(\frac{r}{e}\right)^{r/2} r^{\frac{1}{2}} e^{1/6} \quad \text{for } r \text{ even} \geq 2.$$

[Hint. Define independent random variables $X_1, \ldots, X_n$, where each X_i is $+1$ or -1 with probability $\frac{1}{2}$. X_i is approximated by a normal random variable Y_i with mean 0 and variance 1. If r is even, $EX_i^r = 1 \leq EY_i^r$. Then

$$\mu_r(b) = \frac{1}{n^{r/2}} E\left(\sum_{i=1}^{n} X_i\right)^r \leq \frac{1}{n^{r/2}} E\left(\sum_{i=1}^{n} Y_i\right)^r.]$$

§11. Non-BCH triple-error-correcting codes

In this final section we describe some triple-error-correcting non-BCH* binary codes. These are similar to BCH codes, and have the same parameters, but require a different and interesting technique to find d. They will be used in §7 of Ch. 15 to construct nonlinear Goethals codes.

The block length will be $n = 2^m - 1$, where $m = 2t + 1$, $t \geq 2$. α is a primitive element of $GF(2^m)$ and $M^{(i)}(x)$ is the minimal polynomial of α^i.

Triple-error-correcting binary BCH codes normally have $g(x) = M^{(1)}(x)M^{(3)}(x)M^{(5)}(x)$. But the new codes have $g(x) = M^{(1)}(x)M^{(r)}(x)M^{(s)}(x)$, where $r = 1 + 2^{t-1}$, $s = 1 + 2^t$. By Problem 14 of Ch. 7, $M^{(r)}(x)$ and $M^{(s)}(x)$ have degree m. The first examples of the new codes are:

$m = 5$. A [31, 16, 7] code with $g(x) = M^{(1)}(x)M^{(3)}(x)M^{(5)}(x)$ – this *is* a triple-error-correcting BCH code.

$m = 7$. A [127, 106, 7] code with $g(x) = M^{(1)}(x)M^{(5)}(x)M^{(9)}(x)$.

First we consider the $[n, n - 2m]$ code $\mathcal{B}$ with $g(x) = M^{(r)}(x)M^{(s)}(x)$.

$\mathcal{B}$ is a double-error-correcting code. To show this we use linearized polynomials.

Let $a(x) = x^{l_1} + x^{l_2} + \cdots + x^{l_w}$ be any binary vector of length n and weight w, with locators $X = \{\alpha^{l_1}, \ldots, \alpha^{l_w}\}$. Note that

$$a(\alpha^j) = \sum_{\beta \in X} \beta^j = A_j \tag{41}$$

*But see Research Problem 9.7.

The elements of X may be represented as binary m-tuples. Let V_a be the subspace of GF(2^m) generated by these m-tuples.

Let the *rank* of a, denoted by r_a, be the dimension of this subspace. Naturally $r_a \leqslant w$ with equality iff the elements of X are linearly independent over GF(2).

Each vector of V_a represents an element β of GF(2^m). We set $L(z) = \Pi_{\beta \in V_a}(z - \beta)$. Since $X \subset V_a$,

$$L(\beta) = 0 \quad \text{for } \beta \in X. \tag{42}$$

It follows from §9 of Ch. 4 that $L(z)$ is a linearized polynomial, i.e.

$$L(z) = \sum_{i=0}^{r_a} l_i z^{2^i}, \quad l_i \in \text{GF}(2^m). \tag{43}$$

Lemma 25. *If $a(x)$ is a codeword of $\mathcal{B}$ with rank $r_a \leqslant 4$, then $a(\alpha^{1+2^i}) = 0$ for all $i \geqslant 0$.*

Proof. For this $a(x)$ we have

$$L(z) = l_0 z + l_1 z^2 + l_2 z^4 + l_3 z^8 + l_4 z^{16},$$

where $l_0 \neq 0$. From (42), $L(\beta) = 0$ for all $\beta \in X$ and so

$$0 = \sum_{\beta \in X} \beta L(\beta)^{2^{t-2}} = \sum_{i=0}^{4} l_i^{2^{t-2}} \sum_{\beta \in X} \beta^{1+2^{t+i-2}}$$

$$= \sum_{i=0}^{4} l_i^{2^{t-2}} a(\alpha^{1+2^{t+i-2}}) \quad \text{by (41)}.$$

Rearranging:

$$a(\alpha^{1+2^{t-2}}) = \sum_{i=1}^{4} \left(\frac{l_i}{l_0}\right)^{2^{t-2}} a(\alpha^{1+2^{t+i-2}})$$

Now by definition of $\mathcal{B}$,

$$a(\alpha^{1+2^{t-1}}) = a(\alpha^{1+2^t}) = 0,$$

$$a(\alpha^{1+2^{t+1}}) = a(\alpha^{1+2^t})^{2^{t+1}} = 0,$$

$$a(\alpha^{1+2^{t+2}}) = a(\alpha^{1+2^{t-1}})^{2^{t+2}} = 0,$$

hence $a(\alpha^{1+2^{t-2}}) = 0$. Furthermore,

$$\sum_{\beta \in X} \beta L(\beta)^{2^{t-2-l}} = 0 \quad \text{for } 1 \leqslant l \leqslant t-2,$$

so we can successively show that

$$a(\alpha^{1+2^{t-3}}) = a(\alpha^{1+2^{t-4}}) = \cdots = a(\alpha^2) = 0. \qquad \text{Q.E.D.}$$

Corollary 26. *If $a(x)$ is a codeword of $\mathcal{B}$ with rank ≤ 4, then* $\mathrm{wt}\,(a(x)) \geq 7$.

Proof. From the Lemma, $a(\alpha^2) = a(\alpha^3) = a(\alpha^5) = 0$. Then by the BCH bound, $\mathrm{wt}\,(a(x)) \geq 7$.   Q.E.D.

Theorem 27. $\mathcal{B}$ *has minimum distance at least 5.*

Proof. For $a \in \mathcal{B}$, $\mathrm{wt}\,(a) \leq 4 \Rightarrow r_a \leq 4 \Rightarrow \mathrm{wt}\,(a) \geq 7$ (by Corollary 26), which is a contradiction.   Q.E.D.

Problems. (10) Show that for $a(x) \in \mathcal{B}$, $\mathrm{wt}\,(a(x)) = 5 \Rightarrow r_a = 5$, i.e.

$$a(x) = x^{t_1} + \cdots + x^{t_5},$$

where $\alpha^{t_1}, \ldots, \alpha^{t_5}$ are linearly independent over GF(2).

(11) Let $a(x)$ be an even weight vector of the Hamming code generated by $M^{(1)}(x)$. Suppose $X = \{\beta_1, \beta_2, \ldots, \beta_{2e}\}$, $\bar{X} = \{\beta_1 + \beta_{2e}, \beta_2 + \beta_{2e}, \ldots, \beta_{2e-1} + \beta_{2e}\}$, and let $\bar{a}(x)$ be the vector defined by X. Show that $\bar{a}(x)$ is also in the Hamming code.

Finally we establish the minimum distance of our triple-error-correcting code.

Theorem 28. Let $\mathcal{C}$ be the $[n, n-3m]$ code with generator polynomial $M^{(1)}(x)M^{(r)}(x)M^{(s)}(x)$. Then $\mathcal{C}$ has minimum distance 7.

Proof. Let $\mathcal{A}$ be the Hamming code generated by $M^{(1)}(x)$. Then $\mathcal{C} = \mathcal{A} \cap \mathcal{B}$, and $\mathcal{C}$ has minimum distance ≥ 5.

Suppose $a(x) \in \mathcal{C}$ has weight 5. By Problem 10, $a(\alpha) \neq 0$. So $a(x) \notin \mathcal{A}$, a contradiction.

Now suppose $\mathrm{wt}\,(a(x)) = 6$, and $X = \{\beta_1, \ldots, \beta_6\}$. Since $a(x) \in \mathcal{A}$, $\sum_{i=1}^{6} \beta_i = 0$. Let $\bar{X} = \{\beta_1 + \beta_6, \beta_2 + \beta_6, \ldots, \beta_5 + \beta_6\}$, and let $\bar{a}(x)$ be the vector defined by $\bar{X}$. We show that $\bar{a}(x) \in \mathcal{C}$, a contradiction since $\mathrm{wt}\,(\bar{a}(x)) = 5$. By Problem 11, $\bar{a}(x) \in \mathcal{A}$. Also

$$\bar{a}(\alpha^{1+2^t}) = (\beta_1 + \beta_6)(\beta_1 + \beta_6)^{2^t} + \cdots + (\beta_5 + \beta_6)(\beta_5 + \beta_6)^{2^t}$$
$$= (\beta_1 + \beta_6)(\beta_1^{2^t} + \beta_6^{2^t}) + \cdots + (\beta_5 + \beta_6)(\beta_5^{2^t} + \beta_6^{2^t})$$
$$= \beta_1^{1+2^t} + \cdots + \beta_6^{1+2^t} \quad \text{using } \Sigma\,\beta_i = 0,$$
$$= a(\alpha^{1+2^t}) = 0.$$

Similarly $\bar{a}(\alpha^{1+2^{t-1}}) = a(\alpha^{1+2^{t-1}}) = 0$.   Q.E.D.

Research Problem (9.7). Does $\mathscr{C}$ have the same weight distribution as the triple-error-correcting BCH code? If so, are the codes equivalent? (We conjecture they are not.)

Notes on Chapter 9

§1. The original references to Bose, Ray–Chaudhuri and Hocquenghem were given in Ch. 7. The weight distributions of various BCH codes will be found in Berlekamp [113, Table 16.1], Goldman et al. [518], Kasami [727] and Myrvaagnes [981]. Chien and Frazer [289] apply BCH codes to document retrieval; see also Bose et al. [179].

§§2 and 4. The true minimum distance of BCH codes. A number of results are known besides the ones we give here. For example Kasami and Lin [734] have shown that $d = \delta$ if $n = 2^m - 1$ and $\delta = 2^{m-s-1} - 2^{m-s-i-1} - 1$ for $1 \le i \le m - s - 2$ and $0 \le s \le m - 2i$.

Also Peterson [1039] shows that if the BCH code of length $2^m - 1$ and designed distance δ has $d = \delta$, then the code with designed distance $\delta' = (\delta + 1)2^{m-h} - 1$, where $h \ge \delta$, has minimum distance δ'.

In order to determine the true minimum distance of the codes in Fig. 9.1 we used the following sources: Berlekamp [113, Table 16; 116], Chen [266], Kasami, Lin and Peterson [737], Peterson [1038, 1039], and Theorems 2, 3 and 5. Other results on the exact determination of d will be found in Chen and Lin [274], Hartmann et al. [612], Knee and Goldman [770], Lin and Weldon [839], and Wolf [1428]. Theorem 2 is due to Farr [417] and Theorems 3 and 4 to Peterson [1038, 1039]. See also [932].

We would appreciate hearing from anyone who can remove any of the asterisks from Fig. 9.1. The entry marked with # was found by Kasami and Tokura [744], who give an infinite family of primitive BCH codes with d > δ. Cerveira [255] and Lum and Chien [864] give nonprimitive BCH codes with $d > \delta$.

Stenbit [1276] gives generator polynomials for all the codes in Fig. 9.1.

§3. This section follows Mann [906]. (There is an unfortunate misprint in the abstract of [906]. In line 3, for $i = 1, \ldots, v$ read $i = 1, \ldots, v - 1$.) Berlekamp [112; 113, Ch. 12] gives a general solution to the problem of finding $I(n, \delta)$. See also Peterson [1036].

§4. For Problem 4 see Coxeter and Moser [314, p. 6].

§5. Theorem 13 was discovered by Lin and Weldon [838] and Camion [237]. Berlekamp [122] has proved:

Theorem 29. *For any sequence of primitive binary* BCH *codes of rate R,*

$$d \sim 2n \frac{\log_e R^{-1}}{\log_2 n}, \quad \text{as length } n \to \infty,$$

where d may be interpreted as the designed distance, Bose distance, or true minimum distance.

Unfortunately the proof is too long to give here.

Kasami [728] has given the following generalization of Theorem 13. Any family of cyclic codes is bad if it has the property that the extended codes are invariant under the affine group. By Theorem 16 of Ch. 8 this includes BCH codes. On the other hand McEliece [938] has shown that there exist (possibly nonlinear) codes which are invariant under large groups and meet the Gilbert–Varshamov bound. See also Theorem 31 of Ch. 17.

§6. *Decoding.* The main source is Berlekamp [113]. Excellent descriptions are also given by Peterson and Weldon [1040], and by Gallager [464]. Chien [282] contains a good survey.

Special decoding algorithms for certain BCH codes have been given by Banerji [67], Bartee and Schneider [71], Blokh [164], Cowles and Davida [312], Davida [333], Kasami [725], Matveeva [929, 930], and Polkinghorn [1065].

Theorem 14 is from Peterson [1036]. For the Berlekamp algorithm see the original reference [113, Ch. 7], Massey [922, 922a], Gallager [464], Peterson and Weldon [1040, Ch. 9], and also Ch. 12. Refinements will be found in Burton [216], Ong [1014], Sullivan [1293], and Tzeng et al. [1352]. Bajoga and Walbesser [59] study the complexity of this algorithm.

For the Chien search see Chien [279]. Gore [543, 547] and Mandelbaum [896, 899] describe techniques for speeding up or avoiding the Chien search. The Chien search can be avoided if the zeros of $\sigma(z)$ can be found directly, so the following references on factoring polynomials over finite fields are relevant: Berlekamp [111, 113, 117, 121], Berlekamp et al. [131], Chien et al. [288], Golomb [523], McEliece [937] and Mills and Zierler [961]. Other references on decoding BCH codes are: Berlekamp [110], Chien and Tzeng [293], Davida [334], Forney [435], Laws [795], Massey [920], Michelson [954], Nesenbergs [985], Szwaja [1298], Tanaka et al. [1301, 1304], Ullman [1357], and Wolf [1426].

Correcting more than t errors. The complete decoding of double-error-correcting BCH codes is given by Berlekamp [113, §16.481] and Hartmann [607]. Complete decoding of some (perhaps all) triple-error-correcting BCH codes is given by Van der Horst and Berger [663]. Their algorithm applies to all triple-error-correcting BCH codes if the following problem is settled.

Research Problem (9.8). Show that the maximum weight of a coset leader of any coset of a triple-error-correcting BCH code is 5. (This is known to be true except possibly for $m \geq 12$ and $m \equiv 2$ (mod 4).)

Berger and Van der Horst [106] use their decoding algorithm to apply triple-error-correcting codes to source coding.

Hartmann [609], Hartmann and Tzeng [619], Reddy [1096], and Tzeng and Hartmann [1351] give decoding algorithms for correcting a little beyond the BCH bound in certain cases.

§8. The results of this section are from Gorenstein, Peterson, and Zierler [550]. See also Berlekamp [113, §16.481] and [663]. Leont'ev [816] has partially solved Research Problem 9.4 by showing that a binary BCH code of length $n = 2^m - 1$ and designed distance $\delta = 2t + 1$ is not quasi-perfect if $2 < t < n^{\frac{1}{2}}/\log n$ and $m \geq 7$.

§9. Theorem 19 is from Carlitz and Uchiyama [249], and depends on a deep theorem of Weil [1394]. See also Williams [1419]. Anderson [28] was the first to use Theorem 19 in coding theory; however, our result is slightly stronger than his.

Research Problem (9.9). Can any other bound from number theory, for example those of Lang and Weil [793] or Deligne [342], be used to obtain bounds on the minimum distance of codes?

Theorem 21 is from Sidel'nikov [1208].

§10. Theorem 23 is also due to Sidel'nikov [1208] (except that he has 9 where we have 20). It has been generalized by Delsarte [360]. The proof is modeled on Feller's proof of the Berry–Esséen central limit theorem given in [427, Ch. 16, §5]. Sidel'nikov has also proved that for many BCH codes

$$A_j = \frac{1}{2^k}\binom{n}{j}(1 + \epsilon), \quad \text{where } |\epsilon| < Cn^{-1/10},$$

but the proof is too complicated to give here.

Combining Theorems 21, 23 and 29 we deduce that the weight distributions of primitive binary BCH codes of fixed rate are asymptotically normal.

10

Reed–Solomon and Justesen codes

§1. Introduction

The first part of the chapter deals with Reed–Solomon codes, which are BCH codes over GF(q) with the special property that the length n is $q - 1$. Besides serving as illuminating examples of BCH codes, they are of considerable practical and theoretical importance, as we shall see. They are convenient for building up other codes, either alone (for example by mapping into binary codes, §5) or in combination with other codes, as in concatenated codes (§11).

Justesen codes (§11) are a family of concatenated codes which can be obtained in this way. Justesen codes are distinguished by being the first family of codes we have seen with the property that, as the length increases, both the rate and distance/length remain positive. Thus, unlike BCH codes (see §5 of Ch. 9), asymptotically Justesen codes are *good*.

§2. Reed–Solomon codes

Definition. A *Reed–Solomon* (or RS) code over GF(q) is a BCH code of length $N = q - 1$. Of course q is never 2. Thus the length is the number of nonzero elements in the ground field. We shall use N, K and D to denote the length, dimension, and minimum distance (using capital letters to distinguish them from the parameters of the binary codes which will be constructed later). Figure 10.6 gives a summary of the properties of these codes.

Since $x^{q-1} - 1 = \Pi_{\beta \in GF(q)^*} (x - \beta)$, the minimal polynomial of α^i is simply $M^{(i)}(x) = x - \alpha^i$. Therefore an RS code of length $q - 1$ and designed distance δ has generator polynomial

$$g(x) = (x - \alpha^b)(x - \alpha^{b+1}) \cdots (x - \alpha^{b+\delta-2}). \tag{1}$$

Usually, but not always, $b = 1$.

Examples. (1) As usual take $GF(4) = \{0, 1, \alpha, \beta = \alpha^2\}$ with $\alpha^2 + \alpha + 1 = 0$. An RS code over $GF(4)$ with $N = 3$, designed distance 2 and $b = 2$ has $g(x) = x - \beta$. The 4^2 codewords are shown in Fig. 10.1.

$$
\begin{array}{cccc}
000 & 1\alpha 0 & \beta 0 \alpha & \beta \alpha 1 \\
01\alpha & \alpha \beta 0 & 10\beta & 111 \\
0\alpha\beta & \beta 10 & 1\beta\alpha & \alpha\alpha\alpha \\
0\beta 1 & \alpha 01 & \alpha 1\beta & \beta\beta\beta.
\end{array}
$$

Fig. 10.1. A [3, 2, 2] RS code over $GF(4)$.

(2) The RS code over $GF(5)$ with $N = 4$ and designed distance 3. We take $\alpha = 2$ as the primitive element of $GF(5)$, so that

$$g(x) = (x - \alpha)(x - \alpha^2) = (x - 2)(x - 4) = x^2 + 4x + 3.$$

Some of the 25 codewords are 3410, 2140, 1320, 0341, 1111, ...

The dimension of an RS code is $K = N - \deg g(x) = N - \delta + 1$. The minimum distance D is, by the BCH bound (Theorem 8 of Ch. 7), at least $\delta = N - K + 1$. However, by Theorem 11 of Ch. 1 it can't be greater than this. Therefore

$$D = N - K + 1,$$

and RS codes are maximum distance separable (see §10 of Ch. 1, and the next chapter). It follows that the Hamming weight distribution of any RS code is given by Theorem 6 of Ch. 11.

RS codes are important for several reasons:

(i) They are the natural codes to use when a code is required of length less than the size of the field. For, being MDS, they have the highest possible minimum distance.

(ii) They are convenient for building other codes, as we shall see. For example they can be mapped into binary codes with surprisingly high minimum distance (§5). They are also used in constructing concatenated and Justesen codes (§11).

(iii) They are useful for correcting bursts of errors (§6).

Encoding and decoding are discussed in §7 and §10.

Problem. (1) Show that the dual of an RS code is an RS code.

§3. Extended RS codes

Adding an overall parity check to a code does not always increase the minimum distance, as Problem 42 of Ch. 1 showed. However:

Theorem 1. *Let $\mathscr{C}$ be the $[N = q^m - 1, K, D]$ RS code with generator polynomial*

$$g(x) = (x - \alpha)(x - \alpha^2) \cdots (x - \alpha^{D-1}). \tag{2}$$

Then extending each codeword $c = c_0 c_1 \cdots c_{N-1}$ of $\mathscr{C}$ by adding an overall parity check

$$c_N = -\sum_{i=0}^{N-1} c_i$$

produces an $[N + 1, K, D + 1]$ code.

Proof. Suppose c has weight D. The minimum weight is increased to $D + 1$ provided

$$c(1) = -c_N = \sum_{i=0}^{N-1} c_i \neq 0.$$

But $c(x) = a(x)g(x)$ for some $a(x)$, so $c(1) = a(1)g(1)$. Certainly $g(1) \neq 0$. Furthermore $a(1) \neq 0$, or else $c(x)$ is a multiple of $(x - 1)g(x)$ and has weight $\geq D + 1$ by the BCH bound.   Q.E.D.

Example. The preceding example gives the $[4, 2, 3]$ code shown in Fig. 10.2. In fact a further extension is always possible – see §5 of Ch. 11.

$$
\begin{array}{llll}
0000 & 1\alpha0\beta & \beta0\alpha1 & \beta\alpha10 \\
01\alpha\beta & \alpha\beta01 & 10\beta\alpha & 1111 \\
0\alpha\beta1 & \beta10\alpha & 1\beta\alpha0 & \alpha\alpha\alpha\alpha \\
0\beta1\alpha & \alpha01\beta & \alpha1\beta0 & \beta\beta\beta\beta.
\end{array}
$$

Fig. 10.2. A $[4, 2, 3]$ extended RS code.

§4. Idempotents of RS codes

In this section we assume $q = 2^m$. Minimal RS codes have dimension 1 (hence $D = N$) and are easy to describe. The check polynomial is $h(x) = x + \alpha^i$, for some $i = 0, 1, \ldots, N - 1$, and $g(x) = (x^N + 1)/h(x)$. There are $q - 1$

nonzero codewords, all of weight N, and they consist of all scalar multiples of one codeword.

Suppose

$$E(x) = \sum_{i=0}^{N-1} \beta_i x^i$$

is the corresponding primitive idempotent, with $\beta_0 = 1$. Since the code is cyclic,

$$\beta_1 x E(x) = \beta_1 \beta_{N-1} + \beta_1 x + \sum_{i=1}^{N-2} \beta_1 \beta_i x^{i+1}$$

is also in the code, and must be equal to $E(x)$ (or else $E(x) - \beta_1 x E(x)$ would have weight less than N). Therefore $\beta_2 = \beta_1^2$. Repeating this argument shows that

$$E(x) = 1 + \beta x + (\beta x)^2 + (\beta x)^3 + \cdots + (\beta x)^{N-1} \tag{3}$$

for some β.

For example, when $N = 7$, the primitive idempotents are shown in Fig. 10.3 (taking $\beta = 1, \alpha, \ldots, \alpha^6$), together with the corresponding check polynomials.

$$
\begin{array}{ll}
\qquad\qquad E(x) & h(x) \\
\theta_0 = 1 + x \quad + x^2 \quad + x^3 \quad + x^4 \quad + x^5 \quad + x^6 & x + 1 \\
\theta_1 = 1 + \alpha x \; + \alpha^2 x^2 + \alpha^3 x^3 + \alpha^4 x^4 + \alpha^5 x^5 + \alpha^6 x^6 & x + \alpha^6 \\
\theta_2 = 1 + \alpha^2 x + \alpha^4 x^2 + \alpha^6 x^3 + \alpha x^4 \quad + \alpha^3 x^5 + \alpha^5 x^6 & x + \alpha^5 \\
\theta_3 = 1 + \alpha^3 x + \alpha^6 x^2 + \alpha^2 x^3 + \alpha^5 x^4 + \alpha x^5 \quad + \alpha^4 x^6 & x + \alpha^4 \\
\theta_4 = 1 + \alpha^4 x + \alpha x^2 \quad + \alpha^5 x^3 + \alpha^2 x^4 + \alpha^6 x^5 + \alpha^3 x^6 & x + \alpha^3 \\
\theta_5 = 1 + \alpha^5 x + \alpha^3 x^2 + \alpha x^3 \quad + \alpha^6 x^4 + \alpha^4 x^5 + \alpha^2 x^6 & x + \alpha^2 \\
\theta_6 = 1 + \alpha^6 x + \alpha^5 x^2 + \alpha^4 x^3 + \alpha^3 x^4 + \alpha^2 x^5 + \alpha x^6 & x + \alpha
\end{array}
$$

Fig. 10.3. Idempotents of minimal $[7, 1, 7]$ RS codes over $GF(8)$.

Then the idempotent of the $[2^m - 1, K, 2^m - K]$ RS code (with $b = 1$) is $\sum_{i=1}^{K} \theta_i$. (This is somewhat easier to find than the generator polynomial Equation (1), although both require a table of $GF(2^m)$.)

Problem. (2) Show that the idempotent of the $[7, 2, 6]$ RS code (with $b = 1$) is

$$E(x) = \alpha^4 x + \alpha x^2 + \alpha^4 x^3 + \alpha^2 x^4 + \alpha^2 x^5 + \alpha x^6.$$

Show that the codewords consist of cyclic permutations of the 9 codewords $\theta_1, \theta_2, \alpha^i E(x), 0 \le i \le 6$.

§5. Mapping GF(2^m) codes into binary codes

We know from Ch. 4 that elements of GF(q), where $q = p^m$, can be represented by m-tuples of elements from GF(p). Therefore an [N, K, D] RS code over GF(q) becomes an [$n = mN$, $k = mK$, $d \geqslant D$] code over GF(p). If $q = 2^m$ the binary codes obtained in this way (and others derived from them) often have high minimum distance, as we now see.

Let $\xi_1, \ldots, \xi_m$ be a basis for GF(2^m) over GF(2). Then if $\beta = \Sigma_{i=1}^m b_i \xi_i$ is any element of GF(2^m), $b_i \in$ GF(2), we map β into $b_1, b_2, \ldots, b_m$. This mapping sends linear codes into linear codes (but cyclic codes need not go into cyclic codes).

Examples. (1) Using the basis 1, α for GF(4) over GF(2), 0 maps into 00, 1 into 10, α into 01, α^2 into 11. Then the [3, 2, 2] RS code over GF(4) of Fig. 10.1 becomes the [6, 4, 2] binary code of Fig. 10.4.

$$
\begin{array}{cccc}
000000 & 100100 & 110001 & 110110 \\
001001 & 011100 & 100011 & 101010 \\
000111 & 111000 & 101101 & 010101 \\
001110 & 010010 & 011011 & 111111.
\end{array}
$$

Fig. 10.4. A [6, 4, 2] binary code obtained from Fig. 10.1.

(2) Let $c = (c_0, c_1, \ldots, c_{N-1})$ belong to an [N, K, D] RS code over GF(2^m). Replace each c_i by the corresponding binary m-tuple, and add an overall parity check on each m-tuple. The resulting binary code has parameters

$$
n = (m + 1)(2^m - 1), \qquad k = mK, \qquad d \geqslant 2D = 2(2^m - K), \tag{4}
$$

for any $K = 1, \ldots, 2^m - 2$. The same construction applied to the extended RS code gives

$$
[(m + 1)2^m, mK, d \geqslant 2(2^m - K + 1)] \tag{5}
$$

binary codes, for $K = 1, \ldots, 2^m - 1$.

E.g. From the [15, 10, 6] and [16, 10, 7] codes over GF(2^4) we obtain [75, 40, 12] and [80, 40, 14] binary codes. Even though slightly better codes exist – we shall construct an [80, 40, 16] quadratic residue code in Ch. 16 – nevertheless this construction is impressively simple. (See also §8.1 of Ch. 18.)

(3) Using the basis 1, α, α^6 for GF(2^3) over GF(2), the mapping is

$$
\begin{array}{lll}
0 \to 000, & \alpha^2 \to 101, & \alpha^5 \to 011, \\
1 \to 100, & \alpha^3 \to 110, & \alpha^6 \to 001. \\
\alpha \to 010, & \alpha^4 \to 111, &
\end{array}
$$

Consider the $[7, 5, 3]$ RS code over $GF(2^3)$ with generator polynomial

$$g_1(x) = (x + \alpha^5)(x + \alpha^6)$$
$$= \alpha^4 + \alpha x + x^2.$$

It is surprising that this is mapped onto the $[21, 15, 3]$ binary BCH code with generator polynomial

$$g_2(y) = M^{(1)}(y) = 1 + y + y^2 + y^4 + y^6.$$

For $g_1(x)$ itself is mapped onto the vector

$$111, \ 010, \ 100, \ 000, \ 000, \ 000, \ 000$$

which is $g_2(y)$. Also $\alpha g_1(x)$ is mapped onto $y g_2(y)$, $\alpha^2 g_1(x)$ onto $y^2 g_2(y)$, $x g_1(x)$ onto $y^3 g_2(y)$, and so on.

This is the only known, nontrivial, example of a cyclic code mapping in this way onto a cyclic code!

Problem. (3) Let φ be a linear mapping from $GF(2^m)$ onto $GF(2)^m$, and $\bar{\varphi}$ be the induced mapping from vectors of length $2^m - 1$ over $GF(2^m)$ onto binary vectors of length $m(2^m - 1)$. Suppose $\mathscr{C}_1$ is a cyclic code over $GF(2^m)$ with generator polynomial $g_1(x)$, and $\mathscr{C}_2$ is a binary cyclic code with generator polynomial $g_2(y)$. Show that $\bar{\varphi}$ maps $\mathscr{C}_1$ onto $\mathscr{C}_2$ iff $y^i g_2(y)$ is the image under $\bar{\varphi}$ of some scalar multiple of $g_1(x)$, for all $0 \le i \le m - 1$.

Research Problem (10.1). Find some more examples of cyclic codes mapping onto cyclic codes.

The effect of changing the basis. A change of basis may change the weight distribution and even the minimum weight. For example, consider the $[7, 2, 6]$ MDS code with idempotent $\theta_1 + \theta_6$ and check polynomial $(x + \alpha)(x + \alpha^6)$. (This is not an RS code.) The codewords (over $GF(2^3)$) are cyclic permutations of the 9 vectors in Fig. 10.5.

Using the basis $1, \alpha, \alpha^2$ we obtain the binary weights in column 8. Using the basis $\alpha^3, \alpha^6, \alpha^5$ we have the mapping

$$0 \to 000, \quad \alpha^2 \to 101, \quad \alpha^5 \to 001,$$
$$1 \to 111, \quad \alpha^3 \to 100, \quad \alpha^6 \to 010,$$
$$\alpha \to 011, \quad \alpha^4 \to 110,$$

giving the binary weights in column 9. Notice the minimum weight obtained from the first mapping is 8, but from the second only 6.

			Codeword				Basis #1	Basis #2
1	x	x^2	x^3	x^4	x^5	x^0	Weight	Weight
1	α	α^2	α^3	α^4	α^5	α^6	12	12
1	α^6	α^5	α^4	α^3	α^2	α	12	12
0	α^5	α^3	α^6	α^6	α^3	α^5	14	6
0	α^6	α^4	1	1	α^4	α^6	10	12
0	1	α^5	α	α	α^5	1	10	12
0	α	α^6	α^2	α^2	α^6	α	8	10
0	α^2	1	α^3	α^3	1	α^2	8	12
0	α^3	α	α^4	α^4	α	α^3	10	10
0	α^4	α^2	α^5	α^5	α^2	α^4	12	10

Fig. 10.5. A cyclic code and the weights of two binary codes obtained from it.

Problem. (4) Show that the weight distribution of the first binary code is: $A_0 = 1$, $A_8 = 14$, $A_{10} = A_{12} = 21$, $A_{14} = 7$. For the second: $A_0 = 1$, $A_6 = 7$, $A_{\cdot 0} = 21$, $A_{12} = 35$.

Research Problem (10.2). Given a code $\mathscr{C}$ over $GF(2^m)$, which basis for $GF(2^m)$ over $GF(2)$ maps $\mathscr{C}$ into the binary code $\mathscr{C}^*$ with the greatest minimum distance? How much effect does this have on the minimum distance of $\mathscr{C}^*$? Does it help to use nonlinear mappings?

RS codes contain BCH codes. The codes in Example 2 are so good that it is worth examining their performance for large m. This deteriorates because of

Theorem 2. *The* $[N = 2^m - 1, K, D]$ *RS code with zeros* $\alpha, \alpha^2, \ldots, \alpha^{D-1}$ *contains the primitive binary BCH code of length N and designed distance D. Similarly the extended RS code contains the extended BCH code.*

Proof. If c belongs to the BCH code, c is a binary vector with $c(\alpha) = \cdots = c(\alpha^{D-1}) = 0$, and so also belongs to the RS code.   Q.E.D.

Therefore the minimum distance of the RS code is at most the minimum distance of the BCH code. From this we can show that, just as long BCH codes are bad (§5 of Ch. 9), so are these long binary codes obtained from RS codes.

Theorem 3. *The binary codes obtained from RS codes and having parameters given in (4) and (5) are asymptotically bad. That is, they do not contain an infinite family of codes with both rate and distance/length bounded away from zero.*

Proof. Let $\mathscr{C}$ be an $[N = 2^m - 1, K, D = N - K + 1]$ RS code. From Theorem 2, $\mathscr{C}$ contains the binary BCH code $\mathscr{C}_1$ of length N and designed distance D. Define d_2 by

$$2^{i-1} < D \leqslant d_2 = 2^i - 1,$$

so that $\mathscr{C}_1$ contains the BCH code $\mathscr{C}_2$ of designed distance d_2. By Theorem 5 of Ch. 9, d_2 is the true minimum distance of $\mathscr{C}_2$.

Now let $\mathscr{C}_3$ be the $[(m + 1)N, mK, d_3]$ binary code obtained from $\mathscr{C}$, as in (4). (The proof for (5) is similar.) Then

$$d_3 \leqslant 2d_2 < 4D.$$

Therefore if the rate $= mK/(m + 1)N \approx K/N$ of $\mathscr{C}_3$ is held fixed, the ratio

$$\frac{\text{distance}}{\text{length}} = \frac{d_3}{(m + 1)N} < \frac{4D}{(m + 1)N} = \frac{4(N - K + 1)}{(m + 1)N}$$

approaches zero as $m \to \infty$.   Q.E.D.

However, by using only a slightly more complicated construction, it is possible to get asymptotically good binary codes from RS codes – see §11.

§6. Burst error correction

On many channels the errors are not random but tend to occur in clusters, or bursts.

Definition. A *burst* of length b is a vector whose only nonzeros are among b successive components, the first and last of which are nonzero.

Binary codes obtained from RS codes are particularly suited to correcting several bursts. For a binary burst of length b can affect at most r adjacent symbols from $GF(2^m)$, where r is given by

$$(r - 2)m + 2 \leqslant b \leqslant (r - 1)m + 1.$$

So if D is much greater than r, many bursts can be corrected.

§7. Encoding Reed–Solomon codes

Since RS codes are cyclic, they can be encoded by either of the methods described in §8 of Ch. 7. However, the following simple encoding method (which is in fact the original method of Reed and Solomon) has certain practical advantages.

Let $u = (u_0, u_1, \ldots, u_{K-1})$, $u_i \in GF(q)$, be the message symbols to be encoded, and let

$$u(z) = \sum_{i=0}^{K-1} u_i z^i.$$

Then the codeword corresponding to u is taken to be the vector c whose Mattson–Solomon polynomial is $Nu(z)$, where $N = q - 1$. Thus

$$c = (u(1), u(\alpha), \ldots, u(\alpha^{N-1})). \tag{6}$$

[Or $(u(0), u(1), u(\alpha), \ldots, u(\alpha^{N-1}))$ for the extended code with an overall parity check added.] We show that c *is* in the Reed–Solomon code by verifying that

$$c(x) = \sum_{i=0}^{N-1} c_i x^i$$

has $\alpha, \alpha^2, \ldots, \alpha^{D-1}$ as zeros. In fact the MS polynomial of c, $Nu(z)$, is equal to

$$\sum_{j=0}^{N-1} A_{-j} z^j, \quad \text{where } A_{-j} = c(\alpha^{-j}),$$

by Equation (8) of Ch. 8. Therefore, equating coefficients, and using $N = -1$ in $GF(q)$, we have

$$c(1) = -u_0, \; c(\alpha^{-1}) = -u_1, \ldots, c(\alpha^{-K+1}) = -u_{K-1} \tag{7}$$

and $c(\alpha^j) = 0$ for $1 \leqslant j \leqslant N - K = D - 1$. Thus c is in the RS code, and this is an encoding method for this code. Notice however that this encoder is not systematic.

Definition. Suppose $\mathscr{C}$ is an $(n, M = q^k, d)$ linear or nonlinear code over $GF(q)$, with an encoder which maps a message $u_0, \ldots, u_{k-1}$ onto the codeword $c_0, \ldots, c_{n-1}$. The encoder is called *systematic* if there are coordinates $i_0, \ldots, i_{k-1}$ such that $u_0 = c_{i_0}, \ldots, u_{k-1} = c_{i_{k-1}}$, i.e. if the message is to be found unchanged in the codeword.

For example, the encoder

message	codeword
00	000
01	010
10	101
11	111

is systematic (with $i_0 = 0$, $i_1 = 1$), while the same code with the encoder

message	codeword
00	000
01	111
10	101
11	010

is not. An unsystematic code scrambles the message. Even after the error vector and codeword have been recovered, in an unsystematic code a further computation is necessary to recover the message. In the present example Equation (7) tells how to recover the message from the codeword.

Problem. (5) Show that any linear code has a systematic encoder. What about the nonlinear codes {000, 110, 011, 111} and {000, 100, 010, 001}?

For example, consider the [4, 2, 3] RS code over GF(5) described in Example 2 of §2. Encoder #1 of §8 of Ch. 7 would map the message (u_0, u_1) onto the codeword $(2u_0 + 2u_1, u_0 + 3u_1, u_0, u_1)$. On the other hand the encoder just described maps (u_0, u_1) onto the codeword

$$c = (u_0 + u_1, u_0 + 2u_1, u_0 + 4u_1, u_0 + 3u_1).$$

Problem. (6) Verify the last two statements.

An RS code of length $N = q - 1$ over GF(q) is a cyclic code with generator polynomial $g(x) = (x - \alpha^b)(x - \alpha^{b+1}) \cdots (x - \alpha^{b+\delta-2})$, where α is a primitive element of GF(q). The dimension is $K = N - \delta + 1$ and the minimum distance is δ. (Often $b = 1$.) This is a BCH code, and is MDS (Ch. 11). It may be extended to $[q + 1, K, q - K + 2]$ and (if $q = 2^m$) $[2^m + 2, 3, 2^m]$ and $[2^m + 2, 2^m - 1, 4]$ codes (§5 of Ch. 11). The idempotent is given in §4. RS codes are important in concatenated codes (§11) and burst correction (§6).

Fig. 10.6. Properties of Reed–Solomon codes.

---

*§8. Generalized Reed–Solomon codes

A slightly more general class of codes than RS codes are obtained if Equation (6) is replaced by

$$c = (v_1 u(1), v_2 u(\alpha), \dots, v_N u(\alpha^{N-1})), \tag{8}$$

where the v_i are nonzero elements of GF(q). Equation (6) is the case when all $v_i = 1$. This suggests a further generalization.

Definition. Let $\alpha = (\alpha_1, \dots, \alpha_N)$ where the α_i are distinct elements of GF(q^m), and let $v = (v_1, \dots, v_N)$ where the v_i are nonzero (but not necessarily distinct) elements of GF(q^m). Then the *generalized RS code*, denoted by $\mathrm{GRS}_K(\alpha, v)$, consists of all vectors

$$(v_1 F(\alpha_1), v_2 F(\alpha_2), \dots, v_N F(\alpha_N)) \tag{9}$$

where $F(z)$ ranges over all polynomials of degree $< K$ with coefficients from $\mathrm{GF}(q^m)$. This is an $[N, K]$ code over $\mathrm{GF}(q^m)$. Since F has at most $K - 1$ zeros, the minimum distance is at least $N - K + 1$, and hence is *equal* to $N - K + 1$. The code is MDS.

Theorem 4. *The dual of* $\mathrm{GRS}_K(\alpha, v)$ *is* $\mathrm{GRS}_{N-K}(\alpha, v')$ *for some* v'.

Proof. First suppose $K = N - 1$, and let $\mathscr{D}$ be the dual code to $\mathrm{GRS}_{N-1}(\alpha, v)$. Then $\mathscr{D}$ has dimension 1 and consists of all scalar multiples of some fixed vector $v' = (v'_1, \ldots, v'_N)$. We must show that all $v'_i \neq 0$. v' satisfies

$$v_1 v'_1 + \cdots + v_N v'_N = 0,$$
$$\alpha_1 v_1 v'_1 + \cdots + \alpha_N v_N v'_N = 0,$$
$$\cdots\cdots\cdots\cdots\cdots\cdots\cdots\cdots\cdots$$
$$\alpha_1^{N-2} v_1 v'_1 + \cdots + \alpha_N^{N-2} v_N v'_N = 0,$$

or

$$\begin{bmatrix} 1 & \cdots & 1 \\ \alpha_1 & \cdots & \alpha_N \\ \cdots\cdots\cdots\cdots \\ \alpha_1^{N-2} & \cdots & \alpha_N^{N-2} \end{bmatrix} \begin{bmatrix} v_1 v'_1 \\ \cdot \\ \cdot \\ \cdot \\ v_N v'_N \end{bmatrix} = 0 \qquad (10)$$

If any $v'_i = 0$ then (10) gives a set of simultaneous equations for the other $v_i v'_i$ whose coefficient matrix is Vandermonde. Hence from Lemma 17 of Ch. 4 all $v_i v'_i = 0$ and so all $v'_i = 0$, which is impossible.

Then $\mathrm{GRS}_K(\alpha, v)$ is dual to $\mathrm{GRS}_{N-K}(\alpha, v')$, for all $K < N - 1$, since

$$\sum_{i=1}^{N} (\alpha_i^s v_i)(\alpha_i^t v'_i) = \sum_{i=1}^{N} \alpha_i^{s+t} v_i v'_i = 0$$

for $s \leq K - 1$, $t \leq N - K - 1$, from (10).   Q.E.D.

It follows from Theorem 4 that $\mathrm{GRS}_K(\alpha, v)$ has parity check matrix equal to a generator matrix for $\mathrm{GRS}_{N-K}(\alpha, v')$, which is

$$\begin{bmatrix} v'_1 & \cdots & v'_N \\ \alpha_1 v'_1 & \cdots & \alpha_N v'_N \\ \alpha_1^{N-K-1} v'_1 & \cdots & \alpha_N^{N-K-1} v'_N \end{bmatrix}$$

$$= \begin{bmatrix} 1 & \cdots & 1 \\ \alpha_1 & \cdots & \alpha_N \\ \cdots\cdots\cdots\cdots\cdots \\ \alpha_1^{N-K-1} & \cdots & \alpha_N^{N-K-1} \end{bmatrix} \begin{bmatrix} v'_1 & & 0 \\ & v'_2 & \\ 0 & & v'_N \end{bmatrix} \qquad (11)$$

We shall meet these codes again in §2 of Ch. 12.

Problem. (7) (a) If $\beta_i = c\alpha_i + d$ $(i = 1, \ldots, N)$, for $c \neq 0$, $d \in \mathrm{GF}(q^m)$, show that $\mathrm{GRS}_K(\alpha, v) = \mathrm{GRS}_K(\beta, v)$.

(b) When are $\mathrm{GRS}_K(\alpha, v)$ and $\mathrm{GRS}_K(\alpha, w)$ equivalent?

(c) Show that $\mathrm{GRS}_K(\alpha, v) = \mathrm{GRS}_K(\alpha, w)$ iff $v = \lambda w$, $\lambda \neq 0 \in \mathrm{GF}(q^m)$.

§9. Redundant residue codes

Equation (6) also suggests another way of looking at RS codes. Observe that $u(\alpha^i)$ is the remainder when the message $u(x)$, of degree less than K, is divided by $M^{(i)}(x) = x - \alpha^i$. So we can restate Equation (6) by saying that $u(x)$ is encoded into

$$(r_0, r_1, \ldots, r_{N-1}), \tag{12}$$

where r_i is the residue of $u(x)$ modulo $M^{(i)}(x)$.

$u(x)$ can be recovered from its residues with the aid of:

Theorem 5. (Chinese remainder theorem for polynomials.) *Let* $m_0(x), \ldots, m_{K-1}(x)$ *be polynomials over* $\mathrm{GF}(q)$ *which are pairwise relatively prime, and set* $M(x) = m_0(x)m_1(x) \cdots m_{K-1}(x)$. *If* $r_0(x), \ldots, r_{K-1}(x)$ *are any polynomials over* $\mathrm{GF}(q)$, *there exists exactly one polynomial* $u(x)$ *with* $\deg u(x) < \deg M(x)$ *such that*

$$u(x) \equiv r_i(x) \,(\mathrm{mod}\, m_i(x)), \tag{13}$$

for all $i = 0, \ldots, K - 1$. *In fact, let* $a_i(x)$ *be such that*

$$\frac{M(x)}{m_i(x)} a_i(x) \equiv 1 \,(\mathrm{mod}\, m_i(x)), \quad i = 0, \ldots, K - 1.$$

(Such an $a_i(x)$ *exists by Corollary 15 of Ch. 12.) Then the solution to* (13) *is*

$$u(x) = \sum_{i=0}^{K-1} \frac{M(x)}{m_i(x)} r_i(x) a_i(x) \text{ reduced } \mathrm{mod}\, M(x). \tag{14}$$

Problem (8). State the corresponding theorem for integers.

Theorem 5 shows that $r_0, \ldots, r_{K-1}$ in (12) are enough to reconstruct $u(x)$, in the absence of noise. Thus $r_K(x), \ldots, r_{N-1}(x)$ are redundant residues which are included in the codeword for protection against errors. Any code of this type is called a *redundant residue code*. We have shown that RS codes are redundant residue codes, and in Ch. 12 we shall see that some Goppa codes are also of this type. Other examples are mentioned in the Notes.

§10. Decoding RS codes

Since RS codes are special cases of BCH codes, they can be decoded by the methods of §6 of Ch. 9. The original majority logic decoding method of Reed and Solomon is also worth mentioning because of its considerable theoretical interest, even though it is impractical.

Suppose the codeword (6) has been transmitted, an error vector $e = (e_0, \ldots, e_{N-1})$ occurs, and $y = (y_0, \ldots, y_{N-1})$ is received. Thus the decoder knows

$$y_0 = e_0 + u_0 + u_1 + u_2 + \cdots + u_{K-1},$$
$$y_1 = e_1 + u_0 + \alpha u_1 + \alpha^2 u_2 + \cdots + \alpha^{K-1} u_{K-1}, \tag{15}$$
$$\cdots\cdots\cdots\cdots\cdots\cdots\cdots\cdots\cdots\cdots\cdots\cdots\cdots\cdots\cdots$$
$$y_{N-1} = e_{N-1} + u_0 + \alpha^{N-1} u_1 + \alpha^{2(N-1)} u_2 + \cdots + \alpha^{(K-1)(N-1)} u_{K-1}.$$

If there are no errors, $e = 0$, and any K of these equations can be solved to determine the message $u = (u_0, \ldots, u_{K-1})$, since the coefficient matrix is Vandermonde (Lemma 17 of Ch. 4). Thus there are $\binom{N}{K}$ determinations, or *votes*, for the correct u.

If there are errors, some sets of K equations will give the wrong u. But no incorrect u can receive too many votes.

Theorem 6. *If w errors occur, an incorrect u will receive at most $\binom{w+K-1}{K}$ votes. The correct u will receive at least $\binom{N-w}{K}$ votes.*

Proof. Since the equations in (15) are independent, any K of them have exactly one solution u. To obtain more than one vote, u must be the solution of more than K equations. An incorrect u can be the solution of at most $w + K - 1$ equations, consisting of w erroneous equations and $K - 1$ correct ones. (For if the u is the solution of K correct equations then u is correct.) Thus an incorrect u can be the solution to at most $\binom{w+K-1}{K}$ sets of K equations. Clearly there are $\binom{N-w}{K}$ sets of correct equations giving the correct u. Q.E.D.

Thus the message u will be obtained correctly if $\binom{N-w}{K} > \binom{w+K-1}{K}$, i.e. if $N - w > w + K - 1$, or $D = N - K + 1 > 2w$. So error vectors of weight less than half the minimum distance (and possibly others) can be corrected. Of course if $\binom{N}{K}$ is large this method is impractical.

§11. Justesen codes and concatenated codes

In §5 of Ch. 9 we saw that long BCH codes are bad, and in §5 of this chapter that long binary codes obtained from RS codes are also bad.

However, by a very simple construction it is possible to obtain an infinite family of good binary codes (called Justesen codes) from RS codes, as we now show.

The starting point is an RS code $\mathcal{R}$ over $GF(2^m)$ with parameters $[N = 2^m - 1, K, D = N - K + 1]$. Let α be a primitive element of $GF(2^m)$. Let $a = (a_0, \ldots, a_{N-1})$, $a_i \in GF(2^m)$, be a typical codeword of $\mathcal{R}$. Let b be the vector

$$b = (a_0, a_0; a_1, \alpha a_1; a_2, \alpha^2 a_2; \ldots; a_{N-1}, \alpha^{N-1} a_{N-1}).$$

Finally, replacing each component of b by the corresponding binary m-tuple, we obtain a *binary* vector c of length $2mN$.

Definition. For any N and K, with $0 < K < N$, the Justesen code $\mathcal{J}_{N,K}$ consists of all such vectors c which are obtained from the $[N, K]$ Reed–Solomon code $\mathcal{R}$.

Clearly $\mathcal{J}_{N,K}$ is a binary linear code of length $n = 2mN$ and dimension $k = mK$. The rate of this code is $k/n = K/2N < \frac{1}{2}$. A larger class of Justesen codes which includes codes of rate $\geq \frac{1}{2}$ will be described below.

Justesen codes are an example of *concatenated codes*, which we now describe.

Concatenated codes. Consider the arrangement shown in Fig. 10.7. Suppose the inner encoder and decoder use a code (called the *inner code*) which is an $[n, k, d]$ binary code. The combination of inner encoder, channel, and inner decoder can be thought of as forming a new channel (called a *superchannel*) which transmits binary k-tuples. If these k-tuples are considered as elements of $GF(2^k)$, we can attempt to correct errors on the superchannel by taking the *outer code* to be an $[N, K, D]$ code over $GF(2^k)$. Frequently a Reed–Solomon code is used as the outer code. This combination of codes (or any similar scheme) is called a *concatenated* code. The overall code (the *supercode*) is a **binary code of length nN, dimension kK, and rate $(k/n) \cdot (K/N)$.**

The encoding is done as follows. The kK binary information symbols are divided into K k-tuples, which are thought of as elements of $GF(2^k)$. These are then encoded by the outer encoder into the codeword $a_0 a_1 \cdots a_{N-1}$. Each a_i is now encoded by the inner encoder into a binary n-tuple b_i. Then $b_0 b_1 \cdots b_{N-1}$ is the codeword of the supercode and is transmitted over the channel.

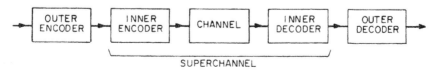

Fig. 10.7. A concatenated code.

Problem. (9) (a) Show that the minimum distance of the concatenated code is at least dD.

(b) For example, use an $[8, 4, 4]$ binary inner code and a $[12, 6, 7]$ outer RS code over $GF(2^4)$ to obtain a $[96, 24, 28]$ binary concatenated code. Obtain a $[104, 24, 32]$ code in the same way.

Justesen codes may be thought of as concatenated codes where the inner encoder uses N distinct codes. Let $\mathscr{C}_i$ be the $[2m, m]$ binary code consisting of the binary representations of the vectors $(u, \alpha^i u)$, $u \in GF(2^m)$. Then the ith symbol a_i of the outer code is encoded by $\mathscr{C}_i$.

Minimum distance of Justesen codes. The key point is that every binary vector (u, v) of length $2m$ with $u \neq 0$, $v \neq 0$ belongs to exactly one of the codes $\mathscr{C}_i$. Therefore a typical codeword c of $\mathscr{J}$ contains at least D *distinct* binary $(2m)$-tuples.

Since there aren't many $(2m)$-tuples of low weight, the total weight of c must be large. Thus if l is as large as possible subject to

$$\sum_{i=1}^{l} \binom{2m}{i} \leq D,$$

then if D is large so is l, and

$$wt(c) \geq \sum_{i=1}^{l} i \binom{2m}{i}$$

is also large.

Estimates of binomial coefficients. Before proving the first theorem we need some estimates of binomial coefficients. These involve the entropy function $H_2(x)$, defined by

$$H_2(x) = -x \log_2 x - (1-x) \log_2 (1-x)$$

where $0 \leq x \leq 1$ (Fig. 10.8). Fano [415] gives a useful table of $H_2(x)$. $H_2(x)$ is

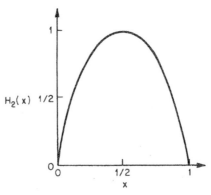

Fig. 10.8. The entropy function $H_2(x)$.

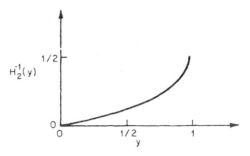

Fig. 10.9. The inverse function $H_2^{-1}(y)$.

used in information theory as a measure of "uncertainty", but to us it is just a convenient function to have because of its role in Lemmas 7 and 8.

We shall also need the inverse function $H_2^{-1}(y)$ (see Fig. 10.9) defined by

$$x = H_2^{-1}(y) \text{ iff } y = H_2(x)$$

for $0 \le x \le \frac{1}{2}$.

Lemma 7. (An estimate for a binomial coefficient.) *Suppose λn is an integer, where $0 < \lambda < 1$. Then*

$$\frac{1}{\sqrt{8n\lambda(1-\lambda)}} 2^{nH_2(\lambda)} \le \binom{n}{\lambda n} \le \frac{1}{\sqrt{2\pi n\lambda(1-\lambda)}} 2^{nH_2(\lambda)}. \qquad (16)$$

Proof. The proof uses *Stirling's formula* for $n!$

$$\sqrt{2\pi}n^{n+1/2} e^{-n+1/12n-1/360n^3} < n!$$
$$< \sqrt{2\pi}n^{n+1/2} e^{-n+1/12n}. \qquad (17)$$

Therefore, with $\mu = 1 - \lambda$,

$$\binom{n}{\lambda n} = \frac{n!}{(\lambda n)!(\mu n)!}$$

$$\ge \frac{1}{\sqrt{2\pi n\lambda\mu}} \frac{1}{\lambda^{\lambda n}\mu^{\mu n}} e^{-1/12n\lambda - 1/12n\mu}.$$

$$= \frac{1}{\sqrt{2\pi n\lambda\mu}} 2^{nH_2(\lambda)} e^{-1/12n\lambda\mu}.$$

The LHS of (16) now follows after a bit of algebra. The proof of the RHS is similar and is left to the reader (Problem 10).

Remark. For most purposes simpler versions of Stirling's formula,

$$\sqrt{2\pi}\, n^{n+1/2}\, e^{-n} < n! < \sqrt{2\pi}\, n^{n+1/2}\, e^{-n+1/12n}, \tag{18}$$

or

$$n! \sim \sqrt{2\pi}\, n^{n+1/2}\, e^{-n} \quad \text{as } n \to \infty$$

are adequate.

Lemma 8. (Estimate for a sum of binomial coefficients.) Suppose λn is an integer, where $\frac{1}{2} < \lambda < 1$. Then

$$\frac{2^{nH_2(\lambda)}}{\sqrt{8n\lambda(1-\lambda)}} \leqslant \sum_{k=\lambda n}^{n} \binom{n}{k} \leqslant 2^{nH_2(\lambda)}. \tag{19}$$

Proof. We first prove the RH inequality. For any positive number r we have

$$2^{r\lambda n} \sum_{k=\lambda n}^{n} \binom{n}{k} \leqslant \sum_{k=\lambda n}^{n} 2^{rk} \binom{n}{k} \leqslant \sum_{k=0}^{n} 2^{rk} \binom{n}{k} = (1+2^r)^n.$$

Thus

$$\sum_{k=\lambda n}^{n} \binom{n}{k} \leqslant \{2^{-r\lambda} + 2^{r(1-\lambda)}\}^n.$$

Choose $r = \log_2(\lambda/(1-\lambda))$. Then this sum is

$$\leqslant 2^{nH_2(\lambda)}(1-\lambda+\lambda)^n = 2^{nH_2(\lambda)}.$$

The LH inequality follows from

$$\sum_{k=\lambda n}^{n} \binom{n}{k} \geqslant \binom{n}{\lambda n}$$

and the previous lemma.   Q.E.D.

Corollary 9. *For* $0 < \mu < \frac{1}{2}$,

$$\frac{2^{nH_2(\mu)}}{\sqrt{8n\mu(1-\mu)}} \leqslant \sum_{k=0}^{\mu n} \binom{n}{k} \leqslant 2^{nH_2(\mu)}. \tag{20}$$

Analysis of Justesen codes.

Lemma 10. *If we are given M distinct nonzero L-tuples, where*

$$M = \gamma(2^{8L} - 1), \qquad 0 < \gamma < 1, \qquad 0 < \delta < 1,$$

then the sum of the weights of these L-tuples is at least

$$L\gamma(2^{\delta L} - 1)(H_2^{-1}(\delta) - o(1)).$$

[Recall that $f(L) = o(g(L))$ means $f(L)/g(L) \to 0$ as $L \to \infty$.]

Proof. The number of these L-tuples having weight $\leqslant \lambda L$ is at most

$$\sum_{i=1}^{\lambda L} \binom{L}{i} \leqslant 2^{L H_2(\lambda)},$$

by Corollary 9, for any $0 < \lambda < \frac{1}{2}$.

So the total weight is at least

$$\lambda L(M - 2^{L H_2(\lambda)}) = \lambda L M(1 - 2^{L H_2(\lambda)}/M).$$

Choose $\lambda = H_2^{-1}(\delta - 1/\log L) = H_2^{-1}(\delta) - o(1)$, with $\lambda < \frac{1}{2}$. Then the total weight is at least

$$L\gamma(2^{\delta L} - 1)(1 - o(1))(H_2^{-1}(\delta) - o(1)),$$
$$= L\gamma(\delta^{\delta L} - 1)(H_2^{-1}(\delta) - o(1)). \qquad \text{Q.E.D.}$$

Theorem 11. (Justesen.) *Let R be fixed, $0 < R < \frac{1}{2}$. For each m choose*

$$K = \lceil R \cdot 2N \rceil = \lceil R2(2^m - 1) \rceil. \qquad (21)$$

Then the Justesen code $\mathcal{J}_{N,K}$ is a binary code of length $n = 2mN$ with rate

$$R_m = \frac{K}{2N} \geqslant R,$$

and the minimum distance d_m satisfying

$$\frac{d_m}{n} \geqslant (1 - 2R)(H_2^{-1}(0.5) - o(1))$$

$$\approx 0.110(1 - 2R).$$

The lower bound on d_m/n is linear in R and is plotted in Fig. 10.10.

Proof. Let c be any nonzero codeword. As we saw earlier, c contains at least $N - K + 1$ distinct nonzero binary $(2m)$-tuples. From (21),

$$R + \frac{1}{2N} > R_m = \frac{K}{2N} \geqslant R,$$

$$N - K + 1 = N\left(1 - \frac{K-1}{N}\right) \geqslant N(1 - 2R).$$

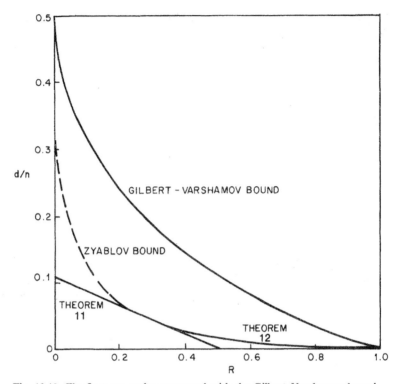

Fig. 10.10. The Justesen codes compared with the Gilbert–Varshamov bound.

We can now apply Lemma 10 with $L = 2m$, $\delta = \frac{1}{2}$, $\gamma = 1 - 2R$, and deduce that

$$\text{wt}\,(c) \geq 2m(1 - 2R)(2^m - 1)(H_2^{-1}(0.5) - o(1)),$$

$$\frac{d_m}{2mN} \geq (1 - 2R)(H_2^{-1}(0.5) - o(1)). \qquad \qquad \text{Q.E.D.}$$

Theorem 11 only gives codes of rate less than $\frac{1}{2}$. A larger class of codes can be obtained by puncturing $\mathcal{J}_{N,K}$ in the following way. Each component of the vector

$$b = (a_0, a_0; a_1, \alpha a_1; a_2, \alpha^2 a_2; \ldots, \alpha^{N-1} a_{N-1})$$

is expressed as a binary m-tuple as before, and then the last s binary digits are punctured from the alternate components

$$a_0 \ \alpha a_1 \ \alpha^2 a_2 \ \cdots \ \alpha^{N-1} a_{N-1}.$$

The set of all such punctured vectors forms the Justesen code $\mathcal{J}_{N,K}^*$. This is a binary linear code with parameters

$$n = (2m - s)N, \qquad k = mK.$$

We shall choose s later. This is a concatenated code for which the inner codes have rate

$$r_m = \frac{m}{2m - s},$$

the outer RS code has rate $R_{\mathrm{RS}} = K/N$, and the overall rate of $\mathscr{G}^*_{N,K}$ is the product

$$R_m = r_m R_{\mathrm{RS}} = \frac{mK}{(2m - s)N}.$$

As before the vector $(a_0, \ldots, a_{N-1})$ has

$$\text{weight} \geq N\left(1 - \frac{K - 1}{N}\right).$$

However, after puncturing each nonzero pair may occur as many as 2^s times. The lowest weight would occur when each nonzero $(2m - s)$-tuples appears exactly 2^s times. The number of distinct nonzero $(2m - s)$-tuples is then at least

$$\frac{1}{2^s} N\left(1 - \frac{K - 1}{N}\right) \geq (2^{m-s} - 1)\left(1 - \frac{K - 1}{N}\right).$$

Given R we choose

$$K = \left\lceil \frac{RN(2m - s)}{m} \right\rceil, \tag{22}$$

so that

$$\frac{K - 1}{N} \leq \frac{R(2m - s)}{m}.$$

Therefore the number of distinct nonzero $(2m - s)$-tuples is

$$\geq (2^{m-s} - 1)\left(1 - \frac{R(2m - s)}{m}\right).$$

From Lemma 10 with $L = 2m - s$, $\delta = (m - s)/L$, $\gamma = 1 - R(2m - s)/m$ we deduce that the weight of the codeword is at least

$$2^s(2m - s)\left(1 - \frac{R(2m - s)}{m}\right)(2^{m-s} - 1)(H^{-1}(\delta) - o(1)),$$

where the initial 2^s is because each $(2m - s)$-tuple occurs 2^s times (in the worst case). So for this code

$$\frac{\text{distance}}{\text{length}} \geq \frac{2^m - 2^s}{2^m - 1}\left(1 - \frac{R(2m - s)}{m}\right)\left(H_2^{-1}\left(\frac{m - s}{2m - s}\right) - o(1)\right)$$

Choose an r such that $\frac{1}{2} \leqslant r < 1$, and set

$$s = \left[\frac{m}{r} (2r - 1) \right], \tag{23}$$

so that

$$r_m = \frac{m}{2m - s} \geqslant r.$$

Then as $m \to \infty$,

$$\frac{s}{m} \to \frac{2r - 1}{r} < 1,$$

$$\frac{2^m - 2^s}{2^m - 1} \to 1,$$

and the lower bound on distance/length approaches

$$\left(1 - \frac{R}{r} \right) H_2^{-1}(1 - r) \quad \text{as } m \to \infty.$$

Finally we shall choose r to maximize this expression. Setting the derivative with respect to r equal to 0 we find that r should be chosen as the solution r_o (say) of

$$R = \frac{r^2}{1 + \log_2 [1 - H_2^{-1}(1 - r)]} \tag{24}$$

provided $r_o \geqslant \frac{1}{2}$. If $r_o < \frac{1}{2}$, which happens when $R < 0.30$, take $r = \frac{1}{2}$. Thus we have proved

Theorem 12. (Justesen.) *Let R be fixed, $0 < R < 1$. The punctured code $\mathcal{J}_{N,K}^*$, with r equal to the maximum of $\frac{1}{2}$ and the solution of (24), and s and K given by (23) and (22), has a lower bound to distance/length equal to*

$$\left(1 - \frac{R}{r} \right) H_2^{-1}(1 - r) \quad \text{as } m \to \infty.$$

For comparison, the asymptotic form of the Gilbert–Varshamov bound (Theorem 12 of Ch. 1, Theorem 30 of Ch. 17) is that codes exist for which distance/length has a lower bound which approaches $H_2^{-1}(1 - R)$ as length $\to \infty$. (See Fig. 10.10.) Consider a point $(r, H_2^{-1}(1 - r))$ on this curve. The straight line joining the projections of this point on the axes, i.e. $(r, 0)$ and $(0, H_2^{-1}(1 - r))$, has equation $y = (1 - R/r)H_2^{-1}(1 - r)$. The Justesen bound of Theorem 12 is the maximum of $(1 - R/r)H_2^{-1}(1 - r)$ for $\frac{1}{2} \leqslant r \leqslant 1$, and so is the envelope of these lines. This bound is also shown in Fig. 10.10.

The bounds of Theorems 11 and 12 meet when $R = 0.30$.

Finally, the following simple argument shows that concatenated codes exist which lie on or above the envelope of these lines for all $0 \leqslant r \leqslant 1$. Consider codes obtained by concatenating an inner $[n, k, d]$ binary code and an $[N, K, D]$ outer code over $\mathrm{GF}(2^k)$. To avoid trivial cases we insist that both n and N approach infinity. The overall rate is $R = kK/nN$.

Theorem 13. (Zyablov.) *There exist concatenated codes with $n \to \infty$, $N \to \infty$, in which the outer code is maximal distance separable, and which satisfy*

$$\frac{\text{distance}}{\text{length}} \geq \max_{0 \leq r \leq 1} \left\{ \left(1 - \frac{R}{r} \right) H_2^{-1}(1 - r) \right\}.$$

Proof. Since the outer code is MDS, $D = N - K + 1$, so

$$\frac{D}{N} = 1 - \frac{K}{N} + \frac{1}{N}.$$

Take the inner code to meet the Garshamov bound, so $d/n \geq H_2^{-1}(1 - r)$, where $r = k/n$. Then for the overall code

$$\frac{\text{distance}}{\text{length}} \geq \frac{dD}{nN} \geq \left(1 - \frac{R}{r} \right) H_2^{-1}(1 - r)$$

as $n \to \infty$, $N \to \infty$.   Q.E.D.

This bound is indicated by the broken line in Fig. 10.10. For $R > 0.30$ Justesen codes meet the bound.

Research Problem (10.3). Give an explicit construction for concatenated codes which meet the bound when $R < 0.30$.

Remark. Blokh and Zyablov have shown [165] that the class of all concatenated codes (with $n \to \infty$ and $N \to \infty$) contains codes meeting the Garshamov bound. As usual this proof is not constructive.

Problem. (10) Prove the RHS inequality of (16).

Notes on Chapter 10

§2. The first time RS codes appear *as codes is* in Reed and Solomon [1106]. However, they had already been explicitly constructed by Bush [220] in 1952, using the language of orthogonal arrays. RS codes are extensively discussed by Forney [436]. Gore [547] has shown that in many situations the binary versions of RS codes have a lower error probability than binary BCH codes with the same parameters.

Solomon [1254] uses codewords of maximum weight in an RS code to encode messages over alphabets which "are not quite the right size, field-wise", i.e. are not of prime power order, such as the English alphabet. Solomon [1250] and Reed et al. [1105, 1107] describe synchronizable codes constructed from RS codes. See also Ebert and Tong [402] and Wolverton [1432].

§3. For extending RS codes see Bush [220], Gross [563], Tanaka and Nishida [1303] and Wolf [1427].

§5. Example 3 was found by Solomon [1252]; see also MacWilliams [879].

§6. Burst error correcting codes. Peterson and Weldon [1040] have an excellent treatment. Some of the most important papers are: Bahl and Chien [57], Bridewell and Wolf [197], Burton [215], Burton et al. [218], Chien et al. [280, 285, 292], Elliott [407], Forney [440], Fujiwara et al. [462], Hagelbarger [573], Iwadare [684, 685], Kasami [723], Kasami and Matora [743], Mandelbaum [897], Pehlert [1033], Posner [1069], Shiva and Sheng [1204], Tanaka and Nishida [1305], Tauglikh [1310], Tavares and Shiva [1314], Tong [1334–1336] and Wainberg and Wolf [1382, 1384].

§7. Other encoding methods for RS codes are described by Stone [1282].

§8. Delsarte [359] has studied generalized RS codes.

§9. For the Chinese remainder theorem see Niven and Zuckerman [995, p. 33] or Uspensky and Heaslet [1359, pp. 189–191]. Bossen [188], Bossen and Yau [189], Mandelbaum [896, 898, 899], and Stone [1282] have described other redundant residue codes, which are also MDS codes over $GF(q^m)$. However, Reed–Solomon codes seem to be the most important codes in this class.

§10. Burton [215], Gore [543, 547], Mandelbaum [896, 898, 899] and Yau and Liu [1446] have studied the decoding of RS codes.

§11. Justesen's original paper is [704]; see also [705]. Sugiyama et al. [1285, 1287] and Weldon [1405] (see also [1406]) have given small improvements on Justesen's construction at very low rates. Concatenated codes were introduced by Forney [436]. See also Blokh and Zyablov [165, 166], Savage [1151], Zyablov [1475, 1476] and §§5 and 8.2 of Ch. 18. Problem 9b is due to Sugiyama and Kasahara [1286].

Lemma 7 is a special case of the Chernoff bound – see Ash [32], or Jelinek [690, p. 117]. Park [1021] gives an inductive proof of the RHS of (16). Massey [923] has given an alternative version of Lemma 10. Theorem 13 is due to Zyablov [1475] (see also [165, 166 and 1476]).

11

MDS codes

§1. Introduction

We come now to one of the most fascinating chapters in all of coding theory: MDS codes. In Theorem 11 of Ch. 1 it was shown that for a linear code over any field, $d \leq n - k + 1$. Codes with $d = n - k + 1$ are called *maximum distance separable*, or MDS for short. The name comes from the fact that such a code has the *maximum* possible *distance* between codewords, and that the codewords may be *separated* into message symbols and check symbols (i.e. the code has a systematic encoder, using the terminology of §7 of Ch. 10). In fact *any* k symbols may be taken as message symbols, as Corollary 3 shows. MDS codes are also called *optimal*, but we prefer the less ambiguous term.

In this chapter various properties of MDS codes will be derived. We shall also see that the problem of finding the longest possible MDS code with a given dimension is equivalent to a surprising list of combinatorial problems, none of which is completely solved – see Research Problem 11.1a to 11.1f. In Fig. 11.2 and Research Problem 11.4 we state what is conjectured to be the solution to some of these problems.

In §2 of the preceding chapter it was shown that there is an $[n = q - 1, k, d = n - k + 1]$ Reed–Solomon (or RS) code over $GF(q)$, for all $k = 1, \ldots, n$, and that these codes are MDS codes. Furthermore in §3 an overall parity check was added producing $[n + 1, k, n - k + 2]$ extended RS codes, also MDS. It is natural to ask if more parity checks can be added, while preserving the property of being MDS. The answer seems to be (§§5, 7 below) that one or two further parity checks can be added, but probably no more. More generally, we state the first version of our problem.

Research Problem (11.1a). Given k and q, find the largest value of n for which an $[n, k, n - k + 1]$ MDS code exists over $GF(q)$. Let $m(k, q)$ denote this largest value of n.

It will turn out that in all the known cases, when an $[n, k, d]$ MDS code exists, then an $[n, k, d]$ RS or extended RS code with the same parameters also exists. Thus as far as is known at present, RS and extended RS codes are the most important class of MDS codes. For this reason we don't give a separate discussion of decoding MDS codes but refer to §10 of Ch. 10.

Problem. (1) Show that $[n, 1, n]$, $[n, n - 1, 2]$ and $[n, n, 1]$ MDS codes exist over any field. These are called *trivial* MDS codes. For a nontrivial code, $2 \leq k \leq n - 2$.

§2. Generator and parity check matrices

Let $\mathscr{C}$ be an $[n, k, d]$ code over $GF(q)$ with parity check matrix H and generator matrix G.

Theorem 1. $\mathscr{C}$ is MDS iff every $n - k$ columns of H are linearly independent.

Proof. $\mathscr{C}$ contains a codeword of weight w iff w columns of H are linearly dependent (Theorem 10 of Ch. 1). Therefore $\mathscr{C}$ has $d = n - k + 1$ iff no $n - k$ or fewer columns of H are linearly dependent.   Q.E.D.

Theorem 2. If $\mathscr{C}$ is MDS so is the dual code $\mathscr{C}^{\perp}$.

Proof. H is a generator matrix for $\mathscr{C}^{\perp}$. From Theorem 1, any $n - k$ columns of H are linearly independent, so only the zero codeword is zero on as many as $n - k$ coordinates. Therefore $\mathscr{C}^{\perp}$ has minimum distance at least $k + 1$, i.e. it has parameters $[n, n - k, k + 1]$.   Q.E.D.

Example.

$$\begin{pmatrix} 1 & 0 & 1 & 1 \\ 0 & 1 & \alpha & \beta \end{pmatrix}$$

is the generator matrix for a $[4, 2, 3]$ MDS code $\mathscr{C}$ over $GF(4) =$

$\{0, 1, \alpha, \beta = \alpha^2\}$. The dual code $\mathscr{C}^\perp$ has generator matrix

$$\begin{pmatrix} 1 & \alpha & 1 & 0 \\ 1 & \beta & 0 & 1 \end{pmatrix}$$

and is also a [4, 2, 3] MDS code.

Corollary 3. *Let $\mathscr{C}$ be an $[n, k, d]$ code over $\mathrm{GF}(q)$. The following statements are equivalent:*

(i) *$\mathscr{C}$ is MDS;*

(ii) *every k columns of a generator matrix G are linearly independent (i.e. any k symbols of the codewords may be taken as message symbols);*

(iii) *every $n - k$ columns of a parity check matrix H are linearly independent.*

Proof. From Theorems 1 and 2.   Q.E.D.

The open problem can now be restated as

Research Problem (11.1b). Given k and q, find the largest n for which there is a $k \times n$ matrix over $\mathrm{GF}(q)$ having every k columns linearly independent.

Equivalently, in vector space terminology:

Research Problem (11.1c). Given a k-dimensional vector space over $\mathrm{GF}(q)$, what is the largest number of vectors with the property that any k of them form a basis for the space?

Problems. (2) Show that the only binary MDS codes are the trivial ones.

(3) The Singleton bound for nonlinear codes. If $\mathscr{C}$ is an (n, M, d) code over $\mathrm{GF}(q)$, show that $d \leq n - \log_q M + 1$.

§3. The weight distribution of an MDS code

Surprisingly, the Hamming weight distribution of an MDS code is completely determined.

Theorem 4. *Let $\mathscr{C}$ be an $[n, k, d]$ code over $\mathrm{GF}(q)$. Then $\mathscr{C}$ is MDS iff $\mathscr{C}$ has a minimum weight codeword in any d coordinates.*

Proof. (Only if.) Given any $n - k + 1$ coordinates, take one of them together with the complementary $k - 1$ coordinates as message symbols (which can be done by Corollary 3). Setting the single coordinate equal to 1 and the $k - 1$ to 0 gives a codeword of weight $n - k + 1$. The proof of the converse is left to the reader.   Q.E.D.

Corollary 5. *The number of codewords in $\mathscr{C}$ of weight $n - k + 1$ is*

$$(q - 1)\binom{n}{n - k + 1}.$$

An MDS code has k distinct nonzero weights, $n - k + 1, \ldots, n$, and the dual code has minimum distance $d' = k + 1$. Therefore by Theorem 29 of Ch. 6, the codewords of weight d form a t-design, which however by Theorem 4 is just a trivial design. Theorem 4 of Ch. 6 also determines the weight distribution, but in this case it is easier to begin from the MacWilliams identities in the form of Problem (6) of Ch. 5, namely

$$\sum_{i=0}^{n-j}\binom{n-i}{j}A_i = q^{k-j}\sum_{i=0}^{j}\binom{n-i}{j-i}A'_i, \quad j = 0, 1, \ldots, n.$$

Since $A_i = 0$ for $1 \leqslant i \leqslant n - k$ and $A'_i = 0$ for $1 \leqslant i \leqslant k$, this becomes

$$\sum_{i=n-k+1}^{n-j}\binom{n-i}{j}A_i = \binom{n}{j}(q^{k-j} - 1), \quad j = 0, 1, \ldots, k - 1.$$

Setting $j = k - 1$ and $k - 2$ gives

$$A_{n-k+1} = \binom{n}{k-1}(q - 1),$$

$$\binom{k-1}{k-2}A_{n-k+1} + A_{n-k+2} = \binom{n}{k-2}(q^2 - 1),$$

$$A_{n-k+2} = \binom{n}{k-2}[(q^2 - 1) - (n - k + 2)(q - 1)].$$

It is not hard to guess (and to verify) that the general solution is

$$A_{n-k+r} = \binom{n}{k-r}\sum_{j=0}^{r-1}(-1)^j\binom{n-k+r}{j}(q^{r-j} - 1).$$

Hence we have

Theorem 6. *The number of codewords of weight w in an $[n, k, d = n - k + 1]$*

MDS *code over* GF(q) *is*

$$A_w = \binom{n}{w} \sum_{j=0}^{w-d} (-1)^j \binom{w}{j} (q^{w-d+1-j} - 1)$$

$$= \binom{n}{w} (q-1) \sum_{j=0}^{w-d} (-1)^j \binom{w-1}{j} q^{w-d-j}. \cdot$$

We note that

$$A_{n-k+2} = \binom{n}{k-2} (q-1)(q-n+k-1).$$

This number must be nonnegative, hence

Corollary 7. *Let $\mathscr{C}$ be an $[n, k, n-k+1]$ MDS code. If $k \geqslant 2$, $q \geqslant n-k+1$. If $k \leqslant n-2$, $q \geqslant k+1$.*

Proof. The second statement follows from examining the weight distribution of $\mathscr{C}^{\perp}$.   Q.E.D.

Problems. (4) Prove the converse part of Theorem 4.

(5) A *real code* consists of all linear combinations with real coefficients of the rows of a generator matrix $G = (q_{ij})$, where the q_{ij} are real numbers. Justify the statement that most real codes are MDS.

Research Problem (11.2). What can be said about the complete weight enumerator (see §6 of Ch. 5) of an MDS code, or even of an RS code?

§4. Matrices with every square submatrix nonsingular

Theorem 8. *An $[n, k, d]$ code $\mathscr{C}$ with generator matrix $G = [I \mid A]$, where A is a $k \times (n-k)$ matrix, is MDS iff every square submatrix (formed from any i rows and any i columns, for any $i = 1, 2, \ldots, \min\{k, n-k\}$) of A is nonsingular.*

Proof. ($\Rightarrow$) Suppose $\mathscr{C}$ is MDS. By Corollary 3, every k columns of G are linearly independent. The idea of the proof is very simple and we shall just illustrate it by proving that the top right 3×3 submatrix A' of A is nonsingular. Take the matrix B consisting of the last $k-3$ columns of I and the first 3 columns of A:

Then det $B = $ det $A' \neq 0$. The general case is handled in the same way. ($\Leftarrow$)
The converse is immediate.   Q.E.D.

Examples. (1) The $[4, 2, 3]$ code over GF(4) shown in Fig. 10.2 has

$$A = \begin{bmatrix} \beta & \alpha \\ \alpha & \beta \end{bmatrix},$$

and indeed every square submatrix of A (of size 1 and 2) is nonsingular.
   (2) The $[5, 2, 4]$ extended RS code over GF(5) has

$$A = \begin{bmatrix} 3 & 4 & 2 \\ 3 & 2 & 4 \end{bmatrix}.$$

From Theorem 8 the next version of our problem is:

Research Problem (11.1d). Given k and q, find the largest r such that there exists a $k \times r$ matrix having entries from GF(q) with the property that every square submatrix is nonsingular.

Problem. (6) (Singleton.) Show that any rectangular submatrix A of the arrays in Fig. 11.1 has the property that any $k \times k$ submatrix of A is nonsingular over GF(q).

$$
\begin{array}{ll}
& \begin{array}{llll} 1 & 1 & 1 & 1 \\ 1 & 2 & 3 & 4 \\ \end{array} \\
q = 5 & \begin{array}{llll} 1 & 3 & 4 \\ 1 & 4 \\ \end{array}
\end{array}
\qquad
\begin{array}{ll}
& \begin{array}{llllll} 1 & 1 & 1 & 1 & 1 & 1 \\ 1 & 3 & 6 & 4 & 2 & 5 \\ \end{array} \\
q = 7 & \begin{array}{llllll} 1 & 6 & 4 & 2 & 5 \\ 1 & 4 & 2 & 5 \\ 1 & 2 & 5 \\ 1 & 5 \\ \end{array}
\end{array}
$$

Fig. 11.1.

Research Problem (11.3). Generalize Fig. 11.1. for larger q.

Problem. (7) (a) Show that any square submatrix of a Vandermonde matrix with real, positive entries is nonsingular. Show that this is not true for Vandermonde matrices over finite fields.

(b) Given $x_1, \ldots, x_n, y_1, \ldots, y_n$ the matrix $C = (c_{ij})$ where $c_{ij} = 1/(x_i + y_j)$ is called a *Cauchy matrix*. Show that

$$\det(C) = \frac{\prod\limits_{1 \leq i < j \leq n} (x_j - x_i)(y_j - y_i)}{\prod\limits_{1 \leq i, j \leq n} (x_i + y_j)}.$$

Hence, provided the x_i are distinct, the y_i are distinct, and $x_i + y_j \neq 0$ for all i, j, it follows that any square submatrix of a Cauchy matrix is nonsingular over any field.

§5. MDS codes from RS codes

Let $\alpha_1, \ldots, \alpha_{q-1}$ be the nonzero elements of $GF(q)$. The $[q, k, d = q - k + 1]$ extended RS code of §3 of Ch. 10 has parity check matrix

$$H_1 = \begin{bmatrix} 1 & \cdots & 1 & 1 \\ \alpha_1 & \cdots & \alpha_{q-1} & 0 \\ \alpha_1^2 & \cdots & \alpha_{q-1}^2 & 0 \\ \cdots\cdots\cdots\cdots\cdots\cdots\cdots \\ \alpha_1^{q-k-1} & \cdots & \alpha_{q-1}^{q-k-1} & 0 \end{bmatrix}.$$

One more parity check can always be added, producing a $[q + 1, k, q - k + 2]$ MDS code, by using the parity check matrix

$$H_2 = \begin{bmatrix} 1 & \cdots & 1 & 1 & 0 \\ \alpha_1 & \cdots & \alpha_{q-1} & 0 & 0 \\ \alpha_1^2 & \cdots & \alpha_{q-1}^2 & 0 & 0 \\ \cdots\cdots\cdots\cdots\cdots\cdots\cdots \\ \alpha_1^{q-k} & \cdots & \alpha_{q-1}^{q-k} & 0 & 1 \end{bmatrix}. \tag{1}$$

To show this, we must verify that any $q - k + 1$ columns of H_2 are linearly independent, i.e. form a nonsingular matrix. In fact, any $q - k + 1$ of the first $q - 1$ columns form a Vandermonde matrix (Lemma 17 of Ch. 4) and are nonsingular. Similarly, given any $q - k + 1$ columns which include one or both of the last two columns, we expand about these columns and again obtain a Vandermonde matrix.

In fact, there exist cyclic codes with the same parameters.

Theorem 9. *For any* k, $1 \leq k \leq q + 1$, *there exists a* $[q + 1, k, q - k + 2]$ *cyclic MDS code over* $GF(q)$.

Proof. We only prove the case $q = 2^m$, the case of odd q being similar. To exclude the trivial cases we assume $2 \leq k \leq q - 1$. Consider the polynomial $x^{2^m+1} + 1$ over $GF(2^m)$. The cyclotomic cosets, under multiplication by 2^m and reduction modulo $2^m + 1$, are

$$\{0\}$$
$$\{1, 2^m\} = \{1, -1\}$$
$$\{2, 2^m - 1\} = \{2, -2\}$$
$$\cdots\cdots\cdots\cdots\cdots\cdots\cdots$$
$$\{2^{m-1}, 2^{m-1} + 1\} = \{2^{m-1}, -2^{m-1}\}.$$

For example, if $2^m + 1 = 33$ we have the cyclotomic cosets

$$\{0\}$$

$\{1, 32\}$	$\{9, 24\}$
$\{2, 31\}$	$\{10, 23\}$
$\{3, 30\}$	$\{11, 22\}$
$\{4, 29\}$	$\{12, 21\}$
$\{5, 28\}$	$\{13, 20\}$
$\{6, 27\}$	$\{14, 19\}$
$\{7, 26\}$	$\{15, 18\}$
$\{8, 25\}$	$\{16, 17\}$

Thus $x^{2^m+1} + 1$ has, besides $x + 1$, only quadratic factors over $GF(2^m)$, and these are of the form

$$x^2 + (\alpha^i + \alpha^{-i})x + 1 = (x + \alpha^i)(x + \alpha^{-i}),$$

where α is a primitive $(2^m + 1)$-st root of unity. Now $\alpha \in GF(2^{2m})$; in fact if ξ is a primitive element of $GF(2^{2m})$ we may take $\alpha = \xi^{2^m-1}$.

Problem. (8) With this value of α show $\alpha^i + \alpha^{-i}$ is in $GF(2^m)$.

Now consider the $[2^m + 1, 2^m + 1 - 2t - 1]$ cyclic code with generator polynomial

$$(x + 1) \prod_{i=1}^{t} (x^2 + (\alpha^i + \alpha^{-i})x + 1).$$

This has $2t + 1$ consecutive zeros

$$\alpha^{-t}, \alpha^{-t+1}, \ldots, \alpha^{-1}, 1, \alpha, \ldots, \alpha^{t-1}, \alpha^t;$$

thus by the BCH bound the minimum distance is at least $2t + 2$. Since $n = 2^m + 1$, $k = 2^m + 1 - 2t - 1$, $n - k + 1 = 2t + 2$, and the code is MDS. This constructs the desired codes for all even values of k.

Similarly the code with generator polynomial

$$\prod_{i=2^{m-1}-t+1}^{2^{m-1}} (x^2 + (\alpha^i + \alpha^{-i})x + 1)$$

has the $2t$ consecutive zeros

$$\alpha^{2^{m-1}-t+1}, \ldots, \alpha^{2^{m-1}}, \alpha^{2^{m-1}+1}, \ldots, \alpha^{2^{m-1}+t},$$

and is a $[2^m + 1, 2^m + 1 - 2t, 2t + 1]$ MDS code. This gives the desired codes for all odd k.   Q.E.D.

Example. Codes of length $n = 9$ over $GF(2^3)$. Let ξ be a primitive element of $GF(2^6)$, $\beta = \xi^9$ a primitive element of $GF(2^3)$, and $\alpha = \xi^7$ a primitive 9th root of unity. Then from Fig. 4.5, $\beta^3 = \beta^2 + 1$, and

$$\alpha + \alpha^{-1} = \xi^7 + \xi^{56} = \beta^5, \qquad \alpha^2 + \alpha^{-2} = \xi^{14} + \xi^{49} = \beta^3,$$
$$\alpha^3 + \alpha^{-3} = \xi^{21} + \xi^{42} = 1, \qquad \alpha^4 + \alpha^{-4} = \xi^{28} + \xi^{35} = \beta^6.$$

Therefore

$$x^9 + 1 = (x + 1)(x^2 + x + 1)(x^2 + \beta^3 x + 1)(x^2 + \beta^5 x + 1)(x^2 + \beta^6 x + 1).$$

The code over GF(8) with check polynomial $x^2 + x + 1$ is degenerate (for $x^2 + x + 1$ divides $x^3 + 1$, see Lemma 8 of Chapter 8). It is a $[9, 2, 6]$ code with idempotent $x + x^2 + x^4 + x^5 + x^7 + x^8$, or 011011011 using an obvious notation.

The other three codes of dimension 2 are $[9, 2, 8]$ codes. Their idempotents are readily found to be:

				Idempotent					*Check polynomial*
1	x	x^2	x^3	x^4	x^5	x^6	x^7	x^8	
0	β^3	β^6	1	β^5	β^5	1	β^6	β^3	$x^2 + \beta^3 x + 1$
0	β^5	β^3	1	β^6	β^6	1	β^3	β^5	$x^2 + \beta^5 x + 1$
0	β^6	β^5	1	β^3	β^3	1	β^5	β^6	$x^2 + \beta^6 x + 1$

The codes with these idempotents are minimal codes (§3 of Ch. 8) and consist of the 9 cyclic shifts of the idempotent and their scalar multiples. The code with generator polynomial $x^2 + \beta^6 x + 1$ and zeros α^4, α^{-4} is a $[9, 7, 3]$ code. The polynomial $(x + 1)(x^2 + \beta^5 x + 1)$ has zeros α^{-1}, 1, α and generates a $[9, 6, 4]$ code; the polynomial $(x^2 + x + 1)(x^2 + \beta^6 x + 1)$ has zeros α^3, α^4, α^5, α^6 and generates a $[9, 5, 5]$ code, and so on.

Problem. (9) Find the idempotents and weight distributions of these codes.

The case $k = 3$ and q even. There are just two known cases when another parity check can be added: when $q = 2^m$ and $k = 3$ or $k = q - 1$.

Theorem 10. *There exist* $[2^m + 2, 3, 2^m]$ *and* $[2^m + 2, 2^m - 1, 4]$ *triply-extended* RS (*and* MDS) *codes.*

Proof. Use the matrix

$$\begin{bmatrix} 1 & \cdots & 1 & 1 & 0 & 0 \\ \alpha_1 & \cdots & \alpha_{q-1} & 0 & 1 & 0 \\ \alpha_1^2 & \cdots & \alpha_{q-1}^2 & 0 & 0 & 1 \end{bmatrix} \tag{2}$$

as either generator or parity check matrix. Any 3 columns are linearly independent, since the α_i^2 are all distinct.   Q.E.D.

§6. *n*-arcs

There is also a connection between MDS codes and finite geometries. From Corollary 3 we see that the problem of finding an $[n, k, n - k + 1]$ MDS code $\mathcal{C}$ can be looked at as the geometric problem of finding a set S of n points in the projective geometry $PG(k - 1, q)$ (see Appendix B) such that every k points of S are linearly independent, i.e. such that no k points of S lie on a hyperplane. The coordinates of the points are the columns of a generator matrix of $\mathcal{C}$.

For example, the columns of (2) comprise $2^m + 2$ points in the projective plane $PG(2, 2^m)$ such that no three points lie on a line.

Definition. An *n*-arc is a set of n points in the geometry $PG(k - 1, q)$ such that no k points lie in a hyperplane $PG(k - 2, q)$, where $n \geq k \geq 3$. E.g. (2) shows a $(2^m + 2)$-arc in $PG(2, 2^m)$.

Thus another version of our problem is:

Research Problem (11.1e). Given k and q, find the largest value of n for which there exists an *n*-arc in $PG(k - 1, q)$.

There is an extensive geometrical literature on this problem, but we restrict ourselves to just one theorem.

Theorem 11. *If $\mathcal{C}$ is a nontrivial $[n, k \geq 3, n - k + 1]$ MDS code over $GF(q)$, q odd, then $n \leq q + k - 2$. Equivalently, for any n-arc in $PG(k - 1, q)$, q odd, $n \leq q + k - 2$.*

Proof. Let $G = (g_{ij})$ be a $k \times n$ generator matrix of $\mathscr{C}$, let $r_1, \ldots, r_k$ denote the rows of G, and let C be the 0-chain in $\mathrm{PG}(k-1, q)$ consisting of the points whose coordinates are the columns of G. A generic point of $\mathrm{PG}(k-1, q)$ will be denoted by $(x_1, \ldots, x_k)$. It is clear that the hyperplane $x_1 = 0$ meets C in those points for which $g_{1i} = 0$. Therefore the weight of the first row, r_1, of G equals n − number of points in which the hyperplane $x_1 = 0$ meets C. Similarly the weight of the codeword $\sum_{i=1}^{k} \lambda_i r_i$ of $\mathscr{C}$ equals n − number of points in which the hyperplane $\sum_{i=1}^{k} \lambda_i x_i = 0$ meets C.

We know from Corollary 7 that $n \le q + k - 1$. Suppose now $n = q + k - 1$, which implies $A_{n-k+2} = 0$. Then C meets the hyperplanes of $\mathrm{PG}(k-1, q)$ in $k - 1, \ k - 3, \ k - 4, \ldots, 1$, or 0 points (but not $k - 2$ since there are no codewords of weight $n - k + 2$).

Pick a hyperplane which contains $k - 3$ points of C, say $P_1, \ldots, P_{k-3}$. Let Σ be a subspace $\mathrm{PG}(k-3, q)$ lying in this hyperplane and containing $P_1, \ldots, P_{k-3}$. Any hyperplane through Σ must meet C in 2 or 0 more points. Let r be the number which meet C in 2 more points. The union of all these hyperplanes is $\mathrm{PG}(k-1, q)$, so certainly contains all the points of C. Therefore

$$2r + k - 3 = q + k - 1,$$

or $2r = q + 2$, which is a contradiction since q is odd. Q.E.D.

§7. The known results

For $k = 3$ and q odd, Theorem 11 says $n \le q + 1$, and so the codes of Theorem 9 have the largest possible n. Thus one case of Research Problem 11.1a is solved: $m(3, q) = q + 1$ if q is odd.

On the other hand for $k = 3$ and q even, $n \le q + 2$ by Corollary 7, and the codes of Theorem 10 show that $m(3, q) = q + 2$ if q is even.

The results for general k are best shown graphically. Figure 11.2 gives the values of k and r for which an $[n = k + r, k]$ MDS code over $\mathrm{GF}(q)$ is known to exist (thus r is the number of parity checks).

By Corollary 2 the figure is symmetric in k and r. Apart from the codes of Theorem 10, no code is known which lies above the broken line $n = k + r = q + 1$ ($r \ge 2, k \ge 2$). Codes above the heavy line are forbidden by Corollary 7.

There is good evidence that the broken line is the true upper bound, and we state this as:

Research Problem (11.4). Prove (or disprove) that all MDS codes, with the exception of those given in Theorem 10, lie beneath the line $n = k + r = q + 1$ in Fig. 11.2.

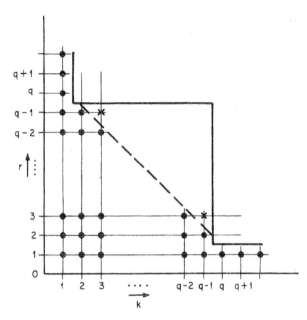

Fig. 11.2. The best $[n = k + r, k]$ MDS codes known. ● means a code exists for all q; × means a code exists iff $q = 2^m$.

This is known to be true for codes with $k \leqslant 5$, or $q \leqslant 11$, or $q > (4k - 9)^2$, and in some other cases.

Stated in terms of the function $m(k, q)$ the conjecture is that

$$m(k, q) = \begin{cases} q + 1 \text{ for } & 2 \leqslant k \leqslant q, \\ k + 1 \text{ for } & q < k, \end{cases} \tag{3}$$

except for

$$m(3, q) = m(q - 1, q) = q + 2 \quad \text{if} \quad q = 2^m. \tag{4}$$

§8. Orthogonal arrays

Definition. An $M \times n$ matrix A with entries from a set of q elements is called an *orthogonal array* of *size M, n constraints, q levels, strength k*, and *index λ* if any set of k columns of A contains all q^k possible row vectors exactly λ times. Such an array is denoted by (M, n, q, k). Clearly $M = \lambda q^k$. The case $q = 2$ was considered in Theorem 8 of Ch. 5.

Examples. The code $\mathcal{A}_{12}$ of Fig. 2.1 is a $(12, 11, 2, 2)$ (see Theorem 8 of Ch. 5). Fig. 11.3 shows a $(4, 3, 2, 2)$, and the codewords in Fig. 10.1 form a $(16, 3, 4, 2)$

with entries from GF(4).

$$\begin{bmatrix} 1 & 1 & 1 \\ - & 1 & - \\ 1 & - & - \\ - & - & 1 \end{bmatrix}$$

Fig. 11.3. A $(4, 3, 2, 2)$ orthogonal array.

Theorem 12. *The rows of a* (q^k, n, q, k) *linear orthogonal array* A *of index unity and symbols from* GF(q) *are the codewords of an* $[n, k]$ *MDS code over* GF(q), *and conversely.*

Proof. Any $q^k \times k$ submatrix of A contains each k-tuple exactly once $\Leftrightarrow$ the corresponding k coordinates can be taken as message symbols $\Leftrightarrow$ the code is MDS, by Corollary 3.   Q.E.D.

Problem. (10) Show that if $H_{4\lambda}$ is a normalized Hadamard matrix of order 4λ (§3 of Ch. 2), then the last $4\lambda - 1$ columns of $H_{4\lambda}$ form a $(4\lambda, 4\lambda - 1, 2, 2)$ orthogonal array of index λ. Fig. 11.3 is the case $\lambda = 1$.

Thus the final version of our problem is:

Research Problem (11.1f). Find the greatest possible n in a (q^k, n, q, k) orthogonal array of index unity.

Notes on Chapter 11

§1. Singleton [1214] seems to have been the first to explicitly study MDS codes. However in 1952 Bush [220] had already discovered Reed–Solomon codes and the extensions given in Theorems 9 and 10, using the language of orthogonal arrays (§8).

Some other redundant residue codes besides RS codes are also MDS – see §9 of Ch. 10.

Assmus and Mattson [41] have shown that MDS codes whose length n is a prime number π are very common, by showing that every cyclic code of length π over GF(p^i) is MDS for all i, for all except a finite number of primes p.

Without giving any details we just mention that an MDS code with $k = 2$ is also equivalent to a set of $n - k$ mutually orthogonal Latin squares of order q (Denes and Keedwell [371, p. 351], Posner [1068], Singleton [1214]). Therefore the more general problem of finding MDS codes over alphabets of size s (i.e. not necessarily over a field) includes the very difficult problem of finding all projective planes! (See Appendix B.)

§3. The weight distribution of MDS codes was found independently by Assmus, Gleason, Mattson and Turyn [50], Forney and Kohlenberg [436], and Kasami, Lin and Peterson [736]. Our derivation follows Goethals [491]. See also [933].

Corollary 7 is due to Bush. It is also given in [1214] and by Borodin [172]. Our proof follows Robillard [1118].

§4. For Problem 7 see Knuth [772, p. 36] and Pólya-Szegö [1067, vol. 2, p. 45].

§5. See the Notes to §3 of Ch. 10.

§6. Segre [1170–1173] and later Thas [1316, 1317], Casse [252], Hirschfeld [655] and many others have studied n-arcs and related problems in finite geometries. Two recent surveys are Barlotti [69, 70]. See also Dowling [385], Gulati et al. [565–569].

Using methods of algebraic geometry Segre [1170; 1173, p. 312] and Casse [252] have improved Theorem 11 as follows:

Theorem 13. *Assume* $q \geq k + 1$.
(i) *If* $k = 3, 4$ *or* 5 *then* (3) *and* (4) *hold.*
(ii) *If* $k \geq 6$ *then* $m(k, q) \leq q + k - 4$.

Thas [1316] has shown:

Theorem 14. *For* q *odd and* $q > (4k - 9)^2$, $m(k, q) = q + 1$.

Maneri, Silverman, and Jurick ([904, 905, 703]) have shown

Theorem 15. (3) *and* (4) *hold for* $q \leq 11$.

Other conditions under which (3) and (4) hold are given by Thas [1318].

It also follows from the geometrical theory that if q is odd then in many (conjecturally all) cases there is an unique $[n = q + 1, k, q - k + 2]$ MDS code. But if q is even this is known to be false.

In a projective plane of order h, a set C of $h + 1$ points no 3 of which are collinear is called an *oval*. Segre [1171; 1173, p. 270] has shown that in a Desarguesian plane of odd order (i.e. $h = q$, an odd prime power) these points form a conic. For example, if we take the columns of the generator matrix ((2)

without the penultimate column) of the $[q+1, 3, q-1]$ MDS code as the points, we see that they satisfy the equation $x_2^2 = x_1 x_3$.

If however $q = 2^m$, all lines which meet the oval C in a single point are concurrent; the point in which they meet is called the *nucleus* or *knot* of C. The points of C together with the nucleus give $2^m + 2$ points no 3 of which are collinear, and give a $[2^m + 2, 3, 2^m]$ MDS code with the same parameters as the code given in Theorem 10.

12

Alternant, Goppa and other generalized BCH codes

§1. Introduction

Alternant codes are a large and powerful family of codes obtained by an apparently small modification of the parity check matrix of a BCH code. Recall from Ch. 7 that a BCH code of length n and designed distance δ over GF(q) has parity check matrix $H = (H_{ij})$ where $H_{ij} = \alpha^{ij}$ ($1 \leq i \leq \delta - 1$, $0 \leq j \leq n - 1$) and $\alpha \in \mathrm{GF}(q^m)$ is a primitive n-th root of unity. By changing H_{ij} to $\alpha_j^{i-1} y_j$, where* $\alpha = (\alpha_1, \ldots, \alpha_n)$ is a vector with distinct components from GF(q^m), and $y = (y_1, \ldots, y_n)$ is a vector with nonzero components from GF(q^m) we get the alternant code $\mathscr{A}(\alpha, y)$. The properties of this code are summarized in Fig. 12.2.

The extra freedom in the definition is enough to ensure that some long alternant codes meet the Gilbert–Varshamov bound (Theorem 3), in contrast to the situation for BCH codes (Theorem 13 of Ch. 9).

In fact, it turns out that alternant codes form a very large class indeed, and a great deal remains to be discovered about them. For example, *which* alternant codes meet the Gilbert–Varshamov bound? How does one find the true dimension and minimum distance?

Of course BCH codes are a special case of alternant codes. In sections 3 to 7 various other subclasses of alternant codes are defined, namely Goppa (§3–§5 – see Fig. 12.3 for a summary), Srivastava (§6, especially Fig. 12.4) and Chien–Choy generalized BCH (§7 and Fig. 12.5) codes. The rather complicated relationship between these codes is indicated in Fig. 12.1 (which is not drawn to scale).

The encoding and decoding of alternant codes is similar to that of BCH

*We apologize for using α in two different ways.

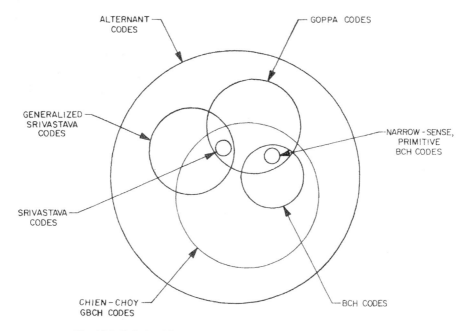

ALTERNANT
CODES

GOPPA CODES

GENERALIZED
SRIVASTAVA
CODES

NARROW - SENSE,
PRIMITIVE
BCH CODES

SRIVASTAVA
CODES

CHIEN − CHOY
GBCH CODES

BCH CODES

Fig. 12.1. Relationship between various subclasses of alternant codes.

codes, and is discussed in §9. The key step in decoding is the use of the Euclidean algorithm to go from the syndrome vector to the error locator and error evaluator polynomials. Of course this method applies equally well to BCH codes, and so we are able to fill in the gap in decoding BCH codes that was left in §6 of Ch. 9.

The Euclidean algorithm, for finding the greatest common divisor of two integers or polynomials, is described in §8. As a consequence of the algorithm we also are finally able to prove the result (Corollary 15), used several times in earlier chapters, which says that if f and g are relatively prime polynomials (or integers), then there exist polynomials (or integers) U and V such that

$$Uf + Vg = 1.$$

§2. Alternant codes

Alternant codes are closely related to the generalized Reed–Solomon codes $\mathrm{GRS}_{k_0}(\alpha, v)$ of §8 of Ch. 10. For convenience we repeat the definition. Let $\alpha = (\alpha_1, \ldots, \alpha_n)$ where the α_i are distinct elements of $\mathrm{GF}(q^m)$, and let $v = (v_1, \ldots, v_n)$ where the v_i are nonzero (but not necessarily distinct) elements of $\mathrm{GF}(q^m)$. Then $\mathrm{GRS}_{k_0}(\alpha, v)$ consists of all vectors

$$(v_1 F(\alpha_1), v_2 F(\alpha_2), \ldots, v_n F(\alpha_n)) \tag{1}$$

where $F(z)$ is any polynomial of degree $< k_0$ with coefficients from $GF(q^m)$. $GRS_{k_0}(\alpha, v)$ is an $[n, k_0, r + 1]$ MDS code over $GF(q^m)$, where $r = n - k_0$, and has parity check matrix

$$H = \begin{bmatrix} y_1 & y_2 & \cdots & y_n \\ \alpha_1 y_1 & \alpha_2 y_2 & \cdots & \alpha_n y_n \\ \alpha_1^2 y_1 & \alpha_2^2 y_2 & \cdots & \alpha_n^2 y_n \\ \cdots\cdots\cdots\cdots\cdots\cdots\cdots \\ \alpha_1^{r-1} y_1 & \cdots & & \alpha_n^{r-1} y_n \end{bmatrix},$$

$$= \begin{bmatrix} 1 & 1 & \cdots & 1 \\ \alpha_1 & \alpha_2 & \cdots & \alpha_n \\ \alpha_1^2 & \alpha_2^2 & \cdots & \alpha_n^2 \\ \cdots\cdots\cdots\cdots\cdots \\ \alpha_1^{r-1} & \alpha_2^{r-1} & \cdots & \alpha_n^{r-1} \end{bmatrix} \begin{bmatrix} y_1 & & 0 \\ & y_2 & \\ & & \ddots & \\ 0 & & & y_n \end{bmatrix} \qquad (2)$$

$$= XY \quad (\text{say}),$$

where $y = (y_1, \ldots, y_n)$, with $y_i \in GF(q^m)$ and $y_i \neq 0$, is such that $GRS_{k_0}(\alpha, v)^\perp = GRS_{n-k_0}(\alpha, y)$.

Definition. The *alternant code* $\mathcal{A}(\alpha, y)$ consists of all codewords of $GRS_{k_0}(\alpha, v)$ which have components from $GF(q)$, i.e. $\mathcal{A}(\alpha, y)$ is the restriction of $GRS_{k_0}(\alpha, v)$ to $GF(q)$. Thus $\mathcal{A}(\alpha, y)$ consists of all vectors a over $GF(q)$ such that $Ha^T = 0$, where H is given by (2).

A parity check matrix $\bar{H}$ with elements from $GF(q)$ can be obtained by replacing each element of H by the corresponding column vector of length m from $GF(q)$, just as was done for BCH codes. Since $\mathcal{A}(\alpha, y)$ is a subfield subcode of $GRS_{k_0}(\alpha, v)$ it follows from §7 of Ch. 7 that $\mathcal{A}(\alpha, y)$ is an $[n, k, d]$ code over $GF(q)$ with

$$n - mr \leq k \leq n - r, \qquad d \geq r + 1.$$

(The properties of $\mathcal{A}(\alpha, y)$ are collected in Fig. 12.2.) It is possible to obtain this estimate on d directly from the parity check matrix.

Theorem 1. $\mathcal{A}(\alpha, y)$ *has minimum distance* $d \geq r + 1$.

Proof. Suppose a is a nonzero codeword of $\mathcal{A}(\alpha, y)$ with weight $\leq r$. Then $Ha^T = XYa^T = 0$. Set $b^T = Ya^T$, then wt $(b) =$ wt (a) since Y is diagonal and invertible. Thus $Xb^T = 0$, which is impossible since X is Vandermonde, by Lemma 17 of Ch. 4.   Q.E.D.

$\mathscr{A}(\alpha, y)$ is defined by the parity check matrix (2) (or (3)), where $\alpha_1, \ldots, \alpha_n$ are distinct elements of $GF(q^m)$ and $y_1, \ldots, y_n$ are nonzero elements of $GF(q^m)$. $\mathscr{A}(\alpha, y)$ is a linear code over $GF(q)$ with

length n, dimension $k \geq n - mr$,
minimum distance $d \geq r + 1$.

$\mathscr{A}(\alpha, y)$ is the restriction of $GRS_{k_0}(\alpha, v)$ to $GF(q)$, where $k_0 = n - r$. The dual $\mathscr{A}(\alpha, y)^\perp = T_m(GRS_{n-k_0}(\alpha, y))$ (Theorem 2). There exist long alternant codes which meet the Gilbert–Varshamov bound (Theorem 3). Important subclasses of alternant codes are BCH, Goppa (§§3–5), Srivastava (§6), and Chien–Choy generalized BCH (§7) codes. For decoding see §9.

Fig. 12.2. Properties of alternant code $\mathscr{A}(\alpha, y)$.

Let $C = (c_{ij})$, $c_{ij} \in GF(q^m)$ be any invertible matrix. Then an equally good parity check matrix for $\mathscr{A}(\alpha, y)$ is (from Problem 31 of Ch. 7).

$H' = CXY$,

$$
= \begin{bmatrix} c_{11} & c_{12} & \cdots & c_{1r} \\ c_{21} & c_{22} & \cdots & c_{2r} \\ \cdots\cdots\cdots\cdots\cdots\cdots \\ c_{r1} & c_{r2} & \cdots & c_{rr} \end{bmatrix} \begin{bmatrix} 1 & 1 & \cdots & 1 \\ \alpha_1 & \alpha_2 & \cdots & \alpha_n \\ \cdots\cdots\cdots\cdots\cdots\cdots \\ \alpha_1^{r-1} & \alpha_2^{r-1} & \cdots & \alpha_n^{r-1} \end{bmatrix} \begin{bmatrix} y_1 & & & 0 \\ & y_2 & & \\ & & \ddots & \\ 0 & & & y_n \end{bmatrix}
$$

$$
= \begin{bmatrix} y_1 g_1(\alpha_1) & y_2 g_1(\alpha_2) & \cdots & y_n g_1(\alpha_n) \\ y_1 g_2(\alpha_1) & y_2 g_2(\alpha_2) & \cdots & y_n g_2(\alpha_n) \\ \cdots\cdots\cdots\cdots\cdots\cdots\cdots\cdots \\ y_1 g_r(\alpha_1) & y_2 g_r(\alpha_2) & \cdots & y_n g_r(\alpha_n) \end{bmatrix} \tag{3}
$$

say, where

$$
g_i(x) = c_{i1} + c_{i2}x + c_{i3}x^2 + \cdots + c_{ir}x^{r-1} (i = 1, \ldots, r) \tag{4}
$$

is a polynomial of degree $< r$ over $GF(q^m)$.

Note from (2) or (3) that it is natural to label the coordinates of the codewords by $\alpha_1, \ldots, \alpha_n$. This is useful for encoding and decoding.

Examples. First put $\alpha = (1, \alpha, \alpha^2, \ldots, \alpha^6)$, $y = (1, 1, \ldots, 1)$ where α is a primitive element of $GF(2^3)$. Then if $r = 2$ the alternant code $\mathscr{A}(\alpha, y)$ has parity check matrix (2) equal to

$$
H = X = \begin{bmatrix} 1 & 1 & 1 & 1 & 1 & 1 & 1 \\ 1 & \alpha & \alpha^2 & \alpha^3 & \alpha^4 & \alpha^5 & \alpha^6 \end{bmatrix}.
$$

Replacing each entry by the corresponding binary vector of length 3 we obtain

$$\bar{H} = \begin{bmatrix} 1111111 \\ 0000000 \\ \dfrac{0000000}{1001011} \\ 0101110 \\ 0010111 \end{bmatrix}.$$

Thus $\mathcal{A}(\alpha, y)$ is a $[7, 3, 4]$ code. On the other hand, if $y = (1, \alpha, \alpha^2, \alpha^3, \alpha^4, \alpha^5, \alpha^6)$ then

$$H = XY = \begin{bmatrix} 1 & \alpha & \alpha^2 & \alpha^3 & \alpha^4 & \alpha^5 & \alpha^6 \\ 1 & \alpha^2 & \alpha^4 & \alpha^6 & \alpha & \alpha^3 & \alpha^5 \end{bmatrix}.$$

The second row of this matrix is redundant, for if

$$\sum_{i=1}^{n} h_i a_i = 0$$

where $h_i \in GF(2^m)$ and $a_i = 0$ or 1 then

$$\sum_{i=1}^{n} h_i^2 a_i = \left(\sum_{i=1}^{n} h_i a_i \right)^2 = 0.$$

Thus we can take

$$\bar{H} = \begin{bmatrix} 1001011 \\ 0101110 \\ 0010111 \end{bmatrix},$$

and $\mathcal{A}(\alpha, y)$ is now a $[7, 4, 3]$ Hamming code. (In this case the effect of y has been to decrease the minimum distance and increase the number of information symbols.)

Other examples of alternant codes are BCH codes. For the parity check matrix of a general BCH code is (Equation (19) of Ch. 7)

$$H = \begin{bmatrix} 1 & \alpha^b & \alpha^{2b} & \cdots & \alpha^{(n-1)b} \\ 1 & \alpha^{b+1} & \alpha^{2(b+1)} & \cdots & \alpha^{(n-1)(b+1)} \\ \cdots \cdots \cdots \cdots \cdots \cdots \cdots \cdots \cdots \\ 1 & \alpha^{b+\delta-2} & \cdots \cdots \cdots \cdots \alpha^{(n-1)(b+\delta-2)} \end{bmatrix}$$

$$= \begin{bmatrix} 1 & 1 & 1 & \cdots & 1 \\ 1 & \alpha & \alpha^2 & \cdots & \alpha^{n-1} \\ 1 & \alpha^2 & \alpha^4 & \cdots & \alpha^{2(n-1)} \\ \cdots \cdots \cdots \cdots \cdots \cdots \\ 1 & \alpha^{\delta-2} & \cdots \cdots \cdots \alpha^{(\delta-2)(n-1)} \end{bmatrix} \begin{bmatrix} 1 & & & 0 \\ & \alpha^b & & \\ & & \alpha^{2b} & \\ & & & \ddots & \\ 0 & & & & \alpha^{(n-1)b} \end{bmatrix},$$

which is an alternant code.

The dual of an alternant code.

Theorem 2. *The dual of the alternant code* $\mathcal{A}(\alpha, y)$ *is the code*

$$T_m(\mathrm{GRS}_{n-k_0}(\alpha, y)) = T_m(\mathrm{GRS}_{k_0}(\alpha, v)^{\perp}). \tag{5}$$

Recall from §7 of Ch. 7 that if $\mathcal{C}^*$ is any code over $\mathrm{GF}(q^m)$, the trace code $T_m(\mathcal{C}^*)$ is the code over $\mathrm{GF}(q)$ consisting of all distinct vectors

$$(T_m(c_1), \ldots, T_m(c_n)) \quad \text{where} \quad (c_1, \ldots, c_n) \in \mathcal{C}^*.$$

Proof. From Theorem 11 of Ch. 7 and Theorem 4 of Ch. 10.   Q.E.D.

An illustration of this theorem will be found following Theorem 11 of Ch. 7.

Long alternant codes are good.

Theorem 3. (a) *Given n, h and δ, let m be any number dividing $n - h$. Then there exists an alternant code* $\mathcal{A}(\alpha, y)$ *over* $\mathrm{GF}(q)$ *with parameters* $[n, k \geq h, d \geq \delta]$ *provided*

$$\sum_{w=1}^{\delta-1} (q-1)^w \binom{n}{w} < (q^m - 1)^{(n-h)/m}. \tag{6}$$

(b) *Hence there exist long alternant codes which meet the Gilbert–Varshamov bound.*

Proof. (i) Let a be any vector over $\mathrm{GF}(q)$. The number of codes $\mathrm{GRS}_{k_0}(\alpha, v)$ which contain a, for fixed k_0 and α, and varying v, is at most $(q^m - 1)^{k_0}$. To show this, observe from (1) that $\mathrm{GRS}_{k_0}(\alpha, v)$ contains a iff

$$\frac{a_i}{v_i} = F(\alpha_i), \quad i = 1, \ldots, n,$$

for some polynomial $F(z)$ of degree $< k_0$. Since $F(z)$ is determined by its values at $\leq k_0$ points, we can choose at most k_0 v_i's before $F(z)$ is determined. There are $q^m - 1$ choices for each of these v_i's.

(ii) Consider the family $\mathcal{F}$ of alternant codes $\mathcal{A}(\alpha, y)$ which are restrictions of some $\mathrm{GRS}_{k_0}(\alpha, v)$ to $\mathrm{GF}(q)$, where $k_0 = n - (n - h)/m$. Then $\mathcal{A}(\alpha, y) \in \mathcal{F}$ has dimension $k \geq n - m(n - k_0) = h$. From (i) the number of $\mathcal{A}(\alpha, y) \in \mathcal{F}$ which contain a is at most $(q^m - 1)^{n-(n-h)/m}$. Therefore the number of $\mathcal{A}(\alpha, y) \in \mathcal{F}$ with minimum distance $< \delta$ is at most

$$(q^m - 1)^{n-(n-h)/m} \sum_{w=1}^{\delta-1} (q-1)^w \binom{n}{w}. \tag{7}$$

(iii) The total number of codes $\mathcal{A}(\alpha, y)$ in $\mathcal{F}$ is equal to the number of choices for v, which is $(q^m - 1)^n$. So if this number exceeds (7), there exists an $\mathcal{A}(\alpha, y)$ with dimension $\geq h$ and minimum distance $\geq \delta$. This proves (a). Asymptotically, when n is large and h/n fixed, (6) is the same as the Gilbert–Varshamov bound (Theorem 12 of Ch. 1, Theorem 30 of Ch. 17).

Q.E.D.

Of course Theorem 3 doesn't say which alternant codes are the best, only that good ones exist. Since the class of alternant codes is so large, it is useful to have names for some subclasses. In the following sections we shall describe the subclasses known as Goppa, Srivastava and Chien–Choy generalized BCH codes. These are obtained by placing restrictions on α or y, or both.

Problem. (1) Consider the binary alternant code with $n = 6$, $\alpha_i = \alpha^{i+1}$ for $i = 0, \ldots, 5$ where α is a primitive element of $GF(2^3)$, all $y_i = 1$, $g_1(x) = 1 + x$ and $g_2(x) = x$. Show that this is a $[6, 2, 4]$ code.

§3. Goppa codes

This is the most interesting subclass of alternant codes. Just as cyclic codes are specified in terms of a generator polynomial (Theorem 1 of Ch. 7), so Goppa codes are described in terms of a Goppa polynomial $G(z)$. In contrast to cyclic codes, where it is difficult to estimate the minimum distance d from the generator polynomial, Goppa codes have the property that $d \geq \deg G(z) + 1$. We first give the definition in terms of Goppa polynomials and then show that these are alternant codes.

The definition of a Goppa code of length n with symbols from $GF(q)$ calls for two things: a polynomial $G(z)$ called the *Goppa polynomial*, having coefficients from $GF(q^m)$, for some fixed m, and a subset $L = \{\alpha_1, \ldots, \alpha_n\}$ of $GF(q^m)$ such that $G(\alpha_i) \neq 0$ for all $\alpha_i \in L$. Usually L is taken to be all the elements of $GF(q^m)$ which are not zeros of $G(z)$.

With any vector $a = (a_1, \ldots, a_n)$ over $GF(q)$ we associate the rational function

$$R_a(z) = \sum_{i=1}^{n} \frac{a_i}{z - \alpha_i}. \tag{8}$$

Definition. The Goppa code $\Gamma(L, G)$ (or Γ) consists of all vectors a such that

$$R_a(z) \equiv 0 \bmod G(z), \tag{9}$$

or equivalently such that $R_a(z) = 0$ in the polynomial ring $GF(q^m)[z]/G(z)$. If $G(z)$ is irreducible then Γ is called an *irreducible* Goppa code.

Figure 12.3 shows the basic properties of these codes. Examples will be given after Theorem 6.

$\Gamma(L, G)$ is a linear code over $GF(q)$, defined by Equation (9).

length $n = |L|$
dimension $k \geq n - mr$, $r = \deg G(z)$
minimum distance $d \geq r + 1$.

$\Gamma(L, G) =$ alternant code $\mathcal{A}(\alpha, y)$ where $y_i = G(\alpha_i)^{-1}$. $\Gamma(L, G)^\perp = T_m(GRS_r(\alpha, y))$. In the binary case if $G(z)$ has no multiple zeros then $d \geq 2r + 1$. There exist long Goppa codes which meet the Gilbert–Varshamov bound. Extended binary double-error-correcting Goppa codes are cyclic (§5).

Fig. 12.3. Properties of the Goppa code $\Gamma(L, G)$.

The parity check matrix of Γ. It is obvious that Γ is a linear code. The parity check matrix can be found from (9). For in the ring of polynomials mod $G(z)$, $z - \alpha_i$ has an inverse (since it does not divide $G(z)$). The inverse is

$$(z - \alpha_i)^{-1} = -\frac{G(z) - G(\alpha_i)}{z - \alpha_i} G(\alpha_i)^{-1},$$

for indeed

$$-(z - \alpha_i) \frac{(G(z) - G(\alpha_i))}{z - \alpha_i} G(\alpha_i)^{-1} \equiv 1 \bmod G(z).$$

Therefore a is in $\Gamma(L, G)$ iff

$$\sum_{i=1}^{n} a_i \frac{G(z) - G(\alpha_i)}{z - \alpha_i} G(\alpha_i)^{-1} = 0 \tag{10}$$

as a polynomial (not mod $G(z)$). If $G(z) = \sum_{i=0}^{r} g_i z^i$, with $g_i \in GF(q^m)$ and $g_r \neq 0$, then

$$\frac{G(z) - G(\alpha_i)}{z - \alpha_i} = g_r(z^{r-1} + z^{r-2}\alpha_i + \cdots + \alpha_i^{r-1}) + g_{r-1}(z^{r-2} + \cdots + \alpha_i^{r-2}) + \cdots$$
$$+ g_2(z + \alpha_i) + g_1.$$

Equating the coefficients of $z^{r-1}, z^{r-2}, \ldots, 1$ to zero in (10) we see that a is in $\Gamma(L, G)$ iff $Ha^{tr} = 0$, where

$$H = \begin{bmatrix} g_r G(\alpha_1)^{-1} & \cdots & g_r G(\alpha_n)^{-1} \\ (g_{r-1} + \alpha_1 g_r) G(\alpha_1)^{-1} & \cdots & (g_{r-1} + \alpha_n g_r) G(\alpha_n)^{-1} \\ \cdots\cdots\cdots\cdots\cdots\cdots\cdots\cdots\cdots & & \cdots\cdots\cdots\cdots\cdots\cdots\cdots\cdots\cdots \\ (g_1 + \alpha_1 g_2 + \cdots + \alpha_1^{r-1} g_r) G(\alpha_1)^{-1} & \cdots & (g_1 + \alpha_n g_2 + \cdots + \alpha_n^{r-1} g_r) G(\alpha_n)^{-1} \end{bmatrix}$$

$$
= \begin{bmatrix} g_r & 0 & 0 & \cdots & 0 \\ g_{r-1} & g_r & 0 & \cdots & 0 \\ g_{r-2} & g_{r-1} & g_r & \cdots & 0 \\ \multicolumn{5}{c}{\cdots\cdots\cdots\cdots\cdots\cdots} \\ g_1 & g_2 & g_3 & \cdots & g_r \end{bmatrix} \begin{bmatrix} 1 & 1 & \cdots & 1 \\ \alpha_1 & \alpha_2 & \cdots & \alpha_n \\ \alpha_1^2 & \alpha_2^2 & \cdots & \alpha_n^2 \\ \multicolumn{4}{c}{\cdots\cdots\cdots\cdots\cdots} \\ \alpha_1^{r-1} & \alpha_2^{r-1} & \cdots & \alpha_n^{r-1} \end{bmatrix} \cdot \begin{bmatrix} G(\alpha_1)^{-1} & & & 0 \\ & G(\alpha_2)^{-1} & & \\ & & \cdot & \\ 0 & & & G(\alpha_n)^{-1} \end{bmatrix}
$$

$$= CXY \quad \text{(say)}, \tag{11}$$

is a parity check matrix for $\Gamma(L, G)$. Since C is invertible, by Problem 31 of Ch. 7 another parity check matrix is

$$H' = XY$$

$$= \begin{bmatrix} G(\alpha_1)^{-1} & \cdots & G(\alpha_n)^{-1} \\ \alpha_1 G(\alpha_1)^{-1} & \cdots & \alpha_n G(\alpha_n)^{-1} \\ \multicolumn{3}{c}{\cdots\cdots\cdots\cdots\cdots\cdots} \\ \alpha_1^{r-1} G(\alpha_1)^{-1} & \cdots & \alpha_n^{r-1} G(\alpha_n)^{-1} \end{bmatrix}, \tag{12}$$

and this is usually the simplest to use.

A parity check matrix with elements from $\mathrm{GF}(q)$ is then obtained by replacing each entry of H (or H') by the corresponding column vector of length m from $\mathrm{GF}(q)$.

Comparing (11) with (3) we see that $\Gamma(L, G)$ is an alternant code $\mathcal{A}(\alpha, y)$ with $\alpha = (\alpha_1, \ldots, \alpha_n)$ and $y = (G(\alpha_1)^{-1}, \ldots, G(\alpha_n)^{-1})$. Therefore $\Gamma(L, G)$ has dimension $k \geq n - rm$ and minimum distance $d \geq r + 1$.

In fact it is easy to find the generalized Reed–Solomon code which produces $\Gamma(L, G)$.

Theorem 4. $\Gamma(L, G)$ *is the restriction to* $\mathrm{GF}(q)$ *of* $\mathrm{GRS}_{n-r}(\alpha, v)$, *where* $v = (v_1, \ldots, v_n)$ *and*

$$v_i = \frac{G(\alpha_i)}{\prod\limits_{j \neq i} (\alpha_i - \alpha_j)}, \quad i = 1, \ldots, n.$$

Proof. (i) Take $u \in \mathrm{GRS}_{n-r}(\alpha, v) \,|\, \mathrm{GF}(q)$. Then

$$u_i = v_i F(\alpha_i) = \frac{F(\alpha_i) G(\alpha_i)}{\prod\limits_{j \neq i} (\alpha_i - \alpha_j)},$$

where $F(z)$ is a polynomial of degree $< n - r$. Thus

$$\sum_{i=1}^{n} \frac{u_i}{z - \alpha_i} = \frac{1}{\prod\limits_{i=1}^{n} (z - \alpha_i)} \sum_{i=1}^{n} F(\alpha_i) G(\alpha_i) \prod_{j \neq i} \frac{(z - \alpha_j)}{(\alpha_i - \alpha_j)}.$$

Let

$$N(z) = \sum_{i=1}^{n} F(\alpha_i)G(\alpha_i) \prod_{j \neq i} (z - \alpha_j)/(\alpha_i - \alpha_j).$$

Then $N(\alpha_i) = F(\alpha_i)G(\alpha_i)$ for $i = 1, \ldots, n$. Also $\deg N(z) \leq n - 1$ and $\deg F(z)G(z) \leq n - 1$. Since the polynomial $N(z) - F(z)G(z)$ is determined by its values at n points, $N(z) = F(z)G(z)$. Therefore

$$\sum_{i=1}^{n} \frac{u_i}{z - \alpha_i} = \frac{F(z)G(z)}{\prod_{i=1}^{n} (z - \alpha_i)}$$

and hence $u \in \Gamma(L, G)$. Thus

$$\Gamma(L, G) \supset \mathrm{GRS}_{n-r}(\alpha, v) \mid \mathrm{GF}(q)$$

(ii) The converse is similar and is left to the reader.   Q.E.D.

From Theorem 2 we obtain:

Theorem 5. *The dual of a Goppa code is given by*

$$\Gamma(L, G)^{\perp} = T_m(\mathrm{GRS}_r(\alpha, y)) \tag{13}$$

where $y_i = G(\alpha_i)^{-1}$.

Problem. (2) Prove directly that $\mathrm{GRS}_r(\alpha, y)$, where $y_i = G(\alpha_i)^{-1}$, and $\mathrm{GRS}_{n-r}(\alpha, y')$, where

$$y'_i = \frac{G(\alpha_i)}{\prod_{j \neq i} (\alpha_j - \alpha_i)},$$

are dual codes.

Binary Goppa codes. Just as for BCH codes one can say a bit more in the binary case (cf. §6 of Ch. 7). Suppose $\Gamma = \Gamma(L, G)$ is a binary Goppa code (with $q = 2$). Let $a = (a_1 \cdots a_n)$ be a codeword of weight w in Γ, with $a_{l_1} = \cdots = a_{l_w} = 1$, and define

$$f_a(z) = \prod_{i=1}^{w} (z - \alpha_{l_i}). \tag{14}$$

Then

$$f'_a(z) = \sum_{i=1}^{w} \prod_{j \neq i} (z - \alpha_{l_i}),$$

$$R_a(z) = \sum_{i=1}^{w} \frac{1}{z - \alpha_{l_i}} = \frac{f'_a(z)}{f_a(z)} \quad \text{from (8)} \tag{15}$$

The α_i's are distinct. from the definition of Γ, so $f'_a(z)$ and $f_a(z)$ have no common factors, and (15) is in lowest terms. Since $G(\alpha_i) \neq 0$, $f_a(z)$ and $G(z)$

are relatively prime, and so from (15)

$$R_a(z) \equiv 0 \bmod G(z) \text{ iff } G(z) \mid f_a'(z).$$

We are working mod 2, so $f_a'(z)$ contains only even powers and is a perfect square. Let $\bar{G}(z)$ be the lowest degree perfect square which is divisible by $G(z)$. Then

$$G(z) \mid f_a'(z) \text{ iff } \bar{G}(z) \mid f_a'(z).$$

We conclude that

$$a \in \Gamma \text{ iff } R_a(z) \equiv 0 \bmod G(z)$$
$$\text{iff } \bar{G}(z) \mid f_a'(z). \tag{16}$$

In particular, if $a \neq 0$, $\deg f_a'(z) \geq \deg \bar{G}(z)$. Hence

$$\text{min. distance of } \Gamma \geq \deg \bar{G}(z) + 1. \tag{17}$$

An important special case is:

Theorem 6. *Suppose $G(z)$ has no multiple zeros, so that $\bar{G}(z) = G(z)^2$. Then*

$$\text{min. distance of } \Gamma \geq 2 \deg G(z) + 1. \tag{18}$$

If $G(z)$ has no multiple zeros then Γ is called a *separable* Goppa code.

Examples of binary Goppa codes. (1) Take $G(z) = z^2 + z + 1$, $L = GF(2^3) = \{0, 1, \alpha, \ldots, \alpha^6\}$ where α is primitive, $q = 2$, and $q^m = 8$. Certainly $G(\beta) \neq 0$ for $\beta \in GF(2^3)$, for the zeros of $z^2 + z + 1$ belong to $GF(2^2)$, $GF(2^4)$, $GF(2^6)$, ... but not to $GF(2^3)$ – see Theorem 8 of Ch. 4. We obtain an irreducible Goppa code Γ of length $n = |L| = 8$, dimension $k \geq 8 - 2.3 = 2$, and minimum distance $d \geq 5$. From (12) a parity check matrix is

$$H = \begin{bmatrix} \dfrac{1}{G(0)} & \dfrac{1}{G(1)} & \dfrac{1}{G(\alpha)} & \cdots & \dfrac{1}{G(\alpha^6)} \\[2mm] \dfrac{0}{G(0)} & \dfrac{1}{G(1)} & \dfrac{\alpha}{G(\alpha)} & \cdots & \dfrac{\alpha^6}{G(\alpha^6)} \end{bmatrix}$$

From Fig. 4.5 we find

$$H = \begin{bmatrix} 1 & 1 & \alpha^2 & \alpha^4 & \alpha^2 & \alpha & \alpha & \alpha^4 \\ 0 & 1 & \alpha^3 & \alpha^6 & \alpha^5 & \alpha^5 & \alpha^6 & \alpha^3 \end{bmatrix}$$

$$= \begin{array}{c} \begin{matrix} 0 & \;1 & \;\alpha & \alpha^2 & \alpha^3 & \alpha^4 & \alpha^5 & \alpha^6 \end{matrix} \\ \left[\begin{array}{cccccccc} 1 & 1 & 0 & 0 & 0 & 0 & 0 & 0 \\ 0 & 0 & 0 & 1 & 0 & 1 & 1 & 1 \\ 0 & 0 & 1 & 1 & 1 & 0 & 0 & 1 \\ \hline 0 & 1 & 1 & 1 & 1 & 1 & 1 & 1 \\ 0 & 0 & 1 & 0 & 1 & 1 & 0 & 1 \\ 0 & 0 & 0 & 1 & 1 & 1 & 1 & 0 \end{array} \right] \end{array}$$

The codewords are

$$0 \ 1 \ \alpha \ \alpha^2 \ \alpha^3 \ \alpha^4 \ \alpha^5 \ \alpha^6$$

$$\begin{bmatrix} 0 & 0 & 0 & 0 & 0 & 0 & 0 & 0 \\ 0 & 0 & 1 & 1 & 1 & 1 & 1 & 1 \\ 1 & 1 & 0 & 0 & 1 & 0 & 1 & 1 \\ 1 & 1 & 1 & 1 & 0 & 1 & 0 & 0 \end{bmatrix},$$

so this is an [8, 2, 5] Goppa code. By adding an overall parity check and reordering the columns the following [9, 2, 6] code is obtained:

$$1 \ \alpha^4 \ \alpha^6 \ \infty \ \alpha^2 \ \alpha^5 \ 0 \ \alpha \ \alpha^3$$

$$\begin{bmatrix} 0 & 0 & 0 & 0 & 0 & 0 & 0 & 0 & 0 \\ 0 & 1 & 1 & 0 & 1 & 1 & 0 & 1 & 1 \\ 1 & 0 & 1 & 1 & 0 & 1 & 1 & 0 & 1 \\ 1 & 1 & 0 & 1 & 1 & 0 & 1 & 1 & 0 \end{bmatrix}. \tag{19}$$

This code is cyclic! An explanation for this phenomenon will be given in §5.

This example can also be used as an illustration of Theorem 4. Here $n - r = 8 - 2 = 6$, and $v_i = G(\alpha_i) = \alpha^{2i} + \alpha^i + 1$, since $\Pi_{j \neq i}(\alpha_j - \alpha_i) = 1$ for all i. Thus Theorem 4 states that the [8, 2, 5] Goppa code is the restriction to GF(2) of the code over GF(2^3) with generator matrix

$$\begin{bmatrix} 1 & 1 & \alpha^5 & \alpha^3 & \alpha^5 & \alpha^6 & \alpha^6 & \alpha^3 \\ 0 & 1 & \alpha^6 & \alpha^5 & \alpha & \alpha^3 & \alpha^4 & \alpha^2 \\ 0 & 1 & 1 & 1 & \alpha^4 & 1 & \alpha^2 & \alpha \\ 0 & 1 & \alpha & \alpha^2 & 1 & \alpha^4 & 1 & 1 \\ 0 & 1 & \alpha^2 & \alpha^4 & \alpha^3 & \alpha & \alpha^5 & \alpha^6 \\ 0 & 1 & \alpha^3 & \alpha^6 & \alpha^6 & \alpha^5 & \alpha^3 & \alpha^5 \end{bmatrix}.$$

It is readily checked that

$$\text{row } 1 + \text{row } 2 + \text{row } 6 = 1 \ 1 \ 1 \ 1 \ 0 \ 1 \ 0 \ 0,$$
$$\text{row } 1 + \text{row } 5 + \text{row } 6 = 1 \ 1 \ 0 \ 0 \ 1 \ 0 \ 1 \ 1,$$
$$\text{row } 2 + \text{row } 5 \qquad\quad = 0 \ 0 \ 1 \ 1 \ 1 \ 1 \ 1 \ 1.$$

(2) Take $G(z) = z^3 + z + 1$ and $L = \mathrm{GF}(2^5)$. Again $G(\beta) \neq 0$ for $\beta \in L$ (using Theorem 8 of Ch. 4). Then $\Gamma(L, G)$ is a [32, 17, 7] irreducible Goppa code, with parity check matrix given by Equation (12) (here α is a primitive element of GF(2^5)):

$$H = \begin{bmatrix} \dfrac{1}{G(0)} & \dfrac{1}{G(1)} & \dfrac{1}{G(\alpha)} & \cdots & \dfrac{1}{G(\alpha^{30})} \\[2ex] \dfrac{0}{G(0)} & \dfrac{1}{G(1)} & \dfrac{\alpha}{G(\alpha)} & \cdots & \dfrac{\alpha^{30}}{G(\alpha^{30})} \\[2ex] \dfrac{0^2}{G(0)} & \dfrac{1^2}{G(1)} & \dfrac{\alpha^2}{G(\alpha)} & \cdots & \dfrac{\alpha^{60}}{G(\alpha^{30})} \end{bmatrix}$$

$$= \begin{bmatrix} 1 & 1 & \alpha^4 & \alpha^8 & \alpha^{14} & \cdots & \alpha^{26} \\ 0 & 1 & \alpha^5 & \alpha^{10} & \alpha^{17} & \cdots & \alpha^{25} \\ 0 & 1 & \alpha^6 & \alpha^{12} & \alpha^{20} & \cdots & \alpha^{24} \end{bmatrix}$$

$$= \begin{bmatrix} 1 & 1 & 0 & 1 & 1 & & 1 \\ 0 & 0 & 0 & 0 & 0 & & 1 \\ 0 & 0 & 0 & 1 & 1 & \cdots & 1 \\ 0 & 0 & 0 & 1 & 1 & & 0 \\ 0 & 0 & 1 & 0 & 1 & & 1 \\ \hline 0 & 1 & 1 & 1 & 1 & & 1 \\ 0 & 0 & 0 & 0 & 1 & & 0 \\ 0 & 0 & 1 & 0 & 0 & \cdots & 0 \\ 0 & 0 & 0 & 0 & 0 & & 1 \\ 0 & 0 & 0 & 1 & 1 & & 1 \\ \hline 0 & 1 & 0 & 0 & 0 & & 0 \\ 0 & 0 & 1 & 1 & 0 & & 1 \\ 0 & 0 & 0 & 1 & 1 & \cdots & 1 \\ 0 & 0 & 1 & 1 & 1 & & 1 \\ 0 & 0 & 0 & 0 & 0 & & 1 \end{bmatrix}$$

where we have used the table of GF(2^5) given in Fig. 4.5. The weight distribution of $\Gamma(L, G)$ was found (by computer) to be:

i:	0	7	8	9	10	11	12	13	14	15
A_i:	1	128	400	800	1903	4072	6876	10360	14420	17448

16	17	18	19	20	21	22	23	24	25	26
18381	17336	14330	10360	6860	4136	2068	760	250	136	47

(3) Of course the coefficients of $G(z)$ need not be restricted to 0's and 1's. For example we could take $G(z) = z^2 + z + \alpha^3$, where α is a primitive element of GF(2^4). From Theorem 15 of Ch. 9 $G(z)$ is irreducible over GF(2^4), since $T_4(\alpha^3) = \alpha^3 + \alpha^6 + \alpha^{12} + \alpha^9 = 1$. Therefore we can take $L = $ GF(2^4), and obtain a [16, 8, 5] irreducible Goppa code.

Problem. (3) Find a parity check matrix for this code.

(4) *Irreducible Goppa codes.* Consider $G(z) = z^3 + z + 1$, which is irreducible over GF(2). The zeros of $G(z)$ lie in GF(2^3) and hence by Theorem 8 of Ch. 4 are in GF(2^6), GF(2^9), ... Provided m is not a multiple of 3 we can take $L = $ GF(2^m), and obtain an

$$[n = 2^m, k \geqslant 2^m - 3m, d \geqslant 7] \quad (\text{for } 3 \nmid m) \tag{20}$$

irreducible Goppa code. When $m = 5$, the bounds for k and d are exact, as we saw in example (2).

Alternatively, taking $G(z)$ to be an irreducible cubic over $GF(2^m)$ we get a code with parameters (20) for any m.

More generally, taking $G(z)$ to be an irreducible polynomial of degree r over $GF(2^m)$ we obtain an

$$[n = 2^m, k \geqslant 2^m - rm, d \geqslant 2r + 1] \qquad (21)$$

irreducible Goppa code for any r and m. The comparable primitive BCH code has parameters

$$[n = 2^m - 1, k \geqslant 2^m - 1 - rm, d \geqslant 2r + 1], \qquad (22)$$

which (if equality holds for k and d in (21) and (22)) has one fewer information symbol.

Problem. (4) Let $G(z)$ have degree r, distinct zeros, coefficients in $GF(2^s)$, and satisfy $G(0) \neq 0$, $G(1) \neq 0$. Let $GF(2^t)$ be the smallest field which contains all the zeros of $G(z)$. Show that we can choose $L = GF(2^m)$ for any m such that $s \mid m$ and $(t, m) = 1$, and obtain a Goppa code with parameters (21).

(5) *BCH codes.* Narrow-sense, primitive BCH codes are a special case of Goppa codes: choose $G(z) = z^r$ and $L = \{1, \alpha, \ldots, \alpha^{n-1}\}$ when $n = q^m - 1$ and α is a primitive element of $GF(q^m)$. Then from Eq. (12),

$$H = \begin{bmatrix} 1 & \alpha^{-r} & \alpha^{-2r} & \cdots & \alpha^{-(n-1)r} \\ 1 & \alpha^{-(r-1)} & \alpha^{-2(r-1)} & \cdots & \alpha^{-(n-1)(r-1)} \\ \cdots\cdots\cdots\cdots\cdots\cdots\cdots\cdots\cdots\cdots\cdots \\ 1 & \alpha^{-1} & \alpha^{-2} & \cdots & \alpha^{-(n-1)} \end{bmatrix}$$

which becomes the parity check matrix of a BCH code (Equation (19) of Ch. 7) when α^{-1} is replaced by β.

To obtain a t-error-correcting *binary* BCH code we take $G(z) = z^{2t}$ and $L = GF(2^m)^*$.

In examples (1) and (2) it turned out that k and d coincided with the bounds given in Fig. 12.3. But this is not always so, as the example of BCH codes shows.

Research Problem (12.1). Find the true dimension and minimum distance of a Goppa code.

Problem. (5) Another form for the parity check matrix. Suppose $G(z)$ has no multiple zeros, say $G(z) = (z - z_1) \cdots (z - z_r)$, where $z_1, \ldots, z_r$ are distinct elements of $GF(2^s)$. Show that $a \in \Gamma(L, G)$ iff $Ha^T = 0$, where $H = (H_{ij})$, $H_{ij} = 1/(z_i - \alpha_j)$, for $1 \leqslant i \leqslant r$, $1 \leqslant j \leqslant n$.

Remark. Note that in this problem H is a Cauchy matrix (see Problem 7 of

Ch. 11). If H is the parity check matrix of a t-error-correcting code then every $2t$ columns of H must be linearly independent, from Theorem 10 of Ch. 1. Now in classical matrix theory there are two *complex* matrices with the property that *every* square submatrix is nonsingular. These are the Vandermonde and Cauchy matrices – see Lemma 17 of Ch. 4 and Problem 7 of Ch. 11. The Vandermonde matrix is the basis for the definition of a BCH code (§6 of Ch. 7), and we have just seen that the Cauchy matrix is the basis for separable Goppa codes.

Problem. (6) A Goppa code with $G(z) = (z - \beta)^r$ for some β is called *cumulative*. Show that there is a weight-preserving one-to-one mapping between $\Gamma(GF(2^m) - \{\beta\}, (z - \beta)^r)$ and the BCH code $\Gamma(GF(2^m)^*, z^r)$.

Example (1) suggests the following problem.

Problem. (7) (Cordaro and Wagner.) Let $\mathscr{C}_n$ be that $[n, 2, d]$ binary code with the highest d and which corrects the most errors of weight $[\frac{1}{2}(d - 1)] + 1$. Set $r = [\frac{1}{3}(n + 1)]$. Show that $d = 2r$ if $n \equiv 0$ or $1 \bmod 3$, and $d = 2r - 1$ if $n \equiv 2 \bmod 3$; and that a generator matrix for $\mathscr{C}_n$ can be taken to consist of r columns equal to $\binom{0}{1}$, r columns equal to $\binom{1}{0}$, and the remaining columns equal to $\binom{1}{1}$.

§4. Further properties of Goppa codes

Adding an overall parity check. Let $\Gamma(L, G)$ be a Goppa code over $GF(q)$ of length $n = q^m$, with $L = GF(q^m) = \{0, 1, \alpha, \ldots, \alpha^{n-2}\}$ and $G(z) = $ a polynomial of degree r with no zeros in $GF(q^m)$. From (12), $a = (a(0), a(1), \ldots, a(a^{n-2}))$ is in $\Gamma(L, G)$ iff

$$\sum_{\beta \in GF(q^m)} \frac{\beta^i a(\beta)}{G(\beta)} = 0 \quad \text{for } i = 0, 1, \ldots, r - 1. \tag{23}$$

$\Gamma(L, G)$ may be extended by adding an overall parity check $a(\infty)$ given by

$$a(\infty) = - \sum_{\beta \in GF(q^m)} a(\beta)$$

or

$$\sum_{\beta \in GF(q^m) \cup \{\infty\}} a(\beta) = 0. \tag{24}$$

With the convention that $1/\infty = 0$, the range of summation in (23) can be changed to $GF(q^m) \cup \{\infty\}$. Finally, combining (23) and (24), we obtain the result that $\hat{a} = (a(0), a(1), \ldots, a(\alpha^{n-2}), a(\infty))$ is in the extended Goppa code iff

$$\sum_{\beta \in GF(q^m) \cup \{\infty\}} \frac{\beta^i a(\beta)}{G(\beta)} = 0 \quad \text{for } i = 0, 1, \ldots, r. \tag{25}$$

The extended code will be denoted by $\hat{\Gamma}(L, G)$.

Problem. (8)(a) Let π be the permutation of $GF(q^m) \cup \{\infty\}$ defined by

$$\pi : \gamma \to \frac{a\gamma + b}{c\gamma + d}, \tag{26}$$

where $a, b, c, d \in GF(q^m)$ satisfy $ad - bc \neq 0$, and $\gamma \in GF(q^m) \cup \{\infty\}$. First check that this *is* a permutation of $GF(q^m) \cup \{\infty\}$. Then show that π sends the extended Goppa code $\hat{\Gamma}(L, G)$ into the equivalent code $\hat{\Gamma}(L, G_1)$, where

$$G_1(z) = (cz + d)^r G \left(\frac{az + b}{cz + d} \right).$$

(b) Suppose $G(z)$ is such that

$$G(z) = e(cz + d)^r G \left(\frac{az + b}{cz + d} \right)$$

for some $e \in GF(q^m)^*$. Show that π fixes $\hat{\Gamma}(L, G)$.

Goppa codes and Mattson–Solomon polynomials. In the special case in which L consists of all n^{th} roots of unity, a nice description of Goppa codes can be given in terms of Mattson–Solomon polynomials (see Cor. 8). Only the binary case will be considered.

Recall from §6 of Ch. 8 that if $a(x) = \sum_{i=0}^{n-1} a_i x^i$, $a_i = 0$ or 1, is any polynomial, its Mattson–Solomon polynomial $A(z)$ is given by

$$A(z) = \sum_{i=0}^{n-1} A_{-i} z^i, \tag{27}$$

where $A_i = a(\alpha^i)$ and $\alpha \in GF(2^m)$ is a primitive n^{th} root of unity. The inverse operation is

$$a(x) = \sum_{i=0}^{n-1} A(\alpha^i) x^i. \tag{28}$$

If $L = \{1, \alpha, \alpha^2, \ldots, \alpha^{n-1}\}$, there is a relation between

$$R_a(z) = \sum_{i=0}^{n-1} \frac{a_i}{z + \alpha^i}$$

of Equation (8) and $A(z)$. This is given by the following theorem. We use the notation that if $f(y)$ is any polynomial, $[f(y)]_n$ denotes the remainder when $f(y)$ is divided by $y^n - 1$.

Theorem 7.

$$A(z) = [z(z^n + 1)R_a(z)]_n, \tag{29}$$

$$R_a(z) = \sum_{i=0}^{n-1} \frac{A(\alpha^i)}{z + \alpha^i}. \tag{30}$$

Proof. Write $R_a(z) = F(z)/(z^n + 1)$, where

$$F(z) = \sum_{i=0}^{n-1} a_i \prod_{j \neq i} (z + \alpha^i).$$

Then

$$z(z^n + 1)R_a(z) = zF(z) = \sum_{i=0}^{n-1} a_i z \prod_{j \neq i} (z + \alpha^i).$$

Also

$$A(z) = \sum_{i=0}^{n-1} a_i \sum_{j=0}^{n-1} \alpha^{-ij} z^j.$$

Then (29) will follow if we show that

$$\left[z \prod_{j \neq i} (z + \alpha^i) \right]_n = \sum_{j=0}^{n-1} \alpha^{-ij} z^j,$$

i.e.

$$z \prod_{j \neq i} (z + \alpha^i) + z^n + 1 = \sum_{j=0}^{n-1} \alpha^{-ij} z^j.$$

To see that this holds, just multiply both sides by $z + \alpha^i$. Finally (30) follows from (28).   Q.E.D.

Corollary 8. *Let $\Gamma(L, G)$ be a binary Goppa code with $L = \{1, \alpha, \ldots, \alpha^{n-1}\}$, where $\alpha \in GF(2^m)$ is a primitive n^{th} root of unity. Then*

$$\Gamma(L, G) = \{a(x): [z^{n-1}A(z)]_n \equiv 0 \bmod G(z)\}. \tag{31}$$

Proof. As in the proof of the theorem, write $R_a(z) = F(z)/(z^n + 1)$. Since $z^n + 1$ and $G(z)$ are relatively prime, $a(x) \in \Gamma(L, G)$ iff $F(z) \equiv 0 \bmod G(z)$. From (29), $A(z) = [zF(z)]_n$ and so $[z^{n-1}A(z)]_n = F(z)$.   Q.E.D.

Note that if $\mathrm{wt}\,(a(x))$ is even, $A_0 = 0$ and $[z^{n-1}A(z)]_n = A(z)/z$. Thus in this case $zG(z)$ divides $A(z)$.

Example. Consider the binary Goppa code $\Gamma(L, G)$ where $G = z^3 + z + 1$, $L =$

$\{1, \alpha, \ldots, \alpha^{14}\}$ and α is a primitive element of $GF(2^4)$. The parity check matrix (12) is

$$
H = \begin{array}{c}
\left[\begin{array}{c}
\begin{array}{cccccccccc}
1\;1 & & & 1\;1\;1\;1\;1 & & & 1\;1 & \\
& 1\;1 & & 1\;1\;1\;1 & & 1\;1 & & 1\;1 \\
& 1 & 1\;1\;1 & & 1 & & 1\;1\;1 & & 1\;1 \\
& 1 & & & 1 & & 1 & & 1 \\
\end{array}
\end{array}\right. \\[2em]
\begin{array}{cccccccccc}
1 & & 1 & 1\;1 & 1\;1\;1 & & & 1\;1 \\
1 & & & & 1\;1 & & 1\;1 & & 1 \\
& & 1 & 1\;1\;1 & & & & 1\;1 \\
1\;1\;1\;1 & & 1 & & 1\;1 & & 1 \\
\end{array} \\[2em]
\begin{array}{cccccccccc}
1\;1 & & 1 & & 1 & & 1\;1\;1 \\
1\;1 & & 1\;1\;1\;1\;1\;1\;1\;1 \\
1\;1\;1\;1\;1 & & & 1 & & 1\;1\;1\;1 \\
& & & 1 & & & 1 & & 1\;1 \\
\end{array}
\end{array}
$$

Then $\Gamma(L, G)$ is a $[15, 3, 7]$ code with generator matrix

$$
\begin{array}{cl}
u_1 & \left[\begin{array}{ccccccccccccccc}
1\;1\;1 & & 1\;1 & & & 1 & & 1 \\
1 & & & 1\;1\;1 & & & 1 & & 1 & & & 1 \\
1\;1\;1\;1\;1\;1\;1\;1\;1\;1\;1\;1\;1\;1\;1
\end{array}\right] \\
u_2 & \\
u_3 &
\end{array}
$$

This is a poor code, but a good illustration of the Corollary. The Mattson–Solomon polynomials corresponding to the codewords u_1, u_3 and u_2 are respectively

$$1 + z + z^2 + z^4 + z^8,$$

$$z + z^2 + z^3 + \cdots + z^{14},$$

$$1 + w^2 z + w z^2 + w z^3 + w^2 z^4 + w z^5 + w^2 z^6 + w^2 z^7$$
$$+ w z^8 + w^2 z^9 + w^2 z^{10} + w z^{11} + w z^{12} + w^2 z^{13} + w z^{14},$$

where w stands for α^5. Applying the Corollary to the first of these we find

$$[z^{14} A(z)]_{15} = 1 + z + z^3 + z^7 + z^{14}$$

which is indeed a multiple of $z^3 + z + 1$. The reader is invited to verify the Corollary for the other two polynomials.

Corollary 9. *Let $\mathscr{C}$ be a binary Goppa code $\Gamma(L, G)$ with $L = \{1, \alpha, \ldots, \alpha^{n-1}\}$, where $\alpha \in GF(2^m)$ is a primitive n^{th} root of unity. If $\mathscr{C}$ is cyclic then $\mathscr{C}$ is a BCH code and $G(z) = z^r$ for some r.*

Proof: Suppose $G(z) \neq \gamma z^l$ for any $\gamma \in GF(2^m)$ and integer l. Then $G(z)$ has a zero $\beta \neq 0$ in some field $GF(2^s)$. Let $a(x)$ be a nonzero codeword in $\mathscr{C}$ with even weight. From Corollary 8, $[z^{n-1}A(z)]_n = A(z)/z$ is a multiple of $G(z)$, hence $A(\beta) = 0$. Since $\mathscr{C}$ is cyclic, by Theorem 22 (iv) of Ch. 8, $A(\alpha^i \beta) = 0$ for $i = 0, \ldots, n-1$. Thus $A(z)$ has at least n distinct zeros, so $\deg A(z) \geq n$, which is impossible. Q.E.D.

Long Goppa codes are good. It was shown in §2 that long alternant codes meet the Gilbert–Varshamov bound. In fact this is true for a much smaller class of codes, as shown by the following problem.

Problem. (9)(a) Show that there exists a Goppa code $\Gamma(L, G)$ over $GF(q)$ with $G(z) =$ an irreducible polynomial over $GF(q^m)$ of degree r and $L = GF(q^m)$, having parameters $[n = q^m, k \geq n - rm, d]$, provided

$$\sum_{w=r+1}^{d-1} \left[\frac{w-1}{r}\right](q-1)^w \binom{q^m}{w} < \frac{1}{r} q^{rm}(1 - (r-1)q^{-rm/2}). \tag{32}$$

[Hint: From Theorem 15 of Ch. 4, show that there are $I_{q^m}(r) \geq (1/r)(q^{rm} - (r-1)q^{rm/2})$ irreducible polynomials of degree r over $GF(q^m)$.]

(b) Hence show that there exist long irreducible Goppa codes which meet the Gilbert–Varshamov bound.

But this doesn't say which $G(z)$ is best.

Research Problem (12.2). Which Goppa polynomial $G(z)$ gives the highest minimum distance (or the lowest redundancy)?

*§5. Extended double-error-correcting Goppa codes are cyclic

In this section certain extended binary double-error-correcting Goppa codes are shown to be cyclic (generalizing Ex. 1 of §3). To do this we find a group of permutations which preserve the code, and then show that one of these permutations consists of a single cycle.

The codes to be considered have Goppa polynomial $G(z)$ which is a quadratic with distinct roots, and L consists of all elements of $GF(2^m)$ that are not zeros of $G(z)$. Thus $\Gamma(L, G)$ is a double-error-correcting code. There are two cases, depending on whether or not $G(z)$ is irreducible. From Theorem 16 and Problem 6 of Ch. 9, $G(z)$ can be taken to be $z^2 + z + \beta$, where β is an element of $GF(2^m)$ with trace 1 if $G(z)$ is irreducible over $GF(2^m)$, or with trace 0 if $G(z)$ is reducible over $GF(2^m)$. Let $\hat{\Gamma}(L, G)$ be the extended code obtained by adding an overall parity check to $\Gamma(L, G)$, with coordinates labeled by $L \cup \{\infty\}$. Thus $\hat{\Gamma}(L, G)$ has length $2^m + 1$ if $G(z)$ is irreducible, or length $2^m - 1$ if $G(z)$ is reducible.

The irreducible case is considered first.

Theorem 10. *Let* $\hat{\Gamma}(L, G)$ *be the* $[2^m + 1, 2^m - 2m, 6]$ *extended double-error-correcting Goppa code just defined, where* $G(z)$ *is irreducible over* GF(2^m). *Then* $\hat{\Gamma}(L, G)$ *is fixed by the group* G_Γ *consisting of the* $2m(2^m + 1)$ *permutations*

$$G_\Gamma = \{C^i, C^i B(\eta), C^i D, C^i DB(\eta); \quad \text{for } i = 0, \dots, m - 1 \text{ and all } \eta \in \mathrm{GF}(2^m)\}$$
(33)

where

$$B(\eta): \gamma \to \frac{\eta\gamma + \beta + \eta}{\gamma + \eta},$$

$$C: \gamma \to \gamma^2 + \beta,$$
(34)

$$D: \gamma \to \gamma + 1,$$

act on all γ *in* GF(2^m) $\cup \{\infty\}$. *(More about* G_Γ *in the next theorem.)*

Proof. (i) First we show that the permutations $B(\eta)$, C and D preserve $\hat{\Gamma}(L, G)$. (Note from Problem 8a that $B(\eta)$ *is* a permutation, for $T_m(\eta^2 + \eta) = 0$, $T_m(\beta) = 1$ hence $\eta^2 + \eta + \beta \neq 0$.) From Problem 8b, $B(\eta)$ and D preserve $\hat{\Gamma}$. To show that C preserves $\hat{\Gamma}$, let $a = (a(0), a(1), \dots, a(\infty))$ be a codeword of $\hat{\Gamma}$ and let $Ca = (a'(0), a'(1), \dots, a'(\infty))$, where $a'(\gamma) = a(\gamma^2 + \beta)$. Then from (25) $Ca \in \hat{\Gamma}$ iff

$$\sum_{\gamma \in \mathrm{GF}(2^m) \cup \{\infty\}} \frac{\gamma^i a'(\gamma)}{\gamma^2 + \gamma + \beta} = 0 \quad \text{for } i = 0, 1, 2.$$
(35)

Set $\alpha = \gamma^2 + \beta$, or $\gamma = \sqrt{}(\alpha + \beta)$ (recall from §6 of Ch. 4 that every element of GF(2^m) has a unique square root). Then (35) becomes

$$\sum_{\alpha \in \mathrm{GF}(2^m) \cup \{\infty\}} \frac{(\alpha + \beta)^{i/2} a(\alpha)}{\alpha + \sqrt{}(\alpha + \beta)}.$$

Squaring this we obtain

$$\sum_{\alpha \in \mathrm{GF}(2^m) \cup \{\infty\}} \frac{(\alpha + \beta)^i a(\alpha)}{\alpha^2 + \alpha + \beta},$$

which from the definition of $\hat{\Gamma}$, Equation (25), is indeed zero for $i = 0, 1, 2$. Therefore C preserves $\hat{\Gamma}$. (ii) Secondly we show that the group generated by

C, D and all $B(\eta)$ is given by (33). Any permutation of the form[†]

$$\gamma \to \frac{a\gamma + b}{c\gamma + d}, \qquad a, b, c, d \in GF(2^m), \quad ad - bc \neq 0, \tag{36}$$

can be represented by the invertible matrix $\binom{ab}{cd}$. For example $B(\eta)$ and D are represented by

$$\begin{pmatrix} \eta & \beta + \eta \\ 1 & \eta \end{pmatrix} \quad \text{and} \quad \begin{pmatrix} 1 & 1 \\ 0 & 1 \end{pmatrix}.$$

Of course $\binom{ab}{cd}$ and $e\binom{ab}{cd}$, for any e in $GF(2^m)^*$, represent the same permutation. If permutations π_1, π_2 are represented by the matrices M_1, M_2, then $\pi_1\pi_2$ (which means first apply π_1, then π_2) is represented by the matrix-product M_2M_1. For example $DB(\eta)$ is represented by

$$\begin{pmatrix} \eta & \beta + \eta \\ 1 & \eta \end{pmatrix}\begin{pmatrix} 1 & 1 \\ 0 & 1 \end{pmatrix} = \begin{pmatrix} \eta & \beta \\ 1 & 1 + \eta \end{pmatrix}. \tag{37}$$

Note that

$$B(\eta)^2 = D^2 = I, \tag{38}$$

$$C^m: \gamma \to \gamma + T_m(\beta) = \gamma + 1 \tag{39}$$

so $C^m = D$ and $C^{2m} = I$.

Let H be the group generated by D and all $B(\eta)$. From (38) and

$$B(\theta)B(\eta) = DB\left(\frac{\eta\theta + \beta + \eta}{\eta + \theta}\right) \quad \text{if } \eta \neq \theta,$$

$$B(\eta)D = DB(1 + \eta),$$

it follows that

$$H = \{I, B(\eta), D, DB(\eta); \quad \text{for all } \eta \in GF(2^m)\},$$

and has order $2(2^m + 1)$. It remains to show that the group generated by H and C is given by (33). This follows from

$$B(\eta)C = CB(\eta^2 + \beta),$$

$$DC = CD. \qquad \qquad \text{Q.E.D.}$$

Problem. (10) Show that $CB(\eta)C^{-1} = B((\eta + \beta)^{2^{m-1}})$.

We shall now prove that $\hat{\Gamma}(L, G)$ is equivalent to a cyclic code.

[†] The group of all such permutations is called $PGL_2(2^m)$ (cf. §5 of Ch. 16).

Theorem 11. (Berlekamp and Moreno.) *Let $\hat{\Gamma}(L, G)$ be an extended irreducible double-error-correcting Goppa code as in Theorem 10. The group G_Γ given by (33), which fixes $\hat{\Gamma}(L, G)$, contains a permutation $DB(\eta_0)$ which is a single cycle consisting of all the elements of $GF(2^m) \cup \{\infty\}$. Thus the coordinate places can be rearranged so that $\hat{\Gamma}(L, G)$ is cyclic.*

Proof. The elements of $GF(2^m) \cup \{\infty\}$ will be represented by nonzero vectors $\binom{r}{s}$ with $r, s \in GF(2^m)$, where $\binom{r}{s}$ represents the element r/s if $s \neq 0$, $\binom{r}{0}$ represents ∞. Also $\binom{r}{s}$ and $e\binom{r}{s}$ represent the same element, for $e \in GF(2^m)^*$. The permutation $DB(\eta)$ given by (37) sends

$$\infty \to \eta \to \eta^2 + \beta \to \frac{\eta^3 + \beta\eta + \beta}{\eta^2 + \eta + 1 + \beta} \to \cdots$$

or equivalently, using the vector notation,

$$\binom{1}{0} \to \begin{pmatrix} \eta & \beta \\ 1 & 1+\eta \end{pmatrix}\binom{1}{0} = \binom{\eta}{1} \to \begin{pmatrix} \eta & \beta \\ 1 & 1+\eta \end{pmatrix}\binom{\eta}{1} = \binom{\eta^2 + \beta}{1} \to \binom{\eta^3 + \beta\eta + \beta}{\eta^2 + \eta + 1 + \beta} \to \cdots$$

We shall find an η_0 such that $DB(\eta_0)$ consists of a single cycle, i.e. such that the least integer n for which

$$\begin{pmatrix} \eta & \beta \\ 1 & 1+\eta \end{pmatrix}^n \binom{1}{0} = k\binom{1}{0}, \quad \text{for some } k \in GF(2^m),$$

is $2^m + 1$. The eigenvalues of $\begin{pmatrix} \eta & \beta \\ 1 & 1+\eta \end{pmatrix}$ are λ and $1 + \lambda$, where

$$\lambda(1 + \lambda) = \eta^2 + \eta + \beta.$$

The eigenvectors are given by

$$\begin{pmatrix} \eta & \beta \\ 1 & 1+\eta \end{pmatrix}\binom{\lambda + \eta}{1} = (\lambda + 1)\binom{\lambda + \eta}{1},$$

$$\begin{pmatrix} \eta & \beta \\ 1 & 1+\eta \end{pmatrix}\binom{\lambda + \eta + 1}{1} = \lambda\binom{\lambda + \eta + 1}{1}.$$

Let us write $\binom{1}{0}$ in terms of the eigenvectors as

$$\binom{1}{0} = \binom{\lambda + \eta}{1} + \binom{\lambda + \eta + 1}{1}.$$

Therefore

$$\begin{pmatrix} \eta & \beta \\ 1 & 1+\eta \end{pmatrix}^n \binom{1}{0} = (\lambda + 1)^n \binom{\lambda + \eta}{1} + \lambda^n \binom{\lambda + \eta + 1}{1}.$$

Now let n be the smallest integer such that

$$\begin{pmatrix} \eta & \beta \\ 1 & 1+\eta \end{pmatrix}^n \binom{1}{0} = k\binom{1}{0}$$

for some $k \in \mathrm{GF}(2^m)$. Then

$$(\lambda + 1)^n(\lambda + \eta) + \lambda^n(\lambda + \eta + 1) = k, \tag{40}$$

$$(\lambda + 1)^n + \lambda^n = 0. \tag{41}$$

Substituting (41) into (40) we obtain

$$k = \lambda^n. \tag{42}$$

Let $\xi \in \mathrm{GF}(2^{2m})$, $\xi \notin \mathrm{GF}(2^m)$. Then

$$(\xi^{2^m} + \xi)^{2^m} = \xi + \xi^{2^m},$$

so by Theorem 8 of Ch. 4,

$$\xi^{2^m} + \xi = \alpha \quad (\text{say}) \in \mathrm{GF}(2^m).$$

Now suppose ξ is a primitive $(2^m + 1)^{\mathrm{st}}$ root of unity. Then $\xi^{2^m} = \xi^{-1}$, so $\xi^{-1} + \xi + \alpha = 0$, or

$$\xi^2 + \alpha\xi + 1 = 0.$$

We choose η_0 by setting $\lambda = \xi/\alpha$, thus

$$\eta_0^2 + \eta_0 + \beta = \lambda(1 + \lambda) = (\xi^2 + \alpha\xi)/\alpha^2 = 1/\alpha^2. \tag{43}$$

Now $\lambda = \xi/\alpha$ is a zero of $x^2 + x + 1/\alpha^2$, so $T_m(1/\alpha^2) = 1$ by Theorem 15 of Ch. 9. Hence $T_m(\beta + 1/\alpha^2) = 0$, and from (43) and Theorem 15 of Ch. 9, η_0 is in $\mathrm{GF}(2^m)$.

From (42), we must have $\lambda^n = (\xi/\alpha)^n \in \mathrm{GF}(2^m)$, i.e.

$$\left(\frac{\xi}{\alpha}\right)^{n(2^m-1)} = (\xi^{(2^m-1)})^n = 1. \tag{44}$$

But $2^m - 1$ and $2^m + 1$ are relatively prime, so ξ^{2^m-1} is also a primitive $(2^m + 1)^{\mathrm{st}}$ root of unity. Therefore $n = 2^m + 1$ is the smallest number for which (44) is possible, and so $DB(\eta_0)$ consists of a single cycle. (Note that Equation (41) holds with $n = 2^m + 1$.) Q.E.D.

From §4 of Ch. 9, any cyclic code of length $2^m + 1$ is reversible (as the code (19) is). So once we know from Theorem 11 that $\hat{\Gamma}(L, G)$ is cyclic, it follows that the automorphism group of $\hat{\Gamma}(L, G)$ contains the reversing permutation R and the permutation σ_2 (which sends i to $2i$ if the coordinates are labeled properly) – see §4 of Ch. 9. By Problem 4 of Ch. 9, the automorphism group of $\hat{\Gamma}(L, G)$ has order at least $2m(2^m + 1)$. In fact R and σ_2 *are* contained in G_Γ, Problem 11, so G_Γ is exactly the group generated by the cyclic permutation $DB(\eta_0)$ and σ_2.

Of course the automorphism group of $\hat{\Gamma}(L, G)$ may be larger than G_Γ, as shown by the examples below.

Problem. (11) (i) Show that the permutation

$$\gamma \to \frac{\beta}{\gamma^2 + \beta + 1} \tag{45}$$

is in G_Γ, fixes the point $\binom{1}{0}$, and sends $(DB(\eta_0))^i \binom{1}{0}$ to $(DB(\eta_0))^{2i} \binom{1}{0}$. So this is the permutation σ_2.

[Hint: Write $\binom{1}{0} = (\lambda + \eta)\binom{\lambda + \eta}{1} + (\lambda + \eta + 1)\binom{\lambda + \eta + 1}{1}$.]

(ii) Show that $R = \sigma_2^m (DB(\eta_0))^{-1} = DB(\eta_0)D$ is the reversing permutation.

Examples. To illustrate Theorems 10 and 11, consider $\hat{\Gamma}(L, G) =$ the $[9, 2, 6]$ code (19) of example 1 in §3. Some of the permutations fixing this code are

$$C: \gamma \to \gamma^2 + 1, \quad \text{which is } (\infty)(01)(\alpha \alpha^6 \alpha^4 \alpha^3 \alpha^2 \alpha^5),$$

$$D: \gamma \to \gamma + 1, \quad \text{which is } (\infty)(01)(\alpha \alpha^3)(\alpha^2 \alpha^6)(\alpha^4 \alpha^5),$$

and

$$DB(\eta_0) = DB(\alpha^2): \gamma \to \frac{\alpha^2 \gamma + 1}{\gamma + \alpha^6}, \quad \text{which is } (1\alpha^4 \alpha^6 \infty \alpha^2 \alpha^5 0 \alpha \alpha^3).$$

As shown in (19), arranging the coordinate places in the order $1, \alpha^4, \alpha^6, \infty, \alpha^2,$ $\alpha^5, 0, \alpha, \alpha^3$ does make the code cyclic. For this code σ_2 (Equation (45)) is $\gamma \to 1/\gamma^2$, which is $(1)(\alpha^4 \alpha^6 \alpha^2 \alpha^3 \alpha \alpha^5)(\infty 0)$.

From Theorem 11 the code is fixed by the 54 permutations of G_Γ. In this case the code is in fact fixed by many additional permutations.

Problem. (12) Show that the automorphism group of the $[9, 2, 6]$ code (19) has order 1296.

Applying Theorem 11 to example 3 of §3 we obtain a $[17, 8, 6]$ cyclic code, which is in fact a quadratic residue code (see Ch. 16) whose automorphism group has order 2448.

As a third example, take $G(z) = z^2 + z + 1$ and $L = \mathrm{GF}(2^5)$. Then $\hat{\Gamma}(L, G)$ is a $[33, 22, 6]$ cyclic code with generator polynomial $1 + x^2 + x^5 + x^6 + x^9 + x^{11}$.

The case when $G(z)$ is reducible. Similar results hold if $G(z)$ is reducible over $\mathrm{GF}(2^m)$, say $G(z) = z^2 + z + \beta = (z + \lambda)(z + \lambda + 1)$, where $\beta, \lambda \in \mathrm{GF}(2^m)$ and $T_m(\beta) = 0$. Now we take $L = \mathrm{GF}(2^m) - \{\lambda, \lambda + 1\}$. The $2^m - 1$ coordinates of $\hat{\Gamma}(L, G)$ are labeled by $L \cup \{\infty\}$. For $\eta \neq \lambda$, $\eta \neq \lambda + 1$ we again define $B(\eta)$, C and D by (34). These are indeed permutations of $L \cup \{\infty\}$, for $B(\eta)$ and D interchange λ and $\lambda + 1$, and C fixes λ and $\lambda + 1$. Then we have:

Theorem 12. *If $G(z)$ is reducible then the $[2^m - 1, 2^m - 2m - 2, 6]$ extended*

Goppa code $\hat{\Gamma}(L, G)$ is fixed by the group G_Γ' consisting of the $2m(2^m - 1)$ permutations

$$\{C^i, C^iB(\eta), C^iD, C^iDB(\eta); \text{ for } i = 0, \ldots, m - 1$$
$$\text{and all } \eta \in GF(2^m) - \{\lambda, \lambda + 1\}\} \qquad (46)$$

acting on $L \cup \{\infty\}$. This group contains a permutation $DB(\eta_0)$ which is a single cycle consisting of all the elements of $L \cup \{\infty\}$. Hence $\hat{\Gamma}(L, G)$ is equivalent to a cyclic code.

Proof. The first statement is proved as in Theorem 10. Note that now $C^m = I$. To prove the second statement we will show that $GF(2^m)$ contains an element $\lambda = \lambda_0$ (say) that does not satisfy the equation

$$\lambda^d + (\lambda + 1)^d = 0 \qquad (47)$$

for any $d < 2^m - 1$. Then choose $\eta_0 \in GF(2^m)$ so that

$$\eta_0^2 + \eta_0 + \beta = \lambda_0(\lambda_0 + 1).$$

This can be done since $T_m(\beta) = T_m(\lambda_0^2 + \lambda_0) = 0$. Then $n = 2^m - 1$ is the smallest number for which Equations (41) and (42) hold, which completes the proof.

It remains to show that there is a λ in $GF(2^m)$ such that (47) does not hold for any $d < 2^m - 1$. The minimum d for which (47) holds must be a divisor of $2^m - 1$. For

$$\lambda^{2^m - 1} = (\lambda + 1)^{2^m - 1} \quad (\text{since } \lambda \in GF(2^m))$$

and so from (47), $\lambda^\delta = (\lambda + 1)^\delta$ where $\delta = \gcd(d, 2^m - 1)$. But d is minimal so $\delta = d$ and $d \mid 2^m - 1$. Let F_d be the number of zeros of $\lambda^d + (\lambda + 1)^d$ in $GF(2^m)$, where $d \mid 2^m - 1$. Now $\lambda^d + (\lambda + 1)^d$ divides $\lambda^{2^m - 1} + (\lambda + 1)^{2^m - 1}$, and the latter equation has $2^m - 2$ zeros in $GF(2^m)$ (namely all of $GF(2^m)$ except 0 and 1). Thus all the zeros of $\lambda^d + (\lambda + 1)^d$ are in $GF(2^m)$, i.e. $F_d = d - 1$.

The number of zeros of $\lambda^{2^m - 1} + (\lambda + 1)^{2^m - 1}$ which are not zeros of (47) for some $d < 2^m - 1$ is

$$\sum_{d \mid 2^m - 1} F_d \mu \left(\frac{2^m - 1}{d} \right), \qquad (48)$$

where μ is the Möbius function defined in Ch. 4. For if ξ is such a zero it will be counted in the term $d = 2^m - 1$ and no other term, and so is counted once. On the other hand, if ξ is a zero of $\lambda^d + (\lambda + 1)^d$ for $d < 2^m - 1$, it is also a zero of $\lambda^D + (\lambda + 1)^D$ whenever $d \mid D$, i.e. the number of times it is counted in the sum is

$$\sum_{d \mid D \mid 2^m - 1} \mu \left(\frac{2^m - 1}{D} \right) = \sum_{d' \mid (2^m - 1)/d} \mu \left(\frac{2^m - 1}{dd'} \right) \quad \text{where } D = dd',$$

$$= 0 \quad \text{by Theorem 14 of Ch. 4.}$$

Finally (48) is equal to

$$\sum_{d\mid 2^m-1} (d-1)\mu\left(\frac{2^m-1}{d}\right) = \sum_{d\mid 2^m-1} d\mu\left(\frac{2^m-1}{d}\right) - \sum_{d\mid 2^m-1} \mu\left(\frac{2^m-1}{d}\right)$$

$$= \varphi(2^m-1) - 0 \quad \text{(by Theorem 14 of Ch. 4)}$$

$$> 0, \quad \text{as required.} \qquad\qquad\qquad \text{Q.E.D.}$$

Problem. (13) Show that this code is also reversible and the reversing permutation may be taken to be $B(\eta_0)$.

Hence in this case $\hat{\Gamma}(L, G)$ is fixed by the group of order $2m(2^m - 1)$ generated by the cyclic permutation $DB(\eta_0)$, σ_2 (Equation (45)), and the reversing permutation $B(\eta_0)$. (See Problem 4 of Ch. 9.)

Research Problem (12.3). What other extended Goppa codes are cyclic?

*§6. Generalized Srivastava codes

Another interesting class of alternant codes are the generalized Srivastava codes. (See Fig. 12.4.)

> The code is defined by the parity check matrix (51), where $\alpha_1, \ldots,$ α_n, $w_1, \ldots, w_s$ are $n + s$ distinct elements of $\mathrm{GF}(q^m)$, and $z_1, \ldots, z_n$ are nonzero elements of $\mathrm{GF}(q^m)$. It is an $[n, k \geqslant n - mst, d \geqslant st + 1]$ code over $\mathrm{GF}(q)$, and is an alternant code. The original Srivastava codes are the case $t = 1$, $z_i = \alpha_i^\mu$ for some μ.

Fig. 12.4. Properties of generalized Srivastava code.

Definition. In the parity check matrix (3) for the alternant code $\mathcal{A}(\alpha, y)$, suppose $r = st$ and let $\alpha_1, \ldots, \alpha_n$, $w_1, \ldots, w_s$ be $n + s$ distinct elements of $\mathrm{GF}(q^m)$, $z_1, \ldots, z_n$ be nonzero elements of $\mathrm{GF}(q^m)$. Also set

$$g_{(l-1)t+k}(x) = \frac{\displaystyle\prod_{j=1}^{s} (x - w_j)^t}{(x - w_l)^k}, \quad l = 1, \ldots, s, \; k = 1, \ldots, t, \tag{49}$$

(note that this *is* a polynomial in x), and

$$y_i = \frac{z_i}{\displaystyle\prod_{j=1}^{s} (\alpha_i - w_j)^t}, \quad i = 1, \ldots, n, \tag{50}$$

so that

$$y_i g_{(l-1)t+k}(\alpha_i) = \frac{z_i}{(\alpha_i - w_l)^{k}}.$$

The resulting code is called a *generalized Srivastava code*. Thus the parity check matrix for a generalized Srivastava code is

$$H = \begin{bmatrix} H_1 \\ H_2 \\ \vdots \\ H_s \end{bmatrix} \tag{51}$$

where

$$H_l = \begin{bmatrix} \dfrac{z_1}{\alpha_1 - w_l} & \dfrac{z_2}{\alpha_2 - w_l} & \cdots & \dfrac{z_n}{\alpha_n - w_l} \\[2mm] \dfrac{z_1}{(\alpha_1 - w_l)^2} & \dfrac{z_2}{(\alpha_2 - w_l)^2} & \cdots & \dfrac{z_n}{(\alpha_n - w_l)^2} \\ \cdots\cdots\cdots\cdots\cdots\cdots\cdots\cdots\cdots\cdots \\ \dfrac{z_1}{(\alpha_1 - w_l)^t} & \dfrac{z_2}{(\alpha_2 - w_l)^t} & \cdots & \dfrac{z_n}{(\alpha_n - w_l)^t} \end{bmatrix} \tag{52}$$

for $l = 1, \ldots, s$.

The original *Srivastava codes* are the special case $t = 1$, $z_i = \alpha_i^{\mu}$ for some μ, and have parity check matrix

$$H = \begin{bmatrix} \dfrac{\alpha_1^{\mu}}{\alpha_1 - w_1} & \cdots & \dfrac{\alpha_n^{\mu}}{\alpha_n - w_1} \\ \cdots\cdots\cdots\cdots\cdots\cdots \\ \dfrac{\alpha_1^{\mu}}{\alpha_1 - w_s} & \cdots & \dfrac{\alpha_n^{\mu}}{\alpha_n - w_s} \end{bmatrix}. \tag{53}$$

Since there are s w_i's, there can be at most $q^m - s$ α_i's, so the length of a generalized Srivastava code is at most $q^m - s$.

If $\alpha_1, \ldots, \alpha_n$ are chosen to be all the elements of $GF(q^m)$ except the w_i's, then $n = q^m - s$ and the codes are called *primitive* (by analogy with BCH codes).

Since it is an alternant code, a generalized Srivastava code has $k \geq n - mst$ and $d \geq st + 1$.

Example. Consider the generalized binary Srivastava code with $m = 6$, $n = 8$, $r = 2$, $s = 1$, $t = 2$, $\alpha_1, \ldots, \alpha_8 =$ the elements of $GF(2^3)$ lying in $GF(2^6)$, i.e.

$$\{\alpha_1, \ldots, \alpha_8\} = \{0, 1, \alpha^9, \alpha^{18}, \alpha^{27}, \alpha^{36}, \alpha^{45}, \alpha^{54}\},$$

where α is a primitive element of $GF(2^6)$. Also

$$w_1 = \alpha, \ z_i = 1 \quad \text{for } i = 1, \ldots, 8.$$

Therefore

$$H = \begin{bmatrix} \dfrac{1}{0-\alpha} & \dfrac{1}{1-\alpha} & \dfrac{1}{\alpha^9-\alpha} & \dfrac{1}{\alpha^{18}-\alpha} & \cdots & \dfrac{1}{\alpha^{54}-\alpha} \\[2mm] \dfrac{1}{(0-\alpha)^2} & \dfrac{1}{(1-\alpha)^2} & \dfrac{1}{(\alpha^9-\alpha)^2} & \dfrac{1}{(\alpha^{18}-\alpha)^2} & \cdots & \dfrac{1}{(\alpha^{54}-\alpha)^2} \end{bmatrix} \tag{54}$$

The second row is the square of the first and is redundant. Expanding the first row in powers of α we find from Fig. 4.5

$$\bar{H} = \begin{bmatrix} 1 & 0 & 0 & 0 & 0 & 0 & 0 & 1 \\ 0 & 1 & 0 & 0 & 1 & 1 & 1 & 1 \\ 0 & 1 & 1 & 0 & 1 & 0 & 0 & 0 \\ 0 & 1 & 0 & 1 & 1 & 1 & 0 & 0 \\ 0 & 1 & 1 & 0 & 0 & 1 & 0 & 0 \\ 1 & 1 & 0 & 1 & 1 & 0 & 1 & 1 \end{bmatrix} \tag{55}$$

The codewords are

$$\begin{bmatrix} 0 & 0 & 0 & 0 & 0 & 0 & 0 & 0 \\ 1 & 1 & 1 & 1 & 0 & 0 & 0 & 1 \\ 0 & 1 & 0 & 1 & 1 & 1 & 1 & 0 \\ 1 & 0 & 1 & 0 & 1 & 1 & 1 & 1 \end{bmatrix}.$$

This is an $[8, 2, 5]$ code which is in fact equivalent to the Goppa code of example 1 of §3.

Problems. (14) Show that a Srivastava code (with $t = 1$) is a Goppa code; hence $d \geq 2s + 1$ for binary Srivastava codes.

(15) Show that the binary Srivastava code (53) with $m = 4$, $s = 2$, $n = 14$, $w_1 = 0$, $w_2 = 1$, $\{\alpha_1, \ldots, \alpha_{14}\} = GF(2^4) - \{0, 1\}$ and $\mu = 0$ is a $[14, 6, 5]$ code, and that the dual code has minimum distance 4.

(16) Show that the binary Srivastava code with $m = 4$, $s = 2$, $n = 13$, $w_1 = \alpha^{-1}$, $w_2 = \alpha^{-3}$, $\{\alpha_1, \ldots, \alpha_{13}\} = GF(2^4) - \{0, \alpha^{-1}, \alpha^{-3}\}$ and $\mu = 1$ is a $[13, 5, 5]$ code, and that the dual code has minimum distance $d' = 5$.

(17) If $m = 1$, show that a generalized Srivastava code is MDS (Ch. 10). These are sometimes called *Gabidulin* codes.

(18) Let $\mathscr{C}$ be a binary primitive generalized Srivastava code with $z_i = 1$ for all i, and $n = 2^m - s$. (i) If $s = 1$, show that $\mathscr{C}$ is unique and is a primitive narrow-sense BCH code. (ii) If $s = 2$, again show that $\mathscr{C}$ is unique and that the extended code $\hat{\mathscr{C}}$ is fixed by a transitive permutation group.

(19) Prove the following properties of the binary non-primitive generalized Srivastava code with parameters:
$m = m_1 m_2$, with $m_1 > 1$, $m_2 > 1$,
$s = m_2$,

$w_i = \alpha^{2^{(i-1)m_1}}$, $i = 1, \ldots, m_2$, where α is a primitive element of $GF(2^m)$,

$\alpha_1, \ldots, \alpha_n$ distinct elements of $GF(2^{m_1}) \subset GF(2^m)$,

$n \leq 2^{m_1}$,

$z_1, \ldots, z_n$ nonzero elements of $GF(2^{m_1})$.

(i) The rows of $H_2, \ldots, H_s$ in (51) are redundant. Hence H can be taken to be the RHS of (52) with w_l replaced by α.

(ii) (54) is a code of this type.

(iii) $k \geq n - mt$ and $d \geq m_2 t + 1$.

(iv) If all $z_i = 1$ then $k \geq n - mt/2$.

(v) If all $z_i = 1$ and $m_2 = 2$, then the extended code is fixed by a transitive permutation group.

*§7. Chien–Choy generalized BCH codes

These are another special class of alternant codes, and are defined in terms of two polynomials $P(z)$ and $G(z)$. (See Fig. 12.5) Let n be relatively prime to q, and let $GF(q^m)$ be the smallest extension field of $GF(q)$ which contains all n^{th} roots of unity.

GBCH(P, G) consists of all $a(x)$ such that (56) holds, and is an $[n, k \geq n - rm, d \geq r + 1]$ code over $GF(q)$, where $P(z)$ and $G(z)$ are polynomials over $GF(q^m)$, $\deg P \leq n - 1$, $r = \deg G \leq n - 1$, which are relatively prime to $z^n - 1$. A parity check matrix is given by (57). These are a special class of alternant codes.

Fig. 12.5. Properties of Chien–Choy generalized BCH code GBCH(P, G).

Definition. The *Chien–Choy generalized BCH code* of length n over $GF(q)$ with associated polynomials $P(z)$ and $G(z)$ – abbreviated GBCH(P, G) – is defined as follows. Let $P(z)$ and $G(z)$ be polynomials with coefficients from $GF(q^m)$ with $\deg P(z) \leq n - 1$ and $r = \deg G(z) \leq n - 1$, which are relatively prime to $z^n - 1$. Then GBCH(P, G) consists of all $a(x)$ with coefficients in $GF(q)$ and degree $\leq n - 1$ for which the MS polynomial $A(z)$ satisfies

$$[A(z)P(z)]_n \equiv 0 \quad \mod G(z). \tag{56}$$

For example, in the binary case GBCH(z^{n-1}, G) is the Goppa code $\Gamma(L, G)$ where L consists of the nth roots of unity, by Corollary 8.

The parity check matrix for GBCH(P, G) is found as follows.
$a(x) \in$ GBCH(P, G)

iff $\exists \, U(z)$ such that $[A(z)P(z)]_n = U(z)G(z)$ and $\deg U(z) \leq n - 1 - r$

iff   (from Theorem 22, Ch. 8) $\exists \, u(x)$ with coefficients from $GF(q^m)$ such that $a(x) * p(x) = u(x) * g(x)$ and $U_1 = \cdots = U_r = 0$, where $u(x)$, $p(x)$ and $g(x)$ are obtained from $U(z)$, $P(z)$ and $G(z)$ by the inverse mapping Equation (9) of Ch. 8.

iff $\exists\, u_0, \ldots, u_{n-1} \in \mathrm{GF}(q^m)$ such that $a_i p_i = u_i g_i$ ($i = 0, \ldots, n-1$) and $U_1 = \cdots = U_r = 0$. (Note that $p_i \neq 0$, $g_i \neq 0$ by hypothesis.)

iff $\displaystyle \sum_{i=0}^{n-1} \frac{a_i p_i}{g_i} \alpha^{ij} (= U_j) = 0$ for $j = 1, \ldots, r$.

In other words $a = (a_0, \ldots, a_{n-1}) \in \mathrm{GBCH}(P, G)$ iff $Ha^T = 0$, where

$$
H = \begin{bmatrix}
p_0/g_0 & p_1\alpha/g_1 & p_2\alpha^2/g_2 & \cdots & p_{n-1}\alpha^{n-1}/g_{n-1} \\
p_0/g_0 & p_1\alpha^2/g_1 & p_2\alpha^4/g_2 & \cdots & p_{n-1}\alpha^{2(n-1)}/g_{n-1} \\
\cdots & \cdots & \cdots & \cdots & \cdots \\
p_0/g_0 & p_1\alpha^r/g_1 & p_2\alpha^{2r}/g_2 & \cdots & p_{n-1}\alpha^{r(n-1)}/g_{n-1}
\end{bmatrix}
$$

$$
= \begin{bmatrix}
1 & 1 & 1 & \cdots & 1 \\
1 & \alpha & \alpha^2 & \cdots & \alpha^{n-1} \\
\cdots & \cdots & \cdots & \cdots & \\
1 & \alpha^{r-1} & & \cdots & \alpha^{(r-1)(n-1)}
\end{bmatrix}
\begin{bmatrix}
p_0/g_0 & & & 0 \\
& p_1\alpha/g_1 & & \\
& & p_2\alpha^2/g_2 & \\
& & & \ddots \\
0 & & & p_{n-1}\alpha^{n-1}/g_{n-1}
\end{bmatrix}
$$

$$(57)$$

Thus H is a parity check matrix for the code. This shows that $\mathrm{GBCH}(P, G)$ is an alternant code with $\alpha_1, \ldots, \alpha_n$ chosen to be the nth roots of unity and $y_i = p_{i-1}\alpha^{i-1}/g_{i-1}$. Therefore $\mathrm{GBCH}(P, G)$ has parameters $[n, k \geq n - rm, d \geq r + 1]$ where $r = \deg G(z)$.

If $P(z) = z^{\delta + b - 2}$, $G(z) = z^{\delta - 1}$ then $\mathrm{GBCH}(P, G)$ is a BCH code, for it is easily seen that the parity check matrix is then

$$
\begin{bmatrix}
1 & \alpha^b & \alpha^{2b} & \cdots & \alpha^{(n-1)b} \\
1 & \alpha^{b+1} & \alpha^{2b+2} & \cdots & \alpha^{(n-1)(b+1)} \\
\cdots & \cdots & \cdots & \cdots & \cdots \\
1 & \alpha^{b+\delta-2} & \alpha^{2(b+\delta-2)} & \cdots & \alpha^{(n-1)(b+\delta-2)}
\end{bmatrix},
$$

which is the parity check matrix of a BCH code (Equation 19 of Ch. 7).

If $P(z) = z^{\delta + b_1 - 2} + z^{\delta + b_2 - 2}$, $G(z) = z^{\delta - 1}$ then $\mathrm{GBCH}(P, G)$ contains the intersection of the corresponding BCH codes, and in fact may equal this intersection.

For example, let $n = 15$ and consider $P(z) = z$, $G(z) = z^2$. The parity check matrix is

$$
\begin{bmatrix}
1 & 1 & 1 & 1 & 1 & 1 & 1 & 1 & 1 & 1 & 1 & 1 & 1 & 1 & 1 \\
1 & \alpha & \alpha^2 & \alpha^3 & \alpha^4 & \alpha^5 & \alpha^6 & \alpha^7 & \alpha^8 & \alpha^9 & \alpha^{10} & \alpha^{11} & \alpha^{12} & \alpha^{13} & \alpha^{14}
\end{bmatrix}
$$

and defines the $[15, 10, 4]$ BCH code with idempotent $\theta_3 + \theta_5 + \theta_7$ (see Problem 7, Ch. 8 for the idempotents of block length 15). If $P(z) = z^5$, $G(z) = z^2$ the parity check matrix is

$$\begin{bmatrix} 1 & \alpha^4 & \alpha^8 & \alpha^{12} & \alpha & \alpha^5 & \alpha^9 & \alpha^{13} & \alpha^2 & \alpha^6 & \alpha^{10} & \alpha^{14} & \alpha^3 & \alpha^7 & \alpha^{11} \\ 1 & \alpha^5 & \alpha^{10} & 1 & ,\alpha^5 & \alpha^{10} & 1 & \alpha^5 & \alpha^{10} & 1 & \alpha^5 & \alpha^{10} & 1 & \alpha^5 & \alpha^{10} \end{bmatrix}.$$

This determines the [15, 9] BCH code with idempotent $\theta_0 + \theta_3 + \theta_7$. GBCH$(z + z^5, z^2)$ has parity check matrix.

$$\begin{bmatrix} 1 & \alpha & \alpha^2 & \alpha^{11} & \alpha^4 & \alpha^{10} & \alpha^7 & \alpha^6 & \alpha^8 & \alpha^{13} & \alpha^5 & \alpha^3 & \alpha^{14} & \alpha^9 & \alpha^{12} \\ 1 & \alpha^2 & \alpha^4 & \alpha^{14} & \alpha^8 & 1 & \alpha^{13} & \alpha^{13} & \alpha & \alpha^7 & 1 & \alpha^{14} & \alpha^{11} & \alpha^7 & \alpha^{11} \end{bmatrix}.$$

When this is written in binary it is seen that one row can be discarded; hence it defines a code of dimension 8 which is in fact the cyclic code with idempotent $\theta_7 + \theta_3$, i.e. with zeros $1, \alpha, \alpha^2, \alpha^4, \alpha^8, \alpha^5, \alpha^{10}$.

Since this example is a cyclic code which is not a BCH code it follows from Corollary 9 that the class of alternant codes includes other codes besides Goppa codes.

Further properties and examples are given in the following problems, and in [286].

Problems. (20) Show that $a(x) \in \mathrm{GBCH}(P, G)$ iff $\deg [A(z)P(z)/G(z)]_n \leq n - 1 - r$.

(21) Show that $\mathrm{GBCH}(P, G) = \mathrm{GBCH}(P^*, z^r)$, where $P(z)^* = [z^r P(z)/G(z)]_n$ and $r = \deg G(z)$. Thus these codes could have been defined just in terms of P^*, without introducing G. However, the definition in terms of two polynomials simplifies the analysis.

§8. The Euclidean algorithm

A crucial step in decoding alternant codes uses the Euclidean algorithm. This is a simple and straightforward algorithm for finding the greatest common divisor (g.c.d.) of two integers or polynomials, or for finding the continued fraction expansion of a real number.

We shall describe the algorithm as it applies to polynomials, since that is the case of greatest interest to us. Only trivial changes are needed to get the algorithm for integers. In this section the coefficients of the polynomials can belong to any field. If $f(z)$ and $g(z)$ are polynomials, by a g.c.d. of $f(z)$ and $g(z)$ we mean a polynomial of highest degree which divides both $f(z)$ and $g(z)$. Any constant multiple of such a polynomial is also a g.c.d. of $f(z)$ and $g(z)$.

Theorem 13. (Euclidean Algorithm for polynomials.) *Given polynomials $r_{-1}(z)$ and $r_0(z)$, with $\deg r_0 \leq \deg r_{-1}$, we make repeated divisions to obtain the series of* *equations*

$$r_{-1}(z) = q_1(z)r_0(z) + r_1(z), \quad \deg r_1 < \deg r_0,$$
$$r_0(z) = q_2(z)r_1(z) + r_2(z), \quad \deg r_2 < \deg r_1,$$
$$\cdots\cdots\cdots\cdots\cdots\cdots\cdots$$
$$r_{j-2}(z) = q_j(z)r_{j-1}(z) + r_j(z), \quad \deg r_j < \deg r_{j-1},$$
$$r_{j-1}(z) = q_{j+1}(z)r_j(z).$$

Then the last nonzero remainder $r_j(z)$ in the division process is a g.c.d. of $r_{-1}(z)$ and $r_0(z)$.

Proof. Clearly $r_j(z)$ divides $r_{j-1}(z)$, hence divides $r_{j-2}(z),\dots$, hence divides both $r_0(z)$ and $r_{-1}(z)$. Conversely, if $h(z)$ divides $r_{-1}(z)$ and $r_0(z)$, then $h(z)$ divides $r_1(z),\dots,r_j(z)$. Therefore $r_j(z)$ is a g.c.d. of $r_{-1}(z)$ and $r_0(z)$. Q.E.D.

Example. We find the g.c.d. of $r_{-1}(z) = z^4 + z^3 + z^2 + 1$ and $r_0(z) = z^3 + 1$ over the field GF(2):

$$z^4 + z^3 + z^2 + 1 = (z + 1)(z^3 + 1) + (z^2 + z),$$
$$z^3 + 1 = (z + 1)(z^2 + z) + (z + 1),$$
$$z^2 + z = z(z + 1).$$

Hence $r_2(z) = z + 1$ is the g.c.d. of $r_{-1}(z)$ and $r_0(z)$.

As a by-product of the Euclidean algorithm we obtain the following useful result.

Theorem 14. *Let $r_{-1}(z)$ and $r_0(z)$ be polynomials with $\deg r_0 \leqslant \deg r_{-1}$, and with g.c.d. $h(z)$. Then there exist polynomials $U(z)$ and $V(z)$ such that*

$$U(z)r_{-1}(z) + V(z)r_0(z) = h(z), \tag{58}$$

where $\deg U$ and $\deg V$ are less than $\deg r_{-1}$.

Proof. Let the polynomials $U_i(z)$ and $V_i(z)$ be defined by

$$U_{-1}(z) = 0, \quad U_0(z) = 1,$$
$$V_{-1}(z) = 1, \quad V_0(z) = 0, \tag{59}$$

and

$$U_i(z) = q_i(z)U_{i-1}(z) + U_{i-2}(z),$$
$$V_i(z) = q_i(z)V_{i-1}(z) + V_{i-2}(z). \tag{60}$$

Then

$$\begin{bmatrix} U_i(z) & U_{i-1}(z) \\ V_i(z) & V_{i-1}(z) \end{bmatrix} = \begin{bmatrix} U_{i-1}(z) & U_{i-2}(z) \\ V_{i-1}(z) & V_{i-2}(z) \end{bmatrix} \begin{bmatrix} q_i(z) & 1 \\ 1 & 0 \end{bmatrix}$$

$$= \cdots$$

$$= \begin{bmatrix} 1 & 0 \\ 0 & 1 \end{bmatrix} \begin{bmatrix} q_1(z) & 1 \\ 1 & 0 \end{bmatrix} \cdots \begin{bmatrix} q_i(z) & 1 \\ 1 & 0 \end{bmatrix} \tag{61}$$

Also

$$\begin{bmatrix} r_{i-2}(z) \\ r_{i-1}(z) \end{bmatrix} = \begin{bmatrix} q_i(z) & 1 \\ 1 & 0 \end{bmatrix} \begin{bmatrix} r_{i-1}(z) \\ r_i(z) \end{bmatrix}$$

$$\begin{bmatrix} r_{i-3}(z) \\ r_{i-2}(z) \end{bmatrix} = \begin{bmatrix} q_{i-1}(z) & 1 \\ 1 & 0 \end{bmatrix} \begin{bmatrix} q_i(z) & 1 \\ 1 & 0 \end{bmatrix} \begin{bmatrix} r_{i-1}(z) \\ r_i(z) \end{bmatrix}$$

$$\cdots\cdots\cdots\cdots\cdots\cdots\cdots\cdots\cdots\cdots\cdots\cdots\cdots$$

$$\begin{bmatrix} r_{-1}(z) \\ r_0(z) \end{bmatrix} = \begin{bmatrix} q_1(z) & 1 \\ 1 & 0 \end{bmatrix} \cdots \begin{bmatrix} q_{i-1} & 1 \\ 1 & 0 \end{bmatrix} \begin{bmatrix} q_i & 1 \\ 1 & 0 \end{bmatrix} \begin{bmatrix} r_{i-1}(z) \\ r_i(z) \end{bmatrix}$$

$$= \begin{bmatrix} U_i(z) & U_{i-1}(z) \\ V_i(z) & V_{i-1}(z) \end{bmatrix} \begin{bmatrix} r_{i-1}(z) \\ r_i(z) \end{bmatrix} \tag{62}$$

The determinant of the RHS of (61) is $(-1)^i$. Hence from (61) and (62)

$$\begin{bmatrix} r_{i-1}(z) \\ r_i(z) \end{bmatrix} = (-1)^i \begin{bmatrix} V_{i-1}(z) & -U_{i-1}(z) \\ -V_i(z) & U_i(z) \end{bmatrix} \begin{bmatrix} r_{-1}(z) \\ r_0(z) \end{bmatrix}. \tag{63}$$

In particular

$$r_i(z) = (-1)^i[-V_i(z)r_{-1}(z) + U_i(z)r_0(z)], \tag{64}$$

which is (58). Also

$$\deg U_i = \sum_{k=1}^{i} \deg q_k,$$

$$\deg r_{i-1} = \deg r_{-1} - \sum_{k=1}^{i} \deg q_k,$$

$$\deg U_i = \deg r_{-1} - \deg r_{i-1} < \deg r_{-1},$$

and similarly for V_i.   Q.E.D.

Corollary 15. *If $r_{-1}(z)$ and $r_0(z)$ are relatively prime then there exist polynomials $U(z)$ and $V(z)$ such that*

$$U(z)r_{-1}(z) + V(z)r_0(z) = 1.$$

Of course similar results hold for integers.

Example. Continuing the previous example, we find

$$U_{-1} = 0, \qquad U_0 = 1, \qquad U_1 = z + 1, \qquad U_2 = z^2,$$
$$V_{-1} = 1, \qquad V_0 = 0, \qquad V_1 = 1, \qquad V_2 = z + 1$$

and Equation (58) states that

$$z + 1 = (z + 1)(z^4 + z^3 + z^2 + 1) + z^2(z^3 + 1).$$

Problem. (22) Show that $U_i(z)$ and $V_i(z)$ are relatively prime.

§9. Decoding alternant codes

In this section we describe an efficient decoding algorithm for alternant codes which makes use of the Euclidean algorithm of the previous section. The decoding is in 3 stages, as for BCH codes: Stage I, find the syndrome. Stage II, find the error locator and error evaluator polynomials using the Euclidean algorithm. Stage III, find the locations and values of the errors, and correct them. Since this decoding method applies equally well to BCH codes, it completes the gap left in the decoding of BCH codes in §6 of ch. 9.

Let $\mathscr{A}$ be an alternant code $\mathscr{A}(\alpha, y)$ over $GF(q)$ with parity check matrix $H = XY$ given by Equation (2), and having minimum distance $d \geq r + 1$, where r is even. Suppose $t \leq r/2$ errors have occurred, in locations

$$X_1 = \alpha_{i_1}, \ldots, X_t = \alpha_{i_t},$$

with error values

$$Y_1 = a_{i_1}, \ldots, Y_t = a_{i_t},$$

as in §6 of Ch. 9.

Stage I of the decoding is to find the syndrome. This is given by

$$S = \begin{bmatrix} 1 & 1 & \cdots & 1 \\ \alpha_1 & \alpha_2 & \cdots & \alpha_n \\ \alpha_1^2 & \alpha_2^2 & \cdots & \alpha_n^2 \\ \cdots & \cdots & \cdots & \cdots \\ \alpha_1^{r-1} & \alpha_2^{r-1} & \cdots & \alpha_n^{r-1} \end{bmatrix} \begin{bmatrix} y_1 & & & 0 \\ & y_2 & & \\ & & y_3 & \\ & & & \ddots \\ 0 & & & y_n \end{bmatrix} \begin{bmatrix} 0 \\ 0 \\ Y_1 \\ \vdots \\ \vdots \end{bmatrix}$$

$$= \begin{bmatrix} S_0 \\ S_1 \\ \vdots \\ S_{r-1} \end{bmatrix} \quad \text{(say)},$$

where

$$S_\mu = \sum_{\nu=1}^{t} \alpha_{i_\nu}^\mu a_{i_\nu} y_{i_\nu}$$

$$= \sum_{\nu=1}^{t} X_\nu^\mu Y_\nu y_{i_\nu}, \tag{65}$$

for $\mu = 0, \ldots, r-1$.

Stage II uses the syndrome to obtain the *error locator polynomial*

$$\sigma(z) = \prod_{i=1}^{t} (1 - X_i z)$$

$$= \sum_{i=0}^{t} \sigma_i z^i, \quad \sigma_0 = 1, \tag{66}$$

and the *error evaluator polynomial*

$$\omega(z) = \sum_{\nu=1}^{t} Y_\nu y_{i_\nu} \prod_{\substack{\mu=1 \\ \mu \neq \nu}}^{t} (1 - X_\mu z). \tag{67}$$

(Note that this is a slightly different definition from that in Equation (17) of Ch. 8). These polynomials are related by

$$\frac{\omega(z)}{\sigma(z)} \equiv S(z) \bmod z^r \tag{68}$$

where

$$S(z) = \sum_{\mu=0}^{r-1} S_\mu z^\mu.$$

(The proof of (68) is the same as that of Theorem 25 of Ch. 8.)

Thus the goal of Stage II is, given $S(z)$, to find $\sigma(z)$ and $\omega(z)$ such that (68) holds and such that the degree of $\sigma(z)$ is as small as possible. There certainly *is* a solution to (68), since it is assumed that $\leq r/2$ errors occurred. Equation (68) is called the *key equation* of the decoding process.

The key equation can be solved by the Euclidean algorithm of the previous section. In fact, Equation (63) implies

$$r_i(z) \equiv (-1)^i U_i(z) r_0(z) \pmod{r_{-1}(z)}. \tag{69}$$

Thus we can use the following algorithm for Stage II.

Algorithm. *Set*

$$r_{-1}(z) = z^r, \quad r_0(z) = S(z),$$

and proceed with the Euclidean algorithm until reaching an $r_k(z)$ such that

$$\deg r_{k-1}(z) \geq \tfrac{1}{2} r \quad \text{and} \quad \deg r_k(z) \leq \tfrac{1}{2} r - 1. \tag{70}$$

Then the error locator and evaluator polynomials are given by

$$\sigma(z) = \delta U_k(z) \tag{71}$$

$$\omega(z) = (-1)^k \delta r_k(z) \tag{72}$$

where δ is a constant chosen to make $\sigma(0) = 1$, *and satisfy*

$$\omega(z) \equiv \sigma(z)S(z) \bmod z^r, \tag{73}$$

$$\deg \sigma(z) \leqslant \tfrac{1}{2}r, \tag{74}$$

$$\deg \omega(z) \leqslant \tfrac{1}{2}r - 1. \tag{75}$$

Proof. (73) follows from (69). Also $\deg \omega = \deg r_k \leqslant \tfrac{1}{2}r - 1$ and $\deg \sigma = \deg U_k = \deg r_{-1} - \deg r_{k-1} \leqslant \tfrac{1}{2}r$ from (70). Q.E.D.

That these are the correct values of $\sigma(z)$ and $\omega(z)$ follows from

Theorem 16. *The polynomials* $\sigma(z)$ *and* $\omega(z)$ *given by* (71), (72) *are the unique solution to* (73) *with* $\sigma(0) = 1$, $\deg \sigma \leqslant \tfrac{1}{2}r$, $\deg \omega \leqslant \tfrac{1}{2}r - 1$, *and* $\deg \sigma$ *as small as possible.*

Proof. (1) We first show that if there are two solutions to (73), say σ, ω and σ', ω', with $\deg \sigma \leqslant \tfrac{1}{2}r$, $\deg \omega \leqslant \tfrac{1}{2}r - 1$, $\deg \sigma' \leqslant \tfrac{1}{2}r$, $\deg \omega' \leqslant \tfrac{1}{2}r - 1$, then

$$\sigma = \mu \sigma' \quad \text{and} \quad \omega = \mu \omega'$$

for some polynomial μ. In fact,

$$\omega \equiv \sigma S \bmod z^r, \qquad \omega' = \sigma' S \bmod z^r,$$

$$\omega \sigma' \equiv \omega' \sigma \bmod z^r.$$

But the degree of each side of this congruence is less than r. Therefore $\omega \sigma' = \omega' \sigma$, and

$$\frac{\omega}{\omega'} = \frac{\sigma}{\sigma'} = \mu \quad \text{(say)}.$$

(2) If the solution σ, ω given by (71), (72) is not that for which $\deg \sigma$ is smallest, then from 1,

$$\sigma = \mu \sigma' \quad \text{and} \quad \omega = \mu \omega'$$

for some μ, where σ', ω' is also a solution. Then from (63) and (72),

$$\omega(z) = (-1)^k \delta r_k(z)$$

$$= -\delta V_k(z)z^r + \sigma(z)S(z),$$

$$\mu(z)\omega'(z) = -\delta V_k(z)z^r + \mu(z)\sigma'(z)S(z). \tag{76}$$

Also from (73), for some $\psi(z)$,

$$\omega'(z) = \sigma'(z)S(z) + \psi(z)z'. \tag{77}$$

(76) and (77) imply $\mu(z) \,|\, V_k(z)$. But $\mu(z)$ also divides $\sigma(z) = \delta U_k(z)$. Since $U_k(z)$ and $V_k(z)$ are relatively prime (Problem 22), $\mu(z)$ is a constant.   Q.E.D.

Stage III is as for BCH codes: the error locators X_i are found as the reciprocals of the roots of $\sigma(z)$, and then the error values are given by

$$Y_\mu = \frac{\omega(X_\mu^{-1})}{y_{i_\mu} \prod_{\nu \neq \mu} (1 - X_\nu X_\mu^{-1})}. \tag{78}$$

Notes on Chapter 12

§2. Alternant codes were defined and studied by Helgert in a series of papers [631–635]. Many important results are given by Delsarte [359]. The name *alternant code* is based on the fact that a matrix or determinant of the form

$$\begin{bmatrix} f_0(x_0) & f_1(x_0) & \cdots & f_{n-1}(x_0) \\ \cdots & & \cdots & \cdots \\ f_0(x_{r-1}) & f_1(x_{r-1}) & \cdots & f_{n-1}(x_{r-1}) \end{bmatrix}$$

is called an alternant – see for example Muir [974, pp. 341, 346].

Other generalizations of Goppa codes have been given by Mandelbaum [903] and Tzeng and Zimmerman [1356].

§3. Goppa introduced his codes in [536] and [537]. See also the short survey article [125] by Berlekamp. §§3,4 are partly based on these three papers. Further properties are given by Goppa [538–539] and Sugiyama et al. [1291]; these papers describe a number of other good codes that can be constructed using Goppa codes as building blocks. For the reference to Gabidulin codes see Goppa [537].

§5. Theorem 11 is due to Berlekamp and Moreno [130], and Theorem 12 to Tzeng and Zimmerman [1355]. Other classes of extended Goppa codes which are cyclic have been given in [1355] and in Tzeng and Yu [1353]. However, Research Problem 12.3 remains unsolved. For $PGL_2(q)$ and related groups see for example Huppert [676, p. 177].

§6. The first published reference to Srivastava codes is Berlekamp [113, §15.1]. These codes and generalized Srivastava codes have been studied by Helgert [631–635] and this section is based on his work. The actual dimension and minimum distance of a number of Srivastava codes are given in [631] and [632]. Sugiyama et al. [1289] have found some good codes by modifying Srivastava codes.

§7. Generalized BCH codes are described by Chien and Choy in [286], and many further properties of these codes are given in that paper.

§8. For the Euclidean algorithm see for example Niven and Zuckerman [995], Stark [1272], and Uspensky and Heaslet [1359].

§9. The idea of expressing the key step in the decoding as finding the shortest recurrence that will produce a given sequence is due to Berlekamp [113, Ch. 7] and Massey [992]. Helgert [635] gave the key equation (67) for decoding alternant codes. See also Retter [1109, 1110] for Goppa codes. The use of the Euclidean algorithm for solving the key equation seems to be due to Sugiyama et al. [1288] (see also [1290]), and our treatment follows that paper. Mills [960] has also noticed the connection between decoding and the continued fraction algorithm. A different decoding algorithm, also based on the Euclidean algorithm, has been given by Mandelbaum [903] (see also [902]). Yet another algorithm has been proposed by Patterson [1030]. See also Sain [1141].

Which decoding algorithm is best? Aho, Hopcroft and Ullman [16, §8.9] describe an algorithm which computes the GCD of two polynomials of degree n in $O(n \log^2 n)$ steps. By using this algorithm instead of the Euclidean algorithm, Sarwate [1145] shows that a t-error-correcting Goppa code of length n, for fixed t/n, can be decoded in $O(n \log^2 n)$ arithmetic operations. (A similar result for RS codes has been obtained by Justesen [706].)

Similarly, using Theorem 29 of Ch. 9, a primitive binary BCH code of length n and rate R can be decoded up to its designed distance in $O(n \log n)$ arithmetic operations. These results are better than those obtained with the Euclidean algorithm, but unfortunately only for excessively large values of n. For practical purposes the original version of the Berlekamp algorithm (Berlekamp [113, Ch. 7]) is probably the fastest, although this depends on the machinery available for the decoding. Nevertheless, decoding using the Euclidean algorithm is by far the simplest to understand, and is certainly at least comparable in speed with the other methods (for $n < 10^6$) and so it is the method we prefer.

Part II

13

Reed–Muller codes

§1. Introduction

Reed–Muller (or RM) codes are one of the oldest and best understood families of codes. However, except for first-order RM codes and codes of modest block lengths, their minimum distance is lower than that of BCH codes. But the great merit of RM codes is that they are relatively easy to decode, using majority-logic circuits (see §§6 and 7).

In fact RM codes are the simplest examples of the class of geometrical codes, which also includes Euclidean geometry and projective geometry codes, all of which can be decoded by majority logic. A brief account of these geometrical codes is given in §8, but regretfully space does not permit a more detailed treatment. In compensation we give a fairly complete bibliography.

§§2, 3 and 4 give the basic properties of RM codes, and Figs. 13.3 and 13.4 give a summary of the properties. Sections 9, 10 and 11 consider the automorphism groups and Mattson–Solomon polynomials of these codes.

The next chapter will discuss 1st order RM codes and their applications, while Chapter 15 studies 2nd order RM codes, and also the general problem of finding weight enumerators of RM codes.

§2. Boolean functions

Reed–Muller codes can be defined very simply in terms of Boolean functions. We are going to define codes of length $n = 2^m$, and to do so we shall need m variables $v_1, \ldots, v_m$ which take the values 0 or 1. Alternatively, let $v = (v_1, \ldots, v_m)$ range over V^m, the set of all binary m-tuples. [In earlier chapters this set has been called F^m, but in Chs. 13 and 14 it will be more convenient to call it V^m.] Any function $f(v) = f(v_1, \ldots, v_m)$ which takes on the values 0 and 1 is called a *Boolean function* (abbreviated B.f.). Such a function can be specified by a *truth table*, which gives the value of f at all of its 2^m arguments. For example when $m = 3$ one such Boolean function is specified by the truth table:

$$v_3 = 0\ 0\ 0\ 0\ 1\ 1\ 1\ 1$$
$$v_2 = 0\ 0\ 1\ 1\ 0\ 0\ 1\ 1$$
$$v_1 = 0\ 1\ 0\ 1\ 0\ 1\ 0\ 1$$
$$f = 0\ 0\ 0\ 1\ 1\ 0\ 0\ 0$$

The last row of the table gives the values taken by f, and is a binary vector of length $n = 2^m$ which is denoted by f. A code will consist of all vectors f, where the function f belongs to a certain class.

The columns of the truth table are for the moment assumed to have the natural ordering illustrated above.

The last row of the truth table can be filled in arbitrarily, so there are 2^{2^m} Boolean functions of m variables.

The usual logical operations may be applied to Boolean functions:

$$f \text{ EXCLUSIVE OR } g = f + g,$$
$$f \text{ AND } g = fg,$$
$$f \text{ OR } g = f + g + fg,$$
$$\text{NOT } f = \bar{f} = 1 + f.$$

(1)

The right-hand side of these equations defines the operations in terms of binary functions.

A Boolean function can be written down immediately from its truth table: in the preceding example,

$$f = v_1 v_2 \bar{v}_3 \text{ OR } \bar{v}_1 \bar{v}_2 v_3,$$

since the right-hand side is equal to 1 exactly when f is. This is called the *disjunctive normal form* for f (see Problem 1).

Using the rules (1) this simplifies to (check!)

$$f = v_3 + v_1 v_2 + v_1 v_3 + v_2 v_3.$$

Notice that $v_i^2 = v_i$. It is clear that in this way any Boolean function can be expressed as a sum of the 2^m functions

$$1, v_1, v_2, \ldots, v_m, v_1 v_2, v_1 v_3, \ldots, v_{m-1} v_m, \ldots, v_1 v_2 \cdots v_m,$$

(2)

with coefficients which are 0 or 1. Since there are 2^{2^m} Boolean functions altogether, all these sums must be distinct.

In other words the 2^m vectors corresponding to the functions (2) are linearly independent.

Problem. (1) (Decomposition of Boolean functions.) If $f(v_1, \ldots, v_m)$ is a B.f., show the following.

(a) $f(v_1, \ldots, v_m) = v_m f(v_1, \ldots, v_{m-1}, 1)$ OR $\bar{v}_m f(v_1, \ldots, v_{m-1}, 0)$, and

(b) $f(v_1, \ldots, v_m) = v_m g(v_1, \ldots, v_{m-1}) + h(v_1, \ldots, v_{m-1})$ where g, h are B.f.'s.

(c) Disjunctive normal form:

$$f(v_1, \ldots, v_m) = \sum_{i_1=0}^{1} \cdots \sum_{i_m=0}^{1} f(i_1, \ldots, i_m) w_1^{i_1} \cdots w_m^{i_m},$$

where

$$w_r^1 = v_r, \qquad w_r^0 = \bar{v}_r.$$

Theorem 1. *Any Boolean function f can be expanded in powers of v_i as*

$$f(v_1, \ldots, v_m) = \sum_{a \in V^m} g(a) v_1^{a_1} \cdots v_m^{a_m}, \tag{3}$$

where the coefficients are given by

$$g(a) = \sum_{b \subset a} f(b_1, \ldots, b_m), \tag{4}$$

and $b \subset a$ means that the 1's in b are a subset of the 1's in a.

Proof. For $m = 1$, the disjunctive normal form for f is

$$f(v_1) = f(0)(1 + v_1) + f(1)v_1$$
$$= f(0)1 + (f(0) + f(1))v_1,$$

which proves (3) and (4). Similarly for $m = 2$ we have

$$f(v_1, v_2) = f(0, 0)(1 + v_1)(1 + v_2) + f(0, 1)(1 + v_1)v_2 + f(1, 0)v_1(1 + v_2)$$
$$+ f(1, 1)v_1 v_2 = f(0, 0)1 + \{f(0, 0) + f(1, 0)\}v_1$$
$$+ \{f(0, 0) + f(0, 1)\}v_2 + \{f(0, 0) + f(0, 1) + f(1, 0) + f(1, 1)\}v_1 v_2,$$

which again proves (3) and (4). Clearly we can continue in this way.   Q.E.D.

Problem. (2) (The randomization lemma.) Let $f(v_1, \ldots, v_{m-1})$ be an arbitrary Boolean function. Show that $v_m + f(v_1, \ldots, v_{m-1})$ takes the values 0 and 1 equally often.

§3. Reed–Muller codes

As in §2, $v = (v_1, \ldots, v_m)$ denotes a vector which ranges over V^m, and f is the vector of length 2^m obtained from a Boolean function $f(v_1, \ldots, v_m)$.

Definition. The rth *order binary Reed–Muller* (or RM) code $\mathcal{R}(r, m)$ of length $n = 2^m$, for $0 \leq r \leq m$, is the set of all vectors f, where $f(v_1, \ldots, v_m)$ is a Boolean function which is a polynomial of degree at most r.

For example, the first order RM code of length 8 consists of the 16 codewords

$$a_0 \mathbf{1} + a_1 v_1 + a_2 v_2 + a_3 v_3, \quad a_i = 0 \text{ or } 1,$$

which are shown in Fig. 13.1.

0	0 0 0 0 0 0 0 0
v_3	0 0 0 0 1 1 1 1
v_2	0 0 1 1 0 0 1 1
v_1	0 1 0 1 0 1 0 1
$v_2 + v_3$	0 0 1 1 1 1 0 0
$v_1 + v_3$	0 1 0 1 1 0 1 0
$v_1 + v_2$	0 1 1 0 0 1 1 0
$v_1 + v_2 + v_3$	0 1 1 0 1 0 0 1
$\mathbf{1}$	1 1 1 1 1 1 1 1
$\mathbf{1} + v_3$	1 1 1 1 0 0 0 0
$\mathbf{1} + v_2$	1 1 0 0 1 1 0 0
$\mathbf{1} + v_1$	1 0 1 0 1 0 1 0
$\mathbf{1} + v_2 + v_3$	1 1 0 0 0 0 1 1
$\mathbf{1} + v_1 + v_3$	1 0 1 0 0 1 0 1
$\mathbf{1} + v_1 + v_2$	1 0 0 1 1 0 0 1
$\mathbf{1} + v_1 + v_2 + v_3$	1 0 0 1 0 1 1 0

Fig. 13.1. The 1ˢᵗ order Reed–Muller code of length 8.

This code is also shown in Fig. 1.14, and indeed we shall see that $\mathcal{R}(1, m)$ is always the dual of an extended Hamming code. $\mathcal{R}(1, m)$ is also the code $\mathscr{C}_{2^m}$ obtained from a Sylvester-type Hadamard matrix in §3 of Ch. 2.

All the codewords of $\mathcal{R}(1, m)$ except $\mathbf{0}$ and $\mathbf{1}$ have weight 2^{m-1}. Indeed any B.f. of degree exactly 1 corresponds to a vector of weight 2^{m-1}, by Problem 2.

In general the rth order RM code consists of all linear combinations of the vectors corresponding to the products

$$1, v_1, \ldots, v_m, v_1 v_2, v_1 v_3, \ldots, v_{m-1} v_m, \ldots \text{ (up to degree } r),$$

which therefore form a basis for the code. There are

$$k = 1 + \binom{m}{1} + \binom{m}{2} + \cdots + \binom{m}{r}$$

such basis vectors vectors, and as we saw in §2 they are linearly independent. So k is the dimension of the code.

For example when $m = 4$ the 16 possible basis vectors for Reed–Muller codes of length 16 are shown in Fig. 13.2 (check!).

1	1 1 1 1 1 1 1 1 1 1 1 1 1 1 1 1
v_4	0 0 0 0 0 0 0 0 1 1 1 1 1 1 1 1
v_3	0 0 0 0 1 1 1 1 0 0 0 0 1 1 1 1
v_2	0 0 1 1 0 0 1 1 0 0 1 1 0 0 1 1
v_1	0 1 0 1 0 1 0 1 0 1 0 1 0 1 0 1
$v_3 v_4$	0 0 0 0 0 0 0 0 0 0 0 0 1 1 1 1
$v_2 v_4$	0 0 0 0 0 0 0 0 0 0 1 1 0 0 1 1
$v_1 v_4$	0 0 0 0 0 0 0 0 0 1 0 1 0 1 0 1
$v_2 v_3$	0 0 0 0 0 0 1 1 0 0 0 0 0 0 1 1
$v_1 v_3$	0 0 0 0 0 1 0 1 0 0 0 0 0 1 0 1
$v_1 v_2$	0 0 0 1 0 0 0 1 0 0 0 1 0 0 0 1
$v_2 v_3 v_4$	0 0 0 0 0 0 0 0 0 0 0 0 0 0 1 1
$v_1 v_3 v_4$	0 0 0 0 0 0 0 0 0 0 0 0 0 1 0 1
$v_1 v_2 v_4$	0 0 0 0 0 0 0 0 0 0 0 1 0 0 0 1
$v_1 v_2 v_3$	0 0 0 0 0 0 0 1 0 0 0 0 0 0 0 1
$v_1 v_2 v_3 v_4$	0 0 0 0 0 0 0 0 0 0 0 0 0 0 0 1

Fig. 13.2. Basis vectors for RM codes of length 16.

The basis vectors for the rth order RM code of length 16, $\mathcal{R}(r, 4)$, are:

Order r	Rows of Fig. 13.2
0	1
1	1–5
2	1–11
3	1–15
4	all

Reed–Muller codes of length 2^{m+1} may be very simply obtained from RM codes of length 2^m using the $|u \mid u + v|$ construction of §9 of Ch. 2.

Theorem 2.

$$\mathcal{R}(r + 1, m + 1) = \{|u \mid u + v|: u \in \mathcal{R}(r + 1, m), v \in \mathcal{R}(r, m)\}.$$

For example $\mathcal{R}(1, 4)$ was constructed in this way in §9 of Ch. 2.

There is an equivalent statement in terms of generator matrices. Let $G(r, m)$ be a generator matrix for $\mathcal{R}(r, m)$. Then the theorem says

$$G(r + 1, m + 1) = \begin{pmatrix} G(r + 1, m) & G(r + 1, m) \\ 0 & G(r, m) \end{pmatrix}.$$

(Indeed, a codeword generated by this matrix has the form $|u \,|\, u + v|$ where $u \in \mathscr{R}(r+1, m)$, $v \in \mathscr{R}(r, m)$.)

Proof. By definition a typical codeword f in $\mathscr{R}(r+1, m+1)$ comes from a polynomial $f(v_1, \ldots, v_{m+1})$ of degree at most $r+1$. We can write (as in Problem 1)

$$f(v_1, \ldots, v_{m+1}) = g(v_1, \ldots, v_m) + v_{m+1} h(v_1, \ldots, v_m),$$

where $\deg(g) \leqslant r+1$ and $\deg(h) \leqslant r$. Let g and h be the vectors (of length 2^m) corresponding to $g(v_1, \ldots, v_m)$ and $h(v_1, \ldots, v_m)$. Of course $g \in \mathscr{R}(r+1, m)$ and $h \in \mathscr{R}(r, m)$. But now consider $g(v_1, \ldots, v_m)$ and $v_{m+1} h(v_1, \ldots, v_m)$ as polynomials in $v_1, \ldots, v_{m+1}$. The corresponding vectors (now of length 2^{m+1}) are $|g \,|\, g|$ and $|0 \,|\, h|$ (see Problem 7). Therefore $f = |g \,|\, g| + |0 \,|\, h|$.   Q.E.D.

Notice the resemblance between the recurrence for RM codes in Theorem 2 and the recurrence for binomial coefficients

$$\binom{m+1}{r+1} = \binom{m}{r+1} + \binom{m}{r}. \tag{5}$$

(See Problem 4.)

The minimum distance of Reed–Muller codes is easily obtained from Theorem 2.

Theorem 3. $\mathscr{R}(r, m)$ *has minimum distance* 2^{m-r}.

Proof. By induction on m, using Theorem 33 of Ch. 2.   Q.E.D.

Figure 13.3 shows the dimensions of the first few $[n, k, d]$ RM codes (check!)

Notice that $\mathscr{R}(m, m)$ contains all vectors of length 2^m; $\mathscr{R}(m-1, m)$ consists of all even weight vectors, and $\mathscr{R}(0, m)$ consists of the vectors 0, 1 (Problem 5).

Theorem 4. $\mathscr{R}(m-r-1, m)$ *is the dual code to* $\mathscr{R}(r, m)$, *for* $0 \leqslant r \leqslant m-1$.

Proof. Take $a \in \mathscr{R}(m-r-1, m)$, $b \in \mathscr{R}(r, m)$. Then $a(v_1, \ldots, v_m)$ is a polynomial of degree $\leqslant m-r-1$, $b(v_1, \ldots, v_m)$ has degree $\leqslant r$, and their product ab has degree $\leqslant m-1$. Therefore $ab \in \mathscr{R}(m-1, m)$ and has even weight. Therefore the dot product $a \cdot b \equiv 0$ (modulo 2). So $\mathscr{R}(m-r-1, m) \subset$

length n	4	8	16	32	64	128	256	512
m	2	3	4	5	6	7	8	9

distance d				dimension k				
1	4	8	16	32	64	128	256	512
2	3	7	15	31	63	127	255	511
4	1	4	11	26	57	120	247	502
8		1	5	16	42	99	219	466
16			1	6	22	64	163	382
32				1	7	29	93	256
64					1	8	37	130
128						1	9	46
256							1	10
512								1

Fig. 13.3. Reed–Muller codes.

$\mathcal{R}(r, m)^{\perp}$. But

$$\dim \mathcal{R}(m - r - 1, m) + \dim \mathcal{R}(r, m)$$

$$= 1 + \binom{m}{1} + \cdots + \binom{m}{m - r - 1} + 1 + \binom{m}{1} + \cdots + \binom{m}{r} = 2^{m},$$

which implies $\mathcal{R}(m - r - 1, m) = \mathcal{R}(r, m)^{\perp}$.   Q.E.D.

These properties are collected in Fig. 13.4.

For any m and any r, $0 \le r \le m$, there is a binary r^{th} order RM code $\mathcal{R}(r, m)$ with the following properties:

$$\text{length } n = 2^{m},$$

$$\text{dimension } k = 1 + \binom{m}{1} + \cdots + \binom{m}{r},$$

$$\text{minimum distance } d = 2^{m-r}.$$

$\mathcal{R}(r, m)$ consists of (the vectors corresponding to) all polynomials in the binary variables $v_1, \ldots, v_m$ of degree $\le r$. The dual of $\mathcal{R}(r, m)$ is $\mathcal{R}(m - r - 1, m)$ (Theorem 4). $\mathcal{R}(r, m)$ is an extended cyclic code (Theorem 11), and can be easily encoded and decoded (by majority logic, §§6, 7). Decoding is especially easy for first-order RM codes (Ch. 14). Aut $(\mathcal{R}(r, m)) = \text{GA}(m)$ (Theorem 24). Good practical codes, also the basis for constructing many other codes (Ch. 15).

Fig. 13.4. Summary of Reed–Muller codes.

Punctured Reed–Muller codes.

Definition. For $0 \leq r \leq m - 1$, the *punctured RM code* $\mathcal{R}(r, m)^*$ is obtained by puncturing (or deleting) the coordinate corresponding to $v_1 = \cdots = v_m = 0$ from all the codewords of $\mathcal{R}(r, m)$.

(In fact we shall see in the next section that an equivalent code is obtained no matter which coordinate is punctured.)

Clearly $\mathcal{R}(r, m)^*$ has length $2^m - 1$, minimum distance $2^{m-r} - 1$, and dimension $1 + \binom{m}{1} + \cdots + \binom{m}{r}$.

Problems. (3) Show that $\mathcal{R}(r, m)$ is a subcode of $\mathcal{R}(r + 1, m)$. In fact show that $\mathcal{R}(r + 1, m) = \{a + b \colon a \in \mathcal{R}(r, m), b$ is zero or a polynomial in $v_1, \ldots, v_m$ of degree exactly $r + 1\}$.

(4) If Theorem 2 is used as a recursive definition of RM codes, use Equation (5) to calculate their dimension. Obtain Fig. 13.3 from Pascal's triangle for binomial coefficients.

(5) Show that $\mathcal{R}(0, m)$ and $\mathcal{R}(0, m)^*$ are repetition codes, $\mathcal{R}(m - 1, m)$ contains all vectors of even weight, and $\mathcal{R}(m, m)$ and $\mathcal{R}(m - 1, m)^*$ contain all vectors, of the appropriate lengths.

(6) Let $|S \mid T| = \{|s \mid t| \colon s \in S, t \in T\}$. Show that

$$\mathcal{R}(r + 1, m + 1) = \cup |S \mid S|$$

where S runs through those cosets of $\mathcal{R}(r, m)$ which are contained in $\mathcal{R}(r + 1, m)$.

(7) Let $f(v_1, \ldots, v_m)$ be a B.f. of m variables, and let f be the corresponding binary vector of length 2^m. Show that the vectors of length 2^{m+1} corresponding to $g(v_1, \ldots, v_{m+1}) = f(v_1, \ldots, v_m)$ and to $h(v_1, \ldots, v_{m+1}) = v_{m+1}f(v_1, \ldots, v_m)$ are $|f \mid f|$ and $|0 \mid f|$ respectively.

§4. RM codes and geometries

Many properties of RM codes are best stated in the language of finite geometries. The *Euclidean geometry* EG$(m, 2)$ of dimension m over GF(2) contains 2^m points, whose coordinates are all the binary vectors $v = (v_1, \ldots, v_m)$ of length m. If the zero point is deleted, the *projective geometry* PG$(m - 1, 2)$ is obtained. (See Appendix B for further details.)

Any subset S of the points of EG$(m, 2)$ has associated with it a binary incidence vector $\chi(S)$ of length 2^m, containing a 1 in those components $s \in S$, and zeros elsewhere.

This gives us another way of thinking about codewords of $\mathcal{R}(r, m)$, namely as (incidence vectors of) subsets of $EG(m, 2)$.

For example, the Euclidean geometry $EG(3, 2)$ consists of 8 points $P_0, P_1, \ldots, P_7$ whose coordinates we may take to be the following column vectors:

	P_0	P_1	P_2	P_3	P_4	P_5	P_6	P_7
$\bar{v}_3$	1	1	1	1	0	0	0	0
$\bar{v}_2$	1	1	0	0	1	1	0	0
$\bar{v}_1$	1	0	1	0	1	0	1	0

The subset $S = \{P_2, P_3, P_4, P_5\}$ has incidence vector

$$\chi(S) = 00111100.$$

This is a codeword of $\mathcal{R}(1, 3)$ – see Fig. 13.1.

For any value of m let us write the complements of the vectors $v_m, \ldots, v_1$, as follows:

$$
\begin{array}{ll}
\bar{v}_m & 1\ 1\ \cdots\ 0\ 0\ 0\ 0 \\
\bar{v}_{m-1} & 1\ 1\ \cdots\ 0\ 0\ 0\ 0 \\
\cdots\cdots\cdots\cdots\cdots\cdots \\
\bar{v}_2 & 1\ 1\ \cdots\ 1\ 1\ 0\ 0 \\
\bar{v}_1 & 1\ 0\ \cdots\ 1\ 0\ 1\ 0.
\end{array}
$$

We take the columns of this array to be the coordinates of the points in $EG(m, 2)$. In this way there is a 1-to-1 correspondence between the points of $EG(m, 2)$ and the components (or coordinate positions) of binary vectors of length 2^m. Any vector x of length 2^m describes a subset of $EG(m, 2)$, consisting of those points P for which $x_P = 1$. Clearly x is the incidence vector of this subset. The number of points in the subset is equal to the weight of x.

For example the vectors v_i themselves are the characteristic vectors of hyperplanes which pass through the origin, the $v_i v_j$ describe subspaces of dimension $m - 2$, and so on. (Of course there are other hyperplanes through the origin besides the v_i. For example no v_i contains the point $11 \cdots 1$. Similarly the $v_i v_j$ are not the only subspaces of dimension $m - 2$, and so on.)

One of the advantages of this geometrical language is that it enables us to say exactly what the codewords of minimum weight are in the rth order Reed–Muller code of length 2^m: they are the $(m - r)$-flats in $EG(m, 2)$. This is proved in Theorems 5 and 7. In Theorem 9 we use this fact to determine the number of codewords of minimum weight. Then in §5 we show that the codewords of minimum weight generate $\mathcal{R}(r, m)$. Along the way we prove that the punctured code $\mathcal{R}(r, m)^*$ is cyclic. The proofs given in §§4, 5 should be omitted on a first reading.

Let H be any hyperplane in $EG(m, 2)$. By definition the incidence vector

$h = \chi(H)$ consists of all points v which satisfy a linear equation in $v_1, \ldots, v_m$. In other words, the Boolean function h is a linear function of $v_1, \ldots, v_m$, and so is a codeword of weight 2^{m-1} in $\mathcal{R}(r, m)$.

We remark that if $f \in \mathcal{R}(r, m)$ is the incidence vector of a set S, then $hf \in \mathcal{R}(r+1, m)$ and is the incidence vector of $S \cap H$. We are now ready for the first main theorem.

Theorem 5. *Let f be a minimum weight codeword of $\mathcal{R}(r, m)$, say $f = \chi(S)$. Then S is an $(m - r)$-dimensional flat in $\mathrm{EG}(m, 2)$ (which need not pass through the origin).*

E.g. the 14 codewords of weight 4 in the $[8, 4, 4]$ extended Hamming code are the 14 planes of Euclidean 3-space $\mathrm{EG}(3, 2)$.

Proof. Let H be any hyperplane $\mathrm{EG}(m - 1, 2)$ in $\mathrm{EG}(m, 2)$ and let H' be the parallel hyperplane, so that $\mathrm{EG}(m, 2) = H \cup H'$.

By the above remark $S \cap H$ and $S \cap H'$ are in $\mathcal{R}(r+1, m)$, and so contain 0 or $\geq 2^{m-r-1}$ points. Since $|S| = 2^{m-r} = |S \cap H| + |S \cap H'|$, $|S \cap H| = 0$, 2^{m-r-1} or 2^{m-r}. The following Lemma then completes the proof of the theorem.

Lemma 6. (Rothschild and Van Lint.) *Let S be a subset of $\mathrm{EG}(m, 2)$ such that $|S| = 2^{m-r}$, and $|S \cap H| = 0$, 2^{m-r-1} or 2^{m-r} for all hyperplanes H in $\mathrm{EG}(m, 2)$. Then S is an $(m - r)$-dimensional flat in $\mathrm{EG}(m, 2)$.*

Proof. By induction on m. The result is trivial for $m = 2$.

Case (i). Suppose for some H, $|S \cap H| = 2^{m-r}$. Then $S \subset H$, i.e. $S \subset \mathrm{EG}(m - 1, 2)$. Let X be any hyperplane in H. There exists another hyperplane H'' of $\mathrm{EG}(m, 2)$ such that $X = H \cap H''$, and $S \cap X = S \cap H''$, i.e. $|S \cap X| = 0$, $2^{m-1-(r-1)-1}$ or $2^{(m-1)-(r-1)}$. By the induction hypothesis S is an $((m - 1) - (r - 1))$-flat in $\mathrm{EG}(m - 1, 2)$ and hence in $\mathrm{EG}(m, 2)$.

Case (ii). If for some H, $|S \cap H| = 0$, then replacing H by its parallel hyperplane reduces this to case (i).

Case (iii). It remains to consider the case when $|S \cap H| = 2^{m-r-1}$ for all H. Consider

$$\sum_{H \subset \mathrm{EG}(m, 2)} |S \cap H|^2 = \sum_{H \subset \mathrm{EG}(m, 2)} \left(\sum_{a \in S} \chi_H(a) \right)^2$$

$$= \sum_{a \in S} \sum_{b \in S} \sum_{H \subset \mathrm{EG}(m, 2)} \chi_H(a) \chi_H(b)$$

$$= |S|(2^m - 1) + |S|(|S| - 1)(2^{m-1} - 1)$$

since there are $2^m - 1$ hyperplanes in $EG(m, 2)$ through a point and $2^{m-1} - 1$ through a line. The LHS is $2^{2m-2r-1}(2^m - 1)$. Substituting $|S| = 2^{m-r}$ on the RHS leads to a contradiction. Q.E.D.

The converse of Theorem 5 is:

Theorem 7. *The incidence vector of any $(m - r)$-flat in $EG(m, 2)$ is in $\mathcal{R}(r, m)$.*

Proof. Any $(m - r)$-flat in $EG(m, 2)$ consists of all points v which satisfy r linear equations over $GF(2)$, say

$$\sum_{j=1}^{m} a_{ij}v_j = b_i, \quad i = 1, \dots, r,$$

or equivalently

$$\sum_{j=1}^{m} a_{ij}v_j + b_i + 1 = 1, \quad i = 1, \dots, r.$$

This can be replaced by the single equation

$$\prod_{i=1}^{r} \left(\sum_{j=1}^{m} a_{ij}v_j + b_i + 1 \right) = 1,$$

i.e., by a polynomial equation of degree $\leq r$ in $v_1, \dots, v_m$. Therefore the flat is in $\mathcal{R}(r, m)$. Q.E.D.

Combining Theorems 5 and 7 we obtain

Theorem 8. *The codewords of minimum weight in $\mathcal{R}(r, m)$ are exactly the incidence vectors of the $(m - r)$-dimensional flats in $EG(m, 2)$.*

Minimum weight codewords in $\mathcal{R}(r, m)^*$ are obtained from minimum weight codewords in $\mathcal{R}(r, m)$ which have a 1 in coordinate 0 (by deleting that 1). Such a codeword is the incidence vector of a subspace $PG(m - r - 1, 2)$ in $PG(m - 1, 2)$. (We remind the reader that in a projective geometry there is no distinction between flats and subspaces – see Appendix B.)

Theorem 9. *The number of codewords of minimum weight in:*
(a) $\mathcal{R}(r, m)^*$ is

$$A_{2^{m-r}-1} = \prod_{i=0}^{m-r-1} \frac{2^{m-i} - 1}{2^{m-r-i} - 1}.$$

(b) $\mathcal{R}(r, m)$ is

$$A_{2^{m-r}} = 2^r \prod_{i=0}^{m-r-1} \frac{2^{m-i}-1}{2^{m-r-i}-1}.$$

Proof. Theorems 3 and 5 of Appendix B.

*§5. The minimum weight vectors generate the code

Theorem 10. *The incidence vectors of the projective subspaces* $PG(\mu-1,2)$ *of* $PG(m-1,2)$ *generate* $\mathcal{R}(r,m)^*$, *where* $\mu = m - r$.

Proof. (The beginner should skip this.) Let α be a primitive element of $GF(2^m)$. Then the points of $PG(m-1,2)$ can be taken to be $\{1, \alpha, \alpha^2, \alpha^3, \ldots, \alpha^{2^m-2}\}$. Let $l = 2^\mu - 2$.

A subset $T = \{\alpha^{d_0}, \ldots, \alpha^{d_u}\}$ of these points will be represented in the usual way by the polynomial

$$w_T(x) = x^{d_0} + \cdots + x^{d_u}.$$

If $T = \{\alpha^{d_0}, \ldots, \alpha^{d_l}\}$ is a $PG(\mu-1,2)$ then the points of T are all nonzero linear combinations over $GF(2)$ of μ linearly independent points $\alpha_0, \ldots, \alpha_{\mu-1}$ (say) of $GF(2^m)$. In other words the points of T are

$$\sum_{j=0}^{\mu-1} a_{ij}\alpha_j = \alpha^{d_i}, \quad i = 0, 1, \ldots, l,$$

where $(a_{i0}, a_{i1}, \ldots, a_{i\mu-1})$ runs through all nonzero binary μ-tuples. Also $xw_T(x)$ represents the $PG(\mu-1,2)$ spanned by $\alpha\alpha_0, \ldots, \alpha\alpha_{\mu-1}$. Thus every cyclic shift of the incidence vector of a $PG(\mu-1,2)$ is the incidence vector of another $PG(\mu-1,2)$.

Let $\mathscr{C}$ be the code generated by all $w_T(x)$, where T is any $PG(\mu-1,2)$. Clearly $\mathscr{C}$ is a cyclic code and is contained in $\mathcal{R}(r,m)^*$; the theorem asserts that in fact $\mathscr{C} = \mathcal{R}(r,m)^*$. We establish this by showing that

$$\dim \mathscr{C} \geq 1 + \binom{m}{1} + \cdots + \binom{m}{r}.$$

The dimension of $\mathscr{C}$ is the number of α^s which are not zeros of $\mathscr{C}$; i.e. the number of α^s such that $w_T(\alpha^s) \neq 0$ for some T. Now

$$w_T(\alpha^s) = \sum_{i=0}^{l} \alpha^{sd_i} = \sum_{i=0}^{l} \left(\sum_{j=0}^{\mu-1} a_{ij}\alpha_j \right)^s$$

$$= \sum_b (b_0\alpha_0 + \cdots + b_{\mu-1}\alpha_{\mu-1})^s$$

where the summation extends over all nonzero binary μ-tuples $b = (b_0, \ldots, b_{\mu-1})$. Call this last expression $F_s(\alpha_0, \ldots, \alpha_{\mu-1})$. Then

$$F_s(\alpha_0, \ldots, \alpha_{\mu-1}) = \sum_{b_0 \cdots b_{\mu-1}} (b_0\alpha_0 + \gamma)^s,$$

where $\gamma = b_1\alpha_1 + \cdots + b_{\mu-1}\alpha_{\mu-1}$,

$$= \sum_{b_1 \cdots b_{\mu-1}} (\gamma^s + (\alpha_0 + \gamma)^s)$$

$$= \sum_{b_1 \cdots b_{\mu-1}} \left(\alpha_0^s + \sum_{j=1}^{s-1} \binom{s}{j} \alpha_0^j \gamma^{s-j} \right)$$

$$= \sum_{j=1}^{s-1} \binom{s}{j} \alpha_0^j F_{s-j}(\alpha_1, \ldots, \alpha_{\mu-1})$$

$$= \cdots$$

$$= \sum_{\substack{\sum j_i = s \\ j_i \geq 1}} \frac{s!}{j_0! \cdots j_{\mu-1}!} \alpha_0^{j_0} \cdots \alpha_{\mu-1}^{j_{\mu-1}}. \qquad (6)$$

This is a homogeneous polynomial of degree s in $\alpha_0, \ldots, \alpha_{\mu-1}$.

Then dim $\mathscr{C}$ is the number of s such that $F_s(\alpha_0, \ldots, \alpha_{\mu-1})$ is not identically zero, when the α_i are linearly independent.

In fact we will just count those $F_s(\alpha_0, \ldots, \alpha_{\mu-1})$ which contain a coefficient which is nonzero modulo 2. We note that such an F

(i) cannot be identically zero, and

(ii) cannot have $\alpha_0, \ldots, \alpha_{\mu-1}$ linearly dependent (Problem 8). From Lucas' theorem, a multinomial coefficient

$$\frac{s!}{j_0! \cdots j_{\mu-1}!}$$

is nonzero modulo 2 iff

$$(j_0)_i + (j_1)_i + \cdots + (j_{\mu-1})_i \leq (s)_i \quad \text{for all } i, \ 0 \leq i \leq m-1,$$

where $(x)_i$ denotes the ith bit in the binary expansion of x.

Therefore (6) contains a nonzero coefficient whenever the binary expansion of s contains $\geq \mu$ 1's. For example if $s = 2^{i_0} + 2^{i_1} + \cdots + 2^{i_{\mu-1}}$, (6) contains a nonzero coefficient corresponding to $j_0 = 2^{i_0}, \ldots, j_{\mu-1} = 2^{i_{\mu-1}}$.

The number of such s in the range $[1, 2^m - 1]$ is

$$\binom{m}{\mu} + \binom{m}{\mu+1} + \cdots + \binom{m}{m} = 1 + \binom{m}{1} + \cdots + \binom{m}{r}. \qquad \text{Q.E.D.}$$

Problem. (8) Show that if $\alpha_0, \ldots, \alpha_{\mu-1}$ are linearly dependent, then $F_s(\alpha_0, \ldots, \alpha_{\mu-1})$ is identically zero modulo 2.

Important Remark. For nonnegative integers s let $w_2(s)$ denote the number of 1's in the binary expansion of s. Then the proof of this theorem has shown that α^s is a nonzero of $\mathcal{R}(r, m)^*$ iff $1 \leq s \leq 2^m - 1$ and $w_2(s) \geq \mu$. Or in other words,

Theorem 11. *The punctured* RM *code* $\mathcal{R}(r, m)^*$ *is a cyclic code, which has as zeros* α^s *for all* s *satisfying*

$$1 \leq w_2(s) \leq m - r - 1 \quad \text{and} \quad 1 \leq s \leq 2^m - 2.$$

The *generator and check polynomials, for the punctured* RM *code* $\mathcal{R}(r, m)^*$ are, for $0 \leq r \leq m - 1$,

$$g(x) = \prod_{\substack{1 \leq w_2(s) \leq m-r-1 \\ 1 \leq s \leq 2^m-2}} M^{(s)}(x) \tag{7}$$

$$h(x) = (x + 1) \prod_{\substack{m-r \leq w_2(s) \leq m-1 \\ 1 \leq s \leq 2^m-2}} M^{(s)}(x) \tag{8}$$

where s runs through representatives of the cyclotomic cosets, and $M^{(s)}(x)$ is the minimal polynomial of α^s. (Remember that an empty product is 1.)

An alternative form is obtained by using α^{-1} instead of α as the primitive element. This has the effect of replacing s by $2^m - 1 - s$ and $w_2(s)$ by $m - w_2(s)$. Then

$$g(x) = \prod_{\substack{r+1 \leq w_2(s) \leq m-1 \\ 1 \leq s \leq 2^m-2}} M^{(s)}(x) \tag{9}$$

$$h(x) = (x + 1) \prod_{\substack{1 \leq w_2(s) \leq r \\ 1 \leq s \leq 2^m-2}} M^{(s)}(x). \tag{10}$$

If the generator polynomial of $\mathcal{R}(r, m)^*$ is given by Equation 7, then by Theorem 4 of Ch. 7, the dual code has generator polynomial (10).

The *idempotent* of $\mathcal{R}(r, m)^*$ is

$$\theta_0 + \sum_{\substack{1 \leq w_2(s) \leq r \\ 1 \leq s \leq 2^m-2}} \theta_s$$

or equivalently (again replacing α by α^{-1}),

$$\theta_0 + \sum_{\substack{m-r \leq w_2(s) \leq m-1 \\ 1 \leq s \leq 2^m-2}} \theta_s$$

where s runs through the representatives of the cyclotomic cosets, and θ_s is defined in Ch. 8.

The dual to $\mathcal{R}(r, m)^*$, by Theorem 5 of Ch. 8, has the idempotent

$$\left(1 + \theta_0 + \sum_{\substack{m-r \leq w_2(s) \leq m-1 \\ 1 \leq s \leq 2^m-2}} \theta_s\right)^* = \sum_{\substack{1 \leq w_2(t) \leq m-r-1 \\ 1 \leq t \leq 2^m-2}} \theta_t^*$$

For example, the idempotents of $\mathcal{R}(1, m)^*$ and $\mathcal{R}(2, m)^*$ may be taken to be

$$\theta_0 + \theta_1^*$$

and

$$\theta_0 + \theta_1^* + \sum_{i=[(m+1)/2]}^{m-1} \theta_{l_i}^*$$

respectively, where $l_i = 1 + 2^i$. Then general forms for codewords in various RM codes of length 2^m are as follows. Here the first part of the codeword is the overall parity check, and the second part is in the cyclic code $\mathcal{R}^*$.

$\mathcal{R}(0, m)$: $|a \mid a\theta_0|$

$\mathcal{R}(1, m)$: $|a \mid a\theta_0 + a_1 x^{i_1}\theta_1^*|$

$\mathcal{R}(2, m)$: $|a \mid a\theta_0 + a_1 x^{i_1}\theta_1^* + \sum_{i=[(m+1)/2]}^{m-1} a_i x^{2^i}\theta_{l_i}^*|$

$\mathcal{R}(m-2, m)$: $|f(1) \mid f(x)(1 + \theta_1)|$

$\mathcal{R}(m-1, m)$: $|f(1) \mid f(x)|$

where $a_1, a_i \in GF(2)$, $i_1, s_i \in \{0, 1, 2, \dots, 2^m - 2\}$, and $f(x)$ is arbitrary.

Nesting habits. Figure 13.5 shows the nesting habits of BCH and punctured RM codes. Here $\mathcal{B}(d)$ denotes the BCH code of designed distance d, and the binary numbers in parentheses are the exponents of the zeros of the codes (one from each cyclotomic coset). The codes get smaller as we move down the page.

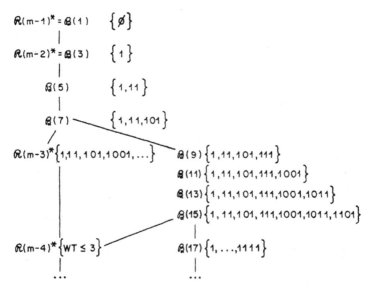

Fig. 13.5. Nesting habits of BCH and punctured RM codes.

We see that

$$\mathscr{R}(r, m)^* \subset \text{BCH code of designed distance } 2^{m-r} - 1,$$
$$\mathscr{R}(r, m) \subset \text{extended BCH code of designed distance } 2^{m-r} - 1.$$

Theorem 12. *The incidence vectors of the* $(m - r)$*-flats in* $\text{EG}(m, 2)$ *generate* $\mathscr{R}(r, m)$.

Proof. We recall that $\mathscr{R}(r, m)$ is $\mathscr{R}(r, m)^*$ with an overall parity check added. By Theorem 10, the incidence vectors of the $(m - r)$-flats with a 1 in coordinate 0 generate $\mathscr{R}(r, m)$. So certainly *all* the $(m - r)$-flats generate $\mathscr{R}(r, m)$. Q.E.D.

We mention without proof the following generalization of this result.

Theorem 13. (MacWilliams and Mann.) *The rank over* $\text{GF}(p)$ *of the incidence matrix of the hyperplanes of an* m*-dimension Euclidean or projective geometry over* $\text{GF}(p^s)$ *is*

$$\binom{m + p - 1}{m}^s + \epsilon$$

where $\epsilon = +1$ *for the projective and* 0 *for the Euclidean geometry.*

Research Problem (13.1). Is there a codeword $a(x) \in \mathscr{R}(r, m)^*$ which is the incidence vector of a $\text{PG}(m - r - 1, 2)$ and generates $\mathscr{R}(r, m)^*$?

§6. Encoding and decoding (I)

There are two obvious ways to encode an RM code. The first uses the generator matrix in the form illustrated in Fig. 13.2 (this is a nonsystematic encoder). The second, which is systematic, makes use of the fact (proved in Theorem 11) that RM codes are extended cyclic codes. In this § we give a decoding algorithm which applies specifically to the first encoder, and then in §7 we give a more general decoding algorithm which applies to any encoder.

We illustrate the first decoder by studying the [16, 11, 4] second-order RM code of length 16, $\mathscr{R}(2, 4)$. As generator matrix G we take the first 11 rows of Fig. 13.2. Thus the message symbols

$$a = a_0 a_4 a_3 a_2 a_1 a_{34} a_{24} a_{14} a_{23} a_{13} a_{12}$$

are encoded into the codeword

$$x = aG = a_0\mathbf{1} + a_4v_4 + \cdots + a_1v_1 + \cdots + a_{12}v_1v_2 \tag{11}$$

$$= x_0x_1 \cdots x_{15} \quad \text{(say).}$$

This is a single-error-correcting code, and we shall show how to correct one error by majority logic decoding. (Unlike the majority logic decoder for RS codes given in Ch. 10, this *is* a practical scheme.) The first step is to recover the 6 symbols $a_{12}, \ldots, a_{34}$. Observe from Fig. 13.2 that if there are no errors,

$$
\begin{aligned}
a_{12} &= x_0 + x_1 + x_2 + x_3 \\
&= x_4 + x_5 + x_6 + x_7 \\
&= x_8 + x_9 + x_{10} + x_{11} \\
&= x_{12} + x_{13} + x_{14} + x_{15},
\end{aligned} \tag{12}
$$

$$
\begin{aligned}
a_{13} &= x_0 + x_1 + x_4 + x_5 \\
&= x_2 + x_3 + x_6 + x_7 \\
&= x_8 + x_9 + x_{12} + x_{13} \\
&= x_{10} + x_{11} + x_{14} + x_{15},
\end{aligned} \tag{13}
$$

. . . .

$$
\begin{aligned}
a_{34} &= x_0 + x_4 + x_8 + x_{12} \\
&= x_1 + x_5 + x_9 + x_{13} \\
&= x_2 + x_6 + x_{10} + x_{14} \\
&= x_3 + x_7 + x_{11} + x_{15}.
\end{aligned}
$$

Equation (12) gives 4 votes for the value of a_{12}, Equation (13) gives 4 votes for a_{13}, and so on. So if one error occurs, the majority vote is still correct, and thus each a_{ij} is obtained correctly.

To find the symbols $a_1, \ldots, a_4$, subtract

$$a_{34}v_3v_4 + \cdots + a_{12}v_1v_2$$

from x, giving say $x' = x_0'x_1' \cdots x_{15}'$. Again from Fig. 13.2 we observe that

$$
\begin{aligned}
a_1 &= x_0' + x_1' \\
&= x_2' + x_3' \\
&\quad \cdots \\
&= x_{14}' + x_{15}', \\
a_2 &= x_0' + x_2' \\
&= \cdots.
\end{aligned}
$$

Now it is easier: there are 8 votes for each a_i, and so if there is one error the majority vote certainly gives each a_i correctly. It remains to determine a_0. We have

$$x'' = x' - a_4 v_4 - \cdots - a_1 v_1$$
$$= a_0 \mathbf{1} + \text{error},$$

and $a_0 = 0$ or 1 according to the number of 1's in x''.

This scheme is called the *Reed decoding algorithm*, and will clearly work for any RM code.

How do we find which components of the codeword x are to be used in the parity checks $(12), (13), \ldots$? To answer this we shall give a geometric description of the algorithm for decoding $\mathcal{R}(r, m)$. We first find a_σ, where $\sigma = \sigma_1 \cdots \sigma_r$ say. The corresponding row of the generator matrix, $v_{\sigma_1} \cdots v_{\sigma_r}$, is the incidence vector of an $(m - r)$-dimensional subspace S of $EG(m, 2)$. For example, the double line in Fig. 13.6 shows the plane S corresponding to a_{12}. Let T be the "complementary" subspace to S with incidence vector $v_{\tau_1} \cdots v_{\tau_{m-r}}$, where $\{\tau_1, \ldots, \tau_{m-r}\}$ is the complement of $\{\sigma_1, \ldots, \sigma_r\}$ in $\{1, 2, \ldots, m\}$. Clearly T meets S in a single point, the origin.

Let $U_1, \ldots, U_{2^{m-r}}$ be all the translates of T in $EG(m, 2)$, including T itself. (These are shaded in Fig. 13.6.) Each U_i meets S in exactly one point.

Theorem 14. *If there are no errors, a_σ is given by*

$$a_\sigma = \sum_{P \in U_i} x_P, \quad i = 1, \ldots, 2^{m-r}$$

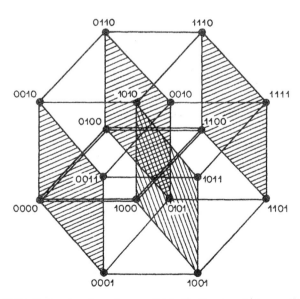

Fig. 13.6. EG(4, 2) showing the subspace S (double lines) and $U_1, \ldots, U_4$ (shaded).

These equations are a generalization of Equations (12), (13), and give 2^{m-r} votes for a_σ.

Proof. Because of the form of the generator matrix, the codeword x is

$$x = \sum_{\rho = \rho_1 \cdots \rho_s} a_\rho v_{\rho_1} \cdots v_{\rho_s}$$

where the sum is over all subsets $\{\rho_1, \ldots, \rho_s\}$ of $\{1, \ldots, m\}$ of size at most r. (This generalizes Equation (11)). Therefore

$$\sum_{P \in U_i} x_P = \sum_\rho a_\rho \sum_{P \in U_i} (v_{\rho_1} \cdots v_{\rho_s})_P$$

$$= \sum_\rho a_\rho N(U_i, \rho),$$

where $N(U_i, \rho)$ is the number of points in the intersection of U_i and the subspace W with incidence vector $v_{\rho_1} \cdots v_{\rho_s}$.

We use the fact that the intersection of two subspaces is a subspace, and all subspaces (except points) contain an even number of points. Now T and W intersect in the subspace

$$v_{\tau_1} \cdots v_{\tau_{m-r}} v_{\rho_1} \cdots v_{\rho_s}.$$

If $s < r$, this subspace has dimension at least 1, and $N(U_i, \rho)$ is even. On the other hand, if $s = r$ but $W \neq S$, then one of the ρ_i must equal one of the τ_j, say $\rho_1 = \tau_1$. Then T and W intersect in

$$v_{\tau_1} \cdots v_{\tau_{m-r}} v_{\rho_2} \cdots v_{\rho_r},$$

which again has dimension at least 1, and $N(U_i, \rho)$ is even. Finally, if $W = S$, $N(U_i, \rho) = 1$. Q.E.D.

This theorem implies that, if no more than $[\frac{1}{2}(2^{m-r} - 1)]$ errors occur, majority logic decoding will recover each of the symbols a_σ correctly, where σ is any string of r symbols. The rest of the a's can be recovered in the same way, as shown in the previous example. Thus the Reed decoding algorithm can correct $[\frac{1}{2}(d - 1)] = [\frac{1}{2}(2^{m-r} - 1)]$ errors.

§7. Encoding and decoding (II)

The Reed decoding algorithm does not apply if the code is encoded systematically as an extended cyclic code, as in §8 of Ch. 7. Fortunately another majority logic decoding algorithm is available, and its description will lead us to a more general class of codes, the finite geometry codes.

If $\mathscr{C}$ is any $[n, k]$ code over GF(q), the rows of the H matrix are parity checks, i.e. define equations

$$\sum_{i=0}^{n-1} h_i x_i = 0$$

which every codeword x must satisfy. Of course any linear combination of the rows of H is also a parity check: so in all there are q^{n-k} parity checks. The art of majority logic decoding is to choose the best subset of these equations.

Definition. A set of parity check equations is called *orthogonal on the ith coordinate* if x_i appears in each equation, but no other x_j appears more than once in the set.

Example. Consider the $[7, 3, 4]$ simplex code, with parity check matrix

$$H = \begin{bmatrix} 1101000 \\ 0110100 \\ 0011010 \\ 0001101 \end{bmatrix}.$$

Seven of the 16 parity checks are shown in Fig. 13.7.

$$\begin{bmatrix} 0 & 1 & 2 & 3 & 4 & 5 & 6 \\ \hline 1 & 1 & 0 & 1 & 0 & 0 & 0 \\ 0 & 1 & 1 & 0 & 1 & 0 & 0 \\ 0 & 0 & 1 & 1 & 0 & 1 & 0 \\ 0 & 0 & 0 & 1 & 1 & 0 & 1 \\ 1 & 0 & 0 & 0 & 1 & 1 & 0 \\ 0 & 1 & 0 & 0 & 0 & 1 & 1 \\ 1 & 0 & 1 & 0 & 0 & 0 & 1 \end{bmatrix}$$

Fig. 13.7. Parity checks on the $[7, 3, 4]$ code.

Rows 1, 5, 7 of Fig. 13.7 are the parity checks

$$x_0 + x_1 + x_3 = 0,$$
$$x_0 + x_4 + x_5 = 0,$$
$$x_0 + x_2 + x_6 = 0,$$

which are orthogonal on coordinate 0. Of course these correspond to the lines through the point 0 in Fig. 2.12. Similarly the lines through 1 give three parity checks orthogonal on coordinate 1, and so on.

Suppose now that an error vector e occurs and $y = x + e$ is received. If there are J parity checks orthogonal on coordinate 0, we define $S_1, \ldots, S_J$ to be the result of applying these parity checks to y. In the above example we have

$$S_1 = y_0 + y_1 + y_3 = e_0 + e_1 + e_3,$$
$$S_2 = y_0 + y_4 + y_5 = e_0 + e_4 + e_5,$$
$$S_3 = y_0 + y_2 + y_6 = e_0 + e_2 + e_6.$$

Theorem 15. *If not more than $[\frac{1}{2}J]$ errors occur, then the true value of e_0 is the value taken by the majority of the S_i's, with the rule that ties are broken in favor of 0.*

Proof. Suppose at most $\frac{1}{2}J$ errors occur. (i) If $e_0 = 0$, then at most $[\frac{1}{2}J]$ equations are affected by the errors. Therefore at least $[\frac{1}{2}J]$ of the S_i's are equal to 0. (ii) If $e_0 = 1$, then less than $[\frac{1}{2}J]$ equations are affected by the other errors. Hence the majority of S_i's are equal to 1.   Q.E.D.

Corollary 16. *If there are J parity checks orthogonal on every coordinate, the code can correct $[\frac{1}{2}J]$ errors.*

Remarks. (i) If the code is cyclic, once a set of J parity checks orthogonal on one coordinate has been found, J parity checks orthogonal on the other coordinates are obtained by cyclically shifting the first set.

(ii) The proof of Theorem 15 shows that some error vectors of weight greater than $[\frac{1}{2}J]$ will cause incorrect decoding. However, one of the nice features of majority logic decoding (besides the inexpensive circuitry) is that often many error vectors of weight *greater* than $[\frac{1}{2}J]$ are also corrected.

(iii) Breaking ties. In case of a tie, the rule is to favor 0 if it is one of the alternatives, but otherwise to break ties in any way. Equivalently, use the majority of $\{0, S_1, S_2, \ldots\}$.

This method of decoding is called *one-step majority logic decoding.* However, usually there are not enough orthogonal parity checks to correct up to one-half of the minimum distance, as the following theorem shows.

Theorem 17. *For a code over $GF(q)$, the number of errors which can be corrected by one-step majority logic decoding is at most*

$$\frac{n-1}{2(d'-1)},$$

where d' is the minimum distance of the dual code.

Proof. The parity checks orthogonal on the first coordinate have the form

$$\begin{array}{ccccccc}
 & \overbrace{}^{\geq d'-1} & & & \\
1 & x\ x\ x & 0\ 0\ 0\ 0 \\
1 & 0\ 0\ 0 & \underbrace{x\ x\ x\ x}_{\geq d'-1} \\
\cdots & &
\end{array}$$

since each corresponds to a codeword in the dual code. Therefore $J \leqslant (n-1)/(d'-1)$. By Remark (ii) above, some error pattern of weight $[\frac{1}{2}(n-1)/(d'-1)] + 1$ will cause the decoder to make a mistake.   Q.E.D.

Examples. (1) For the $[23, 12, 7]$ Golay code, $d' = 8$, and so at most $[22/2.7] = 1$ error can be corrected by one-step decoding.

(2) Likewise most RS codes cannot be decoded by one-step decoding, since $d' = n - d + 2$.

However, there are codes for which one-step majority logic decoding is useful, such as the $[7, 3, 4]$ code of Fig. 13.7 and more generally the difference-set cyclic codes described below.

L-step decoding. Some codes, for example RM codes, can be decoded using several stages of majority logic.

Definition. A set of parity checks $S_1, S_2, \ldots$ is called *orthogonal on coordinates* $a, b, \ldots, c$ if the sum $x_a + x_b + \cdots + x_c$ appears in each S_i, but no other x_j appears more than once in the set.

Example. *2-step decoding of the* $[7, 4, 3]$ *code.* The 7 nonzero parity checks are

$$
\begin{array}{c}
\begin{array}{ccccccc}
0 & 1 & 2 & 3 & 4 & 5 & 6
\end{array} \\
\left[
\begin{array}{ccccccc}
1 & 1 & 1 & 0 & 1 & 0 & 0 \\
0 & 1 & 1 & 1 & 0 & 1 & 0 \\
0 & 0 & 1 & 1 & 1 & 0 & 1 \\
1 & 0 & 0 & 1 & 1 & 1 & 0 \\
0 & 1 & 0 & 0 & 1 & 1 & 1 \\
1 & 0 & 1 & 0 & 0 & 1 & 1 \\
1 & 1 & 0 & 1 & 0 & 0 & 1
\end{array}
\right]
\end{array}
$$

There are two parity checks orthogonal on coordinates 0 and 1, namely

$$S_1 = e_0 + e_1 + e_2 \qquad + e_4,$$
$$S_2 = e_0 + e_1 \qquad + e_3 \qquad\qquad + e_6,$$

two which are orthogonal on coordinates 0 and 2,

$$S_1 = e_0 + e_1 + e_2 \qquad + e_4,$$
$$S_3 = e_0 \qquad + e_2 \qquad\qquad + e_5 + e_6,$$

and so on. Suppose there is one error. Then the majority rule gives the correct value of $e_0 + e_1$ (from the first pair of equations), and of $e_0 + e_2$ (from the

·second pair). Now the equations

$$S_4 = e_0 + e_1$$
$$S_5 = e_0 \qquad + e_2$$

are orthogonal on e_0 and again the majority rule gives e_0. This is a two-step majority logic decoding algorithm to find e_0. A circuit for doing this is shown in Fig. 13.8.

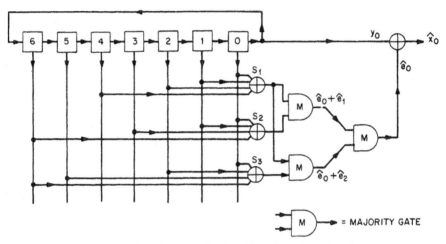

Fig. 13.8. Two-step majority decoding of the $[7, 4, 3]$ code.

Since the code is cyclic, it is enough to design a decoder which corrects the first coordinate. The others are then corrected automatically.

A decoder which has L levels of majority logic is called an *L-step* decoder. The basic idea, as illustrated in the preceding example, is that the number of coordinates in the check sums which are being estimated decreases from level to level, until at the final step we have estimates for the individual coordinates.

Lemma 18. *If there are J checks at each stage of the decoding, then* $[\frac{1}{2}J]$ *errors can be corrected.*

Proof. As for Theorem 15.   Q.E.D.

Even with L-step decoding it may not be possible to correct up to half the minimum distance.

Theorem 19. *For a code over* $GF(q)$, *the number of errors which can be* *corrected by L-step majority decoding is at most*

$$\frac{n}{d'} - \frac{1}{2},$$

where d' is the minimum distance of the dual code.

Proof. Suppose there are a set of J parity checks orthogonal on l coordinates. We shall show that $J \leq 2n/d' - 1$, which by Remark (ii) above proves the theorem. Let the ith parity check involve a_i coordinates besides the l. Since these checks correspond to codewords in the dual code, we have

$$l + a_i \geq d', \qquad a_i + a_j \geq d' (i \neq j). \tag{14}$$

Set $S = \sum_{i=1}^{J} a_i$. Then $S \leq n - l$, and from (14),

$$Jl + S \geq Jd', \qquad (J-1)S \geq \binom{J}{2} d'.$$

Eliminating l gives $n - S \geq (Jd' - S)/J$, and eliminating S then gives

$$(J-1)d' \leq 2n - 2d'. \qquad \text{Q.E.D.}$$

Example. For the $[23, 12, 7]$ Golay code, L-step decoding cannot correct more than 2 errors.

In contrast to this, we have:

Theorem 20. *For the rth .order RM code $\mathcal{R}(r, m)$, $(r + 1)$-step majority decoding can correct $[\frac{1}{2}(d - 1)] = [\frac{1}{2}(2^{m-r} - 1)]$ errors.*

Proof. The dual code is $\mathcal{R}(m - r - 1, m)$ and by Theorem 8 the low weight codewords in the dual code are the incidence vectors of the $(r + 1)$-dimensional flats in $EG(m, 2)$.

Let V be any r-flat. We will find a set of parity checks orthogonal on the coordinates y_P, $P \in V$. In fact, let U be any $(r + 1)$-flat containing V. Now each of the $2^m - 2^r$ points not in V determines a U, and each U is determined by $2^{r+1} - 2^r$ such points. Therefore there are $(2^m - 2^r)/(2^{r+1} - 2^r) = 2^{m-r} - 1$ different U's. Any two such U's meet only in V. Thus we have an estimate for the sum

$$\sum_{P \in V} y_P.$$

This estimate will be correct provided no more than $[\frac{1}{2}(2^{m-r} - 1)]$ errors occur. We repeat this for all r-flats V.

Next, let W be any $(r - 1)$-flat, and let V be any r-flat containing W. There are $2^{m-r+1} - 1$ such V's and from the first stage we know the values of the corresponding sums. Therefore we can obtain an estimate for the value of the sum

$$\sum_{P \in W} y_P.$$

Proceeding in this way, after $r + 1$ steps we finally arrive at an estimate for y_P, for any point P, which will be correct provided no more than $[\frac{1}{2}(d - 1)] = [\frac{1}{2}(2^{m-r} - 1)]$ errors occur.   Q.E.D.

Improvements of the decoding algorithm. A more practical scheme than the preceding is to use the cyclic code $\mathcal{R}(r, m)^*$, since then one need only construct a circuit to decode the first coordinate. The dual code $\mathcal{R}(r, m)^{*\perp}$ now contains the incidence vectors of all $(r + 1)$-flats in EG$(m, 2)$ which do not pass through the origin.

This is illustrated by the $[7, 4, 3]$ code, for which we gave a two-step decoding algorithm in Fig. 13.8. This code is in fact $\mathcal{R}(1, 3)^*$.

The following technique, known as *sequential code reduction*, considerably reduces the number of majority gates, but at the cost of increasing the delay in decoding.

We shall illustrate the technique by applying it to the decoder of Fig. 13.8. The idea is very simple. In Fig. 13.8, let

$$S_4, S_4', S_4'', \ldots$$

denote the output from the first majority gate at successive times. Then

$$S_4 = e_0 + e_1,$$
$$S_4' = e_1 + e_2,$$
$$\cdots$$

Note that

$$S_4 + S_4' = S_5!$$

So if we are willing to wait one clock cycle, S_5 can be obtained without using the second majority gate. The resulting circuit is shown in Fig. 13.9.

In general this technique (when it is applicable) will reduce the number of majority gates, adders, etc., from an exponential function of the number of steps to a linear function, at the cost of a linear delay (as in Fig. 13.10).

The name "sequential code reduction" comes from the fact that at each successive stage of the decoder we have estimates for additional parity checks, and so the codeword appears to belong to a smaller code. In the example, after the first stage we know all the sums $e_i + e_j$, which are in fact parity checks on the $[7, 1, 7]$ repetition code.

Unfortunately, sequential code reduction doesn't apply to all codes (and even when it does apply, it may be a difficult task to find the best decoder).

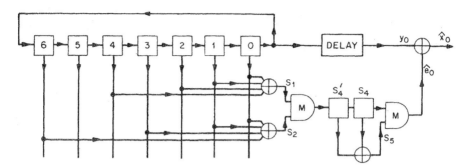

Fig. 13.9. Decoder for [7, 4, 3] code using sequential code reduction.

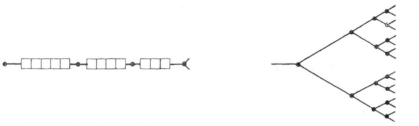

Circuit using sequential code reduction.   Circuit for L-step decoding.

Fig. 13.10.

References to this and to other modifications and generalizations of the basic algorithm are given in the notes.

Threshold decoding. Let us return briefly to the problem of decoding an arbitrary binary linear code. Suppose the codeword x is transmitted and y is received. The decoder decides that the most likely error vector $\hat{e}$ is the coset leader of the coset containing y, and decodes y as $\hat{y} = \hat{x} + \hat{e}$. We saw in Theorem 5 of Ch. 1 that $\hat{e}$ is a function of the syndrome S. More precisely $\hat{e}$ is a binary vector-valued function of the $n - k$ components $S_1, S_2, \ldots, S_{n-k}$ of S (see Fig. 13.11). For small codes we could synthesize $\hat{e}$ by a simple combinational logic circuit using AND's, OR's, etc. (but no delay elements). This can be further simplified if the code is cyclic, in which case the circuit is

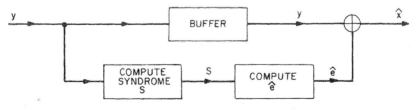

Fig. 13.11.

called a *Meggitt decoder*. Figure 13.8 is a simple example of a Meggitt decoder. See also Problem 36 of Ch. 7.

But larger codes (e.g. BCH codes, Fig. 9.5) require more complicated components to synthesize $\hat{e}$. For example in this chapter we have used majority gates to decode RM codes. A more general notion is that of a *weighted majority gate*, or *threshold gate*, which is a function $\theta(v_1, \ldots, v_m)$ of $v_1, \ldots, v_m$, with real weights $a_1, \ldots, a_m$, defined by

$$\theta(v_1, \ldots, v_m) = \begin{cases} 1 & \text{if } \sum_{i=1}^{m} a_i v_i > \frac{1}{2} \sum_{i=1}^{m} a_i, \\ \\ 0 & \text{if } \sum_{i=1}^{m} a_i v_i \leq \frac{1}{2} \sum_{i=1}^{m} a_i, \end{cases} \tag{15}$$

where these sums are real.

In contrast to majority logic decoding any code can be decoded by a one-step threshold gate decoding algorithm. It is easy to show that this can always be done: what is hard is to find an efficient way of doing it.

Theorem 21. (Rudolph.) *Any binary linear code can be decoded by one-step threshold gate decoding.*

Proof. Write $\hat{e} = (f_1, \ldots, f_n)$, where each component $f_i = f_i(S) = f_i(S_1, \ldots, S_{n-k})$ is a function of the syndrome. Let $F_i(S) = 1 - 2f_i(S)$ be the corresponding real ± 1-valued function. By Equation (11) of Ch. 14, $F_i(S)$ can be written

$$F_i(S) = \frac{1}{2^{n-k}} \sum_{u \in V^{n-k}} \hat{F}_i(u)(-1)^{u \cdot S},$$

where the $\hat{F}_i(u)$ are the Hadamard coefficients of $F_i(S)$ given by Equation (8) of Ch. 14. Then

$$f_i(S) = \frac{1}{2} \left(1 - \frac{1}{2^{n-k}} \sum_{u \in V^{n-k}} \hat{F}_i(u)(-1)^{u \cdot S} \right).$$

If θ is any threshold gate function of the 2^{n-k} inputs $\sum_{i=1}^{n-k} u_i S_i$, $u \in V^{n-k}$, with weights a_u, it is immediate from the definition of θ that

$$\theta \left(\sum_i u_i S_i : u \in V^{n-k} \right) = \frac{1}{2} \left(1 - sgn \sum_{u \in V^{n-k}} a_u (-1)^{u \cdot S} \right),$$

where $sgn(x) = 1$ if $x \geq 0$, $= -1$ if $x < 0$. Therefore

$$f_i(S) = \theta \left(\sum_i u_i S_i : u \in V^{n-k} \right) \tag{16}$$

if we take $a_u = \hat{F}_i(u)/2^{n-k}$. Since we can do this for each i, the theorem follows.   Q.E.D.

Unfortunately Equation (16) represents $\hat{e}$ as a threshold function of *all* the 2^{n-k} parity checks, so this is not a practical algorithm. However in a number of cases it has been possible to find a different one-step threshold gate realization of $\hat{e}$ which involves many fewer parity checks (see Notes).

Research Problem (13.2). Find the most efficient one-step threshold gate realization of a given Boolean function.

§8. Other geometrical codes

(I) *Difference-set cyclic codes.* Let Π be the projective plane $PG(2, p^s)$ of order p^s (see Appendix on Finite Geometries). Π contains $n = p^{2s} + p^s + 1$ points, which can be represented as triples

$$(\beta_1, \beta_2, \beta_3), \qquad \beta_i \in GF(p^s).$$

Note that $(\lambda\beta_1, \lambda\beta_2, \lambda\beta_3)$, $\lambda \in GF(p^s)^*$, is the same point as $(\beta_1, \beta_2, \beta_3)$. Each triple can be regarded as an element of $GF(p^{3s})$, i.e. can be written as a power of α, where α is a primitive element of $GF(p^{3s})$. Some scalar multiple of each triple is equal to α^i for $0 \le i < n$. We label the n points of the plane by these powers of α.

Let $\alpha^{i_1}, \dots, \alpha^{i_l}$, $l = p^s + 1$, be a line of Π. The incidence vector of this line has 1's in exactly the coordinates $i_1, \dots, i_l$. By the proof of Theorem 10, any cyclic shift of this vector is the incidence vector of another line. Since there are n shifts and n lines, every line of Π is obtained in this way.

Let $\mathcal{D}$ be the code generated over $GF(p)$ by these n incidence vectors, and let $\mathcal{C} = \mathcal{D}^\perp$. Clearly $\mathcal{D}$ is a cyclic code of length n. From Theorem 13, $\mathcal{D}$ has dimension $\binom{p+1}{2}^s + 1$.

$\mathcal{C}$ can be decoded by one-step majority logic, as follows. The incidence vectors of the $l = p^s + 1$ lines through a point of Π form a set of orthogonal checks on that coordinate. (They are orthogonal because two lines through a point have no other intersection.) By Corollary 16, one-step majority logic decoding will correct $\frac{1}{2}(p^s + 1)$ errors, and the code has minimum distance at least $p^s + 2$.

Examples. (1) The simplest example is when $p = 2$, $s = 1$. Then $\mathcal{C}$ is the $[7, 3, 4]$ binary simplex code again, as shown in Fig. 13.7.

(2) If $p = s = 2$, $\mathcal{D}$ is generated by the lines of the projective plane of order 4, and $\mathcal{C}$ is a $[21, 11, 6]$ binary code.

These codes are closely related to difference sets.

Definition. A *planar difference set* modulo $n = l(l - 1) + 1$ is a set of l numbers

$d_1, \ldots, d_l$ with the property that the $l(l-1)$ differences $d_i - d_j$ ($i \neq j$), when reduced modulo n, are exactly the numbers $1, 2, \ldots, n-1$ in some order.

Examples. (1) For $l = 3$, the numbers $d_1 = 0$, $d_2 = 1$, $d_3 = 3$ form a planar difference set modulo 7. Indeed, the differences modulo 7 are $1 - 0 = 1$, $3 - 0 = 3$, $3 - 1 = 2$, $0 - 1 = 6$, $0 - 3 = 4$, $1 - 3 = 5$.

(2) $\{0, 1, 3, 9\}$ is a difference set modulo 13, and $\{0, 2, 7, 8, 11\}$ is a difference set modulo 21.

The only known planar difference sets are those obtained from a projective plane Π of order p^s in the following way: let $\alpha^{i_1}, \ldots, \alpha^{i_l}$ be the points of a line of Π. Then $\{i_1, \ldots, i_l\}$ is a planar difference set. For suppose two differences are equal, say

$$i_r - i_s = i_t - i_u, \quad \text{where } i_r \neq i_t.$$

Then the cyclic shift of this line which sends the point α^{i_r} into the point α^{i_t} gives a new line which meets the first in two points, a contradiction. Thus we have proved

Theorem 22. (Singer.) *If $\alpha^{i_1}, \ldots, \alpha^{i_l}$, $l = p^s + 1$, are the points of a line in a projective plane* $\mathrm{PG}(2, p^s)$, *then* $\{i_1, \ldots, i_l\}$ *is a planar difference set modulo* $p^{2s} + p^s + 1$.

Research Problem (13.3). Are there any other planar difference sets?

(II) *Euclidean and projective geometry codes.* Instead of taking $\mathcal{D}$ to be generated by the incidence vectors of the lines of a projective plane, we may use the incidence vectors of the r-dimensional flats of a Euclidean geometry $\mathrm{EG}(m, p^s)$ or a projective geometry $\mathrm{PG}(m, p^s)$. Then $\mathscr{C} = \mathcal{D}^\perp$ is a *Euclidean* or *projective geometry* code. There is no simple formula for the dimension of these codes. Decoding can be done by r-step majority logic decoding as for RM codes. Further details and generalizations can be found in the references given in the notes. All of these codes can be regarded as extensions of RM codes.

§9. Automorphism groups of the RM codes

Let $A = (a_{ij})$ be an invertible $m \times m$ binary matrix and let b be a binary m-tuple. The transformation

$$T: \text{replace} \begin{pmatrix} v_1 \\ \vdots \\ v_m \end{pmatrix} \text{ by } A \begin{pmatrix} v_1 \\ \vdots \\ v_m \end{pmatrix} + b \tag{17}$$

is a permutation of the set of 2^m m-tuples which sends 0 into b.

We may also think of T as permuting Boolean functions:

$$T: \text{replace } f(v_1, \ldots, v_m) \quad \text{by} \quad f\Big(\sum a_{1j}v_j + b_1, \ldots, \sum a_{mj}v_j + b_m\Big). \quad (18)$$

The set of all such transformations T forms a group, with composition as the group operation. The order of this group is found as follows. The first column of A may be chosen in $2^m - 1$ ways, the second in $2^m - 2$, the third in $2^m - 4, \ldots$. Furthermore there are 2^m choices for b. So this group, which is called the *general affine group* and is denoted by GA(m), has order

$$|\text{GA}(m)| = 2^m (2^m - 1)(2^m - 2)(2^m - 2^2) \cdots (2^m - 2^{m-1}). \quad (19)$$

A useful approximation to its order is

$$|\text{GA}(m)| \approx 0.29 \, 2^{m^2+m} \quad \text{for } m \text{ large.}$$

(We encountered another form of this group in §5 of Ch. 8.)

It is clear from (18) that if f is a polynomial of degree r, so is Tf. Therefore the group GA(m) permutes the codewords of the rth order RM code $\mathcal{R}(r, m)$, and

$$\text{GA}(m) \subset \text{Aut } \mathcal{R}(r, m). \quad (20)$$

The subgroup of GA(m) consisting of all transformations

$$T: \text{replace } \begin{pmatrix} v_1 \\ \vdots \\ v_m \end{pmatrix} \text{ by } A \begin{pmatrix} v_1 \\ \vdots \\ v_m \end{pmatrix} \quad (21)$$

(i.e., for which $b = 0$) is the *general linear group* GL($m, 2$) (see §5 of Ch. 8), and has order

$$|\text{GL}(m, 2)| = (2^m - 1)(2^m - 2)(2^m - 2^2) \cdots (2^m - 2^{m-1})$$
$$\approx 0.29 \, 2^{m^2} \quad \text{for } m \text{ large.} \quad (22)$$

Since (21) fixes the zero m-tuple, the group GL($m, 2$) permutes the codewords of the punctured RM code $\mathcal{R}(r, m)^*$:

$$\text{GL}(m, 2) \subset \text{Aut } \mathcal{R}(r, m)^*. \quad (23)$$

Note that GL($m, 2$) is doubly transitive and GA(m) is triply transitive (Problem 9).

Theorem 23. *For* $1 \leq r \leq m - 1$,

(a) Aut $\mathcal{R}(r, m)^* \subset$ Aut $\mathcal{R}(r + 1, m)^*$
(b) Aut $\mathcal{R}(r, m) \subset$ Aut $\mathcal{R}(r + 1, m)$.

Proof of (b). Let $x_1, \ldots, x_B$ be the minimum weight vectors of $\mathcal{R}(r, m)$. For $\pi \in$ Aut $\mathcal{R}(r, m)$, let $\pi x_i = x_{i'}$. Now x_i is an $(m - r)$-flat. If Y is any $(m - r - 1)$-

flat, then for some i, j, $Y = x_i * x_j$. Therefore

$$\pi Y = \pi(x_i * x_j)$$
$$= \pi x_i * \pi x_j = x_{i'} * x_{j'},$$

which is the intersection of two $(m - r)$-flats; and contains 2^{m-r-1} points since π is a permutation. Thus πY is an $(m - r - 1)$-flat. So π permutes the generators of $\mathcal{R}(r + 1, m)$, and therefore preserves the whole code. Part (a) is proved in the same way.   Q.E.D.

It is immediate from Problem 5 that

$$\text{Aut } \mathcal{R}(r, m)^* = \mathcal{S}_{2^m - 1} \quad \text{for } r = 0 \text{ and } m - 1,$$
$$\text{Aut } \mathcal{R}(r, m) = \mathcal{S}_{2^m} \quad \text{for } r = 0, m - 1 \text{ and } m.$$

In the remaining cases we show that equality holds in (20) and (23).

Theorem 24. *For* $1 \leq r \leq m - 2$,
  (a) Aut $\mathcal{R}(r, m)^* = GL(m, 2)$,
  (b) Aut $\mathcal{R}(r, m) = GA(m)$.

Proof. (i) We have

$$\text{simplex code } \mathcal{H}_m^\perp \xrightarrow{\text{add } 1} \mathcal{R}(1, m)^* \xleftarrow{\substack{\text{puncture } 0 \\ \text{coordinate}}} \mathcal{R}(1, m).$$

By Problem 30 of Ch. 8, Aut $\mathcal{H}_m^\perp = $ Aut $\mathcal{R}(1, m)^*$. From (23), and the remark following Theorem 13 of Ch. 8, since $\mathcal{H}_m^\perp$ has dimension m,

$$\text{Aut } \mathcal{H}_m^\perp = \text{Aut } \mathcal{R}(1, m)^* = GL(m, 2).$$

Finally, by Problem 29 of Ch. 8, Aut $\mathcal{H}_m = GL(m, 2)$.
  (ii) Let $G_1 = $ Aut $\mathcal{R}(1, m)^*$, $G_2 = $ Aut $\mathcal{R}(1, m)$. Clearly G_1 is the subgroup of G_2 which fixes the 0 coordinate. Since $GA(m)$ is transitive, so is G_2. Each coset of G_1 in G_2 sends 0 to a different point, so $|G_2| = 2^m |G_1|$. Therefore from (19) and (22) $G_2 = GA(m)$. Again by Problem 29 of Ch. 8, Aut $\mathcal{R}(m - 2, m) = $ Aut $\mathcal{R}(1, m) = GA(m)$.
  (iii) From Theorem 23 and (i), (ii),

$$GA(m) = \text{Aut } \mathcal{R}(1, m) \subseteq \text{Aut } \mathcal{R}(2, m) \subseteq \cdots \subseteq \text{Aut } \mathcal{R}(m - 2, m) = GA(m)$$
$$GL(m, 2) = \text{Aut } \mathcal{R}(1, m)^* \subseteq \text{Aut } \mathcal{R}(2, m)^* \subseteq \cdots \subseteq \text{Aut } \mathcal{R}(m - 2, m)^*$$
$$= GL(m, 2).$$
$$\text{Q.E.D.}$$

Problem. (9) Show that $GL(m, 2)$, $GA(m)$ are in fact groups of the stated orders, and are respectively doubly and triply transitive.

*§10. Mattson–Solomon polynomials of RM codes

In this section we shall show that the Boolean function defining a code-word of an RM code is really the same as the Mattson–Solomon polynomial (Ch. 8) of the codeword.

Let α be a primitive element of GF(2^m). Then $1, \alpha, \ldots, \alpha^{m-1}$ is a basis for GF(2^m). Let $\lambda_0, \ldots, \lambda_{m-1}$ be the complementary basis (Ch. 4). We shall now consider RM codes to be defined by truth tables in which the columns are taken in the order $0, 1, \alpha, \alpha^2, \ldots, \alpha^{2^m-2}$. For example, when $m = 3$, the truth table is shown in Fig. 13.12.

$$0 \ \ 1 \ \ \alpha \ \ \alpha^2 \ \ \alpha^3 \ \ \alpha^4 \ \ \alpha^5 \ \ \alpha^6$$

v_3	0	0	0	1	0	1	1	1
v_2	0	0	1	0	1	1	1	0
v_1	0	1	0	0	1	0	1	1

Fig. 13.12.

Lemma 25. *The MS polynomial of the codeword v_j in $\mathcal{R}(1, m)$ is $T_m(\lambda_j z)$, where T_m is the trace function defined in §8 of Ch. 4.*

Proof. Let $\mathcal{M}$ be the matrix consisting of the rows $v_m, \ldots, v_1$ of the truth table, with the first, or zero, column, deleted. $\mathcal{M}$ is an $m \times 2^m - 1$ matrix $(\mathcal{M}_{ki})$, and

$$\alpha^i = \sum_{k=0}^{m-1} \mathcal{M}_{ki} \alpha^{m-k-1}, \quad 0 \le i \le 2^m - 2.$$

We must show that

$$T_m(\lambda_j \alpha^i) = \mathcal{M}_{m-j-1,i}.$$

In fact,

$$T_m(\lambda_j \alpha^i) = \sum_{k=0}^{m-1} \mathcal{M}_{ki} T_m(\lambda_j \alpha^{m-k-1}) = \mathcal{M}_{m-j-1,i},$$

by the property of a complementary basis.   Q.E.D.

Corollary 26. *If a is any binary vector of length 2^m, corresponding to the Boolean function $a(v_1, \ldots, v_m)$, then the MS polynomial of a is*

$$A(z) = a(T_m(\lambda_0 z), \ldots, T_m(\lambda_{m-1} z)). \tag{24}$$

Notes. (i) $A(0) = \sum_{i=0}^{2^m-2} A(\alpha^i) = a(0, \ldots, 0)$ is an overall parity check on a.

(ii) When evaluating the RHS of (24), high powers of z are reduced by

$z^{2^{m}-1} = 1$. However, once $A(z)$ has been obtained, in order to use the properties of MS polynomials given in Ch. 8, $A(z)$ must be considered as a polynomial in $\mathscr{F}[z]$, $\mathscr{F} = GF(2^m)$.

Conversely, if a is the vector with MS polynomial $A(z)$, the B.f. corresponding to a is

$$\dot{A}(v_1 + v_2\alpha + \cdots + v_m\alpha^{m-1}).$$

Example. Suppose $m = 3$ and the codeword is θ_1^*, so that the MS polynomial is $A(z) = z + z^2 + z^4$. Then the corresponding B.f. is

$$A(v_1 + v_2\alpha + v_3\alpha^2) = v_1 + v_2T_3(\alpha) + v_3T_3(\alpha)$$
$$= v_1 + (v_2 + v_3)(\alpha + \alpha^2 + \alpha^4) = v_1,$$

in agreement with Fig. 13.12.

Problem. (10) Repeat for the codeword θ_3^*.

*§11. The action of the general affine group on Mattson–Solomon polynomials

Definition. An *affine polynomial* is a linearized polynomial plus a constant (see §9 of Ch. 4), i.e. has the form

$$F(z) = \gamma_0 + \sum_{i=0}^{m-1} \gamma_i z^{2^i}, \quad \text{where } \gamma_i \in GF(2^m).$$

Problem. (11) Show that those zeros of an affine polynomial which lie in $GF(2^m)$ form an r-flat in $EG(m, 2)$.

The main result of this section is the following:

Theorem 27. *A transformation of the general affine group $GA(m)$ acts on the MS polynomial $A(z)$ of a vector of length 2^m by replacing z by $F(z)$, where $F(z)$ is an affine polynomial with exactly one zero in $GF(2^m)$. Conversely, any such transformation of MS polynomials arises from a transformation of $GA(m)$.*

Proof. Consider any transformation belonging to $GA(m)$, say

$$v^T \rightarrow Av^T + b,$$

where

$$v = (v_1, \ldots, v_m) = (T_m(\lambda_0 z), \ldots, T_m(\lambda_{m-1} z)).$$

The variable z in the MS polynomial is related to v by

$$z = v_1 + v_2\alpha + \cdots + v_m\alpha^{m-1} = \boldsymbol{\alpha} \cdot \boldsymbol{v}^T,$$

where $\boldsymbol{\alpha} = (1, \alpha, \ldots, \alpha^{m-1})$. Thus z is transformed as follows:

$$z \rightarrow \boldsymbol{\alpha} A \boldsymbol{v}^T + \boldsymbol{\alpha} \cdot \boldsymbol{b}. \tag{25}$$

Now

$$\boldsymbol{\alpha} A \begin{pmatrix} T_m(\lambda_0)z \\ \vdots \\ T_m(\lambda_{m-1}z) \end{pmatrix} = \boldsymbol{\alpha} A \begin{pmatrix} \sum_{j=0}^{m-1} (\lambda_0 z)^{2^j} \\ \vdots \\ \sum_{j=0}^{m-1} (\lambda_{m-1}z)^{2^j} \end{pmatrix}$$

$$= (\boldsymbol{\alpha} A \boldsymbol{\lambda}^T)z + (\boldsymbol{\alpha} A (\boldsymbol{\lambda}^2)^T)z^2 + \cdots + (\boldsymbol{\alpha} A (\boldsymbol{\lambda}^{2^{m-1}})^T)z^{2^{m-1}},$$

where $\boldsymbol{\lambda}^{2^j} = (\lambda_0^{2^j}, \ldots, \lambda_{m-1}^{2^j})$ and the T denotes transpose.

Let $\boldsymbol{\alpha} A = \boldsymbol{\beta} = (\beta_0, \ldots, \beta_{m-1})$ be a new basis for $\mathrm{GF}(2^m)$. Then

$$z \rightarrow \boldsymbol{\alpha} \cdot \boldsymbol{b} + (\boldsymbol{\beta} \cdot \boldsymbol{\lambda}^T)z + (\boldsymbol{\beta} \cdot (\boldsymbol{\lambda}^2)^T)z^2 + \cdots + (\boldsymbol{\beta} \cdot (\boldsymbol{\lambda}^{2^{m-1}})^T)z^{2^{m-1}}, \tag{26}$$

which is an affine polynomial.

Problem. (12) Show that the polynomial (26) has exactly one zero in $\mathrm{GF}(2^m)$. Conversely, any transformation

$$z \rightarrow F(z) = u_0 + f(z), \quad u_0 \in \mathrm{GF}(2^m), \tag{27}$$

when $f(z)$ is a linearized polynomial and $F(z)$ has exactly one zero in $\mathrm{GF}(2^m)$, is in $\mathrm{GA}(m)$. We decompose (27) into

$$z \rightarrow z + u_0,$$

which is clearly in $\mathrm{GA}(m)$, followed by

$$z \rightarrow f(z) = \sum_{j=0}^{m-1} \gamma_j z^{2^j}. \tag{28}$$

Problem. (13) Show that (28) is in $\mathrm{GL}(m, 2)$, i.e. has the form

$$z \rightarrow \sum_{j=0}^{m-1} (\boldsymbol{\beta} \cdot (\boldsymbol{\lambda}^{2^j})^T)z^{2^j}$$

for some basis $\beta_0, \ldots, \beta_{m-1}$ of $\mathrm{GF}(2^m)$.

Notes to Chapter 13.

§1. Reed–Muller codes are named after Reed [1104] and Muller [975] (although Peterson and Weldon [1040, p. 141] attribute their discovery to an earlier, unpublished paper of Mitani [963]).

Nonbinary RM codes have been defined by several authors – see Delsarte et al. [365], Kasami, Lin and Peterson [739–741], Massey et al. [924] and Weldon [1401]. See also [99, 1396].

§2. Truth tables are widely used in switching theory (see for example McCluskey [935, Ch. 3] and in elementary logic, where they are usually written with FALSE instead of 0 and TRUE instead of 1 (see for example Kemeny et al. [755, Ch. 1]). In either form they are of great importance in discrete mathematics. For the disjunctive normal form see Harrison [606, p. 59] or McCluskey [935, p. 78].

§3. Problem 6 is due to S.M. Reddy.

§4. Lemma 6 is from Rothschild and Van Lint [1128].

§5. The proof of Theorem 10 is taken from Delsarte [343].

Lucas' theorem for multinomial coefficients was used in the proof of Theorem 10. This is a straightforward generalization of the usual Lucas theorem for binomial coefficients, which in the binary case is as follows.

Theorem 28. (Lucas [862].) *Let the binary expansions of* n, k *and* $l = n - k$ *be*

$$n = \sum_i n_i 2^i, \qquad k = \sum_i k_i 2^i, \qquad l = \sum_i l_i 2^i,$$

where n_i, k_i, l_i *are* 0 *or* 1. *Then*

$$\binom{n}{k} \equiv \begin{cases} 1 \ (\mathrm{mod}\ 2) & \text{iff } k_i \leq n_i \quad \text{for all } i \\ 0 \ (\mathrm{mod}\ 2) & \text{iff } k_i > n_i \quad \text{for some } i. \end{cases}$$

Equivalently,

$$\binom{n}{k} \equiv \begin{cases} 1 \ (\mathrm{mod}\ 2) & \text{iff } k_i + l_i \leq n_i \quad \text{for all } i \\ 0 \ (\mathrm{mod}\ 2) & \text{iff } k_i + l_i > n_i \quad \text{for some } i. \end{cases}$$

For a proof see for example Berlekamp [113, p. 113]. Singmaster [1216] gives generalizations.

That RM codes are extended cyclic codes (Theorem 11) was simultaneously discovered by Kasami, Lin and Peterson [739, 740] and Kolesnik and Mironchikov (see [774] and the references given there). See also Camion [237].

Theorem 13 is from MacWilliams and Mann [884]. The ranks of the incidence matrices of subspaces of other dimensions have been determined by Goethals and Delsarte [499], Hamada [591], and Smith [1243–1246].

§6. The Reed decoding algorithm was given by Reed [1104], and was the first nontrivial majority logic decoding algorithm. For more about this algorithm see Gore [544] and Green and San Souci [555]. Massey [918] (see also [921]) and Kolesnik and Mironchikov (described in Dobrushin [378]) extensively studied majority logic and threshold decoding. Rudolph [1130, 1131] introduced one-step weighted majority decoding, and this work was extended by Chow [294], Duc [387], Gore [544, 545], Ng [990] and Rudolph and Robbins [1134].

Techniques for speeding up the decoding of Reed–Muller codes were given by Weldon [1403]. See also Peterson and Weldon [1040, Ch. 10]. Decoding Reed–Muller and other codes using a general purpose computer has been investigated by Paschburg et al. [1025, 1026].

Duc [388] has given conditions on a code which must be satisfied if L-step decoding can correct $[\frac{1}{2}(d-1)]$ errors. See also Kugurakov [786].

The Reed decoding algorithm will correct many error patterns of weight greater than $[\frac{1}{2}(d-1)]$. Krichevskii [784] has investigated just how many. Theorems 17 and 19 are given by Lin [835–837].

Other papers on majority-logic decoding are Berman and Yudanina [137], Chen [271], Delsarte [348], Duc and Skattebol [390], Dyn'kin and Tenegol'ts [397], Kasami and Lin [732, 733], Kladov [763], Kolesnik [773], Longobardi et al. [860], Redinbo [1101], Shiva and Tavares [1205], Smith [1245] and Warren [1390].

§7. Decoding by sequential code reduction was introduced by Rudolph and Hartmann [1132]. Some applications are given in [1112]. Meggitt decoders were described in [952]. Theorem 21 is due to Rudolph [1131]. Rudolph and Robbins [1134] show that the same result holds using only positive weights a_i in (15). Longobardi et al. [860] and Robbins [1115] have found efficient threshold decoding circuits for certain codes.

§8. Difference-set cyclic codes were introduced by Weldon [1400]. Their dimension was found by Graham and MacWilliams [553]. For Theorem 22 see Singer [1213].

For more about difference sets see Baumert [82], Hall [589], Mann [908] and Raghavarao [1085, Ch. 10].

Euclidean and projective geometry codes were studied by Prange [1074, 1075], but the first published discussion was given by Rudolph [1130]. There is now an extensive literature on these codes and their generalizations – see for example Chen [267–270], Chen et al. [273, 274, 276], Cooper and Gore [307], Delsarte [343], Goethals [491, 494], Goethals and Delsarte [499], Gore and Cooper [548], Hartmann et al. [611], Kasami et al. [732, 740, 741], Lin et al. [833, 835–837, 840], Rao [1092], Smith [1246] and Weldon [1402].

§9. The class of codes which are invariant under the general affine group has been studied by Kasami, Lin and Peterson [738] and Delsarte [344].

§11. Problem 11 is from Berlekamp [113, Ch. 11].

14

First-order Reed–Muller codes

§1. Introduction

In this chapter we continue the study of Reed–Muller codes begun in the previous chapter, concentrating on the first-order codes $\mathcal{R}(1, m)$. Under the name of pseudo-noise or PN sequences, the codewords of $\mathcal{R}(1, m)$, or more precisely, of the simplex code $\mathcal{S}_m$, are widely used in range-finding, synchronizing, modulation, and scrambling, and in §2 we describe their properties. In §3 the difficult (and essentially unsolved) problem of classifying the cosets of $\mathcal{R}(1, m)$ is investigated. Then §4 describes the encoding and decoding of $\mathcal{R}(1, m)$. This family of codes is one of the few for which maximum likelihood decoding is practical, using the Green Machine decoder. (One of these codes was used to transmit pictures from Mars.) The final section deals with cosets of greatest minimum weight. These correspond to Boolean functions which are least like linear functions, the so-called bent functions.

§2. Pseudo-noise sequences

The codewords (except for $\mathbf{0}$ and $\mathbf{1}$) of the cyclic $[2^m - 1, m, 2^{m-1}]$ simplex code $\mathcal{S}_m$ or the $[2^m, m + 1, 2^{m-1}]$ extended cyclic first-order Reed–Muller code $\mathcal{R}(1, m)$ resemble random sequences of 0's and 1's (Fig. 14.1). In fact we shall see that if c is any nonzero codeword of $\mathcal{S}_m$, then c has many of the properties that we would expect from a sequence obtained by tossing a fair coin $2^m - 1$ times. For example, the number of 0's and the number of 1's in c are as nearly equal as they can be. Also, define a *run* to be maximal string of consecutive identical symbols. Then one half of the runs in c have length 1, one quarter

$$
\begin{array}{ccccccc}
0 & 0 & 0 & 0 & 0 & 0 & 0 \\
1 & 1 & 1 & 0 & 1 & 0 & 0 \\
0 & 1 & 1 & 1 & 0 & 1 & 0 \\
0 & 0 & 1 & 1 & 1 & 0 & 1 \\
1 & 0 & 0 & 1 & 1 & 1 & 0 \\
0 & 1 & 0 & 0 & 1 & 1 & 1 \\
1 & 0 & 1 & 0 & 0 & 1 & 1 \\
1 & 1 & 0 & 1 & 0 & 0 & 1 \\
\end{array}
$$

Fig. 14.1. Codewords of the [7, 3, 4] cyclic simplex code.

have length 2, one eighth have length 3, and so on. In each case the number of runs of 0 is equal to the number of runs of 1. Perhaps the most important property of c is that its auto-correlation function is given by

$$
\rho(0) = 1, \rho(\tau) = -\frac{1}{2^m - 1} \quad \text{for} \quad 1 \le \tau \le 2^m - 2.
$$

(see Fig. 14.4).

This randomness makes these codewords very useful in a number of applications, such as range-finding, synchronizing, modulation, scrambling, etc.

Of course the codewords are not really random, and one way this shows up is that the properties we have mentioned hold for *every* nonzero codeword in a simplex code, whereas in a coin-tossing experiment there would be some variation from sequence to sequence. (For this reason these codewords are unsuitable for serious encryption.)

These codewords can be generated by shift registers, and we now describe how this is done and give their properties.

Let

$$
h(x) = x^m + h_{m-1}x^{m-1} + \cdots + h_1 x + 1
$$

be a primitive irreducible polynomial of degree m over GF(2) (see Ch. 4). As in §3 of Ch. 8, $h(x)$ is the check polynomial of the $[2^m - 1, m, 2^{m-1}]$ simplex code $\mathcal{S}_m$. We construct a linear feedback shift register whose feedback connections are defined by $h(x)$, as in Fig. 14.2.

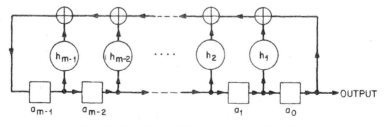

Fig. 14.2. Feedback shift register defined by $h(x)$.

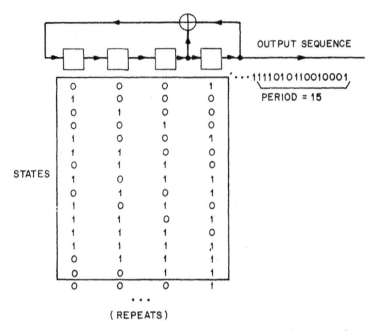

Fig. 14.3. Shift register corresponding to $h(x) = x^4 + x + 1$, showing successive states.

For example, if $h(x) = x^4 + x + 1$, the shift register is shown in Fig. 14.3. Suppose the initial contents (or *state*) of the shift register are $a_{m-1}, \ldots, a_1, a_0$ as in Fig. 14.2. The output is taken from the right-hand end of the register, and is the infinite binary sequence $a = a_0 a_1 a_2 \ldots$.

Definition. For any nonzero initial state, the output a is called a *pseudo-noise (or PN) sequence*. (These sequences are also called pseudo-random sequences, m-sequences, or 'maximal length feedback shift register sequences.) An example is shown in Fig. 14.3, which gives the successive states of the shift register if the initial state is 0001. The output sequence is the 4th column, i.e.,

$$a = a_0 a_1 \cdots = 100\ 010\ 011\ 010\ 111, 100 \ldots \qquad (1)$$

having period 15.

Properties of a PN sequence.

Property I. *A PN sequence a satisfies the recurrence*

$$a_l = a_{l-1} h_{m-1} + \cdots + a_{l-m+1} h_1 + a_{l-m},$$

for $l = m, m+1, \ldots$.

Property II. *For some initial state $a_0 \ldots a_{m-1}$ the sequence a is periodic with period $n = 2^m - 1$, i.e. $a_{n+i} = a_i$ for all $i \geq 0$, and n is the smallest number with this property.*

Proof. Observe that for any a, Property (I) implies

$$
\begin{pmatrix} a_{l-m+1} \\ a_{l-m} \\ \vdots \\ a_{l-1} \\ a_l \end{pmatrix}
=
\begin{pmatrix} 0 & 1 & 0 & \ldots & 0 \\ 0 & 0 & 1 & \ldots & 0 \\ \ldots\ldots\ldots\ldots\ldots \\ 0 & 0 & 0 & \ldots & 1 \\ 1 & h_1 & h_2 & \ldots & h_{m-1} \end{pmatrix}
\begin{pmatrix} a_{l-m} \\ a_{l-m-1} \\ \vdots \\ a_{l-2} \\ a_{l-1} \end{pmatrix}.
$$

Call this $m \times m$ matrix U. Then

$$
\begin{pmatrix} a_{l-m+1} \\ \vdots \\ a_l \end{pmatrix}
= U \begin{pmatrix} a_{l-m} \\ \vdots \\ a_{l-1} \end{pmatrix}
= U^2 \begin{pmatrix} a_{l-m-1} \\ \vdots \\ a_{l-2} \end{pmatrix}
= \cdots
$$

$$
= U^{l-m+1} \begin{pmatrix} a_0 \\ \vdots \\ a_{m-1} \end{pmatrix}.
$$

Now U is the companion matrix of $h(x)$ (see §3 of Ch. 4), and by problem 20 of Ch. 4, $n = 2^m - 1$ is the smallest number such that $U^n = I$. Therefore there is a vector $b = (a_0 \ldots a_{m-1})$ such that $U^n b^T = b^T$, and $U^i b^T \neq b^T$ for $1 \leq i \leq n - 1$. If the initial state is taken to be b, then a has period $2^m - 1$.   Q.E.D.

Property III. *With this initial state, the shift register goes through all possible $2^m - 1$ nonzero states before repeating.*

Proof. Obvious because the period is $2^m - 1$.   Q.E.D.

Note that the zero state doesn't occur, unless a is identically zero. Also $2^m - 1$ is the maximum possible period for an m-stage linear shift register.

Property IV. *For any nonzero initial state, the output sequence has period $2^m - 1$, and in fact is obtained by dropping some of the initial digits from a.*

Proof. This follows from (III).   Q.E.D.

Now let us examine the properties of any segment

$$c = a_l \ldots a_{l+n-1}$$

of length $n = 2^m - 1$ from a.

Property V. *c belongs to the simplex code $\mathscr{S}_m$ with check polynomial $h(x)$.*

Proof. c is clearly in the code with parity check polynomial $h(x)$, since c satisfies Equation (12) of Ch. 7. Since $h(x)$ is a primitive polynomial, this is a simplex code (by §3 of Ch. 8).

Property VI. (The shift-and-add property.) *The sum of any segment c with a cyclic shift of itself is another cyclic shift of c.*

Proof. Immediate from V. 　　　　　　　　　　　　　　　　　　　　　Q.E.D.

Problems. (1) (The window property.) Show that if a window of width m is slid along a PN sequence then each of the $2^m - 1$ nonzero binary m-tuples is seen exactly once in a period.

(2) Consider an m-stage shift register defined by $h(x)$ (as in Fig. 14.2), where $h(x)$ is not necessarily a primitive polynomial. If the initial state is $a_0 = 1, a_1 = \cdots = a_{m-1} = 0$, show that the period of a is p, where p is the smallest positive integer for which $h(x)$ divides $x^p - 1$.

(3) If $h(x)$ is *irreducible*, but not necessarily primitive, then p divides $2^m - 1$.

(4) If $a = a_0 a_1 \ldots$ is a PN sequence, then so is $b = a_0 a_j a_{2j} \ldots$ if j is relatively prime to $2^m - 1$.

(5) Some shift of a, i.e. $b = a_s a_{s+1} a_{s+2} \cdots = b_0 b_1 b_2 \ldots$ (say) has the property that $b_i = b_{2i}$ for all i. (b just consists of repetitions of the idempotent of the simplex code – see §4 of Ch. 8.)

Pseudo-randomness properties.

Property VII. *In any segment c there are 2^{m-1} 1's and $2^{m-1} - 1$ 0's.*

Proof. From (V). 　　　　　　　　　　　　　　　　　　　　　　　　　Q.E.D.

Property VIII. *In c, one half of the runs have length 1, one quarter have length 2, one eighth have length 3, and so on, as long as these fractions give integral numbers of runs. In each case the number of runs of 0's is equal to the number of runs of 1's.*

Problem. (6) Prove this.

Autocorrelation function. We come now to the most important property, the autocorrelation function. The *autocorrelation function* $\rho(\tau)$ of an infinite real or complex sequence $s_0 s_1 s_2 \ldots$ of period n is defined by

$$\rho(\tau) = \frac{1}{n}\sum_{j=0}^{n-1} s_j \bar{s}_{\tau+j} \quad \text{for } \tau = 0, \pm 1, \pm 2, \ldots \tag{2}$$

where the bar denotes complex conjugation. This is a periodic function: $\rho(\tau) = \rho(\tau+n)$. The autocorrelation function of a *binary* sequence $a_0 a_1 \ldots$ of period n is then defined to be the autocorrelation function of the real sequence $(-1)^{a_0}, (-1)^{a_1}, \ldots$ obtained by replacing 1's by -1's and 0's by $+1$'s. Thus

$$\rho(\tau) = \frac{1}{n}\sum_{j=0}^{n-1} (-1)^{a_j + a_{\tau+j}}. \tag{3}$$

Alternatively, let A be the number of places where $a_0 \ldots a_{n-1}$ and the cyclic shift $a_\tau a_{\tau+1} \ldots a_{\tau+n-1}$ agree, and D the number of places where they disagree (so $A + D = n$). Then

$$\rho(\tau) = \frac{A - D}{n}. \tag{4}$$

For example, the PN sequence (1) has autocorrelation function $\rho(0) = 1$, $\rho(\tau) = -\frac{1}{15}$ for $1 \leq \tau \leq 14$, as shown in Fig. 14.4.

Property IX. *The autocorrelation function of a* PN *sequence of period* $n = 2^m - 1$ *is given by*

$$\rho(0) = 1$$

$$\rho(\tau) = -\frac{1}{n} \quad \text{for } 1 \leq \tau \leq 2^m - 2. \tag{5}$$

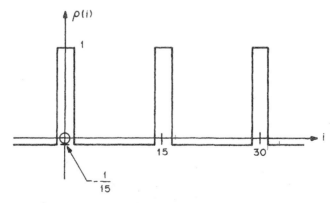

Fig. 14.4. Autocorrelation function of a PN sequence.

Proof. From (4)

$$\rho(\tau) = \frac{n - 2d}{n} \tag{6}$$

where $d = \text{dist}\,(a_0 \ldots a_{n-1}, a_\tau \ldots a_{\tau+n-1}) = \text{wt}\,(a_\sigma \ldots a_{\sigma+n-1})$ for some σ, by (VI). The result then follows from (V) and (VII).   Q.E.D.

Problems. (7) Show that (5) is the best possible autocorrelation function of any binary sequence of period $n = 2^m - 1$, in the sense of minimizing $\max_{0 < i < n} \rho(i)$.

(8) (A test to distinguish a PN sequence from a coin-tossing sequence.)
Let $c_0, \ldots, c_{N-1}$ be N consecutive binary digits from a PN sequence of period $2^m - 1$, where $N < 2^m - 1$, and form the matrix

$$M = \begin{bmatrix} c_0 & c_1 & \cdots & c_{N-b} \\ c_1 & c_2 & \cdots & c_{N-b+1} \\ \cdots\cdots\cdots\cdots\cdots\cdots \\ c_{b-1} & c_b & \cdots & c_{N-1} \end{bmatrix},$$

where $m < b < \frac{1}{2}N$. Show that the rank of M over GF(2) is less than b. [Hint: there are only m linearly independent sequences in $\mathcal{S}_m$.] On the other hand, show that if $c_0, \ldots, c_{N-1}$ is a segment of a coin-tossing sequence, where each c_i is 0 or 1 with probability $\frac{1}{2}$, then the probability that rank $(M) < b$ is at most 2^{2b-N-1}. This is very small if $b \ll \frac{1}{2}N$.

Thus the question "Is rank $(M) = b$?" is a test on a small number of digits from a PN sequence which shows a departure from true randomness. E.g. if $m = 11$, $2^m - 1 = 2047$, $b = 15$, the test will fail if applied to any $N = 50$ consecutive digits of a PN sequence, whereas the probability that a coin-tossing sequence fails is at most 2^{-21}.

§3. Cosets of the first-order Reed–Muller code

The problem of enumerating the cosets of a first-order Reed–Muller code arises in various practical situations. The general problem is unsolved (see Research Problem 1.1 of Chapter 1), although the cosets have been completely enumerated for $n \leqslant 32$. However, there are a few interesting things we can say. For example, finding the weight distribution of the coset containing v amounts to finding the Hadamard transform of v. Also enumerating the cosets of $\mathcal{R}(1, m)$ is equivalent to classifying Boolean functions modulo linear functions. The properties of the cosets described here will also be useful in decoding (see §4). We end this section with a table of cosets of $\mathcal{R}(1, 4)$. Those cosets of greatest minimum weight are especially interesting and will be studied in the final section of the chapter.

Notation. As in §2 of the preceding chapter, $v = (v_1, \ldots, v_m)$ denotes a vector which ranges over V^m, and if $f(v_1, \ldots v_m)$ is any Boolean function, f is the corresponding binary vector of length 2^m.

The first-order Reed–Muller code $\mathcal{R}(1, m)$ consists of all vectors

$$u \underset{0}{\mathbf{1}} + \sum_{i=1}^{m} u_i v_i, \quad u_i = 0 \text{ or } 1, \tag{7}$$

corresponding to linear Boolean functions. Define the *orthogonal code* $\mathcal{O}_m$ to be the $[2^m, m, 2^{m-1}]$ code consisting of the vectors

$$\sum_{i=1}^{m} u_i v_i = u \cdot v, \quad u_i \in \{0, 1\}.$$

Thus

$$\mathcal{R}(1, m) = \mathcal{O}_m \cup (1 + \mathcal{O}_m).$$

Suppose that in the codewords of $\mathcal{R}(1, m)$ we replace 1's by -1's and 0's by 1's. The resulting set of 2^{m+1} real vectors are the coordinates of the vectors of a regular figure in 2^m-dimensional Euclidean space, called a *cross-polytope* (i.e. a generalized octahedron). For example, when $m = 2$, we obtain the 4-dimensional cross-polytope (also called a *16-cell*) shown in Fig. 14.5.

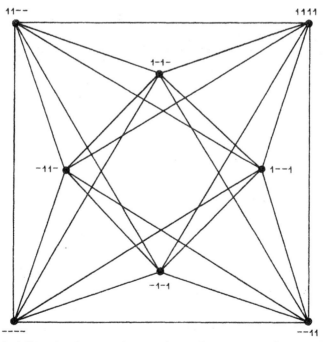

Fig. 14.5. 4-dimensional cross-polytope, with vertices corresponding to codewords of $\mathcal{R}(1, 2)$.

This set of real vectors is also called a biorthogonal signal set – see Problem 43 of Ch. 1.

If the same transformation is applied to the codewords of $\mathcal{O}_m$, we obtain a set of 2^m mutually orthogonal real vectors.

For any vector $u = (u_1, \ldots, u_m)$ in V^m, $f(u)$ will denote the value of f at u, or equally the component of f in the place corresponding to u.

It will be convenient to have a name for the real vector obtained from a binary vector f by replacing 1's by -1's and 0's by $+1$'s – call it F. Thus the component of F in the place corresponding to u is

$$F(u) = (-1)^{f(u)}$$

Hadamard transforms and cosets of $\mathcal{R}(1, m)$. Recall from Ch. 2 that the Hadamard transform of a real vector F is given by

$$\hat{F}(u) = \sum_{v \in V^m} (-1)^{u \cdot v} F(v), \quad u \in V^m,$$

$$= \sum_{v \in V^m} (-1)^{u \cdot v + f(v)} \tag{8}$$

$\hat{F}$ is a real vector of length 2^m. Alternatively

$$\hat{F} = FH,$$

where H is the $2^m \times 2^m$ symmetric Hadamard matrix given by

$$H = (H_{u,v}), \quad H_{u,v} = (-1)^{u \cdot v}, \quad u, v \in V^m. \tag{9}$$

Consequently

$$F = \frac{1}{2^m} \hat{F} H, \tag{10}$$

or

$$F(v) = \frac{1}{2^m} \sum_{u \in V^m} (-1)^{u \cdot v} \hat{F}(u). \tag{11}$$

Observe from (8) that $\hat{F}(u)$ is equal to the number of 0's minus the number of 1's in the binary vector

$$f + \sum_{i=1}^{m} u_i v_i.$$

Thus

$$\hat{F}(u) = 2^m - 2 \, \text{dist} \left\{ f, \sum_{i=1}^{m} u_i v_i \right\}, \tag{12}$$

or

$$\text{dist} \left\{ f, \sum_{i=1}^{m} u_i v_i \right\} = \tfrac{1}{2} \{ 2^m - \hat{F}(u) \}. \tag{13}$$

Also

$$\text{dist}\left\{f, 1 + \sum_{i=1}^{m} u_i v_i\right\} = \tfrac{1}{2}\{2^m + \hat{F}(u)\} \tag{14}$$

Now the weight distribution of that coset (of a code $\mathscr{C}$) which contains f gives the distances of f from the codewords of $\mathscr{C}$. Therefore we have proved:

Theorem 1. *The weight distribution of that coset of $\mathscr{R}(1, m)$ which contains f is*

$$\frac{1}{2}\{2^m \pm \hat{F}(u)\} \quad \text{for } u \in V^m.$$

The weight distribution of the coset containing f is thus determined by the Hadamard transform of F.

Problem. (9) If the coset of $\mathscr{O}_m$ containing f has weight distribution $A'_i(f)$, $0 \le i \le 2^m$, show that the coset of $\mathscr{R}(1, m)$ containing f has weight distribution $A_i(f) = A'_i(f) + A'_{2^m-i}(f)$, $0 \le i \le 2^m$.

Equations 13 and 14 say that the closest codeword of $\mathscr{R}(1, m)$ to f is that $\sum u_i v_i$ for which $|\hat{F}(u)|$ is largest.

Example. For $m = 2$, we have

$$v_2 = 0\ \ 0\ \ 1\ \ 1$$
$$v_1 = 0\ \ 1\ \ 0\ \ 1$$
$$\text{Suppose} \quad f\ = 0\ \ 0\ \ 0\ \ 1 = v_1 v_2.$$
$$\text{Then} \qquad F = 1\ \ 1\ \ 1\ \ - \quad (\text{where} - \text{stands for} - 1).$$

The Hadamard transform coefficients of F are, from (8),

$$\hat{F}(00) = F(00) + F(01) + F(10) + F(11)$$
$$= 1 + 1 + 1 - 1 = 2,$$
$$\hat{F}(01) = F(00) - F(01) + F(10) - F(11) = 2,$$
$$\hat{F}(10) = 2,$$
$$\hat{F}(11) = -2.$$

Indeed,

$$F = \frac{1}{4}\{2 \times (1111) + 2 \times (1\text{-}1\text{-}) + 2 \times (11\text{-}\text{-}\text{-})$$
$$- 2 \times (1\text{-}\text{-}1)\} = 111\text{-},$$

verifying (11).

The code $\mathscr{R}(1, 2)$ consists of the vectors $\{0000, 0101, 0011, 0110, 1111, 1010, 1100, 1001\}$. The weight distribution of the coset containing $f = 0001$ is, according to Theorem 1,

$$A_1(f) = 4, \qquad A_3(f) = 4,$$

which is indeed the case.

The Hadamard transform coefficients of a $(+1, -1)$- vector F satisfy the following orthogonality relation.

Lemma 2.

$$\sum_{u \in V^m} \hat{F}(u)\hat{F}(u + v) = \begin{cases} 2^{2m} & \text{if } v = 0 \\ 0 & \text{if } v \neq 0 \end{cases}$$

Proof.

$$\text{LHS} = \sum_{u \in V^m} \sum_{w \in V^m} (-1)^{u \cdot w} F(w) \cdot \sum_{x \in V^m} (-1)^{(u+v) \cdot x} F(x)$$

$$= \sum_{w,x \in V^m} (-1)^{v \cdot x} F(w) F(x) \sum_{u \in V^m} (-1)^{u \cdot (w+x)}.$$

The inner sum is equal to $2^m \delta_{w,x}$. Therefore

$$\text{LHS} = 2^m \sum_{w \in V^m} (-1)^{v \cdot w} F(w)^2$$

$$= 2^m \sum_{w \in V^m} (-1)^{v \cdot w}, \quad \text{since } F(w) = \pm 1,$$

$$= 2^{2m} \delta_{v,0}. \qquad\qquad\qquad \text{Q.E.D.}$$

Corollary 3. (Parseval's equation.)

$$\sum_{u \in V^m} \hat{F}(u)^2 = 2^{2m}. \tag{15}$$

Note that Corollary 3 and Equation (12) imply that the weight distribution $A_i'(f)$, $0 \le i \le 2^m$, of the coset $f + \mathcal{O}_m$ satisfies

$$\sum_{i=0}^{2^m} (2^m - 2i)^2 A_i'(f) = 2^{2m}.$$

Boolean functions and cosets of $\mathcal{R}(1, m)$. The codewords of $\mathcal{R}(1, m)$ are the linear functions (7) of $v_1, \ldots, v_m$. Let us say that two Boolean functions $f(v_1, \ldots, v_m)$ and $g(v_1, \ldots, v_m)$ are equivalent if the difference

$$f(v_1, \ldots, v_m) - g(v_1, \ldots, v_m)$$

is in $\mathcal{R}(1, m)$. If this is so, f and g are in the same coset of $\mathcal{R}(1, m)$. Equivalence classes of Boolean functions under this definition of equivalence are in 1–1– correspondence with the cosets of $\mathcal{R}(1, m)$.

Theorem 4. *Suppose the Boolean functions f and g are related by*

$$g(v) = f(Av + B), \tag{16}$$

for some invertible m × m binary matrix A and some binary m-tuple B. We say that g is obtained from f by an affine transformation. Then the cosets of $\mathcal{R}(1, m)$ containing f and g have the same weight distribution.

Proof. From Theorem 1 it is enough to show that the sets $\{\pm \hat{G}(u): u \in V^m\}$ and $\{\pm \hat{F}(u): u \in V^m\}$ are equal. In fact,

$$\hat{G}(u) = \sum_{v \in V^m} (-1)^{u \cdot v} G(v)$$

$$= \sum_{v \in V^m} (-1)^{u \cdot v} F(Av + B).$$

Set $v = A^{-1}w + A^{-1}B$, then

$$\hat{G}(u) = \sum_{w \in V^m} (-1)^{u \cdot A^{-1}w} (-1)^{u \cdot A^{-1}B} F(w)$$

$$= \pm \sum_{w \in V^m} (-1)^{u' \cdot w} F(w), \quad \text{where } u' = uA^{-1}$$

$$= \pm \hat{F}(u'). \qquad \text{Q.E.D.}$$

Therefore, in order to lump together cosets with the same weight distribution, we can introduce a stronger definition of equivalence. Namely we define f and g to be equivalent if

$$g(v) = f(Av + B) + a_0 1 + \sum_{i=1}^{m} a_i v_i \tag{17}$$

for some binary invertible matrix A, vector B and constants a_i. Now all Boolean functions in the same equivalence class belongs to cosets with the same weight distribution.

However, the cosets containing f and g may have the same weight distribution even if f and g are not related as in Equation (17). The first time this happens is when $n = 32$, when for example the cosets containing

$$f = v_1 v_2 + v_3 v_4$$

and

$$g = v_1 v_2 v_3 + v_1 v_4 v_5 + v_2 v_3 + v_2 v_4 + v_3 v_5$$

both have the weight distribution

$$A'_{12}(f) = A'_{20}(f) = 16, \qquad A'_{16}(f) = 32.$$

We conclude this section with a table giving the weight distribution of the cosets of $\mathcal{R}(1, 4)$ (Fig. 14.6). The table gives, for each weight distribution, the

Number	Typical Boolean function	Weight Distribution																	Remarks
		0	1	2	3	4	5	6	7	8	9	10	11	12	13	14	15	16	
1	–	1								30								1	The code itself
16	1234		1						15		15						1		
120	123			1				7		16		7				1			
560	1234 + 12				1		3		12		12		3		1				
840	123 + 14					2		8		12		8		2					
35	12					4				24				4					
448	1234 + 12 + 34	–					6		10		10		6						
28	12 + 34							16				16							Bent functions

Fig. 14.6. Cosets of first-order Reed–Muller code of length 16.

simplest corresponding Boolean function f and the number of cosets having this weight distribution. For example, the last line of the table means that there are 28 cosets with weight distribution

$$A_6(f) = A_{10}(f) = 16,$$

where $f = v_1 v_2 + v_3 v_4$ or is obtained from this by a substitution of the form shown in Equation (17). These 28 f's are called bent functions, for reasons which will be given in the final section (see also Problem 16).

§4. Encoding and decoding $\mathcal{R}(1, m)$

General techniques for encoding and decoding Reed–Muller codes were described in Ch. 13. However there are special techniques for first-order Reed–Muller codes which are described in this section.

$\mathcal{R}(1, m)$ is a $[2^m, m + 1, 2^{m-1}]$ code, and so has low rate and can correct many errors. It is therefore particularly suited to very noisy channels. For example $\mathcal{R}(1, 5)$ was succesfully used in the Mariner 9 spacecraft to transmit pictures from Mars, one of which is shown in Fig. 14.7. (Each dot in the picture was assigned one of 64 levels of greyness and encoded into a 32-bit codeword.)

Encoding is especially simple. We describe the method by giving the encoder for $\mathcal{R}(1, 3)$, an $[8, 4, 4]$ code.

A message $u_0 u_1 u_2 u_3$ is to be encoded into the codeword

$$(x_0 x_1 \ldots x_7) = (u_0 u_1 u_2 u_3) \begin{bmatrix} 1 & 1 & 1 & 1 & 1 & 1 & 1 & 1 \\ 0 & 0 & 0 & 0 & 1 & 1 & 1 & 1 \\ 0 & 0 & 1 & 1 & 0 & 0 & 1 & 1 \\ 0 & 1 & 0 & 1 & 0 & 1 & 0 & 1 \end{bmatrix} \tag{18}$$

This is accomplished by the circuit shown in Fig. 14.8. The clock circuit in Fig. 14.8 goes through the successive states $t_1 t_2 t_3 = 000, 001, 010, 011, 100, \ldots$, $111, 000, 001, \ldots$ (i.e., counts from 0 to 7). The circuit forms

$$u_0 + t_1 u_1 + t_2 u_2 + t_3 u_3$$

which, from Equation (18), is the codeword $x_0 x_1 \ldots x_7$. Nothing could be simpler.

Decoding. As we saw in §3 of Ch. 1, maximum likelihood decoding requires comparing the received vector f with every codeword (7) of $\mathcal{R}(1, m)$. I.e., we must find the distance from f to every codeword of $\mathcal{R}(1, m)$, and then decode f as the closest codeword. From Equations (13) and (14) above, this amounts to finding the largest component $|\hat{F}(u)|$, where $\hat{F}$ is the Hadamard transform of F given by Equation (8). Suppose the largest component is $|\hat{F}(u_1, \ldots, u_m)|$.

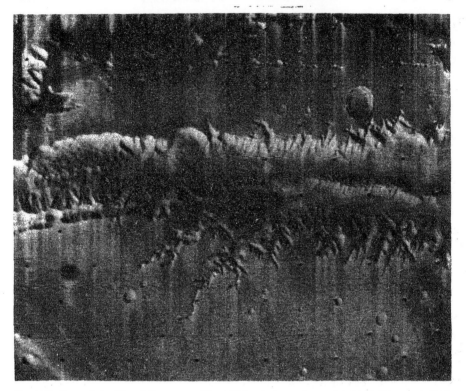

Fig. 14.7. Part of the Grand Canyon on Mars. This photograph was transmitted by the Mariner 9 spacecraft on 19 January 1972 using the first-order Reed–Muller code $\mathscr{R}(1, 5)$. Photograph courtesy of NASA/JPL.

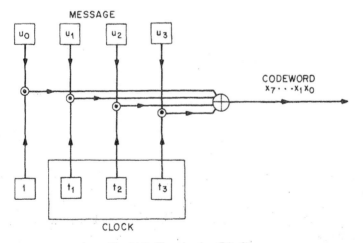

Fig. 14.8. Encoder for $\mathscr{R}(1, 3)$.

If $\hat{F}(u_1, \ldots, u_m) \geqslant 0$ we decode f as

$$\sum_{i=1}^{m} u_i v_i,$$

(from (13)), whereas if $\hat{F}(u_1, \ldots, u_m) < 0$ we decode f as

$$1 + \sum_{i=1}^{m} u_i v_i.$$

Direct calculation of $\hat{F} = FH_{2^m}$ by multiplying F and H_{2^m} would require about $2^m \times 2^m = 2^{2m}$ additions and subtractions. Fortunately there is a faster way to obtain $\hat{F}$, which is a discrete version of the so-called Fast Fourier Transform. This is possible because H_{2^m} can be written as a product of m $2^m \times 2^m$ matrices, each of which has only two non-zero elements per column. Thus only $m2^m$ additions and subtractions are needed to evaluate $\hat{F} = FH_{2^m}$.

In order to explain this decomposition of H_{2^m} we must introduce Kronecker products.

Kronecker product of matrices.

Definition. If $A = (a_{ij})$ is an $m \times m$ matrix and $B = (b_{ij})$ is an $n \times n$ matrix over any field, the *Kronecker product* of A and B is the $mn \times mn$ matrix obtaining from A by replacing every entry a_{ij} by $a_{ij}B$. This product is written $A \otimes B$. Symbolically we have

$$A \otimes B = (a_{ij}B).$$

For example if

$$I_2 = \begin{pmatrix} 1 & 0 \\ 0 & 1 \end{pmatrix} \quad \text{and} \quad H_2 = \begin{pmatrix} 1 & 1 \\ 1 & - \end{pmatrix},$$

$$I_2 \otimes H_2 = \begin{bmatrix} 1 & 1 & 0 & 0 \\ 1 & - & 0 & 0 \\ 0 & 0 & 1 & 1 \\ 0 & 0 & 1 & - \end{bmatrix} \tag{19}$$

$$H_2 \otimes I_2 = \begin{bmatrix} 1 & 0 & 1 & 0 \\ 0 & 1 & 0 & 1 \\ 1 & 0 & - & 0 \\ 0 & 1 & 0 & - \end{bmatrix} \tag{20}$$

This shows that in general $A \otimes B \neq B \otimes A$.

Problem. (10) Prove the following properties of the Kronecker product.
  (i) Associative law:

$$A \otimes (B \otimes C) = (A \otimes B) \otimes C$$

(ii) Distributive law:

$$(A + B) \otimes C = A \otimes C + B \otimes C.$$

(iii)

$$(A \otimes B)(C \otimes D) = (AC) \otimes (BD). \tag{21}$$

Hadamard matrices. Let us define the Hadamard matrix H_{2^m} of order 2^m inductively by

$$H_2 = \begin{pmatrix} 1 & 1 \\ 1 & - \end{pmatrix},$$

$$H_{2^m} = H_2 \otimes H_{2^{m-1}}, \quad \text{for } m \geq 2.$$

H_{2^m} is the Sylvester-type Hadamard matrix of Ch. 2 (see Fig. 2.3; it is also the matrix given by Equation (9) above provided u and v take on values in V^m in the right order.

Theorem 5. (The fast Hadamard transform theorem.)

$$H_{2^m} = M_{2^m}^{(1)} M_{2^m}^{(2)} \cdots M_{2^m}^{(m)}, \tag{22}$$

where

$$M_{2^m}^{(i)} = I_{2^{m-i}} \otimes H_2 \otimes I_{2^{i-1}}, \quad 1 \leq i \leq m,$$

and I_n is an $n \times n$ unit matrix.

Proof. By induction on m. For $m = 1$ the result is obvious. Assume the result is true for m. Then for $1 \leq i \leq m$

$$\begin{aligned} M_{2^{m+1}}^{(i)} &= I_{2^{m+1-i}} \otimes H_2 \otimes I_{2^{i-1}} \\ &= I_2 \otimes I_{2^{m-i}} \otimes H_2 \otimes I_{2^{i-1}} \\ &= I_2 \otimes M_{2^m}^{(i)}, \end{aligned}$$

and

$$M_{2^{m+1}}^{(m+1)} = H_2 \otimes I_{2^m}.$$

Therefore

$$\begin{aligned} M_{2^{m+1}}^{(1)} \cdots M_{2^{m+1}}^{(m+1)} &= (I_2 \otimes M_{2^m}^{(1)}) \cdots (I_2 \otimes M_{2^m}^{(m)})(H_2 \otimes I_{2^m}) \\ &= H_2 \otimes (M_{2^m}^{(1)} \cdots M_{2^m}^{(m)}) \quad \text{by (21)}, \\ &= H_2 \otimes H_{2^m} \quad \text{by the induction hypothesis}, \\ &= H_{2^{m+1}}. \qquad\qquad\qquad\qquad\qquad\qquad \text{Q.E.D.} \end{aligned}$$

Example. For $m = 2$,

$$M_4^{(1)} = I_2 \otimes H_2, \qquad M_4^{(2)} = H_2 \otimes I_2,$$

(see (19) and (20)), and indeed

$$M_4^{(1)} M_4^{(2)} = \begin{pmatrix} 1 & 1 & 0 & 0 \\ 1 & - & 0 & 0 \\ 0 & 0 & 1 & 1 \\ 0 & 0 & 1 & - \end{pmatrix} \begin{pmatrix} 1 & 0 & 1 & 0 \\ 0 & 1 & 0 & 1 \\ 1 & 0 & - & 0 \\ 0 & 1 & 0 & - \end{pmatrix} = \begin{pmatrix} 1 & 1 & 1 & 1 \\ 1 & - & 1 & - \\ 1 & 1 & - & - \\ 1 & - & - & 1 \end{pmatrix} = H_4.$$

For $m = 3$,

$$M_8^{(1)} = I_4 \otimes H_2,$$
$$M_8^{(2)} = I_2 \otimes H_2 \otimes I_2,$$
$$M_8^{(3)} = H_2 \otimes I_4.$$

Decoding circuit: the Green machine. We now give a decoding circuit for $\mathcal{R}(1, m)$ which is based on Theorem 5. This circuit is called the *Green machine* after its discoverer R.R. Green. We illustrate the method by describing the decoder for $\mathcal{R}(1, 3)$.

Suppose $f = f_0 f_1 \cdots f_7$ is the received vector, and let

$$\begin{aligned} F &= F_0 F_1 \cdots F_7 \\ &= ((-1)^{f_0}, (-1)^{f_1}, \ldots, (-1)^{f_7}). \end{aligned}$$

We wish to find

$$\begin{aligned} \hat{F} &= FH_8 \\ &= (F_0 F_1 \cdots F_7) M_8^{(1)} M_8^{(2)} M_8^{(3)}, \quad \text{from (22).} \end{aligned}$$

Now

$$M_8^{(1)} = \begin{bmatrix} 1 & 1 & & & & & & \\ 1 & - & & & & & & \\ & & 1 & 1 & & & & \\ & & 1 & - & & & & \\ & & & & 1 & 1 & & \\ & & & & 1 & - & & \\ & & & & & & 1 & 1 \\ & & & & & & 1 & - \end{bmatrix}$$

So

$$FM_8^{(1)} = (F_0 + F_1, F_0 - F_1, F_2 + F_3, \ldots, F_6 + F_7, F_6 - F_7).$$

The circuit shown in Fig. 14.9 calculates the components of $FM_8^{(1)}$ two at a time. The switches are arranged so that after $F_0 \cdots F_3$ have been read in, the two-stage registers on the right contain $(F_0 - F_1, F_0 + F_1)$ and $(F_2 - F_3, F_2 + F_3)$ respectively. These four quantities are used in the second stage (see Fig. 14.10). Then $F_4 \cdots F_7$ are read in, and $(F_4 - F_5, F_4 + F_5)$ and $(F_6 - F_7, F_6 + F_7)$ are formed in the same pair of two-stage registers.

The second stage calculates

$$(F_0 + F_1, F_0 - F_1, \ldots, F_6 - F_7) M_8^{(2)}, \tag{23}$$

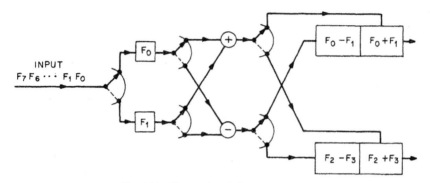

Fig. 14.9. First stage of Green machine.

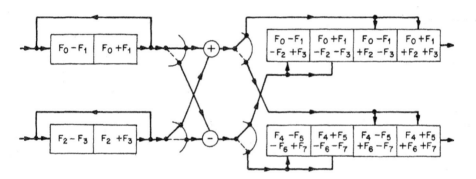

Fig. 14.10. Second stage of Green machine.

where

$$
M_8^{(2)} =
\begin{bmatrix}
1 & & 1 & & & & & \\
 & 1 & & 1 & & & & \\
1 & & - & & & & & \\
 & 1 & & - & & & & \\
 & & & & 1 & & 1 & \\
 & & & & & 1 & & 1 \\
 & & & & 1 & & - & \\
 & & & & & 1 & & - \\
\end{bmatrix}.
$$

So the product (23) is

$$(F_0 + F_1 + F_2 + F_3, \; F_0 - F_1 + F_2 - F_3, \ldots, F_4 - F_5 - F_6 + F_7).$$

This product is formed by the circuit shown in Fig. 14.10. The third stage calculates

$$(F_0 + F_1 + F_2 + F_3, \ldots, F_4 - F_5 - F_6 + F_7)M_8^{(3)}, \tag{24}$$

$$
M_8^{(3)} = \begin{bmatrix}
1 & & & & 1 & & & \\
& 1 & & & & 1 & & \\
& & 1 & & & & 1 & \\
& & & 1 & & & & 1 \\
1 & & & & - & & & \\
& 1 & & & & - & & \\
& & 1 & & & & - & \\
& & & 1 & & & & -
\end{bmatrix}.
$$

So the product (24) is

$$
\begin{aligned}
&= (F_0 + F_1 + \cdots + F_7, \; F_0 - F_1 + F_2 - F_3 + F_4 - F_5 + F_6 - F_7, \ldots, F_0 - F_1 - F_2 \\
&\qquad\qquad\qquad\qquad\qquad\qquad\qquad\qquad\qquad\qquad\quad + F_3 - F_4 + F_5 + F_6 - F_7) \\
&= (\hat{F}_0 \hat{F}_1 \hat{F}_2 \cdots \hat{F}_7)
\end{aligned}
$$

which are desired Hadamard transform components. These are formed by the
circuit shown in Fig. 14.11.

Figures 14.9–14.11 together comprise the Green machine. The final stage is
to find that i for which $|\hat{F}_i|$ is largest. Then f is decoded either as the ith
codeword of $\mathcal{R}(1, 3)$ if $\hat{F}_i \geqslant 0$, or as the complement of the ith codeword if
$\hat{F}_i < 0$.

Note that the Green machine has the useful property that the circuit for
decoding $\mathcal{R}(1, m + 1)$ is obtained from that for $\mathcal{R}(1, m)$ by adding an extra
register to the mth stage and then adding one more stage.

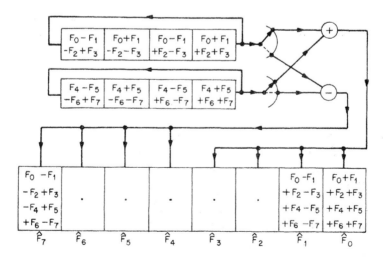

Fig. 14.11. Third stage of Green machine.

Problem. (11) Show that $M_{2^h}^{(j)} M_{2^k}^{(j)} = M_{2^k}^{(j)} M_{2^h}^{(j)}$. (This implies that the order of the stages in the decoder may be changed without altering the final output.)

§5. Bent Functions

Cosets of a first-order Reed–Muller code with the largest minimum weight are especially interesting. When m is even the corresponding Boolean functions are called *bent* functions, because they are in some sense furthest away from linear functions. In this section we study their properties and give several constructions. Bent functions will be used in the next chapter in the construction of the nonlinear Kerdock codes.

Definition. A Boolean function $f(v_1, \ldots, v_m)$ is called *bent* if the Hadamard transform coefficients $\hat{F}(u)$ given by Equation (8) are all $\pm 2^{m/2}$.

Examples. (1) $f(v_1, v_2) = v_1 v_2$ is a bent function, since the $\hat{F}(u)$ are all ± 2 (see the example preceding Lemma 2).

(2) $f(v_1, v_2, v_3, v_4) = v_1 v_2 + v_3 v_4$ is bent, as shown by the last row of Fig. 14.6.

Since $\hat{F}(u)$ is an integer (from Equation (8)) if $f(v_1, \ldots, v_m)$ is bent then m must be even. From now on we assume m is even and ≥ 2.

Theorem 6. *A bent function* $f(v_1, \ldots, v_m)$ *is further away from any linear function*

$$a_0 1 + \sum_{i=1}^{m} a_i v_i$$

than any other Boolean function. More precisely, $f(v_1, \ldots, v_m)$ *is bent iff the corresponding vector f has distance* $2^{m-1} \pm 2^{m/2-1}$ *from every codeword of* $\mathcal{R}(1, m)$. *If f is not bent, f has distance less than* $2^{m-1} - 2^{m/2-1}$ *from some codeword of* $\mathcal{R}(1, m)$.

Proof. If f is not bent, then the $\hat{F}(u)$ are not all $\pm 2^{m/2}$. From Corollary 3, since there are 2^m summands in Equation (15), some $|\hat{F}(u)|$ must be bigger than $2^{m/2}$. Therefore from Equation (13) or (14), the distance between f and some codeword of $\mathcal{R}(1, m)$ is less than $2^{m-1} - 2^{m/2-1}$.   Q.E.D.

Theorem 7. $f(v_1, \ldots, v_m)$ *is bent iff the* $2^m \times 2^m$ *matrix H whose* $(u, v)^{\text{th}}$ *entry is* $(1/2^{m/2}) \hat{F}(u + v)$ *is a Hadamard matrix.*

Proof. From Lemma 2.   Q.E.D.

Note that if $f(v_1, \ldots, v_m)$ is bent, then we may write

$$\frac{\hat{F}(u)}{2^{m/2}} = (-1)^{\hat{f}(u)}, \tag{25}$$

which defines a Boolean function $\hat{f}(u_1, \ldots, u_m)$. The Hadamard transform coefficients of $\hat{f}$ (obtained by setting $f = \hat{f}$ in Equation (8)) are $2^{m/2}(-1)^{f(u)} = \pm 2^{m/2}$. Therefore $\hat{f}$ is also a bent function!

Thus there is a natural pairing $f \leftrightarrow \hat{f}$ of bent functions.

Problem. (12) Show that f is bent iff the matrix whose $(u, v)^{\text{th}}$ entry is $(-1)^{f(u+v)}$, for $u, v \in V^m$, is a Hadamard matrix.

Theorem 8. *If $f(v_1, \ldots, v_m)$ is bent and $m > 2$, then $\deg f \leq \frac{1}{2}m$.*

Proof. Suppose f is bent and $m > 2$. The proof uses the expansion of f given by Theorem 1 of Ch. 13, and requires a lemma.

Let $F(u) = (-1)^{f(u)}$, let $\hat{F}(u)$ be the Hadamard transform of $F(u)$ given by Equation (8), and let $\hat{f}(u)$ be as in Equation (25).

Lemma 9. *If $\mathscr{C}$ is any $[m, k]$ code,*

$$\sum_{u \in \mathscr{C}^\perp} f(u) = 2^{m-k-1} - 2^{\frac{1}{2}m-1} + 2^{\frac{1}{2}m-k} \sum_{u \in \mathscr{C}} \hat{f}(u), \tag{26}$$

where the sums are evaluated as real sums.

Proof. This is just a restatement of Lemma 2 of Ch. 5. We start from

$$\sum_{u \in \mathscr{C}^\perp} F(u) = \frac{1}{2^k} \sum_{u \in \mathscr{C}} \hat{F}(u),$$

and set $F(u) = 1 - 2f(u)$ and $\hat{F}(u) = 2^{m/2}(1 - 2\hat{f}(u))$.   Q.E.D.

To complete the proof of the theorem, we apply the lemma with

$$\mathscr{C}^\perp = \{b \in V^m : b \subset a\},$$
$$\mathscr{C} = \{b \in V^m : b \subset \bar{a}\},$$

where a is some vector of V^m and $\bar{a}$ is its complement. Then $|\mathscr{C}^\perp| = 2^{\text{wt}(a)}$ and (26) becomes

$$\sum_{b \subset a} f(b) = 2^{\text{wt}(a)-1} - 2^{\frac{1}{2}m-1} + 2^{\text{wt}(a)-\frac{1}{2}m} \sum_{b \subset \bar{a}} \hat{f}(b). \tag{27}$$

Now Theorem 1 of Ch. 13 states that

$$f(v_1, \ldots, v_m) = \sum_{a \in V^m} g(a)v_1^{a_1} \cdots v_m^{a_m},$$

where

$$g(a) = \sum_{b \subset a} f(b) \quad (\text{mod } 2).$$

Thus $g(a)$ is given by Equation (27). But if $\text{wt}(a) > \frac{1}{2}m$ and $m > 2$, the RHS of (27) is even, and $g(a)$ is zero. Therefore f has degree at most $\frac{1}{2}m$.   Q.E.D.

Problem. (13) Show that f is bent iff for all $v \neq 0$, $v \in V^m$, the *directional derivative* of f in the direction v, defined by

$$f_v(x) = f(x + v) + f(x), \quad x \in V^m,$$

takes the values 0 and 1 equally often.

We shall now construct some families of bent functions.

Theorem 10.

$$h(u_1, \ldots, u_m, v_1, \ldots, v_n) = f(u_1, \ldots, u_m) + g(v_1, \ldots, v_n)$$

is a bent function (of $m + n$ arguments) iff f is a bent function (of $u_1, \ldots, u_m$) and g is a bent function (of $v_1, \ldots v_n$).

Proof. We shall write $w \in V^{m+n}$ as $w = (u, v)$ where $u \in V^m$ and $v \in V^n$. From Equation (8),

$$\hat{H}(w) = \sum_{t \in V^{m+n}} (-1)^{w \cdot t + h(t)}, \quad \text{where } t = (r, s),$$

$$= \sum_{r \in V^m} \sum_{s \in V^n} (-1)^{u \cdot r + v \cdot s + f(r) + g(s)}$$

$$= \hat{F}(u)\hat{G}(v). \tag{28}$$

If f and g are bent, then from Equation (28) $\hat{H}(w) = \pm 2^{\frac{1}{2}(m+n)}$ and so h is bent. Conversely, suppose h is bent but f is not, so that $|\hat{F}(\lambda)| > 2^{\frac{1}{2}m}$ for some $\lambda \in V^m$. Then with $w = (\lambda, v)$,

$$\pm 2^{\frac{1}{2}(m+n)} = \hat{H}(w) = \hat{F}(\lambda)\hat{G}(v)$$

for all $v \in V^n$. Therefore $|\hat{G}(v)| < 2^{\frac{1}{2}n}$ for all $v \in V^n$, which is impossible by Corollary 3.   Q.E.D.

Theorem 10 enables us to generalize examples (1) and (2):

Corollary 11.

$$v_1 v_2 + v_3 v_4 + \cdots + v_{m-1} v_m \qquad (29)$$

is bent, for any even $m \geq 2$.

It is easy to see that if $f(v)$ is bent then so is any function $f(Av + B)$ obtained from f by an affine transformation.

Problems. (14) Show that if m is even then $v_1 v_2 + v_1 v_3 + \cdots + v_{m-1} v_m$ is bent.

(15) Use Dickson's theorem (Theorem 4 of the next chapter) to show that any quadratic bent function is an affine transformation of (29).

(16) (a) The bent function $v_1 v_2 + v_3 v_4$ can be represented by the graph

$$1 \quad\quad\quad 2$$
$$\bullet\!\!-\!\!-\!\!-\!\!-\!\!-\!\!-\!\!\bullet$$

$$\bullet\!\!-\!\!-\!\!-\!\!-\!\!-\!\!-\!\!\bullet$$
$$4 \quad\quad\quad 3$$

Show that the 28 quadratic bent functions of v_1, v_2, v_3, v_4 are:

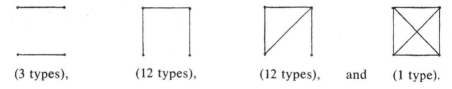

(3 types),   (12 types),   (12 types),   and   (1 type).

(b) Let $f = v_1 v_2 + v_3 v_4$. Show that the vectors of weight 6 in $f + \mathcal{R}(1, 4)$ form a 2-(16, 6, 2) design. (This design is called a *biplane* – see for example Cameron [231]).

(17) Show that $v_1 v_2$ and $v_1 v_2 + v_3 v_4$ are essentially the only bent functions of 2 and 4 arguments.

A Boolean function $f(v) = f(v_1, \ldots, v_m)$ is called *decomposable* if there is a binary matrix A such that

$$f(Av) = f_1(v_1, \ldots, v_l) + f_2(v_{l+1}, \ldots, v_m)$$

for some l. Otherwise f is *indecomposable*.

Problem. (18) Show that if $f(v_1, \ldots, v_m)$ is a bent function of degree $\frac{1}{2} m \geq 3$, then f is indecomposable.

Theorem 12. *For any function* $g(v_1, \ldots, v_m)$, *the function*

$$f(u_1, \ldots, u_m, v_1, \ldots, v_m) = \sum_{i=1}^{m} u_i v_i + g(v_1, \ldots, v_m) \tag{30}$$

is bent.

Proof. From Equations (8) and (13), $f(u_1, \ldots, u_m, v_1, \ldots, v_m)$ is bent iff the number of vectors $(u, v) = (u_1, \ldots, u_m, v_1, \ldots, v_m)$ which are zeros of

$$h = f(u_1, \ldots, u_m, v_1, \ldots, v_m) + \sum_{i=1}^{m} \lambda_i u_i + \sum_{i=1}^{m} \mu_i v_i \tag{31}$$

is $2^{2m-1} \pm 2^{m-1}$, for all $\lambda = (\lambda_1, \ldots, \lambda_m)$, $\mu = (\mu_1, \ldots, \mu_m)$. Substituting (30) into (31) we obtain

$$h = g(v_1, \ldots, v_m) + \sum_{i=1}^{m} (v_i + \lambda_i) u_i + \sum_{i=1}^{m} \mu_i v_i.$$

(i) For any $v \neq \lambda$, the first sum is not identically zero, and h is a linear function of $u_1, \ldots, u_m$. Thus there are 2^{m-1} choices of $u_1, \ldots, u_m$ for which h is zero. The total number of zeros of h of this type is $2^{m-1}(2^m - 1)$.

(ii) Now suppose $v = \lambda$. Then

$$h = g(v_1, \ldots, v_m) + \sum_{i=1}^{m} \mu_i v_i,$$

which is either 0 or 1, independent of $u_1, \ldots, u_m$. If it is 1, the total number of zeros of h is $2^{2m-1} - 2^{m-1}$. But if it is 0, there are 2^m additional zeros (u arbitrary, $v = \lambda$), for a total of $2^{2m-1} + 2^{m-1}$.   Q.E.D.

Problems. (19) Let

$$f_1(u_1, \ldots, u_m, v_1, \ldots, v_m) = \sum_{i=1}^{m} u_i v_i.$$

Show that f_1, $f_1 + u_1 u_2 u_3$, $f_1 + u_1 u_2 u_3 u_4, \ldots,$ and $f_1 + u_1 \cdots u_m$ are $m - 1$ inequivalent bent functions, of degrees $2, 3, \ldots, m$.

(20) Show that

$$f(u_1, \ldots, u_m, v_1, \ldots, v_m) = \quad \varphi_i(u_1, \ldots, u_m) v_i + g(v_1, \ldots, v_m)$$

is bent, where $g(v)$ is arbitrary and $\varphi(u) = (\varphi_1(u), \ldots, \varphi_m(u))$ is a 1-1-mapping of V^m onto itself.

Thus we have constructed a number of bent functions, and many others will be found in the references given in the Notes. But the following problem is unsolved:

Research Problem (14.1). Classify all bent functions of m variables.

Notes on Chapter 14

§2. There is an extensive body of literature on pseudo-noise sequences. See for example Ball et al. [62], Balza et al. [63], Briggs and Godfrey [198], Brusilovskii [207], Carter [251], Chang [262], Chang and Ho [263], Cumming [320], Davies [336], Duvall and Kibler [395], Fillmore and Marx [429], Fredricsson [455], Godfrey [487], Gold [515–516], Goldstein and Zierler [521], Harvey [621], Hurd [677], Kotov [780], Mayo [931], Parsley and Sarwate [1024], Roberts and Davis [1116], Scholefield [1167], Sidel'nikov [1207], Tretter [1339], Tsao [1341], Weathers et al. [1392], Weng [1407], Willett [1416], [1417] and especially Golomb [522, 523], Kautz [750], Selmer [1178] and Zierler [1461, 1465].

For applications of PN sequences to range-finding see Evans and Hagfors [412], Golomb [522, Ch. 6], Gorenstein and Weiss [551], Pettengill [1041]; to synchronizing see Burrowes [213], Golomb [522, Ch. 8], Kilgus [760], and Stiffler [1280]; to modulation see Golomb [522, Ch. 5]; and to scrambling see Feistel et al. [423], Henriksson [641], Nakamura and Iwadare [984] and Savage [1148]. The unsuitability of PN sequences for encryption is mentioned for example by Geffe [469, 470].

Sequences of period 2^m (rather than $2^m - 1$) can be obtained from nonlinear shift registers, and are called *de Bruijn* sequences. See de Bruijn [206], Fredricksen et al. [451–454] and Knuth [772, p. 379].

If $s_0, \ldots, s_{n-1}$ is a sequence of length n, its *periodic* autocorrelation function is defined by (2). Its *aperiodic* autocorrelation function is given by

$$\varphi(\tau) = \sum_{i=0}^{n-\tau-1} s_i \bar{s}_{i+\tau}, \quad 0 \leqslant \tau \leqslant n - 1. \tag{32}$$

The problem of finding binary sequences with small aperiodic autocorrelation function is much harder – see for example Barker [68], Barton and Mallows [73], Caprio [241], Golay [511, 513, 514], Golomb and Scholtz [530], Kruskal [785], Lindner [841], MacWilliams [875], Moharir et al. [968–970], Schneider and Orr [1157], Simmons [1212], Turyn [1343, 1344], Turyn and Storer [1348] and Tseng and Liu [1349]. Problems 7 and 8 are from Golomb [522, p. 48] and Gilbert [480].

§3. For geometric properties of cross-polytopes see Coxeter [313]. Lechner [797–799] has shown that the number of equivalence classes of Boolean functions (with equivalence defined by Equation (17)) is as follows:

$$
\begin{array}{llllll}
m & 1 & 2 & 3 & 4 & 5 \\
\text{number} & 1 & 2 & 3 & 8 & 48.
\end{array}
$$

Research Problem (14.2). Find these numbers when $m \geqslant 6$. How fast do they grow with m?

The cosets of $\mathcal{R}(1,5)$ have been classified by Berlekamp and Welch [133]. Figure 14.6 is based on Sloane and Dick [1233]. See also Hobbs [656], Holmes [660], Kurshan and Sloane [789], and Sarwate [1144]. Berlekamp [119] applied Theorem 1 to study the effect of errors in the transformed vector $\hat{F}$ on the reconstructed vector F. This problem arises when Hadamard transforms are used to transmit pictures. For an application of these cosets in spectroscopy see Tai et al. [1298a].

The cosets of a certain $[n^2, 2n-1, n]$ code arise in studying the Berlekamp–Gale switching game (Brown and Spencer [204], Gordon and Witsenhausen [542]).

§4. A good reference for the encoding and decoding of $\mathcal{R}(1,m)$ and its use in the Mariner '69 spacecraft is Posner [1071]. Detailed descriptions of the encoder and decoder are given by Duffy [392] and Green [557, 558]. For the theory of fast Hadamard and Fourier transforms see Bergland [108], Brigham [199], Cairns [225], Gentleman [472], Green [557, 558], Nicholson [991], Rushforth [1135], Shanks [1187] and Welch [1398]. For the Kronecker product of matrices see Bellman [100] or Marcus [914].

Clark and Davis [295] describe how the Green machine can be modified to apply to other codes.

A coset of $\mathcal{R}(1,5)$ rather than $\mathcal{R}(1,5)$ itself was actually used on the Mariner spacecraft, to facilitate synchronizing the received codewords. (See Posner [1071], or Baumert and Rumsey [90, 91].)

§5. The name "bent function" and most of the results of §5 are due to Rothaus [1127]. Dillon [376] (see also [377]) gives an excellent survey, and constructs several other families of bent functions. He shows that bent functions are equivalent to *elementary Hadamard difference sets*. See also McFarland [950].

15

Second-order Reed–Muller, Kerdock and Preparata codes

§1. Introduction

The key idea in this chapter is that the second-order RM code $\mathcal{R}(2, m)$ is a union of cosets of the first-order RM code $\mathcal{R}(1, m)$, and that these cosets are in 1-1 correspondence with symplectic forms (Theorem 1). A symplectic form is a Boolean function of $2m$ variables $u_1, \ldots, u_m, v_1, \ldots v_m$ given by $\sum_{i,j} u_i B_{ij} v_j$ where $B = (B_{ij})$ is a symmetric binary matrix with zero diagonal. The rank of the matrix is called the rank of the symplectic form, or of the coset, and uniquely determines the weight distribution of the coset (Theorem 5). In §2 we count symplectic forms by rank, and hence find the weight distribution of $\mathcal{R}(2, m)$ (Theorem 8). The problem of finding the weight distribution of a general RM code is unsolved, although in §3 we do give, without proof, two general results. First, Kasami and Tokura's formula for the number of codewords of weight $< 2^{m-r+1}$ in $\mathcal{R}(r, m)$ (Theorem 11). Second, McEliece's theorem that the weight of any codeword in $\mathcal{R}(r, m)$ is divisible by $2^{\lfloor (m-1)/r \rfloor}$ (Corollary 13).

In §4 we obtain the weight distribution of certain interesting small subcodes of $\mathcal{R}(2, m)^*$, including the dual of the double-error-correcting BCH code (Figs. 15.3 and 15.4).

In §5 we build up maximal sets of symplectic forms with the property that the rank of the sum of any two forms in the set is $\geq 2d$. The corresponding subcodes of $\mathcal{R}(2, m)$ have several interesting properties. These subcodes are linear if m is odd; but for even m they are the nonlinear generalized Kerdock codes $\mathcal{DG}(m, d)$. When $d = \frac{1}{2}m$, $\mathcal{DG}(m, \frac{1}{2}m)$ is the Kerdock code $\mathcal{K}(m)$. The first Kerdock code, $\mathcal{K}(4)$, is equivalent to the $(16, 256, 6)$ Nordstrom–Robinson code $\mathcal{N}_{16}$ given in Ch. 2. Shortly after the discovery of $\mathcal{N}_{16}$, Preparata succeeded in generalizing $\mathcal{N}_{16}$ to get an infinite family of codes

$\mathscr{P}(m)$ with parameters $(2^m, 2^{2^m-2m}, 6)$, for all even $m \geq 4$. (See Fig. 15.12 below.) Then in 1972 Kerdock discovered the codes $\mathscr{K}(m)$. These have parameters $(2^m, 2^{2m}, 2^{m-1} - 2^{(m-2)/2})$, and are a kind of dual to $\mathscr{P}(m)$ in the sense that the weight (or distance) distribution of $\mathscr{K}(m)$ is the MacWilliams transform (§5 of Ch. 5) of the weight distribution of $\mathscr{P}(m)$. (See Fig. 15.6.) The codes $\mathscr{P}(m)$ and $\mathscr{K}(m)$ appear to contain at least twice as many codewords as the best linear code with the same length and minimum distance.

The codes $\mathscr{DG}(m, d)$ were discovered by Delsarte and Goethals. In the last paragraph we describe the codes $\mathscr{I}(m)$, found by Goethals, which are a kind of dual to $\mathscr{DG}(m, \frac{1}{2}(m - 2))$.

The construction of the codes $\mathscr{K}(m), \mathscr{P}(m), \mathscr{DG}(m, d), \ldots$ will require a good deal of the algebraic machinery developed in earlier chapters – for example the group algebra and annihilator polynomial from Ch. 5 and idempotents and Mattson–Solomon polynomials from Ch. 8.

§2. Weight distribution of the second-order Reed–Muller code

Codewords of the second-order Reed–Muller code $\mathscr{R}(2, m)$ of length $n = 2^m$ are given by the Boolean functions of degree ≤ 2 in $v = v_1, \ldots, v_m$ (see §3 of Ch. 13). Thus a typical codeword is given by the Boolean function

$$S(v) = \sum_{1 \leq i \leq j \leq m} q_{ij} v_i v_j + \sum_{1 \leq i \leq m} l_i v_i + \epsilon.$$

We may write this as

$$S(v) = vQv^T + Lv^T + \epsilon,$$
$$= Q(v) + L(v) + \epsilon, \tag{1}$$

where

$$Q(v) = vQv^T,$$
$$L(v) = Lv^T, \tag{2}$$

and $Q = (q_{ij})$ is an upper triangular binary matrix, $L = (l_1, \ldots, l_m)$ is a binary vector, and ϵ is 0 or 1. Here $Q(v) = Q(v_1, \ldots, v_m)$ is called a *quadratic form* and $L(v) = L(v_1, \ldots, v_m)$ is a *linear form*. (A *form* is a homogeneous polynomial.)

If $Q(v)$ is fixed and the linear function $L(v) + \epsilon$ varies over the first-order Reed–Muller code $\mathscr{R}(1, m)$, then $S(v)$ runs through a coset of $\mathscr{R}(1, m)$ in $\mathscr{R}(2, m)$. This coset is characterized by Q. Alternatively we can characterize it by the symmetric matrix $B = Q + Q^T$; B has all diagonal elements zero.

Associated with the matrix B is another form $\mathscr{B}(u, v)$, defined by

$$\mathscr{B}(u, v) = uBv^T. \tag{3}$$

Alternatively

$$\mathcal{B}(u, v) = u(Q + Q^T)v^T$$
$$= uQv^T + vQu^T$$
$$= Q(u + v) + Q(u) + Q(v) \tag{4}$$
$$= S(u + v) + S(u) + S(v) + \epsilon \quad \text{by (1)}. \tag{5}$$

$\mathcal{B}(u, v)$ is *bilinear*, i.e. satisfies (from (3))

$$\mathcal{B}(u + v, w) = \mathcal{B}(u, w) + \mathcal{B}(v, w),$$
$$\mathcal{B}(u, v + w) = \mathcal{B}(u, v) + \mathcal{B}(u, w). \tag{6}$$

From Equation (4), $\mathcal{B}(u, v)$ is also *alternating*, i.e. satisfies

$$\mathcal{B}(u, u) = 0 \tag{7}$$

and

$$\mathcal{B}(u, v) = \mathcal{B}(v, u). \tag{8}$$

A binary form which is alternating and bilinear is called *symplectic*. Similarly a binary symmetric matrix with zero diagonal is called a *symplectic matrix*. Thus B is a symplectic matrix.

For example the matrix

$$B = \begin{pmatrix} 0 & 1 \\ 1 & 0 \end{pmatrix}$$

is symplectic, and the corresponding symplectic form is

$$\mathcal{B}(u_1, u_2, v_1, v_2) = (u_1, u_2)\begin{pmatrix} 0 & 1 \\ 1 & 0 \end{pmatrix}\begin{pmatrix} v_1 \\ v_2 \end{pmatrix} = u_1v_2 + u_2v_1.$$

As a concrete example, consider the first- and second-order RM codes of length 16. $\mathcal{R}(1, 4)$ is a $[16, 5, 8]$ extended simplex code, and is generated by the first five rows of Fig. 13.2. $\mathcal{R}(2, 4)$ is a $[16, 11, 4]$ extended Hamming code, and is generated by the first 11 rows of that figure. $\mathcal{R}(2, 4)$ consists of $2^{11-5} = 64$ cosets of $\mathcal{R}(1, 4)$. One coset is $\mathcal{R}(1, 4)$ itself, and the quadratic and symplectic forms corresponding to this coset are zero.

Another coset consists of the 16 vectors $v_1v_2 + v_3v_4 + \mathcal{R}(1, 4)$, four of which are shown in Fig. 15.1.
The Boolean functions $S(v)$ (Equation (1)) corresponding to these four

$$0\ 0\ 0\ 1\ 0\ 0\ 0\ 1\ 0\ 0\ 0\ 1\ 1\ 1\ 1\ 0 = v_1v_2 + v_3v_4$$
$$1\ 1\ 1\ 0\ 1\ 1\ 1\ 0\ 1\ 1\ 1\ 0\ 0\ 0\ 0\ 1 = v_1v_2 + v_3v_4 + 1$$
$$0\ 1\ 0\ 0\ 0\ 1\ 0\ 0\ 0\ 1\ 0\ 0\ 1\ 0\ 1\ 1 = v_1v_2 + v_3v_4 + v_1$$
$$1\ 0\ 1\ 1\ 1\ 0\ 1\ 1\ 1\ 0\ 1\ 1\ 0\ 1\ 0\ 0 = v_1v_2 + v_3v_4 + v_1 + 1$$

Fig. 15.1. Four vectors of the coset $v_1v_2 + v_3v_4 + \mathcal{R}(1, 4)$.

vectors are

$$v_1 v_2 + v_3 v_4,$$
$$v_1 v_2 + v_3 v_4 + 1,$$
$$v_1 v_2 + v_3 v_4 + v_1,$$
$$v_1 v_2 + v_3 v_4 + v_1 + 1.$$

(Incidentally these are bent functions – see §5 of Ch. 14.) They all have the same quadratic part (Equation (2)),

$$Q(v) = v_1 v_2 + v_3 v_4$$

$$= [v_1 v_2 v_3 v_4] \begin{bmatrix} 0100 \\ 0000 \\ 0001 \\ 0000 \end{bmatrix} \begin{bmatrix} v_1 \\ v_2 \\ v_3 \\ v_4 \end{bmatrix}$$

Thus

$$Q = \begin{bmatrix} 0100 \\ 0000 \\ 0001 \\ 0000 \end{bmatrix} \quad \text{and} \quad B = Q + Q^T = \begin{bmatrix} 0100 \\ 1000 \\ 0001 \\ 0010 \end{bmatrix} \tag{9}$$

are the upper triangular and symplectic matrices corresponding to this coset. The symplectic *form* corresponding to this coset (Equation (3)) is

$$\mathcal{B}(u, v) = u B v^T$$
$$= u_1 v_2 + u_2 v_1 + u_3 v_4 + u_4 v_3.$$

Problem. (1) Show that if a function $\mathcal{B}(u, v)$ satisfies Equations (6) and (7) then $\mathcal{B}(u, v) = Q(u + v) + Q(u) + Q(v)$ for some quadratic form $Q(u)$.

Thus we have shown:

Theorem 1. *There is a 1-1-correspondence between symplectic forms and cosets of $\mathcal{R}(1, m)$ in $\mathcal{R}(2, m)$. The zero form corresponds to $\mathcal{R}(1, m)$ itself.*

We shall use the symbol $\mathcal{B}$ to refer to both to the coset and to the corresponding symplectic form.

Clearly the number of distinct symplectic forms or matrices is $2^{\binom{T}{2}}$. But we also need the number of each rank.

Theorem 2. *Let $N(m, r)$ be the number of symplectic $m \times m$ matrices of rank r over GF(2). Then*

$$N(m, 2h + 1) = 0,$$

$$N(m, 2h) = \prod_{i=1}^{h} \frac{2^{2i} - 2}{2^{2i} - 1} \cdot \prod_{i=0}^{2h-1} (2^{m-i} - 1)$$

$$= \frac{(2^m - 1)(2^{m-1} - 1) \cdots (2^{m-2h+1} - 1)}{(2^{2h} - 1)(2^{2h-2} - 1) \cdots (2^2 - 1)} \cdot 2^{h(h-1)}.$$

Proof. Note that $N(1, 0) = 1$, $N(1, 1) = 0$ and $N(2, 0) = 1$, $N(2, 1) = 0$, $N(2, 2) = 1$.

We shall derive a recursion formula for $N(m, r)$. Let A be a fixed $m \times m$ symplectic matrix of rank r, and set

$$B = \begin{pmatrix} A & y^T \\ y & 0 \end{pmatrix},$$

of size $(m + 1) \times (m + 1)$.

Lemma 3. *Of the 2^m different matrices B, $2^m - 2^r$ have rank $r + 2$, and 2^r have rank r.*

Proof of Lemma. If y is independent of the rows of A, which can happen in $2^m - 2^r$ ways, then $\binom{A}{y}$ has rank $r + 1$ and B has rank $r + 2$. On the other hand, if y is dependent on the rows of A, say $y = xA$, then $\binom{A}{y}$ has rank r. To show that $\binom{y^T}{0}$ is dependent on the columns of $\binom{A}{y}$, we observe that

$$\binom{A}{y} x^T = \binom{y^T}{yx^T} = \binom{y^T}{xAx^T} = \binom{y^T}{0}.$$ Q.E.D.

From Lemma 3, the recurrence for $N(m, r)$ is

$$N(m + 1, r) = 2^r N(m, r) + (2^m - 2^{r-2}) N(m, r - 2).$$

The initial value $N(1, 1) = 0$ implies $N(1, 2h + 1) = 0$ for all h. It is straightforward to verify, using the initial value $N(m, 0) = 1$, that the solution of the recurrence is

$$N(m, 2) = \frac{(2^m - 1)(2^{m-1} - 1)}{2^2 - 1},$$

$$N(m, 4) = \frac{(2^m - 1)(2^{m-1} - 1)(2^{m-2} - 1)(2^{m-3} - 1)}{(2^4 - 1)(2^2 - 1)} \cdot 2^2,$$

and in general is the formula stated in the theorem.   Q.E.D.

We next show that the weight distribution of the coset $\mathscr{B}$ depends only on the rank of the matrix B. For this we need a fundamental theorem due to Dickson, which we state as follows:

Theorem 4. (Dickson's theorem.) (1) *If B is a symplectic $m \times m$ matrix of rank $2h$, then there exists an invertible binary matrix R such that RBR^T has zeros everywhere except on the two diagonals immediately above and below the main diagonal, and there has $1010 \ldots 100 \ldots 0$ with h ones.*
Examples of RBR^T are, when $h = 2$,

$$\begin{pmatrix} 0100 \\ 1000 \\ 0001 \\ 0010 \end{pmatrix} \qquad \begin{pmatrix} 01000 \\ 10000 \\ 00010 \\ 00100 \\ 00000 \end{pmatrix}$$

(2) *Any Boolean function of degree ≤ 2,*

$$vQv^T + L(v) + \epsilon,$$

where Q is an upper triangular matrix and L, ϵ are arbitrary, becomes

$$T(y) = \sum_{i=1}^{h} y_{2i-1} y_{2i} + L_1(y) + \epsilon$$

under the transformation of variables $y = vR^{-1}$, where R is given by Part (1) and $B = Q + Q^T$. Moreover $y_1, \ldots, y_{2h}$ are linearly independent.
(3) *If $L_1(y)$ is linearly dependent on $y_1, \ldots, y_{2h}$, we may by an affine transformation of variables write $T(y)$ as*

$$\sum_{i=1}^{h} x_{2i-1} x_{2i} + \epsilon_1, \quad \epsilon_1 = 0 \text{ or } 1, \tag{10}$$

where $x_1, \ldots, x_{2h}$ are linearly independent, and each x_i is a linear form in $y_1, \ldots, y_{2h}, 1$.

Proof. (1) The statement is clearly true for $m = 1, 2$. Suppose it is true for $m \leq t$ and $h \leq [t/2]$. Then a $(t+1) \times (t+1)$ symplectic matrix B may be written as

$$B = \begin{pmatrix} A & y^T \\ y & 0 \end{pmatrix}$$

where A is of the same type. For example, a 3×3 symplectic matrix can be written as

$$\begin{pmatrix} 0 & 0 & a_1 \\ 0 & 0 & a_2 \\ a_1 & a_2 & 0 \end{pmatrix} \quad \text{or} \quad \begin{pmatrix} 0 & 1 & a_1 \\ 1 & 0 & a_2 \\ a_1 & a_2 & 0 \end{pmatrix}.$$

If rank $A < 2[t/2]$, we may, by elementary row and column operations, reduce B to

$$\begin{pmatrix} A & y'^T & 0 & \cdots & 0 \\ y' & 0 & 0 & \cdots & 0 \\ 0 & 0 & 0 & \cdots & 0 \\ & & \cdots\cdots\cdots & & \\ 0 & 0 & 0 & \cdots & 0 \end{pmatrix},$$

where

$$\begin{pmatrix} A & y'^T \\ y' & 0 \end{pmatrix}$$

is of size $\leq t \times t'$. By induction this may be brought into the form described. Thus we may suppose that A has rank $2[t/2]$. If $2[t/2] = t$, then B has rank t, and

$$B = \begin{pmatrix} 0 & 1 & 0 & \cdots & 0 & 0 & a_1 \\ 1 & 0 & 0 & \cdots & 0 & 0 & a_2 \\ & & \cdots\cdots\cdots\cdots\cdots & & \\ 0 & 0 & & \cdots & 0 & 1 & a_{t-1} \\ 0 & 0 & & \cdots & 1 & 0 & a_t \\ a_1 & a_2 & & \cdots & a_{t-1} & a_t & 0 \end{pmatrix}.$$

In this case

$$R = \begin{pmatrix} 1 & 0 & 0 & 0 & \cdots & 0 & 0 & 0 \\ 0 & 1 & 0 & 0 & \cdots & 0 & 0 & 0 \\ a_2 & a_1 & a_4 & a_3 & \cdots & a_t & a_{t-1} & 1 \end{pmatrix}.$$

If $2[t/2] = t - 1$ then

$$B = \begin{pmatrix} 0 & 1 & \cdots & 0 & 0 & 0 & a_1 \\ 1 & 0 & \cdots & 0 & 0 & 0 & a_2 \\ & & \cdots\cdots\cdots\cdots\cdots & & \\ 0 & 0 & \cdots & 0 & 1 & 0 & a_{t-2} \\ 0 & 0 & \cdots & 1 & 0 & 0 & a_{t-1} \\ 0 & 0 & \cdots & 0 & 0 & 0 & \epsilon \\ a_1 & a_2 & \cdots & a_{t-2} & a_{t-1} & \epsilon & 0 \end{pmatrix},$$

where rank $B = t - 1$ if $\epsilon = 0$ and $t + 1$ if $\epsilon = 1$. In either case use

$$R = \begin{pmatrix} 1 & 0 & 0 & \cdots & 0 & 0 & 0 & 0 \\ 0 & 1 & 0 & \cdots & 0 & 0 & 0 & 0 \\ & & \cdots\cdots\cdots\cdots\cdots & & \\ a_2 & a_1 & a_4 & \cdots & a_{t-2} & a_{t-1} & 0 & 1 \end{pmatrix}.$$

(Check this for $m = 4, 5$.)

(2) The transformation $x = uR^{-1}$, $y = vR^{-1}$ changes the bilinear form uBv^T into

$$xRBR^Ty = \sum_{i=1}^{h} (x_{2i-1}y_{2i} + x_{2i}y_{2i-1}).$$

The corresponding quadratic form is

$$\sum_{i=1}^{h} y_{2i-1}y_{2i}.$$

(3) Let

$$T(y) = \sum_{i=1}^{h} y_{2i-1}y_{2i} + L(y) + \epsilon,$$

where $L(y)$ is linearly dependent on the y_i, say

$$L(y) = \sum_{i=1}^{2h} l_iy_i.$$

For example, if $l_1 = 1$, $l_2 = 0$, so that

$$T(y) = y_1y_2 + y_1 + \cdots,$$

the substitution $y_1 = x_1$, $y_2 = x_1 + x_2$ changes this to

$$T(y) = x_1x_2 + \sum_{i=2}^{h} y_{2i-1}y_{2i} + \sum_{i=3}^{2h} l_iy_i.$$

If $l_1 = l_2 = 1$ then $T(y) = y_1y_2 + y_1 + y_2 + \cdots$ is changed by the substitution $y_1 = x_1 + 1$, $y_2 = x_2 + 1$ into

$$T(y) = x_1x_2 + \sum_{i=2}^{h} y_{2i-1}y_{2i} + \sum_{i=3}^{2h} l_iy_i + 1.$$

Clearly we can continue in this way.   Q.E.D.

Remarks. (1) Since $\mathcal{R}(1, m)$ contains the Boolean function 1, the coset $\mathcal{B}$ will contain both

$$\sum_{i=1}^{h} x_{2i-1}x_{2i} \quad \text{and} \quad \sum_{i=1}^{h} x_{2i-1}x_{2i} + 1.$$

(2) The Mattson–Solomon polynomial corresponding to the canonical form

$$\sum_{i=1}^{h} x_{2i-1}x_{2i}$$

is, by Corollary 26 of Ch. 13,

$$\sum_{i=1}^{h} T_m(\alpha_iz)T_m(\beta_iz), \tag{11}$$

where $\alpha_1, \ldots, \alpha_h, \beta_1, \ldots, \beta_h$ are linearly independent elements of $GF(2^m)$.

Problem. (2) In part (3) of the theorem, show that the constant ϵ_1 is given by

$$\epsilon_1 = \sum_{i=1}^{h} l_{2i-1}l_{2i} + \epsilon.$$

Hence show that the number of $L(y)$ for which $\epsilon_1 = 1$ is $2^{2h-1} - 2^{h-1}$.

We now use the canonical expression for a quadratic form given by Theorem 4 to obtain the weight distribution of the corresponding coset, as shown in the following theorem. This result will be used many times in this chapter.

Theorem 5. *If the matrix B has rank $2h$, the weight distribution of the corresponding coset $\mathcal{B}$ of $\mathcal{R}(1, m)$ in $\mathcal{R}(2, m)$ is as follows:*

Weight	Number of Vectors
$2^{m-1} - 2^{m-h-1}$	2^{2h}
2^{m-1}	$2^{m+1} - 2^{2h+1}$
$2^{m-1} + 2^{m-h-1}$	2^{2h}

We shall need two lemmas for the proof.

Lemma 6. *The number of values of $(x_1, \ldots, x_{2h})$ for which*

$$\sum_{i=1}^{h} x_{2i-1}x_{2i}$$

is zero is $2^{2h-1} + 2^{h-1}$.

Proof. If $h = 1$ this number is 3 (they are 00, 10, 01) so the lemma is true. We proceed by induction on h.

$$\sum_{i=1}^{h+1} x_{2i-1}x_{2i} = \sum_{i=1}^{h} x_{2i-1}x_{2i} + x_{2h+1}x_{2h+2}$$

$$= F_1 + F_2 \quad \text{(say)}.$$

There are $3(2^{2h-1} + 2^{h-1})$ cases with $F_1 = F_2 = 0$, and $2^{2h-1} - 2^{h-1}$ cases with $F_1 = F_2 = 1$. The total number is

$$4 \cdot 2^{2h-1} + 2 \cdot 2^{h-1} = 2^{2(h+1)-1} + 2^{(h+1)-1}. \qquad \text{Q.E.D.}$$

Therefore the number of vectors $(x_1, \ldots, x_{2h}, x_{2h+1}, \ldots, x_m)$ for which

$$\sum_{i=1}^{h} x_{2i-1}x_{2i}$$

is zero is

$$2^{m-2h}(2^{2h-1} + 2^{h-1}) = 2^{m-1} + 2^{m-h-1}.$$

Lemma 7. *Let*

$$L_{2h+1} = \sum_{i=2h+1}^{m} a_i x_i$$

where the a_i are not all zero. The number of values of $x_1, \ldots, x_m$ for which

$$\sum_{i=1}^{h} x_{2i-1} x_{2i} + L_{2h+1}$$

is zero is 2^{m-1}.

Proof. Use the randomization lemma of Problem 2 of Ch. 13.   Q.E.D.

Proof of Theorem 5. Let B have rank $2h$. By Dickson's theorem, the quadratic part of any Boolean function in the coset $\mathscr{B}$ can be transformed to

$$Q(x) = \sum_{i=1}^{h} x_{2i-1} x_{2i}.$$

Let the linear part be $L(x)$. Suppose $L(x)$ has the form

$$L(x) = \sum_{i=1}^{2h} a_i x_i,$$

which can happen in 2^{2h} ways. Using Part 3 of Dickson's theorem $Q(x) + L(x)$ becomes either

$$\sum_{i=1}^{h} y_{2i-1} y_{2i} \quad \text{or} \quad \sum_{i=1}^{h} y_{2i-1} y_{2i} + 1.$$

By the remark following Lemma 6 these have weights $2^{m-1} - 2^{m-h-1}$ and $2^{m-1} + 2^{m-h-1}$ respectively. On the other hand if $L(x)$ is not dependent on $x_1, \ldots, x_{2h}$, which happens in $2^{m+1} - 2^{2h+1}$ ways, by Lemma 7 the codeword has weight 2^{m-1}.   Q.E.D.

Remark. Theorem 5 shows that the larger h is, the greater is the minimum weight in the coset. $h = 0$ corresponds to $\mathscr{R}(1, m)$ itself. The largest possible minimum weight occurs when $2h = m$ (and so m must be even). The Boolean function corresponding to such a coset is a quadratic bent function (see §5 of Ch. 14, especially Problem 15).

We illustrate Theorem 5 by finding the weight distribution of the coset $v_1 v_2 + v_3 v_4 + \mathscr{R}(1, 4)$. The matrix B is given in Equation (9) and has rank 4. Thus $h = 2$, and the coset has weight distribution $A_6 = A_{10} = 16$ (in agreement with the last row of Fig. 14.6).

An immediate consequence of Theorem 5 is:

Theorem 8. (Weight distribution of second-order Reed–Muller code.) *Let A_i be the number of codewords of weight i in $\mathcal{R}(2, m)$. Then $A_i = 0$ unless $i = 2^{m-1}$ or $i = 2^{m-1} \pm 2^{m-1-h}$ for some h, $0 \le h \le [\frac{1}{2}m]$. Also $A_0 = A_{2^m} = 1$ and*

$$A_{2^{m-1} \pm 2^{m-1-h}} = 2^{h(h+1)} \cdot \frac{(2^m - 1)(2^{m-1} - 1) \cdots (2^{m-2h+1} - 1)}{(2^{2h} - 1)(2^{2h-2} - 1) \cdots (2^2 - 1)}$$

$$\textit{for } 1 \le h \le [\tfrac{1}{2}m]. \qquad (12)$$

There is no simple formula for $A_{2^{m-1}}$, but of course

$$A_{2^{m-1}} = 2^{1+m+\binom{m}{2}} - \sum_{i \ne 2^{m-1}} A_i.$$

Proof. From Theorems 2 and 5   Q.E.D.

Example. The weight distribution of the [32, 16, 8] Reed–Muller code $\mathcal{R}(2, 5)$ is

i			A_i
0	or	32	1
8	or	24	620
12	or	20	13888
		16	36518

Remarks. (1) Putting $h = 1$ in (12) gives $A_{2^{m-2}} = \frac{4}{3}(2^m - 1)(2^{m-1} - 1)$ for the number of codewords of minimum weight, in agreement with Theorem 9(b) of Ch. 13.

(2) Equation (12) is rather complicated, but can be simplified by writing it in terms of Gaussian binomial coefficients.

Definition. For a real number $b \ne 1$, and all nonnegative integers k, the b-ary *Gaussian binomial coefficients* $\begin{bmatrix} x \\ k \end{bmatrix}$ are defined by

$$\begin{bmatrix} x \\ 0 \end{bmatrix} = 1,$$

$$\begin{bmatrix} x \\ k \end{bmatrix} = \frac{(b^x - 1)(b^{x-1} - 1) \cdots (b^{x-k+1} - 1)}{(b^k - 1)(b^{k-1} - 1) \cdots (b - 1)}, \quad k = 1, 2, \dots. \qquad (13)$$

(Here x is a real number, usually an integer.) For example

$$\begin{bmatrix} 3 \\ 2 \end{bmatrix} = \frac{(b^3 - 1)(b^2 - 1)}{(b^2 - 1)(b - 1)} = b^2 + b + 1.$$

There are many similarities between Gaussian binomial coefficients and ordinary binomial coefficients, as the following problem shows (cf. Problem 18 of Ch. 1).

Problem. (3) *Properties of Gaussian binomial coefficients.*

(a)
$$\lim_{b \to 1} \begin{bmatrix} x \\ k \end{bmatrix} = \binom{x}{k}.$$

(b)
$$\begin{bmatrix} n \\ k \end{bmatrix} = \begin{bmatrix} n \\ n - k \end{bmatrix}.$$

(c) Define [n] by
$$[0] = 1,$$
$$[n] = (b^n - 1)(b^{n-1} - 1) \cdots (b - 1), \quad \text{for } n = 1, 2, \ldots.$$

Then

(d)
$$\begin{bmatrix} n \\ k \end{bmatrix} = \frac{[n],}{[k][n - k]}.$$

(e)
$$\begin{bmatrix} n \\ k \end{bmatrix} = \begin{bmatrix} n - 1 \\ k - 1 \end{bmatrix} + b^k \begin{bmatrix} n - 1 \\ k \end{bmatrix}.$$

$$(y + 1)(y + b)(y + b^2) \cdots (y + b^{n-1}) = \sum_{k=0}^{n} \begin{bmatrix} n \\ k \end{bmatrix} b^{(n-k)(n-k-1)/2} y^k.$$

(f)
$$y^n = \sum_{k=0}^{n} (-1)^{n-k} \begin{bmatrix} n \\ k \end{bmatrix} (y + 1)(y + b) \cdots (y + b^{k-1}).$$

(g)
$$\sum_{k=i}^{j} (-1)^{k-i} b^{(k-i)(k-i-1)/2} \begin{bmatrix} k \\ i \end{bmatrix} \begin{bmatrix} j \\ k \end{bmatrix} = \delta_{i,j}.$$

In terms of Gaussian binomial coefficients with $b = 4$, Equation (12) becomes

$$A_{2^{m-1} \pm 2^{m-1-h}} = 2^{h(h+1)} [h] \begin{bmatrix} \frac{1}{2} m \\ h \end{bmatrix} \begin{bmatrix} \frac{1}{2}(m - 1) \\ h \end{bmatrix}. \tag{14}$$

Another useful property of these coefficients is:

Theorem 9. *The number of distinct (although not necessarily inequivalent)* [n, k] *codes over* GF(q) *is the q-ary Gaussian binomial coefficient* $\begin{bmatrix} n \\ k \end{bmatrix}$ *(with* $b = q$).

Proof. The number of ways of choosing k linearly independent vectors is

$$(q^n - 1)(q^n - q) \cdots (q^n - q^{k-1}).$$

Each such set is the basis for an [n, k] code $\mathscr{C}$. But the number of bases in $\mathscr{C}$

is

$$(q^k - 1)(q^k - q) \cdots (q^k - q^{k-1}).$$

Therefore the number of distinct $\mathcal{C}$'s is

$$\frac{(q^n - 1)(q^n - q) \cdots (q^n - q^{k-1})}{(q^k - 1)(q^k - q) \cdots (q^k - q^{k-1})} = \begin{bmatrix} n \\ k \end{bmatrix}. \qquad \text{Q.E.D.}$$

This is easy to remember: the number of $[n, k]$ codes is $\begin{bmatrix} n \\ k \end{bmatrix}$.

Problems. (4) (Sarwate.) Let $W(x, y)$ be the weight enumerator of $\mathcal{R}(2, m)$, and let $w(z; m) = W(1, z)$. Show that

$$w(z; m) = w(z^2; r \cdot - 1) + z^{2^{m-1}} w(1; m - 1)$$
$$+ 2^{m-1}(2^m - 2)z^{2^{m-2}} w(z^2; m - 2)$$

[Hint: the three terms correspond to the codewords in $\mathcal{R}(1, m - 1)$ of weights 0, 2^{m-1} and 2^{m-2} respectively.]

(5) Use the methods of this section to prove the following theorem. Let $ns = 2^{2rl} - 1$, $s > 1$, $l > 1$, and s is a divisor of $2^r + 1$. Let $\mathcal{C}$ be an $[n, 2rl]$ minimal cyclic code. Show that $\mathcal{C}$ consists of $0, n$ codewords of weight $(2^{2rl-1} + \epsilon(s - 1)2^{rl-1})/s$, and $n(s - 1)$ codewords of weight $(2^{2rl-1} - \epsilon 2^{rl-1})/s$, where $\epsilon = (-1)^l$. [Hint: Let $\beta \in GF(2^{2rl})$ be a nonzero of $\mathcal{C}$. If $(a_0 \cdots a_{n-1}) \in \mathcal{C}$, then $a_i = T_{2rl}(\gamma \beta^i)$ for some $\gamma \in GF(2^{2rl})$ (Theorem 9 of Ch. 8). First take $s = 2^r + 1$, and $\beta = \xi^{2^r+1}$ where ξ is a primitive element of $GF(2^{2rl})$. Show that

$$Q(\xi) = T_{2rl}(\gamma \xi^{2^r+1})$$

is a quadratic form in ξ, and that the rank of the corresponding symplectic form is either $2rl - 2r$ if $\gamma = \eta^{2^r+1}$ for some $\eta \in GF(2^{2rl})$, or $2rl$ otherwise. Secondly, if s is a proper divisor of $2^r + 1$, divide the group generated by ξ^s into cosets of the subgroup generated by ξ^{2^r+1}.]

*§3. Weight distribution of arbitrary Reed–Muller codes

The first-order RM code $\mathcal{R}(1, m)$ has weight distribution

$$A_0 = A_{2^m} = 1, \qquad A_{2^{m-1}} = 2^{m+1} - 2; \qquad (15)$$

and the weight distribution of the second-order is given in Theorem 8. Now $\mathcal{R}(m - r - 1, m)$ is dual to $\mathcal{R}(r, m)$, so from Theorem 1 of Ch. 5 the weight distributions of $\mathcal{R}(m - 2, m)$ (the extended Hamming code) and $\mathcal{R}(m - 3, m)$ are also known. But so far there is no formula for any other general class of Reed–Muller codes.

Research Problem (15.1). Find the weight distribution of $\mathcal{R}(3, m), \ldots$.

Nevertheless there are some general results about the weight enumerator of $\mathcal{R}(r, m)$, which we now state without proof.

Theorem 8 shows that in $\mathcal{R}(2, m)$ with minimum distance $d = 2^{m-2}$, the only weights occurring in the range d to $2d$ are of the form

$$w = 2d - 2^i \quad \text{for some } i.$$

This property holds for all RM codes, and in fact Kasami and Tokura have found the number of codewords of weight w in $\mathcal{R}(r, m)$ for all $w < 2d = 2^{m-r+1}$ (actually for all $w < 2\frac{1}{2}d$ – see below). To do this they found a canonical form for all the relevant Boolean functions, as follows:

Theorem 10. *Let $f(v_1, \ldots, v_m)$ be a Boolean function of degree at most r, where $r \geq 2$, such that*

$$\text{wt } (f) < 2^{m-r+1}.$$

Then f can be transformed by an affine transformation into either
  (i) $f = v_1 \cdots v_{r-\mu}(v_{r-\mu+1} \cdots v_r + v_{r+1} \cdots v_{r+\mu})$,

where μ satisfies $3 \leq \mu \leq r$ and $\mu \leq m - r$, or
  (ii) $f = v_1 \cdots v_{r-2}(v_{r-1}v_r + v_{r+1}v_{r+2} + \cdots + v_{r+2\mu-3}v_{r+2\mu-2})$,

where μ satisfies $2 \leq 2\mu \leq m - r + 2$.

From this result and Theorem 8 Kasami and Tokura [745] proved:

Theorem 11. (Weight distribution of $\mathcal{R}(r, m)$ in the range d to $2d$.) *Let A_w be the number of codewords of weight w in $\mathcal{R}(r, m)$, where $r \geq 2$, and suppose*

$$d = 2^{m-r} \leq w < 2d.$$

Define $\alpha = \min(m - r, r)$ and $\beta = \frac{1}{2}(m - r + 2)$. Then
  (i) *$A_w = 0$ unless $w = w(\mu) = 2^{m-r+1} - 2^{m-r+1-\mu}$ for some μ in the range $1 \leq \mu \leq \max(\alpha, \beta)$. The case $\mu = 1$ corresponds to $w = d$ and is taken care of by Theorem 9(b) of Ch. 13.*
  (ii) *If $\mu = 2$ or $\max(\alpha, 2) < \mu \leq \beta$ then*

$$A_{w(\mu)} = \frac{2^{r+\mu^2+\mu-2} \prod_{i=0}^{r+2\mu-3} (2^{m-i} - 1)}{\prod_{i=0}^{r-3} (2^{r-2-i} - 1) \prod_{i=0}^{\mu-1} (4^{i+1} - 1)}. \tag{16}$$

  (iii) *If $\max(\beta, 2) < \mu \leq \alpha$ then*

$$A_{w(\mu)} = \frac{2^{r+\mu^2+\mu-1} \displaystyle\prod_{i=0}^{r+\mu-1} (2^{m-i}-1)}{\displaystyle\prod_{i=0}^{r-\mu-1} (2^{r-\mu-i}-1) \prod_{i=0}^{\mu-1} (2^{\mu-i}-1)^2}. \tag{17}$$

(iv) *If* $3 \leq \mu \leq \min(\alpha, \beta)$ *then* $A_{w(\mu)}$ *is equal to the sum of* (16) *and* (17).

Theorem 11 has been extended to weights $\leq 2\frac{1}{2}d$ by Kasami, Tokura and Azumi [746], but the algebra becomes very complicated and it would seem that a different approach is needed to go further.

The weight enumerators of all Reed–Muller codes of lengths ≤ 256 are now known – see Kasami et al. [746], Sarwate [1144], Sugino et al. [1284], and Van Tilborg [1325]. The smallest Reed–Muller codes for which the weight distributions are not presently known (in 1977) are $\mathcal{R}(3,9)$, $\mathcal{R}(4,9)$ and $\mathcal{R}(5,9)$.

The second general result is McEliece's theorem that the weight of every codeword in $\mathcal{R}(r,m)$ is divisible by $2^{\lfloor (m-1)/r \rfloor}$. This follows from:

Theorem 12. (McEliece [939, 941].) *If $\mathscr{C}$ is a binary cyclic code then the weight of every codeword in $\mathscr{C}$ is divisible by 2^{l-1}, where l is the smallest number such that l nonzeros of the code (with repetitions allowed) have product 1.*

The proof of this theorem is difficult and is omitted.

Corollary 13. *All weights in $\mathcal{R}(r,m)$ are multiples of*

$$2^{\lfloor (m-1)/r \rfloor} = 2^{\lceil m/r \rceil - 1}$$

Proof. From Equation (10) of Ch. 13, α^s for $1 \leq s \leq 2^m - 2$ is a nonzero of the punctured cyclic code $\mathcal{R}(r,m)^*$ iff the binary expansion of s has from 1 to r ones in it. Now the product

$$\alpha^{s_1} \alpha^{s_2} \cdots \alpha^{s_l} = 1$$

iff

$$s_1 + s_2 + \cdots + s_l = 2^m - 1. \tag{18}$$

Looking at Equation (18) in binary, we see that each term on the left has at most r ones, whereas the RHS contains m ones. From this it follows that the smallest l for which (18) has a solution is $\lceil m/r \rceil$. The result then follows from Theorem 12.   Q.E.D.

Example. The weights in $\mathcal{R}(2,5)$ are divisible by 2^2, as we saw in the example following Theorem 8.

*§4. Subcodes of dimension 2m of $\mathcal{R}(2, m)^*$ and $\mathcal{R}(2, m)$

We begin by proving a simple theorem which makes it easy to find the weight distribution of many small subcodes of $\mathcal{R}(2, m)^*$ and $\mathcal{R}(2, m)$, by finding the rank of the corresponding symplectic forms.

Theorem 14. *Let B be a symplectic matrix of rank 2h. Then the set of binary m-tuples v such that $uBv^T = 0$ for all $u \in V^m$ is a space of dimension $m - 2h$.*

Proof. Let R be the matrix described in Theorem 4 and set $u' = uR^{-1} = (u'_1, \ldots u'_m)$, $v' = vR^{-1} = (v'_1, \ldots, v'_m)$. Then $uBv^T = u'RBR^Tv'$.

Now RBR^T is of the form

$$
2h \updownarrow \left[\begin{array}{c|c} \begin{matrix} 01 \cdots 00 \\ 10 \cdots 00 \\ \\ 00 \cdots 01 \\ 00 \cdots 10 \end{matrix} & 0 \\ \hline 0 & 0 \end{array} \right]
$$

$$\xleftarrow{\hspace{1em}} 2h \xrightarrow{\hspace{1em}}$$

Thus

$$u'RBR^Tv' = u'_1v'_2 + u'_2v'_1 + u'_3v'_4 + u'_4v'_3 + \cdots$$
$$+ u'_{2h-1}v'_{2h} + u'_{2h}v'_{2h-1}.$$

This is identically zero for all u' iff

$$v' = (0 \cdots 0v'_{2h+1} \cdots v'_m).$$

Such v' form a space of dimension $m - 2h$.   Q.E.D.

Now let $S(v)$ (Equation (1)) be the Boolean function describing a codeword c of $\mathcal{R}(2, m)$. The components of c are obtained from $S(v)$ by letting $(v_1 \cdots v_m)$ run through all binary m-tuples. We can also consider these components to be the values of $S(\xi)$, for $\xi \in GF(2^m)$. Define $Q(\xi)$ for $\xi \in GF(2^m)$ by

$$Q(\xi) = \bar{\xi}Q\bar{\xi}^T,$$

where $\bar{\xi}$ is the m-tuple corresponding to $\xi \in GF(2^m)$. The values of the corresponding symplectic form are, by Equation (4),

$$\mathcal{B}(\xi, \eta) = Q(\xi + \eta) + Q(\xi) + Q(\eta) \tag{19}$$

We shall find that $\mathcal{B}(\xi, \eta)$ can usually be written as

$$\mathcal{B}(\xi, \eta) = T_m(\xi L_B(\eta)),$$

where $L_B(x)$ is a linearized polynomial. Then from Theorem 14

$$\text{rank } B = m - \text{dimensional (kernel } L_B), \tag{20}$$

where kernel L_B is the subspace of η in $GF(2^m)$ for which $L_B(\eta) = 0$.

Small subcodes of $\mathcal{R}(2, m)^*$. In this section we shall find the weight distribution of the $[2^m - 1, 2m]$ cyclic code with idempotent $\theta_1^* + \theta_{l_i}^*$, where $l_i = 1 + 2^i$, for all i. This is a subcode of $\mathcal{R}(2, m)^*$. In particular the code with idempotent $\theta_1^* + \theta_3^*$ is the dual of the double-error-correcting BCH code.

Our method for finding the weight distribution is as follows. First the Mattson–Solomon polynomial of a typical codeword is obtained, and from this the Boolean function $S(\xi)$ describing the codeword. This is a quadratic since the code is a subcode of $\mathcal{R}(2, m)^*$. Then we find the corresponding symplectic form $\mathcal{B}(\xi, \eta)$, the rank of this form from Equation (20), and use Theorem 5 to find the possible weights in the code. Finally the methods of Chapter 6 are used to obtain the weight distribution.

From Equation (8) of Ch. 13 the nonzeros of $\mathcal{R}(2, m)^*$ can be taken to be α^s with $w_2(s) > m - 3$; i.e. those powers of α with exponents belonging to the cyclotomic cosets

$$C_0, C_{-1} \text{ and } C_{-l_i}, \quad \text{for all } l_i = 1 + 2^i,$$

$$1 \le i \le \left[\frac{m}{2} \right].$$

The idempotent of $\mathcal{R}(2, m)^*$ is

$$\theta_0 + \theta_1^* + \sum_{i=1}^{[m/2]} \theta_{l_i}^*.$$

Case (I). m odd, $m = 2t + 1$. This case is easy because of the following result, whose proof we leave to the reader.

Problem (6) (a) Show that if m is odd, and $1 \le i \le t$, then $2^m - 1$ and $2^i + 1$ are relatively prime. (b) Show that $2^i + 1$ and $2^{m-i} + 1$ are the only odd numbers in the cyclotomic coset C_{l_i}, and that this coset has m elements.

We shall find the weight distribution of the $[2^m - 1, 2m]$ subcode $\mathscr{C}_i$ of $\mathcal{R}(2, m)^*$ with idempotent $\theta_1^* + \theta_{l_i}^*$. A nonzero codeword of $\mathscr{C}_i$ has one of the forms $x^j \theta_1^*$, $x^j \theta_{l_i}^*$; or $x^j \theta_1^* + x^k \theta_{l_i}^*$, for some integers j and k. The first two codewords have weight 2^{m-1}, and give us no trouble.

The third type of codeword is a cyclic shift of the codeword $a =$

$x^{l-k}\theta_1^* + \theta_{l_i}^*$. The MS polynomial of a is

$$A(z) = \sum_{l \in C_1} (\gamma z)^l + \sum_{l \in C_{l_i}} z^l$$

$$= T_m(\gamma z) + T_m(z^{1+2^i}),$$

see §6 of Ch. 8. Let $S(\xi)$ be the Boolean function describing a. Then from Theorem 20 of Ch. 8,

$$S(\xi) = T_m(\gamma\xi) + T_m(\xi^{1+2^i}), \quad \text{for all } \xi \in \mathrm{GF}(2^m)^*. \tag{21}$$

The constant term ϵ (see Equation (1)) is $\epsilon = S(0) = 0$. Let a' be the vector obtained from a by adding an overall parity check (which is zero).

We use Equation (21) to find the symplectic form corresponding to that coset of $\mathcal{R}(1, m)$ in $\mathcal{R}(2, m)$ which contains a'. This is, from (5) and (21),

$$\mathcal{B}(\xi, \eta) = T_m(\gamma(\xi + \eta)) + T_m((\xi + \eta)^{1+2^i})$$
$$+ T_m(\gamma\xi) + T_m(\xi^{1+2^i}) + T_m(\gamma\eta) + T_m(\eta^{1+2^i}).$$

Since

$$(\xi + \eta)^{1+2^i} = \xi^{1+2^i} + \xi^{2^i}\eta + \xi\eta^{2^i} + \eta^{1+2^i},$$

$\mathcal{B}(\xi, \eta)$ becomes

$$T_m(\xi^{2^i}\eta + \xi\eta^{2^i}) = T_m(\xi(\eta^{2^{m-i}} + \eta^{2^i})), \quad \text{since } T_m(x) = T_m(x^2).$$

Next we use Equation (20) to find the rank of $\mathcal{B}$. $\mathcal{B}(\xi, \eta)$ is zero for all values of ξ

$$\text{iff} \qquad \eta^{2^{m-i}} + \eta^{2^i} = 0$$
$$\text{iff} \quad (\eta^{2^{m-2i}} + \eta)^{2^i} = 0$$
$$\text{iff} \qquad\qquad \eta = \eta^{2^{m-2i}}$$
$$\text{iff } \eta \in \mathrm{GF}(2^{m-2i}) \cap \mathrm{GF}(2^m) = \mathrm{GF}(2^s),$$

where $s = (m - 2i, m) = (i, m)$ (by Theorem 8 of Ch. 4). Thus the space of η for which $\mathcal{B}(\xi, \eta) = 0$ for all ξ has dimension s. From Equation (20), the rank of B is $m - s$. We conclude from Theorem 5 that the codewords of the form $x^l\theta_1^* + x^k\theta_{l_i}^*$ have weights 2^{m-1} and $2^{m-1} \pm 2^{(m+s-2)/2}$.

Finally the dual code $\mathscr{C}_i^\perp$ is a subcode of the Hamming code, so has $d' \geq 3$. We can therefore use Theorem 2 of Ch. 6 to find the weight distribution of $\mathscr{C}_i$, which is given in Fig. 15.2. Figure 15.2 extends the result of Th. 34 of Ch. 8.

i	A_i
0	1
$2^{m-1} - 2^{(m+s-2)/2}$	$(2^m - 1)(2^{m-s-1} + 2^{(m-s-2)/2})$
2^{m-1}	$(2^m - 1)(2^m - 2^{m-s} + 1)$
$2^{m-1} + 2^{(m+s-2)/2}$	$(2^m - 1)(2^{m-s-1} - 2^{(m-s-2)/2})$

Fig. 15.2. Weight distribution of the code with idempotent $\theta_1^* + \theta_{l_i}^*$, where $l_i = 1 + 2^i$, $s = (i, m)$, m odd.

Note that $\mathcal{C}_3$, with idempotent $\theta_1^* + \theta_3^*$, is a $[2^m - 1, 2m, 2^{m-1} - 2^{(m-1)/2}]$ code which is the dual of the double-error-correcting BCH code. In this case $i = 1$, $s = 1$ and the weight distribution is shown in Fig. 15.3. (This was used in §8 of Ch. 9 to show that the double-error-correcting BCH code is quasi-perfect.)

i	A_i
0	1
$2^{m-1} - 2^{(m-1)/2}$	$(2^m - 1)(2^{m-2} + 2^{(m-3)/2})$
2^{m-1}	$(2^m - 1)(2^{m-1} + 1)$
$2^{m-1} + 2^{(m-1)/2}$	$(2^m - 1)(2^{m-2} - 2^{(m-3)/2})$

Fig. 15.3. Weight distribution of dual of double-error-correcting BCH code of length $2^m - 1$, m odd.

Examples. For $n = 31$, $\mathcal{C}_1$ and $\mathcal{C}_2$ are $[31, 10, 12]$ codes; for $n = 127$, $\mathcal{C}_1$, $\mathcal{C}_2$ and $\mathcal{C}_3$ are $[127, 14, 56]$ codes; and for $n = 511$, $\mathcal{C}_1$, $\mathcal{C}_2$ and $\mathcal{C}_4$ are $[511, 18, 240]$ codes. The weight distribution of these codes are the same as those given in Fig. 8.7. For $n = 511$, $\mathcal{C}_3$ is a $[511, 18, 224]$ code with the weight distribution

Weight	Number of Words
0	1
224	36.511
256	449.511
288	28.511

Case (II). m *even.* This case is more difficult because $l_i = 1 + 2^i$ is frequently not prime to $2^m - 1$.

We begin by doing a special case, the $[2^m - 1, 2m, 2^{m-1} - 2^{m/2}]$ code $\mathcal{C}_3$ with idempotent $\theta_1^* + \theta_3^*$, which is the dual of the double-error-correcting BCH code.

Now 3 divides $2^m - 1$ since m is even, so the general form of a codeword of $\mathcal{C}_3$ is $a_0 x^{i_1} \theta_1^* + b(x) \theta_3^*$. This has MS polynomial

$$\sum_{j \in C_1} (\beta_1 z)^j + \sum_{j \in C_3} (\beta_2 z)^j,$$

(§6 of Ch. 8). The corresponding Boolean function has values

$$S(\xi) = T_m(\beta_1 \xi + (\beta_2 \xi)^3) \quad \text{for } \xi \in GF(2^m)^*,$$

from Theorem 20 of Ch. 8. The symplectic form is

$$\mathcal{B}(\xi, \eta) = T_m(\beta_2^3(\xi^2\eta + \xi\eta^2))$$
$$= T_m(\xi(\beta_2^3\eta^2 + \beta_2^{1+2^{m-1}}\eta^{2^{m-1}})).$$

So we must find the dimension of the space of zeros of the expression

$$\beta_2^3 \eta^2 + \beta_2^{1+2^{m-1}} \eta^{2^{m-1}} = (\gamma^2 \eta + \gamma \eta^{2^{m-2}})^2, \tag{22}$$

where $\gamma^2 = \beta^{2^{m-1}+1}$, $\gamma^4 = \beta^3$. Now η is a zero of (22) iff η is zero or

$$\eta^{2^{m-2}-1} = \gamma. \tag{23}$$

How many different γ's are there in $GF(2^m)^*$ which satisfy this equation? If ω is a primitive element of $GF(2^2) \subset GF(2^m)$, then

$$\eta^{2^{m-2}-1} = (\omega\eta)^{2^{m-2}-1} = (\omega^2\eta)^{2^{m-2}-1},$$

for all $\eta \in GF(2^m)$. Conversely, if

$$\eta^{2^{m-2}-1} = \eta_1^{2^{m-2}-1},$$

then $\eta/\eta_1 \in GF(2^{m-2}) \cap GF(2^m) = GF(2^2)$. Therefore there are $\frac{1}{3}(2^m - 1)$ γ's in $GF(2^m)^*$ which satisfy (23).

If γ has the form $\eta^{2^{m-2}-1}$ for some $\eta \in GF(2^m)^*$, then Equation (22) has four zeros in $GF(2^m)$, and the rank of B is $m - 2$. Therefore the weight of the corresponding codeword is 2^{m-1} or $2^{m-1} \pm 2^{m/2}$. If γ is not of this form, (22) has only the zero $\eta = 0$, and the codeword has weight $2^{m-1} \pm 2^{(m-2)/2}$.

The minimum distance of the dual code is 5, and again the weight distribution of $\mathscr{C}_3$ may be obtained from Theorem 2 of Ch. 6, and is shown in Fig. 15.4.

i	A_i
0	1
$2^{m-1} - 2^{m/2}$	$\frac{1}{32}2^{(m-2)/2-1}(2^{(m-2)/2} + 1)(2^m - 1)$
$2^{m-1} - 2^{m/2-1}$	$\frac{1}{32}2^{(m+2)/2-1}(2^{m/2} + 1)(2^m - 1)$
2^{m-1}	$(2^{m-2} + 1)(2^m - 1)$
$2^{m-1} + 2^{m/2-1}$	$\frac{1}{32}2^{(m+2)/2-1}(2^{m/2} - 1)(2^m - 1)$
$2^{m-1} + 2^{m/2}$	$\frac{1}{32}2^{(m-2)/2-1}(2^{(m-2)/2} - 1)(2^m - 1)$

Fig. 15.4. Weight distribution of the dual of the double-error-correcting BCH code for even m.

Now we do the general case, the code $\mathscr{C}_i$ with idempotent $\theta_1^* + \theta_{l_i}^*$, $l_i = 1 + 2^i$, where $1 < i < \frac{1}{2}m$. (The case $i = \frac{1}{2}m$ was treated in Ch. 8.)

The codewords of $\mathscr{C}_i$ are of one of the forms $x^i\theta_1^*$, $a(x)\theta_{l_i}^*$, or $x^i\theta_1^* + a(x)\theta_{l_i}^*$. The first of these always has weight 2^{m-1}; while for the others we obtain a symplectic form

$$\mathscr{B}(\xi, \eta) = T_m(\xi(\beta\eta^{2^i} + \beta^{2^{m-i}}\eta^{2^{m-i}})).$$

Hence we need to find the number of η such that

$$0 = \beta\eta^{2^i} + \beta^{2^{m-i}}\eta^{2^{m-i}} = (\gamma^{2^i}\eta + \gamma\eta^{2^{m-2i}})^{2^i}, \tag{24}$$

or equivalently the number of nonzero η such that

$$\eta^{2^{m-2i}-1} = \gamma^{2^i-1}. \tag{25}$$

The number of solutions to (25) is given by Problem 7 below. From this, if $(m, i) = (m, 2i) = s$, (24) has 2^s solutions in $GF(2^m)$ for any choice of γ.

Therefore $\mathscr{C}_i$ contains just 3 weights, 2^{m-1} and $2^{m-1} \pm 2^{(m+s-2)/2}$, and the weight distribution is the same as if m were odd and is shown in Fig. 15.2.

On the other hand, if $(m, 2i) = 2(m, i) = 2s$, Equation (24) has either 1 or 2^{2s} solutions depending on the choice of γ, and therefore $\mathscr{C}_i$ contains 5 weights, namely 2^{m-1}, $2^{m-1} \pm 2^{(m+2s-2)/2}$, and $2^{m-1} \pm 2^{m/2-1}$. By Problem 8 $\mathscr{C}_i^{\perp}$ has minimum distance at least 5, and again we can use Theorem 2 of Ch. 6 to obtain the weight distribution, which is shown in Fig. 15.5.

i	A_i
0	1
2^{m-1}	$(2^m - 1)\{(2^s - 1)2^{m-2s} + 1\}$
$2^{m-1} \pm 2^{(m+2s-2)/2}$	$2^{(m-2s-2)/2}(2^m - 1)(2^{(m-2s)/2} \mp 1)/(2^s + 1)$
$2^{m-1} \pm 2^{m/2-1}$	$2^{(m+2s-2)/2}(2^m - 1)(2^{m/2} \mp 1)/(2^s + 1).$

Fig. 15.5. Weight distribution of code with idempotent $\theta_1^* + \theta_{l_i}^*$, where $l_i = 1 + 2^i$, $1 < i < \frac{1}{2}m$, $(m, 2i) = 2(m, i) = 2s$.

Problems. (7) Let m be even and g arbitrary. Show that the number of integers x in the range $0 \le x \le 2^m - 2$ which satisfy

$$(2^{m-2i} - 1)x \equiv (2^i - 1)g \bmod 2^m - 1$$

is

(i) $2^s - 1$ if $(m, 2i) = (m, i) = s$,

(ii) $2^s - 1$ if $(m, 2i) = 2(m, i) = 2s$ and $(2^s + 1) \mid g$,

(iii) 0   if $(m, 2i) = 2(m, i) = 2s$ and $(2^s + 1) \nmid g$.

(8) Use the Hartmann–Tzeng generalized BCH bound, Problem 24 of Ch. 7, to show that the minimum distance of $\mathscr{C}_i^{\perp}$ is at least 5.

(9) Show that $2^m - 1$ and $2^i + 1$ are relatively prime iff $(m, i) = (m, 2i)$. If $(m, 2i) = 2(m, i)$ then $2^i + 1$ and $2^m - 1$ have a nontrivial common factor.

(10) Show that dual of the $[2^m, 2^m - 1 - 2m, 6]$ extended d.e.c. BCH code has parameters

$$[2^m, 2m + 1, 2^{m-1} - 2^{\lfloor m/2 \rfloor}].$$

*§5. The Kerdock code and generalizations

In this rather long section we are going to study a special kind of subcode of the second-order RM code $\mathscr{R}(2, m)$. Let $\mathscr{S}$ be the set of symplectic forms $\mathscr{B}(\xi, \eta)$ associated with the codewords of this subcode (see Equations (4) and (19)). Then $\mathscr{S}$ has the property that the rank of every nonzero form in $\mathscr{S}$ is at least $2d$, and the rank of the sum of any two distinct forms in $\mathscr{S}$ is also at least $2d$. Here d is some fixed number in the range $1 \le d \le [m/2]$. In Ch. 21 we shall see that the maximum size of such a set $\mathscr{S}$ is

$$
\begin{cases}
2^{(2t+1)(t-d+1)} & \text{if } m = 2t+1, \tag{26} \\
2^{(2t+1)(t-d+2)} & \text{if } m = 2t+2. \tag{27}
\end{cases}
$$

Hence the maximum size of the corresponding subcode of $\mathscr{R}(2, m)$ is $2^{1+m}|\mathscr{S}|$. In Ch. 21 we shall also find the number of symplectic forms of each rank in a maximal size set, and hence the distance distribution of the corresponding code.

For odd m these maximum codes turn out to be linear, but for even m they are nonlinear codes which include the Nordstrom–Robinson and Kerdock codes as a special case. The material in this section is due to Delsarte and Goethals.

Case (I). *m odd, $m = 2t+1$.* The general codeword of $\mathscr{R}(2, m)^*$ is

$$
b\theta_0 + a_0 x^{l_0}\theta_1^* + \sum_{j=1}^{t} a_j x^{l_j}\theta_{l_j}^*, \quad l_j = 1 + 2^j.
$$

This has the MS polynomial

$$
\sum_{s \in C_1} (\gamma_0 z)^s + \sum_{j=1}^{t} \sum_{s \in C_{l_j}} (\gamma_j z)^s,
$$

so the corresponding Boolean function and symplectic form are

$$
S(\xi) = \sum_{j=0}^{t} T_{2t+1}(\gamma_j \xi^{1+2^j})
$$

and

$$
\mathscr{B}(\xi, \eta) = \sum_{j=1}^{t} T_{2t+1}(\gamma_j(\xi\eta^{2^j} + \xi^{2^j}\eta))
$$

$$
= T_{2t+1}(\xi L_B(\eta)), \tag{28}
$$

where

$$
L_B(\eta) = \sum_{j=1}^{t} (\gamma_j\eta^{2^j} + (\gamma_j\eta)^{2^{2t+1-j}}). \tag{29}
$$

By setting either the first $d-1$ or the last $d-1$ γ_i's equal to zero in (28) we get a symplectic form of rank $\geqslant 2d$.

Clearly the sum of two such symplectic forms also has $\gamma_i = 0$ for the same values of i. (This is why the corresponding code is linear.)

Theorem 15. *If $\gamma_1 = \gamma_2 = \cdots = \gamma_{d-1} = 0$ then the symplectic form (28) has rank $\geqslant 2d$.*

Proof. If $\gamma_1 = \cdots = \gamma_{d-1} = 0$ then

$$L_B(\eta) = \sum_{j=d}^{t} (\gamma_j \eta^{2^j} + (\gamma_j \eta)^{2^{2t+1-j}})$$

$$= L'_B(\eta)^{2^d},$$

where degree $L'_B(\eta) \leqslant 2^{2(t-d)+1}$. Thus the dimension of the space of η for which $L_B(\eta) = 0$ is at most $2(t-d)+1$, and so rank $B \geqslant 2t+1-2(t-d)-1 = 2d$.
                                                                        Q.E.D.

Theorem 16. *If* $\gamma_{t-d+2} = \gamma_{t-d+3} = \cdots \gamma_t = 0$ *then again the symplectic form* (28) *has rank* $\geqslant 2d$.

Proof. The exponents of η which occur in $L_B(\eta)$ are now $2, 2^2, \ldots, 2^{t-d+1}, 2^{t+d}, 2^{t+d+1}, \ldots, 2^{2t}$. Set $\beta = \eta^{2^{t+d}}$. Then the exponents of β in $L_B(\eta)$ are $1, 2, \ldots, 2^{t-d}, 2^{t-d+2}, \ldots, 2^{2(t-d)+2}$. The highest power of β is at most $2(t-d)+2$, so the dimension of the space of β for which $L_B(\eta) = 0$ is at most $2(t-d)+2$. Since m is odd, this dimension must be odd (Equation (20)), and is at most $2(t-d)+1$. Hence the rank of B is at least $2t+1-2(t-d)-1 = 2d$.   Q.E.D.

By setting $\gamma_i = 0$ we are removing the idempotent $\theta_{l_i}^*$ from the code. So as a corollary to Theorems 15 and 16 we have:

Corollary 17. *Let* $m = 2t+1$ *be odd, and let d be any number in the range* $1 \leqslant d \leqslant t$. *Then there exist two*

$$[2^m, m(t-d+2)+1, 2^{m-1}-2^{m-d-1}],$$

subcodes of $\mathcal{R}(2, m)$. *These are obtained by extending the cyclic subcodes of* $\mathcal{R}(2, m)^*$ *having idempotents*

$$\theta_0 + \theta_1^* + \sum_{j=d}^{t} \theta_{l_j}^*, \tag{30}$$

and

$$\theta_0 + \theta_1^* + \sum_{j=1}^{t-d+1} \theta_{l_j}^*. \tag{31}$$

These codes have weights 2^{m-1} *and* $2^{m-1} \pm 2^{m-h-1}$ *for all h in the range* $d \leqslant h \leqslant t$.

Proof. We have seen that the nonzero symplectic forms associated with the codewords have rank $\geqslant 2d$. The result then follows from Theorem 5.   Q.E.D.

Remarks. From Equation (26) these codes have the maximum possible size. The weight distribution can be obtained from the results given in §8 of Ch. 21.

Case (II). *m even, $m = 2t + 2$.* Again we are looking for maximal sets of symplectic forms such that every form has rank $\geq 2d$ and the sum of every two distinct forms also has rank $\geq 2d$. But now for even m the resulting codes turn out to be nonlinear. We begin with the case $d = \frac{1}{2}m$, in which case we get the Kerdock codes. Before giving the definition, we summarize in Fig. 15.6 the properties of this family of codes. (For $m = 4$, see Fig. 2.19.)

For even $m \geq 4$, $\mathcal{K}(m)$ is a nonlinear code with

　　　length $n = 2^m$,

　　　contains 2^{2m} codewords,

　　　minimum distance $2^{m-1} - 2^{(m-2)/2}$.

The first few codes are $(16, 256, 6)$, $(64, 2^{12}, 28)$, $(256, 2^{16}, 120)$ codes, and the first of these is equivalent to the Nordstrom–Robinson code $\mathcal{N}_{16}$. The general form for a codeword of $\mathcal{K}(m)$ is Equation (34). $\mathcal{K}(m)$ is systematic. The weight and distance distributions coincide and are given in Fig. 15.7. Also $\mathcal{R}(1, m) \subset \mathcal{K}(m) \subset \mathcal{R}(2, m)$. The codewords of each weight in $\mathcal{K}(m)$ form a 3-design.

Fig. 15.6. Summary of properties of Kerdock code $\mathcal{K}(m)$.

---

i	A_i
0	1
$2^{m-1} - 2^{(m-2)/2}$	$2^m(2^{m-1} - 1)$
2^{m-1}	$2^{m+1} - 2$
$2^{m-1} + 2^{(m-2)/2}$	$2^m(2^{m-1} - 1)$
2^m	1

Fig. 15.7. Weight (or distance) distribution of Kerdock code $\mathcal{K}(m)$.

Definition of Kerdock code $\mathcal{K}(m)$. $\mathcal{K}(m)$ consists of $\mathcal{R}(1, m)$ together with $2^{m-1} - 1$ cosets of $\mathcal{R}(1, m)$ in $\mathcal{R}(2, m)$. The cosets are chosen to be of maximum rank, m, and the sum of any two cosets is also of rank m. Alternatively, the Boolean functions associated with these cosets are quadratic bent functions (§5 of Ch. 14), with the property that the sum of any two of them is again a bent function. Since there are a total of 2^{m-1} cosets (or associated symplectic forms), by Equation (27) $\mathcal{K}(m)$ is as large as it can be.

　　We shall write the codewords of $\mathcal{R}(1, m)$ in the form $|u|u + v|$, where $u \in \mathcal{R}(1, m - 1)$ and $v \in \mathcal{R}(0, m - 1)$ (see Theorem 2 of Ch. 13). With this notation $\mathcal{R}(1, m)$ consists of the codewords

$$|a|a\theta_0 + a_1 x^{i}\theta_1^* | b | b\theta_0 + a_1 x^{i}\theta_1^* |, \tag{32}$$

where $a, a_1, b, b_1 \in GF(2)$ and $0 \leqslant i \leqslant 2^{m-1} - 2$.

The *Kerdock code* $\mathcal{K}(m)$, for $m = 2t + 2 \geqslant 4$, consists of $\mathcal{R}(1, m)$ together with $\mu = 2^{2t+1} - 1$ cosets of $\mathcal{R}(1, m)$ in $\mathcal{R}(2, m)$ having coset representatives

$$w_j = \left| 0 | x^j \sum_{i=1}^{t} \theta_{l_i}^* | 0 | x^j \left(\theta_1^* + \sum_{i=1}^{t} \theta_{l_i}^* \right) \right|, \tag{33}$$

for $j = 1, \ldots, \mu$ where as usual $l_i = 1 + 2^i$. Let $w_0 = 0$.

Thus the general codeword of $\mathcal{K}(m)$ is

$$\left| a | a\theta_0 + a_1 x^{i}\theta_1^* + \epsilon x^j \sum_{i=1}^{t} \theta_{l_i}^* | b | b\theta_0 + a_1 x^{i}\theta_1^* + \epsilon x^j \left(\theta_1^* + \sum_{i=1}^{t} \theta_{l_i}^* \right) \right|, \tag{34}$$

where $\epsilon = 0$ or 1 and $1 \leqslant j \leqslant \mu$.

Problems. (11) Let $\mathscr{C}$ be any code which consists of a linear code $\mathscr{H}$ together with l cosets of $\mathscr{H}$ with coset representatives w_j:

$$\mathscr{C} = \mathscr{H} \cup \bigcup_{j=1}^{l} (w_j + \mathscr{H}).$$

Thus $|\mathscr{C}| = (l + 1)|\mathscr{H}|$. $\mathscr{H}$ and $\mathscr{C}$ are represented by the elements

$$H = \sum_{v \in \mathscr{H}} z^v \quad \text{and} \quad \mathscr{C} = H + \sum_{j=1}^{l} z^{w_j} H$$

of the group algebra (see §5 of Ch. 5). Show that the distance distribution of $\mathscr{C}$ is equal to the weight distribution of

$$H + \frac{2}{l+1} \sum_{i=1}^{l} z^{w_i} H + \frac{2}{l+1} \sum_{i=1}^{l-1} \sum_{j=i+1}^{l} z^{w_i + w_j} H.$$

(12) Show that the linear code generated by $\mathcal{K}(m)$ is $\mathcal{R}(2, m)$.

The MS polynomials for the left and right halves of w_j are

$$L(z) = \sum_{i=1}^{t} \sum_{s \in C_{l_i}} (\gamma z)^s,$$

and

$$R(z) = \sum_{i=1}^{t} \sum_{s \in C_{l_i}} (\gamma z)^s + \sum_{s \in C_1} (\gamma z)^s,$$

respectively, where $\gamma \in GF(2^{2t+1})^*$ depends on j.

Let the elements of $GF(2^{2t+1})$ be $(\xi_0 = 0, \xi_1, \ldots, \xi_\mu)$, in some fixed order. Then

$$w_j = |L(\xi_0)|L(\xi_1), \ldots, L(\xi_\mu)|R(\xi_0)|R(\xi_1), \ldots, R(\xi_\mu)|,$$

where

$$L(\xi_\nu) = \sum_{i=1}^{t} T_{2t+1}(\gamma\xi_\nu)^{2^{i+1}},$$

$$R(\xi_\nu) = \sum_{i=1}^{t} T_{2t+1}(\gamma\xi_\nu)^{2^{i+1}} + T_{2t+1}(\gamma\xi_\nu),$$

for $0 \le \nu \le \mu$. If we write the elements of $GF(2^{2t+2})$ as pairs (ξ, ϵ), where $\xi \in GF(2^{2t+1})$ and $\epsilon = 0$ or 1, then we can use these pairs to index the components of w_j (or other vectors of $\mathcal{R}(2, m)$), where $\epsilon = 0$ on the LHS of w_j, and $\epsilon = 1$ on the RHS of w_j.

For example, if $t = 1$ (so that w_j has length 16), we write the elements of $GF(2^4)$ as pairs (ξ, ϵ), where $\xi \in GF(2^3)$ and $\epsilon = 0$ or 1, as shown in Fig. 15.8. Here y is a primitive element of $GF(2^4)$ with $y^4 + y + 1 = 0$, and α is a primitive element of $GF(2^3)$ with $\alpha^3 + \alpha + 1 = 0$.

y^i	4-tuple	ξ, ϵ
0	0000	$0, 0$
1	1000	$1, 0$
y	0100	$\alpha, 0$
y^2	0010	$\alpha^2, 0$
y^3	0001	$0, 1$
y^4	1100	$\alpha^3, 0$
y^5	0110	$\alpha^4, 0$
y^6	0011	$\alpha^2, 1$
y^7	1101	$\alpha^3, 1$
y^8	1010	$\alpha^6, 0$
y^9	0101	$\alpha, 1$
y^{10}	1110	$\alpha^5, 0$
y^{11}	0111	$\alpha^4, 1$
y^{12}	1111	$\alpha^5, 1$
y^{13}	1011	$\alpha^6, 1$
y^{14}	1001	$1, 1$

Fig. 15.8. The field $GF(2^4)$ in the (ξ, ϵ) notation, $\xi \in GF(2^3)$.

For concreteness Fig. 15.9 shows w_7 and some codewords of $\mathcal{R}(1, 4)$, first (Fig. 15.9a) as codewords of an extended cyclic code with the overall parity check in the component labeled ∞, and second (Fig. 15.9b) indexed by the pairs (ξ, ϵ), $\xi \in GF(2^3)$, $\epsilon = 0$ or 1. Figure 15.8 is used to convert from Fig. 15.9a to 15.9b. Note that in Fig. 15.9 $w_7 = x^4\theta_1^* + \theta_3^* + \theta_5^*$ and hence is in $\mathcal{R}(2, 4)$.

(a) as codewords of an extended cyclic code.

∞	1	y	y^2	y^3	y^4	y^5	y^6	y^7	y^8	y^9	y^{10}	y^{11}	y^{12}	y^{13}	y^{14}	name
0	1	1	1	0	0	1	1	1	0	1	0	1	1	1	0	$w_7 = x^4\theta_1^* + \theta_3^* + \theta_5^*$
0	0	0	0	1	0	0	1	1	0	1	0	1	1	1	1	$a = \theta_1^*$
0	1	1	1	1	0	0	0	1	0	0	1	1	0	1	0	$b = y^4\theta_1^*$
1	1	1	1	0	1	1	0	0	1	0	0	1	0	0	0	$c = \theta_0 + \theta_1^*$

(Here θ_i^* is an idempotent of block length 15 – see Problem 7 of Ch. 8.)

(b) indexed by (ξ, ϵ), $\xi \in \mathrm{GF}(2^3)$, $\epsilon = 0$ or 1.

∞	1,0	$\alpha,0$	$\alpha^2,0$	$\alpha^3,0$	$\alpha^4,0$	$\alpha^5,0$	$\alpha^6,0$	0,1	1,1	$\alpha,1$	$\alpha^2,1$	$\alpha^3,1$	$\alpha^4,1$	$\alpha^5,1$	$\alpha^6,1$
0	1	1	1	0	1	0	0	0	0	1	1	1	1	1	1
0	0	0	0	0	0	0	0	1	1	1	1	1	1	1	1
0	1	1	1	0	0	1	0	1	0	0	0	1	1	0	1
1	1	1	1	1	1	1	1	0	0	0	0	0	0	0	0

name

$$w_7 = |0|\theta_3^*|0|\theta_1^* + \theta_3^*|$$
$$a = |0|0|1|\theta_0|$$
$$b = |0|x^2\theta_1^*|1|\theta_0 + x^2\theta_1^*|$$
$$c = |1|\theta_0|0|0|$$

(Here θ_i^* is an idempotent of block length 7 – see Problem 5 of Ch. 8.)

Fig. 15.9. w_7 and some codewords of $\mathcal{R}(1, 4)$.

Next we want a Boolean function $f_\gamma(v) = f_\gamma(v_1, \ldots, v_{2t+2})$ which represents w_j, i.e. such that

$$f_\gamma = w_j \quad \text{(equality of vectors of length } 2^{2t+2}).$$

We set ξ equal to the element of $\mathrm{GF}(2^{2t+1})$ represented by the binary vector $v_1, \ldots, v_{2t+1}$ and set $\epsilon = v_{2t+2}$. Then f_γ is given by

$$f_\gamma(v_1, \ldots, v_{2t+2}) = f_\gamma(\xi, \epsilon)$$

$$= \sum_{i=1}^{t} T_{2t+1}(\gamma\xi)^{1+2^i} + \epsilon T_{2t+1}(\gamma\xi).$$

Then indeed

$$f_\gamma = w_j.$$

The crucial result is the following.

Theorem 18. $f_\gamma(v)$ *is a bent function, and for* $\gamma \neq \delta$, $f_\gamma(v) + f_\delta(v)$ *is also a bent function.*

Proof. The proof is by showing that the symplectic forms corresponding to $f_\gamma(v)$ and to $f_\gamma(v) + f_\delta(v)$ have rank $2t + 2$. The symplectic form corresponding to $f_\gamma(\xi, \epsilon)$ is, from Equation (5),

$$\mathcal{B}_\gamma((\xi, \epsilon_1), (\eta, \epsilon_2)) = \sum_{i=1}^{t} T_{2t+1}(\gamma^{2^{i+1}}(\xi^{2^i}\eta + \xi\eta^{2^i}))$$
$$+ T_{2t+1}(\gamma(\epsilon_1\eta + \epsilon_2\xi))$$
$$= T_{2t+1}(\gamma^2 \xi \eta) + T_{2t+1}(\gamma\xi)T_{2t+1}(\gamma\eta)$$
$$+ T_{2t+1}(\gamma(\epsilon_1\eta + \epsilon_2\xi)). \tag{35}$$

To find the rank of $\mathcal{B}_\gamma$ we must find the dimension of the space of (η, ϵ_2) such that $\mathcal{B}_\gamma((\xi, \epsilon_1), (\eta, \epsilon_2)) = 0$ for all (ξ, ϵ_1). By choosing $(\xi, \epsilon_1) = (0, 1)$ we deduce that $T_{2t+1}(\gamma\eta) = 0$, hence $\mathcal{B}_\gamma = T_{2t+1}((\gamma\xi)(\gamma\eta + \epsilon_2))$. So we must have $\gamma\eta + \epsilon_2 = 0$,

$$\therefore\ T_{2t+1}(\gamma\eta + \epsilon_2) = 0,$$
$$\therefore\ T_{2t+1}(\epsilon_2) = 0,$$
$$\therefore\ \epsilon_2 = 0 \quad \text{(because } 2t + 1 \text{ is odd).}$$
$$\therefore\ \mathcal{B}_\gamma = T_{2t+1}(\gamma^2\xi\eta)$$

and so η must be 0.

Therefore $(\eta, \epsilon_2) = (0, 0)$ is the only pair for which $\mathcal{B}_\gamma((\xi, \epsilon_1), (\eta, \epsilon_2)) = 0$ for all (ξ, ϵ_1). Hence the rank of $\mathcal{B}_\gamma$ is $2t + 2$.

It remains to show that $\mathcal{B}_\gamma + \mathcal{B}_\delta$ has rank $2t + 2$ if $\gamma \neq \delta$. The proof is similar to the preceding and is left to the reader.   Q.E.D.

Weight distribution of $\mathcal{K}(m)$. From Theorem 5 a coset of $\mathcal{R}(1, m)$ of rank m contains 2^m vectors of weight $2^{m-1} - 2^{m/2-1}$ and 2^m vectors of weight $2^{m-1} + 2^{m/2-1}$. The first-order *RM* code itself contains $2^m - 2$ codewords of weight 2^{m-1} and the words $\mathbf{0}$ and $\mathbf{1}$. Hence the weight distribution of $\mathcal{K}(m)$ is as given in Fig. 15.7. From Problem 11 and Theorem 18 the distance distribution is also given by Fig. 15.7. In particular, $\mathcal{K}(4)$ is equivalent to $\mathcal{N}_{16}$, since $\mathcal{N}_{16}$ is unique (§8 of Ch. 2).

In §6 we shall describe another nonlinear code, the Preparata code $\mathcal{P}(m)$, which has the surprising property that its weight (or distance) distribution $\{B_i\}$ is the MacWilliams transform of the weight distribution of $\mathcal{K}(m)$. Thus from Equation (13) of Ch. 5,

$$2^{2m}B_i = P_i(0) + 2^m(2^{m-1} - 1)\{P_i(2^{m-1} - 2^{(m-2)/2}) + P_i(2^{m-1} + 2^{(m-2)/2})\}$$
$$+ (2^{m+1} - 2)P_i(2^{m-1}) + P_i(2^m), \tag{36}$$

where $P_i(x)$ is a Krawtchouk polynomial (§2 of Ch. 5). It is readily checked that the minimum distance in $\mathscr{P}(m)$ is $d' = 6$. It then follows from Theorem 9 of Ch. 6 that the codewords of each weight in $\mathscr{K}(m)$ form 3-designs (since $\bar{s} = 3$).

Case II. *m even, $m = 2t + 2$ (continued)*. We now study the general case, where d is any number in the range $1 \leqslant d \leqslant \frac{1}{2}m$. The nonlinear codes we are about to describe were discovered by Delsarte and Goethals, and are denoted by $\mathscr{DG}(m, d)$. Here is the definition.

If $d = \frac{1}{2}m$ let $\mathscr{C} = \mathscr{R}(1, 2t + 1)$, while if $1 \leqslant d < \frac{1}{2}m$ let $\mathscr{C}$ be the

$$[2^{2t+1}, (2t + 1)(t - d + 2) + 1, 2^{2t} - 2^{2t-d}]$$

code defined by Equation (31) of Corollary 17. Let $w_0 = 0, w_1, \ldots, w_\mu$ be the vectors defined by Equation (33), and let $v_0 = 0 \cdots 01 \cdots 1$.

Definition. For $1 \leqslant d \leqslant \frac{1}{2}m = t + 1$ the code $\mathscr{DG}(m, d)$ consists of all vectors

$$|c(x)|c(x)| + w_j + \epsilon v_0, \tag{37}$$

for $c(x) \in \mathscr{C}, 0 \leqslant j \leqslant \mu, \epsilon = 0$ or 1.

Theorem 19. $\mathscr{DG}(m, d)$, *where $m = 2t + 2 \geqslant 4$, is a code of length 2^{2t+2} containing $2^{(2t+1)(t-d+2)+2t+3}$ codewords and having minimum distance $2^{2t+1} - 2^{2t+1-d}$. If $d = 1$, $\mathscr{DG}(m, 1) = \mathscr{R}(2, m)$, while for $2 \leqslant d \leqslant \frac{1}{2}m$, $\mathscr{DG}(m, d)$ is a nonlinear subcode of $\mathscr{R}(2, m)$.*

Proof. There are $2^{(2t+1)(t-d+2)+2t+3}$ codewords of the form (37). That these are all distinct and in fact have distance at least $2^{2t+1} - 2^{2t+1-d}$ apart will follow from Theorem 20 below.   Q.E.D.

Note that $\mathscr{DG}(m, d)$ contains the Kerdock code $\mathscr{K}(m)$ as a subcode. This is so because $|c(x)|c(x)| + \epsilon v_0$ includes all codewords of $\mathscr{R}(1, 2t + 2)$. Also $\mathscr{DG}(m, \frac{1}{2}m)$ is equal to $\mathscr{K}(m)$. Thus these codes are a generalization of the Kerdock codes. From Problem 12 the linear code generated by $\mathscr{DG}(m, d)$ is $\mathscr{R}(2, m)$.

The first few codes $\mathscr{DG}(m, d)$ are shown in Fig. 15.10, where $k = \log_2$ (number of codewords) and $\delta = $ minimum distance.

$\mathscr{DG}(m, d)$ is contained in $\mathscr{R}(2, m)$ and is a union of cosets of $\mathscr{R}(1, m)$. Consider the coset of $\mathscr{R}(1, m)$ which contains the codeword (37). The symplectic form corresponding to this coset is

		$\mathscr{D}\mathscr{G}(m, d)$			
m	d	n	k	δ	name
4	1	16	11	4	$\mathscr{R}(2, 4)$
	2	16	8	6	$\mathscr{K}(4) = \mathscr{N}_{16}$
6	1	64	22	16	$\mathscr{R}(2, 6)$
	2	64	17	24	
	3	64	12	28	$\mathscr{K}(6)$
8	1	256	37	64	$\mathscr{R}(2, 8)$
	2	256	30	96	
	3	256	23	112	
	4	256	16	120	$\mathscr{K}(8)$

Fig. 15.10. The parameters $(n, 2^k, \delta = \text{minimum distance})$ of the first $\mathscr{D}\mathscr{G}(m, d)$ codes.

$$
\mathscr{B}_\gamma((\xi, \epsilon_1), (\eta, \epsilon_2)) = T_{2t+1}\left(\xi \sum_{i=1}^{t-d+1} (\gamma_i \eta^{2^i} + (\gamma_i \eta)^{2^{2t+1-i}})\right)
$$
$$
+ T_{2t+1}(\gamma^2 \xi \eta) + T_{2t+1}(\gamma \xi) T_{2t+1}(\gamma \eta)
$$
$$
+ T_{2t+1}(\gamma(\epsilon_1 \eta + \epsilon_2 \xi)). \tag{38}
$$

The first term comes from $|c(x)| c(x)|$ via Equations (28), (29), and the rest from w_j via Equation (35). γ corresponds to w_j. The number of distinct symplectic forms of this type is $(2^{2t+1})^{t-d+2}$. From the following theorem and Equation (27) this number is as large as it can be.

The crucial property of these codes is given by:

Theorem 20. *The rank of any symplectic form (38) is at least 2d, and the rank of the sum of any two such forms is also at least 2d.*

Proof. We write (38) as

$$
\mathscr{B}_\gamma((\xi, \epsilon_1), (\eta, \epsilon_2)) = T_{2t+1}(\xi L(\eta))
$$
$$
+ T_{2t+1}(\xi L_\gamma(\eta)) + T_{2t+1}(\gamma(\epsilon_1 \eta + \epsilon_2 \xi)),
$$

where

$$
L(\eta) = \sum_{i=1}^{t-d+1} (\gamma_i \eta^{2^i} + (\gamma_i \eta)^{2^{2t+1-i}}),
$$
$$
L_\gamma(\eta) = \gamma^2 \eta + \gamma T_{2t+1}(\gamma \eta). \tag{39}
$$

It suffices to show that the sum $\mathscr{B}_\gamma + \mathscr{B}_\delta$, $\gamma \neq \delta$, of two such forms has rank $\geq 2d$. Since $\mathscr{C}$ is a linear code $T_{2t+1}(\xi L(\eta)) + T_{2t+1}(\xi L'(\eta)) = T_{2t+1}(\xi L''(\eta))$, where L'' is given by (39) for suitable γ_i's. Therefore

$$\mathcal{B}_\gamma((\xi, \epsilon_1), (\eta, \epsilon_2)) + \mathcal{B}_\delta((\xi, \epsilon_1), (\eta, \epsilon_2))$$
$$= T_{2t+1}(\xi L''(\eta)) + T_{2t+1}(\xi L_\gamma(\eta)) + T_{2t+1}(\xi L_\delta(\eta)) + T_{2t+1}((\gamma + \delta)(\epsilon_1 \eta + \epsilon_2 \xi)).$$

$$(40)$$

It is convenient to define

$$\mathcal{B}'(\xi, \eta) = \mathcal{B}_\gamma((\xi, 0), (\eta, 0)) + \mathcal{B}_\delta((\xi, 0), (\eta, 0)),$$

which is a symplectic form with $\xi, \eta \in \mathrm{GF}(2^{2t+1})$. Then

$$\mathcal{B}'(\xi, \eta) = T_{2t+1}(\xi L_B(\eta)),$$

$$(41)$$

where

$$L_B(\eta) = \sum_{i=1}^{t-d+1} (\gamma_i \eta^{2^i} + (\gamma_i \eta)^{2^{2t+1-i}})$$
$$+ (\gamma^2 + \delta^2)\eta + \gamma T_{2t+1}(\gamma \eta) + \delta T_{2t+1}(\delta \eta).$$

Let the rank of $\mathcal{B}'(\xi, \eta)$ be $2h$. We shall show that $2h \geq 2d - 2$. Expand

$$L_B(\eta) = \sum_{i=0}^{2t+1} \lambda_i \eta^{2^i}.$$

Then

$$\lambda_i = \gamma^{1+2^i} + \delta^{1+2^i} \quad \text{for} \quad t - d + 2 \leq i \leq t + d - 1.$$

$$(42)$$

Let us assume that the underlying Boolean function is in the canonical form given by Theorem 4. Its MS polynomial is given by Equation (11). It is readily seen that the values of the corresponding symplectic form are $T_{2t+1}(\xi L_B(\eta))$, where

$$L_B(\eta) = \sum_{j=1}^{h} (\alpha_j T_{2t+1}(\beta_j \eta) + \beta_j T_{2t+1}(\alpha_j \eta))$$

$$(43)$$

and $\alpha_1, \ldots, \alpha_h, \beta_1, \ldots, \beta_h$ are linearly independent elements of $\mathrm{GF}(2^{2t+1})$. Comparing (42) and (43) we see that, for $t - d + 2 \leq i \leq t + d - 1$,

$$\gamma^{1+2^i} + \delta^{1+2^i} = \sum_{j=1}^{h} (\alpha_j \beta^{2^i} + \alpha_j^{2^i} \beta_j)$$
$$= \sum_{j=1}^{h} (\alpha_j^{1+2^i} + \beta_j^{1+2^i} + (\alpha_j + \beta_j)^{1+2^i}).$$

If we write

$$s_i = \gamma^{1+2^i} + \delta^{1+2^i} + \sum_{j=1}^{h} (\alpha_j^{1+2^i} + \beta_j^{1+2^i} + (\alpha_j + \beta_j)^{1+2^i}),$$

then clearly

$$s_i = 0 \quad \text{for} \quad t - d + 2 \leq i \leq t + d - 1.$$

$$(44)$$

Let U be the set $\{\gamma, \delta, \alpha_1, \beta_1, \alpha_1 + \beta_1, \ldots, \alpha_h, \beta_h, \alpha_h + \beta_h\}$, so that we can write

$$s_i = \sum_{\alpha \in U} \alpha^{1+2^i}. \tag{45}$$

Let $V(U)$ be the smallest subspace of $GF(2^{2t+1})$ which contains all of U. Clearly $\dim V(U) \leqslant 2h + 2$. Let $\sigma(z)$ be the linearized polynomial corresponding to U (see §9 of Ch. 4), given by

$$\sigma(z) = \prod_{\lambda \in V(U)} (z + \lambda) = \sum_{i=0}^{2h+2} \sigma_i z^{2^i}.$$

Note that $\sigma_0 \neq 0$ (for otherwise $\sigma(z)$ is a perfect square). Then

$$\begin{aligned}
0 &= \sum_{\alpha \in U} \alpha \sigma^{2^i}(\alpha) \\
&= \sigma_0^{2^i} \sum_{\alpha \in U} \alpha^{1+2^i} + \sigma_1^{2^i} \sum_{\alpha \in U} \alpha^{1+2^{i+1}} + \cdots \\
&\quad + \sigma_{2h+2}^{2^i} \sum_{\alpha \in U} \alpha^{1+2^{i+2h+2}} \\
&= \sigma_0^{2^i} s_i + \sum_{j=1}^{2h+2} \sigma_j^{2^i} s_{i+j} \quad \text{by (45).} \tag{46}
\end{aligned}$$

Now we assume that $2h + 2 \leqslant 2d - 2$ and arrive at a contradiction. This will prove that $2h + 2 > 2d - 2$ or $2h \geqslant 2d - 2$, as claimed.

Setting $i = t - d + 1$ in (46) and using (44) we deduce that $s_{t-d+1} = 0$. Then setting $i = t - d, \ldots$ implies $s_i = 0$ for all $0 \leqslant i \leqslant t + d - 1$. But $s_0 = \gamma^2 + \delta^2 \neq 0$, a contradiction.

Therefore the restricted form $\mathcal{B}'$ has rank at least $2d - 2$. If the rank of $\mathcal{B}'$ is $2d$ or more, so is the rank of $\mathcal{B}_\gamma + \mathcal{B}_\delta$ and we are done. Suppose then that the rank of $\mathcal{B}'$ is exactly $2h = 2d - 2$. We need a lemma.

Lemma 21. *Let $\mathcal{B}((\xi, \epsilon_1), (\eta, \epsilon_2))$ be the symplectic form*

$$\mathcal{B}((\xi, \epsilon_1), (\eta, \epsilon_2)) = T_{2t+1}(\xi L(\eta)) + \beta(\epsilon_2 \xi + \epsilon_1 \eta)),$$

where $L(\eta)$ is a linearized polynomial over $GF(2^{2t+1})$ and $\beta \in GF(2^{2t+1})$. Let

$$\begin{aligned}
\mathcal{B}'(\xi, \eta) &= \mathcal{B}((\xi, 0), (\eta, 0)) \\
&= T_{2t+1}(\xi L(\eta)).
\end{aligned}$$

Then if the equation $L(\eta) = \beta$ has no solutions in $GF(2^{2t+1})$,

$$\text{rank } \mathcal{B} = \text{rank } \mathcal{B}' + 2.$$

Proof of Lemma. To find the rank of $\mathcal{B}$ we need to know the number of (η, ϵ_2) such that $\mathcal{B}((\xi, \epsilon_1), (\eta, \epsilon_2)) = 0$ for all (ξ, ϵ_1). Setting $(\xi, \epsilon_1) = (0, 1)$ we see that $T_{2t+1}(\beta \eta) = 0$. Thus $\mathcal{B} = T_{2t+1}(\xi(L(\eta) + \beta \epsilon_2))$. So we must have $L(\eta) + \beta \epsilon_2 = 0$.

If $L(\eta) = \beta$ has no solution, the only zeros have $\epsilon_2 = 0$. In fact the zeros are

$$\{(\eta, 0): L(\eta) = T_{2t+1}(\beta\eta) = 0\}.$$

Now from (20),

$$\text{rank } \mathcal{B}' = 2t + 1 - \dim (\text{kernel } L(\eta)),$$

and hence

$$\text{rank } \mathcal{B} \geqslant 2t + 2 - \dim (\text{kernel } L(\eta)).$$

Thus rank $\mathcal{B} >$ rank $\mathcal{B}'$, i.e. rank $\mathcal{B} =$ rank $\mathcal{B}' + 2$ (since it cannot be more than this). This completes the proof of the lemma.   Q.E.D.

For the proof of the theorem, $\mathcal{B}$ and $\mathcal{B}'$ are given by (40) and (41), and $\beta = \gamma + \delta$. To complete the proof we must show that if rank $\mathcal{B}' = 2d - 2$ then the equation $L_B(\eta) = \gamma + \delta$ has no solution in $GF(2^{2t+1})$.

Suppose then that $\gamma + \delta = L_B(\eta)$ for some $\eta \in GF(2^{2t+1})$. From Equation (43) $\gamma + \delta$ is linearly dependent on $\alpha_1, \ldots, \alpha_h, \beta_1, \ldots, \beta_h$, i.e.,

$$\gamma + \delta = \sum_{j=1}^{h} (a_j\alpha_j + b_j\beta_j), \quad \text{for } a_j, b_j \in GF(2),$$

$$= \sum_{j=1}^{h} (a_j(\alpha_j + b_j\gamma) + b_j(\beta_j + a_j\gamma)).$$

Let U' be the set

$$U' = \{\gamma + \delta, \alpha_1 + b_1\gamma, \beta_1 + a_1\gamma, \alpha_1 + \beta_1 + (a_1 + b_1)\gamma, \ldots,$$
$$\alpha_h + b_h\gamma, \beta_h + a_h\gamma, \alpha_h + \beta_h + (a_h + b_h)\gamma\},$$

and $V(U')$ the smallest subspace of $GF(2^{2t+1})$ containing all of U'. Clearly $\dim V(U') \leqslant 2h = 2d - 2$. Now defining

$$s_i' = \sum_{\alpha \in U'} \alpha^{1+2^i},$$

it is straightforward to verify that $s_i' = s_i$. Then by the same argument as before, $s_i = 0$ for $i \leqslant t + d - 1$. But again $s_0 \neq 0$, a contradiction. Hence our assumption that $\gamma + \delta = L_B(\eta)$ is false, thus completing the proof of the theorem.   Q.E.D.

From Theorem 5 it follows that $\mathcal{DG}(m, d)$ has distances $0, 2^{m-1}, 2^m$ and $2^{m-1} \pm 2^{m-h-1}$ for $d \leqslant h \leqslant t + 1$. In §8 of Ch. 21 we will show how to find the distance distribution of this code. The distance distributions in the special cases $d = t + 1$ and $d = t$ are given in Figs. 15.7 and 15.13.

Problem. (13) Show that $\mathcal{DG}(m, d + 1)$ is a union of disjoint translates of $\mathcal{DG}(m, d)$.

*§6. The Preparata code

We now describe another nonlinear code, the Preparata code $\mathscr{P}(m)$, which has the property that its weight (or distance) distribution $\{B_i\}$ is the Mac-Williams transform of the weight distribution of the Kerdock code $\mathscr{K}(m)$, and is given by Equation (36). Of course the Preparata code is not the dual of the Kerdock code in the usual sense, since both codes are nonlinear.

$\mathscr{P}(m)$ can be constructed in a similar way to the Kerdock code, as the union of a linear code Π and $2^{m-1} - 1$ cosets of Π in $\mathscr{R}(m-2, m)$. Recall that $\mathscr{R}(m-2, m)$ is the extended Hamming code and is the dual code to $\mathscr{R}(1, m)$.

From Theorem 2 of Ch. 13, $\mathscr{R}(m-2, m)$ consists of the codewords

$$|u|u+v|: u \text{ even weight}, \quad v \in \mathscr{R}(m-3, m-1);$$

i.e. consists of the codewords

$$|h(1)|h(x)|h(1) + k(1)|h(x) + k(x)(1 + \theta_1)|$$

where $h(x)$ and $k(x)$ are arbitrary.

Definition. For m even, $m = 2t + 2 \geqslant 4$, the *Preparata code* $\mathscr{P}(m)$ consists of a linear code Π (which is contained in $\mathscr{R}(m-2, m)$), together with $\mu = 2^{m-1} - 1$ cosets of Π in $\mathscr{R}(m-2, m)$ having coset representatives $w_j, j = 1, \ldots, \mu$. Π consists of the codewords

$$|g(1)|g(x)(1 + \theta_1)|f(1) + g(1)|g(x)(1 + \theta_1) + f(x)(1 + \theta_1 + \theta_3)|, \qquad (47)$$

where $f(x)$ and $g(x)$ are arbitrary, and

$$w_j = |1|x^j|0|x^j\theta_1|, \quad j = 1, \ldots, \mu. \qquad (48)$$

Thus Π has dimension $2^{m-1} - 1 - (m-1) + 2^{m-1} - 1 - 2(m-1) = 2^m - 3m + 1$. Π is a $[16, 5, 8]$ code if $m = 4$, and a $[2^m, 2^m - 3m + 1, 6]$ code if $m \geqslant 6$.

$\mathscr{P}(m)$ consists of the following 2^k codewords, where $k = 2^m - 2m$:

$$|g(1) + a|g(x)(1 + \theta_1) + ax^j|f(1) + g(1)|g(x)(1 + \theta_1)$$
$$+ f(x)(1 + \theta_1 + \theta_3) + ax^j\theta_1|, \qquad (49)$$

where $a = 0$ or 1 and $1 \leqslant j \leqslant \mu$. (See Fig. 15.12.)

The first Preparata code coincides with the first Kerdock code (and hence is equivalent to the Nordstrom–Robinson code $\mathscr{N}_{16}$):

Lemma 22. $\mathscr{P}(4) = \mathscr{K}(4)$, *i.e. the Preparata and Kerdock codes of length 16 are the same.*

Proof. The primitive idempotents for length 7 are $\theta_0, \theta_1, \theta_3$ with $\theta_1^* = \theta_3, \theta_3^* = \theta_1, \theta_0 + \theta_1 + \theta_3 = 1$. Thus Π consists of the codewords

$$|g(1)|g(x)(\theta_0 + \theta_3)|f(1) + g(1)|g(x)(\theta_0 + \theta_3) + f(x)\theta_0|$$
$$= |g(1)|g(1)\theta_0 + \alpha_1 x^i \theta_1^*|f(1) + g(1)|f(1)\theta_0 + g(1)\theta_0$$
$$+ \alpha_1 x^{i_1} \theta_1^*|.$$

Therefore $\Pi = \mathcal{R}(1, 4)$.

From Equations (33) and (48) the coset representatives for $\mathcal{K}(4)$ and $\mathcal{P}(4)$ are respectively

$$w_i(\mathcal{K}) = |0|x^j \theta_3^*|0|x^j(\theta_1^* + \theta_3^*)|,$$
$$w_j(\mathcal{P}) = |1|x^j|0|x^j \theta_1|.$$

Their sum is

$$|1|x^j(\theta_0 + \theta_1^*)|0|x^j \theta_1^*|,$$

which is in $\mathcal{R}(1, 4)$. Hence $w_i(\mathcal{K})$ and $w_i(\mathcal{P})$ define the same coset of $\mathcal{R}(1, 4)$. Therefore $\mathcal{P}(4) = \mathcal{K}(4)$.   Q.E.D.

Problem. (14) Show that the linear code generated by $\mathcal{P}(m)$ is a $[2^m, 2^m - 2m + 1, 4]$ subcode of $\mathcal{R}(m - 2, m)$ consisting of all codewords of the form

$$|F(1)|F(x)|\epsilon + F(1) + G(1)|\epsilon\theta_0 + F(x) + G(x)(1 + \theta_1 + \theta_3)|.$$

Lemma 23. $\Pi^{\perp}$ *consists of the codewords*

$$|c|c\theta_0 + c_1 x^{i_1}\theta_0^* + c_3 x^{i_3}\theta_3^*|d|d\theta_0 + d_1 x^{i_1}\theta_1^* + c_3 x^{i_3}\theta_3^*|,$$

where c, c_1, c_3, d, d_1 *are* 0 *or* 1 *and* $i_1, i_3, j_1 \in \{0, \ldots, 2^{m-1} - 2\}$.

Proof. We may write $\Pi = \mathcal{A}_1 + \mathcal{A}_2$, where $\mathcal{A}_1$ and $\mathcal{A}_2$ consist of the vectors

$$|0|0|f(1)|f(x)(1 + \theta_1 + \theta_3)|,$$
$$|g(1)|g(x)(1 + \theta_1)|g(1)|g(x)(1 + \theta_1)|$$

respectively. Then $\mathcal{A}_1^{\perp}$ and $\mathcal{A}_2^{\perp}$ consist of the vectors

$$|c|h_1(x)|d|d\theta_0 + d_1 x^{i_1}\theta_1^* + c_3 x^{i_3}\theta_3^*|,$$
$$|e_1|(e_1 + e_2)\theta_0 + h(x) + c_1 x^{i_1}\theta_1^*|e_2|h(x)|$$

respectively. Therefore in $\Pi^{\perp} = \mathcal{A}_1^{\perp} \cap \mathcal{A}_2^{\perp}$ (see Problem 33 of Ch. 1) we must have $c = e_1$, $h_1(x) = (e_1 + e_2)\theta_0 + h(x) + c_1 x^{i_1}\theta_1^*$, etc. These equations imply that $\Pi^{\perp}$ is as stated.   Q.E.D.

Note that $\mathcal{R}(1, m) \subset \Pi^{\perp} \subset \mathcal{R}(2, m)$ (see Fig. 15.11).

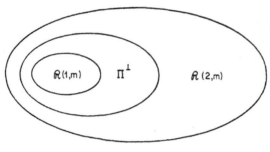

Fig. 15.11.

We come now to the main theorem of this section.

Theorem 24. *The weight (or distance) distribution $\{B_i\}$ of $\mathcal{P}(m)$ is the MacWilliams transform of the weight distribution of $\mathcal{K}(m)$, and has the generating function*

$$2^{2m} \sum_{i=0}^{2^m} B_i z^i = (1 + z)^{2^m} + 2^m (2^{m-1} - 1)$$
$$\cdot \{(1 + z)^{2^{m-1}+2^{(m-2)/2}}(1 - z)^{2^{m-1}-2^{(m-2)/2}}$$
$$+ (1 + z)^{2^{m-1}-2^{(m-2)/2}}(1 - z)^{2^{m-1}+2^{(m-2)/2}}\}$$
$$+ (2^{m+1} - 2)(1 - z^2)^{2^{m-1}} + (1 - z)^{2^m}. \tag{50}$$

Alternatively B_i is given by Equation (36).

Proof. The distance distribution of $\mathcal{P}(m)$ is (from Problem 11) the weight distribution of the element

$$D = H + \frac{1}{2^{m-2}} \sum_{i=1}^{2^{m-1}-1} z^{w_i} H + \frac{1}{2^{m-2}} \sum_{1 \le i < j \le 2^{m-1}-1} z^{w_i + w_j} H$$

of the group algebra, where

$$H = \sum_{v \in \Pi} z^v.$$

The MacWilliams transform of this weight distribution is the set of numbers

$$A_l' = \frac{1}{|\mathcal{P}(m)|} \sum_{\text{wt}(u)=l} \chi_u(D), \quad 0 \le l \le 2^m,$$

where the sum is over all vectors u of length 2^m and weight l (from Theorem 5 of Ch. 5).

We show that $A_l' = 0$ unless l is one of the numbers $0, 2^{m-1} \pm 2^{(m-2)/2}, 2^{m-1}$ or 2^m; i.e. one of the weights of $\mathcal{K}(m)$; and that in fact the set of numbers A_l' is the weight distribution of $\mathcal{K}(m)$.

We consider separately the contribution to A_i' from $u \in \mathcal{R}(1, m)$, $u \in \Pi^\perp - \mathcal{R}(1, m)$ and $u \notin \Pi^\perp$.

Case (I). $u \in \mathcal{R}(1, m)$.

Lemma 25. *If* $u \in \mathcal{R}(1, m)$, $\chi_u(D) = |\mathcal{P}(m)|$.

Proof. $\mathcal{P}(m) \subset \mathcal{R}(1, m)^\perp$, thus $(-1)^{u \cdot v} = 1$ for all $v \in \mathcal{P}(m)$. Q.E.D.

Corollary 26. *The codewords of* $\mathcal{R}(1, m)$ *contribute 1 to* A_0', *1 to* A_{2^m}' *and* $2^{m+1} - 2$ *to* $A_{2^{m-1}}'$.

Case (II). $u \notin \Pi^\perp$.

Lemma 27. *If* $u \notin \Pi^\perp$, $\chi_u(D) = 0$.

Proof. We have $\chi_u(H) = 0$, hence $\chi_u(z^{w_i} H) = (-1)^{u \cdot w_i} \chi_u(H) = 0$, and $\chi_u(z^{w_i + w_j} H) = (-1)^{u \cdot (w_i + w_j)} \chi_u(H) = 0$. Q.E.D.

Case (III). $u \in \Pi^\perp - \mathcal{R}(1, m)$. This case is more complicated. We must calculate

$$\chi_u(D) = \chi_u(H)\left(1 + \frac{1}{2^{m-2}} \sum_{i=1}^{2^{m-1}-1} (-1)^{u \cdot w_i} + \frac{1}{2^{m-2}} \sum_{1 \leqslant i < j \leqslant 2^{m-1}-1} (-1)^{u \cdot (w_i + w_j)}\right).$$

Suppose $u \cdot w_i$ is even for α values of i and odd for β values of i. Then clearly

$$2^{m-1} - 1 = \alpha + \beta,$$

$$\sum_{i=1}^{2^{m-1}-1} (-1)^{u \cdot w_i} = \alpha - \beta.$$

Also $u \cdot (w_i + w_j)$ is even for $\binom{\alpha}{2} + \binom{\beta}{2}$ pairs w_i, w_j and odd for $\alpha\beta$ pairs. Thus

$$\sum_{1 \leqslant i < j \leqslant 2^{m-1}-1} (-1)^{u \cdot (w_i + w_j)} = \binom{\alpha}{2} + \binom{\beta}{2} - \alpha\beta$$
$$= \tfrac{1}{2}\{(\alpha - \beta)^2 - (\alpha + \beta)\},$$

and

$$\frac{\chi_u(D)}{|\mathcal{P}(m)|} = \frac{1}{2^{2m-2}} (\alpha - \beta + 1)^2. \tag{51}$$

We now find the possible values of β.

Lemma 28. *Let* $\mathcal{R}(1, m) + s$ *be a coset of* $\mathcal{R}(1, m)$ *in* $\Pi^{\perp}$. *Then for any u in the coset,* $(-1)^{w_j \cdot u} = (-1)^{w_j \cdot s}$.

Proof. $u = s + r$, $r \in \mathcal{R}(1, m)$, and $(-1)^{w_j \cdot r} = 1$.   Q.E.D.

The dimension of $\Pi^{\perp}$ is $3m - 1$, and the dimension of $\mathcal{R}(1, m)$ is $m + 1$, so there are 2^{2m-2} cosets. We take the coset representatives to be

$$s = |0|c_1 x^{i_1}\theta_1^* + c_3 x^{i_3}\theta_3^*|0|c_3 x^{i_3}\theta_3^*|,$$

where $c_1, c_3 \in GF(2)$, $0 \leqslant i_1, i_3 \leqslant 2^{m-1} - 1$.

Consider the scalar product of s and $w_j = |1|x^j|0|x^j\theta_1|$. Now $x^j\theta_1 \cdot x^{i_3}\theta_3^*$ is always even (since the dual of the code generated by θ_1 is generated by $1 + \theta_1^* = \sum_{r \neq 1} \theta_r^*$, and this contains the code generated by θ_3^*). The scalar product is 1 if x^j has a nonzero coefficient in $c_1 x^{i_1}\theta_1^* + c_3 x^{i_3}\theta_3^*$, and is otherwise zero, hence $\beta = \mathrm{wt}\,(c_1 x^{i_1}\theta_1^* + c_3 x^{i_3}\theta_3^*)$. Figure 15.3 shows that this vector has weight 0, 2^{m-2}, or $2^{m-2} \pm 2^{(m-2)/2}$.

If $\beta = 2^{m-2}$, then $\alpha = 2^{m-2} - 1$ and from (51) $\chi_u(D)/|\mathcal{P}(m)| = 0$.

If $\beta = 2^{m-2} \pm 2^{(m-2)/2}$, then $\alpha = 2^{m-2} \mp 2^{(m-2)} - 1$ and $\chi_u(D)/|\mathcal{P}(m)| = 1/2^{m-2}$.

If $c_1 x^{i_1}\theta_1^* + c_3 x^{i_3}\theta_3^*$ has weight $2^{m-2} \pm 2^{(m-2)/2}$, the coset representative s has weight $2^{m-1} \pm 2^{(m-2)/2}$ since $x^{i_3}\theta_3^*$ has weight 2^{m-2}. Thus the coset of $\mathcal{R}(1, m)$ in $\Pi^{\perp}$ is a maximum rank coset in $\mathcal{R}(2, m)$; and contains 2^m vectors of weight $2^{m-1} + 2^{(m-2)/2}$ and 2^m vectors of weight $2^{m-1} - 2^{(m-2)/2}$. This coset contributes $2^m/2^{m-2} = 4$ to A'_l for $l = 2^{m-1} \pm 2^{(m-2)/2}$. Now from Fig. 15.3 the total number of s of weights $2^{m-1} + 2^{(m-2)/2}$ and $2^{m-1} - 2^{(m-2)/2}$ is $2^{m-2}(2^{m-1} - 1)$. This completes Case III.

The contributions to A'_l from the three cases are as follows:

Case:	I	II	III
A'_0	1	0	0
$A'_{2^{m-1}}$	$2^{m+1} - 2$	0	0
A'_{2^m}	1	0	0
$A'_{2^{m-1} \pm 2^{(m-2)/2}}$	0	0	$2^m(2^{m-1} - 1)$

Adding the three cases we obtain the weight distribution of the Kerdock code as shown in Fig. 15.7. This proves that the distance distribution of $\mathcal{P}(m)$ is the transform of the weight distribution of $\mathcal{K}(m)$. It remains to show that the weight and distance distributions of $\mathcal{P}(m)$ coincide.

It is readily checked from (50) that $\mathcal{P}(m)$ has minimum distance 6. Since $\mathcal{K}(m)$ has $\bar{s} = 3$ it follows from Theorem 3 of Ch. 6 that $\mathcal{P}(m)$ is distance invariant.   Q.E.D.

Corollary 29. $\mathcal{P}(m)$ *has minimum distance* 6.

Remark. $\mathcal{P}(m)$ has the same length and minimum distance as the $[2^m, 2^m - 2m - 1, 6]$ extended BCH code, but contains twice as many codewords. We shall see in Chapter 17 that $\mathcal{P}(m)$ has the greatest possible number of codewords for this minimum distance.

The properties of $\mathcal{P}(m)$ are summarized in Fig. 15.12.

For even $m \geqslant 4$, $\mathcal{P}(m)$ is a nonlinear code with

> length $n = 2^m$,
> contains 2^{n-2m} codewords,
> minimum distance 6.

The general form for a codeword is Equation (49). $\mathcal{P}(m)$ is systematic and quasi-perfect, contains twice as many codewords as an extended double-error-correcting BCH code of the same length, and has the greatest possible number of codewords for this minimum distance. The weight and distance distributions coincide and are given by Equations (36) or (50). $\mathcal{P}(4) = \mathcal{K}(4) = (16, 256, 6)$ Nordstrom–Robinson code. Also $\mathcal{R}(m-3, m) \subset \mathcal{P}(m) \subset \mathcal{R}(m-2, m)$. The codewords of each weight in $\mathcal{P}(m)$ form a 3-design (Theorem 33).

Fig. 15.12. Properties of Preparata code $\mathcal{P}(m)$.

The shortened Preparata code $\mathcal{P}(m)^*$. Let $\mathcal{P}(m)^*$ be the code obtained from $\mathcal{P}(m)$ by deleting one coordinate place (any one). In this section we show:

(1) The weight and distance distributions of $\mathcal{P}(m)^*$ do not depend on which coordinate is deleted (Theorem 32).

(2) The codewords of each weight in $\mathcal{P}(m)^*$ form a 2-design (Theorem 33).

(3) The whole space F^{2^m-1} is the union of disjoint translates of $\mathcal{P}(m)^*$ by vectors of weights 1, 2, and 3; i.e. $\mathcal{P}(m)^*$ is quasi-perfect (Theorem 34).

Let $\mathscr{C}$ be any code containing $\mathbf{0}$, of even block length n, and such that all codewords have even weight. Let $\mathscr{C}^*$ be the code obtained from $\mathscr{C}$ by deleting a fixed coordinate place, which for convenience we suppose to be the last. Let $\{A_i\}$ and $\{a_i\}$ be the distance distributions of $\mathscr{C}$ and $\mathscr{C}^*$. As usual the transformed distributions are denoted by primes.

Theorem 30.
$$A_i' = a_i' + a_{n-i}'.$$

Proof. Let $\sigma_i = \{u \in F^n: \mathrm{wt}\,(u) = i\}$, $\hat{\sigma}_i = \{u \in F^{n-1}: \mathrm{wt}\,(u) = i\}$. The transforms of the distance distributions of $\mathscr{C}^*$ and $\mathscr{C}$ are, by Theorem 5 of Ch. 5,

$$|\mathscr{C}|^2 a_i' = \sum_{\hat{w}\in\hat{\sigma}_i}\sum_{\hat{a},\,\hat{\theta}\in\mathscr{C}^*}(-1)^{\hat{w}\cdot(\hat{a}+\hat{\theta})},$$

$$|\mathscr{C}|^2 A_i' = \sum_{w\in\sigma_i}\sum_{u,\,v\in\mathscr{C}}(-1)^{w\cdot(u+v)},$$

$$= \sum_{\hat{w}\in\hat{\sigma}_i}\sum_{\hat{a},\,\hat{\theta}\in\mathscr{C}^*}(-1)^{\hat{w}\cdot(\hat{a}+\hat{\theta})}$$

$$+ \sum_{\hat{w}\in\hat{\sigma}_{i-1}}\sum_{\hat{a},\,\hat{\theta}\in\mathscr{C}^*}(-1)^{\hat{w}\cdot(\hat{a}+\hat{\theta})+u_n+v_n},$$

where we have written $w = |\hat{w}|w_n|$, $u = |\hat{u}|u_n|$ etc., with

$$u_n = \sum_{i=1}^{n-1}\hat{u}_i, \qquad v_n = \sum_{i=1}^{n-1}\hat{v}_i$$

since $\mathscr{C}$ is an even weight code. Therefore

$$|\mathscr{C}|^2 A_i' = |\mathscr{C}|^2 a_i' + \sum_{\hat{w}\in\hat{\sigma}_{i-1}}\sum_{\hat{a},\,\hat{\theta}\in\mathscr{C}^*}(-1)^{(\hat{w}+1)\cdot(\hat{a}+\hat{\theta})}$$

But the last term is $|\mathscr{C}|^2 a_{n-i}'$.   Q.E.D.

Corollary 31. $A_i' = A_{n-i}'$; in particular $A_0' = A_n' = 1$.

Remark. Theorem 30 and Corollary 31 are trivial for linear codes, since $\{A_i'\}$ and $\{a_i'\}$ are the weight distributions of $\mathscr{C}^\perp$ and $(\mathscr{C}^*)^\perp$.

Problem. (15) For linear codes show that $(\mathscr{C}^*)^\perp$ is obtained from $\mathscr{C}^\perp$ by deleting the fixed coordinate place and discarding all resulting odd weight vectors.

We now apply Theorem 30 to $\mathscr{P}(m)$; viz.

$$A_{2^{m-1}+2^{(m-2)/2}}' = A_{2^{m-1}-2^{(m-2)/2}}' = a_{2^{m-1}+2^{(m-2)/2}}' + a_{2^{m-1}-2^{(m-2)/2}}',$$

$$A_{2^{m-1}}' = 2a_{2^{m-1}}'.$$

Thus the transform of the distance distribution of $\mathscr{P}(m)^*$ contains $s' = 3$ nonzero terms besides $a_0' = 1$. Since $\mathscr{P}(m)^*$ has minimum distance d at least 5, we can find $a_{2^{m-1}\pm 2^{(m-2)/2}}'$, $a_{2^{m-1}}'$ from Theorem 2 of Ch. 6. Since $s' < d$, by Theorem 6 of Ch. 6 this is also the transform of the weight distribution. The result is given in the following theorem.

Theorem 32. Let $\mathscr{P}(m)^*$ be the punctured Preparata code of length $n - 1$, where $n = 2^m$, m even ≥ 4.

(1) *The transform of the distance (and weight) distribution of $\mathscr{P}(m)^*$ is:*

$$a_0' = 1, \; a_{n/2}' = n - 1,$$

$$a_{(n-\sqrt{n})/2}' = \tfrac{1}{4}(n-2)(n+\sqrt{n}), \qquad a_{(n+\sqrt{n})/2}' = \tfrac{1}{4}(n-2)(n-\sqrt{n}).$$

(2) *The distance (and weight) distribution of $\mathcal{P}(m)^*$ is given by the generating function*

$$\frac{2^{n-1}}{|\mathcal{P}(m)|} \sum_{i=0}^{n-1} a_i z^i = (1+z)^{n-1}$$
$$+ \tfrac{1}{4}(n + \sqrt{n})(n-2)(1+z)^{(1/2)(n+\sqrt{n})-1}(1-z)^{(1/2)(n-\sqrt{n})}$$
$$+ \tfrac{1}{4}(n - \sqrt{n})(n-2)(1+z)^{(1/2)(n-\sqrt{n})-1}(1-z)^{(1/2)(n+\sqrt{n})}$$
$$+ (n-1)(1+z)^{n/2-1}(1-z)^{n/2}.$$

where $|\mathcal{P}(m)| = |\mathcal{P}(m)^*| = 2^{n-2m-2}$.

(3) *The distance (and weight) distribution of $\mathcal{P}(m)^*$ does not depend on which coordinate of $\mathcal{P}(m)$ is deleted.*

(4) *The minimum distance of $\mathcal{P}(m)^*$ is 5.*

Theorem 33. *The codewords of any fixed weight in $\mathcal{P}(m)^*$ form a 2-design. In particular the*

$$a_5 = \tfrac{1}{60}(2^m - 1)(2^m - 2)(2^m - 4) \tag{52}$$

codewords of weight 5 in $\mathcal{P}(m)^*$ form a

$$2 - (2^m - 1, 5, \tfrac{1}{3}(2^m - 4)) \ design. \tag{53}$$

The codewords of any fixed weight in $\mathcal{P}(m)$ form a 3-design. In particular the

$$A_6 = \tfrac{1}{360}2^m(2^m - 1)(2^m - 2)(2^m - 4) \tag{54}$$

codewords of weight 6 in $\mathcal{P}(m)$ form a

$$3 - (2^m, 6, \tfrac{1}{3}(2^m - 4)) \ design. \tag{55}$$

Proof. For $\mathcal{P}(m)^*$, $d = 5$ and $s' = 3$, so the first statement follows from Theorem 24 of Ch. 6. Similarly for the third statement. To find the value of λ in (53), observe that the annihilator polynomial of $\mathcal{P}(m)^*$ is

$$\alpha(x) = 2^{2m-1}\left(1 - \frac{x}{2^{m-1} - 2^{(1/2)(m-2)}}\right)\left(1 - \frac{x}{2^{m-1} + 2^{(1/2)(m-2)}}\right)\left(1 - \frac{x}{2^{m-1}}\right)$$

$$= P_0(x) + P_1(x) + \frac{1}{r}(P_2(x) + P_3(x)), \tag{56}$$

where $r = \tfrac{1}{3}(2^m - 1)$. From Theorem 23 of Ch. 6,

$$\lambda = \frac{1 - (1/r)}{(1/r)} = \frac{1}{3}(2^m - 4).$$

a_5 is the number of blocks in this design and from Equation (21) of Ch. 2 is given by Equation (52). A_6 then follows from Theorem 14 of Ch. 8.   Q.E.D.

Problem. (16) Show that there are

$$a_6 = \tfrac{1}{360}(2^m - 1)(2^m - 2)(2^m - 4)(2^m - 6)$$

codewords of weight 6 in $\mathcal{P}(m)^*$.

The last two theorems in this section show that, first, the whole space, and second, the Hamming code, is a union of disjoint translates of $\mathcal{P}(m)^*$.

Theorem 34. *The whole space* F^{n-1}, $n = 2^m$, *is the union of disjoint translates of* $\mathcal{P}(m)^*$ *by all vectors of weights* $0, 1, 2$, *and* $2^{m-1} - 1$ *vectors* $g_1, \ldots, g_{2^{m-1}-1}$ *of weight* 3.

Proof. Let $\{a_i(f)\}$ be the weight distribution of the translate $f + \mathcal{P}(m)^*$. We shall use Equation (19) of Ch. 6, where (from Equation (56) above) $\alpha_0 = \alpha_1 = 1$, $\alpha_2 = \alpha_3 = 1/r$. Therefore wt $(f) = 1$ implies $a_0(f) = 0$, $a_1(f) = 1$, and $a_2(f) = a_3(f) = 0$. Also wt $(f) = 2$ implies $a_0(f) = a_1(f) = 0$, $a_2(f) = 1$ and $a_3(f) = r - 1 = \tfrac{1}{3}(2^m - 4)$.
The

$$1 + (2^m - 1) + \binom{2^m - 1}{2}$$

translates of $\mathcal{P}(m)^*$ by vectors of weights 0, 1 and 2 are clearly disjoint. The number of remaining vectors in F^{n-1} is $|\mathcal{P}(m)^*|(2^{m-1} - 1)$. These must lie in translates with $a_0(f) = a_1(f) = a_2(f) = 0$, $a_3(f) = r$.
Let T be the set of vectors of weight 3 in F^{n-1}. Write $T = T_2 \cup T_3$, where

$$T_2 = \{f \in T : \text{dist} (f, u) = 2 \quad \text{for some } u \in \mathcal{P}(m)^*\},$$
$$T_3 = \{f \in T : \text{dist} (f, u) \geqslant 3 \quad \text{for all } u \in \mathcal{P}(m)^*\}.$$

We shall show that the remaining translates of $\mathcal{P}(m)^*$ are $g + \mathcal{P}(m)^*$ for those $g \in T_3$ which have a 1 in any fixed coordinate.

Problems. (17) If $g \in T_3$, show that the translate $g + \mathcal{P}(m)^*$ is disjoint from all translates $f + \mathcal{P}(m)^*$, where wt $(f) = 0$, 1 or 2.
(18) If $g_1, g_2 \in T_3$ have a nonempty intersection, show that the translates $g_1 + \mathcal{P}(m)^*$ and $g_2 + \mathcal{P}(m)^*$ are disjoint.

By Theorem 33 there are $\tfrac{1}{3}(2^m - 4)$ codewords of weights 5 in $\mathcal{P}(m)^*$ which have 1's in any two fixed coordinates. Each of these codewords contains 3 vectors of T_2 with 1's in these two coordinates. The total number of weight 3 vectors in F^{n-1} with 1's in these two coordinates is $2^m - 3$. Hence the number in T_3 is $(2^m - 3) - (2^m - 4) = 1$.

Thus the vectors of T_3 form a $2-(2^m-1,3,1)$ design; hence also a 1-$(2^m-1,3,\lambda)$ design with $\lambda = 2^{m-1}-1$ from Theorem 9 of Ch. 2. Therefore there are $2^{m-1}-1$ vectors $g_1,\ldots,g_{2^{m-1}-1}$ in T_3 with a 1 in a fixed coordinate. The translates of $\mathscr{P}(m)^*$ by these vectors are disjoint and exactly cover the remaining vectors in F^{n-1}.   Q.E.D.

Corollary 35. $\mathscr{P}(m)^*$ *is quasi-perfect.*

Theorem 36. *The union of $\mathscr{P}(m)^*$ and the translates $g_i + \mathscr{P}(m)^*$, $i = 1,\ldots,2^{m-1}-1$, forms a linear $[2^m-1, 2^m-1-m, 3]$ Hamming code.*

Proof. (I) We first show that for $j = 0, 1,\ldots,2^{m-1}-2$

$$|x^j|1|x^j| \tag{57}$$

and

$$|x^j\theta_3|0|0|, \tag{58}$$

define the same coset of $\mathscr{P}(m)^*$. In fact, their sum is $|x^j(1+\theta_3)|1|x^j|$, which is indeed in $\mathscr{P}(m)^*$ — set $g(x) = x^j\theta_3$, $f(x) = x^j$ and $a = 1$ in Equation (49).

(II) We next show that the coset of $\mathscr{P}(m)^*$ defined by (58) has minimum weight 3; thus the vectors (57) are in fact the same as the vectors $g_1,\ldots,g_\mu$ of Theorem 34. A typical vector of such a coset is

$$|g(x)(1+\theta_1) + ax^i + x^i\theta_3|f(1)|g(x)(1+\theta_1) + f(x)(1+\theta_1+\theta_3) + ax^i\theta_1|. \tag{59}$$

If $a = 0$, then the RHS is in the Hamming code, so has weight at least 3 unless it is 0. If it is 0, then $f(x) = g(x) = 0$ and the LHS has weight 2^{m-2}. If $a = 1$, the LHS and the RHS are both nonzero. Including the parity check in the middle, the weight is ≥ 3.

(III) Therefore the translates of $\mathscr{P}(m)^*$ by the vectors (57) (or 58)) form a code with 2^{2^m-1-m} vectors and minimum weight 3. It remains to show that this code is linear; by Problem 28 of Chapter 1 it is therefore equivalent to a Hamming code. We must show that the sum of two vectors of the form (59),

$$|(g_1(x) + g_2(x))(1+\theta_1) + a_1 x^{i_1} + a_2 x^{i_2} + (x^{i_1} + x^{i_2})\theta_3|$$
$$|f_1(1) + f_2(1)|(g_1(x) + g_2(x))(1+\theta_1) + (f_1(x) +$$
$$+ f_2(x))(1+\theta_1+\theta_3) + (a_1 x^{i_1} + a_2 x^{i_2})\theta_1|, \tag{60}$$

is again of the form (59). Only the case $a_1 = a_2 = 1$ gives any trouble. Since the codewords of weight 3 in a Hamming code form a 2-design (Theorem 15 of Ch. 2), there is a $g_3(x)$ and a j_3 such that

$$x^{i_1} + x^{i_2} + x^{i_3} = g_3(x)(1+\theta_1), \tag{61}$$

and, multiplying by θ_1,

$$(x^{i_1} + x^{i_2})\theta_1 = x^{i_3}\theta_1. \tag{62}$$

Note that the $x^j\theta_3$ are coset representatives for cosets of the double-error-correcting BCH code in the Hamming code (of length $2^{m-1} - 1$). Hence we can write

$$g_3(x)(1 + \theta_1) = x^{j_4}\theta_3 + \beta(x)(1 + \theta_1 + \theta_3), \tag{63}$$

for suitable j_4 and $\beta(x)$. Also

$$(1 + \theta_1)(1 + \theta_1 + \theta_3) = 1 + \theta_1 + \theta_3. \tag{64}$$

From Equations (61)–(64), we see that (60) is of the form of Equation (59) with $g(x) = g_1(x) + g_2(x) + \beta(x)(1 + \theta_1 + \theta_3)$, $a = 1$, $j = j_3$, $x^j\theta_3 = (x^{j_1} + x^{j_2} + x^{j_4})\theta_3$ (this can be done since θ_3 is the idempotent of a simplex code), and $f(x) = f_1(x) + f_2(x) + \beta(x)(1 + \theta_1 + \theta_3)$.   Q.E.D.

Problems. (19) Preparata defined $\mathscr{P}(m)$ to consist of all codewords

$$|m(1) + q(1)|m(x) + q(x)|i|m(x) + (m(1) + i)\theta_0 + s(x) + q(x)\theta_1|$$

where $m(x)$ is in the Hamming code generated by $M^{(1)}(x)$, $s(x)$ is in the distance 6 BCH code generated by $M^{(0)}(x)M^{(1)}(x)M^{(3)}(x)$, $q(x) \in \{0, 1, x, \ldots, x^{2^{m-1}-2}\}$, and $i = 0$ or 1. Show that this defines the same code as Equations (47) and (48).

(20) (a) Show that the first $2^m - 2m$ bits of $\mathscr{P}(m)^*$ can be taken to be information symbols; hence the code is systematic.

(b) Show that the remaining symbols of $\mathscr{P}(4)^*$ are quadratic functions of the information symbols.

(21) In the notation of Problem 16 of Ch. 14, show that the Nordstrom–Robinson code consists of the $[16, 5, 8]$ Reed–Muller code and the seven cosets corresponding to the bent functions

(22) Show that the nonlinear code of Fig. 5.1 consists of the vectors $(x_1, \ldots, x_8)$ where the x_i satisfy five quadratic equations.

*§7. Goethals' generalization of the Preparata codes

We saw in §5 that the codes $\mathscr{D}\mathscr{G}(m, d)$ generalize the Kerdock codes, for $\mathscr{D}\mathscr{G}(m, m/2) = \mathscr{K}(m)$ (for m even). The Preparata codes, as we have just seen, are a kind of dual to the Kerdock code, in the sense that the weight distribution of $\mathscr{P}(m)$ is the MacWilliams transform of that of $\mathscr{K}(m)$. In this section we construct a nonlinear triple-error-correcting code $\mathscr{I}(m)$ which is the same kind of dual of $\mathscr{D}\mathscr{G}(m, \frac{1}{2}(m - 2))$.

The construction is due to Goethals [495] and [496], and we refer the reader to his papers for a proof of the following facts.

For $m = 2t + 2 \geqslant 6$, $\mathscr{I}(m)$ is a nonlinear code of length 2^m, containing 2^k codewords where $k = 2^m - 3m + 1$, and with minimum distance 8. The first few codes have parameters

$$(64, 2^{47}, 8), (256, 2^{233}, 8), (1024, 2^{995}, 8), \ldots .$$

$\mathscr{I}(m)$ contains 4 times as many codewords as the extended triple-error-correcting BCH code of the same length.

As usual, $\mathscr{I}(m)$ consists of a linear code π together with $2^{m-1} - 1$ cosets of π. π consists of the codewords

$$|g(1)|g(x)(1 + \theta_1)|f(1) + g(1)|g(x)(1 + \theta_1) + f(x)(1 + \theta_1 + \theta_r + \theta_s)|, \quad (65)$$

where $r = 1 + 2^{t-1}$ and $s = 1 + 2^t$. Hence π has dimension $2^m - 4m + 2$. The cosets representatives w_i are given by Equation (48).

The weight and distance distribution of $\mathscr{I}(m)$ coincide, and are equal to the MacWilliams transform of the weight distribution of $\mathscr{DG}(m, \frac{1}{2}(m - 2))$. The latter distribution is given in Fig. 15.13 (although this won't be proved until Ch. 21).

i	A_i
0 or 2^{2t+2}	1
$2^{2t+1} \pm 2^{t+1}$	$2^{2t}(2^{2t+1} - 1)(2^{2t+2} - 1)/3$
$2^{2t+1} \pm 2^t$	$2^{2t+2}(2^{2t+1} - 1)(2^{2t+1} + 4)/3$
2^{2t+1}	$2(2^{2t+2} - 1)(2^{4t+1} - 2^{2t} + 1)$

Fig. 15.13. Weight (or distance) distribution of $\mathscr{DG}(m, \frac{1}{2}(m - 2))$.

Research Problems (15.2). Let $\mathscr{P}'(m)$ be obtained from $\mathscr{P}(m)$ by changing θ_3 to θ_s in the linear subcode. Does $\mathscr{P}'(m)$ have the same properties as $\mathscr{P}(m)$?

(15.3) Find a code whose distance distribution is the transform of that of $\mathscr{DG}(m, d)$, for any d.

(15.4) Show that $\mathscr{P}(m)$ and $\mathscr{K}(m)$ contain at least twice as many codewords as any linear code of the same length and minimum distance. What about $\mathscr{DG}(m, d)$ and $\mathscr{I}(m)$?

Notes on Chapter 15

§2. Theorem 2 is due to Albert [18], see also MacWilliams [878]. For Theorem 4 see Dickson [374, p. 197]. Theorem 8 is from Sloane and Berlekamp [1232]; see also McEliece [936]. Properties of the codewords of minimum weight in $\mathscr{R}(2, m)$ have been studied by Berman and Yudanina [138].

For Gaussian binomial coefficients see Berman and Fryer [134], Goldman and Rota [519] or Pólya and Alexanderson [1066]. For Problem 4 see Sarwate [1144]; Problem 5 is due to Delsarte and Goethals [362] and Wolfmann [1431].

§3. Theorem 10 is from Kasami and Tokura [745], and extends the results of Berlekamp and Sloane [132].

Theorem 12 extends an earlier result of Solomon and McEliece [1256]; the proof is given in [939] and [941]. This result has been extended to Abelian group codes – see Delsarte [347] and Delsarte and McEliece [369]. Van Lint [848] has given a direct proof of Corollary 13.

§4. The weight distributions given in §4 are due to Kasami [727, 729].

§5. For odd values of m the linear codes described in this section were analyzed by Berlekamp [118] and Delsarte and Goethals [364]. The weight distributions of other subcodes of $\mathcal{R}(2, m)$ can be found in [118, 729]. See also Dowling [384].

Research Problem. (15.5) $\mathcal{K}(4) = \mathcal{P}(4)$ is unique. Are either $\mathcal{K}(m)$ or $\mathcal{P}(m)$ unique for $m > 4$?

We mention (without giving any more explanation) the following result:

Theorem 38. (Berlekamp [120], Goethals [493], Snover [1247].) *The automorphism groups of the Nordstrom–Robinson codes $\mathcal{K}(4)^*$ and $\mathcal{K}(4)$ are respectively the alternating group $\mathcal{A}_7$, and $\mathcal{A}_7$ extended by the elementary abelian group of order* 16. *The latter is a triply transitive group of order* 16.15.14.12.

The Kerdock codes were discovered in 1972 – see Kerdock [758]. The description given here follows N. Patterson (unpublished) and Delsarte and Goethals [364]. The generalizations to the codes $\mathcal{DG}(m, d)$ were also given in [364]. Another way of looking at the Kerdock codes has been given by Cameron and Seidel [235]. For a possible generalization to GF(3) see Patterson [1031].

§6. The Preparata codes were introduced in [1081]; (see also [1080, 1082]). In this paper Preparata also shows that these codes are systematic, and gives algebraic encoding and decoding methods for them. Mykkeltveit [978] has shown that the Kerdock codes are systematic. The weight distribution of $\mathcal{P}(m)$ was first obtained by Semakov and Zinov'ev [1181]. In Ch. 18 we shall use the Preparata codes to construct a number of other good nonlinear double-error-correcting codes.

The Preparata codes are the chief example of a class of codes called *nearly perfect* codes (the definition will be given in Ch. 17). A more general class of codes are *uniformly packed codes*. See Bassalygo et al. [78, 79], Goethals and Snover [504], Goethals and Van Tilborg [505], Lundström et al. [844a, 844b], Semakov et al. [1183] and Van Tilborg [1326, 1327]. The proof of Theorem 36 follows Zaitsev et al. [1451].

§7. The material in this section is due to Goethals [495] and [496]. Research problems 15.2 and 15.3 have been solved by Goethals [498].

16

Quadratic-residue codes

§1. Introduction

The quadratic-residue (QR) codes $\mathcal{Q}$, $\bar{\mathcal{Q}}$, $\mathcal{N}$, $\bar{\mathcal{N}}$ are cyclic codes of prime block length p over a field GF(l), where l is another prime which is a quadratic residue modulo p. (Only the cases $l = 2$ and 3 have been studied to any extent.) $\mathcal{Q}$ and $\mathcal{N}$ are equivalent codes with parameters $[p, \frac{1}{2}(p + 1), d \geqslant \sqrt{p}]$, while $\bar{\mathcal{Q}}$ and $\bar{\mathcal{N}}$ are equivalent codes with parameters $[p, \frac{1}{2}(p - 1), d \geqslant \sqrt{p}]$. Also $\mathcal{Q} \supset \bar{\mathcal{Q}}$ and $\mathcal{N} \supset \bar{\mathcal{N}}$. These codes are defined in §2 and a summary of their properties is given in Fig. 16.1.

Examples of quadratic-residue codes are the binary [7, 4, 3] Hamming code, and the binary [23, 12, 7] and ternary [11, 6, 5] perfect Golay codes $\mathcal{G}_{23}$ and $\mathcal{G}_{11}$ (see §10 of Ch. 6 and Ch. 20). Other examples are given in Fig. 16.2.

Thus QR codes have rate close to $\frac{1}{2}$, and tend to have high minimum distance (at least if p is not too large, but see Research Problem 16.1). Several techniques for decoding QR and other cyclic codes are described in §9. The most powerful method is permutation decoding, which makes use of the fact that these codes have large automorphism groups.

Other properties of QR codes discussed in this chapter are idempotents (§3), dual codes and extended codes of length $p + 1$ (§4), and automorphism groups (§5). Extended QR codes are fixed by the group $PSL_2(p)$ if $l = 2$, or by a slight generalization of this group if $l > 2$ (Theorem 12), and in many cases this is the full group (Theorem 13).

Further properties of binary QR codes are described in §6, including methods for finding the true minimum distance. It is also shown in §6 that some QR codes have a generator matrix of the form $[I \mid A]$, where A is a circulant or bordered circulant matrix. Such codes are called *double circulant codes*. Double circulant codes over GF(2) and GF(3) for a particular choice of

A are studied in §§7,8. These have properties similar to QR codes, although less is known about them. The codes over GF(3) are the Pless *symmetry codes*.

In this chapter, proofs are usually given only for the case $p = 4k - 1$; the case $p = 4k + 1$ being similar and left to the reader.

§2. Definition of quadratic-residue codes

We are going to define quadratic-residue (QR) codes of prime length p over GF(l), where l is another prime which is a quadratic residue mod p. In the important case of binary quadratic residue codes ($l = 2$), this means that p has to be a prime of the form $8m \pm 1$ (by Theorem 23 of the Notes).

Let Q denote the set of quadratic residues modulo p and N the set of nonresidues. (See §3 of Ch. 2 and §6 of Ch. 4). If ρ is a primitive element of the field GF(p), then $\rho^e \in Q$ iff e is even, while $\rho^e \in N$ iff e is odd. Thus Q is a cyclic group generated by ρ^2. Since $l \in Q$, the set Q is closed under multiplication by l. Thus Q is a disjoint union of cyclotomic cosets mod p. Hence

$$q(x) = \prod_{r \in Q} (x - \alpha^r) \quad \text{and} \quad n(x) = \prod_{n \in N} (x - \alpha^n) \tag{1}$$

have coefficients from GF(l), where α is a primitive p^{th} root of unity in some field containing GF(l). Also

$$x^p - 1 = (x - 1)q(x)n(x). \tag{2}$$

Let R be the ring GF(l)$[x]/(x^p - 1)$.

Definition. The *quadratic-residue* codes $\mathcal{Q}, \bar{\mathcal{Q}}, \mathcal{N}, \bar{\mathcal{N}}$ are cyclic codes (or ideals) of R with generator polynomials

$$q(x), \qquad (x - 1)q(x), \qquad n(x), \qquad (x - 1)n(x) \tag{3}$$

respectively. Sometimes $\mathcal{Q}$ and $\mathcal{N}$ are called *augmented* QR codes, and $\bar{\mathcal{Q}}$ and $\bar{\mathcal{N}}$ *expurgated* QR codes.

Clearly $\mathcal{Q} \supset \bar{\mathcal{Q}}$ and $\mathcal{N} \supset \bar{\mathcal{N}}$. In the binary case $\bar{\mathcal{Q}}$ is the even weight subcode of $\mathcal{Q}$, and $\bar{\mathcal{N}}$ is the even weight subcode of $\mathcal{N}$.

The permutation of coordinates in R induced by $x \to x^n$ for a fixed nonresidue n interchanges $\mathcal{Q}$ and $\mathcal{N}$, and also $\bar{\mathcal{Q}}$ and $\bar{\mathcal{N}}$, so that these codes are equivalent. $\mathcal{Q}$ and $\mathcal{N}$ have dimension $\frac{1}{2}(p + 1)$, and $\bar{\mathcal{Q}}$ and $\bar{\mathcal{N}}$ have dimension $\frac{1}{2}(p - 1)$. (See Fig. 16.1.)

Example. If $l = 2$ and $p = 7$, then $\mathcal{Q}$ has generator polynomial $(x + \alpha)(x + \alpha^2)(x + \alpha^4) = x^3 + x + 1$, where $\alpha \in$ GF(2^3), and is the [7, 4, 3] Hamming code.

$\bar{\mathcal{2}}$ is the [7, 3, 4] subcode with generator polynomial $(x + 1)(x^3 + x + 1)$. $\mathcal{N}$ is the equivalent [7, 4, 3] Hamming code with generator polynomial $(x + \alpha^3)(x + \alpha^5)(x + \alpha^6) = x^3 + x^2 + 1$. (Note that a different choice of α would interchange $\mathcal{2}$ and $\mathcal{N}$.)

The Golay codes. It has already been mentioned in §6 of Ch. 2 and §10 of Ch. 6 that there are two perfect Golay codes, namely a [23, 12, 7] binary code $\mathcal{G}_{23}$ and an [11, 6, 5] ternary code $\mathcal{G}_{11}$. One definition of $\mathcal{G}_{23}$ was given in Ch. 2. We now give a second definition and also define $\mathcal{G}_{11}$, as QR codes.

Definition. The *Golay code* $\mathcal{G}_{23}$ is the QR code $\mathcal{2}$ (or $\mathcal{N}$) for $l = 2$ and $p = 23$. The *Golay code* $\mathcal{G}_{11}$ is the code $\mathcal{2}$ (or $\mathcal{N}$) for $l = 3$ and $p = 11$. These have parameters [23, 12, 7] and [11, 6, 5] respectively. (It will be shown in §7 that the two definitions of $\mathcal{G}_{23}$ are equivalent. This will also follow from Corollary 16 of Ch. 20.)

Over GF(2) (see Fig. 7.1)

$$x^{23} + 1 = (x + 1)(x^{11} + x^{10} + x^6 + x^5 + x^4 + x^2 + 1)(x^{11} + x^9 + x^7 + x^6 + x^5 + x + 1),$$

so the generator polynomial of $\mathcal{G}_{23}$ can be taken to be either

$$x^{11} + x^{10} + x^6 + x^5 + x^4 + x^2 + 1 \quad \text{or} \quad x^{11} + x^9 + x^7 + x^6 + x^5 + x + 1. \quad (4)$$

Over GF(3),

$$x^{11} - 1 = (x - 1)(x^5 + x^4 - x^3 + x^2 - 1)(x^5 - x^3 + x^2 - x - 1),$$

so the generator polynomial of $\mathcal{G}_{11}$ can be taken to be either

$$x^5 + x^4 - x^3 + x^2 - 1 \quad \text{or} \quad x^5 - x^3 + x^2 - x - 1. \quad (5)$$

Other examples are shown in Fig. 16.2. (Upper and lower bounds on d are given in some cases.)

They are defined by (3) and are codes over GF(l) with

       length p = prime,
       dimension $\frac{1}{2}(p + 1)$ for $\mathcal{2}, \mathcal{N}$; $\frac{1}{2}(p - 1)$ for $\bar{\mathcal{2}}, \bar{\mathcal{N}}$,
       minimum distance $d \geq \sqrt{p}$,

where l is a prime which is a quadratic residue mod p. If $p = 4k - 1$, $d^2 - d + 1 \geq p$ (Theorem 1). Idempotents are given by Theorems 2 and 4. For generator matrices see Equations (22), (23), (28), (31), (39), (44), (45). If $p = 4k - 1$, $\mathcal{2}^{\perp} = \bar{\mathcal{2}}$, $\mathcal{N}^{\perp} = \bar{\mathcal{N}}$; if $p = 4k + 1$, $\mathcal{2}^{\perp} = \bar{\mathcal{N}}$, $\mathcal{N}^{\perp} = \bar{\mathcal{2}}$. The extended codes $\hat{\mathcal{2}}, \hat{\mathcal{N}}$ are defined by adding the overall parity check (27). If $p = 4k - 1$, $\hat{\mathcal{2}}$ and $\hat{\mathcal{N}}$ are self-dual; if $p = 4k + 1$, $(\hat{\mathcal{2}})^{\perp} = \hat{\mathcal{N}}$. Aut $(\hat{\mathcal{2}})$ contains $PSL_2(p)$ (Theorems 10 and 12). Figure 16.2 gives examples and Fig. 16.3 properties of the binary codes.

Fig. 16.1. Properties of quadratic-residue codes $\mathcal{2}, \bar{\mathcal{2}}, \mathcal{N}, \bar{\mathcal{2}}$.

(a) Over GF(2)

n	k	d	n	k	d	n	k	d
8	4	4	74	37	14	138	69	14–22
18	9	6	80	40	16	152	76	20
24	12	8	90	45	18	168	84	16–24
32	16	8	98	49	16	192	96	16–28
42	21	10	104	52	20	194	97	16–28
48	24	12	114	57	12–16	200	100	16–32
72	36	12	128	64	20			

(b) Over GF(3)

n	k	d	n	k	d	n	k	d
12	6	6	48	24	15	74	37	?
14	7	6	60	30	18	84	42	?
24	12	9	62	31	?	98	49	?
38	19	?	72	36	?			

Fig. 16.2. A table of extended quadratic-residue codes $\hat{\mathcal{Q}}$.

The square root bound on the minimum distance.

Theorem 1. *If d is the minimum distance of* $\mathcal{Q}$ *or* $\mathcal{N}$, *then* $d^2 \geq p$. *Furthermore, if* $p = 4k - 1$, *then this can be strengthened to*

$$d^2 - d + 1 \geq p. \tag{6}$$

Proof. Let $a(x)$ be a codeword of minimum nonzero weight d in $\mathcal{Q}$. If n is a nonresidue, $\bar{a}(x) = a(x^n)$ is a codeword of minimum weight in $\mathcal{N}$. Then $a(x)\bar{a}(x)$ must be in $\mathcal{Q} \cap \mathcal{N}$, i.e. is a multiple of

$$\prod_{r \in Q} (x - \alpha^r) \prod_{n \in N} (x - \alpha^n) = \prod_{j=1}^{p-1} (x - \alpha^j) = \sum_{j=0}^{p-1} x^j.$$

Thus $a(x)\bar{a}(x)$ has weight p. Since $a(x)$ has weight d, the maximum number of nonzero coefficients in $a(x)\bar{a}(x)$ is d^2, so that $d^2 \geq p$.

If $p = 4k - 1$, we may take $n = -1$ by property (Q2) of Ch. 2. Now in the product $a(x)a(x^{-1})$ there are d terms equal to 1, so the maximum weight of the product is $d^2 - d + 1$.   Q.E.D.

Example. The $[7, 4, d]$ quadratic residue code over GF(l) has $d \geq 3$ (and if $l = 2$, $d = 3$).

As the codes in Fig. 16.2 show, d is often greater than the bound given by Theorem 1.

Research Problems (16.1). Fill in the gaps in Fig. 16.2, extend these tables, and compute similar tables for other primes l.
  (16.2) How does the minimum distance of QR codes behave as $p \to \infty$?

§3. Idempotents of quadratic-residue codes

The case $l = 2$ is considered first.

Theorem 2. *If $l = 2$ and $p = 4k - 1$, then α can be chosen so that the idempotents of $\mathcal{Q}, \bar{\mathcal{Q}}, \mathcal{N}, \bar{\mathcal{N}}$ are*

$$E_q(x) = \sum_{r \in Q} x^r, \qquad F_q(x) = 1 + \sum_{n \in N} x^n,$$

$$E_n(x) = \sum_{n \in N} x^n, \qquad F_n(x) = 1 + \sum_{r \in Q} x^r, \tag{7}$$

respectively.

Proof. Since 2 is a quadratic residue mod p (Theorem 23), $(E_q(x))^2 = E_q(x)$, etc., so these polynomials are idempotent. Thus $E_q(\alpha^i) = 0$ or 1 by Lemma 2 of Ch. 8. For any quadratic residue s,

$$E_q(\alpha^s) = \sum_{r \in Q} \alpha^{rs} = \sum_{r' \in Q} \alpha^{r'} = E_q(\alpha),$$

independent of s. Similarly

$$E_q(\alpha^t) = \sum_{r \in Q} \alpha^{rt} = \sum_{r \in Q} \alpha^{-r} = E_q(\alpha^{-1}),$$

for any nonresidue t. Since $E_q(\alpha) + E_q(\alpha^{-1}) = 1$, either

$$E_q(\alpha^s) = 0 \quad \text{for all } s \in Q \quad \text{and} \quad E_q(\alpha^t) = 1 \quad \text{for all } t \in N \tag{8}$$

or
$$E_q(\alpha^s) = 1 \quad \text{for all } s \in Q \quad \text{and} \quad E_q(\alpha^t) = 0 \quad \text{for all } t \in N,$$

depending on the choice of α. Let us choose α so that (8) holds, then $E_q(x)$ is the idempotent of $\mathcal{Q}$. Also

$$E_n(\alpha^t) = \sum_{n \in N} \alpha^{nt} = \sum_{r \in Q} \alpha^r = 0 \quad \text{for } t \in N,$$

and $E_n(\alpha^s) = 1$ for $s \in Q$. Thus $E_n(x)$ is the idempotent of $\mathcal{N}$.

Finally, $F_q(\alpha^s) = 0$ for $s \in Q$ and $F_q(1) = 0$, so $F_q(x)$ is the idempotent of $\bar{\mathcal{Q}}$. Similarly for $\bar{\mathcal{Q}}$.   Q.E.D.

Note. If $l = 2$ and $p = 4k + 1$, the idempotents of $\mathcal{Q}$, $\bar{\mathcal{Q}}$, $\mathcal{N}$, $\bar{\mathcal{N}}$ may be taken to be

$$1 + \sum_{r \in Q} x^r, \ \sum_{n \in N} x^n, 1 + \sum_{n \in N} x^n, \ \sum_{r \in Q} x^r \qquad (9)$$

respectively.

Example. If $l = 2$ and $p = 23$, then $Q = \{1, 2, 3, 4, 6, 8, 9, 12, 13, 16, 18\}$ and the idempotents of $\mathcal{Q}$ and $\mathcal{N}$ are

$$\sum_{r \in Q} x^r \quad \text{and} \quad \sum_{r \in Q} x^{-r} \qquad (10)$$

respectively. Both generate a code equivalent to the Golay code $\mathcal{G}_{23}$.

Idempotents if $l > 2$. In describing the idempotents for $l > 2$, the following number-theoretic result will be useful. Recall from property (Q3) of Ch. 2 that the Legendre symbol $\chi(i)$ is defined by

$$\chi(i) = \begin{array}{ll} 0 & \text{if } i \text{ is a multiple of } p, \\ 1 & \text{if } i \text{ is a quadratic residue mod } p, \\ -1 & \text{if } i \text{ is a nonresidue mod } p. \end{array}$$

Also $\chi(i)\chi(j) = \chi(ij)$. Define the *Gaussian sum*

$$\theta = \sum_{i=1}^{p-1} \chi(i)\alpha^i. \qquad (11)$$

Since $\theta^l = \theta$, $\theta \in \mathrm{GF}(l)$.

Theorem 3. *If $p = 4k - 1$, then $\theta^2 = -p$. (This result holds if α is a primitive p^{th} root of unity over any field.)*

Proof.

$$\theta^2 = \sum_{i=1}^{p-1} \sum_{j=1}^{p-1} \chi(i)\chi(j)\alpha^{i+j}$$

The $p - 1$ terms in the sum with $i + j = p$ all have coefficient -1, for one of i and j is a residue and the other is a nonresidue mod p (since -1 is a nonresidue). Hence

$$\theta^2 = -(p - 1) + \sum_{i+j \neq p} \sum \chi(i)\chi(j)\alpha^{i+j}.$$

The terms with $i = j$ contribute

$$\sum_{i=1}^{p-1} \chi(i)^2 \alpha^{2i} = \sum_{i=1}^{p-1} \alpha^{2i} = \sum_{i=1}^{p-1} \alpha^i = -1.$$

Thus

$$\theta^2 = -p + \sum_{k=1}^{p-1} \alpha^k \psi(k),$$

where

$$\psi(k) = \sum_{\substack{i=1 \\ i \neq k, 2i \neq k}}^{p-1} \chi(i(k-i)).$$

It remains to show that $\psi(k) = 0$. Let $M_k = \{t: t = i(k-i)$ for some i with $1 \leq i \leq p-1$, $i \neq k$, $2i \neq k\}$. Then $|M_k| = \frac{1}{2}(p-3)$. We show that M_k contains $\frac{1}{4}(p-3)$ residues and $\frac{1}{4}(p-3)$ nonresidues mod p, so that indeed $\psi(k) = 0$. If $t = i(k-i)$ then $i^2 - ki + t = 0$, and $k^2 - 4t = ((i^2 - t)/i)^2$. Since $i^2 \neq t$, $k^2 - 4t \in Q$. Thus $-4t = r - k^2$ for some $r \in Q$, $r \neq k^2$; or $\{-4M_k\} = \{r - k^2: r \in Q, r \neq k^2\}$. By Perron's Theorem (Theorem 24 of the Notes), this set contains equal numbers of residues and nonresidues. Q.E.D.

Note. If $p = 4k + 1$, then $\theta^2 = +p$.

Example. When $p = 7$, for any l,

$$\theta = \alpha + \alpha^2 - \alpha^3 + \alpha^4 - \alpha^5 - \alpha^6,$$

$$\theta^2 = -6 + \sum_{i=1}^{6} \alpha^i = -7.$$

Theorem 4. *If $l > 2$ and $p = 4k \pm 1$ the idempotents of $\mathcal{Q}, \bar{\mathcal{Q}}, \mathcal{N}, \bar{\mathcal{N}}$ are*

$$E_q(x) = \frac{1}{2}\left(1 + \frac{1}{p}\right) + \frac{1}{2}\left(\frac{1}{p} - \frac{1}{\theta}\right) \sum_{r \in Q} x^r + \frac{1}{2}\left(\frac{1}{p} + \frac{1}{\theta}\right) \sum_{n \in N} x^n, \qquad (12)$$

$$F_q(x) = \frac{1}{2}\left(1 - \frac{1}{p}\right) - \frac{1}{2}\left(\frac{1}{p} + \frac{1}{\theta}\right) \sum_{r \in Q} x^r - \frac{1}{2}\left(\frac{1}{p} - \frac{1}{\theta}\right) \sum_{n \in N} x^n, \qquad (13)$$

$$E_n(x) = \frac{1}{2}\left(1 + \frac{1}{p}\right) + \frac{1}{2}\left(\frac{1}{p} + \frac{1}{\theta}\right) \sum_{r \in Q} x^r + \frac{1}{2}\left(\frac{1}{p} - \frac{1}{\theta}\right) \sum_{n \in N} x^n, \qquad (14)$$

$$F_n(x) = \frac{1}{2}\left(1 - \frac{1}{p}\right) - \frac{1}{2}\left(\frac{1}{p} - \frac{1}{\theta}\right) \sum_{r \in Q} x^r - \frac{1}{2}\left(\frac{1}{p} + \frac{1}{\theta}\right) \sum_{n \in N} x^n \qquad (15)$$

respectively.

Proof. It is easy to check that for $r \in Q$,

$$E_q(\alpha') = F_q(\alpha') = 0, \qquad E_n(\alpha') = F_n(\alpha') = 1,$$

while if $n \in N$,

$$E_q(\alpha^n) = F_q(\alpha^n) = 1, \qquad E_n(\alpha^n) = F_n(\alpha^n) = 0.$$

Also $E_q(1) = E_n(1) = 1$, $F_q(1) = F_n(1) = 0$. This proves that $E_q(x)$, $F_q(x)$, $E_n(x)$, $F_n(x)$ are the desired idempotents, from Lemma 2 of Ch. 8.   Q.E.D.

Example. If $l = 3$ and $p = 11$, then $Q = \{1, 3, 4, 5, 9\}$. Also $\alpha \in GF(3^5)$ is a primitive 11th root of unity. Rather than calculating θ from (11) we shall use Theorem 3: $\theta^2 = -11 \equiv 1 \pmod{3}$, so $\theta = \pm 1$. We choose α so that $\theta = 1$. Then from (12) and (14) the idempotents of $\mathcal{Q}$ and $\mathcal{N}$ are

$$-\sum_{r \in Q} x^r \quad \text{and} \quad -\sum_{r \in Q} x^{-r} \tag{16}$$

respectively. Both generate a code equivalent to the Golay code $\mathcal{G}_{11}$. Note that if $l = 3$ and $p = 4k - 1$ the idempotents will always be of this form, since θ^2 is a quadratic residue modulo 3, and so $\theta^2 = 1$.

It may be helpful to indicate how these idempotents were discovered. A plausible guess is that they have the form

$$a + b \sum_{r \in Q} x^r + c \sum_{n \in N} x^n. \tag{17}$$

To square such an expression we use:

Lemma 5. *Suppose $p = 4k - 1$. Then in R,*

$$\left(\sum_{r \in Q} x^r \right)^2 = \tfrac{1}{4}(p - 3) \sum_{r \in Q} x^r + \tfrac{1}{4}(p + 1) \sum_{n \in N} x^n,$$

$$\left(\sum_{n \in N} x^n \right)^2 = \tfrac{1}{4}(p + 1) \sum_{r \in Q} x^r + \tfrac{1}{4}(p - 3) \sum_{n \in N} x^n,$$

$$\sum_{r \in Q} x^r \sum_{n \in N} x^n = \tfrac{1}{2}(p - 1) + \tfrac{1}{4}(p - 3) \sum_{i=1}^{p-1} x^i.$$

Proof. From Perron's Theorem 24.   Q.E.D.

It follows that (17) is an idempotent iff

$$a^2 + (p - 1)bc = a,$$
$$2ab + \tfrac{1}{4}(p - 3)b^2 + \tfrac{1}{4}(p + 1)c^2 + \tfrac{1}{2}(p - 3)bc = b,$$
$$2ac + \tfrac{1}{4}(p + 1)b^2 + \tfrac{1}{4}(p - 3)c^2 + \tfrac{1}{2}(p - 3)bc = c.$$

The solution of these equations gives the idempotents of Theorem 4.

§4. Extended quadratic-residue codes

We first find the duals of the QR codes.

Theorem 6.

$$\mathcal{Q}^{\perp} = \bar{\mathcal{Q}}, \qquad \mathcal{N}^{\perp} = \bar{\mathcal{N}} \quad \text{if } p = 4k - 1, \tag{18}$$

$$\mathcal{Q}^{\perp} = \bar{\mathcal{N}}, \qquad \mathcal{N}^{\perp} = \bar{\mathcal{Q}} \quad \text{if } p = 4k + 1. \tag{19}$$

In both cases,

$$\mathcal{Q} \text{ is generated by } \bar{\mathcal{Q}} \text{ and } \mathbf{1}, \tag{20}$$

$$\mathcal{N} \text{ is generated by } \bar{\mathcal{N}} \text{ and } \mathbf{1}. \tag{21}$$

Proof. Suppose $p = 4k - 1$. The zeros of $\mathcal{Q}$ are α^r for $r \in Q$. Hence by Theorem 4 of Ch. 7, the zeros of $\mathcal{Q}^{\perp}$ are 1 and α^{-n} for $n \in N$. But $-n \in Q$, so $\mathcal{Q}^{\perp} = \bar{\mathcal{Q}}$. From Theorems 2 and 4,

$$E_q(x) = F_q(x) + \frac{1}{p} \sum_{i=0}^{p-1} x^i,$$

which implies (20). Similarly for the other cases.   Q.E.D.

Next we find generator matrices for these codes. Let

$$F_q(x) = \sum_{i=0}^{p-1} f_i x^i$$

be the idempotent of $\bar{\mathcal{Q}}$, given by Theorem 2 or 4. Then a generator matrix for $\bar{\mathcal{Q}}$ is the $p \times p$ circulant matrix

$$\bar{G} = \begin{bmatrix} f_0 & f_1 & \cdots & f_{p-1} \\ f_{p-1} & f_0 & \cdots & f_{p-2} \\ \cdots\cdots\cdots\cdots\cdots \\ f_1 & f_2 & \cdots & f_0 \end{bmatrix} \tag{22}$$

$$= (g_{ij}), \quad 0 \le i, j \le p - 1, \quad \text{with } g_{ij} = f_{j-i},$$

and with subscripts taken mod p. A generator matrix for $\mathcal{Q}$ is

$$\left[\begin{array}{c} \bar{G} \\ \hline 1 \; 1 \cdots 1 \end{array} \right] \tag{23}$$

and similarly for $\mathcal{N}$ and $\bar{\mathcal{N}}$. Of course (22) has rank $\frac{1}{2}(p - 1)$.

Examples. If $l = 2$, $p = 7$, a generator matrix for the $[7, 4, 3]$ Hamming code $\mathcal{Q}$

is

$$
\begin{array}{c}
\begin{array}{ccccccc} 0 & 1 & 2 & 3 & 4 & 5 & 6 \end{array}\\
\boxed{\begin{array}{ccccccc}
1 & & & 1 & & 1 & 1\\
1 & 1 & & & 1 & & 1\\
1 & 1 & 1 & & & 1 &\\
& 1 & 1 & 1 & & & 1\\
1 & & 1 & 1 & 1 & &\\
& 1 & & 1 & 1 & 1 &\\
& & 1 & & 1 & 1 & 1
\end{array}}\\[2pt]
\boxed{\begin{array}{ccccccc}
1 & 1 & 1 & 1 & 1 & 1 & 1
\end{array}}
\end{array}
\tag{24}
$$

If $l = 3$, $p = 11$, a generator matrix for the $[11, 6, 5]$ Golay code $\mathscr{G}_{11}$ is

$$
\begin{array}{c}
\begin{array}{ccccccccccc} 0 & 1 & 2 & 3 & 4 & 5 & 6 & 7 & 8 & 9 & 10 \end{array}\\
\boxed{\begin{array}{ccccccccccc}
1 & 1 & & & & 1 & 1 & 1 & & & 1\\
1 & 1 & 1 & & & & 1 & 1 & 1 & &\\
& 1 & 1 & 1 & & & & 1 & 1 & 1 &\\
1 & & 1 & 1 & 1 & & & & 1 & & 1\\
1 & 1 & & 1 & 1 & 1 & & & & & 1\\
1 & 1 & 1 & & 1 & 1 & 1 & & & &\\
& 1 & 1 & 1 & & 1 & 1 & 1 & & &\\
& & 1 & 1 & 1 & & 1 & 1 & 1 & &\\
& & & 1 & 1 & 1 & & 1 & 1 & 1 &\\
1 & & & & 1 & 1 & 1 & & 1 & 1 &\\
1 & & & & & 1 & 1 & 1 & & 1 & 1
\end{array}}\\[2pt]
\boxed{\begin{array}{ccccccccccc}
1 & 1 & 1 & 1 & 1 & 1 & 1 & 1 & 1 & 1 & 1
\end{array}}
\end{array}
\tag{25}
$$

Extended QR codes. QR codes may be extended by adding an overall parity check in such a way that

$$
\begin{aligned}
(\hat{\mathscr{Q}})^{\perp} = \hat{\mathscr{Q}},\ (\hat{\mathscr{N}})^{\perp} = \hat{\mathscr{N}}, &\quad \text{if } p = 4k - 1,\\
(\hat{\mathscr{Q}})^{\perp} = \hat{\mathscr{N}}, &\quad \text{if } p = 4k + 1,
\end{aligned}
\tag{26}
$$

where $\hat{\ }$ denotes the extended code. If $a = (a_0, \dots, a_{p-1})$ is a codeword of $\mathscr{Q}$ (or $\mathscr{N}$), and $p = 4k - 1$, the extended code is formed by appending

$$
a_\infty = -y \sum_{i=0}^{p-1} a_i,
\tag{27}
$$

where $1 + y^2 p = 0$. Since $(yp)^2 = -p = \theta^2$, it follows that $y = \epsilon\theta/p$, where $\epsilon = \pm 1$ (either choice of sign will do). Note that y is chosen so that the codeword $(1, 1, \dots, 1, -yp)$ of $\hat{\mathscr{Q}}$ (or $\hat{\mathscr{N}}$) is orthogonal to itself. If $l = 2$ or 3, y may be taken to be 1. $\hat{\mathscr{Q}}$ and $\hat{\mathscr{N}}$ are $[p + 1, \frac{1}{2}(p + 1)]$ codes.

Theorem 7. *If* $p = 4k - 1$, *the extended* QR *codes* $\hat{\mathscr{Q}}$ *and* $\hat{\mathscr{N}}$ *are self-dual.*

Proof. A generator matrix for $\hat{\mathscr{Q}}$ is obtained by adding a column to (23), and is given by

$$\hat{G} = \left[\begin{array}{ccc|c} & & & 0 \\ & \bar{G} & & 0 \\ & & & \vdots \\ & & & 0 \\ \hline 1 \; 1 \; \cdots \; 1 & & & -yp \end{array} \right]. \tag{28}$$

Since $\bar{\mathscr{Q}} \subset (\bar{\mathscr{Q}})^{\perp}$, every row of $\bar{G}$ is orthogonal to itself and to every other row. Thus every row of $\hat{G}$ is orthogonal to all the rows of $\hat{G}$, and so $\hat{\mathscr{Q}} \subset (\hat{\mathscr{Q}})^{\perp}$. But $\hat{G}$ has rank $\frac{1}{2}(p + 1)$, so $\hat{\mathscr{Q}} = (\hat{\mathscr{Q}})^{\perp}$.

Thus the extended Golay codes $\mathscr{G}_{24} = \hat{\mathscr{G}}_{23}$ and $\mathscr{G}_{12} = \hat{\mathscr{G}}_{11}$ are self-dual.

Note. If $p = 4k + 1$, the extended codes may be defined so that $(\hat{\mathscr{Q}})^{\perp} = \hat{\mathscr{N}}$.

Theorem 8. (i) *If* $l = 2$ *and* $p = 4k - 1$, *the weight of every codeword in* $\hat{\mathscr{Q}}$ *is divisible by* 4, *and the weight of every codeword in* $\mathscr{Q}$ *is congruent to* 0 *or* 3 mod 4. (ii) *If* $l = 3$ *(and without restriction on* p) *the weight of every codeword in* $\hat{\mathscr{Q}}$ *is divisible by* 3, *and the weight of every codeword in* $\mathscr{Q}$ *is congruent to* 0 *or* 2 mod 3.

Proof. (i) If 2 is a quadratic residue mod p, then by Theorem 23 $p = 8m \pm 1$. So we may assume $p = 8m - 1$. The number of residues or nonresidues is $4m - 1$, and so the weight of each row of $\hat{G}$ is a multiple of 4. The result then follows from Problem 38 of Ch. 1. (ii) If $a \in \hat{\mathscr{Q}}$ then $a_0^2 + a_1^2 + \cdots + a_{\infty}^2 \equiv 0$ mod 3, by Theorem 7. But $a_i = \pm 1$, $a_i^2 = 1$, and the result follows.   Q.E.D.

Note. If $l = 2$ and $p = 4k + 1$, all we can say is that the weights of $\hat{\mathscr{Q}}$ are even.

Application. Theorems 1, 8 establish that $\mathscr{G}_{23}$ and $\mathscr{G}_{11}$ have minimum distances 7 and 5 respectively. Thus $\mathscr{G}_{24}$ only contains codewords of weights 0, 8, 12, 16 and 24 (as we saw in Ch. 2), and $\mathscr{G}_{12}$ only contains weights 0, 6, 9 and 12 (see Ch. 19).

Problem. (1) Show that the weight distribution of a binary quadratic residue code is asymptotically normal. [Hint: Theorem 23 of Ch. 9].

§5. The automorphism group of QR codes

In this section it is shown that the extended QR code $\hat{\mathcal{Q}}$ is fixed by the large permutation group $PSL_2(p)$.

Definition of $PSL_2(p)$. Let p be a prime of the form $8m \pm 1$. The set of all permutations of $\{0, 1, 2, \ldots, p - 1, \infty\}$ of the form

$$y \to \frac{ay + b}{cy + d}, \tag{29}$$

where $a, b, c, d \in GF(p)$ are such that $ad - bc = 1$, forms a group called the *projective special linear group* $PSL_2(p)$ (sometimes also called the *linear fractional* group).

The properties of $PSL_2(p)$ that we need are given in:

Theorem 9. (a) *$PSL_2(p)$ is generated by the three permutations*

$$\begin{aligned}
S&: \ y \to y + 1 \\
V&: \ y \to \rho^2 y \\
T&: \ y \to -\frac{1}{y}
\end{aligned} \tag{30}$$

where ρ is a primitive element of $GF(p)$.
   (b) *In fact $PSL_2(p)$ consists of the $\frac{1}{2}p(p^2 - 1)$ permutations*

$$\begin{aligned}
V^i S^j&: \ y \to \rho^{2i} y + j \\
V^i S^j TS^k&: \ y \to k - (\rho^{2i} y + j)^{-1}
\end{aligned}$$

where $0 \le i < \frac{1}{2}(p - 1), \ 0 \le j, \ k < p$.

   (c) *If $p = 8m - 1$, the generators S, V, T satisfy $S^p = V^{\frac{1}{2}(p-1)} = T^2 = (VT)^2 = (ST)^3 = 1$, and $V^{-1}SV = S^{\rho^2}$.*
   (d) *$PSL_2(p)$ is doubly transitive.*

Proof. (a), (b) A typical element of $PSL_2(p)$

$$y \to \frac{ay + b}{cy + d}, \quad ad - bc = 1$$

can be written either as

$$y \to a^2 y + ab \quad \text{if } c = 0 \quad \text{(for then } d = 1/a\text{)},$$

or

$$y \to \frac{a}{c} - \frac{1}{c^2 y + cd} \quad \text{if } c \neq 0 \quad \left(\text{for then } b = \frac{ad}{c} - \frac{1}{c}\right),$$

and these are respectively $V^i S^{ab}$ (where $a = \rho^i$), and $V^i S^{cd} T S^{a/c}$ (where $c = \rho^i$). (c) and (d) are left as straight-forward exercises for the reader.

$$\text{Q.E.D.}$$

The binary case of the main theorem is:

Theorem 10. (Gleason and Prange.) *If $l = 2$ and $p = 8m \pm 1$, the extended quadratic-residue code $\hat{\mathcal{Q}}$ is fixed by $\mathrm{PSL}_2(p)$.*

Proof. S is a cyclic shift, and V fixes the idempotent, so $\mathcal{Q}$ and hence $\hat{\mathcal{Q}}$ are fixed by S and V. From Theorem 9(a) it remains to show that $\hat{\mathcal{Q}}$ is also fixed by T. Only the case $p = 8m - 1$ is treated. We shall show that each row of the generator matrix (28) of $\hat{\mathcal{Q}}$, i.e.

$$\hat{G} = \left[\begin{array}{c|c} \bar{G} & \begin{matrix} 0 \\ 0 \\ \vdots \\ 0 \end{matrix} \\ \hline 1\ 1\ \cdots\ 1 & 1 \end{array} \right], \tag{31}$$

is transformed by T into another codeword of $\hat{\mathcal{Q}}$. (i) The first row of $\hat{G}$, R_0 say, is

$$\left| 1 + \sum_{n \in N} x^n \, | \, 0 \right|.$$

Then

$$T(R_0) = \left| \sum_{r \in Q} x^r \, | \, 1 \right| = R_0 + 1 \in \hat{\mathcal{Q}}.$$

(ii) Suppose $s \in Q$; the $(s + 1)^{\text{th}}$ row of $\hat{G}$ is

$$R_s = \left| x^s + \sum_{n \in N} x^{n+s} \, | \, 0 \right|.$$

We shall show that $T(R_s) = R_{-1/s} + R_0 + 1 \in \hat{\mathcal{Q}}$. It is just a matter of keeping track of where the 1's go. $T(R_s)$ has 1's in coordinate places $-1/s$ and $-1/(n + s)$ for $n \in N$, which comprise ∞ (if $n = -s$), $2m - 1$ residues and $2m$ nonresidues (by Perron's Theorem 24). Also

$$R_{-1/s} = \left| x^{-1/s} + \sum_{n \in N} x^{n-1/s} \, | \, 0 \right|$$

has 1's in places $-1/s$ and $n - 1/s$ $(n \in N)$, which comprise $2m$ residues and $2m$ nonresidues. Therefore the sum $T(R_s) + R_{-1/s}$ has a 1 in place ∞ and a 0 in place $-1/s$ (a nonresidue). If $-1/(n+s) \in N$, then $-1/(n+s) = n' - (1/s)$ for some $n' \in N$, and the 1's in the sum cancel. Thus the nonresidue coordinate places in the sum always contain 0. On the other hand, if $-1/(n+s) \in Q$, then $-1/(n+s) \neq n' - 1/s$ for all $n' \in N$, and so the sum contains 1 in coordinate places which are residues. I.e.,

$$T(R_s) + R_{-1/s} = \left| \sum_{r \in Q} x^r \, | 1 \right| = R_0 + 1.$$

Similarly if $t \in N$,

$$T(R_t) = R_{-1/t} + R_0. \qquad \text{Q.E.D.}$$

From Theorem 18 of Ch. 8, the codewords of each weight in $\hat{\mathcal{Q}}$ form a 2-design. In some cases much stronger statements can be made – see the end of §8 below.

Corollary 11. *An equivalent code $\mathcal{Q}$ is obtained no matter which coordinate place of $\hat{\mathcal{Q}}$ is deleted.*

Proof. From Corollary 15 of Ch. 8, since $PSL_2(p)$ is transitive. Q.E.D.

The automorphism group of nonbinary QR codes. If $l > 2$ then (as defined in §5 of Ch. 8) $\text{Aut}(\hat{\mathcal{Q}})$ consists of all monomial matrices which preserve $\hat{\mathcal{Q}}$. Since $\hat{\mathcal{Q}}$ is linear it is fixed by the $l - 1$ scalar matrices

$$\begin{pmatrix} \beta & & & 0 \\ & \beta & & \\ & & \ddots & \\ 0 & & & \beta \end{pmatrix}$$

for $\beta \in GF(l)^*$. It is convenient to ignore these and to work instead with $\text{Aut}(\hat{\mathcal{Q}})\dagger = \text{Aut}(\hat{\mathcal{Q}})/\text{scalar matrices}$. Clearly $|\text{Aut}(\hat{\mathcal{Q}})| = (l-1)|\text{Aut}(\hat{\mathcal{Q}})\dagger|$.

Theorem 12. (Gleason and Prange.) *If $l > 2$, $\text{Aut}(\hat{\mathcal{Q}})\dagger$ contains the group isomorphic to $PSL_2(p)$ which is generated by S, V and T', where T' is the following monomial generalization of T: the element moved from coordinate place i is multiplied by $\chi(i)$ for $1 \leq i \leq p - 1$; the element moved from place 0 is multiplied by ϵ; and the element moved from ∞ by $\pm \epsilon$; where ϵ was defined in the previous section, and the $+$ sign is taken if $p = 4k + 1$, the $-$ sign if $p = 4k - 1$.*

Proof. Similar to that of Theorem 10.

Theorems 10 and 12 only state that Aut $(\hat{\mathcal{Q}})$† *contains* $\mathrm{PSL}_2(p)$. There are three* known cases in which Aut $(\hat{\mathcal{Q}})$† is larger than this. For $l = 2$ and $p = 7$, $\hat{\mathcal{Q}}$ is the [8, 4, 4] Hamming code and $|\mathrm{Aut}(\hat{\mathcal{Q}})| = 1344$, by Theorem 24 of Ch. 13. For $l = 2$, $p = 23$ and $l = 3$, $p = 11$ $\hat{\mathcal{Q}}$ is an extended Golay code and Aut $(\hat{\mathcal{Q}})$† is a Mathieu group (see Chapter 20).

But for all other values of l and p, Aut $(\hat{\mathcal{Q}})$† is probably equal to $\mathrm{PSL}_2(p)$. This is known to be true in many cases, for example:

Theorem 13. (Assmus and Mattson.) *If $\frac{1}{2}(p-1)$ is prime and $5 < p \leqslant 4079$ then apart from the three exceptions just mentioned, Aut $(\hat{\mathcal{Q}})$† is (isomorphic to)* $\mathrm{PSL}_2(p)$.

The proof is omitted.

Problems. (2) Show that $\hat{\mathcal{Q}}$ and $\bar{\mathcal{Q}}$ have the same minimum distance, while that of $\mathcal{Q}$ is 1 less.

(3) Show that (29) belongs to $\mathrm{PSL}_2(p)$ iff $ad - bc$ is a quadratic residue mod p.

(4) Show that the generator V in (30) is redundant: S and T alone generate $\mathrm{PSL}_2(p)$.

(5) If $p = 8m - 1$, show that the relations given in Theorem 9(c) define an abstract group of order $\frac{1}{2}p(p^2 - 1)$, which is therefore isomorphic to $\mathrm{PSL}_2(p)$.

Research Problem (16.3). Show that the conclusion of Theorem 13 holds for all $p > 23$.

§6. Binary quadratic residue codes

Figure 16.3 gives a summary of the properties of binary QR codes.

Examples of binary quadratic residue codes $\mathcal{Q}$.

(i) The [7, 4, 3] Hamming code (see p. 481).

(ii) The [17, 9, 5] code with generator polynomial $x^8 + x^5 + x^4 + x^3 + 1$ and idempotent $x^{16} + x^{15} + x^{13} + x^{11} + x^8 + x^4 + x^2 + x + 1$.

(iii) The [23, 12, 7] Golay code $\mathcal{G}_{23}$ (see p. 482).

(iv) The [31, 16, 7] code with generator polynomial $x^{15} + x^{12} + x^7 + x^6 + x^2 + x + 1$. The quadratic residues mod 31 are $C_1 \cup C_5 \cup C_7$.

(v) The [47, 24, 11] code with generator polynomial $x^{23} + x^{19} + x^{18} + x^{14} + x^{13} + x^{12} + x^{10} + x^9 + x^7 + x^6 + x^5 + x^3 + x^2 + x + 1$.

---

*Four if the [6, 3, 4] code over GF(4) is counted – see p. 520 below.

$\mathcal{Q}$, $\bar{\mathcal{Q}}$, $\mathcal{N}$, $\bar{\mathcal{N}}$ have length $p = 8m \pm 1$ (Theorem 23) and generator polynomials

$$q(x) = \prod_{r \in Q} (x + \alpha^r), (x + 1)q(x), \qquad n(x) = \prod_{n \in N} (x + \alpha^n), (x + 1)n(x)$$

$$(32)$$

respectively, where α is a primitive p^{th} root of unity over GF(2). $\mathcal{Q}$, $\mathcal{N}$ have dimension $\frac{1}{2}(p + 1)$; $\bar{\mathcal{Q}}$, $\bar{\mathcal{N}}$ consist of the even weight codewords of $\mathcal{Q}$, $\mathcal{N}$ and have dimension $\frac{1}{2}(p - 1)$. $\mathcal{Q}$ and $\mathcal{N}$ are equivalent codes. For the minimum distance see Theorem 1. Examples are given in Fig. 16.2.

For $p = 8m - 1$ the idempotents of $\mathcal{Q}$, $\bar{\mathcal{Q}}$, $\mathcal{N}$, $\bar{\mathcal{N}}$ may be taken to be

$$\sum_{r \in Q} x^r, \qquad 1 + \sum_{n \in N} x^n, \qquad \sum_{n \in N} x^n, \qquad 1 + \sum_{r \in Q} x^r. \qquad (33)$$

For $p = 8m + 1$ the idempotents may be taken to be

$$1 + \sum_{r \in Q} x^r, \qquad \sum_{n \in N} x^n, \qquad 1 + \sum_{n \in N} x^n, \qquad \sum_{r \in Q} x^r. \qquad (34)$$

If $p = 8m - 1$, adding an overall parity check to $\mathcal{Q}$, $\mathcal{N}$ gives self-dual codes $\hat{\mathcal{Q}}$, $\hat{\mathcal{N}}$ with parameters $[p + 1, \frac{1}{2}(p + 1)]$. All weights in $\hat{\mathcal{Q}}$, $\hat{\mathcal{N}}$ are divisible by 4. If $p = 8m + 1$, $(\hat{\mathcal{Q}})^{\perp} = \hat{\mathcal{N}}$. In both cases $\hat{\mathcal{Q}}$, $\hat{\mathcal{N}}$ are invariant under $\text{PSL}_2(p)$. $\hat{\mathcal{Q}}$ is a quasi-cyclic code with generator matrix (31), (44), and sometimes (39) or (48).

Fig. 16.3. Properties of binary QR codes.

The first two codes with $p = 8m - 1$ are both perfect codes. If $\mathscr{C}$ is any binary self-dual code with all weights divisible by 4, we will see in Ch. 19 that $d \leq 4[n/24] + 4$. This (upper) bound is considerably larger than the square root (lower) bound of Theorem 1, and is attained by the QR codes $\hat{\mathcal{Q}}$ of lengths 8, 24, 32, 48, 80, 104 (and possibly others). This phenomenon has led to a considerable amount of work on determining the actual minimum distance of QR codes, some of which we now describe.

Other forms for the generator matrix of $\hat{\mathcal{Q}}$. The generator matrix $\hat{G}$ for $\hat{\mathcal{Q}}$ given in Eq. (31) is a $(p + 1) \times (p + 1)$ matrix which only has rank $\frac{1}{2}(p + 1)$. Sometimes it is easy to find a generator matrix for $\hat{\mathcal{Q}}$ in the form $[I \,|\, A]$, where I is a unit matrix and A is either a circulant matrix or a bordered circulant.

Canonical form 1 – *two circulants.*

Lemma 14. *For any prime* $p > 3$, $\text{PSL}_2(p)$ *contains a permutation* π_1 *consisting of two cycles of length* $\frac{1}{2}(p + 1)$.

Proof. As in §5 of Ch. 12 we represent the coordinates $\{0, 1, \ldots, p-1, \infty\}$ on which $PSL_2(p)$ acts by nonzero vectors $\binom{x}{y}$, where $x, y \in GF(p)$. Thus $\binom{x}{y}$ represents the coordinate x/y. We also represent an element $i \rightarrow (ai+b)/(ci+d)$ of $PSL_2(p)$ by the matrix $\binom{ab}{cd}$. We must find $\pi_1 \in PSL_2(p)$ such that for all nonzero $\binom{x}{y}$,

$$\pi_1^r \binom{x}{y} = k \binom{x}{y} \quad \text{for some } k \in GF(p)^* \text{ iff } \tfrac{1}{2}(p+1)|r. \tag{35}$$

Then π_1 does indeed consist of two cycles of length $\tfrac{1}{2}(p+1)$.

Let $\alpha \in GF(p^2)$ be a primitive $(p^2-1)^{\text{th}}$ root of unity. Then $\eta = \alpha^{2p-2}$ is a primitive $\tfrac{1}{2}(p+1)^{\text{th}}$ root of unity. Set $\lambda = 1/(1+\eta)$. Then it is immediate that $\lambda^r = (1-\lambda)^r$ iff $\tfrac{1}{2}(p+1)|r$.

Now define $a = \lambda^2 - \lambda$. Then $-a$ is a quadratic residue mod p. For $-a = \alpha^{2p-2}/(1+\eta)^2 = b^2$ where $b = \alpha^{p-1}/(1+\eta)$, and $b \in GF(p)$ since

$$b^p = \frac{\alpha^{p^2-p}}{1+\eta^p} = \frac{\alpha^{1-p}}{1+\eta^{-1}} = b.$$

Now we take $\pi_1 = \binom{1\ 1}{a\ 0} \in PSL_2(p)$. The eigenvectors of π_1 are given by

$$\pi_1 \binom{1}{\lambda-1} = \lambda \binom{1}{\lambda-1}, \quad \pi_1 \binom{1}{-\lambda} = (1-\lambda) \binom{1}{-\lambda},$$

so

$$\pi_1 = B \binom{\lambda \quad 0}{0 \quad 1-\lambda} B^{-1}, \quad \text{where } B = \binom{1 \qquad 1}{\lambda-1 \quad -\lambda},$$

$$\pi_1^r = B \binom{\lambda^r \qquad 0}{0 \quad (1-\lambda)^r} B^{-1}.$$

Therefore (35) holds iff $\lambda^r = (1-\lambda)^r = k$, which from the definition of λ is true iff $\tfrac{1}{2}(p+1)|r$.   Q.E.D.

For example, if $p = 7$, π_1 is $i \rightarrow (i+1)/3i$, which consists of the cycles $(\infty 560)(1324)$.

In general, let π_1 consist of the cycles

$$(l_1 l_2 \cdots l_{\frac{1}{2}(p+1)})(r_1 r_2 \cdots r_{\frac{1}{2}(p+1)}) \tag{36}$$

We take any codeword $c \in \hat{\mathcal{Q}}$ and arrange the coordinates in the order $l_1 \cdots l_{\frac{1}{2}(p+1)} r_1 \cdots r_{\frac{1}{2}(p+1)}$ given by (36). Then the codewords c, $\pi_1 c$, $\pi_1^2 c, \ldots, \pi_1^{(p-1)/2} c$ form an array

$$[L \,|\, R], \tag{37}$$

where L and R are $\tfrac{1}{2}(p+1) \times \tfrac{1}{2}(p+1)$ circulant matrices. For example, if $c = (c_0 c_1 \cdots c_6 c_\infty) = (01101001)$ is the extended idempotent of the $[8, 4, 4]$ code

$\hat{\mathcal{D}}$ then $[L|R]$ is

$$
\begin{array}{r|ccccccc}
 & \infty\ 5 & 6\ 0\ 1\ 3\ 2\ 4 \\
\hline
c: & 1 & 1\ 0\ 1\ 1 \\
\pi_1 c: & 1 & 1\ 1\ 0\ 1 \\
\pi_1^2 c: & 1 & 1\ 1\ 1\ 0 \\
\pi_1^3 c: & & 1\ 0\ 1\ 1\ 1 \\
\end{array}
\tag{38}
$$

Of course (37) always generates a subcode of $\hat{\mathcal{D}}$. But if we can find a codeword c such that either L or R has rank $\frac{1}{2}(p+1)$, we can invert it (see Problem 7) and obtain a generator matrix for $\hat{\mathcal{D}}$ in the first canonical form

$$
\hat{G}_1 = [I \,|\, A], \tag{39}
$$

where A is a $\frac{1}{2}(p+1) \times \frac{1}{2}(p+1)$ circulant matrix. For example, (38) is already in this form.

A code with generator matrix of the form (39) is called a *double circulant code*. We also apply the same name to codes with generator matrices (45), (46), etc., differing from (39) only in the presence of borders around various parts of the matrix. It will be shown in §7 that such codes are examples of quasi-cyclic codes. Further properties of these codes are given there.

Example. The $[24, 12, 8]$ Golay code $\mathcal{G}_{24}$. If $p = 23$ the permutation $\pi_1: i \to (i+1)/5i$ consists of the cycles

$$(21, 7, 16, 12, 19, 22, 0, \infty, 14, 15, 18, 2)(20, 17, 4, 6, 1, 5, 3, 11, 9, 13, 8, 10).$$
$$\tag{40}$$

For $c(x)$ we take the sum of the idempotent

$$
E(x) = \sum_{r \in Q} x^r
$$

of G_{23} and three cyclic shifts of $E(x)$ (with an overall parity check added):

	∞	0	1	2	3	4	5	6	7	8	9	10	11	12	13	14	15	16	17	18	19	20	21	22
$E(x) = 1$		1	1	1	1		1		1	1				1	1				1		1			
$xE(x) = 1$			1	1	1	1		1		1	1				1	1				1		1		
$x^2E(x) = 1$				1	1	1	1		1		1	1				1	1				1		1	
$x^3E(x) = 1$					1	1	1	1		1		1	1				1	1				1		1
$c(x) =$		1		1		1	1			1									1				1	1

Arranging the coordinates of $c(x)$ in the order (40) we obtain the generator matrix for $\mathcal{G}_{24}$ shown in Fig. 16.4.

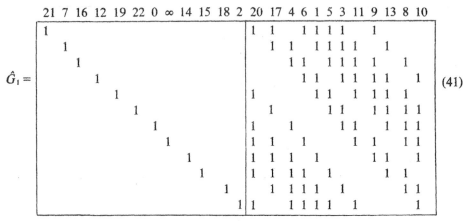

$$\hat{G}_1 = \qquad\qquad\qquad\qquad\qquad\qquad\qquad\qquad\qquad\qquad (41)$$

Fig. 16.4. A double circulant generator matrix for the Golay code $\mathcal{G}_{24}$.

Research Problem (16.4). Is it always possible to find c such that either L or R in (37) is invertible? Find a general method for getting the canonical form (39) for an extended QR code.

Canonical form 2: Two bordered circulants. The second canonical form arises because $\mathrm{PSL}_2(p)$ contains the permutation π_2: $i \to ri$, where $r = \rho^2$ is a generator of the cyclic group of quadratic residues mod p. The cycle structure of π_2 is

$$(\infty)(n, nr, nr^2, \ldots, nr^{(p-3)/2})(1, r, r^2, \ldots, r^{(p-3)/2})(0), \qquad (42)$$

where n is any nonresidue. (For $p = 8m - 1$ we may take n to be -1). Let $\hat{G}_2$ be obtained from $\hat{G}$ by arranging the rows and columns of $\hat{G}$ in the order shown in (42).

Example. The generator matrix for the $[8, 4, 4]$ code ((24) with an overall parity check) becomes

$$\hat{G}_2 = \qquad\qquad\qquad\qquad\qquad\qquad\qquad\qquad\qquad (43)$$

	∞	6	5	3	1	2	4	0
∞	1	1	1	1	1	1	1	1
6		1	1			1	1	
5			1	1	1			1
3	1			1	1	1		
1	1				1		1	1
2		1			1	1		1
4				1		1	1	1
0	1	1	1					1

In general we write $\hat{G}_2$ as

$$
\hat{G}_2 =
\begin{array}{c|ccc|ccc|c}
 & \infty & n\ nr\cdots & 1\ r\cdots & & & & 0 \\
\hline
\infty & 1 & 1\cdots\cdots 1 & 1\cdots\cdots 1 & & & & 1 \\
\hline
n & 0 & & & & & & \epsilon+1 \\
nr & \vdots & \multicolumn{2}{c}{A} & \multicolumn{2}{c}{B} & & \vdots \\
\vdots & 0 & & & & & & \epsilon+1 \\
\hline
1 & 0 & & & & & & \epsilon \\
r & \vdots & \multicolumn{2}{c}{C} & \multicolumn{2}{c}{D} & & \vdots \\
\vdots & 0 & & & & & & \epsilon \\
\hline
0 & 0 & 1\cdots\cdots 1 & 0\cdots\cdots 0 & & & & 1 \\
\end{array}
\tag{44}
$$

where

$$
\epsilon =
\begin{cases}
1 & \text{if } p = 8m - 1, \\
0 & \text{if } p = 8m + 1.
\end{cases}
$$

Problem. (6) (a) Show that A, B, C, D are circulant matrices.

(b) If $p = 8m - 1$, show that $C = C^T = B + J$ and $D = A^T = A + C + J$, where J is a matrix all of whose entries are 1. Also the rows of A, B and D have weight $2m$ and the rows of C have weight $2m - 1$.

(c) If $p = 8m + 1$, show that $A = A^T$, $B = C^T$, $D = D^T$, the rows of B, C and D have weight $2m$, and the rows of A have weight $2m - 1$.

Now if $p = 8m - 1$ C may be invertible (A, B and D are certainly *not* since they have even weight), and if $p = 8m + 1$, A may be. If so we readily arrive at the following canonical form for the generator matrix:

$$
\hat{G}_3 =
\left[
\begin{array}{c|cc}
 & a\ a\ \cdots\ a & b \\
\cline{2-3}
I & & c \\
 & T & c \\
 & & \vdots \\
 & & c \\
\end{array}
\right]
\tag{45}
$$

where I is a $\tfrac{1}{2}n \times \tfrac{1}{2}n$ unit matrix, T is a $(\tfrac{1}{2}n - 1) \times (\tfrac{1}{2}n - 1)$ circulant matrix (here $n = p + 1$), and a, b, c are 0 or 1. This is called a (bordered) double circulant code.

Examples. (1) For $p = 7$, C in (43) is already an identity matrix and we obtain

$$
\hat{G}_3 =
\left[
\begin{array}{cccc|ccc|c}
1 & & & & 1 & 1 & 1 & 0 \\
 & 1 & & & 1 & & 1 & 1 \\
 & & 1 & & 1 & 1 & & 1 \\
 & & & 1 & & 1 & 1 & 1 \\
\end{array}
\right].
\tag{46}
$$

(2) The Golay code $\mathcal{G}_{24}$ again. The first two rows of the generator matrix $\hat{G}$, Equation (31), are:

∞	0	1	2	3	4	5	6	7	8	9	10	11	12	13	14	15	16	17	18	19	20	21	22	
	1			1	1		1	1					1	1		1		1	1	1	1			
1	1			1		1			1	1				1	1	1				1	1	1	$\cdots (*)$

Using (42), we arrange the rows and columns in the following order:

$$\left|\,\infty\,\left|\,14\ 5\ 10\ 20\ 17\ 11\ 22\ 21\ 19\ 15\ 7\,\right|\,1\ 2\ 4\ 8\ 16\ 9\ 18\ 13\ 3\ 6\ 12\,\right|\,0\,\right|$$
$$\qquad\qquad\quad 1\qquad 1\ 1\ 1\qquad 1\qquad\quad 1\qquad 1\ 1\qquad 1\qquad\qquad 1\ 1\quad|1|$$

The first rows of C and D are obtained from $(*)$ and are shown above. To find the inverse of C (see Problem 7) we write the first row of C as $a(x) = x^3 + x^5 + x^6 + x^7 + x^9$ and find the inverse of $a(x) \bmod x^{11} + 1$. This is $1 + x + x^2 + x^5 + x^8 + x^9 + x^{10}$. To find $C^{-1}D$ we compute

$$(1 + x + x^2 + x^5 + x^8 + x^9 + x^{10})(1 + x^3 + x^4 + x^6 + x^9 + x^{10}) \bmod x^{11} + 1$$

$$= 1 + x^2 + x^6 + x^7 + x^8 + x^{10}$$

$$= 1 + \sum_{n \in N} x^n, \quad N = \text{nonresidues mod 11}. \qquad (47)$$

Thus $\mathcal{G}_{24}$ has a generator matrix

$$\hat{G}_3 = \left[\begin{array}{c|cc} & \begin{matrix} 1\ 1\ \cdots\ 1 \end{matrix} & 0 \\ & & 1 \\ I_{12} & T & 1 \\ & & \vdots \\ & & 1 \end{array}\right] \qquad (48)$$

where $T = C^{-1}D$ is the circulant with first row given by (47):

$$\begin{array}{l} 0\ 1\ 2\ 3\ 4\ 5\ 6\ 7\ 8\ 9\ 10 \\ 1\ 0\ 1\ 0\ 0\ 0\ 1\ 1\ 1\ 0\ 1 \end{array} \qquad (49)$$

This is equivalent to the definition of $\mathcal{G}_{24}$ given in Fig. 2.13 (see Problem 9). Thus we have shown that some binary QR codes are equivalent to (possibly bordered) double circulant codes. (The same procedure can be applied to non-binary codes.)

Research Problem (16.5). Sometimes it is not possible to invert C or A (e.g. for $p = 127$ – see Karlin [715]). In this case is it still possible to find a generator matrix of the form (45)?

Problems. (7) – Circulant matrices. (i) Show that the algebra of $n \times n$ circulant matrices over a field F is isomorphic to the algebra of polynomials in the ring

$F[x]/(x^n - 1)$, if the circulant matrix

$$A = \begin{bmatrix} a_0 & a_1 & \cdots & a_{n-1} \\ a_{n-1} & a_0 & \cdots & a_{n-2} \\ \cdots\cdots\cdots\cdots\cdots \\ a_1 & a_2 & \cdots & a_0 \end{bmatrix} \tag{50}$$

is mapped onto the polynomial

$$a(x) = a_0 + a_1 x + \cdots + a_{n-1} x^{n-1}. \tag{51}$$

(ii) The sum and product of two circulants is a circulant. In particular $AB = C$ where $c(x) = a(x)b(x) \bmod x^n - 1$.

(iii) A is invertible iff $a(x)$ is relatively prime to $x^n - 1$. The inverse, if it exists, is B where $a(x)b(x) = 1 \bmod x^n - 1$.

(iv) A^T is a circulant corresponding to the polynomial $a_0 + a_{n-1}x + \cdots + a_1 x^{n-1}$.

(v) Let T be the circulant with $a_1 = 1$, $a_i = 0$ for $i \neq 1$. Show that $A = \sum_{i=0}^{n-1} a_i T^i$.

(vi) Let β be a primitive n^{th} root of unity in some field containing F. Show that $a(1), a(\beta), \ldots, a(\beta^{n-1})$ are the eigenvalues of A (see Problem 20 of Ch. 7).

(8) Let $\mathcal{A}$ (resp. $\mathcal{B}$) be a binary $[2m, m]$ code with generator matrix $[I | A]$ (resp. $[I | B]$) where A and B are circulants with first rows

$$a(x) = \sum_{i=0}^{m-1} a_i x^i, \qquad b(x) = \sum_{i=0}^{m-1} b_i x^i.$$

(i) Show that $\mathcal{A}$ and $\mathcal{B}$ are equivalent (a) if $B = A^T$, or (b) if $a_i = b_{m-i-1}$ for all i (i.e. if $b(x)$ is the reciprocal of $a(x)$), or (c) if $a(x)b(x) \equiv 1 \bmod x^m - 1$, or (d) if $b(x) = a(x)^2$ and m is odd, or (e) if $b(x) = a(x^u)$ where u and m relatively prime.

(ii) Show that $\mathcal{A}^\perp$ is equivalent to $\mathcal{A}$.

(iii) Show that $\mathcal{A}^\perp = \mathcal{A}$ iff $AA^T = I$.

(9) Show that the generator matrix (48), (49) for $\mathcal{G}_{24}$ is equivalent to the one given in Fig. 2.13.

(10) Let $\mathcal{C}$ have generator matrix (45) with $a = c = 1$, $b = 0$. Show that $\mathcal{C}^\perp = \mathcal{C}$ iff $TT^T = I + J$.

A method for finding low weight codewords in $\mathcal{Q}$ (Karlin and & MacWilliams [718]). The generator matrix (45) makes it simpler to find the minimum distance of $\mathcal{Q}$ or $\hat{\mathcal{Q}}$. Many of the results shown in Fig. 16.2 were found in this way.

In some cases it is possible to find low weight codewords directly, as we now show.

The permutation $V: y \to \rho^2 y$, where ρ is a primitive element of $GF(p)$, is in Aut $(\mathcal{Q})$. The order of this permutation is $\frac{1}{2}(p - 1)$; if this is a composite number, say $\frac{1}{2}(p - 1) = sf$, then $\mathcal{Q}$ contains codewords which are fixed by the permutation $U: y \to \rho^{2s} y$. Let $\mathcal{U}$ be the subcode consisting of such codewords. We describe how to find $\mathcal{U}$ for $p = 8m - 1$. Let $e = 2s$, $p - 1 = ef$, where s and f are odd. Divide the integers mod p into e classes

$$T_i = \{\rho^{ej+i} : j = 0, 1, \ldots, f - 1\}, \quad i = 0, \ldots, e - 1.$$

For example, if $p = 71$, $p - 1 = 70 = 14.5$, $e = 14$, $\rho = 7$:

$$T_0 = \{1, 7^{14}, 7^{28}, 7^{42}, 7^{56}\} = \{1, 54, 5, 57, 25\}$$
$$T_1 = \{7, 7^{15}, 7^{29}, 7^{43}, 7^{57}\} = \{7, 23, 35, 44, 33\}$$

Define the polynomials

$$X_i = \sum_{t \in T_i} x^t \quad \text{for } i = 0, \ldots, e - 1. \tag{52}$$

Problem 11. (i) Show that X_i is fixed by U, and that any polynomial which is fixed by U is a linear combination of $1, X_0, \ldots, X_{e-1}$.

(ii) Show that $E_q(x) = X_0 + X_2 + \cdots + X_{e-2}$.

(iii) Show that $X_i E_q(x) = \eta_i + \sum_{j=0}^{e-1} \epsilon_{ij} X_j$ for some η_i, ϵ_{ij} in GF(2), for $i = 0, \ldots, e - 1$.

(iv) Thus $\mathcal{U}$, the subcode of $\mathcal{Q}$ fixed by U, consists of linear combinations of $E_q(x)$ and $X_i E_q(x)$ for $i = 0, \ldots, e - 1$.

The coefficients η_i and ϵ_{ij} are shown in Fig. 16.5.

	1	X_0 X_2 $\cdots$ X_{e-2}	X_s X_{2+s} $\cdots$ X_{e-2+s}
$E_q(x)$	0	1　1 $\cdots$ 1	0　0 $\cdots$ 0
$X_0 E_q(x)$ $\vdots$ $X_{e-2} E_q(x)$	0 $\vdots$ 0	A	B
$X_s E_q(x)$ $\vdots$ $X_{e-2+s} E_q(x)$	1 $\vdots$ 1	$B + J$	$A + I + J$

Fig. 16.5.

(These are not the same as the matrices A and B in (44).) Then

$$G = \begin{array}{|c|c|c|}
\hline
0 & 1 \cdots 1 & 0 \cdots 0 \\
\hline
\begin{array}{c}0\\ \vdots\\ 0\end{array} & A & B \\
\hline
\begin{array}{c}1\\ \vdots\\ 1\end{array} & B+J & A+I+J \\
\hline
\end{array} \qquad (53)$$

is a generator matrix for $\mathcal{U}$. We state the following results without proof. G is a $(2s+1) \times (2s+1)$ matrix with rank $s+1$. A, B are circulant matrices, and the first rows are given by the following easy formula:

Let

$$[m] = \begin{cases} 1 & \text{if } m \text{ is a quadratic residue mod } p, \\ 0 & \text{if } m \text{ is a nonresidue or } 0. \end{cases}$$

Then

$$a_i = [\rho^{2if} - 1], \qquad b_i = [-\rho^{2if} - 1], \quad i = 1, \ldots, s-1,$$
$$a_0 = \tfrac{1}{2}(s+1) \bmod 2, \qquad b_0 = 0.$$

Thus $\mathcal{U}$ may be regarded as a $[2s+1, s+1]$ code. The connection with $\mathcal{Q}$ is that if $\eta + X_{i_1} + \cdots + X_{i_s}$ is a minimum weight codeword in $\mathcal{U}$, then $\mathcal{Q}$ contains codewords of weight $\eta + sf$. Thus the minimum weight of $\mathcal{U}$ provides an upper bound to the minimum weight of $\mathcal{Q}$. In fact in all cases for which we know the minimum weight of $\mathcal{Q}$, this weight is found in $\mathcal{U}$.

Example. $p = 71, e = 14, f = 5, \rho = 7$.

To find the first rows of A and B we calculate, using Vinogradov [1371, pp. 220–225],

$$\begin{array}{ll}
49^5 - 1 = 44 \in N & -49^5 - 1 = 25 \in Q \\
49^{10} - 1 = 36 \in Q & -49^{10} - 1 = 33 \in N \\
49^{15} - 1 = 31 \in N & -49^{15} - 1 = 39 \in Q \\
49^{20} - 1 = 19 \in Q & -49^{20} - 1 = 50 \in Q \\
49^{25} - 1 = 17 \in N & -49^{25} - 1 = 52 \in N \\
49^{30} - 1 = 29 \in Q & -49^{30} - 1 = 40 \in Q,
\end{array}$$

and the first rows of A, B are therefore

$$\begin{array}{|c|c|c|c|c|c|c|}
\hline
0 & 1 & 2 & 3 & 4 & 5 & 6 \\
\hline
0 & 0 & 1 & 0 & 1 & 0 & 1 \\
\hline
\end{array}, \qquad
\begin{array}{|c|c|c|c|c|c|c|}
\hline
0 & 1 & 2 & 3 & 4 & 5 & 6 \\
\hline
0 & 1 & 0 & 1 & 1 & 0 & 1 \\
\hline
\end{array}.$$

Notice that all that is needed to write down the matrix G is a table of powers of ρ. Further the same matrix G will appear for many different primes p.

The matrix A is frequently invertible, in which case one finds a generator matrix for $\mathcal{U}$ in the form

$$
\left[
\begin{array}{c|ccc|ccc}
0 & 1 & 1 & \cdots & 1 & 0 & 0 & \cdots & 0 \\
\hline
0 & & & & & & \\
\vdots & & I & & & R+J & \\
0 & & & & & & \\
\hline
1 & 1 & 1 & \cdots & 1 & 1 & 1 & \cdots & 1
\end{array}
\right]
\tag{54}
$$

where R is an orthogonal circulant matrix.

Example. (Cont.) The inverse of the polynomial $x^2 + x^4 + x^6 \bmod x^7 + 1$ is $x + x^2 + x^3 + x^4 + x^5$. Thus G becomes

$$
\left[
\begin{array}{c|ccccccc|ccccccc}
0 & 1 & 1 & 1 & 1 & 1 & 1 & 1 & 0 & 0 & 0 & 0 & 0 & 0 & 0 \\
\hline
0 & 1 & & & & & & & 1 & 1 & 1 & 0 & 1 & 1 & 1 \\
0 & & 1 & & & & & & 1 & 1 & 1 & 1 & 0 & 1 & 1 \\
0 & & & 1 & & & & & 1 & 1 & 1 & 1 & 1 & 0 & 1 \\
0 & & & & 1 & & & & 1 & 1 & 1 & 1 & 1 & 1 & 0 \\
0 & & & & & 1 & & & 0 & 1 & 1 & 1 & 1 & 1 & 1 \\
0 & & & & & & 1 & & 1 & 0 & 1 & 1 & 1 & 1 & 1 \\
0 & & & & & & & 1 & 1 & 1 & 0 & 1 & 1 & 1 & 1 \\
\hline
1 & 1 & 1 & 1 & 1 & 1 & 1 & 1 & 1 & 1 & 1 & 1 & 1 & 1 & 1
\end{array}
\right]
$$

The sum of the first, second and last rows of this matrix is $1 + X_0 + X_{13}$, so $\mathcal{Q}$ is a [71, 36] code which contains codewords of weight $1 + 2 \cdot 5 = 11$.

The procedure for $p = 8m + 1$ is similar.

Some negative results obtained in this way are shown in Fig. 16.6.

	p	bound
$p = 8m - 1$	631	83
	727	67
	751	99
	991	91
$p = 8m + 1$	233	25
	241	41
	257	33
	761	75
	1361	135

Fig. 16.6. Upper bounds on the minimum distance of $\mathcal{Q}$.

Research Problem (16.6). If $\mathscr{Q}$ contains nontrivial subcodes $\mathscr{U}$, is a codeword of minimum weight always contained in one of these subcodes?

Another method of using the group to find weights in $\mathscr{Q}$. A powerful method for finding the weight enumerator of QR and other codes has been used by Mykkeltveit, Lam and McEliece [979]. It is based on the following.

Lemma 15. *Let $\mathscr{C}$ be a code and H any subgroup of* Aut $(\mathscr{C})$. *If A_i is the total number of codewords in $\mathscr{C}$ of weight i, and $A_i(H)$ the number which are fixed by some element of H, then*

$$A_i \equiv A_i(H) \bmod |H|.$$

Proof. The codewords of weight i can be divided into two classes, those fixed by some element of H, and the rest. If $a \in \mathscr{C}$ is not fixed by any element of H, then the $|H|$ codewords ga for $g \in H$ must all be distinct. Thus $A_i = A_i(H) +$ a multiple of $|H|$.

For a prime q dividing $|\text{Aut}(\mathscr{C})|$, let S_q be a maximal subgroup of Aut $(\mathscr{C})$ whose order is a power of q (S_q is a Sylow subgroup of Aut $(\mathscr{C})$; these groups are especially simple if Aut $(\mathscr{C}) = \text{PSL}_2(p)$.) By letting H run through the groups S_q the lemma determines $A_i \bmod q^a$ and hence determines $A_i \bmod |\text{Aut}(\mathscr{C})|$, from the Chinese Remainder Theorem (Problem 8 of Ch. 10). If $|\text{Aut}(\mathscr{C})|$ is large this often enough to determine A_i exactly. See [979] for examples.

Problem. (12) For some values of p in Fig. 16.2, $\hat{\mathscr{Q}}$ has $d' > \bar{s}$, in the notation of Ch. 6. Find these values and investigate the designs formed by the codewords.

§7. Double circulant and quasi-cyclic codes

A *double circulant* code was defined in the previous section to be a code with generator matrix of the form

$$
\left[\begin{array}{c|cc}
 & a\ a\ \cdots\ a & b \\
\hline
 & & c \\
I & A & c \\
 & & \vdots \\
 & & c
\end{array}\right],
$$

where A is a circulant matrix with first row $a(x) = a_0 + a_1x + a_2x^2 + \cdots$, and a, b, c are 0 or 1. We use the same name for various codes obtained from (55), for example that obtained by deleting the first row and the first and last columns. The latter transformation changes (55) to

$$[I \,|\, A] \tag{56}$$

where I and A are $k \times k$ matrices: (56) generates a $[2k, k]$ double circulant code which we denote in this section by $\mathcal{D}_a$. Some properties of $\mathcal{D}_a$ are given in Problem 8.

Problem. (13) Show that $\mathcal{D}_a$ is particularly simple to encode: the message $u(x)$ becomes the codeword $|u(x)|u(x)a(x)|$.

Quasi-cyclic codes. A code is called *quasi-cyclic* if there is some integer s such that every cyclic shift of a codeword by s places is again a codeword. Of course a cyclic code is a quasi-cyclic code with $s = 1$.

By taking the columns of (56) in the order $1, k + 1, 2, k + 2, \ldots$ we see that a double circulant code $\mathcal{D}_a$ is equivalent to a quasi-cyclic code with $s = 2$. For example (46) gives

$$\begin{bmatrix} 1 & 1 & 0 & 0 & 0 & 1 \\ 0 & 1 & 1 & 1 & 0 & 0 \\ 0 & 0 & 0 & 1 & 1 & 1 \end{bmatrix},$$

which is indeed quasi-cyclic with $s = 2$. Also Lemma 14 and Equation (44) show that any binary QR code $\mathcal{Q}$ with one coordinate deleted, or any extended code $\hat{\mathcal{Q}}$, is equivalent to a quasi-cyclic code with $s = 2$.

Since many of the codes in this Chapter have the form (56), we know that there are good codes of this type. In fact some long double circulant (or quasi-cyclic) codes $\mathcal{D}_a$ meet the Gilbert–Varshamov bound. The following theorem is a slightly weaker result.

Theorem 16. *If k is a prime such that 2 is a primitive element of* GF(k), *and if*

$$1 + \binom{2k}{2} + \binom{2k}{4} + \cdots + \binom{2k}{2r-2} < 2^{k-1}, \quad \text{where } 2r < k,$$

then there is a $[2k, k]$ double circulant binary code $\mathcal{D}_a$ with minimum distance at least $2r$.

Proof. Choose $a(x)$ to have odd weight less than k. The number of such codes $\mathcal{D}_a$ is $2^{k-1} - 1$. From the assumption about k, the complete factorization of

$x^k + 1$ over GF(2) is $(x + 1) \sum_{i=0}^{k-1} x^i$. Hence $a(x)$ has an inverse in the ring GF(2)$[x]/(x^k + 1)$.

Let $|l(x)|r(x)|$ be a typical codeword of $\mathscr{D}_a$; by Problem 13, $r(x) = l(x)a(x)$. Hence wt $(l(x))$ and wt $(r(x))$ have the same parity. Note also that $\mathscr{D}_a$ contains the codeword **1** (the sum of the rows of (56)).

If $l(x)$ has odd weight less than k, then $|l(x)|r(x)|$ for a given $r(x)$ is in exactly one code $\mathscr{D}_a$, namely the one for which $a(x) = l(x)^{-1}r(x)$.

If $l(x)$ has even weight then again $|l(x)|r(x)|$ is in exactly one $\mathscr{D}_a$, namely that for which

$$a(x) = (1 + l(x))^{-1}(1 + r(x)).$$

Therefore each $|l(x)|r(x)|$ with weight less than k is in exactly one code $\mathscr{D}_a$. The remainder of the proof is as usual – cf. Theorem 31 of Ch. 17.   Q.E.D.

Unfortunately it is at present only a conjecture that there are infinitely many primes k such that 2 is a primitive element of GF(k). If this were proved, then we could simply take k to be arbitrarily large in Theorem 16 and deduce that some very large codes $\mathscr{D}_a$ meet the Gilbert–Varshamov bound. But even without this result Kasami has used Theorem 16 to show that some codes $\mathscr{D}_a$ meet the Gilbert–Varshamov bound – see [730] for the proof.

The double circulant codes $\mathscr{B}$. We now define a certain family of binary double circulant codes of length $8m$, where $q = 4m - 1$ is a prime of the form $8k + 3$ (but now $8m - 1$ need not be a prime).

Definition. Let $\mathscr{B}$ be the $[8m = 2q + 2, q + 1]$ code with generator matrix

$$
\begin{array}{cccccccccc}
 & l_\infty & l_0 & l_1 & \cdots & l_{q-1} & r_\infty & r_0 & r_1 & \cdots & r_{q-1}
\end{array}
$$

$$
M = \left[
\begin{array}{c|c|c|c}
\begin{matrix} 1 \\ 1 \\ \vdots \\ 1 \end{matrix} & I & \begin{matrix} 0 \\ 0 \\ \vdots \\ 0 \end{matrix} & A \\
\hline
0 & 0\ 0\ \cdots\ 0 & 1 & 1\ 1\ \cdots\ 1
\end{array}
\right]
\tag{57}
$$

where A is a circulant matrix which in the first row has a 1 in position r_i iff $i = 0$ or i is a quadratic residue mod q, i.e.

$$a(x) = 1 + \sum_{r \in Q} x^r.$$

For example, if $q = 3$,

$$
M = \begin{array}{c@{\hspace{0.5em}}c@{\hspace{0.5em}}c@{\hspace{0.5em}}c@{\hspace{0.5em}}c@{\hspace{0.5em}}c@{\hspace{0.5em}}c@{\hspace{0.5em}}c}
l_\infty & l_0 & l_1 & l_2 & r_\infty & r_0 & r_1 & r_2
\end{array}
$$

$$
M = \left[\begin{array}{c|ccc|c|ccc}
1 & 1 & 0 & 0 & 0 & 1 & 1 & 0 \\
1 & 0 & 1 & 0 & 0 & 0 & 1 & 1 \\
1 & 0 & 0 & 1 & 0 & 1 & 0 & 1 \\
\hline
0 & 0 & 0 & 0 & 1 & 1 & 1 & 1
\end{array} \right]
\tag{58}
$$

which of course is equivalent to the extended Hamming code (46). For $q = 11$ we obtain the [24, 12, 8] Golay code, as we saw in Problem 8.

Problem. (14) Show that $\mathcal{B}$ is self-dual and has all weights divisible by 4.

If $q = 19$ $\mathcal{B}$ is a [40, 20, 8] code, and if $q = 43$ $\mathcal{B}$ is a [88, 44, 16] code, both of which have the greatest possible minimum distance for self-dual codes with weights divisible by 4, as will be shown in Ch. 19. The weight distributions of these codes will also be given there.

$\mathcal{B}$ can also be considered as an extended cyclic code over GF(4). Let $\omega \in$ GF(4) be a primitive cube root of unity, with $\omega^2 + \omega + 1 = 0$, and change l_i, r_i into $v_i = l_i\omega + r_i\omega^2$; i.e.

$$
\begin{array}{ccc}
l_i & r_i & v_i \\
0 & 0 & \leftrightarrow 0 \\
1 & 0 & \leftrightarrow \omega \\
0 & 1 & \leftrightarrow \omega^2 \\
1 & 1 & \leftrightarrow 1
\end{array}
$$

Problem. (15) Show that this mapping sends $\mathcal{B}$ into an extended cyclic code $\mathcal{D}$ over GF(4). The idempotent of $\mathcal{D}$ is

$$
\omega^2 \sum_{r \in Q} x^r + \omega \sum_{n \in N} x^n,
$$

where Q and N are the quadratic residues and nonresidues mod q. Show that $\mathcal{D}$ is obtained by extending the cyclic code with generator polynomial $\Pi_{r \in Q}(x + \alpha^r)$, where α is a primitive q^{th} root of unity in GF(4'). (Thus we may say that $\mathcal{D}$ is an extended quadratic-residue code over GF(4).)

Example. Consider a code in which the first row of the generator matrix is

1 1 0 0 0 0 0 0 0 0 0 0 0 1 1 0 1 1 1 0 0 0 1 0.

(This is the Golay code.) This maps into the GF(4) vector

$$\omega \ 1 \ \omega^2 \ 0 \ \omega^2 \ \omega^2 \ \omega^2 \ 0 \ 0 \ 0 \ \omega^2 \ 0$$

By adding the vector **1** and multiplying by ω we obtain the (extended) idempotent

$$1 \ 0 \ \omega^2 \ \omega \ \omega^2 \ \omega^2 \ \omega^2 \ \omega \ \omega \ \omega \ \omega^2 \ \omega.$$

Problems. (16) Show that Aut $(\mathscr{B})$ contains (i) $PSL_2(q)$ applied simultaneously to both sides of (57), and (ii) the permutation

$$(l_\infty \ r_\infty)(l_0 \ r_0)(l_1 \ r_{q-1})(l_2 \ r_{q-2}) \ \cdots \ (l_{q-1} \ r_1) \tag{59}$$

which interchanges the two sides.

(17) Show that $\mathscr{D}$, the code over GF(4), is fixed by $PSL_2(q)$ and by the generalized permutation $(v_\infty)(v_0)(v_1 v_{q-1})(v_2 v_{q-2}) \cdots$ with ω and ω^2 interchanged.

Research Problem (16.7) Is there a square root bound for the minimum distance of $\mathscr{B}$, analogous to Theorem 1?

We conclude this section with a table (Fig. 16.7) showing some examples of good double circulant codes. The first row of A, $a(x) = a_0 + a_1 x + \cdots$ is given in octal as $/a_0 a_1 a_2 / a_3 a_4 a_5 / \cdots$ E.g. 426 stands for $1 + x^4 + x^6 + x^7$.

n	k	d	Definition	$a(x)$	n	k	d	Definition	$a(x)$
6	3	3	(56)	3	28	14	8	(56)	727
8*	4	4	(38)	7	30	15	8	(56)	2167
10	5	4	(56)	7	31	16	8	(57)*	23642
12	6	4	(56)	6	32	16	8	(56)	557
14	7	4	(56)	77	39	20	8	(57)*	236503
16	8	5	(56)	426	40*	20	8	(57)	636503
18	9	6	(56)	362	40	20	9	(56)	5723
20	10	6	(56)	75	42	21	10	(56)	14573
22	11	7	(56)	355	56*	28	12	(57)	
24*	12	8	(41)	675	64*	32	12	(57)	
24*	12	8	(48)	5072 or 6704					
26	13	7	(56)	653	88*	44	16	(57)	§
28	14	8	(57)	26064	108	54	20	(57)	

*Self-dual, weight divisible by 4.

*With first column deleted.

§$a(x) = 62473705602153$.   Fig. 16.7. Good double circulant codes.

§8. Quadratic-residue and symmetry codes over GF(3)

Examples of extended QR codes over GF(3) are given in Fig. 16.2. These codes have length $12m$ or $12m + 2$, by Problem 25 below, and have all weights divisible by 3 (Theorem 8). The first is the Golay code $\mathcal{G}_{12}$ (see p. 482). The minimum distances of the $[24, 12, 9]$, $[48, 24, 15]$, and $[60, 30, 18]$ codes were found with the aid of a computer. As will be shown in Ch. 19, these are the largest possible minimum distances for self-dual codes over GF(3).

Pless symmetry codes. These are double circulant codes over GF(3) (see Fig. 16.8).

Definition. Let p be a prime $\equiv -1 \pmod 6$, and let S be the following $(p + 1) \times (p + 1)$ matrix (with rows and columns labeled $\infty, 0, 1, \ldots, p-1$):

$$
\begin{array}{c}
\quad\quad\quad \infty\ \ 0\ \ 1\ \cdots\ p-1 \\[4pt]
S =
\begin{array}{c}
\infty \\
0 \\
1 \\
\vdots \\
p-1
\end{array}
\left[
\begin{array}{c|ccc}
0 & 1\ 1\ \cdots\ \ 1 \\
\hline
\epsilon & \\
\epsilon & \\
\vdots & \quad\ C \\
\epsilon & \\
\end{array}
\right]
\end{array}
\tag{60}
$$

where $\epsilon = 1$ if $p = 4k + 1$, $\epsilon = -1$ if $p = 4k - 1$, and C is a circulant matrix in which the first row has a 0 in column 0, a 1 in columns which are quadratic residues mod p, and a -1 for the nonresidues. Then the *Pless symmetry code* S_{2p+2} is the $[2p + 2, p + 1]$ code over GF(3) with generator matrix $[I \,|\, S]$, where I is a $(p + 1) \times (p + 1)$ unit matrix.

Example. For $p = 5$, S_{12} has generator matrix

$$
\begin{array}{ccc}
& \infty\ \ 0\ 1\ 2\ 3\ 4 \\[4pt]
\left[
\begin{array}{ccccc|cccccc}
1 & & & & & 0 & 1\ 1\ 1\ 1\ 1 \\
\hline
& 1 & & & & 1 & 0\ 1\ -\ -\ 1 \\
& & 1 & & & 1 & 1\ 0\ 1\ -\ - \\
& & & 1 & & 1 & -\ 1\ 0\ 1\ - \\
& & & & 1 & 1 & -\ -\ 1\ 0\ 1 \\
& & & & & 1 & 1\ 1\ -\ -\ 1\ 0 \\
\end{array}
\right]
\end{array}
\tag{61}
$$

Problems. (18) Show that S_{12} is equivalent to the Golay code $\mathcal{G}_{12}$.

(19) (i) Show that $SS^T = -I$ (over GF(3)).

(ii) If $p = 4k + 1$, $S = S^T = -S^{-1}$; if $p = 4k - 1$, $S = -S^T = S^{-1}$.

Theorem 17. S_{2p+2} is a self-dual code with all weights divisible by 3.

Proof. By Problem 19.   Q.E.D.

Examples. The first five symmetry codes have parameters

$$[12, 6, 6], [24, 12, 9], [36, 18, 12], [48, 24, 15], [60, 30, 18] \qquad (62)$$

S_{24}, S_{48} and S_{60} have the same Hamming weight enumerator as the extended QR codes with the same parameters (see Ch. 19), but are not equivalent to these codes (Problem 20). Note that these 5 codes have $d = \frac{1}{4}n + 3$, which is the largest it can be (see Ch. 19). Unfortunately the next code is an [84, 42] code for which it is known that $d \leq 21$.

Theorem 18. The automorphism group of S_{2p+2} contains the following monomial transformations: if a codeword $|L|R|$ is in S_{2p+2} so are

$$-|L|R|, \qquad |R| - \epsilon L|, \qquad |S(L)|S(R)|,$$
$$|V(L)|V(R)| \quad \text{and} \quad |T'(L)|T'(R)|,$$

where S, V, T' are as in Theorem 9. Hence Aut (S_{2p+2}) contains a subgroup isomorphic to PSL$_2(p)$.

> For $p \equiv -1 \pmod 6$, S_{2p+2} is a $[2p + 2, p + 1]$ double circulant, self-dual code over GF(3) with generator matrix $[I\,|\,S]$, S given by (60). Aut (S_{2p+2}) contains PSL$_2(p)$, and if $|L|R|$ is a codeword, so is $|R| - \epsilon L|$. (62) gives examples.
>
> Fig. 16.8. Properties of Pless symmetry codes.

Problems. (20) Use Theorems 12 and 13 to show that $\hat{\mathcal{D}}$ and S_{2p+2} are not equivalent for $2p + 2 = 24$, 48 and 60 [Hint: PSL$_2(p) \not\subset$ PSL$_2(2p + 1)$.]

(21) Show that any extended QR or symmetry code of length $n = 12m$ over GF(3) contains at least $2n$ codewords of weight n. In fact there is an equivalent code which contains the rows of a Hadamard matrix of order n and their negatives.

Research Problem (16.8). How does d grow with p in S_{2p+2}?

5-designs. When the results of Ch. 6, especially Theorem 9 and Corollary 30, are applied to the codes in Fig. 16.2 and (62), a number of 5-designs are obtained. The parameters are determined by the weight enumerators of these codes (derived in Ch. 19) and are shown in Fig. 16.9. For example the second line of the figure means that the codewords of weights 8, 12, 16 and 24 in the [24, 12, 8] extended binary Golay code form 5-designs, and the first of these is a 5-(24, 8, 1) design (in agreement with Theorems 24–26 of Ch. 2). See also (67) below.

(i) From QR codes

$[n, k, d]$	field	design from min. wt. words	these weights also give 5-designs
[12, 6, 6]	GF(3)	5-(12, 6, 1)	9
[24, 12, 8]	GF(2)	5-(24, 8, 1)	12, 16, 24
[24, 12, 9]	GF(3)	5-(24, 9, 6)	12, 15
[48, 24, 12]	GF(2)	5-(48, 12, 8)	16, 20, 24, ..., 36, 48
[48, 24, 15]	GF(3)	5-(48, 12, 364)	18, 21, 24, 27
[60, 30, 18]	GF(3)	5-(60, 18, 1530)	21, 24, 27, 30, 33

(ii) From symmetry codes

$[n, k, d]$	field	design from min. wt. words	these weights also give 5-designs
[24, 12, 9]	GF(3)	5-(24, 9, 6)	12, 15
[36, 18, 12]	GF(3)	5-(36, 12, 45)	15, 18, 21
[48, 24, 15]	GF(3)	5-(48, 12, 364)	18, 21, 24, 27
[60, 30, 18]	GF(3)	5-(60, 18, 1530)	21, 24, 27, 30, 33

Fig. 16.9. 5-Designs from codes.

§9. Decoding of cyclic codes and others

Encoding of cyclic codes was described in §8 of Ch. 7, and specific techniques for decoding BCH codes were given in §6 of Ch. 9, alternant codes in §9 of Ch. 12, and Reed–Muller codes in §§6, 7 of Ch. 13. In this section several decoding methods are described which may be applicable to cyclic or double circulant codes.

Let $\mathscr{C}$ be an $[n, k, d]$ binary code. Suppose the codeword $c = c_0 c_1 \cdots c_{n-1}$ is transmitted, an error vector $e = e_0 e_1 \cdots e_{n-1}$ occurs with weight t where $d \geqslant 2t + 1$, and the vector $y = c + e = y_0 y_1 \cdots y_{n-1}$ is received. For simplicity we assume that $\mathscr{C}$ has generator matrix $G = [A \,|\, I]$ and parity check matrix $H = [I \,|\, A^T]$. Thus $c_0 \cdots c_{r-1}$ are the $r = n - k$ check symbols in the codeword and $c_r \cdots c_{n-1}$ are the k information symbols. For example we might be using Encoder #1 of §8 of Ch. 7.

The following theorem is the basis for these decoding methods.

Theorem 19. *Suppose an error vector $e = e_0 e_1 \cdots e_{n-1}$ of weight t occurs, where $2t + 1 \leq d$. Let y be the received vector, with syndrome $S = Hy^T$. If wt $(S) \leq t$, then the information symbols $y_r \cdots y_{n-1}$ are correct, and $S = (e_0 \cdots e_{r-1})^T$ gives the errors. If wt $(S) > t$, then at least one information symbol $y_i (r \leq i \leq n - 1)$ is incorrect.*

Proof. (i) Suppose the information symbols are correct, i.e. $e_i = 0$ for $r \leq i \leq n - 1$. Then $S = [I \,|\, A^T] e^T = (e_0 \cdots e_{r-1})^T$ and wt $(S) \leq t$. (ii) Let $e_{(1)} = e_0 \cdots e_{r-1}$, $e_{(2)} = e_r \cdots e_{n-1}$, and suppose $e_{(2)} \neq 0$. Consider the codeword $c' = e_{(2)} G = e_{(2)}[A \,|\, I] = |e_{(2)} A | e_{(2)}|$. Since $c' \neq 0$, wt $(c') \geq 2t + 1$ by hypothesis. Therefore wt $(e_{(2)}) + $ wt $(e_{(2)} A) \geq 2t + 1$. Now $S = He^T = e_{(1)}^T + A^T e_{(2)}^T$, so wt $(S) \geq$ wt $(A^T e_{(2)}^T) -$ wt $(e_{(1)}^T) \geq 2t + 1 -$ wt $(\dot{e}_{(1)}) -$ wt $(e_{(2)}) \geq t + 1$.   (Q.E.D.

Decoding method I: permutation decoding. This method applies to codes with a fairly large automorphism group, for example cyclic codes or, even better, quadratic-residue codes.

Suppose we wish to correct all error vectors e of weight $\leq t$, for some fixed $t \leq [\frac{1}{2}(d - 1)]$. In order to carry out the decoding, we must find a set $P = \{\pi_1 = 1, \pi_2, \ldots, \pi_s\}$ of permutations which preserve $\mathscr{C}$ and which have the property that for any error vector e of weight $\leq t$, some $\pi_i \in P$ moves all the 1's in e out of the information places.

Example. Let $\mathscr{C}$ be the $[7, 4, 3]$ Hamming code, with $t = 1$. Then a suitable set of permutations is $P = \{1, S^3, S^6\}$, where S is the cyclic permutation, as shown by Fig. 16.10. E.g. if $e = 0001000$, S^6 moves the 1 out of the information places: $S^6 e = 0010000$.

Information
Places

$\overbrace{}$

$1 \cdot c = c_0 c_1 c_2$	$c_3 c_4 c_5 c_6$
$S^3 \cdot c = c_4 c_5 c_6$	$c_0 c_1 c_2 c_3$
$S^6 \cdot c = c_1 c_2 c_3$	$c_4 c_5 c_6 c_0$

Fig. 16.10. The three permutations for permutation decoding of $[7, 4, 3]$ code.

The decoding method is as follows. When y is received, each $\pi_i y$ and its syndrome $S^{(i)} = H(\pi_i y)^T$ in turn is computed, until an i is found for which wt $(S^{(i)}) \leq t$. Then from Theorem 19, the errors are all in the first r places of $\pi_i y$, and are given by $e_0 \cdots e_{r-1} = S^{(i)T}$. Therefore we decode y as

$$c = \pi_i^{-1}(\pi_i y + e_0 \cdots e_{r-1} 0 \cdots 0). \tag{63}$$

If wt $(S^{(i)}) > t$ for all i, we conclude that more than t errors have occurred. Very little work has been done on finding minimal sets P.

Error trapping (Rudolph and Mitchell [1133]). An example of permutation decoding in which only the cyclic permutation S is used. Thus $P = \{1, S, S^2, \ldots, S^{n-1}\}$. This method will correct all error vectors e which contain a circular run of at least k zeros, i.e. contain a string of at least k consecutive zeros if e is written in a circle.

Problem. (22) Show that all errors of weight $\leq t$ will be corrected, where $t = [(n-1)/k]$.

So this method is only useful for correcting error vectors with a big gap between the 1's – i.e. for correcting a small number of random errors, or a burst of errors. For example if applied to the [23, 12, 7] Golay code, $t = [\frac{22}{12}] = 1$, so some errors of weights 2 and 3 will not be corrected.

However, the method is very easy to implement. All we do is keep shifting y until the syndrome of some shift $S^i y$ has weight $\leq t$, and then decode using (63). Note from Problem 36(b) of Ch. 7 that if the division circuit of Encoder #1 is used to calculate the syndrome of y, it is easy to get the syndromes corresponding to cyclic shifts of y. These are simply the contents of the shift register at successive clock cycles.

Examples. (1) The [31, 21, 5] double-error-correcting BCH code. Here $[\frac{30}{21}] = 1$, so error trapping will not correct some double 'errors. However, suppose we are willing to use the permutation σ_2 which sends $y_0 y_1 \cdots y_{30}$ to $z_0 z_1 \cdots z_{30}$ where $z_{2i} = y_i$ (subscript mod 31). σ_2 preserves the code, as shown in §5 of Ch. 8. Also $\sigma_2^5 = 1$. It is not difficult to show that the set of 155 permutations $P = \{S^i \sigma_2^j, 0 \leq i \leq 30, 0 \leq j \leq 4\}$ will move any error pattern of weight 2 out of the information places, and hence can be used for permutation decoding.

(2) The [23, 12, 7] Golay code. Again error trapping can only correct 1 error. However the set of 92 permutations $P = \{S^i \sigma_2^j, 0 \leq i \leq 22, j = 0, 1, 2 \text{ or } 10\}$ will move any error pattern of weight ≤ 3 out of the information places (see MacWilliams [873]), and can be used for permutation decoding.

Neither of these sets of permutations is minimal.

Research Problem (16.9). Find a minimal set P of permutations for decoding (a) the Golay codes $\mathcal{G}_{23}$ and $\mathcal{G}_{24}$, (b) the [31, 21, 5] BCH code, (c) the [48, 24, 12] code $\hat{\mathcal{D}}$, etc.

Decoding method II: covering polynomials (Kasami [725]). This is a modification of error trapping which permits more errors to be corrected.

Suppose we wish to correct all error patterns of weight $\leq t$, where now t is any integer such that $2t + 1 \leq d$.

To do this we choose a set of polynomials

$$Q_0(x) = 0, Q_1(x), \ldots, Q_a(x),$$

called *covering polynomials*, with the following property. For any error vector $e(x)$ of weight $\leq t$, there is a $Q_i(x)$ such that some cyclic shift of $e(x)$ agrees with $Q_i(x)$ in the information symbols. For example, if $e(x)$ contains a circular run of k consecutive zeros, there is some cyclic shift of $e(x)$ which agrees with $Q_0(x) = 0$ in the information symbols.

Example. For the $[23, 12, 7]$ Golay code $\mathcal{G}_{23}$, with $t = 3$, we may choose $Q_0(x) = 0$, $Q_1(x) = x^{16}$, $Q_2(x) = x^{17}$. For it is easy to check that for any error vector $e(x)$ of weight ≤ 3, drawn as a wheel on the left of Fig. 16.11, we can turn the wheel so that it agrees with one of the three wheels on the right in the information places 11–22. E.g. if $e(x) = x^8 + x^{15} + x^{22}$, the cyclic shift $x^{17}e(x)$ has just one nonzero symbol in the information places, in x^{16}, and agrees with $Q_1(x)$.

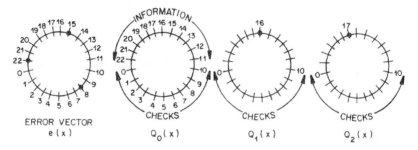

Fig. 16.11. Covering polynomials for Golay code.

Let $q_j = HQ_j^T$ be the syndrome of Q_j, $j = 0, \ldots, a$. The proof of the following theorem is parallel to that of Theorem 19.

Theorem 20. *Let $S = Hy^T$. The information symbols (i.e. the last k symbols) of $y(x) + Q_i(x)$ are correct iff* $\operatorname{wt}(S^T + q_i) \leq t - \operatorname{wt}(Q_i(x))$. *If this is the case then the errors in $y(x) + Q_i(x)$, which are in the first r symbols, are given by* $e_0 \cdots e_{r-1} = S^T + q_i$, *and the errors in $y(x)$ itself are given by* $Q_i + S^T + q_i$.

In view of this theorem we have the following decoding algorithm. Keep shifting $y(x)$ until some shift, $x^i y(x)$ say, has syndrome $S^{(i)}$ such that

$$\operatorname{wt}(S^{(i)} + q_j) \leq t - \operatorname{wt}(Q_j(x))$$

for some $j = 0, 1, \ldots, a$. Then decode $y(x)$ as

$$c(x) = x^{n-i}(x^i y(x) + Q_j^T + S^T + q_i). \tag{64}$$

Problem. (23) Show that $Q_0(x) = 0$, $Q_1(x) = x^{22}$ and $Q_2(x) = x^{23}$ are covering polynomials for the $[31, 16, 7]$ BCH code with $t = 3$.

Very similar decoding methods have been proposed by Karlin [716], Tzeng and Zimmerman [1354], and Berlekamp [123]. For example instead of shifting the received vector and adding the syndromes of the covering polynomials, one can add all vectors of weight $\leq t/2$ to the information places or the check places until a syndrome is found with weight $\leq t$.

As an illustration we give Berlekamp's decoding method for the $[24, 12, 8]$ Golay code $\mathscr{G}_{24}$, using the generator matrix (48), which we write as

$$G = \left[\begin{array}{c|c} I_{12} & \begin{array}{c|c} 1\ 1\ \cdots\ 1 & 0 \\ \hline & 1 \\ A & \vdots \\ & 1 \end{array} \end{array} \right] = [I_{12} | B]$$

$$= (u_1, \ldots, u_{12}, b_{13}, \ldots, b_{24}),$$

giving names to the columns. Thus u_i is a vector with a 1 in coordinate i. Since $\mathscr{G}_{24}$ is self-dual, $GG^T = 0$, hence $I + BB^T = 0$ or $B^{-1} = B^T$.

Also let

$$B = \begin{pmatrix} b_1^T \\ b_2^T \\ \vdots \\ b_{12}^T \end{pmatrix}$$

giving names to the rows.

Suppose the codeword $c = (c_1, \ldots, c_{24})$ is transmitted, the error vector $e = (e_1, \ldots, e_{24})$ occurs, and $y = c + e$ is received. The syndrome is

$$S = Hy^T = He^T$$

$$= \sum_{i=1}^{12} e_i u_i + \sum_{i=13}^{24} e_i b_i. \tag{65}$$

Also

$$B^T S = B^T He^T = [B^T | I] e^T$$

$$= \sum_{i=1}^{12} e_i b_i + \sum_{i=13}^{24} e_i u_{i-12}. \tag{66}$$

Suppose that wt $(e) \leq 3$. Then at least one of the following conditions must hold:

Case I: $\mathrm{wt}\,(e_{13}, \ldots, e_{24}) = 0$, $\mathrm{wt}\,(S) \leqslant 3, \displaystyle\sum_{i=1}^{12} e_i u_i = S.$

Case II: $\mathrm{wt}\,(e_1, \ldots, e_{12}) = 0$, $\mathrm{wt}\,(B^T S) \leqslant 3,$ $\displaystyle\sum_{i=13}^{24} e_i u_{i-12} = B^T S.$

Case III: $\mathrm{wt}\,(e_{13}, \ldots, e_{24}) = 1$,   so for some j, $13 \leqslant j \leqslant 24$,

$\mathrm{wt}\,(S + b_j) \leqslant 2$ and $\displaystyle\sum_{i=1}^{12} e_i u_i = S + b_j.$

Case IV: $\mathrm{wt}\,(e_1, \ldots, e_{12}) = 1$,   so for some j, $1 \leqslant j \leqslant 12$,

$\mathrm{wt}\,(B^T S + b_j) \leqslant 2$ and $\displaystyle\sum_{i=13}^{24} e_i u_{i-12} = B^T S + b_j.$

Thus the decoding can be done by computing the weights of the 26 vectors S, $S + b_j$ $(1 \leqslant j \leqslant 12)$, $B^T S$, $B^T S + b_j$ $(1 \leqslant j \leqslant 12)$.

Example. $S^T = 11100 \cdots 0$, $\mathrm{wt}\,(S) = 3$, so case I applies. $S^T = u_1 + u_2 + u_3$, so $e = 11100 \cdots 0$.

Decoding method III: using t-designs (Assmus et al. [33, 42, 43], Goethals [492, 493]). Threshold decoding (see §§6, 7 of Ch. 13) is often possible if the dual code contains a t-design. The idea is as follows. Suppose the codewords of some weight w in the dual code form a t-*design*, for some t. Let $h_1, \ldots, h_a$ be the blocks of this design (= codewords of weight w in $\mathscr{C}^\perp$) which have a 1 in the first coordinate. Form the a parity checks $y \cdot h_i$, where y is the received vector, and let ν be the number of times $y \cdot h_i = 1$.

The decoding algorithm we seek is of the following form:

(i) If $\nu = 0$, no error has occurred;

(ii) If $\nu < \theta_1$ or if ν belongs to some set of exceptional values Ψ_1, then the first coordinate is correct;

(iii) If $\nu \geqslant \theta_2$ or if $\nu \in \Psi_2$, then the first coordinate is incorrect;

(iv) For all other values of ν, decide that an uncorrectable error has occurred.

Repeat for all coordinates of the received vector.

We illustrate by giving a threshold decoding algorithm for the Golay code $\mathscr{G}_{24}$ (from [493]). Since $\mathscr{G}_{24}$ is self-dual, any codeword can be used as parity check. Let us see how to decode one coordinate, say the first. Take as parity checks the 253 codewords of weight 8 containing the first coordinate (from Fig. 2.14). If there is exactly one error, the number ν of parity checks which fail is either 253 or 77, depending on whether or not the first coordinate is in error. If two errors occur, ν is either 176 or $2 \cdot 56 = 112$ in the two cases. Finally if there are 3 errors, ν is either $120 + 21 = 141$ or $3 \cdot 40 + 5 = 125$ in the two cases. To summarize:

number of errors	ν if first coordinate in error	not in error
1	253	77
2	176	112
3	141	125

Therefore there is a simple threshold test: if more than 133 out of the 253 parity checks fail, the first coordinate is in error, if less than 133 fail this coordinate is correct.

This test can then be applied successively to all the coordinates (using the fact that Aut $(\mathcal{G}_{24})$ is transitive, Theorems 9 and 10).

A more complicated algorithm for the [48, 24, 12] QR code is given in [43].

It appears to be difficult to find a general threshold decoding algorithm of this type, or even to decide how many errors such an algorithm can correct. Rahman and Blake [1086] have given a lower bound on this number in terms of the parameters of the t-design.

Research Problem (16.10). If the codewords of each weight in $\mathcal{C}^{\perp}$ form a t-design, how many errors can be corrected by threshold decoding?

Of course if we allow arbitrary weights to be given to the parity checks we know from Theorem 21 of Ch. 13 that in theory $[\frac{1}{2}(d-1)]$ errors can be corrected for any code (but we don't know how to do it).

Notes on Chapter 16

Further properties of quadratic residues (cf. §3 of Ch. 2) General references are LeVeque [825, Vol. I] and Ribenboim [1111]. We begin by proving the fundamental theorem that 2 is a quadratic residue mod p if $p = 8m \pm 1$, and a nonresidue if $p = 8m \pm 3$. First some lemmas

Lemma 21. (Euler's criterion.) *For any integer a, $\chi(a) \equiv a^{(p-1)/2}$ mod p if p is odd.*

Proof. The result is obvious if a is a multiple of p, so assume $a \not\equiv 0$ mod p. Let ρ be a primitive element of GF(p); then $\rho^{(p-1)/2} \equiv -1$ mod p. Now $a \equiv \rho^i$ mod p for some i and so $a^{(p-1)/2} \equiv (-1)^i$ mod p. From §6 of Ch. 4, a is a quadratic residue iff i is even.   Q.E.D.

Lemma 22. (Gauss' criterion.) *Let p be an odd prime and let μ be the number*

of elements of the set $a, 2a, \ldots, \frac{1}{2}(p-1)a$ *whose numerically least residue* mod p *is negative. Then* $\chi(a) = (-1)^\mu$.

Example. $p = 7$, $a = 2$. The set $2, 4, 6$ becomes $2, -3, -1$, and $\mu = 2$.

Proof. Let $r_1, r_2, \ldots, -r'_1, -r'_2, \ldots$ be the values of the numerically least residues of $a, 2a, \ldots, \frac{1}{2}(p-1)a$. Clearly $r_i \neq r_j$, $r'_i \neq r'_j$ if $i \neq j$, and $r_i \neq r'_j$; for if $m_1 a \equiv r_i \bmod p$ and $m_2 a \equiv -r'_j \bmod p$ then $r_i = r'_j$ implies $(m_1 + m_2)a \equiv 0 \bmod p$, or $m_1 + m_2$ is divisible by p. This is impossible since $m_1 + m_2 \leqslant p - 1$. Hence the $\frac{1}{2}(p-1)$ numbers r_i, r'_i are distinct integers between 1 and $\frac{1}{2}(p-1)$, so are the numbers $1, 2, \ldots, \frac{1}{2}(p-1)$ in some order. Thus

$$a \cdot 2a \cdots \tfrac{1}{2}(p-1)a \equiv (-1)^\mu \left(\frac{p-1}{2} \right)! \bmod p$$

or

$$a^{(p-1)/2} \equiv (-1)^\mu \bmod p.$$

The result now follows from Lemma 21. Q.E.D.

Theorem 23. *2 is a quadratic residue* mod p *if* $p = 8m \pm 1$, *and a nonresidue if* $p = 8m \pm 3$.

Proof. Set $a = 2$ in Lemma 22. μ is the number of integers $2m(1 \leqslant m \leqslant \frac{1}{2}(p-1))$ such that $2m > \frac{1}{2}p$, or $m > p/4$. Thus $\mu = \frac{1}{2}(p-1) - [p/4]$. The four cases are

p	μ
$8m - 3$	$4m - 2 - (2m - 1) = 2m - 1$
$8m - 1$	$4m - 1 - (2m - 1) = 2m$
$8m + 1$	$4m - 2m = 2m$
$8m + 3$	$4m + 1 - 2m = 2m + 1$

Q.E.D.

Theorem 24. (Perron.) (i) *Suppose* $p = 4k - 1$. *Let* $r_1, \ldots, r_{2k}$ *be the $2k$ quadratic residues* mod p *together with 0, and let a be a number relatively prime to p. Then among the $2k$ numbers* $r_i + a$ *there are k residues (possibly including 0) and k nonresidues.*

(ii) *Suppose* $p = 4k - 1$. *Let* $n_1, \ldots, n_{2k-1}$ *be the $2k - 1$ nonresidues, and let a be prime to p. Then among the $2k - 1$ numbers* $n_i + a$ *there are k residues (possibly including 0) and $k - 1$ nonresidues.*

(iii) *Suppose* $p = 4k + 1$. *Among the $2k + 1$ numbers* $r_i + a$ *are, if a is itself a* *residue, $k + 1$ residues (including 0) and k nonresidues; and, if a is a nonresidue, k residues (not including 0) and $k + 1$ nonresidues.*

(iv) *Suppose $p = 4k + 1$. Among the $2k$ numbers $n_i + a$ are, if a is itself a residue, k residues (not including 0) and k nonresidues; and, if a is a nonresidue, $k + 1$ residues (including 0) and $k - 1$ nonresidues.*

For the proof see Perron [1035].

Problems. (24) Gauss' quadratic reciprocity law. Set

$$\left(\frac{a}{p}\right) = \begin{cases} 1 & \text{when } a \text{ is a quadratic residue mod } p \\ -1 & \text{when } a \text{ is a nonresidue mod } p. \end{cases}$$

If p, q are distinct odd primes, show that

$$\left(\frac{p}{q}\right)\left(\frac{q}{p}\right) = (-1)^{\frac{1}{2}(p-1)\cdot\frac{1}{2}(q-1)}.$$

(25) Show that 3 is a quadratic residue mod p iff $p \equiv \pm 1 \pmod{12}$.

§2. Quadratic-residue codes have been extensively studied by Assmus and Mattson in a long series of reports and papers [34], [37–39, 41–47, 49–53, 927] and a great deal of this chapter is based on their work. An alternative method of defining and generalizing QR codes has recently been given by Ward [1387]. His approach definitely merits further study.

It is possible to use (3) to define QR codes over any field GF(q), where q is not necessarily prime, in which case most of the properties given in this chapter still hold. QR codes over GF(4) were mentioned briefly in Problem 15. Since the cases $q = 2$ and 3 are the most important we have restricted ourselves to $q = l =$ prime. But there are some interesting QR codes over GF(4) – see Assmus and Mattson [42, 44]. One example is the [6, 3, 4] code over GF(4) given in Example E9 of Ch. 19, §1. The group of this code is three times the alternating group on 6 letters. With this additional exception, Theorem 13 still holds even if l is allowed to be a prime power. See also Stein and Bhargava [1274]. Another example is a [30, 15, 12] code over GF(4) – see [41]. The weight distribution of the latter is found in Ch. 19, and hence from Corollary 30 of Ch. 6 gives 5-designs with parameters

$$5\text{-}(30, 12, 220), \quad 5\text{-}(30, 14, 5390) \quad \text{and} \quad 5\text{-}(30, 16, 123000), \qquad (67)$$

which could be added to Fig. 16.9. For more about GF(4) codes see [1478].

The sources for Fig. 16.2 are Assmus and Mattson (references as above), Berlekamp [113, p. 360], Karlin [715, 717], Karlin and MacWilliams [718], Mykkeltveit et al. [979], and Rosenstark [1124]. See also Pless [1052], Solomon [1248].

Van Tilborg [1325] has obtained the following strengthening of Theorem 1.

Theorem 25. *If $\mathcal{Q}$ is a binary* QR *code of length* $p = 8m - 1$, *and* $d^2 - d + 1 = p$, *then* (i) $p = 64i^2 + 40i + 7 \geqslant 2551$ *and* $d = 8i + 3$, *for some i, and* (ii) *there exists a projective plane of order* $d - 1$. *If* $d^2 - d + 1 > p$, *then* $d^2 - d - 11 \geqslant p$.

This shows that the [47, 24] QR code has $d \geqslant 11$. Now $d \geqslant 15$ is impossible by the sphere-packing bound (Ch. 1), hence $d = 11$ or 12 by Theorem 8. But d must be odd by Corollary 15 of Ch. 8, so we conclude that $d = 11$.

Assmus, Mattson and Sachar [49] have generalized Theorem 1 as follows.

Theorem 26. *Suppose $\mathscr{C}$ is an* $[n, \frac{1}{2}(n + 1), d]$ *cyclic code over* GF(q) *such that* $\mathscr{C} \supseteq \mathscr{C}^{\perp}$. *If the supports of the minimum weight codewords form a 2-design then* $d^2 - d + 1 \geqslant n$.

§3. For Theorem 3 see Ribenboim [1111]. Theorem 4 is due to N. Patterson, unpublished.

§5. For properties of $\mathrm{PSL}_2(p)$ see for example Conway [306], Coxeter and Moser [314] or Huppert [676]. Theorems 10, 12, 13 are given by Assmus and Mattson [41, 47, 927]. See also Shaughnessy [1193, 1194]. We mention without giving any further details the following theorem of Assmus and Mattson [47, 53], and Shaughnessy [1193]:

Theorem 27. *Suppose* Aut $(\hat{\mathcal{Q}})\dagger$ *properly contains* $\mathrm{PSL}_2(p)$. *Then*

(i) Aut $(\mathcal{Q})\dagger$ *is isomorphic to a nonsolvable transitive permutation group on* p *letters.*

(ii) *If* $p = 2m + 1 \geqslant 23$, *and m is a prime, then* Aut $(\hat{\mathcal{Q}})\dagger$ *is simple and is isomorphic to a 5-fold transitive group.*

The proof of Theorem 13 is based on Theorem 27 and a computer search made by Parker and Nikolai [1023] for groups with certain properties.

Research Problem (16.11). Extend Parker and Nikolai's search to larger values of p.

Some further results about permutation groups which are applicable to QR codes are given in Neumann [988, 989].

It is worth mentioning that Kantor [714] has shown that $\mathrm{PSL}_2(p)$ is the full automorphism group of $(p + 1) \times (p + 1)$ Hadamard matrices of Paley type for all $p > 11$. (See also Assmus and Mattson [39], Hall [588, 590].) Problem 2 is from [47].

§§6, 7. Leech [803] seems to have been the first to obtain the generator matrix

(48) for $\mathcal{G}_{24}$. Karlin [715] studied the decomposition of QR codes into double circulants. For properties and tables of double circulant and quasi-cyclic codes see Bhargava et al. [142], Chen et al. [275], Hoffner and Reddy [659], Karlin [715], Kasami [730], Peterson and Weldon [1040, §8.14], Stein and Bhargava [1273], Tavares et al. [1311], and Townsend and Weldon [1338]. Another proof of Lemma 14 follows from Huppert [676, Theorem 8.4(d), p. 192].

The techniques for finding low weight codewords are based on Karlin [715], Karlin and MacWilliams [718], Mykkeltveit, Lam and McEliece [979]. Assmus and Mattson [44] describe another method using cyclotomy.

That 2 is a primitive element of infinitely many fields GF(k), k = prime, is known as Artin's conjecture; see Shanks [1186, p. 81].

For the double circulant codes $\mathcal{B}$ see Karlin [715, 717]. The entries in Fig. 16.7 are taken from [275, 715, 717, 1273, 1311]. The weight distributions of some of these codes are given in [1273] and [1311]. These references also give some good multiple circulant codes, with generator matrices of the form $[I|A|B|\cdots]$ where $A, B, \ldots$ are circulants. But even less is known about codes of this type. See also Bhargava and Stein [141]. For circulant and other matrices over finite fields see Bellman [100], Berlekamp [109], Carlitz [247–248], Daykin [339], MacWilliams [878, 881] and Muir [974].

§8. For QR codes over GF(3) see the papers of Assmus and Mattson. The Pless symmetry codes have been extensively studied by Pless [1055–1057, 1060]; see also Blake [160, 161]. Figure 16.9 is from Assmus and Mattson [41]. Theorems 17, 18 and Problem (20) are due to Pless [1055].

§9. Permutation decoding was introduced by Prange [1078], but our treatment follows MacWilliams [873]. General properties of cyclic codes which can be permutation decoded have been studied by Shiva et al. [1201, 1202], and Yip et al. [1447].

Berlekamp (unpublished) has found a set of 40 permutations which can be used to decode the Golay code $\mathcal{G}_{23}$.

Kasami [725] sketches circuits for implementing Decoding Method II, and gives covering polynomials for several other codes.

Research Problem (16.12). Give a systematic method of finding covering polynomials.

Berlekamp [123] also considers which error patterns of weight 4 can be corrected. Further papers on decoding cyclic and other codes are: Ahamed [8–11], Chase [264], Chien and Lum [291], Ehrich and Yau [405], Hartmann and Rudolph [614], Kautz [749], Mitchell [964], Miyakawa and Kaneko [966], Nili [994], Omura [1012, 1013], Pierce [1046], Solomon [1255], Tanaka and Kaneku [1299], Weldon [1404], Zierler [1462, 1463] and Zlotnikov and Kaiser [1473].

17

Bounds on the size of a code

§1. Introduction

Probably the most basic problem in coding theory is to find the largest code of a given length and minimum distance. Let

$$A(n, d) = \text{maximum number of codewords in any}$$
$$\text{binary code (linear or nonlinear) of length } n$$
$$\text{and minimum distance } d \text{ between codewords.}$$

In this chapter upper and lower bounds are given for $A(n, d)$, both for small n (see Figs. 1, 2 of Appendix A) and large n (see Fig. 17.7 below). For large n the best *upper* bound is the McEliece–Rodemich–Rumsey–Welch bound (Theorems 35 and 36), which is obtained via the important linear programming approach (see §4). This is a substantial improvement on the Elias bound, the old record-holder (Theorem 34). The best *lower* bound for large n is the Gilbert-Varshamov bound (Theorem 30), but there is still a considerable gap between the upper and lower bounds (see Research Problem 17.9).

One way of tackling $A(n, d)$ is by studying sets of codewords having constant weight. To this end we define

$$A(n, d, w) = \text{maximum number of binary vectors of}$$
$$\text{length } n, \text{ Hamming distance at least } d \text{ apart,}$$
$$\text{and constant weight } w.$$

Fig. 3 of Appendix A gives a table of small values of $A(n, d, w)$. The best upper bounds on $A(n, d, w)$ are the Johnson bounds (§2), which are used in §3 to obtain bounds on $A(n, d)$. The linear programming approach, which applies to both large and small values of n, and to both $A(n, d)$ and $A(n, d, w)$, is described in §4.

Another important function is

$B(n, d)$ = greatest number of codewords in any
linear binary code of length n and minimum distance d.

Of course $B(n, d) \leq A(n, d)$. An upper bound which applies specifically to $B(n, d)$ is the Griesmer bound, described in §5. An interesting technique for constructing linear codes which sometimes meet the Griesmer bound is presented in §6. The construction uses *anticodes*, which are codes which have an *upper* bound on the distance between codewords (and may contain repeated codewords). An excellent table of $B(n, d)$ covering the range $n \leq 127$ has been given by Helgert and Stinaff [636], so is not given here. For large n the best bounds known for $B(n, d)$ are the same as those for $A(n, d)$.

Instead of asking for the largest code of given length and minimum distance, one could look for the code of shortest length having M codewords and minimum distance d. Call this shortest length $N(M, d)$.

Problem. (1) Show that the solution to either problem gives the solution to the other. More precisely, show that if the values of $A(n, d)$ are known then so are the values of $N(M, d)$ and vice versa. State a similar result for linear codes.

Most of this chapter can be read independently of the others. The following papers dealing with bounds are not mentioned elsewhere in the chapter, but are included for completeness: Bambah et al. [64], Bartow [74, 75], Bassalygo [76], Berger [105], Chakravarti [259], Freiman [457], Hatcher [626], Joshi [700], Levenshtein [823, 824], Levy [830], MacDonald [868], McEliece and Rumsey [948], Myravaagnes [980], Peterson [1037], Sacks [1140], Sidelnikov [1209, 1210], Strait [1283] and Welch et al. [1399].

§2. Bounds on $A(n, d, w)$

We begin by studying $A(n, d, w)$, the maximum number of binary vectors of length n, distance at least d apart, and constant weight w. This is an important function in its own right, and gives rise to bounds on $A(n, d)$ via Theorems 13 and 33. In this section we give Johnson's bounds (Theorems 2–5) on $A(n, d, w)$, and then a few theorems giving the exact value in some special cases. A table of small values of $A(n, d, w)$ is given in Fig. 3 of Appendix A. (Another bound on $A(n, d, w)$, using linear programming, is described in §4.)

Theorem 1.

(a) $\qquad\qquad A(n, 2\delta - 1, w) = A(n, 2\delta, w).$
(b) $\qquad\qquad A(n, 2\delta, w) = A(n, 2\delta, n - w).$
(c) $\qquad\qquad A(n, 2\delta, w) = 1 \quad \text{if } w < \delta.$
(d) $\qquad\qquad A(n, 2\delta, \delta) = [\frac{n}{\delta}].$

Proof (a) follows because the distance between vectors of equal weight is even. To get (b) take complements. (c) is obvious. (d) follows because the codewords must have disjoint sets of 1's.   Q.E.D.

The next result generalizes (d):

Theorem 2. (Johnson [693]; see also [694–697].)

$$A(n, 2\delta, w) \leqslant \left[\frac{\delta n}{w^2 - wn + \delta n} \right] \qquad (1)$$

provided the denominator is positive.

Proof. Let $\mathscr{C}$ be an $(n, M, 2\delta)$ constant weight code which attains the bound $A(n, 2\delta, w)$; thus $M = A(n, 2\delta, w)$. Let $X = (x_{ij})$ be an $M \times n$ array of the codewords of $\mathscr{C}$; each row has weight w. Evaluate in two ways the sum of the inner products of the rows:

$$\sum_{i=1}^{M} \sum_{\substack{j=1 \\ j \neq i}}^{M} \sum_{\nu=1}^{n} x_{i\nu} x_{j\nu}.$$

Since the distance between any two rows is at least 2δ, their inner product is at most $w - \delta$. Hence the sum is $\leqslant (w - \delta)M(M - 1)$.

On the other hand, the sum is also equal to

$$\sum_{\nu=1}^{n} \sum_{i=1}^{M} \sum_{\substack{j=1 \\ j \neq i}}^{M} x_{i\nu} x_{j\nu}.$$

If k_ν is the number of 1's in the νth column of X, this column contributes $k_\nu(k_\nu - 1)$ to the sum. Hence

$$\sum_{\nu=1}^{n} k_\nu(k_\nu - 1) \leqslant (w - \delta)M(M - 1). \qquad (2)$$

But

$$\sum_{\nu=1}^{n} k_\nu = wM$$

(the total number of 1's in X), and $\Sigma_{\nu=1}^{n} k_\nu^2$ is minimized if all $k_\nu = wM/n$, in which case

$$\sum_{\nu=1}^{n} k_{\nu}^2 = \frac{w^2 M^2}{n}.$$

Therefore from (2)

$$\frac{w^2 M^2}{n} - wM \leq (w - \delta)M(M - 1).$$

Solving for M gives (1).   Q.E.D.

Since the k_{ν}'s must be integers this can be strengthened slightly.

Theorem 3. *Suppose* $A(n, 2\delta, w) = M$, *and define k and t by*

$$wM = nk + t, \quad 0 \leq t < n.$$

(This is the total number of 1's in all codewords.) Then

$$nk(k - 1) + 2kt \leq (w - \delta)M(M - 1). \tag{3}$$

Proof. The minimum of $\Sigma_{\nu=1}^{n} k_{\nu}^2$, subject to

$$\sum_{\nu=1}^{n} k_{\nu} = wM$$

and the k_{ν}'s being integers, is attained when $k_1 = \cdots = k_t = k + 1$, $k_{t+1} = \cdots = k_n = k$. This minimum is

$$t(k + 1)^2 + (n - t)k^2,$$

so from (2)

$$t(k + 1)^2 + (n - t)k^2 - (nk + t) \leq (w - \delta)M(M - 1).   \text{Q.E.D.}$$

Examples. From Theorem 2

$$A(9, 6, 4) \leq \left[\frac{27}{16 - 36 + 27}\right] = 3,$$

and the code $\{111100000, 100011100, 010010011\}$ shows that $A(9, 6, 4) = 3$. Again from Theorem 2,

$$A(8, 6, 4) \leq \left[\frac{24}{16 - 32 + 24}\right] = 3.$$

But if $A(8, 6, 4) = 3$, then applying Theorem 3 we find

$$3.4 = 1.8 + 4,$$

so $k = 1$, $t = 4$ and Equation (3) is violated. Hence $A(8, 6, 4) \leq 2$. The code $\{11110000, 10001110\}$ shows that $A(8, 6, 4) = 2$.

The following recurrence is useful in case Theorem 2 doesn't apply (and in some cases when it does).

Theorem 4.

$$A(n, 2\delta, w) \leq \left\lfloor \frac{n}{w} A(n - 1, 2\delta, w - 1) \right\rfloor. \tag{4}$$

Proof. Consider the codewords which have a 1 in the ith coordinate. If this coordinate is deleted, a code is obtained with length $n - 1$, distance $\geq 2\delta$, and constant weight $w - 1$. Therefore the number of such codewords is $\leq A(n - 1, 2\delta, w - 1)$. The total number of 1's in the original code thus satisfies

$$wA(n, 2\delta, w) \leq nA(n - 1, 2\delta, w - 1). \qquad \text{Q.E.D.}$$

Corollary 5.

$$A(n, 2\delta, w) \leq \left\lfloor \frac{n}{w} \left\lfloor \frac{n - 1}{w - 1} \cdots \left\lfloor \frac{n - w + \delta}{\delta} \right\rfloor \cdots \right\rfloor \right\rfloor. \tag{5}$$

Proof. Iterate Theorem 4 and then use Theorem 1(d). Q.E.D.

In practice Theorem 4 is applied repeatedly until a known value of $A(n, d, w)$ is reached.

Example.

$$A(20, 8, 7) \leq \left\lfloor \frac{20}{7} A(19, 8, 6) \right\rfloor \qquad \text{from (4)},$$

$$\leq \left\lfloor \frac{20}{7} \left\lfloor \frac{19}{6} A(18, 8, 5) \right\rfloor \right\rfloor \qquad \text{from (4)}.$$

Now Theorem 2 applies:

$$A(18, 8, 5) \leq \left\lfloor \frac{72}{25 - 90 + 72} \right\rfloor = \left\lfloor \frac{72}{7} \right\rfloor = 10.$$

But $A(18, 8, 5) = 10$ is impossible from Theorem 3, since $50 = 18 \cdot 2 + 14$ but $18 \cdot 2 + 2 \cdot 2 \cdot 14 > (5 - 4) \cdot 10 \cdot 9$. Hence $A(18, 8, 5) \leq 9$, and so

$$A(20, 8, 7) \leq \left\lfloor \frac{20}{7} \left\lfloor \frac{19 \cdot 9}{6} \right\rfloor \right\rfloor = \left\lfloor \frac{20 \cdot 28}{7} \right\rfloor = 80.$$

In fact equality holds in both of these (see Fig. 3 of Appendix A).

Problem. (2) Show

$$A(n, 2\delta, w) \leqslant \left[\frac{n}{n-w} A(n-1, 2\delta, w)\right].$$ (6)

[Hint: use Theorem 1(b).]

Sometimes (6) gives better results than (4).

Research Problem (17.1). Given n, 2δ, w, in which order should (4), (6) and (1) be applied to get the smallest bound on $A(n, 2\delta, w)$?

Problems. (3) Some easy values. Show

$$A(10, 6, 3) = 3, \quad A(10, 6, 4) = 5, \quad A(10, 6, 5) = 6,$$
$$A(11, 6, 4) = 6, \quad A(11, 6, 5) = 11,$$
$$A(12, 6, 4) = 9.$$

(4) Assuming a $4r \times 4r$ Hadamard matrix exists, show that

$$A(4r - 2, 2r, 2r - 1) = 2r,$$ (7)

$$A(4r - 1, 2r, 2r - 1) = 4r - 1,$$ (8)

$$A(4r, 2r, 2r) = 8r - 2.$$ (9)

[Hint: use the codes $\mathscr{A}_n$, $\mathscr{C}_n$, $\mathscr{A}'_n$ of page 49 of Ch. 2.]
(5) Show

$$A(n + 1, 2\delta, w) \leqslant A(n, 2\delta, w) + A(n, 2\delta, w - 1).$$

(6) Show that

$$A(n, 2\delta, w) \leqslant \frac{n(n - 1) \cdots (n - w + \delta)}{w(w - 1) \cdots \delta}$$ (10)

with equality iff a Steiner system $S(w - \delta + 1, w, n)$ exists (cf. §5 of Ch. 2).

Thus known facts about Steiner systems (see Chen [277], Collens [300], Dembowski [370], Denniston [372], Doyen and Rosa [386], Hanani [595–600], Di Paola et al. [1020], Witt [1423, 1424]) imply corresponding results about $A(n, d, w)$. For example, the Steiner systems $S(3, q + 1, q^2 + 1)$ exist for any prime power q [370, Ch. 6], and so

$$A(q^2 + 1, 2q - 2, q + 1) = q(q^2 + 1), \quad q = \text{prime power}.$$ (11)

Similarly, from *unitals* [370, p. 104],

$$A(q^3 + 1, 2q, q + 1) = q^2(q^2 - q + 1), \quad q = \text{prime power}.$$ (12)

Problem 6 also shows that the determination of $A(n, 2\delta, w)$ is in general a very difficult problem. For example $A(111, 20, 11) \leqslant 111$ from (10), with equality iff there exists a projective plane of order 10. The following results are quoted without proof.

Theorem 6. (Schönheim [1158]; see also Fort and Hedlund [446], Schönheim [1161], Spencer [1259] and Swift [1294].)

$$A(n, 4, 3) = \begin{cases} \left[\dfrac{n}{3}\left[\dfrac{n-1}{2}\right]\right] & for\ n \not\equiv 5 (\mathrm{mod}\ 6), \\[2em] \left[\dfrac{n}{3}\left[\dfrac{n-1}{2}\right]\right] - 1 & for\ n \equiv 5 (\mathrm{mod}\ 6). \end{cases}$$

Theorem 7. (Kalbfleisch and Stanton [708], Brouwer [201a].)

$$A(n, 4, 4) = \begin{cases} \dfrac{n(n-1)(n-2)}{4 \cdot 3 \cdot 2} & if\ n \equiv 2\ or\ 4 (\mathrm{mod}\ 6), \\[1.5em] \dfrac{n(n-1)(n-3)}{4 \cdot 3 \cdot 2} & if\ n \equiv 1\ or\ 3 (\mathrm{mod}\ 6). \\[1.5em] \dfrac{n(n^2 - 3n - 6)}{4 \cdot 3 \cdot 2} & if\ n \equiv 0 (\mathrm{mod}\ 6) \end{cases}$$

Research Problem (17.2). Find $A(n, 4, 4)$ in the remaining cases of Theorem 7.

Theorem 8. (Hanani [595–598].)

$$A(n, 6, 4) = \frac{n(n-1)}{4 \cdot 3} \quad \textit{iff}\ n \equiv 1\ or\ 4 (\mathrm{mod}\ 12),$$

$$A(n, 8, 5) = \frac{n(n-1)}{5 \cdot 4} \quad \textit{iff}\ n \equiv 1\ or\ 5 (\mathrm{mod}\ 20).$$

The proofs of Theorems 6–8 give explicit constructions for codes attaining these bounds.

Theorem 9. (Erdös and Hanani [411], Wilson [1420–1422].)
   (a) *For each fixed* $k \geqslant 2$, *there is an* $n_o(k)$ *such that for all* $n > n_o(k)$,

$$A(n, 2k - 2, k) = \frac{n(n-1)}{k(k-1)} \quad \textit{iff}\ n \equiv 1\ or\ k (\mathrm{mod}\ k(k-1)).$$

Also

$$\lim_{n \to \infty} \frac{k(k-1)}{n(n-1)} A(n, 2k-2, k) = 1.$$

(b) *If p is a prime power then*

$$\lim_{n \to \infty} \frac{(p+1)p(p-1)}{n(n-1)(n-2)} A(n, 2p-2, p+1) = 1.$$

(See also the papers of Alltop [21–24] and Rokowska [1122].)

But even when none of the above theorems apply, there is still a good method for getting lower bounds on $A(n, 2\delta, w)$: we find a code of length n and minimum distance 2δ, and use the set of all codewords of weight w. E.g. using the 2576 codewords of weight 12 in the [24, 12, 8] Golay code $\mathcal{G}_{24}$ shows $A(24, 8, 12) \geq 2576$. (Actually equality holds.)

More generally, one can use the vectors of weight w in any *coset* of a code of length n and distance 2δ. E.g. using the cosets of $\mathcal{G}_{24}$ given in Problem 13 of Ch. 2 shows $A(24, 8, 6) \geq 77$, $A(24, 8, 10) \geq 960$, etc.

Constant weight codes have a number of practical applications – see Cover [310], Freiman [456], Gilbert [483], Hsiao [666], Kautz and Elspas [752], Maracle and Wolverton [911], Nakamura et al. [983], Neumann [986, 987], Schwartz [1169] and Tang and Liu [1307]. But little is known in general about:

Research Problem (17.3). Find good methods of encoding and decoding constant weight codes (cf. Cover [310]).

Problem. (7) Equations (4) and (1) imply $A(12, 6, 5) \leq [12A(11, 6, 4)/5] = 14$. Show that $A(12, 6, 5) = 12$.
[Hint. (i) By trial and error find an array showing that $A(12, 6, 5) \geq 12$.

(ii) In any array realizing $A(12, 6, 5)$, the number of 1's in a column is $\leq A(11, 6, 4) = 6$.

(iii) The number of pairs of 1's in any two columns is $\leq A(10, 6, 3) = 3$.

(iv) Suppose columns 2, 3 contain 3 11's. Without loss of generality the first 3 rows are

```
0 1 1 1 1 1 0 0 0 0 0 0
0 1 1 0 0 0 1 1 1 0 0 0
0 1 1 0 0 0 0 0 0 1 1 1
```

Let b, c, d, e be the number of rows beginning 110, 101, 100, 000 respectively. Then $b + c + d + e \geq 9$, $b \leq 3$, $c \leq 3$, $b + c + d \leq 6$ hence $e \geq 3$. But $e \leq A(9, 6, 5) = 3$, hence $e = 3$ and there are ≤ 12 rows.

(v) Suppose no pair of columns contain 3 11's. Then the column sums are ≤ 5

and again there are ≤ 12 rows. For if column 1 has weight ≥ 6, look at the rows with 1 in column 1, and show that some column has inner product ≥ 3 with column 1.]

Problem. (8) (Hard.) Show that $A(13, 6, 5) = 18$.

Some other papers dealing with $A(n, d, w)$ are Niven [996] and Stanton et al. [1270].

§3. Bounds on $A(n, d)$

The bounds on $A(n, d, w)$ obtained in §2 can now be used to bound $A(n, d)$. First we quote some results established in §§2, 3 of Ch. 2. Theorem 11 gives $A(n, d)$ exactly if $n \leq 2d$.

Theorem 10.
(a) $$A(n, 2\delta) = A(n - 1, 2\delta - 1)$$
(b) $$A(n, d) \leq 2A(n - 1, d)$$

Theorem 11. (Plotkin and Levenshtein.) *Provided suitable Hadamard matrices exist*:
(a)
$$A(n, 2\delta) = 2 \left[\frac{2\delta}{4\delta - n} \right] \quad \textit{if } 4\delta > n \geq 2\delta,$$
$$A(4\delta, 2\delta) = 8\delta.$$

(b)
$$A(n, 2\delta + 1) = 2 \left[\frac{2\delta + 2}{4\delta + 3 - n} \right] \quad \textit{if } 4\delta + 3 > n \geq 2\delta + 1,$$
$$A(4\delta + 3, 2\delta + 1) = 8\delta + 8.$$

The following result, proved in Theorem 6 of Ch. 1, is useful if d is small.

Theorem 12. (The sphere-packing or Hamming bound.)
$$A(n, 2\delta + 1) \left(1 + \binom{n}{1} + \cdots + \binom{n}{\delta} \right) \leq 2^n. \tag{13}$$

This theorem can be strengthened by introducing the function $A(n, d, w)$.

Theorem 13. (Johnson.)

$$A(n, 2\delta + 1)\left\{1 + \binom{n}{1} + \cdots + \binom{n}{\delta} + \frac{\binom{n}{\delta+1} - \binom{2\delta+1}{\delta} A(n, 2\delta+2, 2\delta+1)}{\left[\frac{n}{\delta+1}\right]}\right\} \leq 2^n.$$

(14)

Proof. (i) Let $\mathscr{C}$ be an (n, M, d) code with $M = A(n, d)$, $d = 2\delta + 1$, which contains the 0 vector. Let S_i be the set of vectors at distance i from $\mathscr{C}$, i.e.

$$S_i = \{u \in F^n: \text{dist}(u, v) \geq i \text{ for all } v \in \mathscr{C}, \text{ and}$$

$$\text{dist}(u, v) = i \text{ for some } v \in \mathscr{C}\}.$$

Thus $S_0 = \mathscr{C}$. Then

$$S_0 \cup S_1 \cup \cdots \cup S_{d-1} = F^n,$$

for if there were a vector at distance $\geq d$ from $\mathscr{C}$ we could add it to $\mathscr{C}$ and get a larger code. The sphere-packing bound (Theorem 12) follows because $S_0, \ldots, S_\delta$ are disjoint. To obtain (14) we estimate $S_{\delta+1}$.

(ii) Pick an arbitrary codeword P and move it to the origin. The codewords of weight $d = 2\delta + 1$ form a constant weight code with distance $\geq 2\delta + 2$, i.e. the number of codewords of weight d is $\leq A(n, 2\delta + 2, 2\delta + 1)$.

(iii) Let $W_{\delta+1}$ be the set of vectors in F^n of weight $\delta + 1$. Any vector in $W_{\delta+1}$ belongs to either S_δ or $S_{\delta+1}$. Corresponding to each codeword Q of weight d there are $\binom{d}{\delta}$ vectors of weight $\delta + 1$ at distance δ from Q. These vectors are in $W_{\delta+1} \cap S_\delta$, and are all distinct. Therefore

$$|W_{\delta+1} \cap S_{\delta+1}| = |W_{\delta+1}| - |W_{\delta+1} \cap S_\delta|$$

$$\geq \binom{n}{\delta+1} - \binom{d}{\delta} A(n, 2\delta + 2, 2\delta + 1).$$

(iv) A vector R in $W_{\delta+1} \cap S_{\delta+1}$ is at distance $\delta + 1$ from at most $[n/(\delta + 1)]$ codewords. For move the origin to R and consider how many codewords can be at distance $\delta + 1$ from R and have mutual distance d (really $d + 1 = 2\delta + 2$). Such codewords must have disjoint sets of 1's, so their number is $\leq [n/(\delta + 1)]$.

(v) Now let P vary over all the codewords, and count the points in $S_0 \cup \cdots \cup S_{\delta+1}$; (14) follows.   Q.E.D.

Example.

$$A(12, 5) \leq \frac{2^{12}}{1 + \binom{12}{1} + \binom{12}{2} + \{\binom{12}{3} - 10A(12, 6, 5)\}/4}$$

But $A(12, 6, 5) = 12$ from Problem 7. Therefore $A(12, 5) \leq 39$. For comparison the sphere-packing bound only gives $A(12, 5) \leq 51$. We shall see in the next section that in fact $A(12, 5) = 32$.

Corollary 14.

$$A(n, 2\delta + 1) \left\{ 1 + \binom{n}{1} + \cdots + \binom{n}{\delta} + \frac{\binom{n}{\delta}\left(\frac{n - \delta}{\delta + 1} - \left[\frac{n - \delta}{\delta + 1}\right]\right)}{\left[\frac{n}{\delta + 1}\right]} \right\} \leq 2^n. \quad (15)$$

Proof. From Corollary 5,

$$A(n, 2\delta + 2, 2\delta + 1) \leq \frac{n(n - 1) \cdots (n - \delta + 1)}{(2\delta + 1)2\delta \cdots (\delta + 2)} \left[\frac{n - \delta}{\delta + 1}\right].$$

Substituting this in (14) gives (15).   Q.E.D.

If equality holds in (13) the code is called perfect (see Ch. 6). A code for which equality holds in (15) is called *nearly perfect* (Goethals & Snover [504]).

Problems. (9) (a) Show that a perfect code is nearly perfect.

(b) Show that a nearly perfect code is quasi-perfect.

(10) Show that the $[2^r - 2, 2^r - r - 2, 3]$ shortened Hamming code, and the punctured Preparata code $\mathcal{P}(m)^*$ (p. 471 of Ch. 15) are nearly perfect. Hence $A(2^r - 2, 3) = 2^{2r-r-2}$ and $A(2^m - 1, 5) = 2^{2^m-2m}$ for even $m \geq 4$. This shows that the Preparata code has the greatest possible number of codewords for this length and distance.

It has recently been shown by K. Lindström [844] that there are no other nearly perfect binary codes. The proof is very similar to that of Theorem 33 of Ch. 6.

Problems. (11) For even n show that $A(n, 3) \leq 2^n/(n + 2)$ [Hint: Combine Theorems 6 and 13. For $n \equiv 4 \pmod 6$ one can say a bit more.]

(12) (Johnson.) Prove the following refinement of (14):

$$A(n, 2\delta + 1) \left\{ 1 + \binom{n}{1} + \cdots + \binom{n}{\delta} + \frac{C_{\delta+1}}{A(n, 2\delta + 2, \delta + 1)} + \frac{C_{\delta+2}}{A(n, 2\delta + 2, \delta + 2)} \right\} \leq 2^n,$$

$$(16)$$

where

$$C_{\delta+1} = \binom{n}{\delta+1} - \binom{2\delta+1}{\delta} A(n, 2\delta+2, 2\delta+1),$$

$$C_{\delta+2} = \binom{n}{\delta+2} - \binom{2\delta+1}{\delta-1} A(n, 2\delta+2, 2\delta+1)$$

$$- \binom{2\delta+1}{\delta}\binom{n-2\delta-1}{1} A(n, 2\delta+2, 2\delta+1)$$

$$- \binom{2\delta+2}{\delta} A(n, 2\delta+2, 2\delta+2)$$

$$- \binom{2\delta+3}{\delta+1} A(n, 2\delta+2, 2\delta+3).$$

[Hint: Any vector at distance $\delta+2$ from a codeword belongs to one of $S_{\delta-1}$, S_δ, $S_{\delta+1}$ or $S_{\delta+2}$.]

(13) Generalize (16).

(14) Show that if $n \leq 2d$ then $A(n, d)$ is either 1 or an even number, provided suitable Hadamard matrices exist. [Hint: Theorem 11.] Use Fig. 1 of Appendix A to show that if $A(n, d)$ is odd and greater than 1 then $A(n, d) \geq 37$.

Research Problem (17.4). Is it true that if $A(n, d) > 1$ then $A(n, d)$ is even, for all n and d? (Elspas [409].)

Just as the bound for $A(n, 2\delta+1)$ given in (14) depends on $A(n, 2\delta+2, 2\delta+1)$, so it is possible to give bounds for $A(n, d, w)$ which depend on a function denoted by $T(w_1, n_1, w_2, n_2, d)$. These bounds are rather complicated and we do not give them here (see [697]). But the function T is of independent interest.

Definition. $T(w_1, n_1, w_2, n_2, d)$ is the maximum number of binary vectors of length $n_1 + n_2$, having Hamming distance at least d apart, where each vector has exactly w_1 1's in the first n_1 coordinates and exactly w_2 1's in the last n_2 coordinates.

Example. $T(1, 3, 2, 4, 6) = 2$, as illustrated by the vectors

$$\begin{array}{ll} 1\ 0\ 0 & 1\ 1\ 0\ 0 \\ 0\ 1\ 0 & 0\ 0\ 1\ 1 \end{array}$$

(the 1's must be disjoint!).

Problem. (15) Prove:

(a) $T(w_1, n_1, w_2, n_2, 2\delta) = T(n_1 - w_1, n_1, w_2, n_2, 2\delta)$.

(b) If $2w_1 + 2w_2 = 2\delta$ then

$$T(w_1, n_1, w_2, n_2, 2\delta) = \min\left\{\left[\frac{n_1}{w_1}\right], \left[\frac{n_2}{w_2}\right]\right\}.$$

(c) $T(w_1, n_1, w_2, n_2, 2\delta) \leq \left[\frac{n_1}{w_1} T(w_1 - 1, n_1 - 1, w_2, n_2, 2\delta)\right]$.

(d) $T(w_1, n_1, w_2, n_2, 2\delta) \leq \left[\frac{n_1}{n_1 - w_1} T(w_1, n_1 - 1, w_2, n_2, 2\delta)\right]$.

(e) $T(w_1, n_1, w_2, n_2, 2\delta) \leq \left[\dfrac{\delta}{\dfrac{w_1^2}{n_1} + \dfrac{w_2^2}{n_2} + \delta - w_1 - w_2}\right]$.

provided the denominator is positive.

(f) $T(0, n_1, w_2, n_2, 2\delta) = A(n_2, 2\delta, w_2)$.

(g) $T(w_1, n_1, w_2, n_2, 2\delta) \leq A(n_2, 2\delta - 2w_1, w_2)$.

(h) (A generalization of Theorem 3). Suppose $T(w_1, n_1, w_2, n_2, 2\delta) = M$, and for $i = 1, 2$ write $Mw_i = q_i n_i + r_i$, $0 \leq r_i < n_i$. Then

$$M^2(w_1 + w_2) \geq r_1(q_1 + 1)^2 + (n_1 - r_1)q_1^2$$
$$+ r_2(q_2 + 1)^2 + (n_2 - r_2)q_2^2 + \delta M(M - 1).$$

(i)

$$T(1, 9, 4, 9, 8) = 3, \qquad T(1, 9, 4, 9, 10) = 2,$$
$$T(1, 9, 5, 13, 10) = 3, \qquad T(1, 9, 6, 14, 10) = 7,$$
$$T(1, 8, 7, 15, 10) \leq 11, \qquad T(1, 9, 7, 15, 10) \leq 12.$$

Tables of $T(w_1, n_1, w_2, n_2, d)$ are given by Best et al. [140].

§4. Linear programming bounds

It is sometimes possible to use linear programming to obtain excellent bounds on the size of a code with a given distance distribution. This section begins with a brief general treatment of linear programming. Then the applications are given, first to arbitrary binary codes, and then to constant weight codes. Asymptotic results, valid when n is large, are postponed to §7.

Linear programming (see for example Simonnard [1211] or Ficken [428]). This is a technique for maximizing (or minimizing) a linear form, called the *objective function*, subject to certain linear constraints. A typical problem is the following:

Problem. (I) *The primal linear programming problem.* Choose the real variables $x_1, \ldots, x_s$ so as to maximize the objective function

$$\sum_{j=1}^{s} c_j x_j \tag{17}$$

subject to the inequalities

$$x_j \geq 0, \quad j = 1, \ldots, s, \tag{18}$$

$$\sum_{j=1}^{s} a_{ij} x_j \geq -b_i, \quad i = 1, \ldots, n. \tag{19}$$

Associated with this is the dual problem, which has as many variables as the primal problem has constraints, and as many constraints as there are variables in the primal problem.

Problem. (II) *The dual linear programming problem.* Choose the real variables $u_1, \ldots, u_n$ so as to

$$\text{minimize } \sum_{i=1}^{n} u_i b_i \tag{20}$$

subject to the inequalities

$$u_i \geq 0, \quad i = 1, \ldots, n, \tag{21}$$

$$\sum_{i=1}^{n} u_i a_{ij} \leq -c_j, \quad j = 1, \ldots, s. \tag{22}$$

It is convenient to restate these in matrix notation.

    (I)′ Maximize cx^T subject to $x \geq 0$, $Ax^T \geq -b^T$,

    (II)′ Minimize ub^T subject to $u \geq 0$, $uA \leq -c$.

A vector x is called a *feasible solution* to (I) or (I)′ if it satisfies the inequalities, and an *optimal solution* if it also maximizes cx^T. Similarly for (II). The next three theorems give the basic facts that we shall need.

Theorem 15. *If x and u are feasible solutions to* (I) *and* (II) *respectively then*

$$cx^T \leq ub^T.$$

Proof.

$$\sum_{j=1}^{s} a_{ij} x_j \geq -b_i \quad \text{and} \quad u_i \geq 0$$

together imply

$$u_i \sum_{j=1}^{s} a_{ij} x_j \geq -u_i b_i \quad \text{or} \quad uAx^T \geq -ub^T.$$

Similarly

$$\sum_{i=1}^{n} u_i a_{ij} \leq - c_j \quad \text{and} \quad x_j \geq 0$$

imply $uAx^T \leq - cx^T$. Then $cx^T \leq - uAx^T \leq ub^T$.   Q.E.D.

Theorem 16. (The duality theorem.) *Let x and u be feasible solutions to* (I) *and* (II) *respectively. Then x and u are both optimal iff* $cx^T = ub^T$.

Proof. ($\Leftarrow$) Suppose $cx^T = ub^T$ but x is not optimal. Then there is a feasible solution y to (I) such that $cy^T > cx^T = ub^T$, contradicting the previous theorem. The second half of the proof is omitted (see [1211]).   Q.E.D.

Theorem 17. (The theorem of complementary slackness.) *Let x and u be feasible solutions to* (I) *and* (II) *respectively. Then x and u are both optimal iff for every* $j = 1, \ldots, s$, *either*

$$x_j = 0 \quad or \quad \sum_{i=1}^{n} u_i a_{ij} = - c_j;$$

and for every $i = 1, \ldots, n$, *either*

$$u_i = 0 \quad or \quad \sum_{j=1}^{s} a_{ij} x_j = - b_i.$$

In words, if a primal constraint is not met with equality, the corresponding dual variable must be zero, and vice versa.

The proof may be left to the reader.

Applications to codes (Delsarte [350–352]). Let $\mathscr{C}$ be an (n, M, d) binary code in which the distances between codewords are $\tau_0 = 0 < \tau_1 < \tau_2 < \cdots < \tau_s$. Let $\{B_i\}$ be the distance distribution of $\mathscr{C}$, i.e. B_i is the average number of codewords at distance i from a fixed codeword (§1 of Ch. 2). Thus $B_0 = 1$, $B_{\tau_j} \geq 0$ $(j = 1, \ldots, s)$ and $B_i = 0$ otherwise. Also $M = 1 + \sum_{j=1}^{s} B_{\tau_j}$. The transformed distribution $\{B_k'\}$ is given by (Theorem 5 of Ch. 5)

$$B_k' = \frac{1}{M} \sum_{i=0}^{n} B_i P_k(i)$$

$$= \frac{1}{M} \sum_{j=0}^{s} B_{\tau_j} P_k(\tau_j), \quad k = 0, \ldots, n, \tag{23}$$

where $P_k(x)$ is a Krawtchouk polynomial (§2 of Ch. 5). Also $B_0' = 1$ and $\sum_{k=0}^{n} B_k' = 2^n / M$. We now make good use of Theorem 6 of Ch. 5 (or Theorem

12 of Ch. 21), which states that

$$B'_k \geqslant 0 \quad \text{for} \quad k = 0, \ldots, n. \tag{24}$$

Thus if $\mathscr{C}$ is *any* code with distances $\{\tau_j\}$, $j = 0, \ldots, s$, between codewords, then $B_{\tau_1}, \ldots, B_{\tau_s}$ is a feasible solution to the following linear programming problem.

Problem. (III) Choose $B_{\tau_1}, \ldots, B_{\tau_s}$ so as to

$$\text{maximize} \sum_{j=1}^{s} B_{\tau_j}$$

subject to

$$B_{\tau_j} \geqslant 0, \quad j = 1, \ldots, s, \tag{25}$$

$$\sum_{j=1}^{s} B_{\tau_j} P_k(\tau_j) \geqslant -\binom{n}{k}, \quad k = 1, \ldots, n. \tag{26}$$

Therefore we have the following theorem.

Theorem 18. (The first version of the *linear programming bound* for codes.) *If* $B_{\tau_1}^*, \ldots, B_{\tau_s}^*$ *is an optimal solution to* (III), *then* $1 + \Sigma_{i=1}^s B_{\tau_i}^*$ *is an upper bound on the size of* $\mathscr{C}$.

Note that (III) certainly has a feasible solution: $B_{\tau_i} = 0$ for all i.

It is often possible to invent other constraints that the B_{τ_j}'s must satisfy, besides (25) and (26). We illustrate by an example.

Example. *The Nadler code is optimal.* Let us find the largest possible double-error-correcting code of length 12. Adding an overall parity check to such a code gives a code $\mathscr{C}$ of length 13, minimum distance 6, in which the only nonzero distances are 6, 8, 10 and 12; suppose $\mathscr{C}$ has M codewords. Let $A_i(u)$ be the number of codewords in $\mathscr{C}$ at distance i from the codeword $u \in \mathscr{C}$. The distance distribution of $\mathscr{C}$ is

$$B_i = \frac{1}{M} \sum_{u \in \mathscr{C}} A_i(u), \quad i = 0, \ldots, 13.$$

Then $B_0 = 1$, and the B_i's are zero except (possibly) for B_6, B_8, B_{10} and B_{12}.

The inequalities (26) become

$$B_6 - 3B_8 - 7B_{10} - 11B_{12} \geq -13,$$

$$-6B_6 - 2B_8 + 18B_{10} + 54B_{12} \geq -\binom{13}{2},$$

$$-6B_6 + 14B_8 - 14B_{10} - 154B_{12} \geq -\binom{13}{3}, \qquad (27)$$

$$15B_6 - 5B_8 - 25B_{10} + 275B_{12} \geq -\binom{13}{4},$$

$$15B_6 - 25B_8 + 63B_{10} - 297\,B_{12} \geq -\binom{13}{5},$$

$$-20B_6 + 20B_8 - 36B_{10} + 132B_{12} \geq -\binom{13}{6}.$$

(There are only 6 distinct inequalities – see Problem 43 of Ch. 5.)

Another inequality may be found. Clearly $A_{12}(u) \leq 1$ (for taking $u = 0$ we see that the number of $v \in \mathscr{C}$ of weight 12 is at most 1). Also

$$A_{10}(u) \leq A(13, 6, 10)$$

$$= A(13, 6, 3) \quad \text{by Theorem 1,}$$

$$= 4 \qquad \qquad \text{from Fig. 3 of Appendix A.}$$

(It is easy to prove this directly.) Finally, if $A_{12}(u) = 1$ then $A_{10}(u) = 0$. So certainly

$$A_{10}(u) + 4A_{12}(u) \leq 4 \quad \text{for all } u \in \mathscr{C}.$$

Summing on $u \in \mathscr{C}$ gives the new inequality

$$B_{10} + 4B_{12} \leq 4. \qquad (28)$$

Actually (28) and the first two constraints of (27) turn out to be enough, and so we consider the problem:

$$\text{Maximize } B_6 + B_8 + B_{10} + B_{12}$$

subject to

$$B_6 \geq 0, \qquad B_8 \geq 0, \qquad B_{10} \geq 0, \qquad B_{12} \geq 0$$

and

$$B_6 - 3B_8 - 7B_{10} - 11B_{12} \geq -13,$$
$$-6B_6 - 2B_8 + 18B_{10} + 54B_{12} \geq -78, \qquad (29)$$
$$- B_{10} - 4B_{12} \geq -4.$$

The dual problem is:

$$\text{Minimize } 13u_1 + 78u_2 + 4u_3$$

subject to

$$u_1 \geq 0, \qquad u_2 \geq 0, \qquad u_3 \geq 0$$

and

$$
\begin{aligned}
u_1 - 6u_2 & \leq -1, \\
-3u_1 - 2u_2 & \leq -1, \\
-7u_1 + 18u_2 - u_3 & \leq -1, \\
-11u_1 + 54u_2 - 4u_3 & \leq -1.
\end{aligned}
\tag{30}
$$

Feasible solutions to these two problems are

$$B_6 = 24, \qquad B_8 = 3, \qquad B_{10} = 4, \qquad B_{12} = 0, \tag{31}$$

$$u_1 = u_2 = \tfrac{1}{3}, \qquad u_3 = \tfrac{16}{3}. \tag{32}$$

In fact since the corresponding objective functions are equal:

$$24 + 3 + 4 + 0 = 13 \cdot \tfrac{1}{3} + 78 \cdot \tfrac{1}{3} + 4 \cdot \tfrac{16}{3} = 31,$$

it follows from Theorem 16 that (31) and (32) are optimal solutions. (These solutions are easily obtained by hand using the simplex method – see [428] or [1211].)

The following argument shows that (31) is the unique optimal solution. Let x_6, x_8, x_{10}, x_{12} be any optimal solution to the primal problem. The u_i of (32) are all positive and satisfy the first three constraints of (30) with equality but not the fourth. Hence from Theorem 17, the x_i must satisfy the primal constraints (29) with equality, and $x_{12} = 0$. These equations have the unique solution

$$x_6 = 24, \qquad x_8 = 3, \qquad x_{10} = 4.$$

Thus (31) is the unique optimal solution.

The other constraints in (27) are also satisfied by (31), hence from Theorem 18, $\mathscr{C}$ has at most 32 codewords. So we have proved that any double-error-correcting code of length 12 has at most 32 codewords. This bound is attained by the Nadler code of Ch. 2, §8.

Note that this method has told us the distance distribution of $\mathscr{C}$. Also, since the first two of the constraints (27) are met with equality, in the transformed distribution $B_1' = B_2' = 0$, and $d' = 3$. But there are only 3 nonzero distances in $\mathscr{C}$. Hence by Theorem 6 of Ch. 6, $\mathscr{C}$ is distance invariant, and $A_i(u) = B_i$ for all $u \in \mathscr{C}$. (These remarks apply only to the extended code $\mathscr{C}$, not to the original code of length 12.)

If the bound obtained by the linear programming method is an odd number, say $|\mathscr{C}| \leq b$, b odd, then the bound can sometimes be reduced by the following argument (see Best et al. [140]). For suppose $|\mathscr{C}| = b$, then from (36) of Ch. 5,

$$bB'_k = \sum_{j=0}^{n} B_j P_k(j)$$

$$= \frac{1}{b} \sum_{\text{wt}(u)=k} \left(\sum_{v \in \mathscr{C}} (-1)^{u \cdot v} \right)^2.$$

The inner sum contains an odd number of ± 1's, hence is nonzero. Therefore $bB'_k \geq (1/b)\binom{n}{k}$, i.e. we can replace (26) by the stronger condition

$$\sum_{j=1}^{s} B_{\tau_j} P_k(\tau_j) \geq -\left(1 - \frac{1}{b}\right)\binom{n}{k}. \tag{33}$$

The following example shows how this is used.

Example. $A(8, 3) = 20$. Let $\mathscr{C}$ be a largest single-error-correcting code of length 8, and let $\hat{\mathscr{C}}$ be the extended code of length 9 and distance 4 with distance distribution $B_0 = 1$, B_4, B_6, B_8. The conditions (26) are:

$$\begin{aligned}
B_4 - 3B_6 - 7B_8 &\geq - \quad 9, \\
-4B_4 \quad\quad + 20B_8 &\geq - \quad 36, \\
-4B_4 + 8B_6 - 28B_8 &\geq - \quad 84, \\
6B_4 - 6B_6 + 14B_8 &\geq -126,
\end{aligned} \tag{34}$$

and we can also impose the conditions

$$\begin{aligned}
B_8 &\leq 1, \\
B_6 + 4B_8 &\leq 12,
\end{aligned} \tag{35}$$

which are proved in the same way as (28). By linear programming we find that the largest value of $B_4 + B_6 + B_8$ subject to (34) and (35) is $20\frac{1}{3}$, so $|\mathscr{C}| \leq 21$. Suppose $|\mathscr{C}| = 21$. Then (33) applies, and the RHS's of (34) can be multiplied by 20/21. Linear programming now gives $B_4 + B_6 + B_8 \leq 19.619\ldots$, so $|\mathscr{C}| \leq 20$. Since an $(8, 20, 3)$ code was given in Ch. 2, it follows that $A(8, 3) = 20$. Many other bounds in Fig. 1 of Appendix A were obtained in this way.

We now return to the general case and consider the dual problem to (III). This will enable us to obtain analytical bounds. The dual to (III) is:

Problem. (IV) Choose $\beta_1, \ldots, \beta_n$ so as to

$$\text{minimize} \quad \sum_{k=1}^{n} \beta_k \binom{n}{k} \tag{36}$$

subject to

$$\beta_k \geq 0, \quad k = 1, \ldots, n, \tag{37}$$

$$\sum_{k=1}^{n} \beta_k P_k(\tau_j) \leq -1, \quad j = 1, \ldots, s. \tag{38}$$

The advantage of using the dual formulation is that *any* feasible solution to

(IV) is, by Theorem 15, an upper bound to the size of the code, whereas only the *optimal* solution to (III) gives an upper bound. The easiest way to specify a feasible solution to (IV) is in terms of a polynomial we shall call $\beta(x)$.

Lemma 19. *Let*

$$\beta(x) = 1 + \sum_{k=1}^{n} \beta_k P_k(x).$$

Then $\beta_1, \ldots, \beta_n$ *is a feasible solution to* (IV) *iff* $\beta_k \geq 0$ *for* $k = 1, \ldots, n$, *and* $\beta(\tau_j) \leq 0$ *for* $j = 1, \ldots, s$.

Proof.

$$\beta(\tau_j) = 1 + \sum_{k=1}^{n} \beta_k P_k(\tau_j). \qquad\qquad \text{Q.E.D.}$$

From the preceding remark we have immediately:

Theorem 20. (The second version of the linear programming bound for codes.) (Delsarte [350].) *Suppose a polynomial* $\beta(x)$ *of degree at most n can be found with the following properties. If the Krawtchouk expansion* (p. 168 *of Ch.* 6) *of* $\beta(x)$ *is*

$$\beta(x) = \sum_{k=0}^{n} \beta_k P_k(x), \qquad\qquad (39)$$

then $\beta(x)$ *should satisfy*

$$\beta_0 = 1, \qquad\qquad (40)$$

$$\beta_k \geq 0 \text{ for } k = 1, \ldots, n, \qquad\qquad (41)$$

$$\beta(\tau_j) \leq 0 \text{ for } j = 1, \ldots, s. \qquad\qquad (42)$$

Then if $\mathscr{C}$ *is any code of length n and distances* $\{\tau_j\}$, $j = 1, \ldots, s$, *between codewords,*

$$|\mathscr{C}| \leq \beta(0) = 1 + \sum_{k=1}^{n} \beta_k \binom{n}{k}. \qquad\qquad (43)$$

Corollary 21. *If*

$$\beta(x) = \sum_{k=0}^{n} \beta_k P_k(x)$$

satisfies (i) $\beta_0 = 1$, $\beta_k \geq 0$ *for* $k = 1, \ldots, n$, *and* (ii) $\beta(j) \leq 0$ *for* $j = d$, $d + 1, \ldots, n$, *then*

$$A(n, d) \leq \beta(0). \qquad\qquad (44)$$

From Theorem 17, any code which meets (43) with equality must satisfy certain conditions:

Theorem 22. *Let $B_{\tau_1}, \ldots, B_{\tau_s}$ and $\beta_1, \ldots, \beta_n$ be feasible solutions to* (III) *and* (IV) *respectively. They are optimal solutions iff*

(i) $$\beta_k B'_k = 0 \quad for\ 1 \leq k \leq n,$$

and

(ii) $$\beta(\tau_j) B_{\tau_j} = 0 \quad for\ j = 1, \ldots, s.$$

Examples. (1) We begin with a very simple example, and show that the dual of a Hamming code is optimal. That is, we use Corollary 21 to show that a code $\mathscr{C}$ of length $2^m - 1$ and minimum distance 2^{m-1} can have at most 2^m codewords. (Of course this also follows from Theorem 11a.) We choose

$$\beta(x) = 2(2^{m-1} - x)$$
$$= P_0(x) + P_1(x),$$

thus $\beta_0 = \beta_1 = 1$, $\beta_k = 0$ for $k > 1$. Since

$$\beta(j) \leq 0 \quad for\ j = 2^{m-1},\ 2^{m-1} + 1, \ldots, 2^m - 1,$$

the hypotheses of Corollary 21 are satisfied, and so from (44)

$$|\mathscr{C}| \leq \beta(0) = 2^m.$$

Thus the dual of a Hamming code is optimal.

Furthermore, from Theorem 22(ii), $B_i = 0$ in $\mathscr{C}$ for $i > 2^{m-1}$; thus $\mathscr{C}$ has just one non-zero distance, 2^{m-1}.

(2) The Hadamard code $\mathscr{C}_{4m}$ of Ch. 2 with parameters $(4m, 8m, 2m)$ is optimal: $A(4m, 2m) = 8m$. (Another special case of Theorem 11a.) To show this, use Corollary 21 with

$$\beta(x) = \frac{1}{m}(2m - x)(4m - x)$$

$$= P_0(x) + P_1(x) + \frac{1}{2m} P_2(x).$$

(3) Let us find a bound on the size of a code with minimum distance d, using a $\beta(x)$ which is *linear*, say $\beta(x) = 1 + \beta_1 P_1(x) = 1 + \beta_1(n - 2x)$, where $\beta_1 > 0$. We want $\beta(d)$, $\beta(d+1), \ldots, \beta(n) \leq 0$ and $\beta(0) = 1 + \beta_1 n$ as small as possible. The best choice is to set $\beta(d) = 0$, i.e. $\beta_1 = 1/(2d - n)$. Then (44) gives $|\mathscr{C}| \leq \beta(0) = 2d/(2d - n)$. This is a weaker version of the Plotkin bound, Theorem 11.

Problem. (16) (Delsarte.) Obtain the sphere packing bound from Corollary 21.

[Hint: Let $d = 2e + 1$. Take

$$\beta_k = \left\{ L_e(k) \middle/ \sum_{i=0}^{e} \binom{n}{i} \right\}^2, \quad 0 \leq k \leq n,$$

where $L_e(x)$ is the Lloyd polynomial of Ch. 6, Theorem 32. Use Corollary 18 of Ch. 5 to show

$$\beta(i) = 0 \quad \text{for} \quad 2e + 1 \leq i \leq n, \; \beta(0) = 2^n \middle/ \sum_{i=0}^{e} \binom{n}{i}.]$$

Example. (4) *The Singleton bound* (Delsarte). One way of satisfying the hypotheses of Corollary 21 is to make $\beta(j) = 0$ for $j = d, \ldots, n$. Thus we take

$$\beta(x) = 2^{n-d+1} \prod_{j=d}^{n} \left(1 - \frac{x}{j} \right).$$

The coefficients β_k are given by (Theorem 20 of Ch. 5)

$$\beta_k = \frac{1}{2^n} \sum_{i=0}^{n} \beta(i) P_i(k)$$

$$= \frac{1}{2^{d-1}} \sum_{i=0}^{n} \binom{n-i}{n-d+1} P_i(k) \middle/ \binom{n}{d-1}$$

$$= \binom{n-k}{d-1} \middle/ \binom{n}{d-1} \geq 0,$$

by Problem 41 of Ch. 5. Note $\beta_k = 0$ for $n - d + 2 \leq k \leq n$. Therefore:

if $\mathscr{C}$ is any (n, M, d) binary code,

$$M \leq \beta(0) = 2^{n-d+1}. \tag{45}$$

This is a generalization of the Singleton bound (Ch. 1, Theorem 11) to nonlinear codes.

Problems. (17) Show that if $\mathscr{C}$ is an (n, M, d) code over $GF(q)$, $M \leq q^{n-d+1}$.

(18) (Delsarte.) Let $\mathscr{C}$ be an (n, M, d) binary code with the property that if $u \in \mathscr{C}$ then the complementary vector $\bar{u}$ is also in $\mathscr{C}$. (Example: any linear code containing **1**.) Prove that

$$M \leq \frac{8d(n-d)}{n - (n-2d)^2}, \tag{46}$$

provided the denominator is positive. Equation (46) is known as the *Grey-Rankin* bound, after Grey [561] and Rankin [1090, 1091]. [Hint: Form an $(n, \frac{1}{2}M, d)$ code $\mathscr{C}'$ by taking one codeword of $\mathscr{C}$ from each complementary pair. The distances in $\mathscr{C}'$ are in the range $[d, n-d]$. Use Corollary 18 of Ch. 5 $\beta(x) = a(d-x)(n-d-x)$ where a is a suitable constant.]

Note that the $(n - 1, 2n, \frac{1}{2}(n - 2))$ conference matrix codes of Ch. 2, §4 satisfy (46) with equality, yet are not closed under complementation. See Research Problem 2.2.

(19) Show that every feasible solution to Problem (III) satisfies $B_{\tau_i} \leq \binom{n}{\tau_i}$.

(20) (McEliece [944].) Another version of Corollary 21: if

$$\delta(x) = \frac{1}{2^n} \sum_{k=0}^{n} \delta_k P_k(x)$$

satisfies (i) $\delta(0) = 1$, $\delta(j) \geq 0$ for $j = 1, \ldots, n$, and (ii) $\delta_k \leq 0$ for $k = d, d + 1, \ldots, n$, then

$$A(n, d) \leq \delta_0 = \sum_{j=0}^{n} \binom{n}{j} \delta(j). \tag{47}$$

[Hint: Set $\beta_k = \delta(k)$ and use Corollary 18 of Ch. 5.]

(21) ([944].) If n is even, choose $\delta(x)$ to be a quadratic with $\delta(0) = 1$, $\delta(n/2 + 1) = 0$, and hence rederive Problem 11.

(22) ([944].) If $n \equiv 1 \pmod 4$, choose $\delta(x) = \{(x - a)^2 - \frac{1}{2} - \frac{1}{2}P_n(x)\}/(a^2 - 1)$, where $a = (n + 1)/2$, and show $A(n, 3) \leq 2^n/(n + 3)$. In particular, show $A(2^m - 3, 3) = 2^{2^m - m - 3}$.

Research Problem (17.5). Give a combinatorial proof that $A(2^m - 3, 3) = 2^{2^m - m - 3}$.

The linear programming bound for constant weight codes. (Delsarte [352].) There is also a linear programming bound for constant weight codes. Let $\mathcal{D}$ be a binary code of length n, distance 2δ, and constant weight w. Then $|\mathcal{D}| \leq A(n, 2\delta, w)$. Let $\{B_{2i}\}$, $i = 0, 1, \ldots, w$, be the distance distribution of $\mathcal{D}$. It follows from the theory of association schemes (see Theorem 12 of Ch. 21) that the transformed quantities B'_{2k} are nonnegative, where now

$$B'_{2k} = \frac{1}{|\mathcal{D}|} \sum_{i=0}^{w} B_{2i} Q_k(i), \quad k = 0, \ldots, w, \tag{48}$$

and the coefficients $Q_k(i)$ are given by

$$Q_k(i) = \frac{n - 2k + 1}{n - k + 1} E_i(k) \binom{n}{k} \bigg/ \binom{w}{i}\binom{n - w}{i}, \tag{49}$$

and

$$E_i(x) = \sum_{j=0}^{i} (-1)^{i-j} \binom{w - j}{i - j}\binom{w - x}{j}\binom{n - w + j - x}{j} \tag{50}$$

is an Eberlein polynomial (see §6 of Ch. 21).

So we can get bounds on $A(n, 2\delta, w)$ by linear programming: maximize $B_0 + B_2 + \cdots + B_{2w}$ subject to $B_{2i} \geq 0$ and $B'_{2k} \geq 0$ for all i and k. As before it is often possible to impose additional constraints.

Example. $A(24, 10, 8) \leqslant 69$. When $n = 24$, $2\delta = 10$ and $w = 8$ the variables B_{10}, B_{12}, B_{14}, B_{16} must satisfy an additional constraint, namely $B_{16} \leqslant A(16, 10, 8) = 4$ (from Appendix A). (For B_{16} is bounded by the number of vectors of weight 8 which are disjoint from a fixed vector of weight 8 and have mutual distance at least 10 apart). By linear programming the maximum of $B_{10} + B_{12} + B_{14} + B_{16}$ subject to these constraints is 68, and is attained when

$$B_{10} = 56, \qquad B_{12} = 0, \qquad B_{14} = 8, \qquad B_{16} = 4.$$

Thus $A(24, 10, 8) \leqslant 69$.

Research Problem (17.6). Is there a code with this distance distribution? The answer is no. (R. E. Kibler, written communication.)

The entries in Fig. 3 of Appendix A which are marked with an L were obtained in this way. It is also possible to give a linear programming bound for $T(w_1, n_1, w_2, n_2, d)$ – see Best et al. [140].

§5. The Griesmer bound

For an $[n, k, d]$ linear code the Singleton bound (45) says $n \geqslant d + k - 1$. The Griesmer bound, Equation (52), increases the RHS of this inequality. The bound is best stated in terms of

$$N(k, d) = \text{length of the shortest binary}$$
$$\text{linear code of dimension } k \text{ and}$$
$$\text{minimum distance } d.$$

Theorem 23.

$$N(k, d) \geqslant d + N\left(k - 1, \left\lceil \frac{d}{2} \right\rceil\right). \tag{51}$$

Proof. Let $\mathscr{C}$ be an $[N(k, d), k, d]$ code, with generator matrix G. Without loss of generality,

G_1 has rank $k - 1$, or else we could make the first row of G_1 zero and $\mathscr{C}$ would have minimum distance less than d. Let G_1 generate an $[N(k, d) - d, k - 1, d_1]$

code. Suppose $|u \mid v| \in \mathscr{C}$ where $\mathrm{wt}(u) = d_1$. Since $|u \mid \bar{v}| \in \mathscr{C}$, we have

$$d_1 + \mathrm{wt}(v) \geq d,$$
$$d_1 + d - \mathrm{wt}(v) \geq d,$$

and, adding, $2d_1 \geq d$ or $d_1 \geq \lceil d/2 \rceil$. Therefore

$$N(k - 1, \lceil d/2 \rceil) \leq N(k, d) - d. \qquad \text{Q.E.D.}$$

Theorem 24. (The *Griesmer bound* [562]; see also Solomon and Stiffler [1257].)

$$N(k, d) \geq \sum_{i=0}^{k-1} \left\lceil \frac{d}{2^i} \right\rceil. \tag{52}$$

Proof. Iterating Theorem 23 we find:

$$N(k, d) \geq d + N\left(k - 1, \left\lceil \frac{d}{2} \right\rceil \right)$$

$$\geq d + \left\lceil \frac{d}{2} \right\rceil + N\left(k - 2, \left\lceil \frac{d}{4} \right\rceil \right)$$

$$\dots\dots\dots\dots\dots\dots\dots$$

$$\geq \sum_{i=0}^{k-2} \left\lceil \frac{d}{2^i} \right\rceil + N\left(1, \left\lceil \frac{d}{2^{k-1}} \right\rceil \right)$$

$$= \sum_{i=0}^{k-1} \left\lceil \frac{d}{2^i} \right\rceil. \qquad \text{Q.E.D.}$$

Examples. (1) Let us find $N(5, 7)$, i.e. the shortest triple-error-correcting code of dimension 5. From Theorem 24,

$$N(5, 7) \geq 7 + \lceil \tfrac{7}{2} \rceil + \lceil \tfrac{7}{4} \rceil + \lceil \tfrac{7}{8} \rceil + \lceil \tfrac{7}{16} \rceil$$
$$= 7 + 4 + 2 + 1 + 1 = 15.$$

In fact there is a $[15, 5, 7]$ BCH code, so $N(5, 7) = 15$.

(2) If $d = 2^{k-1}$, then from Theorem 24

$$N(k, 2^{k-1}) \geq 2^{k-1} + 2^{k-2} + \cdots + 2 + 1 = 2^k - 1.$$

In fact the $[2^k - 1, k, 2^{k-1}]$ simplex code (see Ch. 1) shows that this bound is realized.

§6. Constructing linear codes; anticodes

Let S_k be the generator matrix of a $[2^k - 1, k, 2^{k-1}]$ simplex code; thus the columns of S_k consist of all distinct nonzero binary k-tuples arranged in some order. As the preceding example showed, this code meets the Griesmer bound, as does the code whose generator matrix consists of several copies of S_k placed side-by-side.

An excellent way of constructing good linear codes is to take S_k, or several copies of S_k, and delete certain columns. (See Figs. 17.1a, 17.1b.) The columns to be deleted themselves form the generator matrix of what is called an *anticode* (after Farrell et al. [418], [420], [421]; see also [889], [1098]). This is a code which has *upper* bound on the distance between its codewords. Even a distance of *zero* between codewords is allowed, and thus an anticode may contain repeated codewords.

Definition. If G is any $k \times m$ binary matrix, then all 2^k linear combinations of its rows form the codewords of an *anticode* of *length m*. The *maximum distance* δ of the anticode is the maximum weight of any of its codewords. If rank $G = r$, each codeword occurs 2^{k-r} times.

Example. (1)

$$G = \begin{bmatrix} 0 & 1 & 1 \\ 1 & 0 & 1 \\ 1 & 1 & 0 \end{bmatrix}$$

generates the anticode

$$\begin{array}{ccc} 0 & 0 & 0 \\ 0 & 1 & 1 \\ 1 & 0 & 1 \\ 1 & 1 & 0 \\ 1 & 1 & 0 \\ 1 & 0 & 1 \\ 0 & 1 & 1 \\ 0 & 0 & 0 \end{array}$$

of length $m = 3$, with 2^3 codewords and maximum distance $\delta = 2$. Each codeword occurs twice. Similarly one finds an anticode of length 3, with 2^k codewords and maximum distance 2, for any $k \geq 2$.

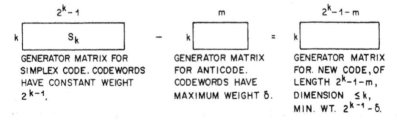

GENERATOR MATRIX FOR SIMPLEX CODE. CODEWORDS HAVE CONSTANT WEIGHT 2^{k-1}.

GENERATOR MATRIX FOR ANTICODE. CODEWORDS HAVE MAXIMUM WEIGHT δ.

GENERATOR MATRIX FOR NEW CODE, OF LENGTH 2^k-1-m, DIMENSION $\leq k$, MIN. WT. $2^{k-1}-\delta$.

Fig. 17.1a. Using an anticode to construct new codes.

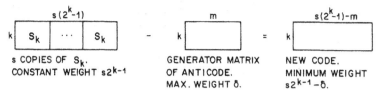

Fig. 17.1b. If the anticode has repeated columns.

Using an anticode to construct new codes. First suppose the $k \times m$ generator matrix of the anticode has distinct columns. Then (see Fig. 17.1a) the columns of this matrix can be deleted from S_k, leaving a generator matrix for the new code. Clearly the new code has length $2^k - 1 - m$, dimension $\leqslant k$, and minimum distance $2^{k-1} - \delta$.

Example. (2) Deleting the columns of the anticode in Example 1 from S_k we obtain a $[2^k - 4, k, 2^{k-1} - 2]$ code, provided $k \geqslant 3$.

Let the new code $\mathscr{C}$ have dimension r, and suppose $r \leqslant k - 1$. A generator matrix for $\mathscr{C}$ then contains $2^k - 1 - m$ distinct nonzero columns, no k of which are linearly independent. Thus $2^k - 1 - m \leqslant 2^{k-1} - 1$, so $m \geqslant 2^{k-1}$. Hence if $m < 2^{k-1}$ the dimension of $\mathscr{C}$ is k.

More generally, suppose s is the maximum number of times any column occurs in the generator matrix of the anticode. Then the new code is formed by deleting the columns of this matrix from s copies of S_k placed side-by-side (Fig. 17.1b), and has length $s(2^k - 1) - m$, dimension $\leqslant k$, and minimum distance $s2^{k-1} - \delta$.

Further examples will be specified using the language of projective geometry (see Appendix B). The *points* of the *projective geometry* $PG(k - 1, 2)$ consist of all $2^k - 1$ nonzero binary k-tuples, and so are exactly the columns of S_k. A subset of these columns is thus just a subset of the points of the geometry.

Examples. (3) Delete a line $(= 3$ points) from S_3:

$$\begin{bmatrix} 0 & 0 & 0 & 1 & 1 & 1 & 1 \\ 0 & 1 & 1 & 0 & 0 & 1 & 1 \\ 1 & 0 & 1 & 0 & 1 & 0 & 1 \end{bmatrix} - \begin{bmatrix} 0 & 1 & 1 \\ 1 & 0 & 1 \\ 1 & 1 & 0 \end{bmatrix} = \begin{bmatrix} 0 & 0 & 1 & 1 \\ 0 & 1 & 0 & 1 \\ 1 & 0 & 0 & 1 \end{bmatrix}$$

$$\quad\quad S_3 \quad\quad\quad\quad \text{Anticode} \quad \text{New code}$$
$$\quad\quad\quad\quad\quad\quad\quad \text{(a line)} \quad \text{with } n = 4,$$
$$\quad\quad\quad\quad\quad\quad\quad\quad\quad\quad\quad k = 3, d = 2.$$

(4) For $k = 5$, form the anticode whose columns consist of the 15 points of a projective subspace of dimension 3 and a single point not on this subspace, as shown in Fig. 17.2.

```
 0 1 2 3 4 5 6 7 8 9 10 11 12 13 14
┌─────────────────────────────────────┐
│ 1       1 1 1       1  1  1     1     │
│   1       1     1 1   1  1        1  1 │
│     1       1 1 1 1      1  1  1     │
│       1       1   1 1      1  1  1  1 │
│                                    1 │
└─────────────────────────────────────┘
```

Fig. 17.2. Generator matrix for an anticode with $m = 16$, $k = 5$, $\delta = 9$.

Deleting these columns from S_5 gives a $[15, 5, 7]$ code meeting the Griesmer bound.

(5) In S_6 choose a projective subspace B_5 of dimension 4 (a 4-flat) containing 31 points, and a 3-flat B_4 (15 points). These meet in a plane C_3 (7 points). Choose a plane B_3 which does not meet C_3. (B_3 intersects B_5 in a line C_2 and B_4 in a point C_1.) Choose a line B_2 which does not meet B_4 or B_3 (and meets B_5 in a point C'_1). Finally choose a point B_1 which does not lie on any previous B_i. The generator matrix of the anticode has as columns the $31 + 15 + 7 + 3 + 1 = 57$ points of B_5, B_4, B_3, B_2 and B_1. The columns of C_3, C_2, C_1 and C'_1 appear twice in this matrix. The maximum weight is $16 + 8 + 4 + 2 + 1 = 31$, and $k = 6$. We delete this anticode from two copies of S_6, as follows:

S_6	S_6
Delete B_5, C_1, B_1	Delete B_4, $B_3 - C_1$, B_2

Note that the points of C_3, C_2, C_1 and C'_1 get deleted *twice*, once from each copy of S_6. The resulting code has length $n = 2 \cdot 63 - 57 = 69$, $k = 6$, $d = 2 \cdot 32 - 31 = 33$, and meets the Griesmer bound.

Problems. (23) Use a line and a 3-flat in $PG(5, 2)$ to get an anticode with $m = 18$, $k = 6$, $\delta = 10$, and hence find a $[45, 6, 22]$ code.

(24) Construct a $[53, 6, 26]$ code.

A General Construction (Solomon and Stiffler [1257], Belov et al. [102]; see also Alltop [25].) The following technique for forming anticodes from subspaces generalizes the preceding examples and gives a large class of codes which meet the Griesmer bound.

Given d and k, define

$$s = \left\lceil \frac{d}{2^{k-1}} \right\rceil \quad \text{and} \quad d = s2^{k-1} - \sum_{i=1}^{p} 2^{u_i - 1}, \tag{53}$$

where $k > u_1 > \cdots > u_p \geq 1$.

Suppose we can find p projective subspaces $B_1, \ldots, B_p$ of S_k, where B_i has

$2^{u_i} - 1$ points, with the property that

$$\text{every } s + 1 \ B_i\text{'s have an empty intersection.} \tag{54}$$

This means that any point of S_k belongs to at most s B_i's. The anticode consists of the columns of $B_1, \ldots$ and B_p; no column occurs more than s times. The new code is obtained by deleting the anticode from s copies of S_k. It has length

$$n = s(2^k - 1) - \sum_{i=1}^{p} (2^{u_i} - 1),$$

dimension k, minimum distance given by (53), and meets the Griesmer bound. These codes are denoted by BV in Fig. 2 of Appendix A.

Theorem 25. (Belov et al. [102].) *The construction is possible* $\Leftrightarrow$

$$\sum_{i=1}^{\min(s+1, p)} u_i \leq sk.$$

Proof. ($\Leftarrow$) Let $f_i(x)$ be a binary irreducible polynomial of degree $k - u_i$, for $i = 1, \ldots, p$, and let A_i consist of all multiples of $f_i(x)$ having degree less than k. Then A_i is a vector space of dimension u_i, spanned by $f_i(x)$, $xf_i(x), \ldots, x^{u_i-1}f_i(x)$. B_i can be identified with the nonzero elements of this space in the obvious way. Suppose

$$\sum_{i=1}^{\min(s+1, p)} u_i \leq sk.$$

We must show that if I is any $(s + 1)$-subset of $\{1, 2, \ldots, p\}$ (if $s + 1 \leq p$) or if $I = \{1, \ldots, p\}$ (if $s + 1 > p$) then $\cap_{i \in I} B_i = \emptyset$. In fact,

$$\bigcap_{i \in I} B_i = \emptyset \Leftrightarrow \bigcap_{i \in I} A_i = 0$$

$$\Leftrightarrow \deg \operatorname*{lcm}_{i \in I} f_i(x) \geq k$$

$$\Leftrightarrow \sum_{i \in I} (k - u_i) \geq k$$

$$\Leftrightarrow \sum_{i \in I} u_i \leq sk$$

$$\Leftarrow \sum_{i=1}^{\min(s+1, p)} u_i \leq sk,$$

(since $u_1 > u_2 > \cdots$), as required.

($\Rightarrow$) Suppose the construction is possible, i.e. (54) holds. If $s + 1 > p$ there is nothing to prove. Assume $s + 1 \leq p$. From (54), for any $(s + 1)$-set I we have $\bigcap_{i \in I} B_i = \emptyset$. But by Problem 27,

$$\bigcap_{i \in I} B_i \geq \left\lceil \exp_2 \left(\sum_{i \in I} u_i - sk \right) \right\rceil - 1.$$

Hence $\Sigma_{i \in I} u_i \leq sk$.   Q.E.D.

Corollary 26. *Given d and k, there is an [n, k, d] code meeting the Griesmer bound if*

$$\sum_{i=1}^{\min(s+1, p)} u_i \leq sk,$$

where $s, u_1, \ldots, u_p$ are defined by (53).

Note that the [5, 4, 2] even weight code meets the Griesmer bound. In this case $s = 1$, $d = 2 = 8 - 4 - 2$, $u_1 = 3$, $u_2 = 2$, $p = 2$, and $\Sigma_{i=1}^{2} u_i > sk$. So the conditions given in the Corollary are not necessary for a code to meet the Griesmer bound. Baumert and McEliece [87] have shown that, given k, the Griesmer bound is attained for all sufficiently large d.

Research Problem (17.7). Find necessary and sufficient conditions on k and d for the Griesmer bound (52) to be tight.

The above construction can be used to find the largest linear codes in the region $n \leq 2d$.

Theorem 27. [Venturini [1369]; rediscovered by Patel [1027]. See Patel [1028] for the generalization to GF(q).] *Given n and d, let*

$$d = x_j 2^j - \sum_{i=0}^{j-1} a_{ij} 2^i, \quad \text{for } j = 0, 1, \ldots \tag{55}$$

where x_j is a positive integer and $a_{ij} = 0$ or 1, and let

$$t_j = x_j - \sum_{i=0}^{j-1} a_{ij}. \tag{56}$$

There is a unique value of j, say $j = k$, for which $t_{k-1} \geq 2d - n$ and $t_k < 2d - n$. Then the largest linear code of length n and minimum distance d contains $B(n, d) = 2^k$ codewords.

The proof is omitted. To construct such an [n, k, d] code we proceed as follows. Set $t = 2d - n$. Then

$$t_{k-1} = x_{k-1} - \sum_{i=0}^{k-2} a_{i,k-1} \geq t, \tag{57}$$

$$n = 2d - t$$

$$= (t_{k-1} - t) + x_{k-1}(2^k - 1) - \sum_{i=0}^{k-2} a_{i,k-1}(2^{i+1} - 1), \tag{58}$$

from (55). Form the anticode consisting of projective subspaces B_{i+1} (with $2^{i+1} - 1$ points) for every i for which $a_{i,k-1} = 1$. Delete this anticode from x_{k-1} copies of S_k, which is possible by (57). Finally, add $t_{k-1} - t$ arbitrary columns. The resulting code has length n (from (58)), dimension k, and minimum distance (at least)

$$x_{k-1}2^{k-1} - \sum_{i=0}^{k-2} a_{i,k-1}2^i = d, \quad \text{from (55)}.$$

Problem. (25) Show $A(100, 50) = 200$, $B(100, 50) = 32$. Thus $B(n, d)$ can be much smaller than $A(n, d)$.

Theorem 28. *If $d \leqslant 2^{k-1}$, and $\mathscr{C}$ is an $[n, k, d]$ code which attains the Griesmer bound, then $\mathscr{C}$ has no repeated columns.*

Proof. Suppose $\mathscr{C}$ has a repeated column. Then the generator matrix of $\mathscr{C}$ may be taken to be

$$G = \begin{bmatrix} 1 & 1 & \cdots \\ \hline 0 & 0 \\ & \cdot & \cdot & G_1 \\ 0 & 0 \end{bmatrix}.$$

Let $\mathscr{C}'$ be the $[n - 2, k - 1, d']$ code generated by G_1. Then from (52)

$$n - 2 \geqslant \sum_{i=0}^{k-2} \left\lceil \frac{d'}{2^i} \right\rceil \geqslant \sum_{i=0}^{k-2} \left\lceil \frac{d}{2^i} \right\rceil,$$

$$n > \sum_{i=0}^{k-2} \left\lceil \frac{d}{2^i} \right\rceil + 1 = \sum_{i=0}^{k-1} \left\lceil \frac{d}{2^i} \right\rceil,$$

and so $\mathscr{C}$ does not meet the Griesmer bound.   Q.E.D.

Definition. An *optimal* anticode is one with the largest length m for given values of k and maximum distance δ.

Suppose an anticode is deleted from the appropriate number of copies of a simplex code. If the resulting code meets the Griesmer bound, the anticode is certainly optimal. Thus the above construction gives many optimal anticodes formed from subspaces. E.g. the anticodes of Examples 3, 4, 5 are optimal. Farrell and Farrag [420, 421] give tables of optimal anticodes.

Theorem 29. *Let $\mathcal{A}$ be an optimal anticode of length m, with 2^k codewords and maximum distance δ, such that no column appears in $\mathcal{A}$ more than s times. Suppose further that the code obtained by deleting $\mathcal{A}$ from s copies of S_k meets the Griesmer bound. Then the new anticode $\mathcal{A}'$ obtained by taking every codeword of $\mathcal{A}$ twice is also optimal.*

Proof. By hypothesis

$$s(2^k - 1) - m = \sum_{i=0}^{k-1} \left\lceil \frac{s2^{k-1} - \delta}{2^i} \right\rceil. \tag{59}$$

From (59) and

$$\left\lceil \frac{s2^k - \delta}{2^i} \right\rceil = s2 - \left\lceil \frac{\delta}{2^i} \right\rceil \quad \text{for } i \leq k$$

it follows that

$$s(2^{k+1} - 1) - m = \sum_{i=0}^{k} \left\lceil \frac{s2^k - \delta}{2^i} \right\rceil,$$

hence $sS_{k+1} - \mathcal{A}'$ meets the Griesmer bound.   Q.E.D.

This theorem often enables us to get an optimal anticode with parameters m, $k + 1$, δ from one with parameters m, k, δ.

Not all optimal anticodes have a simple geometrical description. For example, the $[t + 1, t, 2]$ even weight code $\mathscr{C}$ meets the Griesmer bound. Thus $S_t - \mathscr{C}$ is an optimal anticode with length $2^t - t - 2$, 2^t codewords and maximum distance $2^{t-1} - 2$. From Theorem 29 we get optimal anticodes with 2^k codewords for all $k \geq t$, and hence new codes meeting the Griesmer bound, as follows.

Example. (6)

$$t = 3, \, m = 3, \, \delta = 2 \begin{cases} [\ 15, 4, \ \ 8] \rightarrow [\ 12, 4, \ \ 6] & (k = 4) \\ [\ 31, 5, 16] \rightarrow [\ 28, 5, 14] & (k = 5) \end{cases}$$

$$t = 4, \, m = 10, \, \delta = 6 \begin{cases} [\ 31, 5, 16] \rightarrow [\ 21, 5, 10] & (k = 5) \\ [\ 63, 6, 32] \rightarrow [\ 53, 6, 26] & (k = 6) \end{cases}$$

$$t = 5, \, m = 25, \, \delta = 14 \begin{cases} [\ 63, 6, 32] \rightarrow [\ 38, 6, 18] & (k = 6) \\ [127, 7, 64] \rightarrow [102, 7, 50] & (k = 7). \end{cases}$$

The first two anticodes consists of the 3 points of a line. However the others are not of the geometrical type constructed above. E.g. the second example consists of 10 points and 10 lines in PG(3, 2) with 3 points on each line and 3 lines through each point, as shown in Fig. 17.3. Figure 17.3 is called the *Desargues configuration*, since it arises in proving Desargues' theorem – see Hilbert and Cohn–Vossen [654, Fig. 135, p. 123].

Research Problem (17.8). Is there a natural description of the geometrical configurations corresponding to the other optimal anticodes?

Problem. (26) ([420].) Since an anticode can contain repeated codewords, anticodes may be "stacked" to get new ones.

(a) Construct anticodes $\mathscr{A}$ and $\mathscr{B}$ with parameters $m = 10$, $k = 5$, $\delta = 6$, and $m = 16$, $k = 5$, $\delta = 9$, respectively.

(b) Show that the stacks

$$\begin{matrix} \mathbf{0} & \mathscr{A} & \mathscr{A} \\ \mathbf{1} & \mathscr{A} & \bar{\mathscr{A}} \end{matrix} \quad \text{and} \quad \begin{matrix} \mathbf{0} & \mathscr{B} & \mathscr{B} \\ \mathbf{1} & \mathscr{B} & \bar{\mathscr{B}} \end{matrix}$$

(where $\mathbf{0}$ and $\mathbf{1}$ denote columns of 0's and 1's) form anticodes with $m = 21$, $k = 6$, $\delta = 12$; and $m = 33$, $k = 6$, $\delta = 18$.

(c) Hence obtain $[42, 6, 20]$ and $[30, 6, 14]$ codes.

As a final example, Fig. 17.4 lists some of the optimal linear codes with $k = 6$ (thus giving values of $N(6, d)$).

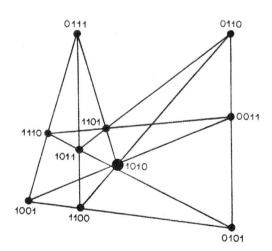

Fig. 17.3. The Desargues configuration.

Code	anticode
[63, 6, 32]	—
[60, 6, 30]	a line
[56, 6, 28]	a plane
[53, 6, 26]	plane + line, or Fig. 17.3
[48, 6, 24]	3-flat
[45, 6, 22]	3-flat + line
[42, 6, 20]	see Problem 26
[38, 6, 18]	see Example 6
[32, 6, 16]	4-flat
[30, 6, 14]	see Problem 26
[26, 6, 12]	see Baumert and McEliece [87]
[23, 6, 10]	see Baumert and McEliece [87]
[18, 6, 8]	see Baumert and McEliece [87]
[15, 6, 6]	(A BCH code)
[11, 6, 4]	(A shortened Hamming code)
[7, 6, 2]	(Even weight code)

Fig. 17.4. Optimal codes with $k = 6$.

Helgert and Stinaff [636] give a much more extensive table.

Problems. (27) ([102]) For $i = 1, \ldots, t$ let $\mathcal{C}_i$ be an $[n, k_i]$ binary linear code. Show that

$$\dim \bigcap_{i=1}^{t} \mathcal{C}_i \geqslant \max \left\{ 0, \sum_{i=1}^{t} k_i - (t-1)n \right\}.$$

[Hint: induction on t.]

(28) Show that if $\mathcal{C}$ is an optimal anticode with δ even and $m < 2^k - 1$, then an extra column can be added to the generator matrix to give an $m + 1, k, \delta + 1$ optimal anticode.

(29) ([889].) Let $\mathcal{C}$ have generator matrix $[I_k \mid A]$ where A is a $k \times (n - k)$ matrix whose rows have weight $\leqslant 1$ and columns have even weight > 0. This implies $3k \geqslant 2n$. Show that $\mathcal{C}$ is an $[n, k]$ code with maximum weight k.

§7. Asymptotic bounds

This section treats asymptotic bounds, applicable when n is large. In this case it turns out that the simplest results are obtained if the *rate* of the largest code,

$$R = \frac{1}{n} \log_2 A(n, d)$$

is plotted as a function of d/n. Figure 17.7 shows the best bounds presently known. *All* codes lie on or below the McEliece–Rodemich–Rumsey–Welch upper bound (Theorems 35, 37), while the *best* codes lie on or above the Gilbert–Varshamov lower bound. The latter was given in Ch. 1, Theorem 12, and asymptotically takes the following form.

Theorem 30. (The Gilbert–Varshamov lower bound.) *Suppose $0 \leq \delta < \frac{1}{2}$. Then there exists an infinite sequence of $[n, k, d]$ binary linear codes with $d/n \geq \delta$ and rate $R = k/n$ satisfying*

$$R \geq 1 - H_2\left(\frac{d}{n}\right), \quad \text{for all } n.$$

Notation. $f(n) \lesssim g(n)$ as $n \to \infty$ means $f(n) \leq g(n)(1 + \epsilon(n))$, where $|\epsilon(n)| \to 0$ as $n \to \infty$. Also $H_2(x) = -x \log_2 x - (1-x) \log_2(1-x)$. (See §11 of Ch. 10).

Proof. From Theorem 12 of Ch. 1 and Corollary 9 of Ch. 10.   Q.E.D.

There are alternant codes (Theorem 3 of Ch. 12), Goppa codes (Problem 9 of Ch. 12), double circulant or quasi-cyclic codes (§7 of Ch. 16) and self-dual codes (§6 of Ch. 19) which meet the Gilbert–Varshamov bound. (See also Research Problem 9.2.) Often the following very simple argument is enough to show that a family of codes contains codes which meet the Gilbert–Varshamov bound.

Theorem 31. *Suppose there is an infinite family of binary codes $\Phi_1, \Phi_2, \ldots$, where Φ_i is a set of $[n_i, k_i]$ codes such that (i) $k_i/n_i > R$ and (ii) each nonzero vector of length n_i belongs to the same number of codes in Φ_i. Then there are codes in this family which asymptotically meet the Gilbert–Varshamov bound, or more precisely, are such that*

$$R \geq 1 - H_2\left(\frac{d}{n}\right) \quad \text{for all } n.$$

Proof. Let N_0 be the total number of codes in Φ_i, and N_1 the number which contain a particular nonzero vector v. By hypothesis N_1 is independent of the choice of v, hence

$$(2^{n_i} - 1)N_1 = (2^{k_i} - 1)N_0.$$

The number of vectors v of weight less than d is

$$\sum_{j=1}^{d-1} \binom{n_i}{j},$$

hence the number of codes in Φ_i with minimum distance less than d is at most

$$N_1 \sum_{j=1}^{d-1} \binom{n_i}{j}.$$

Provided this is less than N_0, i.e. provided

$$\sum_{j=1}^{d-1} \binom{n_i}{j} < (2^{n_i} - 1)/(2^{k_i} - 1),$$

Φ_i contains a code of minimum distance $\geq d$. The result now follows from Corollary 9 of Ch. 10. Q.E.D.

Problems. (30) Show that there are linear codes of any given rate R which meet the Gilbert–Varshamov bound. [Hint: take Φ_i to be the set of linear codes of length i and dimension $k_i = [Ri]$.]

(31) (Kosholev [779], Kozlov [781], Pierce [1044]; see also Gallager [464, 465].) Let G be a $k \times n$ binary matrix whose entries are chosen at random, and let $\mathscr{C}(G)$ be the code with generator matrix G. Show that if k/n is fixed and $n \to \infty$, $\mathscr{C}(G)$ meets the Gilbert–Varshamov bound with probability approaching 1.

If $d/n \geq 1/2$ then it follows from the Plotkin bound that $(1/n) \log_2 A(n, d) \to 0$ as $n \to \infty$, so in what follows we assume $d/n < 1/2$.

We shall give a series of upper bounds, of increasing strength (and difficulty). The first is the sphere-packing bound, Theorem 12, which asymptotically becomes:

Theorem 32. (The sphere-packing or Hamming upper bound.) *For any* (n, M, d) *code,*

$$R \leq 1 - H_2\left(\frac{d}{2n}\right) \quad as \ n \to \infty.$$

The next is the Elias bound, which depends on the following simple result ($A(n, d, w)$ was defined in §1):

Theorem 33.

$$A(n, 2\delta) \leq \frac{2^n A(n, 2\delta, w)}{\binom{n}{w}} \quad \text{for all } 0 \leq w \leq n.$$

Proof. Let $\mathscr{C}$ be an $(n, M, 2\delta)$ code with the largest possible number of codewords, i.e. $M = A(n, 2\delta)$. For $u, v \in F^n$ define

$$\chi(u, v) = \begin{cases} 1 & \text{if } u \in \mathscr{C} \text{ and } \operatorname{dist}(u, v) = w, \\ 0 & \text{otherwise.} \end{cases}$$

We evaluate in two ways the following sum.

$$\sum_{u \in F^n} \sum_{v \in F^n} \chi(u, v) = \sum_{u \in \mathscr{C}} \binom{n}{w} = \binom{n}{w} A(n, 2\delta)$$

$$= \sum_{v \in F^n} \sum_{u \in F^n} \chi(u, v)$$

$$\leq \sum_{v \in F^n} A(n, 2\delta, w) = 2^n A(n, 2\delta, w).$$

Q.E.D.

Theorem 34. (The Elias upper bound [1192].) *For any* (n, M, d) *code,*

$$R \leq 1 - H_2\left(\frac{1}{2} - \frac{1}{2}\sqrt{\left(1 - \frac{2d}{n}\right)}\right), \quad \text{as } n \to \infty. \tag{60}$$

Proof. From Theorem 2, $A(n, 2\delta, w) \leq \delta n/(w^2 - wn + \delta n)$. Set

$$w = w_0 = \left[\frac{n}{2} - \frac{n}{2}\sqrt{\left(1 - \frac{4(\delta - 1)}{n}\right)}\right].$$

Then $A(n, 2\delta, w_0) \leq \delta$, and from Theorem 33, $A(n - 1, 2\delta - 1) = A(n, 2\delta) \leq 2^n \delta/\binom{n}{w_0}$. Now (60) follows from Ch. 10, Lemma 7.   Q.E.D.

Finally we come to the strongest upper bound presently known. This is in two parts; we begin with the simpler.

*Theorem 35. (The McEliece–Rodemich–Rumsey–Welch upper bound I [947].) *For any* (n, M, d) *code,*

$$R \leq H_2\left(\frac{1}{2} - \sqrt{\frac{d}{n}\left(1 - \frac{d}{n}\right)}\right). \tag{61}$$

Proof. This will follow from the second linear programming bound (Corollary 21), with the proper choice of $\beta(x)$. Set

$$\alpha(x) = \frac{1}{a - x}\{P_{t+1}(x)P_t(a) - P_t(x)P_{t+1}(a)\}^2, \tag{62}$$

where a and t will be chosen presently. From the Christoffel–Darboux formula (Ch. 5, Problem 47),

$$P_{t+1}(x)P_t(a) - P_t(x)P_{t+1}(a) = \frac{2(a-x)}{t+1}\binom{n}{t}\sum_{i=0}^{t}\frac{P_i(x)P_i(a)}{\binom{n}{i}},$$

thus

$$\alpha(x) = \frac{2}{t+1}\binom{n}{t}\{P_{t+1}(x)P_t(a) - P_t(x)P_{t+1}(a)\}\sum_{i=0}^{t}\frac{P_i(x)P_i(a)}{\binom{n}{i}}. \tag{63}$$

Since $\alpha(x)$ is a polynomial in x of deg $2t+1$, it may be expanded as

$$\alpha(x) = \sum_{k=0}^{2t+1} \alpha_k P_k(x).$$

We will then choose

$$\beta(x) = \alpha(x)/\alpha_0.$$

In order to apply Corollary 21 it must be shown that

(i) $\alpha(i) \leq 0$ for $i = d, \ldots, n$,

(ii) $\alpha_i \geq 0$ for $i = 1, \ldots, n$,

(iii) $\alpha_0 > 0$ (so that $\beta_0 = 1$).

From (62), $\alpha(i) \leq 0$ if $x > a$, so (i) holds if $a < d$. Since the Krawtchouk polynomials form an orthogonal family (Ch. 5, Theorem 16) it follows from Theorem 3.3.1 of Szegö [1297] that

(a) $P_t(x)$ has t distinct real zeros in the open interval $(0, n)$; and

(b) if $x_1^{(t)} < \cdots < x_t^{(t)}$ are the zeros of $P_t(x)$, and we set $x_0^{(t)} = 0$, $x_{t+1}^{(t)} = n$, then $P_{t+1}(x)$ has exactly one zero in each open interval $(x_i^{(t)}, x_{i+1}^{(t)})$, $i = 0, \ldots, t$ (see Fig. 17.5). To make (ii) and (iii) hold, we choose a in the range $x_1^{(t+1)} < a < x_1^{(t)}$.

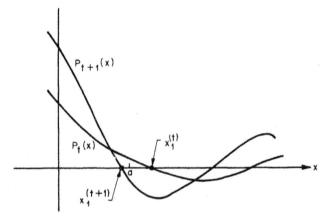

Fig. 17.5. The zeros of Krawtchouk polynomials are interlaced.

Then for $0 \leqslant i \leqslant t$

$$P_t(a)P_i(a) > 0, \qquad P_{t+1}(a)P_i(a) < 0. \tag{64}$$

Now from Corollary 4 of Ch. 21, it follows that, for $k + l < n$,

$$P_k(x)P_l(x) = \sum_m c_{klm}P_m(x), \tag{65}$$

where the coefficients c_{klm} are nonnegative integers. From (63)–(65) the coefficients α_i are nonnegative, proving (ii).

From Ch. 5, Theorem 20,

$$\alpha_0 = \frac{1}{2^n}\sum_{i=0}^{n}\binom{n}{i}\alpha(i),$$

which simplifies to (using (63))

$$\alpha_0 = -\frac{2}{t+1}\binom{n}{t}P_{t+1}(a)P_t(a) > 0,$$

and proves (iii). So provided

$$a < d \quad \text{and} \quad x_1^{(t+1)} < a < x_1^{(t)},$$

Corollary 21 applies, giving

$$A(n, d) \leqslant \beta(0) = \frac{\alpha(0)}{\alpha_0}$$

$$= \frac{\dfrac{1}{a}\left\{\binom{n}{t+1}P_t(a) - \binom{n}{t}P_{t+1}(a)\right\}^2}{-\dfrac{2}{t+1}\binom{n}{t}P_{t+1}(a)P_t(a)}$$

$$= \frac{\binom{n}{t}\{n - t - (t+1)Q\}^2}{-2a(t+1)Q}, \tag{66}$$

where

$$Q = \frac{P_{t+1}(a)}{P_t(a)} < 0.$$

Next we need a result about the asymptotic behaviour of $x_1^{(t)}$ as $n \to \infty$ and the ratio t/n is held (approximately) fixed.

Lemma 36. *If $t = [\lambda n]$, where $0 < \lambda < 1$, then*

$$\frac{1}{n}x_1^{[\lambda n]} \leqslant \frac{1}{2} - \sqrt{\lambda(1-\lambda)}, \quad \text{as } n \to \infty. \tag{67}$$

Proof of Lemma. Suppose (67) is false. Then there is a fixed ϵ with $0 < \epsilon < \sqrt{\lambda(1-\lambda)}$ and an infinite sequence of n's such that $x_1^{[\lambda n]} \geq n(r + 2\epsilon)$, where $r = \frac{1}{2} - \sqrt{\lambda(1-\lambda)}$. For each n in the sequence set $t = t(n) = [\lambda n]$. Then

$$P_t(x) = (-2)^t (x - x_1^{(t)}) \cdots (x - x_t^{(t)})/t!,$$

$$\log_e \frac{P_t(i+1)}{P_t(i)} = \sum_{k=1}^{t} \log_e \left(1 + \frac{1}{i - x_k^{(t)}}\right).$$

Set $i = i(n) = [n(r + \epsilon)]$, so that $|i - x_k^{(t)}| \geq |i - x_1^{(t)}| \geq \epsilon n$, and

$$\log_e \left(1 + \frac{1}{i - x_k^{(t)}}\right) = \frac{1}{i - x_k^{(t)}} + O\left(\frac{1}{n^2}\right),$$

$$\log_e \frac{P_t(i+1)}{P_t(i)} = \sum_{k=1}^{t} \frac{1}{i - x_k^{(t)}} + O\left(\frac{1}{n}\right).$$

Similarly

$$\log_e \frac{P_t(i-1)}{P_t(i)} = -\sum_{k=1}^{t} \frac{1}{i - x_k^{(t)}} + O\left(\frac{1}{n}\right). \tag{68}$$

Hence

$$\log_e \frac{P_t(i+1)}{P_t(i)} - \log_e \frac{P_t(i)}{P_t(i-1)} = O\left(\frac{1}{n}\right),$$

$$\frac{P_t(i+1)}{P_t(i)} = \frac{P_t(i)}{P_t(i-1)} \left(1 + O\left(\frac{1}{n}\right)\right). \tag{69}$$

Now from Problem 46 of Ch. 5,

$$(n - i)P_t(i+1) = (n - 2t)P_t(i) - iP_t(i-1),$$

$$(n - i)\frac{P_t(i+1)}{P_t(i)} \cdot \frac{P_t(i)}{P_t(i-1)} - (n - 2t)\frac{P_t(i)}{P_t(i-1)} + i = 0. \tag{70}$$

If we set $\rho = P_t(i)/P_t(i-1)$ then from (69), (70)

$$(n - i)\rho^2 - (n - 2t)\rho + i + \rho^2 O(1) = 0.$$

From (68), $\rho \leq e^{1/\epsilon}(1 + O(1/n))$, so

$$(n - i)\rho^2 - (n - 2t)\rho + i + O(1) = 0. \tag{71}$$

Since ρ is real, we must have

$$(n - 2t)^2 - 4i(n - i) + O(n) \geq 0,$$

i.e.,

$$(1 - 2\lambda)^2 - 4(r + \epsilon)(1 - r - \epsilon) + O\left(\frac{1}{n}\right) \geq 0.$$

But $(1 - 2\lambda)^2 = 4r(1 - r)$, so

$$\epsilon^2 - \epsilon(1 - 2r) + O\left(\frac{1}{n}\right) \geq 0.$$

But this is false for large n, since

$$\epsilon < \sqrt{\lambda(1 - \lambda)} = (1 - 2r)/2.$$

This completes the proof of the Lemma.

It is possible to show by contour integration that $(1/n)x_1^{[\lambda n]} \sim \frac{1}{2} - \sqrt{\lambda(1 - \lambda)}$ as $n \to \infty$, but to prove Theorem 35 a crude lower bound is sufficient. Suppose $t < n/2$. Since $P_t(0) = \binom{n}{t} > 0$ and $P_t(1) = 1 - (2t/n)\binom{n}{t} > 0$, it follows from Theorem 3.41.2 of Szegö that $P_t(x) > 0$ for $0 < x < 1$. Hence

$$x_1^{(t)} \geq 1. \tag{72}$$

We can now choose a and t in the proof of the theorem. From Fig. 17.5 there is a value of a in the range $x_1^{(t+1)} < a < x_1^{(t)}$, say $a = a_0$, for which $Q = P_{t+1}(a_0)/P_t(a_0) = -1$. Set $a = a_0$. From the lemma if we choose $t = [\lambda n]$ where λ is such that

$$\frac{1}{2} - \sqrt{\lambda(1 - \lambda)} < \frac{d}{n},$$

i.e. if

$$\frac{1}{2} - \sqrt{\frac{d}{n}\left(1 - \frac{d}{n}\right)} < \lambda < \frac{1}{2}, \tag{73}$$

then $x_1^{(t)} \leq d$ for n large. Using these values we get, from (66),

$$A(n, d) \leq \frac{(n + 1)^2 \binom{n}{t}}{2(t + 1)x_1^{(t+1)}}$$

so

$$R = \frac{1}{n}\log_2 A(n, d) \leq H_2\left(\frac{t}{n}\right), \text{ using (72) and Ch. 10, Lemma 7,}$$

$$\leq H_2\left(\frac{1}{2} - \sqrt{\frac{d}{n}\left(1 - \frac{d}{n}\right)}\right) \quad \text{from (73).} \qquad \text{Q.E.D.}$$

The same kind of argument can be applied to the linear programming bound for $A(n, d, w)$ (see p. 545). Combining this with Theorem 33 gives:

Theorem 37. (The McEliece–Rodemich–Rumsey–Welch upper bound II [947].) *For any* (n, M, d) *code,*

$$R \leq B\left(\frac{d}{n}\right), \quad \text{as } n \to \infty, \tag{74}$$

where

$$B(\delta) = \min B(u, \delta),$$
$$0 < u \leqslant 1 - 2\delta$$

and

$$B(u, \delta) = 1 + h(u^2) - h(u^2 + 2\delta u + 2\delta),$$
$$h(x) = H_2(\tfrac{1}{2} - \tfrac{1}{2}\sqrt{(1-x)}).$$

The proof is omitted.

Notice that $B(1 - 2\delta, \delta) = H_2(\tfrac{1}{2} - \sqrt{(\delta(1-\delta))})$, so

$$B\left(\frac{d}{n}\right) \leqslant H_2\left(\frac{1}{2} - \sqrt{\left(\frac{d}{n}\left(1 - \frac{d}{n}\right)\right)}\right),$$

and Theorem 37 is never weaker than Theorem 35. In fact it turns out that for $d/n \geqslant 0.273$, $B(d/n)$ is actually *equal* to

$$H_2\left(\frac{1}{2} - \sqrt{\left(\frac{d}{n}\left(1 - \frac{d}{n}\right)\right)}\right)$$

and in this range Theorems 35 and 37 coincide. For $d/n < 0.273$, Theorem 37 is slightly stronger, as shown by the table in Fig. 17.6. The best upper and lower bounds are plotted in Fig. 17.7.

$\dfrac{d}{n}$	Gilbert–Varshamov lower bound, Theorem 30	Sphere-packing upper bound, Theorem 32	Elias upper bound, Theorem 34	McEliece–Rodemich–Rumsey–Welch upper bounds	
				Theorem 35	Theorem 37
0	1	1	1	1	1
0.1	0.531	0.714	0.702	0.722	0.693
0.2	0.278	0.531	0.492	0.469	0.461
0.3	0.119	0.390	0.312	0.250	0.250
0.4	0.029	0.278	0.150	0.081	0.081
0.5	0	0.189	0	0	0

Fig. 17.6. Bounds on the rate R of the best binary code as a function of d/n, for n large.

Research Problem (17.9). What is the true upper bound on $(1/n) \log_2 A(n, d)$ as a function of d/n, as $n \to \infty$?

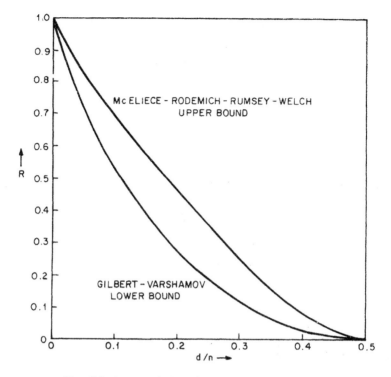

Fig. 17.7. Asymptotic bounds on the best binary codes.

(At present it is not even known if the limit

$$\lim_{n\to\infty} \frac{1}{n} \log_2 A(n, d)$$

exists, when d/n is fixed and between 0 and $\frac{1}{2}$.)

It is possible to obtain an asymptotic bound for Corollary 21 which applies just outside the region where the Plotkin bound is tight.

Theorem 38. (McEliece [944].) *For any positive j satisfying $j = o(\sqrt{d})$, we have*

$$A(2d + j, d) \leq 2d(j + 2), \quad as \ d \to \infty. \tag{75}$$

Proof. This time we choose $\alpha(x)$ to be a cubic, with roots at d, l and $l + 1$, where

$$l \sim \left(d + \frac{j}{2}\right)\left(\frac{j + 2}{j + 1}\right).$$

Then

$$\alpha(x) = (d - x)(l - x)(l + 1 - x)$$
$$= \alpha_0 + \alpha_1 P_1(x) + \alpha_2 P_2(x) + \alpha_3 P_3(x),$$

where

$$\alpha_0 \sim \frac{d^2(j + 2)}{2(j + 1)^2}, \qquad \alpha_1 \sim \frac{d^2}{2(j + 1)^2},$$

$$\alpha_2 \sim \frac{d}{j + 1}, \qquad \alpha_3 = \frac{3}{4}.$$

We take $\beta(x) = \alpha(x)/\alpha_0$. Since the α_i are positive, and $\beta(i) \leq 0$ for $i = d, \ldots, 2d + j$, Corollary 21 applies, and so

$$A(2d + j, d) < \alpha(0)/\alpha_0 \sim 2d(j + 2) \quad \text{as } d \to \infty. \qquad \text{Q.E.D.}$$

For $j = 1$ (75) gives $A(2d + 1, d) \leq 6d$, weaker than Theorem 11(b). But for $j = 2, 3, \ldots$ (75) is the best asymptotic result known.

Notes on Chapter 17

§1. $B(n, d)$ is related to the *critical problem* of combinatorial geometry – see Crapo and Rota [315], Dowling [385], Kelly and Rota [754] and White [1411].

§3. The purported proof that $A(9, 3) \leq 39$, $A(10, 3) \leq 78$ and $A(11, 3) \leq 154$ given by Wax [1391] (based on density arguments of Blichfeldt [163] and Rankin [1089]) is incorrect – see Best et al. [140].

§4. McEliece et al. (unpublished) and Best and Brouwer [139] have independently used the linear programming method to show that

$$A(2^m - 4, 3) = 2^{2^m - m - 4}.$$

18

Methods for combining codes

§1. Introduction

This chapter describes methods for combining codes to get new ones. One of the simplest ways to combine two codes is to form their direct product (see Fig. 18.1), and the first part of this chapter studies product codes and their generalizations. We use the informal term "product code" whenever the codewords are rectangles. After defining the direct product (in §2) we give a necessary and sufficient condition for a cyclic code to be direct product of cyclic codes (§3). Since this is not always possible, §§4, 5 and 6 give some other ways of factoring cyclic codes.

The construction in §4 takes the direct product of codes over a larger field and then obtains a binary code by the trace mapping. We saw in §11 of Ch. 10 that concatenated codes can be very good. Of course a concatenated code is also a kind of product code. §5 studies concatenated codes in the special case when the inner code is an irreducible cyclic code – this is called the ∗ construction. Finally §6 gives yet another method of factoring, due to Kasami, which applies to *any* cyclic code. This works by expressing the Mattson–Solomon (MS) polynomials of the codewords in terms of the MS polynomials of shorter codes.

The second part of the chapter gives a number of other powerful and ingenious ways to combine codes; a summary of these techniques will be found at the beginning of that Part (on page 581).

PART I: Product codes and generalizations

§2. Direct product codes

Let $\mathcal{A}$ and $\mathcal{B}$ be respectively $[n_1, k_1, d_1]$ and $[n_2, k_2, d_2]$ linear codes over GF(q). Suppose for simplicity that the information symbols are the first k_1 symbols of $\mathcal{A}$ and the first k_2 symbols of $\mathcal{B}$.

Definition. The *direct product* $\mathcal{A} \otimes \mathcal{B}$ is an $[n_1n_2, k_1k_2, d_1d_2]$ code whose codewords consist of all $n_1 \times n_2$ arrays constructed as follows (see Fig. 18.1).

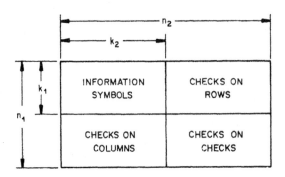

Fig. 18.1. Codewords of the direct product code $\mathcal{A} \otimes \mathcal{B}$.

The top left corner contains the k_1k_2 information symbols. The first k_2 columns are chosen so as to belong to $\mathcal{A}$, and then the rows are completed so as to belong to $\mathcal{B}$. This is also called the *Kronecker* product of $\mathcal{A}$ and $\mathcal{B}$, and is the simplest kind of product code. The columns are codewords of $\mathcal{A}$, and the rows are codewords of $\mathcal{B}$.

Example. (1) The direct product of the $[3, 2, 2]$ binary code with itself is the $[9, 4, 4]$ code consisting of the 16 arrays shown in Fig. 18.2.

Problems. (1) Verify that $\mathcal{A} \otimes \mathcal{B}$ *is* a linear code over GF(q) with parameters $[n_1n_2, k_1k_2, d_1d_2]$.

(2) The codewords could be also formed by first completing the top k_1 rows, then all the columns. Show that this gives the same code.

More generally, let $\mathcal{A}$ and $\mathcal{B}$ be arbitrary $[n_1, k_1, d_1]$ and $[n_2, k_2, d_2]$ linear codes, without assuming that the initial symbols are the information symbols.

$[3, 2, 2] \otimes [3, 2, 2] = [9, 4, 4]$ code

0 0 0	0 0 0	0 0 0	0 0 0
0 0 0	0 1 1	1 0 1	1 1 0
0 0 0	0 1 1	1 0 1	1 1 0
0 1 1	0 1 1	0 1 1	0 1 1
0 0 0	0 1 1	1 0 1	1 1 0
0 1 1	0 0 0	1 1 0	1 0 1
1 0 1	1 0 1	1 0 1	1 0 1
0 0 0	0 1 1	1 0 1	1 1 0
1 0 1	1 1 0	0 0 0	0 1 1
1 1 0	1 1 0	1 1 0	1 1 0
0 0 0	0 1 1	1 0 1	1 1 0
1 1 0	1 0 1	0 1 1	0 0 0

Fig. 18.2.

Then the direct product $\mathcal{A} \otimes \mathcal{B}$ is defined to be the $[n_1 n_2, k_1 k_2, d_1 d_2]$ code whose codewords consist of all $n_1 \times n_2$ arrays in which the columns belong to $\mathcal{A}$ and the rows to $\mathcal{B}$.

Unfortunately direct product codes usually have poor minimum distance (but are easy to decode). Sometimes the rectangular form of the codewords makes them useful in certain applications – see Notes.

Problem. (3) Let G_1 and G_2 be generator matrices for $\mathcal{A}$ and $\mathcal{B}$ respectively. Show that the Kronecker product $G_1 \otimes G_2$ (defined in §4 of Ch. 14) is a generator matrix for $\mathcal{A} \otimes \mathcal{B}$.

Suppose $\mathcal{A}$ and $\mathcal{B}$ are cyclic codes, as in Example 1. Then the direct product $\mathcal{A} \otimes \mathcal{B}$ is invariant under cyclic permutation of all the rows simultaneously, or all the columns simultaneously. We shall represent a typical codeword

$$
\begin{bmatrix}
a_{00} & a_{01} & \cdots & a_{0, n_2-1} \\
a_{10} & a_{11} & \cdots & a_{1, n_2-1} \\
\cdots\cdots\cdots\cdots\cdots\cdots\cdots \\
a_{n_1-1, 0} & & \cdots & a_{n_1-1, n_2-1}
\end{bmatrix}
\tag{1}
$$

of $\mathcal{A} \otimes \mathcal{B}$ by the polynomial

$$
f(x, y) = \sum_{i=0}^{n_1-1} \sum_{j=0}^{n_2-1} a_{ij} x^i y^j.
\tag{2}
$$

If we assume $x^{n_1} = 1$ and $y^{n_2} = 1$, then $xf(x, y)$ and $yf(x, y)$ represent cyclic shifts of the rows and of the columns, and also belong to $\mathcal{A} \otimes \mathcal{B}$. In other words, $\mathscr{A} \otimes \mathscr{B}$ is an ideal in the group algebra $\mathscr{G}$ of the abelian group generated by x and y with $x^{n_1} = y^{n_2} = 1$.

Suppose n_1 and n_2 are relatively prime. Then by the Chinese remainder theorem (§9 of Ch. 10) for each pair i_1, i_2 with $0 \leqslant i_1 < n_1$, $0 \leqslant i_2 < n_2$ there is a unique integer $I(i_1, i_2)$ in the range $0 \leqslant I(i_1, i_2) < n_1 n_2$ such that

$$I(i_1, i_2) \equiv i_1 \quad (\mathrm{mod}\ n_1),$$
$$I(i_1, i_2) \equiv i_2 \quad (\mathrm{mod}\ n_2). \tag{3}$$

This implies that we can rewrite $f(x, y)$ in terms of the single variable $z = xy$, by replacing each term $x^{i_1} y^{i_2}$ by $z^{I(i_1, i_2)}$. (In this case $\mathscr{G}$ is the group algebra of the cyclic group of order $n_1 n_2$ generated by z.)

For example if $n_1 = 3$, $n_2 = 5$ the values of $I(i_1, i_2)$ are:

i_1 \ i_2	0	1	2	3	4	
0	0	6	12	3	9	
1	10	1	7	13	4	$I(i_1, i_2)$
2	5	11	2	8	14	

The codeword $f(x, y)$ becomes

$$
\begin{aligned}
f(x, y) = \ & a_{00} && + a_{01}z^6 + a_{02}z^{12} + a_{03}z^3 + a_{04}z^9 \\
& + a_{10}z^{10} + a_{11}z + a_{12}z^7 + a_{13}z^{13} + a_{14}z^4 \\
& + a_{20}z^5 + a_{21}z^{11} + a_{22}z^2 + a_{23}z^8 + a_{24}z^{14}.
\end{aligned}
\tag{4}
$$

Theorem 1. *If $\mathscr{A}$ and $\mathscr{B}$ are cyclic codes and $(n_1, n_2) = 1$ then $\mathscr{C} = \mathscr{A} \otimes \mathscr{B}$ is also cyclic.*

Proof. As shown above, a codeword $f(x, y) \in \mathscr{C}$ can be written as $g(z)$ where $z = xy$. Also if $f(x, y) \in \mathscr{C}$, $yf(x, y) \in \mathscr{C}$, hence $xyf(x, y) = zg(z) \in \mathscr{C}$. Therefore $\mathscr{C}$ is an ideal in the group algebra of the cyclic group generated by z.   Q.E.D.

Example. (2) Let $\mathscr{A}$ and $\mathscr{B}$ be the $[3, 2, 2]$ and $[5, 4, 2]$ even weight codes. Since $(3, 5) = 1$, $\mathscr{C} = \mathscr{A} \otimes \mathscr{B}$ is a cyclic $[15, 8, 4]$ code. Some typical codewords of $\mathscr{C}$ are

$$
u = \begin{bmatrix} 0 & 1 & 0 & 1 & 0 \\ 1 & 0 & 0 & 1 & 0 \\ 1 & 1 & 0 & 0 & 0 \end{bmatrix}, \quad
v = \begin{bmatrix} 0 & 0 & 0 & 0 & 0 \\ 0 & 1 & 1 & 1 & 1 \\ 0 & 1 & 1 & 1 & 1 \end{bmatrix}, \quad
w = \begin{bmatrix} 1 & 1 & 0 & 0 & 0 \\ 0 & 1 & 1 & 0 & 0 \\ 1 & 0 & 1 & 0 & 0 \end{bmatrix}.
$$

The first four columns are arbitrary codewords of $\mathscr{A}$ and the last column is their sum. Equation (4) gives the cyclic representation of $\mathscr{C}$. For example, the idempotent of $\mathscr{C}$ is (cf. Problem 7 of Ch. 8)

$$\theta_1 + \theta_7 = z + z^2 + z^4 + z^7 + z^8 + z^{11} + z^{13} + z^{14},$$

and the generator polynomial is $1 + z + z^2 + z^5 + z^6 + z^7$. These are the codewords v and w respectively.

Problems. (4) Suppose $(n_1, n_2) = 1$ and $E_1(x)$, $E_2(y)$ are the idempotents of $\mathcal{A}$
and $\mathcal{B}$. Show that the idempotent of $\mathcal{A} \otimes \mathcal{B}$ is obtained from $E_1(x)E_2(y)$ by
replacing xy by z.

(5) Suppose $(n_1, n_2) = 1$. If $\beta_1, \ldots, \beta_{k_1}$ and $\gamma_1, \ldots, \gamma_{k_2}$ are respectively the
nonzeros of $\mathcal{A}$ and $\mathcal{B}$, show that the $k_1 k_2$ nonzeros of $\mathcal{A} \otimes \mathcal{B}$ are $\beta_i \gamma_j$, for
$1 \le i \le k_1$, $1 \le j \le k_2$.

(6) Take $\mathcal{A}$ and $\mathcal{B}$ to be respectively $[3, 2, 2]$ and $[7, 3, 4]$ codes. Find the
parameters, weight distribution, idempotent, and generator polynomial of
$\mathcal{A} \otimes \mathcal{B}$.

(7) Other choices for z besides $z = xy$ may make $\mathcal{A} \otimes \mathcal{B}$ into a cyclic code.
When do different choices give a different cyclic code? [Hint: for the code of
Problem 6 show that $z = xy$ and $z = xy^3$ give different results.]

Sometimes knowing that a code is a direct product enables one to find its
weight distribution.

(8) (Hard.) Let $\mathcal{A}$ and $\mathcal{B}$ be $[2^{m_1} - 1, m_1, 2^{m_1-1}]$ and $[2^{m_2} - 1, m_2, 2^{m_2-1}]$ simplex
codes, where $(m_1, m_2) = 1$ and $m_1 > m_2$. Show that $\mathcal{A} \otimes \mathcal{B}$ has m_2 distinct
nonzero weights, namely $w_i = (2^{m_2} - 2^{m_2-i})2^{m_1-1}$ for $i = 1, \ldots, m_2$. Show that the
number of codewords of weight w_i is

$$F(m_2, i)(2^{m_1} - 1)(2^{m_1} - 2) \cdots (2^{m_1} - 2^{m_1-i}),$$

where $F(r, k)$, $r \ge k$, is given by the recurrence

$$F(r, 0) = 1, F(r, r) = 1,$$
$$F(r + 1, k) = F(r, k - 1) + 2^k F(r, k).$$

*§3. Not all cyclic codes are direct products of cyclic codes

For example, suppose $\mathcal{C}$ is a nondegenerate irreducible $[n, k]$ binary cyclic
code with $n = n_1 n_2$, $(n_1, n_2) = 1$ and $n_1 > 1$, $n_2 > 1$. We shall show that only in
certain cases is it possible to write $\mathcal{C} = \mathcal{A} \otimes \mathcal{B}$ where $\mathcal{A}$ and $\mathcal{B}$ are cyclic
codes of lengths n_1 and n_2.

Let $\alpha, \alpha^2, \alpha^4, \ldots, a^{2^{k-1}}$ be the nonzeros of $\mathcal{C}$, α a primitive n^{th} root of unity in
$GF(2^k)$. Since $(n_1, n_2) = 1$, by §8 of Ch. 12 there exist integers a, b such that
$an_1 + bn_2 = 1$. Define $\beta = \alpha^{bn_2}$ and $\gamma = \alpha^{an_1}$, and let k_1, k_2 be the least integers for
which $\beta \in GF(2^{k_1})$, $\gamma \in GF(2^{k_2})$.

Theorem 2. *There exist cyclic codes $\mathcal{A}$ and $\mathcal{B}$, of lengths n_1 and n_2, such that
$\mathcal{C} = \mathcal{A} \otimes \mathcal{B}$ iff $(k_1, k_2) = 1$. (Then the dimensions of $\mathcal{A}$ and $\mathcal{B}$ are k_1 and k_2.)*

Proof. Making the substitution $z = xy$ puts any codeword $g(z) \in \mathscr{C}$ into an $n_1 \times n_2$ array described by $f(x, y)$. Note that

$$z^{an_1} = x^{an_1} y^{an_1} = y^{an_1} = y^{bn_2} y^{an_1} = y,$$

and similarly $z^{bn_2} = x$. Then $z^{an_1} g(z) = yf(x, y) \in \mathscr{C}$, and similarly $xf(x, y) \in \mathscr{C}$. Therefore the columns of the arrays obtained in this way form a cyclic code $\mathscr{A}$ (say), and the rows form a cyclic code $\mathscr{B}$.

Now $\beta^{n_1} = \gamma^{n_2} = 1$, and $\alpha = \beta\gamma$. So the nonzeros of $\mathscr{C}$ are

$$\beta\gamma, (\beta\gamma)^2, \dots, (\beta\gamma)^{2^{k-1}}.$$

Lemma 3. *The nonzeros of $\mathscr{A}$ and $\mathscr{B}$ are $\beta, \beta^2, \dots, \beta^{2^{k_1-1}}$ and $\gamma, \gamma^2, \dots, \gamma^{2^{k_2-1}}$ respectively.*

Proof. Let $\gamma_j \in GF(2^k)$ be an n_2^{th} root of unity not in the set $\{\gamma, \gamma^2, \dots, \gamma^{2^{k_2-1}}\}$, and let $\beta_l \in GF(2^k)$ be *any* n_1^{th} root of unity. Thus $\beta_l\gamma_j$ is a zero of $\mathscr{C}$. Let

$$f(x, y) = r_0(y) + xr_1(y) + \cdots + x^{n_1-1}r_{n_1-1}(y)$$

be a nonzero codeword of $\mathscr{C}$, where $r_i(y)$ is the ith row. Then

$$f(\beta_l, \gamma_j) = \sum_{i=0}^{n_1-1} \beta_l^i r_i(\gamma_j) = 0$$

holds for all n_1 choices of β_l, giving n_1 linear homogeneous equations for the n_1 quantities $r_i(\gamma_j)$, $0 \le i \le n_1 - 1$. Since the coefficient matrix is Vandermonde (Lemma 17 of Ch. 4), hence nonsingular, $r_i(\gamma_j) = 0$ for all i. Therefore γ_j is a zero of the row code $\mathscr{B}$. The nonzeros of $\mathscr{B}$ must therefore be as stated. Similarly for $\mathscr{A}$. Q.E.D.

Clearly k_1 and k_2 divide k, and in fact if s is the g.c.d. of k_1 and k_2 then

$$k = \frac{k_1 k_2}{s} = \text{l.c.m.} \{k_1, k_2\}.$$

The theorem now follows easily. If $(k_1, k_2) = 1$ then $k = k_1 k_2$ and $\mathscr{C} = \mathscr{A} \otimes \mathscr{B}$. On the other hand if $\mathscr{C} = \mathscr{A} \otimes \mathscr{B}$ then $k = k_1 k_2$, so $s = 1$. Q.E.D.

Examples. (3) Let $\mathscr{C}$ be the $[15, 4, 8]$ cyclic code with idempotent $z + z^2 + z^3 + z^4 + z^6 + z^8 + z^9 + z^{12}$. Let $n_1 = 3$, $n_2 = 5$. The nonzeros of $\mathscr{C}$ are $\alpha, \alpha^2, \alpha^4, \alpha^8$ with $\alpha^{15} = 1$. Then $a = 2$, $b = -1$, $\beta = \alpha^{10}$, $\gamma = \alpha^6$, $\beta^4 = \beta$, $\gamma^6 = \gamma$, and $k_1 = 2$, $k_2 = 4$. Since 2 and 4 are not relatively prime, $\mathscr{C} \ne \mathscr{A} \otimes \mathscr{B}$. Indeed, $\mathscr{A}$ and $\mathscr{B}$ are respectively $[3, 2, 2]$ and $[5, 4, 2]$ codes, and $\mathscr{A} \otimes \mathscr{B}$ is a $[15, 8, 4]$ code.

(4) Let $\mathscr{C}$ be the $[21, 6, 8]$ code of Problem 6, with nonzeros $\alpha, \alpha^2, \alpha^4, \alpha^8, \alpha^{16}, \alpha^{11}$ where $\alpha^{21} = 1$. Then $n_1 = 3$, $n_2 = 7$, $-2.3 + 7 = 1$, $\beta = \alpha^7$, $\gamma = \alpha^{15}$, $k_1 = 2$, $k_2 = 3$, and indeed $\mathscr{C} = \mathscr{A} \otimes \mathscr{B}$.

§4. Another way of factoring irreducible cyclic codes

With the same notation as in §3, if $(k_1 k_2) = s > 1$ then $\mathscr{C} \neq \mathscr{A} \otimes \mathscr{B}$. However, $\mathscr{C}$ can be factored in a different way:

Theorem 4. *It is always possible to find irreducible cyclic codes $\hat{\mathscr{A}}$ and $\hat{\mathscr{B}}$ over* $GF(2^s)$ *such that*

$$\mathscr{A} = T_s(\hat{\mathscr{A}}),$$
$$\mathscr{B} = T_s(\hat{\mathscr{B}}), \tag{5}$$

and

$$\mathscr{C} = T_s(\hat{\mathscr{A}} \otimes \hat{\mathscr{B}}). \tag{6}$$

As in Ch. 10, if $\mathscr{D}$ is an $[n, k, d]$ code over $GF(2^s)$, *the trace of $\mathscr{D}$, $T_s(\mathscr{D})$, is the* $[n, k' \leq sk, d' \leq d]$ *binary code consisting of all distinct codewords*

$$(T_s(u_1), \ldots, T_s(u_n)) \quad \text{where } (u_1, \ldots, u_n) \in \mathscr{D}.$$

Proof. To find $\hat{\mathscr{A}}$ and $\hat{\mathscr{B}}$ we proceed as follows. The nonzeros of $\mathscr{C}$ are $\alpha, \alpha^2, \ldots, \alpha^{2^{k-1}}$, corresponding to the cyclotomic coset $C_1 = \{1, 2, 2^2, \ldots, 2^{k-1}\} \bmod n$. Over $GF(2^s)$ C_1 splits into the s cyclotomic cosets

$$\hat{C}_1 = 1, 2^s, \ldots, 2^{s(k/s-1)},$$
$$\hat{C}_2 = 2, 2^{s+1}, \ldots, 2^{s(k/s-1)+1},$$
$$\cdots\cdots\cdots\cdots\cdots\cdots\cdots\cdots\cdots$$
$$\hat{C}_{2^{s-1}} = 2^{s-1}, 2^{2s-1}, \ldots, 2^{k-1}.$$

The idempotents over $GF(2^s)$ associated with these cosets are defined by

$$\hat{\theta}_{2^i}(\alpha^l) = \begin{cases} 1, & l \in \hat{C}_{2^i}, \\ 0, & \text{otherwise}, \end{cases}$$

for $i = 0, \ldots, s-1$. Then by Theorem 6 of Ch. 8,

$$\hat{\theta}_{2^i}(z) = \sum_{j=0}^{n-1} \epsilon_{ij} z^j, \tag{7}$$

where

$$\epsilon_{ij} = \sum_{l \in \hat{C}_{2^i}} \alpha^{-lj}. \tag{8}$$

The idempotent of $\mathscr{C}$ itself is

$$\theta(z) = \sum_{j=0}^{n-1} \epsilon_j z^j, \tag{9}$$

where

$$\epsilon_j = \sum_{l \in C_1} \alpha^{-lj}. \tag{10}$$

Then from the above definitions

$$\theta(z) = \sum_{i=0}^{s-1} \hat{\theta}_{2^i}(z)$$

$$= \sum_{j=0}^{n-1} T_s(\epsilon_{ij}) z^j \quad \text{for all } i = 0, \ldots, s-1. \tag{11}$$

Thus $\mathscr{C} = T_s(\hat{\mathscr{C}}_{2^i})$, $0 \leq i \leq s-1$. Now $\hat{\mathscr{C}}_1$ is an $[n_1 n_2, k/s]$ code over $GF(2^s)$ with nonzeros $\alpha, \alpha^{2^s}, \ldots, \alpha^{2^{k-s}}$. We apply the technique of the previous section to $\hat{\mathscr{C}}_1$. As before, $\beta = \alpha^{bn_2}$, $\gamma = \alpha^{an_1}$, where $\beta \in GF(q^{k_1/s})$, $\gamma \in GF(q^{k_2/s})$ and $q = 2^s$. Then a codeword of $\hat{\mathscr{C}}_1$ can be written as an $n_1 \times n_2$ array, in which each column belongs to an $[n_1, k_1/s]$ cyclic code $\hat{\mathscr{A}}_1$ with nonzeros $\beta, \beta^{2^s}, \ldots, \beta^{2^{k_1-s}}$, and each row to an $[n_2, k_2/s]$ cyclic code $\hat{\mathscr{B}}_1$, with nonzeros $\gamma, \gamma^{2^s}, \ldots, \gamma^{2^{k_2-s}}$. Since k_1/s and k_2/s are relatively prime, $\hat{\mathscr{C}}_1 = \hat{\mathscr{A}}_1 \otimes \hat{\mathscr{B}}_1$, and

$$\mathscr{C} = T_s(\hat{\mathscr{A}}_1 \otimes \hat{\mathscr{B}}_1). \qquad \text{Q.E.D.}$$

Example. (3) (cont.) Let $\mathscr{C}$ be the $[15, 4, 8]$ code discussed above, with $s = 2$. Set $\omega = \alpha^5$; ω is a primitive element of $GF(2^2)$. Then $\hat{\theta}_1(z)$, $\hat{\theta}_2(z)$ are as shown in Fig. 18.3

	0	1	2	3	4	5	6	7	8	9	10	11	12	13	14
$\hat{\theta}_1$	0	ω	ω^2	ω	ω	0	ω^2	1	ω^2	ω^2	0	1	ω	1	1
$\hat{\theta}_2$	0	ω^2	ω	ω^2	ω^2	0	ω	1	ω	ω	0	1	ω^2	1	1
sum = trace	0	1	1	1	1	0	1	0	1	1	0	0	1	0	0

Fig. 18.3.

Arrange the idempotent $\hat{\theta}_1$ in a 3×5 array

	1	y	y^2	y^3	y^4
1	0	ω^2	ω	ω	ω^2
x	0	ω	1	1	ω
x^2	0	1	ω^2	ω^2	1

Thus $\hat{\theta}_1(x, y) = (1 + \omega^2 x + \omega x^2)(\omega^2 y + \omega y^2 + \omega y^3 + \omega^2 y^4)$, which is the product of the idempotents of $\hat{\mathscr{A}}_1$ and $\hat{\mathscr{B}}_1$. These are $[3, 1, 3]$ and $[5, 2, 4]$ codes over

$GF(2^2)$. Similarly $\hat{\theta}_2$ gives the array

	1	y	y^2	y^3	y^4
1	0	ω	ω^2	ω^2	ω
x	0	ω^2	1	1	ω^2
x^2	0	1	ω	ω	1

and $\hat{\theta}_2(x, y) = (1 + \omega x + \omega^2 x^2)(\omega y + \omega^2 y^2 + \omega^2 y^3 + \omega y^4)$.   Clearly $\mathscr{C} = T_2(\hat{\mathscr{A}}_1 \otimes \hat{\mathscr{B}}_1) = T_2(\hat{\mathscr{A}}_2 \otimes \hat{\mathscr{B}}_2)$.

Problems. (9) Show that $T_2(\hat{\mathscr{A}}_1 \otimes \hat{\mathscr{B}}_2)$ and $T_2(\hat{\mathscr{A}}_2 \otimes \hat{\mathscr{B}}_1)$ are both equal to the other $[15, 4, 8]$ code.

(10) Use the above method to factor the nondegenerate irreducible codes of length $63 = 7.9$. (The necessary idempotents are given in Fig. 8.1.)

§5. Concatenated codes: the * construction

Any concatenated code (Ch. 10) is a kind of product code. If the inner code is irreducible and n_1, n_2 are relatively prime then we shall see that the concatenated code is cyclic. We call this the * construction.

Two codes are needed. (i) An $[n_1, k_1, d_1]$ irreducible cyclic binary code $\mathscr{A}$ with (say) nonzeros $\beta, \beta^2, \beta^4, \ldots, \beta^{2^{k_1-1}}$, where $\beta \in GF(2^{k_1})$ is a primitive n_1^{th} root of unity; and (ii) an $[n_2, k_2, d_2]$ cyclic code $\mathscr{B}$ over $GF(2^{k_1})$. Recall from §3 of Ch. 8 that $\mathscr{A}$ is isomorphic to $GF(2^{k_1})$. There is an isomorphism φ from $\mathscr{A}$ to $GF(2^{k_1})$ given by

$$a(x)^\varphi = a(\beta),$$

with inverse map ψ which sends $\delta \in GF(2^{k_1})$ into the binary n_1-tuple

$$(\delta)^\psi = (T_{k_1}(\delta), T_{k_1}(\delta\beta^{-1}), T_{k_1}(\delta\beta^{-2}), \ldots, T_{k_1}(\delta\beta^{-(n_1-1)})) \in \mathscr{A}. \tag{12}$$

*The * construction.* The new code, which is denoted by $\mathscr{A} * \mathscr{B}$, is formed by taking each codeword $(\delta_0, \ldots, \delta_{n_2-1})$ *of* $\mathscr{B}$ and replacing every δ_i by the binary column vector obtained by transposing $(\delta_i)^\psi$. Thus $\mathscr{A} * \mathscr{B}$ is an $[n_1 n_2, k_1 k_2]$ binary code with minimum distance $\geq d_1 d_2$, whose codewords are $n_1 \times n_2$ arrays. In other words, $\mathscr{A} * \mathscr{B}$ is formed by concatenating the inner code $\mathscr{A}$ with the outer code $\mathscr{B}$.

Example. (5) Let $\mathscr{A}$ be the $[3, 2, 2]$ code which is isomorphic to $GF(2^2)$ under the mapping

$$
\begin{array}{ccc}
\mathrm{GF}(2^2) & \overset{\psi}{\longrightarrow} & \mathscr{A} \\
0 & & 0\ 0\ 0 \\
1 & & 0\ 1\ 1 \\
\omega & & 1\ 0\ 1 \\
\omega^2 & & 1\ 1\ 0
\end{array}
\tag{13}
$$

Let $\mathscr{B}$ be the $[5, 2, 4]$ code over $\mathrm{GF}(2^2)$ with idempotent

$$\omega^2 y + \omega y^2 + \omega y^3 + \omega^2 y^4.$$

Then $\mathscr{B}$ consists of the codewords

$$
\begin{array}{ccccc}
0 & 0 & 0 & 0 & 0 \\
0 & \omega^2 & \omega & \omega & \omega^2 \\
\omega^2 & 0 & \omega^2 & \omega & \omega \\
\omega & \omega^2 & 0 & \omega^2 & \omega \\
\omega & \omega & \omega^2 & 0 & \omega^2 \\
\omega^2 & \omega & \omega & \omega^2 & 0
\end{array}
$$

and their multiples by ω and ω^2. Then $\mathscr{A} * \mathscr{B}$ is found as follows:

$$
0\ \omega^2\ \omega\ \omega\ \omega^2 \rightarrow
\begin{bmatrix}
0 & 1 & 1 & 1 & 1 \\
0 & 1 & 0 & 0 & 1 \\
0 & 0 & 1 & 1 & 0
\end{bmatrix}
$$

$$
1\ 0\ 1\ \omega^2\ \omega^2 \rightarrow
\begin{bmatrix}
0 & 0 & 0 & 1 & 1 \\
1 & 0 & 1 & 1 & 1 \\
1 & 0 & 1 & 0 & 0
\end{bmatrix}
$$

....

and is our old friend the $[15, 4, 8]$ code.

Recall that the idempotent $E(x)$ of $\mathscr{A}$ maps onto the element $1 \in \mathrm{GF}(2^{k_1})$. If $\mathscr{A}$ is a simplex code then a typical codeword $x^i E(x)$ is mapped onto β^i. But in general we also need to know $(\xi)^\psi$, where ξ is a primitive element of $\mathrm{GF}(2^{k_1})$.

Problem. (11) Show that the idempotent of $\mathscr{B}$ maps onto the idempotent of $\mathscr{A} * \mathscr{B}$.

Theorem 5. *If* $(n_1, n_2) = 1$ *then* $\mathscr{A} * \mathscr{B}$ *can be made into a cyclic code by the transformation used in Theorem 1 (i.e. by setting* $z = xy$*).*

Proof. Let (1) be a typical codeword of $\mathscr{A} * \mathscr{B}$, obtained from $(\delta_0, \ldots, \delta_{n_2-1}) \in \mathscr{B}$. As in Theorem 1 it will follow that $\mathscr{A} * \mathscr{B}$ is a cyclic code if the cyclic shifts

$$\begin{bmatrix} a_{0,n_2-1} & \cdots & a_{0,n_2-2} \\ \cdots\cdots\cdots\cdots\cdots \\ a_{n_1-1,n_2-1} & \cdots & a_{n_1-1,n_2-2} \end{bmatrix} \quad \text{and} \quad \begin{bmatrix} a_{n_1-1,0} & \cdots & a_{n_1-1,n_2-1} \\ a_{00} & \cdots & a_{0,n_2-1} \\ \cdots\cdots\cdots\cdots\cdots\cdots \\ a_{n_1-2,0} & \cdots & a_{n_1-2,n_2-1} \end{bmatrix}$$

are in $\mathcal{A} * \hat{\mathcal{B}}$. In fact these are obtained from the codewords $(\delta_{n_2-1}, \delta_0, \ldots, \delta_{n_2-2})$ and $(\beta\delta_0, \ldots, \beta\delta_{n_2-1})$ (by Lemma 10 of Ch. 8).   Q.E.D.

Example. We choose $\hat{\mathcal{B}}$ to be a cyclic MDS code, for example the $[9, 4, 6]$ code over $GF(2^3)$ with zeros $\alpha^{-2}, \alpha^{-1}, 1, \alpha, \alpha^2$, where α is a primitive element of $GF(2^6)$. This has weight distribution (see Ch. 11)

i	0	6	7	8	9
A_i	1	588	504	1827	1176

Take $\mathcal{A}$ to be the $[7, 3, 4]$ simplex code. Then $\mathcal{A} * \hat{\mathcal{B}}$ is a cyclic $[63, 12, 24]$ code with weight distribution

i	0	24	28	32	36
A_i	1	588	504	1827	1176

Theorem 6. *Assume* $(n_1, n_2) = 1$. *The* $k_1 k_2$ *nonzeros of* $\mathcal{A} * \hat{\mathcal{B}}$ *are*

$$z = \beta^{2^j} \gamma_i^{2^j}, \quad \text{for } j = 0, 1, \ldots, k_1 - 1, \ i = 1, \ldots, k_2,$$

where $\beta, \beta^2, \ldots, \beta^{2^{k_1-1}}$ *and* $\gamma_1, \ldots, \gamma_{k_2}$ *are the nonzeros of* $\mathcal{A}$ *and* $\hat{\mathcal{B}}$ *respectively.*

Proof. A typical codeword of $\mathcal{A} * \hat{\mathcal{B}}$ is

$$r_0(y) + x r_1(y) + \cdots + x^{n_1-1} r_{n_1-1}(y).$$

Now

$$r_0(y) + \beta^{2^j} r_1(y) + \cdots + (\beta^{2^j})^{n_1-1} r_{n_1-1}(y) = \delta_0^{2^j} + \delta_1^{2^j} y + \cdots + \delta_{n_2-1}^{2^j} y^{n_2-1},$$

where $(\delta_0, \delta_1, \ldots, \delta_{n_2-1}) \in \hat{\mathcal{B}}$. Also $(\delta_0^{2^j}, \ldots, \delta_{n_2-1}^{2^j})$ is a codeword of the $[n_2, k_2]$ cyclic code $\hat{\mathcal{B}}'$ with nonzeros $\gamma_1^{2^j}, \ldots, \gamma_{k_2}^{2^j}$. Thus $\beta^{2^j} \gamma_i^{2^j}$ is a nonzero of $\mathcal{A} * \hat{\mathcal{B}}$.   Q.E.D.

Example. (5) (cont.) Let α be a primitive element of $GF(2^4)$ and set $\omega = \alpha^5$. Then the nonzeros of $\mathcal{A}$, $\hat{\mathcal{B}}$ and $\mathcal{A} * \hat{\mathcal{B}}$ are $\beta = \alpha^5$, $\beta^2 = \alpha^{10}$; $\gamma_1 = \alpha^3$, $\gamma_2 = \alpha^{12}$; and $\alpha^5 \cdot \alpha^3 = \alpha^8$, $\alpha^5 \cdot \alpha^{12} = \alpha^2$, $\alpha^{10} \cdot \alpha^6 = \alpha$, $\alpha^{10} \cdot \alpha^9 = \alpha^4$. α, α^2, α^4, α^8 are indeed the nonzeros of the $[15, 4, 8]$ code.

Problem. (12) Show that any irreducible $[n_1 n_2, k]$ binary code $\mathscr{C}$ with $(n_1, n_2) =$

1, $n_1 > 1$, $n_2 > 1$, can be factored as $\mathscr{C} = \mathscr{A} * \mathscr{B}$. [Hint: Take $\mathscr{B}$ to be the $[n_2, k/k_1]$ irreducible code over $\mathrm{GF}(2^{k_1})$ with nonzeros $\gamma^{2^{ik_1}}$, $i = 0, 1, \ldots, k/k_1 - 1$.]

*§6. A general decomposition for cyclic codes

This section gives a method of Kasami [731] which expresses the Mattson–Solomon polynomials of any cyclic code of composite length in terms of MS polynomials of shorter codes. (The methods of §§3–5 do not always apply to reducible codes.)

Let $\mathscr{C}$ be an $[n = n_1 n_2, k, d]$ binary cyclic code, where $(n_1, n_2) = 1$, $n_1 > 1$, $n_2 > 1$. (The same procedure works over any finite field.)

Notation. As before, $a n_1 + b n_2 = 1$, $\alpha^n = 1$, $\beta = \alpha^{b n_2}$, $\gamma = \alpha^{a n_1}$. Also $I(i, j)$ is defined by (3).

Case (I). $\mathscr{C}$ is irreducible, with say nonzeros α^i, $i \in C_1$, where C_1 is the cyclotomic coset $\{1, 2, 2^2, 2^3, \ldots, 2^{k-1}\} \bmod n_1 n_2$. Write

$$C_1 = \{I(\mu_1, \nu_1), \ldots, I(\mu_k, \nu_k)\}.$$

Then $\mu_1, \ldots, \mu_k$ consists of several repetitions of the cyclotomic coset

$$C_1' = \{1, 2, 2^2, \ldots, 2^{r-1}\} \bmod n_1.$$

The set $\{\nu : I(1, \nu) \in C_1\}$ consists of $\{1, 2^r, 2^{2r}, \ldots, 2^{k-r}\} \bmod n_2$.

For example, if $\mathscr{C}$ is a $[63, 6, 32]$ code then $n = 63$, $n_1 = 7$, $n_2 = 9$, $k = 6$, and

$$C_1 = \{1, 2, 4, 8, 16, 32\}$$
$$= \{I(1, 1), I(2, 2), I(4, 4), I(1, 8), I(2, 7), I(4, 5)\},$$
$$C_1' = \{1, 2, 4\}, \, r = 3,$$
$$\{\nu : I(1, \nu) \in C_1\} = \{1, 8\}.$$

Let $\mathscr{A}$ be the $[n_1, r]$ irreducible binary cyclic code with nonzeros $\beta, \beta^2, \ldots, \beta^{2^{r-1}}$, and $\mathscr{B}$ the $[n_2, k - r + 1]$ irreducible cyclic code over $\mathrm{GF}(2^r)$ with nonzeros $\gamma, \gamma^{2^r}, \ldots, \gamma^{2^{k-r}}$.

The MS polynomial of a codeword of $\mathscr{A}$ is

$$F(Y) = T_r(\xi Y^{-1}),$$

where ξ is an arbitrary element of $\mathrm{GF}(2^r)$. The MS polynomial of a codeword of $\mathscr{B}$ is

$$\Phi(Z) = c Z^{-1} + (c Z^{-1})^{2^r} + \cdots + (c Z^{-1})^{2^{k-r}},$$

where c is an arbitrary element of $\mathrm{GF}(2^k)$.

Theorem 7. (Kasami.) *The MS polynomial of any codeword of $\mathscr{C}$ is obtained from $F(Y)$ by replacing ξ by $\Phi(Z)$ and setting $X = YZ$ and $Y^{n_1} = Z^{n_2} = 1$.*

Proof. Let $F(Y, Z)$ be the polynomial obtained in this way. Then

$$F(Y, Z) = T_r(\Phi(Z)Y^{-1})$$
$$= T_k(cX^{-I(1, 1)}) = T_k(cX^{-1})$$

after some algebra. But this *is* the MS polynomial of a codeword of $\mathscr{C}$.

Q.E.D.

For the preceding example,

$$F(Y, Z) = T_3(cY^{-1}Z^{-1} + c^8 Y^{-1}Z^{-8})$$
$$= T_3(cX^{-I(1, 1)} + c^8 X^{-I(1, 8)})$$
$$= T_6(cX^{-1}), \qquad\qquad c \in GF(2^6).$$

Case (II). In general, let $\mathscr{C}$ be any $[n, k]$ cyclic code, with nonzeros α^j, $j \in \{C_{a_i} : i = 1, \ldots, s\}$. Suppose $|C_{a_i}| = k_i$. For each $i = 1, \ldots, s$, let

$$C'_i = \{\mu : I(\mu, \nu) \in C_{a_i} \quad \text{for some } \nu\},$$

let b_i be the smallest number in C'_i, $r_i = |C'_i|$ and

$$K_i = \{\nu : I(b_i, \nu) \in C_{a_i}\}.$$

(It may happen that $C'_i = C'_j$ while $K_i \neq K_j$, or vice-versa.) Let $\mathscr{A}$ be the binary cyclic code of length n_1 with nonzeros β^j, $j \in \{C'_i, i = 1, \ldots, s\}$, and let $\mathscr{B}_i$ be the irreducible cyclic code over $GF(2^{r_i})$ of length n_2 with nonzeros γ^μ, $\mu \in K_i$. The MS polynomial of a codeword of $\mathscr{A}$ is

$$F(Y) = \sum_{i=1}^{s} T_{r_i}(\xi_i Y^{-b_i}),$$

for $\xi_i \in GF(2^{r_i})$. Note that, if $C'_i = C'_j$, this cyclotomic coset appears twice in the sum, once with ξ_i and once with ξ_j. The MS polynomial of a codeword of $\mathscr{B}_i$ is

$$\Phi_i(Z) = c_i Z^{-\mu} + (c_i Z^{-\mu})^{2^{r_i}} + \cdots + (c_i Z^{-\mu})^{2^{k_i - r_i}},$$

where $\mu \in K_i$, $c_i \in GF(2^{k_i})$.

Theorem 8. (Kasami.) *The MS polynomial of any codeword of $\mathscr{C}$ is obtained from $F(Y)$ by replacing ξ_i by $\Phi_i(Z)$, and setting $X = YZ$, $Y^{n_1} = Z^{n_2} = 1$.*

For example let $\mathscr{C}$ be the $[63, 18]$ code with nonzeros α^j, $j \in C_1 \cup C_5 \cup C_{15}$.

Then $n = 63$, $n_1 = 7$, $n_2 = 9$,

$$C_1 = \{I(1, 1), I(2, 2), I(4, 4), I(1, 8), I(2, 7), I(4, 5)\},$$
$$C_5 = \{I(5, 5), I(3, 1), I(6, 2), I(5, 4), I(3, 8), I(6, 7)\},$$
$$C_{15} = \{I(1, 6), I(2, 3), I(4, 6), I(1, 3), I(2, 6), I(4, 3)\},$$
$$C_1' = C_3' = \{1, 2, 4\}, b_1 = b_3 = 1, r_1 = r_3 = 3,$$
$$C_2' = \{5, 3, 6\}, b_2 = 3, r_2 = 3,$$
$$K_1 = K_2 = \{1, 8\}, K_3 = \{6, 3\},$$
$$\beta = \alpha^{36}, \gamma = \alpha^{28}.$$

$\mathcal{A}$ has nonzeros β^j, $j = 0, \ldots, 6$ and is the $[7, 6, 2]$ even weight code. $\mathcal{B}_1 = \mathcal{B}_2$ has nonzeros γ, γ^8, check polynomial $(x + \gamma)(x + \gamma^8) = x^2 + \alpha^{54}x + 1$, and $\mathcal{B}_3$ has nonzeros γ^3, γ^6 and check polynomial $(x + \gamma^3)(x + \gamma^6) = x^2 + x + 1$.

The MS polynomials of $\mathcal{A}, \mathcal{B}_1, \mathcal{B}_2, \mathcal{B}_3$ are respectively

$$F(Y) = T_3(\xi_1 Y^{-1}) + T_3(\xi_2 Y^{-3}) + T_3(\xi_3 Y^{-1}),$$
$$\Phi_1(Z) = c_1 Z^{-1} + c_1^8 Z^{-8},$$
$$\Phi_2(Z) = c_2 Z^{-1} + c_2^8 Z^{-8},$$
$$\Phi_3(Z) = c_3 Z^{-3} + c_3^8 Z^{-6}.$$

After replacing ξ_i by $\Phi_i(Z)$ we obtain

$$T_6(c_1 X^{-1}) + T_6(c_2 X^{-5}) + T_6(c_3 X^{-15}),$$

which is the MS polynomial of a codeword of $\mathcal{C}$.

PART II: Other methods of combining codes

We have already seen a few ways of putting codes together to get new ones:

(i) The direct sum of two codes (Problem 17 of Ch. 1, §9 of Ch. 2);

(ii) The technique of pasting two codes side by side (Problem 17 of Ch. 1, Fig. 2.6);

(iii) The $|u|u+v|$ construction (§9 of Ch. 2);

(iv) Concatenating two codes (§11 of Ch. 10);

and of course the various product constructions described in Part I of this chapter. It is very easy to invent other constructions, and indeed the literature contains a large number (see the papers mentioned in the Notes). What makes the constructions given below special is that they have all been used to find exceptionally good codes. They are record holders. We usually speak of one code as being better than another if it contains more codewords (and has the same length and minimum distance), or alternatively if it has a greater length (and the same redundancy and minimum distance). This neglects all problems of encoding and decoding, but we justify this by arguing that once good codes, by this definition, have been found, decoding algorithms will follow. Appendix A contains a table giving the best codes presently known by length up to 512; many of these were found by the methods described here.

We begin (in §7) by describing techniques for increasing the length of a code in an efficient way. The *tail constructions* X and $X4$ are given in §7.1 and §7.2. Section 7.3 is a digression, giving a brief summary of what is known about the largest single- and double-error-correcting codes; even these comparatively simple problems are unsolved. §7.4 describes the $|a+x|b+x|a+b+x|$ construction, important because it is one of the simplest ways to get the Golay code. Then a construction of Piret for doubling the length of an irreducible code is given in §7.5.

§8 gives some constructions related to concatenated codes. In particular, §8.2 presents a powerful technique of Zinov'ev for generalizing concatenated codes.

Finally, section 9 gives some clever methods for shortening a code.

§7. Methods which increase the length

§7.1. Construction X: adding tails to the codewords. This combines three codes to produce a fourth. Suppose $\mathscr{C}_1$ and $\mathscr{C}_2$ are (n_1, M_1, d_1) and $(n_1, M_2 = bM_1, d_2)$ codes, with the property that $\mathscr{C}_2$ is the union of b disjoint translates

of $\mathscr{C}_1$:

$$\mathscr{C}_2 = (x_1 + \mathscr{C}_1) \cup (x_2 + \mathscr{C}_1) \cup \cdots \cup (x_b + \mathscr{C}_1), \tag{14}$$

for suitable vectors $x_1, x_2, \ldots, x_b$.

Example. (6) Let $\mathscr{C}_1$ be the $(4, 2, 4)$ repetition code $\{0000, 1111\}$, and $\mathscr{C}_2$ the $(4, 8, 2)$ even weight code $\{0000, 1111, 0011, 1100, 0101, 1010, 1001, 0110\}$. Then $b = 4$ and

$$\mathscr{C}_2 = \mathscr{C}_1 \cup (0011 + \mathscr{C}_1) \cup (0101 + \mathscr{C}_1) \cup (1001 + \mathscr{C}_1).$$

Let $\mathscr{C}_3 = \{y_1, \ldots, y_b\}$ be any (n_3, b, d_3) code. In the example we could take $\mathscr{C}_3$ to be the $(3, 4, 2)$ code $\{000, 011, 101, 110\}$. Then the new code $\mathscr{C}$ is defined to be

$$|x_1 + \mathscr{C}_1|y_1| \cup |x_2 + \mathscr{C}_1|y_2| \cup \cdots \cup |x_b + \mathscr{C}_1|y_b|. \tag{15}$$

In other words $\mathscr{C}$ consists of the vectors $|x_i + u|y_i|$ for $u \in \mathscr{C}_1$, $i = 1, \ldots, b$.

Simply stated, $\mathscr{C}_2$ is divided into cosets of $\mathscr{C}_1$ and a different tail (y_i) is attached to each coset – see Fig. 18.4.

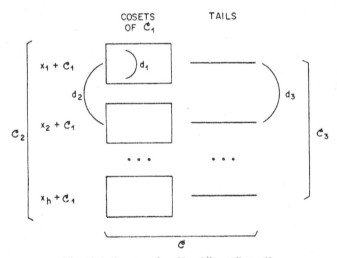

Fig. 18.4. Construction X: adding tails to $\mathscr{C}_2$.

Theorem 9. *The new code $\mathscr{C}$ has parameters*

$$(n_1 + n_3, M_2, \min \{d_1, d_2 + d_3\}).$$

Proof. Let $X = |x|y|$ and $X' = |x'|y'|$ be distinct codewords of $\mathscr{C}$. If x and x' belong to the same coset of $\mathscr{C}_1$ in $\mathscr{C}_2$, dist $(X, X') \geq$ dist $(x, x') \geq d_1$. If x and x' belong to different cosets, then dist $(x, x') \geq d_2$ and dist $(y, y') \geq d_3$.　　　Q.E.D.

Example. (6) (cont.) Attaching the tails $\mathscr{C}_3$ we find that $\mathscr{C}$ is the $(7, 8, 4)$ linear code {0000000, 1111000, 0011011, 1100011, 0101101, 1010101, 1001110, 0110110}.

As this example shows, if $\mathscr{C}_1$, $\mathscr{C}_2$ and $\mathscr{C}_3$ are linear then we can arrange the x_i and y_i so that $\mathscr{C}$ is linear.

Construction X can be applied whenever nested codes are found. For example, the primitive BCH code of designed distance d_2 is a union of cosets of the BCH code of designed distance $d_1 > d_2$ (§6 of Ch. 7). So the construction applies to any pair of primitive BCH codes. E.g. if $d_1 \geqslant d_2 + 2$ and $b = 2^k$, then $\mathscr{C}_3$ can be taken to be the $(k + 1, 2^k, 2)$ code. In this case we are combining $[n_1, k_1, d_1 \geqslant d_2 + 2]$ and $[n_1, k + k_1, d_2]$ codes to obtain an $[n_1 + k + 1, k + k_1, d_2 + 2]$ code. (This generalizes the Andryanov–Saskovets construction [113, p. 333].) Some examples are

$\mathscr{C}_1$	$\mathscr{C}_2$	$\mathscr{C}$
$[31, 6, 15]$	$[31, 11, 11]$	$[37, 11, 13]$
$[63, 36, 11]$	$[63, 39, 9]$	$[67, 39, 11]$

Codes formed by applying construction X to BCH codes are indicated by XB in Fig. 2 of Appendix A.

Other nested codes to which the construction may be applied are cyclic codes (indicated by XC); Preparata codes, §6 of Ch. 15 (XP); and the Delsarte–Goethals codes, Fig. 15.10 (DG). E.g. in view of Theorem 36 of Ch. 15, we can use $\mathscr{C}_1 = \mathscr{P}(m)^*$, $\mathscr{C}_2 = $ a Hamming code, and obtain an infinite family of nonlinear

$$(2^m + m - 1, 2^{2^m - m - 1}, 5), \quad m = 4, 6, 8, \ldots \tag{16}$$

codes. If $m = 4$ this is a $(19, 2^{11}, 5)$ code.

Problems. (13) (a) Apply Construction X to the $\mathscr{D}\mathscr{G}(m, d)$ codes (§5 of Ch. 15) to obtain $(74, 2^{17}, 28)$ and $(74, 2^{12}, 32)$ codes.

(b) Apply Construction X with $\mathscr{C}_1 = \mathscr{R}(1, 6)$, $\mathscr{C}_2 = \mathscr{K}(6)$ to obtain a $(70, 2^{12}, 30)$ code.

(14) (a) *Construction X3.* Let the $(n_1, M_3 = cM_2 = bcM_1, d_3)$ code $\mathscr{C}_3$ be a union of c disjoint translates of the $(n_1, M_2 = bM_1, d_2)$ code $\mathscr{C}_2$, say

$$\mathscr{C}_3 = \bigcup_{i=1}^{c} (u_i + \mathscr{C}_2);$$

where $\mathscr{C}_2$ is the union of b disjoint translates of the (n_1, M_1, d_1) code $\mathscr{C}_1$, say

$$\mathscr{C}_2 = \bigcup_{j=1}^{b} (v_j + \mathscr{C}_1).$$

Each codeword of $\mathscr{C}_3$ can be written uniquely as $u_i + v_j + c$, $c \in \mathscr{C}_1$. Also let

$\mathscr{C}_4 = \{y_1, \ldots, y_b\}$ and $\mathscr{C}_5 = \{z_1, \ldots, z_c\}$ be (n_4, b, Δ) and (n_5, c, δ) codes. Show that the code consisting of all vectors $|u_i + v_j + c \,|\, y_i \,|\, z_i|$ has parameters

$$(n_1 + n_4 + n_5, M_3, d = \min\{d_1, d_2 + \Delta, d_3 + \delta\}).$$

(b) For example, take $\mathscr{C}_1 = [255, 179, 21]$, $\mathscr{C}_2 = [255, 187, 19]$, $\mathscr{C}_3 = [255, 191, 17]$ BCH codes, and $\mathscr{C}_4 = [9, 8, 2]$, $\mathscr{C}_5 = [8, 4, 4]$ codes to obtain a $[272, 191, 21]$ code.

(c) The following *double tail* construction has been successfully used by Kasahara et al. [721]. Let α be a primitive element of $GF(2^m)$, and define three BCH codes:

Code	Zeros	Parameters
$\mathscr{B}_1$	$1, \alpha, \alpha^2, \ldots, \alpha^{2i}$	$[2^m - 1, k, 2i + 2]$
$\mathscr{B}_2$	$\alpha^{-2}, \alpha^{-1}, \ldots, \alpha^{2i}$	$[2^m - 1, k - m, 2i + 4]$
$\mathscr{B}_3$	$1, \alpha, \alpha^2, \ldots, \alpha^{2i+2}$	$[2^m - 1, k - m', 2i + 4]$

where m' is the degree of the minimal polynomial of α^{2i+1}. Now $\mathscr{B}_1$ is the union of 2^m cosets of $\mathscr{B}_2$. So we can append a tail $r(x)$ of length m to $u(x) \in \mathscr{B}_1$ such that vectors in the same coset have the same tail, but vectors in different cosets have different tails. Again $\mathscr{B}_1$ is the union of $2^{m'}$ cosets of $\mathscr{B}_3$, and we add a tail $s(x)$ of length m' in a similar way. Then the new code $\mathscr{C}$ consists of the vectors

$$|u(x) \,|\, r(x) \,|\, s(x) \,|\, s(1)|.$$

Show that $\mathscr{C}$ has parameters

$$[2^m + m + m', k, 2i + 5]. \tag{17}$$

(d) Use this construction to obtain $[22, 6, 9]$ $(m = 4)$ and $[76, 50, 9]$ $(m = 6)$ codes.

For other examples and generalizations see [721]. Codes obtained in this way are denoted by XQ in Fig. 2 of Appendix A.

§7.2. *Construction X4: combining four codes.* Suppose we are given four codes: an (n_1, M_1, d_1) code $\mathscr{C}_1$, an $(n_1, M_2 = bM_1, d_2)$ code $\mathscr{C}_2$, an $(n_3, M_3, d_3 \geq d_1)$ code $\mathscr{C}_3$, and an $(n_3, M_4 = bM_3, d_4)$ code $\mathscr{C}_4$, with the properties that (i) $\mathscr{C}_2$ is a union of b disjoint cosets of $\mathscr{C}_1$:

$$\mathscr{C}_2 = (x_1 + \mathscr{C}_1) \cup (x_2 + \mathscr{C}_1) \cup \cdots \cup (x_b + \mathscr{C}_1),$$

and (ii) $\mathscr{C}_4$ is a union of b disjoint cosets of $\mathscr{C}_3$:

$$\mathscr{C}_4 = (y_1 + \mathscr{C}_3) \cup (y_2 + \mathscr{C}_3) \cup \cdots \cup (y_b + \mathscr{C}_3);$$

for suitable vectors $x_1, \ldots, x_b$ and $y_1, \ldots, y_b$. Then the new code $\mathscr{C}$ consists of all the vectors

$$|x_i + u \,|\, y_i + v|, \quad i = 1, \ldots, b, u \in \mathscr{C}_1, v \in \mathscr{C}_3. \tag{18}$$

Simply stated, the vectors of the ith coset of $\mathscr{C}_1$ in $\mathscr{C}_2$ are appended to the vectors of the ith coset of $\mathscr{C}_3$ in $\mathscr{C}_4$ in all possible ways – see Fig. 18.5.

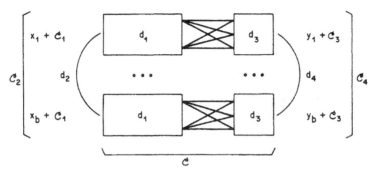

Fig. 18.5. Construction X4, combining four codes.

Theorem 10. *The new code $\mathscr{C}$ has parameters*

$$(n_1 + n_3, M_2 M_3, d \geq \min \{d_1, d_2 + d_4\}). \tag{19}$$

The proof is immediate. If $\mathscr{C}_1 - \mathscr{C}_4$ are linear then $\mathscr{C}$ can be made linear.

Example. (7) Take $\mathscr{C}_1 = $ the $(15, 256, 5)$ Nordstrom–Robinson code (Lemma 22 of Ch. 15), $\mathscr{C}_2 = $ the $(15, 1280, 3)$ code consisting of $\mathscr{C}_1$ together with any four of the eight cosets of $\mathscr{C}_1$ in the $[15, 11, 3]$ Hamming code, $\mathscr{C}_3 = $ the $(5, 2, 5)$ code $\{00000, 11111\}$, and $\mathscr{C}_4 = $ the $(5, 10, 2)$ code $\{00000, 11111, 11000, 00111, 10100, 01011, 10010, 01101, 10001, 01110\}$. Then $\mathscr{C}$ is a $(20, 2560, 5)$ code.

Further examples are given in §7.3 and in Fig. 2 of Appendix A (indicated by $X4$).

Sugiyama *et al.* [1291] have successfully applied Constructions X and $X4$ and modifications of them to Goppa codes. A large number of good codes found in this way are denoted by GP in Fig. 2 of Appendix A.

Problem. (15) Apply Construction $X4$ to extend e-error-correcting BCH codes of length $n_1 = 2^m - 1$. Let $m = \alpha e - \beta$, where $0 \leq \beta < e$, and take $\mathscr{C}_1, \mathscr{C}_2, \mathscr{C}_3$ equal to $[n_1, n_1 - em, 2e + 1]$, $[n_1, n_1 - (e - 1)m, 2e - 1]$ and $[2^\alpha, 2^\alpha - e\alpha - 1, 2e + 2]$ BCH codes. Take $\mathscr{C}_4$ equal to $\mathscr{C}_3$ plus $2^m - 1$ other cosets of $\mathscr{C}_3$ in the $[2^\alpha, 2^\alpha - 1, 2]$ even weight code. Show that $\mathscr{C}$ is a $[2^m + 2^\alpha - 1, 2^m + 2^\alpha - em - \beta - 2, 2e + 1]$ code. In the most favorable case, when m is divisible by e and

$\beta = 0$, the number of information symbols in $\mathscr{C}_1$ has been increased by $2^{m/e} - 1 \approx n^{1/e}$ at the cost of adding one check symbol.

§7.3. *Single- and Double-Error-Correcting Codes*

Single-error-correcting codes. The best binary linear single-error-correcting codes are known. They are Hamming and shortened Hamming codes, with parameters $[n, n - m, 3]$, where $2^{m-1} \leq n \leq 2^m - 1$.

Problem. (16) Prove this statement!

For at least half the values of n there exist nonlinear codes with more codewords – see Theorem 34 of Ch. 2 and Fig. 2 of Appendix A.

Research Problems (18.1). Determine $A(n, 3)$ for $n \geq 9$. (For $n \leq 8$ see Fig. 1 of Appendix A.)

Double-error-correcting (d.e.c.) codes.
  Much less is known.
  (18.2) What is the longest binary linear double-error-correcting code of redundancy r, for $r = 4, 5, \ldots$? If this length is denoted by $n_L(r)$, the following values are known (from Helgert and Stinaff [636]):

r:	4	5	6	7	8	9	10	$\cdots$
$n_L(r)$:	5	6	8	11	17	23–29	32–42	$\cdots$

$$\tag{20}$$

Now let us study the best linear *or* nonlinear d.e.c. codes. We have already seen the following codes.
  (i) Primitive d.e.c. BCH codes with parameters $[2^m - 1, 2^m - 1 - 2m, 5]$. However, these are always inferior to:
  (ii) Nonprimitive d.e.c. BCH codes with parameters $[2^m, 2^m - 2m, 5]$, obtained by shortening the code of length $2^m + 1$ and zeros α^i, $i \in C_0 \cup C_1$.
  (iii) The Preparata codes $\mathscr{P}(m)^*$ with parameters $(2^m - 1, 2^{2^m - 2m}, 5)$ for even $m \geq 4$.
  (iv) The extended Preparata codes (16).
  (v) The $(11, 24, 5)$ Hadamard code of §3 of Ch. 2, and the $(20, 2560, 5)$ code of Example 7 above.
  The following d.e.c. codes are also known:
  (vi) $[2^m + 2^{[(m+1)/2]} - 1, 2^m + 2^{[(m+1)/2]} - 2m - 2, 5]$ shortened BCH codes, given in Problem 28 below.
  (vii) A $[74, 61, 5]$ shortened alternant code (Helgert [633]) and a $[278, 261, 5]$ shortened generalized BCH code (Chien and Choy [286]).
  (viii) Wagner's $[23, 14, 5]$ quasiperfect code ([1378]).

Research Problem (18.3). Give a simple construction of the latter code.

Let $n^*(r)$ be the length of the longest binary (linear or nonlinear) d.e.c. code of redundancy r. Of course $n^*(r) \geq n_L(r)$. From (ii) and (iii) we have

$$n^*(4a + 2) \geq 2^{2a+1} \quad \text{and} \quad n^*(4a + 3) \geq 2^{2a+2} - 1. \tag{21}$$

Problem. (17) Show $n^*(4a + 3) = 2^{2a+2} - 1$ [Hint: Corollary 14 of Ch. 17.]

Construction $X4$ can be used to get further lower bounds on $n^*(r)$. Take $\mathscr{C}_1 = \mathscr{P}(m)^*$, $\mathscr{C}_2 =$ Hamming code $\mathscr{H}_m$, $\mathscr{C}_3 =$ the longest distance 6 code of redundancy m, of length $n^*(m - 1) + 1$, and $\mathscr{C}_4 =$ an even weight code of the same length. The new code $\mathscr{C}$ has length $2^m + n^*(m - 1)$, redundancy $2m$, and so

$$n^*(2m) \geq n^*(m - 1) + 2^m, \quad \text{for even } m \geq 4. \tag{22}$$

Thus the Preparata code $\mathscr{P}(m)^*$ has been extended by the addition of $\sqrt{(n + 1)}$ information symbols at the cost of adding one check symbol.

Problems. (18) Show that, for even $m \geq 4$, $n^*(2m + 1) \geq n^*(m) + 2^m$.
(19) Show $n^*(12) \geq 70$, $n^*(16) \geq 271$, $n^*(20) \geq 1047$ and $n^*(21) \geq 1056$.

The best lower bounds on $n^*(r)$ known to us are shown in Fig. 2 of Appendix A (and Table II of [1237]). The only exact values known are $n^*(6) = 8$, $n^*(7) = 15$, and $n^*(4a + 3)$ from Problem 17. Also $n^*(8) = 19$ or 20.

Research Problem (18.4). Determine $n^*(r)$.

Even less is known about triple-error-correcting codes!

§7.4. The $|a + x|b + x|a + b + x|$ Construction. Let $\mathscr{C}_1$ and $\mathscr{C}_2$ be (n, M_1, d_1) and (n, M_2, d_2) binary codes. The new code $\mathscr{C}$ consists of all vectors

$$|a + x|b + x|a + b + x|, \quad a, b \in \mathscr{C}_1, x \in \mathscr{C}_2. \tag{23}$$

Clearly $\mathscr{C}$ is a code of length $3n$ containing $M_1^2 M_2$ codewords, and is linear if $\mathscr{C}_1$ and $\mathscr{C}_2$ are. No simple formula is known for the minimum distance of $\mathscr{C}$, although a lower bound is given by:

Theorem 11. *For any binary vectors a, b, x,*

$$\text{wt}|a+x|b+x|a+b+x| = 2\text{wt}(a.\text{OR}.b) - \text{wt}(x) + 4s,$$
$$\geq 2\text{wt}(a.\text{OR}.b) - \text{wt}(x), \qquad (24)$$

where s is the number of times a is 0, b is 0 and x is 1.

Problem. (20) Prove Theorem 11.

Theorem 12. (Turyn.) *Let $\mathscr{B}_1$ be the $[7, 4, 3]$ Hamming code with nonzeros $\alpha^i, i \in C_1$, and let $\mathscr{B}_2$ be formed by reversing the codewords of $\mathscr{B}_1$. Let $\mathscr{C}_1$ and $\mathscr{C}_2$ be $[8, 4, 4]$ codes obtained by adding overall parity checks to $\mathscr{B}_1$ and $\mathscr{B}_2$. Then the code $\mathscr{C}$ given by (23) is the $[24, 12, 8]$ Golay code $\mathscr{G}_{24}$.*

Proof. Only the minimum distance needs to be checked. That this is the Golay code then follows from Theorem 14 of Ch. 20. Let $u = |a+x|b+x|a+b+x|$ be a nonzero codeword of $\mathscr{C}$. Each of a, b, x has weight 0, 4 or 8. (i) If at most one of a, b, x has weight 4, wt $(u) \geq 8$. (ii) If two of a, b, x have weight 4, then wt $(u) \geq 8$. (For if wt $(a) = 4$, wt $(x) = 4$ then wt $(a+x) \geq 2$, etc.) (iii) Suppose wt $(a) = $ wt $(b) = $ wt $(x) = 4$. If $a \neq b$, wt $(a.\text{OR}.b) \geq 6$ and Theorem 11 implies wt $(u) \geq 8$. Hence in all cases wt $(u) \geq 8$.   Q.E.D.

Problem. (21) Generalizing Theorem 12, let $\mathscr{C}_1$ and $\mathscr{C}_2$ be $[2^m, m+1, 2^{m-1}]$ first-order Reed–Muller codes one of which is the reverse of the other (except for the overall parity check). Show that $\mathscr{C}$ has parameters

$$[3 \cdot 2^m, 3m+3, 2^m], m \geq 3.$$

[Hint: see [1237].] When $m = 4$ this is a $[48, 15, 16]$ code.

Research Problem (18.5). Find other applications of (23). Are there other constructions like (23) which give good codes?

§7.5. Piret's construction. This is a technique for doubling the length of an irreducible code. Let $\mathscr{C}$ be an $[n, k, d]$ irreducible cyclic binary code, with idempotent $\theta_1(x)$ (see Ch. 8). Here k is the smallest integer such that n divides $2^k - 1$. Put $N = (2^k - 1)/n$, let ξ be a primitive element of $\text{GF}(2^k)$, and let $\alpha = \xi^N$. Thus $\alpha^n = 1$, and the nonzeros of $\mathscr{C}$ may be taken to be $\alpha, \alpha^2, \alpha^{2^2}, \ldots, \alpha^{2^{k-1}}$. Also $\mathscr{C}$ consists of the codewords $i(x)\theta_1(x) \pmod{x^n - 1}$, where deg $i(x) < k$.

We recall from Ch. 8 that $\mathscr{C}$ is isomorphic to $GF(2^k)$, with

Element of $\mathscr{C}$	$\leftrightarrow$	Element of $GF(2^k)$
$\theta_1(x)$	$\leftrightarrow$	1
$x\theta_1(x)$	$\leftrightarrow$	α
$\gamma(x)\theta_1(x)$	$\leftrightarrow$	ξ (this defines $\gamma(x)$)

Then any nonzero codeword of $\mathscr{C}$ can be written as

$$\gamma(x)^j\theta_1(x) \leftrightarrow \xi^j,$$

$0 \leqslant j \leqslant 2^k - 2$. Let $w_j = \text{wt}\,(\gamma(x)^j\theta_1(x))$. Then $w_{j+N} = w_j$. (For $\gamma(x)^{j+N}\theta_1(x) \leftrightarrow \xi^{j+N} = \alpha\xi^j \leftrightarrow x\gamma(x)^j\theta_1(x)$.) We now choose the integer a so as to maximize

$$d' = \min_{0 \leqslant j \leqslant N-1} (w_j + w_{j+a}). \tag{24a}$$

The new code $\mathscr{D}$ consists of 0 and the vectors

$$u_j = |\gamma(x)^j\theta_1(x)|\gamma(x)^{j+a}\theta_1(x)|, \tag{25}$$

$0 \leqslant j \leqslant 2^k - 2$. Then $\mathscr{D}$ is a $[2n, k, d']$ code where d' is given by (24a).

Often it is possible to enlarge $\mathscr{D}$ by adding one or both of the generators $11 \cdots 100 \cdots 0$ and $00 \cdots 011 \cdots 1$, and adding one or two overall parity checks.

For example, let $\mathscr{C}$ be the $[9, 6, 2]$ code with idempotent $\theta_1(x) = x^3 + x^6$. Here $N = 7$, ξ is a primitive element of $GF(2^6)$ (see Fig. 4.5), $\alpha = \xi^7$, $\gamma(x) = x^3 + x^4 + x^6 + x^8$, and $(w_0, w_1, \ldots, w_6) = (2, 6, 6, 4, 6, 4, 4)$. The best choice for a is 1, and $\mathscr{D}$ is an $[18, 6, 6]$ code which can be enlarged to a $[20, 8, 6]$ code.

Codes obtained in this way are indicated by the symbol PT in Fig. 2 of Appendix A.

Problems. (22) Let $\mathscr{C}$ be a $[17, 8, 6]$ code, with $N = 15$. Obtain $[34, 8, 14]$ and $[35, 9, 14]$ codes.

(23) Let $\mathscr{C}$ be a $[21, 6, 8]$ code, with $N = 3$. Obtain a $[44, 8, 18]$ code.

Research Problem (18.6). Let $\mathscr{C}$ be a cyclic code of length n. How should $a(x)$ be chosen so that the minimum distance of the code $\{|u(x)|a(x)u(x)(\text{mod } x^n - 1)|: u(x) \in \mathscr{C}\}$ is as large as possible?

§8. Constructions related to concatenated codes

§8.1. A method for improving concatenated codes. Let $\mathscr{C}'$ be an $[n', k', d' = n' - k' + 1]$ MDS code over $GF(2^k)$ (see Ch. 11), with codewords written as column vectors. If each symbol in a codeword is replaced by the corresponding binary k-tuple (as in Example 2, §5 of Ch. 10), an $n' \times k$ binary

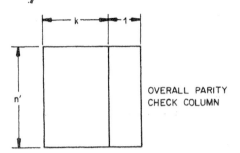

Fig. 18.6. A typical codeword of $\mathscr{C}$.

array is obtained. Adding an overall parity check to each row gives an $[n'(k+1), kk', 2d']$ binary code $\mathscr{C}$ – see Fig. 18.6. (This is a simple form of concatenated code.) The minimum distance of $\mathscr{C}$ is at least $2d'$ because there are at least d' nonzero rows, each having weight ≥ 2.

Kasahara et al. [722] have pointed out that it is often possible to increase the dimension of $\mathscr{C}$ by 1. The new code $\mathscr{D}$ consists of the codewords of $\mathscr{C}$ together with the codewords formed by complementing the last column of Fig. 18.6.

Theorem 13. *The new code $\mathscr{D}$ has parameters*

$$[n'(k+1), kk'+1, 2d'], \tag{26}$$

provided $n' \geq 2d'$.

Problem. (24) Prove this.

For example, let $\mathscr{C}'$ be a $[17, 11, 7]$ Reed–Solomon code over $GF(2^4)$. Then $\mathscr{D}$ is an $[85, 45, 14]$ code. For further examples and a generalization see [722]. Codes obtained in this way are denoted by KS in Fig. 2 of Appendix A.

Problem. (25) Construct $[156, 76, 24]$ and $[168, 76, 28]$ codes.

§8.2. Zinov'ev's generalized concatenated codes. This is a generalization of concatenated codes, and produces a number of good codes (denoted by ZV in Fig. 2 of Appendix A). The construction calls for the following codes. (1) A collection of r codes $\mathscr{A}_1, \ldots, \mathscr{A}_r$, where $\mathscr{A}_i$ is an (n, N_i, δ_i) code over $GF(q_i)$ and q_i is a power of 2. (2) An $(m, q_1 q_2 \ldots q_r, d_1)$ binary code $\mathscr{B}^{(0)}$ which is the union of q_1 disjoint codes $\mathscr{B}_{i_1}^{(1)} (0 \leq i_1 \leq q_1 - 1)$, where $\mathscr{B}_{i_1}^{(1)}$ is an $(m, q_2 \ldots q_r, d_2)$ code. Each $\mathscr{B}_{i_1}^{(1)}$ must be the union of q_2 disjoint codes $\mathscr{B}_{i_1 i_2}^{(2)} (0 \leq i_2 \leq q_2 - 1)$, where $\mathscr{B}_{i_1 i_2}^{(2)}$ has parameters $(m, q_3 \ldots q_r, d_3)$, and so on. Finally, each $\mathscr{B}_{i_1 i_2 \ldots i_{r-2}}^{(r-2)}$

must be the union of q_r disjoint codes $\mathscr{B}^{(r-1)}_{i_1 i_2 \ldots i_{r-1}} (0 \leqslant i_{r-1} \leqslant q_{r-1} - 1)$, each with parameters (m, q_r, d_r). Thus

$$\mathscr{B}^{(0)} = \bigcup_{i_1=0}^{q_1-1} \mathscr{B}^{(1)}_{i_1}, \quad \mathscr{B}^{(1)}_{i_1} = \bigcup_{i_2=0}^{q_2-1} \mathscr{B}^{(2)}_{i_1 i_2}, \ldots,$$

and $d_1 < d_2 < \cdots < d_r$. A typical codeword of $\mathscr{B}^{(0)}$ will be denoted by $b_{i_1 i_2 \ldots i_r} (0 \leqslant i_1 \leqslant q_1 - 1, \ldots, 0 \leqslant i_r \leqslant q_r - 1)$, with subscripts chosen so that $b_{i_1 i_2 \ldots i_r}$ belongs to $\mathscr{B}^{(1)}_{i_1}$ and to $\mathscr{B}^{(2)}_{i_1 i_2}$, etc., and is the i_r^{th} codeword of $\mathscr{B}^{(r-1)}_{i_1 i_2 \ldots i_{r-1}}$.

The construction. Form an $n \times r$ array

$$\begin{bmatrix} a_1^{(1)} & a_1^{(2)} & \cdots & a_1^{(r)} \\ a_2^{(1)} & a_2^{(2)} & \cdots & a_2^{(r)} \\ \multicolumn{4}{c}{\cdots\cdots\cdots\cdots\cdots} \\ a_n^{(1)} & a_n^{(2)} & \cdots & a_n^{(r)} \end{bmatrix} \tag{27}$$

where the first column is in $\mathscr{A}_1$, the second is in $\mathscr{A}_2, \ldots$. Replace each row $a_i^{(1)}, a_i^{(2)}, \ldots, a_i^{(r)}$ by the binary vector

$$b_{a_i^{(1)}, a_i^{(2)}, \ldots, a_i^{(r)}}$$

(To do this, label the elements of each field $GF(q_i)$ by the numbers $0, 1, \ldots, q_i - 1$ in some arbitrary but fixed way.) The resulting $n \times m$ binary arrays (considered as binary vectors of length nm) form the new code $\mathscr{L}$, called a *generalized concatenated code*.

Theorem 14. (Zinov'ev [1470].) *$\mathscr{L}$ is an*

$$(nm, N_1 N_2 \ldots N_r, d \geqslant \min \{\delta_1 d_1, \ldots, \delta_r d_r\})$$

binary code. (Of course $\mathscr{L}$ is not necessarily linear).

Example. Take $\mathscr{B}^{(0)} = (4, 16, 1)$ binary code. $\mathscr{B}^{(0)}$ is the union of the $(4, 8, 2)$ even weight code $\mathscr{B}^{(1)}_0$ and the $(4, 8, 2)$ odd weight code $\mathscr{B}^{(1)}_1$. Both $\mathscr{B}^{(1)}_0$ and $\mathscr{B}^{(1)}_1$ are the union of 4 translates of the $(4, 2, 4)$ code $\{0000, 1111\}$. Thus $r = 3$, $q_1 = 2$, $q_2 = 4$, $q_3 = 2$. We also take

$$\mathscr{A}_1 = (8, 2, 8) \text{ code over } GF(2),$$
$$\mathscr{A}_2 = (8, 4^4, 4) \text{ code over } GF(4) \text{ (generated by (43) of Ch. 1)},$$
$$\mathscr{A}_3 = (8, 2^7, 2) \text{ code over } GF(2).$$

Then $\mathscr{L}$ is a $(32, 2^{16}, 8)$ code.

Proof of Theorem. Only the minimum distance is not obvious. Suppose $(a_j^{(i)})$ and $(c_j^{(i)})$ are two arrays (27), which first differ in the νth column. Then they differ in at least δ_ν places in the νth column. By definition $b_{i_1 i_2 \ldots i_{\nu-1} \alpha \ldots}$ and

$b_{i_1 i_2 \ldots i_{\nu-1} \beta} \ldots (\alpha \neq \beta)$ both belong to $\mathcal{B}^{(\nu-1)}_{i_1 i_2 \ldots i_{\nu-1}}$, and so differ in at least d_ν places. Therefore the corresponding binary codewords of $\mathcal{X}$ differ in at least $\delta_\nu d_\nu$ places. Q.E.D.

Problems. (26) Use $r = 2$, $q_1 = 8$, $q_2 = 2$, $\mathcal{B}^{(0)} = [8, 4, 4]$, $\mathcal{B}^{(1)}_1 = [8, 1, 8]$ binary codes, $\mathcal{A}_1 = [7, 2, 6]$ code over GF(8), $\mathcal{A}_2 = [7, 4, 3]$ code over GF(2) to obtain a [56, 10, 24] binary code $\mathcal{X}$.

(27) Use $r = 4$, $\mathcal{B}^{(0)} = [16, 11, 4]$, $\mathcal{B}^{(1)}_i = (16, 2^8, 6)$, $\mathcal{B}^{(2)}_{ij} = [16, 5, 8]$, $\mathcal{B}^{(3)}_{ijk} = [16, 1, 16]$, $\mathcal{A}_1 = [6, 1, 6]$ over GF(8), $\mathcal{A}_2 = [6, 3, 4]$ over GF(8), $\mathcal{A}_3 = [6, 4, 3]$ over GF(16), $\mathcal{A}_4 = [6, 5, 2]$ over GF(2) to obtain a $(96, 2^{33}, 24)$ binary code $\mathcal{X}$.

§9. Methods for shortening a code

§9.1. Constructions Y1–Y4.

Let $\mathcal{C}$ be an $[n, k, d]$ binary code with parity check matrix H. If i information symbols are set to zero an $[n - i, k - i, d]$ code is obtained. We can do better if we know the minimum distance d' of the dual code $\mathcal{C}^\perp$. For then we can assume that the first row of H has weight d'. If we delete the columns of H where the first row is 1, the remaining matrix H_s has a row of zeros and is the parity check matrix of a code $\mathcal{C}_1$ of length $n - d'$, dimension $(n - d') - (n - k - 1) = k - d' + 1$, and minimum distance at least d (by Theorem 10 of Ch. 1). In fact $\mathcal{C}_1$ consists of exactly those codewords of $\mathcal{C}$ which have zeros under the 1's of the first row of H (with these zeros deleted). In short:

$[n, k, d]$ code, $d' =$ dist. of dual

$$\Rightarrow [n - d', k - d' + 1, \geqslant d] \text{ code.} \tag{28}$$

An example was given in §8 of Ch. 1, where this method was used to construct the Nordstrom–Robinson code from the Golay Code. Further examples are given in Hashim [623], Helgert and Stinaff [636, 637] and Sloane et al. [1237]. Codes obtained in this way are indicated by $Y1$ in Fig. 2 of Appendix A.

Problems. (28) Take $\mathcal{C}$ to be a $[2^{m+1}, 2^{m+1} - 2m - 3, 6]$ extended BCH code. From Problem 10 of Ch. 15, the dual code $\mathcal{C}^\perp$ has minimum distance $d' = 2^m - 2^{\lfloor (m+1)/2 \rfloor}$. Hence obtain a d.e.c. code of length $2^m + 2^{\lfloor (m+1)/2 \rfloor} - 1$ and redundancy $2m + 1$.

(b) The dual of the [48, 15, 16] code of Problem 21 has minimum distance 4. Obtain a [44, 12, 16] code.

(29) *Construction Y2.* Suppose the first row of H is $11 \ldots 100 \ldots 0$, of weight d', and let S consist of the codewords of $\mathcal{C}$ beginning with d' zeros. Thus $\mathcal{C}_1$ is S with the first d' coordinates deleted. Let T be the union of S and all of $d' - 1$ cosets of S in $\mathcal{C}$ with coset leaders $110^{n-2}, 1010^{n-3}, \ldots, 10^{d'-2}10^{n-d'}$.

Show that if the first d' coordinates of T are deleted the result is an

$$(n - d', d'2^{k-d'+1}, \geqslant d - 2) \tag{29}$$

nonlinear code. For example, use the fact that the dual of the [31, 10, 12] BCH code has minimum distance $d' = 5$ to obtain a (26, 320, 10) code.

Construction Y3. Similarly use all cosets with coset representatives of weight 2 to obtain an

$$\left(n - d', \left(1 + \binom{d'}{2}\right)2^{n-d'+1}, \geqslant d - 4\right) \tag{30}$$

nonlinear code. For applications see [1237].

Problem. (30) *Construction Y4.* Let d'' be the minimum weight of the vector u OR v, for $u, v \in \mathscr{C}^{\perp}$, $u \neq v$. Show that deleting *these* columns gives an

$$[n - d'', k - d'' + 2, \geqslant d]$$

code. Applications are given in [637].

§9.2. A construction of Helgert and Stinaff. Suppose $\mathscr{C}$ is an $[n, k, d]$ code with $k \times n$ generator matrix G. The first row of G can be assumed to be $u = 1^d 0^{n-d}$, thus

$$G = \left[\begin{array}{c|c} 11 \cdots 1 & 00 \cdots 0 \\ \hline G_1 & G_2 \end{array}\right].$$

Then G_2 is the generator matrix of an $[n - d, k - 1, d_2]$ code $\mathscr{C}_2$, where $d_2 \geqslant \lceil d/2 \rceil$. For let v be any nonzero codeword of $\mathscr{C}_2$ of weight w, corresponding to a codeword t of $\mathscr{C}$.

$$
\begin{array}{cccc}
\leftarrow \lambda \rightarrow & \leftarrow d - \lambda \rightarrow & \leftarrow w \rightarrow & \\
u = 11 \cdots 1 & 11 \cdots 1 & 00 \cdots 0 & 00 \cdots 0 \\
t = 11 \cdots 1 & 00 \cdots 0 & 11 \cdots 1 & 00 \cdots 0 \\
u + t = 00 \cdots 0 & 11 \cdots 1 & 11 \cdots 1 & 00 \cdots 0
\end{array}
$$

Then $\lambda + w \geqslant d$ and $d - \lambda + w \geqslant d$, hence $2w \geqslant d$ and $w \geqslant \lceil d/2 \rceil$. In short (Helgert and Stinaff [636])

$$[n, k, d] \Rightarrow \left[n - d, k - 1, \left\lceil \frac{d}{2} \right\rceil\right]. \tag{31}$$

For example,

$$[24, 12, 8] \text{ Golay} \Rightarrow [16, 11, 4] \text{ Hamming},$$
$$[51, 8, 24] \text{ cyclic (Ref. [266])} \Rightarrow [27, 7, 12].$$

Codes obtained in this way are denoted by HS in Fig. 2 of Appendix A.

Notes on Chapter 18

§1. General references on product codes and generalizations are Berlekamp and Justesen [127], Burton and Weldon [219], Elias [406], Forney [436], Goethals [490], Gore [546], Kasami [726, 731], Lee and Cheng [802] and Weng [1407]. See also Hartmann et al. [616–618]. Applications of product codes are given by Bahl and Chien [58], Chien and Ng [292], Goethals [489], Kuroda [787], Tang and Chien [1306], Weng and Sollman [1408] and Wolf and Elspas [1430].

For decoding product codes see: Abramson [4], Duc [389], Duc and Skattebol [391], Kasami and Lin [733], Lin and Weldon [839], Reddy [1097, 1099], Reddy and Robinson [1100], Wainberg [1381] and Weldon [1404].

Some applications use codewords which are rectangles (rather than vectors), and here product codes and their generalizations are very useful. The theory of two- and higher-dimensional codes is described by Calabro and Wolf [229], Gordon [541], Ikai and Kojima [681], MacWilliams and Sloane [885], Nomura et al. [998–1001], Reed and Stewart [1106a] and Spann [1258]. For applications see Calingaert [230], Imai [682], Knowlton [771], Nakamura and Iwadare [984], Patel and Hong [1029], Schroeder [1168], Sloane et al. [1230], [1234–1236] and especially [885].

Goldberg [517] and Rao and Reddy [1094] have found several good codes by taking the union of a product code and certain of its cosets. This is a promising technique which needs further study.

Research Problem (18.7). Generalize the constructions of [517] and [1094].

§2. For the theory of ideals in the group algebra $\mathscr{G}$ see Ikai and Kojima [681] and Nomura et al. [998–1001]. ·

§7. The following papers give other techniques for constructing codes: Alltop [25], Assmus and Mattson [40], Bambah et al. [64], Bell [98], Blake [159], Bobrow et al. [168–170], Bredeson [195], Campopiano [240], Dagnino [326], Dênes and Keedwell [371, Ch. 10], Goppa [538], Hakimi et al. [575–577], Hashim and Constantinides [624], Hsaio et al. [668–670], Lecointe [800], Levitt and Wolf [828], Marchukov [912], Massey et al. [924], Olderogge [1010], Salzer [1142], Sauer [1147], Shiva [1198], Shrikhande [1206], Wallis [1385] and Wolf [1425, 1428].

Another class of codes, abelian group codes, show promise, but so far have not led to any really good codes. See Berman [135, 136], Camion [238], Delsarte [345], and MacWilliams [877, 880].

A number of interesting computer searches have been made for good codes. See for example Chen [266], Fontaine and Peterson [433, 434], Hashim and Constantinides [625], Tokura et al. [1331], and Wagner [1377–1380].

Constructions X and $X4$ are described by Sloane et al. [1237], which also gives encoding and decoding methods. See also Zaitsev et al. [1451] and Miyakawa et al. [965]. Theorem 12 is from [53]. Curtis [322] gives a different and very interesting proof of this theorem. Problem 13(b) is due to Ohno et al. [1009], Problems 14(a), (b) and 28(b) to Tezuka and Nakanishi [1315], and Problems 14(c), (d) to Kasahara et al. [721].

§7.5. For Piret's construction see [1047], and also [1048–1050].

19

Self-dual codes and invariant theory

§1. Introduction

A linear code $\mathscr{C}$ is *self-dual* if $\mathscr{C} = \mathscr{C}^\perp$ (§8 of Ch. 1). We have seen that many good codes are self-dual, among them the extended Golay codes and certain quadratic residue codes. The most important property of self-dual codes is given by Gleason's theorem (Theorem 3c), which imposes strong constraints on the weight enumerator. This theorem is proved by a powerful nineteenth-century technique known as invariant theory (§§2, 3). The same technique proves numerous similar theorems, dealing with other kinds of weight enumerators (§4). One corollary of Gleason's theorem is an upper bound on the minimum distance of a self-dual code (Theorems 13, 17). However (Corollary 16, Theorem 17) this upper bound can only be attained for small values of n. Nevertheless self-dual codes exist which meet the Gilbert–Varshamov bound (Theorems 21, 24).

Most of the results in this chapter apply equally well to a larger class of codes, namely linear or nonlinear codes with the property that the transform of the distance distribution is equal to the distance distribution (e.g. the Nordstrom–Robinson code). Such codes are called *formally self-dual*. The formally self-dual codes encountered in this chapter have the same distance distribution as weight distribution, and so we work throughout with the weight distribution.

We remind the reader that a self-dual code has even length n and dimension $\frac{1}{2}n$.

The most interesting self-dual or formally self-dual codes have the property that all distances are multiples of some constant $t > 1$. There are just four nontrivial cases when this can happen.

Theorem 1. (Gleason, Pierce and Turyn [53].) *Suppose $\mathscr{C}$ is a formally self-dual code over* $GF(q)$ *in which the distance between codewords is always a multiple of* $t > 1$. *Then either* $\mathscr{C}$ *is a trivial* $[n, \frac{1}{2}n, 2]$ *code with Hamming weight enumerator*

$$((q - 1)x^2 + y^2)^{n/2}, \tag{1}$$

for any q, or else one of the following holds:

(i)	$q = 2, \ t = 2,$
(ii)	$q = 2, \ t = 4,$
(iii)	$q = 3, \ t = 3,$
(iv)	$q = 4, \ t = 2.$

The proof is omitted. Codes of type (ii) are sometimes called *even* self-dual codes.

The following are some examples of self-dual codes.

Codes of type (i). (E1) The binary code $\{00, 11\}$ is the simplest self-dual code, and has weight enumerator (w.e.)

$$W_1(x, y) = x^2 + y^2. \tag{2}$$

More generally the code over $GF(q)$ with generator matrix $[ab]$, a, $b \neq 0$, is formally self-dual and has w.e. $x^2 + (q - 1)y^2$. The direct sum of $n/2$ such codes is formally self-dual and has w.e. given by Equation (1).

(E2) The $(16, 256, 6)$ Nordstrom-Robinson code $\mathscr{N}_{16}$ (§8 of Ch. 2, Ch. 15).

(E3) The $[18, 9, 6]$ extended QR code (§6 of Ch. 16).

(E4) The $(8, 16, 2)$ formally self-dual code of Fig. 5.1.

Codes of type (ii). (E5) The $[8, 4, 4]$ Hamming code with w.e.

$$W_2(x, y) = x^8 + 14x^4y^4 + y^8. \tag{3}$$

(E6) The $[24, 12, 8]$ Golay code $\mathscr{G}_{24}$ with w.e.

$$W_3(x, y) = x^{24} + 759x^{16}y^8 + 2576x^{12}y^{12}$$
$$+ 759x^8y^{16} + y^{24}. \tag{4}$$

More generally any extended QR code of length $n = 8m$ is of type (ii) (Theorems 7, 8 of Ch. 16).

Recall from Ch. 5 that a code over $GF(3)$ has two weight enumerators, the complete w.e. $\mathscr{W}(x, y, z) = \sum A_{ijk}x^iy^jz^k$, where A_{ijk} is the number of codewords containing i 0's, j 1's, and k 2's; and the Hamming w.e. $W(x, y) = \mathscr{W}(x, y, y)$.

Codes of type (iii). (E7) The $[4, 2, 3]$ ternary code $\#6$ of Ch. 1, with Hamming w.e.

$$W_4(x, y) = x^4 + 8xy^3. \tag{5}$$

(E8) The $[12, 6, 6]$ ternary Golay code $\mathscr{G}_{12}$ (defined by (25) of Ch. 16), with w.e.'s

$$\mathscr{W}_5(x, y, z) = x^{12} + y^{12} + z^{12} + 22(x^6 y^6 + x^6 z^6 + y^6 z^6)$$
$$+ 220(x^6 y^3 z^3 + x^3 y^6 z^3 + x^3 y^3 z^6),$$
$$W_6(x, y) = x^{12} + 264 x^6 y^6 + 440 x^3 y^9 + 24 y^{12}. \tag{6}$$

More generally, any extended ternary QR or symmetry code is of type (iii) (by Equation (26) and Theorems 7, 8 and 17 of Ch. 16).

Codes of type (iv). (E9) The $[6, 3, 4]$ formally self-dual QR code over GF(4) (see p. 520 of Ch. 16). This is obtained by adding an overall parity check to the cyclic code with generator polynomial $(x - \beta)(x - \beta^4)$, where $\beta \in \mathrm{GF}(16)$ is a fifth root of unity. A generator matrix is

$$\begin{bmatrix} 1 & \omega^2 & 1 & 0 & 0 & \omega^2 \\ 0 & 1 & \omega^2 & 1 & 0 & \omega^2 \\ 0 & 0 & 1 & \omega^2 & 1 & \omega^2 \end{bmatrix},$$

where $\omega \in \mathrm{GF}(4)$, $\omega^2 + \omega + 1 = 0$.

Problems. (1) Let $\mathscr{C}$ be a cyclic $[2n - 1, n]$ code. Find a necessary and sufficient condition on the zeros of $\mathscr{C}$ for the extended code $\mathscr{C}^*$ to be self-dual. [Hint: study the cases $n = 8, 16$.]

(2) Let q be even. Show that there exists an RS code $\mathscr{C}$ over GF(q) such that $\mathscr{C}^*$ is self-dual. Show this is impossible if q is odd.

Remark. It follows from Theorem 3c below that the $[32, 16, 8]$ QR and second-order RM codes have the same weight distribution. However, they are not equivalent because: (i) They have different groups. The group of the QR code is at least PSL$_2$(31), of order $\frac{1}{2} \cdot 31(31^2 - 1) = 14880$, by Theorem 13 of Ch. 16. However, since 29 and 31 are twin primes, a theorem of Neumann [988, 989] implies that the group is no bigger than this. On the other hand, the group of the RM code is the general affine group GA(5), of order $32 \cdot 31 \cdot 30 \cdot 28 \cdot 24 \cdot 16 = 319979520$, from Theorem 24 of Ch. 13. (ii) The cosets have different weight distributions (Berlekamp and Welch [133]). (iii) They have different biweight enumerators (cf. Problem 36 of Ch. 5).

Two inequivalent $[14, 3, 5]$ codes with the same weight distribution are given by Fontaine and Peterson ([434]).

§2. An introduction to invariant theory

This section is an elementary introduction to invariant theory, showing how it is used to study weight enumerators.

Suppose $\mathscr{C}$ is a binary self-dual code with all weights divisible by 4, or an even self-dual code for short, and let $W(x, y)$ be its weight enumerator. Since $\mathscr{C}$ is self-dual, Theorem 1 of Ch. 5 implies

$$W(x, y) = \frac{1}{2^{n/2}} W(x + y, x - y)$$

$$= W\left(\frac{x + y}{\sqrt{2}}, \frac{x - y}{\sqrt{2}}\right) \tag{7}$$

(for $W(x, y)$ is homogeneous of degree n). Since all weights are divisible by 4, $W(x, y)$ only contains powers of y^4. Therefore

$$W(x, y) = W(x, iy), \tag{8}$$

where $i = \sqrt{(-1)}$. The problem we wish to solve is to find all polynomials $W(x, y)$ satisfying (7) and (8).

Invariants. Equation (7) says that $W(x, y)$ is unchanged, or *invariant*, under the linear transformation

$$T_1: \qquad \begin{aligned} &\text{replace } x \text{ by } \frac{x + y}{\sqrt{2}}, \\ &\text{replace } y \text{ by } \frac{x - y}{\sqrt{2}}, \end{aligned}$$

or, in matrix notation,

$$T_1: \quad \text{replace } \begin{pmatrix} x \\ y \end{pmatrix} \text{ by } \frac{1}{\sqrt{2}} \begin{pmatrix} 1 & 1 \\ 1 & -1 \end{pmatrix} \begin{pmatrix} x \\ y \end{pmatrix}.$$

Similarly Equation (8) says that $W(x, y)$ is also invariant under the transformation

$$T_2: \qquad \begin{aligned} &\text{replace } x \text{ by } x \\ &\text{replace } y \text{ by } iy \end{aligned}$$

or

$$T_2: \quad \text{replace } \begin{pmatrix} x \\ y \end{pmatrix} \text{ by } \begin{pmatrix} 1 & 0 \\ 0 & i \end{pmatrix} \begin{pmatrix} x \\ y \end{pmatrix}.$$

Of course $W(x, y)$ must therefore be invariant under any combination T_1^2, T_1T_2, $T_1T_2T_1, \ldots$ of these transformations. It is not difficult to show (as we shall see in §3) that the matrices

$$\frac{1}{\sqrt{2}} \begin{pmatrix} 1 & 1 \\ 1 & -1 \end{pmatrix} \quad \text{and} \quad \begin{pmatrix} 1 & 0 \\ 0 & i \end{pmatrix}$$

when multiplied together in all possible ways produce a *group* $\mathscr{G}_1$ containing 192 matrices.

So our problem now says: find the polynomials $W(x, y)$ which are invariant under all 192 matrices in the group $\mathcal{G}_1$.

How many invariants? The first thing we want to know is how many invariants there are. This isn't too precise, because of course if f and g are invariants, so is any constant multiple cf and also $f + g$, $f - g$ and the product fg. Also it is enough to study the *homogeneous* invariants (in which all terms have the same degree).

So the right question to ask is: how many linearly independent, homogeneous invariants are there of each degree d? Let's call this number a_d.

A convenient way to handle the numbers $a_0, a_1, a_2, \ldots$ is by combining them into a power series or generating function

$$\Phi(\lambda) = a_0 + a_1\lambda + a_2\lambda^2 + \cdots.$$

Conversely, if we know $\Phi(\lambda)$, the numbers a_d can be recovered from the power series expansion of $\Phi(\lambda)$.

At this point we invoke a beautiful theorem of T. Molien, published in 1897 ([971]; Bourbaki [190, p. 110], Burnside [211, p. 301], Miller et al. [955, p. 259]).

Theorem 2. *For any finite group $\mathcal{G}$ of complex $m \times m$ matrices, $\Phi(\lambda)$ is given by*

$$\Phi(\lambda) = \frac{1}{|\mathcal{G}|} \sum_{A \in \mathcal{G}} \frac{1}{\det(I - \lambda A)} \tag{9}$$

where $|\mathcal{G}|$ is the number of matrices in $\mathcal{G}$, det stands for determinant, and I is a unit matrix. In words, $\Phi(\lambda)$ is the average, taken over all matrices A in the group, of the reciprocal of the polynomial $\det(I - \lambda A)$.

We call $\Phi(\lambda)$ the *Molien series* of $\mathcal{G}$. The proof of this theorem is given in §3.

For our group $\mathcal{G}_1$, from the matrices corresponding to $I, T_1, T_2, \ldots$ we get

$$\Phi(\lambda) = \frac{1}{192} \left\{ \frac{1}{(1-\lambda)^2} + \frac{1}{1-\lambda^2} + \frac{1}{(1-\lambda)(1-i\lambda)} + \cdots \right\}. \tag{10}$$

There are shortcuts, but it is quite feasible to work out the 192 terms directly (many are the same) and add them. The result is a surprise: everything collapses to give

$$\Phi(\lambda) = \frac{1}{(1-\lambda^8)(1-\lambda^{24})}. \tag{11}$$

Interpreting $\Phi(\lambda)$. The very simple form of (11) is trying to tell us something.

Expanding in powers of λ, we have

$$\Phi(\lambda) = a_0 + a_1\lambda + a_2\lambda^2 + \cdots$$
$$= (1 + \lambda^8 + \lambda^{16} + \lambda^{24} + \cdots)(1 + \lambda^{24} + \lambda^{48} + \cdots). \qquad (12)$$

We can deduce one fact immediately: a_d is zero unless d is a multiple of 8. I.e., the degree of a homogeneous invariant must be a multiple of 8. (This already proves that the block length of an even self-dual code is multiple of 8.) But we can say more. The RHS of (12) is exactly what we would find if there were two "basic" invariants, of degrees 8 and 24, such that all invariants are formed from sums and products of them.

This is because two invariants, θ, of degree 8, and φ, of degree 24, would give rise to the following invariants.

degree d	invariants	number a_d
0	1	1
8	θ	1
16	θ^2	1
24	θ^3, φ	2
32	$\theta^4, \theta\varphi$	2
40	$\theta^5, \theta^2\varphi$	2
48	$\theta^6, \theta^3\varphi, \varphi^2$	3
$\cdots$	$\cdots$	$\cdots$

$$(13)$$

Provided all the products $\theta^i\varphi^j$ are linearly independent – which is the same thing as saying that θ and φ are algebraically independent – the numbers a_d in (13) are exactly the coefficients in

$$1 + \lambda^8 + \lambda^{16} + 2\lambda^{24} + 2\lambda^{32} + 2\lambda^{40} + 3\lambda^{48} + \cdots$$
$$= (1 + \lambda^8 + \lambda^{16} + \lambda^{24} + \cdots)(1 + \lambda^{24} + \lambda^{48} + \cdots)$$
$$= \frac{1}{(1 - \lambda^8)(1 - \lambda^{24})}, \qquad (14)$$

which agrees with (11). So if we can find two algebraically independent invariants of degrees 8 and 24, we will have solved our problem. The answer will be that any invariant of this group is a polynomial in θ and φ. Now $W_2(x, y)$ and $W_3(x, y)$, the weight enumerators of the Hamming and Golay codes, have degrees 8 and 24 and are invariant under the group. So we can take $\theta = W_2(x, y)$ and $\varphi = W_3(x, y)$. (It's not difficult to verify that they are algebraically independent.) Actually, it is easier to work with

$$\varphi = W'_3(x, y) = \frac{W_2(x, y)^3 - W_3(x, y)}{42} = x^4y^4(x^4 - y^4)^4 \qquad (15)$$

rather than $W_3(x, y)$ itself. So we have proved the following theorem, discovered by Gleason in 1970.

Theorem 3a. *Any invariant of the group $\mathcal{G}_1$ is a polynomial in $W_2(x, y)$ (Equation (3)) and $W_3'(x, y)$ (Equation (15)).*

This also gives us the solution to our original problem:

Theorem 3b. *Any polynomial which satisfies Equations (7) and (8) is a polynomial in $W_2(x, y)$ and $W_3'(x, y)$.*

Finally, we have characterized the weight enumerator of an even self-dual code.

Theorem 3c. (Gleason [486].) *The weight enumerator of any even self-dual code is a polynomial in $W_2(x, y)$ and $W_3'(x, y)$.*

Alternative proofs of this theorem are given by Berlekamp et al. [129], Broué and Enguehard [201], and Feit [424] (see also Assmus and Mattson [47]). But the proof given here seems to be the most informative, and the easiest to understand and to generalize.

Notice how the exponents 8 and 24 in the denominator of (11) led us to guess the degrees of the basic invariants.

This behavior is typical, and is what makes the technique exciting to use. One starts with a group of matrices $\mathcal{G}$, computes the complicated-looking sum shown in Equation (9), and simplifies the result. Everything miraculously collapses, leaving a final expression resembling Equation (11) (although not always quite so simple – the precise form of the final expression is given in §3). This expression then tells the degrees of the basic invariants to look for.

Problem. (3) (Gleason [486; 129, 883].) Use the above technique to show that the weight enumerator of any binary self-dual code is a polynomial in $W_1(x, y) = x^2 + y^2$ and

$$W_2'(x, y) = \frac{W_1(x, y)^4 - W_2(x, y)}{4} = x^2 y^2 (x^2 - y^2)^2. \tag{16}$$

[Hint: By problem 38 of Ch. 1, all weights are divisible by 2. The group is generated by T_1 and

$$\begin{pmatrix} 1 & 0 \\ 0 & -1 \end{pmatrix},$$

has order 16 and Molien series $1/(1 - \lambda^2)(1 - \lambda^4)$.]

Finding the basic invariants. In general, finding the basic invariants simpler problem than finding $\Phi(\lambda)$. Either one can use the weight enumer of codes having the appropriate properties, as in the above example, or basic invariants can be found by *averaging*, using the following simple result (proved in §3).

Theorem 4. *If $f(x) = f(x_1, \ldots, x_m)$ is any polynomial in m variables, and $\mathcal{G}$ is a finite group of $m \times m$ matrices, then*

$$\bar{f}(x) = \frac{1}{|\mathcal{G}|} \sum_{A \in \mathcal{G}} A \circ f(x)$$

is an invariant, where $A \circ f(x)$ denotes the polynomial obtained by applying the transformation A to the variables in f.

Of course $\bar{f}(x)$ may be zero. An example of the use of this theorem is given below.

An application of Theorem 3. To show the power of Theorem 3c, we use it to find the w.e. of the [48, 24] QR code. From Theorems 7, 8, 25 of Ch. 16, this is an even self-dual code with minimum distance 12. Therefore the weight enumerator of the code, which is a homogeneous polynomial of degree 48, has the form

$$W(x, y) = x^{48} + A_{12}x^{36}y^{12} + \cdots. \tag{17}$$

The coefficients of $x^{47}y$, $x^{46}y^2, \ldots, x^{37}y^{11}$ are zero. Here A_{12} is the unknown number of codewords of weight 12. It is remarkable that, once we know Equation (17), the weight enumerator is completely determined by Theorem 3c. For Theorem 3c says that $W(x, y)$ must be a polynomial in $W_2(x, y)$ and $W_3'(x, y)$. Since $W(x, y)$ is homogeneous of degree 48, W_2 is homogeneous of degree 8, and W_3' is homogeneous of degree 24, this polynomial must be a linear combination of W_2^6, $W_2^3 W_3'$ and $W_3'^2$.

Thus Theorem 3c says that

$$W(x, y) = a_0 W_2^6 + a_1 W_2^3 W_3' + a_2 W_3'^2 \tag{18}$$

for some real numbers a_0, a_1, a_2. Expanding Equation (18) we have

$$\begin{aligned} W(x, y) = {} & a_0(x^{48} + 84x^{44}y^4 + 2946x^{40}y^8 + \cdots) \\ & + a_1(x^{44}y^4 + 38x^{40}y^8 + \cdots) \\ & + a_2(x^{40}y^8 - \cdots), \end{aligned} \tag{19}$$

and equating coefficients in Equations (17), (19) we get

$$a_0 = 1, \qquad a_1 = -84, \qquad a_2 = 246.$$

Therefore $W(x, y)$ is uniquely determined. When the values of a_0, a_1, a_2 are substituted in (18) it is found that

$$W(x, y) = x^{48} + 17296x^{36}y^{12} + 535095x^{32}y^{16}$$
$$+ 3995376x^{28}y^{20} + 7681680x^{24}y^{24} + 3995376x^{20}y^{28}$$
$$+ 535095x^{16}y^{32} + 17296x^{12}y^{36} + y^{48}. \tag{20}$$

Since the minimum distance is as large as it can be, this is called an *extremal weight enumerator* – see §5. Direct calculation of this weight enumerator would require finding the weight of each of the $2^{24} \approx 1.7 \times 10^7$ codewords, a respectable job even for a computer.

Of course there is also a fair amount of algebra involved in the invariant theory method, although in the preceding example it can be done by hand. The reader may find it helpful if we give a second example, in which the algebra can be shown in full.

A very simple example. The weight enumerator of a self-dual code over GF(q) by Theorem 13 of Ch. 5 satisfies the equation

$$W\left(\frac{x+(q-1)y}{\sqrt{q}}, \frac{x-y}{\sqrt{q}}\right) = W(x, y). \tag{21}$$

Problem: find all polynomials which satisfy Equation (21).

The solution proceeds as before. Equation (21) says that $W(x, y)$ must be invariant under the transformation

$$T_3: \quad \text{replace } \binom{x}{y} \text{ by } A\binom{x}{y},$$

where

$$A = \frac{1}{\sqrt{q}} \begin{pmatrix} 1 & q-1 \\ 1 & -1 \end{pmatrix}. \tag{22}$$

Now $A^2 = I$, so $W(x, y)$ must be invariant under the group $\mathcal{G}_2$ consisting of the two matrices I and A.

To find how many invariants there are, we compute the Molien series $\Phi(\lambda)$ from Equation (9). We find

$$\det(I - \lambda I) = (1-\lambda)^2,$$

$$\det(I - \lambda A) = \det \begin{pmatrix} 1 - \dfrac{\lambda}{\sqrt{q}} & -\dfrac{q-1}{\sqrt{q}}\lambda \\ -\dfrac{\lambda}{\sqrt{q}} & 1 + \dfrac{\lambda}{\sqrt{q}} \end{pmatrix} = 1 - \lambda^2,$$

$$\Phi(\lambda) = \frac{1}{2}\left(\frac{1}{(1-\lambda)^2} + \frac{1}{1-\lambda^2}\right)$$

$$= \frac{1}{(1-\lambda)(1-\lambda^2)}. \tag{23}$$

which is even simpler than Equation (11). Equation (23) suggests that there might be two basic invariants, of degrees 1 and 2 (the exponents in the denominator). If algebraically independent invariants of degrees 1 and 2 can be found, say g and h, then Equation (23) implies that any invariant of $\mathcal{G}_2$ is a polynomial in g and h.

This time we shall use the method of averaging to find the basic invariants. Let us average x over the group – i.e., apply Theorem 4 with $f(x, y) = x$. The matrix I leaves x unchanged, of course, and the matrix A transforms x into $(1/\sqrt{q})(x + (q - 1)y)$. Therefore the average,

$$\bar{f}(x, y) = \frac{1}{2}\left[x + \frac{1}{\sqrt{q}}\{x + (q - 1)y\}\right]$$
$$= \frac{(\sqrt{q} + 1)\{x + (\sqrt{q} - 1)y\}}{2\sqrt{q}},$$

is an invariant. Of course any scalar multiple of $\bar{f}(x, y)$ is also an invariant, so we may divide by $(\sqrt{q} + 1)/2\sqrt{q}$ and take

$$g = x + (\sqrt{q} - 1)y \tag{24}$$

to be the basic invariant of degree 1. To get an invariant of degree 2 we average x^2 over the group, obtaining

$$\frac{1}{2}\left[x^2 + \frac{1}{q}\{x + (q - 1)y\}^2\right].$$

This can be cleaned up by subtracting $((q + 1)/2q)g^2$ (which of course is an invariant), and dividing by a suitable constant. The result is

$$h = y(x - y),$$

the desired basic invariant of degree 2.

Finally g and h must be shown to be algebraically independent: it must be shown that no sum of the form

$$\sum_{i,j} c_{ij}g^i h^j, \quad c_{ij} \text{ complex and not all zero}, \tag{25}$$

is identically zero when expanded in powers of x and y. This can be seen by looking at the leading terms. (The leading term of a polynomial is the first one to be written down when using the natural ordering illustrated in Equations (15), (20), (24).) Thus the leading term of g is x, the leading term of h is xy, and the leading term of $g^i h^j$ is $x^{i+j}y^j$. Since distinct summands in Equation (25) have distinct leading terms, (25) can only add to zero if all the c_{ij} are zero. Therefore g and h re algebraically independent. So we have proved:

Theorem 5. *Any invariant of the group $\mathcal{G}_2$, or equivalently any polynomial satisfying (21), or equivalently, the weight enumerator of any self-dual code over* GF(q), *is a polynomial in $g = x + (\sqrt{q} - 1)y$ and $h = y(x - y)$.*

At this point the reader should cry Stop!, and point out that a self-dual code must have even length and so every term in the weight enumerator must have even degree. But in Theorem 5 g has degree 1.

Thus we haven't made use of everything we know about the code. $W(x, y)$ must also be invariant under the transformation

$$\text{replace } \binom{x}{y} \text{ by } B \binom{x}{y},$$

where

$$B = \begin{pmatrix} -1 & 0 \\ 0 & -1 \end{pmatrix} = -I.$$

This rules out terms of odd degree. So $W(x, y)$ is invariant under the group $\mathscr{G}_3$ generated by A and B, which consists of

$$I, \quad A, \quad -I, \quad -A.$$

The reader can easily work out that the new Molien series is

$$\Phi_{\mathscr{G}_3}(\lambda) = \frac{1}{2} \{ \Phi_{\mathscr{G}_2}(\lambda) + \Phi_{\mathscr{G}_2}(-\lambda) \}$$

$$= \frac{1}{2} \left\{ \frac{1}{(1-\lambda)(1-\lambda^2)} + \frac{1}{(1+\lambda)(1-\lambda^2)} \right\}$$

$$= \frac{1}{(1-\lambda^2)^2}. \tag{26}$$

There are now two basic invariants, both of degree 2 (matching the exponents in the denominator of (26)), say g^2 and h, or the equivalent and slightly simpler pair $g^* = x^2 + (q-1)y^2$ and $h = y(x-y)$. Hence:

Theorem 6. *The weight enumerator of any self-dual code over* GF(q) *is a polynomial in* g^* *and* h.

The general plan of attack. As these examples have illustrated, there are two stages in using invariant theory to solve a problem.

Stage I. Convert the assumptions about the problem (e.g. the code) into algebraic constraints on polynomials (e.g. weight enumerators).

Stage II. Use the invariant theory to find all possible polynomials satisfying these constraints.

§3. The basic theorems of invariant theory

Groups of matrices. Given a collection $A_1, \ldots, A_r$ of $m \times m$ invertible matrices we can form a group $\mathcal{G}$ from them by multiplying them together in all possible ways. Thus $\mathcal{G}$ contains the matrices $I, A_1, A_2, \ldots, A_1 A_2, \ldots, A_2 A_1^{-1} A_2^{-1} A_3, \ldots$. We say that $\mathcal{G}$ is *generated* by $A_1, \ldots, A_r$. Of course $\mathcal{G}$ may be infinite, in which case the theory of invariants described here doesn't directly apply. (But see Dieudonnè and Carrell [375], Rallis [1087] and Weyl [1410].)

Example. Let us show that the group $\mathcal{G}_1$ generated by the matrices

$$M = \frac{1}{\sqrt{2}} \begin{pmatrix} 1 & 1 \\ 1 & -1 \end{pmatrix} \quad \text{and} \quad J = \begin{pmatrix} 1 & 0 \\ 0 & i \end{pmatrix}$$

that was encountered in §2 does indeed have order 192. The key is to discover (by randomly multiplying matrices together) that $\mathcal{G}_1$ contains

$$J^2 = \begin{pmatrix} 1 & 0 \\ 0 & -1 \end{pmatrix}, \quad E = (MJ)^3 = \frac{1+i}{\sqrt{2}} \begin{pmatrix} 1 & 0 \\ 0 & 1 \end{pmatrix},$$

$$E^2 = i \begin{pmatrix} 1 & 0 \\ 0 & 1 \end{pmatrix}, \quad R = MJ^2 M = \begin{pmatrix} 0 & 1 \\ 1 & 0 \end{pmatrix}$$

So $\mathcal{G}_1$ contains the matrices

$$\alpha \begin{pmatrix} 1 & 0 \\ 0 & \pm 1 \end{pmatrix}, \quad \alpha \begin{pmatrix} 0 & 1 \\ \pm 1 & 0 \end{pmatrix}, \quad \alpha \in \{1, i, -1, -i\},$$

which form a subgroup $\mathcal{H}_1$ of order 16. From this it is easy to see that $\mathcal{G}_1$ consists of the union of 12 cosets of $\mathcal{H}_1$:

$$\mathcal{G}_1 = \bigcup_{k=1}^{12} a_k \mathcal{H}_1, \tag{27}$$

where $a_1, \ldots, a_6$ are respectively

$$\begin{pmatrix} 1 & 0 \\ 0 & 1 \end{pmatrix}, \begin{pmatrix} 1 & 0 \\ 0 & i \end{pmatrix}, \frac{1}{\sqrt{2}} \begin{pmatrix} 1 & 1 \\ 1 & -1 \end{pmatrix}, \frac{1}{\sqrt{2}} \begin{pmatrix} 1 & 1 \\ i & -i \end{pmatrix}, \frac{1}{\sqrt{2}} \begin{pmatrix} 1 & i \\ 1 & -i \end{pmatrix}, \frac{1}{\sqrt{2}} \begin{pmatrix} 1 & i \\ i & 1 \end{pmatrix},$$

$a_7 = \eta a_1, \ldots, a_{12} = \eta a_6$, and $\eta = (1+i)/\sqrt{2}$. Thus $\mathcal{G}_1$ consists of the 192 matrices

$$\eta^\nu \begin{pmatrix} 1 & 0 \\ 0 & \alpha \end{pmatrix}, \quad \eta^\nu \begin{pmatrix} 0 & 1 \\ \alpha & 0 \end{pmatrix}, \quad \eta^\nu \frac{1}{\sqrt{2}} \begin{pmatrix} 1 & \beta \\ \alpha & -\alpha\beta \end{pmatrix}, \tag{28}$$

for $0 \le \nu \le 7$ and $\alpha, \beta \in \{1, i, -1, -i\}$.

As a check, one verifies that every matrix in (28) can be written as a product of M's and J's; that the product of two matrices in (28) is again in (28); and that the inverse of every matrix in (28) is in (28). Therefore (28) *is a group*, and is the group generated by M and J. Thus $\mathcal{G}_1$ is indeed equal to (28).

We have gone into this example in some detail to emphasize that it is important to begin by understanding the group thoroughly. (For an alternative way of studying $\mathcal{G}_1$ see [201; pp. 160–161].

Invariants. To quote Hermann Weyl [1409], "the theory of invariants came into existence about the middle of the nineteenth century somewhat like Minerva: a grown-up virgin, mailed in the shining armor of algebra, she sprang forth from Cayley's Jovian head." Invariant theory become one of the main branches of nineteenth century mathematics, but dropped out of fashion after Hilbert's work: see Fisher [431] and Reid [1108]. Recently, however, there has been a resurgence of interest, with applications in algebraic geometry (Dieudonnè and Carrell [375], Mumford [976]), physics (see for example Agrawala and Belinfante [7] and the references given there), combinatorics (Doubilet et al. [381], Rota [1125]), and coding theory ([883, 894, 895]).

There are several different kinds of invariants, but here an invariant is defined as follows.

Let $\mathcal{G}$ be a group of g $m \times m$ complex matrices $A_1, \ldots, A_g$, where the $(i, k)^{\text{th}}$ entry of A_α is $a_{ik}^{(\alpha)}$. In other words $\mathcal{G}$ is a group of linear transformations on the variables $x_1, \ldots, x_m$, consisting of the transformations

$$T^{(\alpha)}: \text{replace } x_i \text{ by } x_i^{(\alpha)} = \sum_{k=1}^{m} a_{ik}^{(\alpha)} x_k, \quad i = 1, \ldots, m \qquad (29)$$

for $\alpha = 1, 2, \ldots, g$. It is worthwhile giving a careful description of how a polynomial $f(x) = f(x_1, \ldots, x_m)$ is transformed by a matrix A_α in $\mathcal{G}$. The transformed polynomial is

$$A_\alpha \circ f(x) = f(x_1^{(\alpha)}, \ldots, x_m^{(\alpha)})$$

where each $x_i^{(\alpha)}$ is replaced by $\sum_{k=1}^{m} a_{ik}^{(\alpha)} x_k$. Another way of describing this is to think of $x = (x_1, \ldots, x_m)^T$ as a column vector. Then $f(x)$ is transformed into

$$A_\alpha \circ f(x) = f(A_\alpha x), \qquad (30)$$

where $A_\alpha x$ is the usual product of a matrix and a vector. One can check that

$$B \circ (A \circ f(x)) = (AB) \circ f(x) = f(ABx). \qquad (31)$$

For example,

$$A = \begin{pmatrix} 1 & 2 \\ 0 & -1 \end{pmatrix}$$

transforms $x_1^2 + x_2$ into $(x_1 + 2x_2)^2 - x_2$.

Definition. An *invariant* of $\mathcal{G}$ is a polynomial $f(x)$ which is unchanged by every linear transformation in $\mathcal{G}$. In other words, $f(x)$ is an invariant of $\mathcal{G}$ if

$$A_\alpha \circ f(x) = f(A_\alpha x) = f(x)$$

for all $\alpha = 1, \ldots, g$.

Example. Let

$$\mathcal{G}_4 = \left\{ \begin{pmatrix} 1 & 0 \\ 0 & 1 \end{pmatrix}, \begin{pmatrix} -1 & 0 \\ 0 & -1 \end{pmatrix} \right\},$$

a group of order $g = 2$. Then x^2, xy and y^2 are homogeneous invariants of degree 2.

Even if $f(x)$ isn't an invariant, its average over the group is, as was mentioned in §2.

Theorem 4. *Let $f(x)$ be any polynomial. Then*

$$\bar{f}(x) = \frac{1}{g} \sum_{\alpha=1}^{g} A_\alpha \circ f(x) \tag{32}$$

is an invariant of $\mathcal{G}$.

Proof. Any $A_\beta \in \mathcal{G}$ transforms the right-hand side of (32) into

$$\frac{1}{g} \sum_{\alpha=1}^{g} (A_\alpha A_\beta) \circ f(x). \quad \text{by (31).} \tag{33}$$

As A_α runs through $\mathcal{G}$, so does $A_\alpha A_\beta$, if A_β is fixed. Therefore (33) is equal to

$$\frac{1}{g} \sum_{\gamma=1}^{g} A_\gamma \circ f(x),$$

which is $\bar{f}(x)$. Therefore $\bar{f}(x)$ is an invariant.   Q.E.D.

More generally, any symmetric function of the g polynomials $A_1 \circ f(x), \ldots, A_g \circ f(x)$ is an invariant of $\mathcal{G}$.

Clearly if $f(x)$ and $h(x)$ are invariants of $\mathcal{G}$, so are $f(x) + h(x)$, $f(x)h(x)$, and $cf(x)$ (c complex). This is equivalent to saying that the set of invariants of $\mathcal{G}$, which we denote by $\mathcal{J}(\mathcal{G})$, forms a *ring* (see p. 189).

One of the main problems of invariant theory is to describe $\mathcal{J}(\mathcal{G})$. Since the transformations in $\mathcal{G}$ don't change the degree of a polynomial, it is enough to describe the homogeneous invariants (for any invariant is a sum of homogeneous invariants).

Basic invariants. Our goal is to find a "basis" for the invariants of $\mathcal{G}$, that is, a set of basic invariants such that any invariant can be expressed in terms of this set. There are two different types of bases one might look for.

Definition. Polynomials $f_1(x), \ldots, f_r(x)$ are called *algebraically dependent* if there is a polynomial p in r variables with complex coefficients, not all zero, such that $p(f_1(x), \ldots, f_r(x)) \equiv 0$. Otherwise $f_1(x), \ldots, f_r(x)$ are called *algebraically independent*. A fundamental result from algebra is:

Theorem 7. (Jacobson [687], vol. 3, p. 154.) *Any $m + 1$ polynomials in m variables are algebraically dependent.*

The first type of basis we might look for is a set of m algebraically independent invariants $f_1(x), \ldots, f_m(x)$. Such a set is indeed a "basis," for by Theorem 7 any invariant is algebraically dependent on $f_1, \ldots, f_m$ and so is a root of a polynomial equation in $f_1, \ldots, f_m$. The following theorem guarantees the existence of such a basis.

Theorem 8. (Burnside [211, p. 357].) *There always exist m algebraically independent invariants of $\mathcal{G}$.*

Proof. Consider the polynomial

$$\prod_{\alpha=1}^{g} (t - A_\alpha \circ x_1)$$

in the variables $t, x_1, \ldots, x_m$. Since one of the A_α is the identity matrix, $t = x_1$ is a zero of this polynomial. When the polynomial is expanded in powers of t, the coefficients are invariants by the remark immediately following the proof of Theorem 4. Therefore x_1 is an algebraic function of invariants. Similarly each of $x_2, \ldots, x_m$ is an algebraic function of invariants. Now if the number of algebraically independent invariants were m' ($< m$), the m independent variables $x_1, \ldots, x_m$ would be algebraic functions of the m' invariants, a contradiction. Therefore the number of algebraically independent invariants is at least m. By Theorem 7 this number cannot be greater than m.   Q.E.D.

Example. For the preceding group $\mathcal{G}_4$, we may take $f_1 = (x + y)^2$ and $f_2 = (x - y)^2$ as the algebraically independent invariants. Then any invariant is a root of a polynomial equation in f_1 and f_2. For example,

$$x^2 = \tfrac{1}{4}(\sqrt{f_1} + \sqrt{f_2})^2,$$
$$xy = \tfrac{1}{4}(f_1 - f_2),$$

and so on.

However, by far the most convenient description of the invariants is a set $f_1, \ldots, f_l$ of invariants with the property that any invariant is a *polynomial* in $f_1, \ldots, f_l$. Then $f_1, \ldots, f_l$ is called a *polynomial basis* (or an *integrity basis*) for the invariants of $\mathcal{G}$. Of course if $l > m$ then by Theorem 7 there will be polynomial equations, called *syzygies*, relating $f_1, \ldots, f_l$.

For example, $f_1 = x^2$, $f_2 = xy$, $f_3 = y^2$ form a polynomial basis for the invariants of $\mathcal{G}_4$. The syzygy relating them is

$$f_1 f_3 - f_2^2 = 0.$$

The existence of a polynomial basis, and a method of finding it, is given by the next theorem.

Theorem 9. (Noether [997]; see also Weyl [1410, p. 275].) *The ring of invariants of a finite group $\mathcal{G}$ of complex $m \times m$ matrices has a polynomial basis consisting of not more than $\binom{m+g}{m}$ invariants, of degree not exceeding g, where g is the order of $\mathcal{G}$. Furthermore this basis may be obtained by taking the average over $\mathcal{G}$ of all monomials*

$$x_1^{b_1} x_2^{b_2} \cdots x_m^{b_m}$$

of total degree $\Sigma\, b_i$ not exceeding g.

Proof. Let the group $\mathcal{G}$ consist of the transformations (29). Suppose

$$f(x_1, \ldots, x_m) = \sum_e c_e x_1^{e_1} \cdots x_m^{e_m},$$

c_e complex, is any invariant of $\mathcal{G}$. (The sum extends over all $e = e_1 \cdots e_m$ for which there is a nonzero term $x_1^{e_1} \cdots x_m^{e_m}$ in $f(x_1, \ldots, x_m)$.) Since $f(x_1, \ldots, x_m)$ is an invariant, it is unchanged when we average it over the group, so

$$f(x_1, \ldots, x_m) = \frac{1}{g} \{ f(x_1^{(1)}, \ldots, x_m^{(1)}) + \cdots + f(x_1^{(g)}, \ldots, x_m^{(g)}) \}$$

$$= \frac{1}{g} \sum_e c_e \{ (x_1^{(1)})^{e_1} \cdots (x_m^{(1)})^{e_m} + \cdots + (x_1^{(g)})^{e_1} \cdots (x_m^{(g)})^{e_m} \}$$

$$= \frac{1}{g} \sum_e c_e J_e \quad \text{(say)}.$$

Every invariant is therefore a linear combination of the (infinitely many) special invariants

$$J_e = \sum_{\alpha=1}^{g} (x_1^{(\alpha)})^{e_1} \cdots (x_m^{(\alpha)})^{e_m}.$$

Now J_e is (apart from a constant factor) the coefficient of $u_1^{e_1} \cdots u_m^{e_m}$ in

$$P_e = \sum_{\alpha=1}^{g} (u_1 x_1^{(\alpha)} + \cdots + u_m x_m^{(\alpha)})^e, \tag{34}$$

where $e = e_1 + \cdots + e_m$. In other words the P_e are the power sums of the g

quantities

$$u_1 x_1^{(1)} + \cdots + u_m x_m^{(1)}, \ldots, u_1 x_1^{(g)} + \cdots + u_m x_m^{(g)}.$$

Any power sum P_e, $e = 1, 2, \ldots$, can be written as a polynomial with rational coefficients in the first g power sums $P_1, P_2, \ldots, P_g$ (Problem 3). Therefore any J_e for

$$e = \sum_{i=1}^{m} e_i > g$$

(which is a coefficient of P_e) can be written as a polynomial in the special invariants

$$J_e \quad \text{with } e_1 + \cdots + e_m \leq g$$

(which are the coefficients of $P_1, \ldots, P_g$). Thus any invariant can be written as a polynomial in the J_e with $\sum_{i=1}^{m} e_i \leq g$. The number of such J_e is the number of $e_1, e_2, \ldots, e_m$ with $e_i \geq 0$ and $e_1 + \cdots + e_m \leq g$, which is $\binom{m+g}{m}$. Finally $\deg J_e \leq g$, and J_e is obtained by averaging $x_1^{e_1} \cdots x_m^{e_m}$ over the group.

Q.E.D.

Molien's theorem. Since we know from Theorem 9 that a polynomial basis always exists, we can go ahead with confidence and try to find it, using the methods described in §2. To discover when a basis has been found, we use Molien's theorem (Theorem 2 above). This states that if a_d is the number of linearly independent homogeneous invariants of $\mathcal{G}$ with degree d, and

$$\Phi_{\mathcal{G}}(\lambda) = \sum_{d=0}^{\infty} a_d \lambda^d,$$

then

$$\Phi_{\mathcal{G}}(\lambda) = \frac{1}{g} \sum_{\alpha=1}^{g} \frac{1}{\det (I - \lambda A_\alpha)}. \tag{35}$$

The proof depends on the following theorem.

Theorem 10. (Miller et al. [955, p. 258], Serre [1185, p. 29].) *The number of linearly independent invariants of $\mathcal{G}$ of degree 1 is*

$$a_1 = \frac{1}{g} \sum_{\alpha=1}^{g} trace\ (A_\alpha).$$

Proof. Let

$$S = \frac{1}{g} \sum_{\alpha=1}^{g} A_\alpha.$$

Changing the variables on which $\mathcal{G}$ acts from $x_1, \ldots, x_m$ to $y_1, \ldots, y_m$, where $(y_1, \ldots, y_m) = (x_1, \ldots, x_m)T^{tr}$, changes S to $S' = TST^{-1}$. We may choose T so that S' is diagonal (see Burnside [211, p. 252]). Now $S^2 = S$, $(S')^2 = S'$, hence the diagonal entries of S' are 0 or 1. So with a change of variables we may assume

$$S = \begin{bmatrix} 1 & & & & & & 0 \\ & \ddots & & & & & \\ & & 1 & 0 & & & \\ & & & 0 & & & \\ & & & & \ddots & & \\ 0 & & & & & 0 \end{bmatrix},$$

with say r 1's on the diagonal. Thus $S \circ y_i = y_i$ if $1 \leq i \leq r$, $S \circ y_i = 0$ if $r+1 \leq i \leq m$.

Any linear invariant of $\mathcal{G}$ is certainly fixed by S, so $a_1 \leq r$. On the other hand, by Theorem 4,

$$S \circ y_i = \frac{1}{g} \sum_{\alpha=1}^{g} A_\alpha \circ y_i$$

is an invariant of $\mathcal{G}$ for any i, and so $a_1 \geq r$.   Q.E.D.

Before proving Theorem 2 let us introduce some more notation. Equation (29) describes how A_α transforms the variables $x_1, \ldots, x_m$. The dth *induced matrix*, denoted by $A_\alpha^{[d]}$, describes how A_α transforms the products of the x_i taken d at a time, namely $x_1^d, x_2^d, \ldots, x_1^{d-1}x_2, \ldots$ (Littlewood [857, p. 122]). E.g.

$$A_\alpha = \begin{pmatrix} a & b \\ c & d \end{pmatrix}$$

transforms x_1^2, $x_1 x_2$ and x_2^2 into

$$a^2 x_1^2 + 2ab x_1 x_2 + b^2 x_2^2,$$
$$ac x_1^2 + (ad + bc) x_1 x_2 + bd x_2^2,$$
$$c^2 x_1^2 + 2cd x_1 x_2 + d^2 x_2^2$$

respectively. Thus the 2nd induced matrix is

$$A_\alpha^{[2]} = \begin{bmatrix} a^2 & 2ab & b^2 \\ ac & ad+bc & bd \\ c^2 & 2cd & d^2 \end{bmatrix}.$$

Proof of Theorem 2. To prove Equation (35), note that a_d is equal to the number of linearly independent invariants of degree 1 of $\mathcal{G}^{[d]} = \{A_\alpha^{[d]}: \alpha =$

$1, \ldots, g$. By Theorem 10,

$$a_d = \frac{1}{g} \sum_{\alpha=1}^{g} \text{trace } A_\alpha^{[d]}.$$

Therefore to prove Theorem 2 it is enough to show that the trace of $A_\alpha^{[d]}$ is equal to the coefficient of λ^d in

$$\frac{1}{\det(I - \lambda A_\alpha)} = \frac{1}{(1 - \lambda \omega_1) \cdots (1 - \lambda \omega_m)}, \tag{36}$$

where $\omega_1, \ldots, \omega_m$ are the eigenvalues of A_α. By a suitable change of variables we can make

$$A_\alpha = \begin{bmatrix} \omega_1 & & 0 \\ & \ddots & \\ 0 & & \omega_m \end{bmatrix}, \qquad A_\alpha^{[d]} = \begin{bmatrix} \omega_1^d & & & 0 \\ & \omega_d^2 & \ddots & \\ & & \omega_1^{d-1}\omega_2 & \\ 0 & & & \ddots \end{bmatrix},$$

and trace $A_\alpha^{[d]} = $ sum of the products of $\omega_1, \ldots, \omega_m$ taken d at a time. But this is exactly the coefficient of λ^d in the expansion of (36).   Q.E.D.

It is worth remarking that the Molien series does not determine the group. For example there are two groups of 2×2 matrices with order 8 having

$$\Phi(\lambda) = \frac{1}{(1 - \lambda^2)(1 - \lambda^4)}$$

(namely the dihedral group $\mathcal{D}_4$ and the abelian group $\mathcal{Z}_2 \times \mathcal{Z}_4$). In fact there exist abstract groups $\mathcal{A}$ and $\mathcal{B}$ whose matrix representations can be paired in such a way that every representation of $\mathcal{A}$ has the same Molien series as the corresponding representation of $\mathcal{B}$ (Dade [324]).

A standard form for the basic invariants. The following notation is very useful in describing the ring $\mathcal{J}(\mathcal{G})$ of invariants of a group $\mathcal{G}$. The complex numbers are denoted by C, and if $p(x)$, $q(x), \ldots$ are polynomials $C[p(x),$ $q(x), \ldots]$ denotes the set of all polynomials in $p(x)$, $q(x)$ with complex coefficients. For example Theorem 3a just says that $\mathcal{J}(\mathcal{G}_1) = C[\theta, \varphi]$.

Also $\oplus$ will denote the usual direct sum operation. For example a statement like $\mathcal{J}(\mathcal{G}) = R \oplus S$ means that every invariant of $\mathcal{G}$ can be written uniquely in the form $r + s$ where $r \in R$, $s \in S$. (Theorem 12 below illustrates this.)

Using this notation we can now specify the most convenient form of polynomial basis for $\mathcal{J}(\mathcal{G})$.

Definition. *A good polynomial basis* for $\mathscr{I}(\mathscr{G})$ consists of homogeneous invariants $f_1, \ldots, f_l (l \geq m)$ where $f_1, \ldots, f_m$ are algebraically independent and

$$\mathscr{I}(\mathscr{G}) = \mathbb{C}[f_1, \ldots, f_m] \quad \text{if } l = m, \tag{37a}$$

or, if $l > m$,

$$\mathscr{I}(\mathscr{G}) = \mathbb{C}[f_1, \ldots, f_m] \oplus f_{m+1} \mathbb{C}[f_1, \ldots, f_m] \oplus \cdots \oplus f_l \mathbb{C}[f_1, \ldots, f_m]. \tag{37b}$$

In words, this says that any invariant of $\mathscr{G}$ can be written as a polynomial in $f_1, \ldots, f_m$ (if $l = m$), or as such a polynomial plus f_{m+1} times another such polynomial plus $\cdots$ (if $l > m$). Speaking loosely, this says that, to describe an arbitrary invariant, $f_1, \ldots, f_m$ are "free" invariants and can be used as often as needed, while $f_{m+1}, \ldots, f_l$ are "transient" invariants and can each be used at most once.

For a good polynomial basis $f_1, \ldots, f_l$ we can say exactly what the syzygies are. If $l = m$ there are no syzygies. If $l > m$ there are $(l - m)^2$ syzygies expressing the products $f_i f_j$ ($i \geq m$, $j \geq m$) in terms of $f_1, \ldots, f_l$.

It is important to note that the Molien series can be written down by inspection from the degrees of a good polynomial basis. Let $d_1 = \deg f_1, \ldots, d_l = \deg f_l$. Then

$$\Phi_{\mathscr{G}}(\lambda) = \frac{1}{\prod_{i=1}^{m}(1 - \lambda^{d_i})}, \quad \text{if } l = m, \tag{38a}$$

or

$$\Phi_{\mathscr{G}}(\lambda) = \frac{1 + \sum_{j=l+1}^{m} \lambda^{d_j}}{\prod_{i=1}^{m}(1 - \lambda^{d_i})}, \quad \text{if } l > m. \tag{38b}$$

(This is easily verified by expanding (38a) and (38b) in powers of λ and comparing with (37b).)

Some examples will make this clear.

(1) For the group $\mathscr{G}_1$ of §2, $f_1 = W_2(x, y)$ and $f_2 = W_3'(x, y)$ form a good polynomial basis, with degrees $d_1 = 8$, $d_2 = 24$. Indeed, from Theorem 3a and Equation (11),

$$\mathscr{I}(\mathscr{G}_1) = \mathbb{C}[W_2(x, y), W_3'(x, y)]$$

and

$$\Phi_{\mathscr{G}_1}(\lambda) = \frac{1}{(1 - \lambda^8)(1 - \lambda^{24})}.$$

(2) For the group $\mathscr{G}_4$ defined above, $f_1 = x^2$, $f_2 = y^2$, $f_3 = xy$ is a good polynomial basis, with $d_1 = d_2 = d_3 = 2$. The invariants can be described as

$$\mathscr{I}(\mathscr{G}_4) = \mathbb{C}[x^2, y^2] \oplus xy\, \mathbb{C}[x^2, y^2]. \tag{39}$$

In words, any invariant can be written uniquely as a polynomial in x^2 and y^2 plus xy times another such polynomial. E.g.

$$(x + y)^4 = (x^2)^2 + 6x^2y^2 + (y^2)^2 + xy(4x^2 + 4y^2).$$

The Molien series is

$$\Phi_{\mathscr{G}_4}(\lambda) = \frac{1}{2}\left\{\frac{1}{(1 - \lambda)^2} + \frac{1}{(1 + \lambda)^2}\right\}$$

$$= \frac{1 + \lambda^2}{(1 - \lambda^2)^2}$$

in agreement with (38b) and (39). The single syzygy is $x^2 \cdot y^2 = (xy)^2$. Note that $f_1 = x^2$, $f_2 = xy$, $f_3 = y^2$ is not a good polynomial basis, for the invariant y^4 is not in the set $\mathbb{C}[x^2, xy] \oplus y^2\mathbb{C}[x^2, xy]$.

Fortunately the following result holds.

Theorem 11. (Hochster and Eagon [657, Proposition 13]; independnetly proved by Dade [325].) *A good polynomial basis exists for the invariants of any finite group of complex $m \times m$ matrices.*

(The proof is too complicated to give here.)

So we know that for any group the Molien series can be put into the standard form of Equations (38a), (38b) (with denominator consisting of a product of m factors $(1 - \lambda^{d_i})$ and numerator consisting of sum of powers of λ with positive coefficients); and that a good polynomial basis Equations (37a), (37b) can be found whose degrees match the powers of λ occurring in the Molien series.

On the other hand the converse is not true. It is not always true that when the Molien series has been put into the form (38a), (38b) (by cancelling common factors and multiplying top and bottom by new factors), then a good polynomial basis for $\mathscr{J}(\mathscr{G})$ can be found whose degrees match the powers of λ in $\Phi(\lambda)$. This is shown by the following example, due to Stanley [1262].

Let $\mathscr{G}_6$ be the group of order 8 generated by the matrices

$$\begin{pmatrix} -1 & 0 & 0 \\ 0 & -1 & 0 \\ 0 & 0 & -1 \end{pmatrix} \quad \text{and} \quad \begin{pmatrix} 1 & 0 & 0 \\ 0 & 1 & 0 \\ 0 & 0 & i \end{pmatrix}.$$

The Molien series is

$$\Phi_{\mathscr{G}_6}(\lambda) = \frac{1}{(1 - \lambda^2)^3} \tag{40}$$

$$= \frac{1 + \lambda^2}{(1 - \lambda^2)^2(1 - \lambda^4)}. \tag{41}$$

A good polynomial basis exists corresponding to Equation (41), namely

$$\mathcal{J}(\mathcal{G}_6) = \mathbb{C}[x^2, y^2, z^4] \oplus xy\mathbb{C}[x^2, y^2, z^4],$$

but there is no good polynomial basis corresponding to (40).

Research Problem (19.1). Which forms of $\Phi(\lambda)$ correspond to a good polynomial basis and which do not?

One important special case has been solved. Shephard and Todd [1196] have characterized those groups for which (37a) and (38a) hold, i.e., for which a good polynomial basis exists consisting only of algebraically independent invariants. These are the groups known as unitary groups generated by reflections. A complete list of the 37 irreducible groups of this type is given in [1196].

Problem. (4) Let

$$P_e = \sum_{i=1}^{n} y_i^e, \quad e = 0, 1, 2, \dots,$$

where $y_1, \dots, y_n$ are indeterminates. Show that P_e is a polynomial in $P_1, \dots, P_n$ with rational coefficients. [Hint: Problem 52 of Ch. 8; see also [756]].

*§4. Generalizations of Gleason's theorem

All of the generalized weight enumerators of self-dual codes can be characterized by invariant theory. We work out one further example in detail, illustrating the general plan of attack described in §2 in a situation where it is more difficult to find a good polynomial basis. Several other generalizations are given as problems.

The complete weight enumerator of a ternary self-dual code. Let $\mathcal{C}$ be an $[n, \frac{1}{2}n, d]$ ternary self-dual code which contains some codeword of weight n. By suitably multiplying columns by -1 we can assume that $\mathcal{C}$ contains the codeword $\mathbf{1} = 111 \cdots 1$.

The goal of this section is to characterize the complete weight enumerator of $\mathcal{C}$ by proving:

Theorem 12. *If $\mathcal{W}(x, y, z)$ is the complete weight enumerator of a ternary self-dual code which contains $\mathbf{1}$, then*

$$\mathcal{W}(x, y, z) \in \mathbb{C}[\alpha_{12}, \beta_6^2, \delta_{36}] \oplus \beta_6 \gamma_{18} \mathbb{C}[\alpha_{12}, \beta_6^2, \delta_{36}]$$

(i.e., $W(x, y, z)$ can be written uniquely as a polynomial in $\alpha_{12}, \beta_6^2, \delta_{36}$ plus $\beta_6\gamma_{18}$ times another such polynomial), where

$$\alpha_{12} = a(a^3 + 8p^3),$$
$$\beta_6 = a^2 - 12b,$$
$$\gamma_{18} = a^6 - 20a^3p^3 - 8p^6,$$
$$\delta_{36} = p^3(a^3 - p^3)^3,$$

and

$$a = x^3 + y^3 + z^3,$$
$$p = 3xyz,$$
$$b = x^3y^3 + x^3z^3 + y^3z^3.$$

Note that

$$\gamma_{18}^2 = \alpha_{12}^3 - 64\delta_{36}.$$

(The subscript of a polynomial gives its degree.)

Proof. The proof follows the two stages described in §2.

Stage I. Let a typical codeword $u \in \mathscr{C}$ contain a 0's, b 1's, and c 2's. Then since $\mathscr{C}$ is self-dual and contains 1

$$u \cdot u = 0 \,(\text{mod. } 3) \Rightarrow 3 \,|\, (b + c)$$
$$\text{(the Hamming weight is divisible by 3)},$$
$$u \cdot 1 = 0 \,(\text{mod. } 3) \Rightarrow 3 \,|\, (b - c) \Rightarrow 3 \,|\, b \text{ and } 3 \,|\, c,$$
$$1 \cdot 1 = 0 \,(\text{mod. } 3) \Rightarrow 3 \,|\, (a + b + c) \Rightarrow 3 \,|\, a.$$

Therefore $W(x, y, z)$ is invariant under the transformations

$$\begin{pmatrix} \omega & 0 & 0 \\ 0 & 1 & 0 \\ 0 & 0 & 1 \end{pmatrix}, \quad J_3 = \begin{pmatrix} 1 & 0 & 0 \\ 0 & \omega & 0 \\ 0 & 0 & 1 \end{pmatrix}, \begin{pmatrix} 1 & 0 & 0 \\ 0 & 1 & 0 \\ 0 & 0 & \omega \end{pmatrix}, \quad \omega = e^{2\pi i/3}.$$

Also $-u$ contains a 0's, c 1's, b 2's, and $1+u$ contains c 0's, a 1's, b 2's. Therefore $W(x, y, z)$ is invariant under

$$\begin{pmatrix} 1 & 0 & 0 \\ 0 & 0 & 1 \\ 0 & 1 & 0 \end{pmatrix}, \begin{pmatrix} 0 & 1 & 0 \\ 0 & 0 & 1 \\ 1 & 0 & 0 \end{pmatrix},$$

i.e., under any permutation of its arguments.

Finally, from Equation (43) of Ch. 5, $W(x, y, z)$ is invariant under

$$M_3 = \frac{1}{\sqrt{3}} \begin{pmatrix} 1 & 1 & 1 \\ 1 & \omega & \omega^2 \\ 1 & \omega^2 & \omega \end{pmatrix}.$$

These 6 matrices generate a group $\mathscr{G}_7$, of order 2592, consisting of 1944 matrices of the type

$$s^{\nu} \begin{pmatrix} 1 & & \\ & \omega^a & \\ & & \omega^b \end{pmatrix} M_3^e \begin{pmatrix} 1 & & \\ & \omega^c & \\ & & \omega^d \end{pmatrix}, \quad s = e^{2\pi i/12},$$

and 648 matrices of the type

$$s^{\nu} \begin{pmatrix} 1 & & \\ & \omega^a & \\ & & \omega^b \end{pmatrix} P,$$

where $0 \le \nu \le 11$, $0 \le a, b, c, d \le 2$, $e = 1$ or 3, and P is any 3×3 permutation matrix.

Thus Stage I is completed: the assumptions about the code imply that $\mathscr{W}(x, y, z)$ is invariant under the group $\mathscr{G}_7$.

Stage II. consists of showing that the ring of invariants of $\mathscr{G}_7$ is equal to $\mathbb{C}[\alpha_{12}, \beta_6^2, \delta_{36}] \oplus \beta_6 \gamma_{18} \mathbb{C}[\alpha_{12}, \beta_6^2, \delta_{36}]$. First, since we have a list of the matrices in $\mathscr{G}_7$, it is a straightforward hand calculation to obtain the Molien series, Eq. (9). As usual everything collapses and the final expression is

$$\Phi_{\mathscr{G}_7}(\lambda) = \frac{1 + \lambda^{24}}{(1 - \lambda^{12})^2 (1 - \lambda^{36})}.$$

This suggests the degrees of a good polynomial basis that we should look for.

Next, $\mathscr{G}_7$ is generated by J_3, M_3, and all permutation matrices P. Obviously the invariants must be symmetric functions of x, y, z having degree a multiple of 3. So we take the algebraically independent symmetric functions a, p, b, and find functions of them which are invariant under J_3 and M_3. For example, β_6 is invariant under J_3, but is sent into $-\beta_6$ by M_3. We denote this by writing

$$\beta_6 \xleftrightarrow{\;J_3\;} \beta_6, \qquad \beta_6 \xleftrightarrow{\;M_3\;} -\beta_6.$$

Therefore β_6^2 is an invariant. Again

$$a \xleftrightarrow{\;M_3\;} \frac{1}{\sqrt{3}}(a + 2p) \xrightarrow{\;J_3\;} \frac{1}{\sqrt{3}}(a + 2\omega p) \xleftarrow{\;M_3\;} \frac{i}{\sqrt{3}}(a + 2\omega^2 p),$$

so another invariant is

$$\alpha_{12} = a(a + 2p)(a + 2\omega p)(a + 2\omega^2 p) = a(a^3 + 8p^3).$$

Again

$$\gamma_{18} \xleftrightarrow{\;J_3\;} \gamma_{18}, \qquad \gamma_{18} \xleftrightarrow{\;M_3\;} -\gamma_{18},$$

so $\beta_6\gamma_{18}$ is an invariant. Finally

$$p \xleftarrow{\ M_3\ } \frac{1}{\sqrt{3}}(a-p) \xrightarrow{\ J_3\ } \frac{1}{\sqrt{3}}(a-\omega p) \xleftarrow{\ M_3\ } \frac{s}{\sqrt{3}}(a-\omega^2 p)$$

gives the invariant

$$\delta_{36} = p^3(a-p)^3(a-\omega p)^3(a-\omega^2 p)^3 = p^3(a^3-p^3)^3.$$

The syzygy $\gamma_{18}^2 = \alpha_{12}^3 - 64\delta_{36}$ is easily verified, and one can show that $\alpha_{12}, \beta_6^2, \delta_{36}$ are algebraically independent. Thus $f_1 = \alpha_{12}$, $f_2 = \beta_6^2$, $f_3 = \delta_{36}$, $f_4 = \beta_6\gamma_{18}$ is a good polynomial basis for $\mathscr{J}(\mathscr{G}_7)$, and the theorem is proved.   Q.E.D.

Remark. Without the assumption that the code contains the all-ones vector the theorem (due to R. J. McEliece) becomes much more complicated ([883, §4.7], Mallows et al. [892]). See also Problem 5.

Applications of Theorem 12. For the ternary Golay code (Equation (6)) $\mathscr{W} = \frac{1}{6}(5\alpha_{12} + \beta_6^2)$. For the [24, 12, 9] symmetry code of Ch. 16,

$$\mathscr{W} = \tfrac{67}{144}\alpha_{12}^2 + \tfrac{1}{8}\alpha_{12}\beta_6^2 + \tfrac{1}{432}\beta_6^4 + \tfrac{11}{27}\beta_6\gamma_{18}.$$

The complete weight enumerator of the symmetry codes of lengths 36, 48 and 60 have also been obtained with the help of Theorem 12 (see Mallows et al. [892]).

Other generalizations. The following problems contain further theorems of the same type. Some results not given here are the Lee enumerator of a self-dual code over GF(7) (MacWilliams et al. [883, §5.3.2], Mallows and Sloane [894]); other split w.e.'s – see Mallows and Sloane [895]; and the biweight enumerator and joint w.e. (§6 of Ch. 5) of binary self-dual codes [883, §4.9, §5.4.1]. Rather less is known about the complete and Lee w.e.'s of a self-dual code over GF(q) – see [883, §§5.3.1, 5.3.2] and Theorem 6 above.

The Hamming w.e.'s of the codes of types (i) and (ii) of Theorem 1 are described in Theorem 3c and Problem 3. The other two types are as follows.

Problems. (5) (Gleason [486], Berlekamp et al. [129], Feit [425, 883].) The Hamming w.e. of a self-dual code over GF(3) is a polynomial in $W_4(x, y)$ (Equation (5)) and

$$W_6'(x, y) = \tfrac{1}{24}\{W_4(x, y)^3 - W_6(x, y)\} = y^3(x^3 - y^3)^3. \qquad (42)$$

[Hint: The group is generated by

$$\frac{1}{\sqrt{3}}\begin{pmatrix} 1 & 2 \\ 1 & -1 \end{pmatrix} \quad \text{and} \quad \begin{pmatrix} 1 & 0 \\ 0 & \omega \end{pmatrix}, \quad \omega = e^{2\pi i/3},$$

has order 48 and Molien series $1/(1-\lambda^4)(1-\lambda^{12})$.]

(6) ([883].) (a) The Hamming w.e. of a self-dual code over GF(4) with all weights divisible by 2 is a polynomial in $f = x^2 + 3y^2$ and $g = y^2(x^2 - y^2)^2$. [Hint: The group has order 12.] The Hamming w.e. of the [6, 3, 4] code of example (E9) is $f^3 - 9g$. The [30, 15, 12] code mentioned on p. 76 of Ch. 16 has Hamming w.e. $f^{15} - 45f^{12}g + 585f^9g^2 - 2205f^6g^3 + 1485f^3g^4 - 3249g^5$.

(b) A [2m, m] code over GF(4) with even weights is formally self-dual. If it is self-dual, then it has binary basis.

(7) Lee enumerator for GF(5). Let $\mathscr{C}$ be a self-dual code over GF(5), with Lee enumerator

$$\mathscr{L}(x, y, z) = \sum_{u \in \mathscr{C}} x^{l_0(u)} y^{l_1(u)} z^{l_2(u)},$$

where $l_0(u)$ is the number of 0's in u, $l_1(u)$ is the number of ± 1's, and $l_2(u)$ is the number of ± 2's (see p. 145 of Ch. 5). Then $\mathscr{L}(x, y, z)$ is a polynomial in A, B and C, where

$$A = x^2 + YZ,$$
$$B = 8x^4YZ - 2x^2Y^2Z^2 + Y^3Z^3 - x(Y^5 + Z^5),$$
$$C = 320x^6Y^2Z^2 - 160x^4Y^3Z^3 + 20x^2Y^4Z^4 + 6Y^5Z^5$$
$$\qquad - 4x(Y^5 + Z^5)(32x^4 - 20x^2YZ + 5Y^2Z^2)$$
$$\qquad + Y^{10} + Z^{10}, \tag{43}$$

and $Y = 2y$, $Z = 2z$. [Hint: the group has order 120. See [883, §5.3.2] and Klein [768, pp. 236–243].] The following examples illustrate this result.

Generators for code	Lee weight enumerator
{12}	A
{100133, 010313, 001331}	$A^3 - \frac{3}{8}B$
{1122000000, 0000100122, 0000010213, 1414141414, 2420430100}	$A^5 - \frac{5}{8}A^2B + \frac{1}{256}C$

(8) (Feit [424], Mallows and Sloane [895].) Suppose n is odd and let $\mathscr{C}$ be an $[n, \frac{1}{2}(n-1)]$ binary code which is contained in its dual and has all weights divisible by 4. Examples are the little [7, 3, 4] Hamming and [23, 11, 8] Golay codes, with w.e.'s

$$\varphi_7 = x^7 + 7x^3y^4, \tag{44}$$

$$\gamma_{23} = x^{23} + 506x^{15}y^8 + 1288x^{11}y^{12} + 253x^7y^{16}. \tag{45}$$

Then n must be of the form $8m \pm 1$. If $n = 8m - 1$, the w.e. of $\mathscr{C}$ is an element of

$$\varphi_7\mathbb{C}[W_2(x, y), W_3'(x, y)] \oplus \gamma_{23}\mathbb{C}[W_2(x, y), W_3'(x, y)], \tag{46}$$

while if $n = 8m + 1$, the w.e. is an element of

$$x\mathbb{C}[W_2(x, y), W_3'(x, y)] \oplus \psi_{17}\mathbb{C}[W_2(x, y), W_3'(x, y)], \tag{47}$$

where

$$\psi_{17} = x^{17} + 17x^{13}y^4 + 187x^9y^8 + 51x^5y^{12} \tag{48}$$

is the w.e. of the $[17, 8, 4]$ code $\bar{I}_{17}^{(3)}$ found by Pless [1058]. Hence show that the w.e.'s of the $[31, 15, 8]$ and $[47, 23, 12]$ QR codes are respectively

$$-14\varphi_7 W_3'(x, y) + \gamma_{23} W_2(x, y),$$

$$\tfrac{1}{7}\{-253\varphi_7 W_2(x, y)^2 W_3'(x, y) + \gamma_{23}(7 W_2(x, y)^3 - 41 W_3'(x, y))\}.$$

For the corresponding theorems for binary self-dual codes with weights divisible by 2, and for ternary self-dual codes, see Mallows et al. [892], [895].

(9) Split weight enumerators. Suppose $\mathscr{C}$ is a $[2m, m]$ even self-dual code which contains the codewords $0^m 1^m$ and $1^m 0^m$, and has the property that the number of codewords with $(w_L, w_R) = (j, k)$ is equal to the number with $(w_L, w_R) = (k, j)$. (Here w_L and w_R denote the left and right weights – see p. 149 of Ch. 5.) Such a code is "balanced" about its midpoint, and the division into two halves is a natural one. Then the split w.e. $\mathscr{S}_{\mathscr{C}}(x, y, X, Y)$ of $\mathscr{C}$ (p. 149 of Ch. 5) is a polynomial in η_8, θ_{16} and γ_{24}, where

$$\eta_8 = x^4X^4 + x^4Y^4 + y^4X^4 + y^4Y^4 + 12x^2y^2X^2Y^2, \tag{49}$$

$$\theta_{16} = (x^2X^2 - y^2Y^2)^2(x^2Y^2 - y^2X^2)^2, \tag{50}$$

$$\gamma_{24} = x^2y^2X^2Y^2(x^4 - y^4)^2(X^4 - Y^4)^2. \tag{51}$$

(10) Examples of split w.e.'s. Use a detached-coefficient notation for $\mathscr{S}$, and instead of the terms

$$\alpha(x^ay^bX^cY^d + x^ay^bX^dY^c + x^by^aX^cY^d + x^by^aX^dY^c)$$

write a row of a table:

$$\begin{array}{ccccc} c/o & x & y & X & Y & \# \\ \alpha & a & b & c & d & 4 \end{array}$$

giving respectively the coefficient, the exponents, and the number of terms of this type. The sum of the products of the first and last columns is the total number of codewords. Use the preceding problem to obtain the split w.e.'s shown in Fig. 19.1 (taking the generator matrices for these codes in the canonical form of Equation (57) of Ch. 16).

(11) Suppose Π is a projective plane of order n, where $n \equiv 2 \pmod 4$ (see p. 59 of Ch. 2). Let $A = (a_{ij})$ be the $(n^2 + n + 1) \times (n^2 + n + 1)$ adjacency matrix of Π, where $a_{ij} = 1$ if the ith line passes through the jth point, and $a_{ij} = 0$ otherwise. Let $\mathscr{C}$ be the binary code generated by the rows of A, and let $\mathscr{C}^*$ be obtained by adding an overall parity check to $\mathscr{C}$. (i) Show that if $n = 2$, $\mathscr{C}$ is the $[7, 4, 3]$ Hamming code. (ii) Show in general that

$$\mathscr{C}^* \text{ is an } [n^2 + n + 2, \tfrac{1}{2}(n^2 + n + 2), n + 2]$$

even self-dual code. (iii) Hence show that there is no projective plane of order

Code	$\mathscr{S}$	c/o	x	y	X	Y	$\#$
[8, 4, 4] Hamming	η_8	1	4	0	4	0	4
		12	2	2	2	2	1
	θ_{16}	1	8	0	4	4	4
		-2	6	2	6	2	4
		4	4	4	4	4	1
	γ_{24}	1	10	2	10	2	4
		-2	10	2	6	6	4
		4	6	6	6	6	1
[24, 12, 8] Golay code $\mathscr{G}_{24}$		1	12	0	12	0	4
		132	10	2	6	6	4
		495	8	4	8	4	4
		1584	6	6	6	6	1
[48, 24, 12] QR code		1	24	0	24	0	4
		276	22	2	14	10	8
		3864	20	4	16	8	8
		13524	20	4	12	12	4
		9016	18	6	18	6	4
		125580	18	6	14	10	8
		256335	16	8	16	8	4
		950544	16	8	12	12	4
		1835400	14	10	14	10	4
		3480176	12	12	12	12	1

Fig. 19.1. Split w.e.'s of even self-dual codes.

6 [Hint: from Theorem 3c, since $8 \nmid 44$]. Note that the results of problem 6 apply to the w.e. of $\mathscr{C}$. The w.e. of the hypothetical plane of order 10 is discussed by Assmus et al. [43, 48], MacWilliams et al. [888] and Mallows and Sloane [895].

Research Problem (19.2). Let $\mathscr{C}$ be a binary code with weights divisible by 4, having parameters $[n, \frac{1}{2}n]$ if n is even or $[n, \frac{1}{2}(n-1)]$ if n is odd, with $\mathscr{C} \subseteq \mathscr{C}^\perp$. Characterize the biweight enumerator of $\mathscr{C}$.

§5. The nonexistence of certain very good codes

In this section Theorem 3c is used to obtain an upper bound (Theorem 13) on the minimum distance of an even self-dual code. It is then shown (Corollary 16) that this upper bound can only be attained for small values of n. The method generalizes the calculation of the w.e. of the $[48, 24, 12]$ QR code given in §2. Similar results hold for the other types of self-dual codes given in Theorem 1.

Let $\mathscr{C}$ be an $[n, \frac{1}{2}n, d]$ even self-dual code, having w.e. $W(x, y) = x^n + A_d x^{n-d} y^d + \cdots$. From Theorem 3c, $W(x, y)$ is a polynomial in $W_2(x, y)$ and $W_3'(x, y)$, and can be written as

$$W(x, y) = \sum_{r=0}^{\mu} a_r W_2(x, y)^{j-3r} W_3'(x, y)^r, \tag{52}$$

where $n = 8j = 24\mu + 8\nu$, $\nu = 0$, 1 or 2.

Suppose the $\mu + 1 = [n/24] + 1$ coefficients a_i in (52) are chosen so that

$$W(x, y) = x^n + A^*_{4\mu+4} x^{n-4\mu-4} y^{4\mu+4} + \cdots$$

$$= W(x, y)^* \quad \text{(say)}. \tag{53}$$

I.e., the a_i are chosen so that $W(x, y)$ has as many leading coefficients as possible equal to zero. It will be shown below that this determines the a_i uniquely. The resulting $W(x, y)^*$ given by (53) is the weight enumerator of that even self-dual code with the greatest minimum weight we could hope to attain, and is called an *extremal* weight enumerator.

If a code exists with weight enumerator $W(x, y)^*$, it has minimum distance $d^* = 4\mu + 4$, unless it should happen that $A^*_{4\mu+4}$ in (53) is accidentally zero, in which case $d^* \geq 4\mu + 8$. But this doesn't happen.

Theorem 13. (Mallows and Sloane [893].) $A^*_{4\mu+4}$, *the number of codewords of minimum nonzero weight in the extremal weight enumerator, is given by:*

$$\binom{n}{5}\binom{5\mu - 2}{\mu - 1} \Big/ \binom{4\mu + 4}{5}, \quad \text{if } n = 24\mu, \tag{54}$$

$$\frac{1}{4} n(n - 1)(n - 2)(n - 4) \frac{(5\mu)!}{\mu!(4\mu + 4)!}, \quad \text{if } n = 24\mu + 8, \tag{55}$$

$$\frac{3}{2} n(n - 2) \frac{(5\mu + 2)!}{\mu!(4\mu + 4)!}, \quad \text{if } n = 24\mu + 16, \tag{56}$$

and is never zero. Therefore the minimum distance of an even self-dual code of length n is at most $4[n/24] + 4$.

Proof. Here we give an elementary proof for the case $n = 24\mu$; the other cases are similar. A second proof which applies to all cases simultaneously will be given below.

By Theorem 9 of Ch. 6 the codewords of weight $4\mu + 4$ form a 5-design. It is easiest to calculate the parameter $\lambda^{(4)}_{4\mu+4}$ of the consequent 4-design. From Equation (9) of Ch. 6 this is given by

$$\lambda^{(4)}_{4\mu+4} 4^{4\mu-2}(4\mu - 2)! = -\frac{4^{4\mu-2}(5\mu - 2)!}{\mu!} + \frac{1}{2^{12\mu}} \sum_{r=4}^{24\mu} \binom{24\mu - 4}{r - 4} \frac{S(r)}{4\mu + 4 - r},$$

where

$$S(x) = \prod_{j=1}^{4\mu-1} (4\mu + 4j - x).$$

The following identities are easily verified.

$$\frac{S(r-4) - S(r)}{16\mu - 4} = \frac{S(r)}{4\mu + 4 - r},$$

$$\sum_{r=4}^{24\mu} \binom{24\mu - 4}{r - 4} S(r-4) = -\sum_{l=4}^{24\mu} \binom{24\mu - 4}{l - 4} S(l),$$

$$\frac{1}{2^{12\mu}} \sum_{r=4}^{24\mu} \binom{24\mu - 4}{r - 4} S(r) = S(24\mu), \quad \text{(Equation (14) of Ch. 6)}.$$

Hence

$$\lambda^{(4)}_{4\mu+4} 4^{4\mu-2}(4\mu - 2)! = -\frac{4^{4\mu-2}(5\mu - 2)!}{\mu!} + \frac{2 \cdot 4^{4\mu-2}(5\mu - 1)!}{(4\mu - 1)\mu!},$$

$$\lambda^{(4)}_{4\mu+4} = \frac{(6\mu - 1)(5\mu - 2)!}{\mu!(4\mu - 1)!},$$

and so

$$\lambda^{(5)}_{4\mu+4} = \frac{4\mu}{24\mu - 4} \lambda^{(4)}_{4\mu+4}$$

$$= \binom{5\mu - 2}{\mu - 1}.$$

Thus the number of codewords of weight $4\mu + 4$ is given by (54).

Q.E.D.

The known codes which attain the bound in Theorem 13 are shown in Fig. 19.2. For the double circulant codes on this list see Fig. 16.7.

The extremal weight enumerators for $n \le 200$ and $n = 256$ (and in particular the w.e.'s of all codes in Fig. 19.2 are given in [893]. These were obtained from Theorem 15.)

[n, k, d]	Code
[8, 4, 4]	Hamming code.
[16, 8, 4]	Direct sum of two Hamming codes.
[24, 12, 8]	Golay code.
[32, 16, 8]	2nd order RM or QR code.
[40, 20, 8]	Double circulant code (Fig. 16.7)
[48, 24, 12]	QR code.
[56, 28, 12]	Double circulant code.
[64, 32, 12]	Double circulant code.
[80, 40, 16]	QR code.
[88, 44, 16]	Double circulant code.
[104, 52, 20]	QR code.

Fig. 19.2. Even self-dual codes with $d = 4[n/24] + 4$

Feit [426] has found a [96, 48, 16] even self-dual code using the $|a + x | b + x | a + b + x|$ construction of Ch. 18. All even self-dual codes of length ≤ 24 are given by Pless [1058], Pless and Sloane [1062, 1063]. The first gap in Fig. 19.2 is at $n = 72$.

Research Problem (19.3). ([1227].) Is there a [72, 36, 16] even self-dual code?

Problem. (12) Suppose $\mathscr{C}$ is a binary self-dual code (of type (i)), with length $n = 8m$ and minimum distance $2m + 2$. Show that the number of codewords of weight $2m + 2$ is

$$\frac{8m(8m - 1)(8m - 2)}{(2m + 2)(2m + 1)2m} \binom{3m - 2}{m - 1}.$$

But if n is very large then the extremal w.e. contains a negative coefficient, and the bound of Theorem 13 cannot be attained. This can be proved by the method used to prove Theorem 13, as shown in the following problem.

Problem. (13) Suppose $\mathscr{C}$ is an even self-dual code of length $n = 24\mu$ whose w.e. is the extremal w.e. $W^*(x, y)$. The codewords of each weight form a 5-design. (a) Show that the parameter $\lambda^{(4)}_{4\mu+8}$ of the 4-design formed by the codewords of weight $4\mu + 8$ is

$$-\frac{(4\mu - 1)(6\mu - 1)}{\mu} \binom{5\mu - 2}{\mu - 1} + \binom{5\mu - 1}{\mu + 1} - \binom{5\mu - 3}{\mu - 1}$$
$$+ \frac{6\mu - 1}{\mu} \binom{5\mu - 2}{\mu - 1}\binom{20\mu - 4}{4} \bigg/ \binom{4\mu + 4}{4}.$$

(b) Hence show that $A^*_{4\mu+8} < 0$ for sufficiently large μ.

An alternative, more powerful proof makes use of:

Theorem 14. Bürmann–Lagrange (Whittaker and Watson [1414, p. 133], Goldstein [520], Good [533], Sack [1137–1139]). *Let $f(x)$ and $\Phi(x)$ be analytic functions near $x = 0$, with $\Phi(0) \neq 0$. Provided that the equation*

$$x = \epsilon\Phi(x)$$

defines x uniquely in some neighborhood of the origin, then $f(x)$ can be expanded in powers of ϵ as follows:

$$f(x) = f(0) + \sum_{r=1}^{\infty} \frac{\epsilon^r}{r!} \frac{d^{r-1}}{da^{r-1}} [f'(a)\Phi(a)^r]_{a=0}, \tag{57}$$

valid for sufficiently small ϵ. (The prime denotes differentiation.)

To apply this theorem to the extremal weight enumerator, we first replace x by 1 and y^4 by x in (52). Then $W_2(x, y)$ becomes $f(x) = 1 + 14x + x^2$ and $W_3'(x, y)$ becomes $g(x) = x(1 - x)^4$. For simplicity suppose that n is a multiple of 24, so $\nu = 0$ and $j = 3\mu$; the other two cases are handled similarly. Equating (52) and (53) gives

$$W^* = \sum_{r=0}^{\mu} a_r f(x)^{3\mu-3r} g(x)^r = 1 + \sum_{r=\mu+1}^{6\mu} A^*_{4r} x^r. \tag{58}$$

Divide by $f(x)^{3\mu}$:

$$f(x)^{-3\mu} = \sum_{r=0}^{\mu} a_r \left(\frac{g(x)}{f(x)^3}\right)^r - f(x)^{-3\mu} \sum_{r=\mu+1}^{6\mu} A^*_{4r} x^r, \tag{59}$$

where

$$\frac{g(x)}{f(x)^3} = \frac{x(1-x)^4}{(1+14x+x^2)^3} \sim x \text{ for } x \text{ small.}$$

Let us expand $f(x)^{-3\mu}$ by Theorem 14 in powers of $\epsilon = g(x)/f(x)^3$, with

$$\Phi(x) = \frac{x}{\epsilon} = \frac{(1+14x+x^2)^3}{(1-x)^4}.$$

$$f(x)^{-3\mu} = \sum_{r=0}^{\infty} \alpha_r \epsilon^r, \tag{60}$$

where $\alpha_0 = 1$ and

$$\alpha_r = \frac{1}{r!} \frac{d^{r-1}}{dx^{r-1}} \left[\frac{d}{dx}(1+14x+x^2)^{-3\mu} \cdot \frac{(1+14x+x^2)^{3r}}{(1-x)^{4r}} \right]_{x=0}$$

$$= \frac{-3\mu}{r!} \frac{d^{r-1}}{dx^{r-1}} \left[\frac{(14+2x)(1+14x+x^2)^{3r-3\mu-1}}{(1-x)^{4r}} \right]_{x=0} \tag{61}$$

for $r = 1, 2, \ldots$ Comparing (59) and (60) we see that

$$a_r = \alpha_r \quad \text{for } r = 0, 1, \ldots, \mu, \tag{62}$$

and

$$\sum_{r=\mu+1}^{\infty} \alpha_r f(x)^{3\mu-3r} g(x)^r = -\sum_{r=\mu+1}^{6\mu} A_{4r}^* x^r. \tag{63}$$

So far we have proved:

Theorem 15. ([893].) *The extremal weight enumerator of an even self-dual code of length $n = 24\mu$ is given by (58), where the coefficients a_r are given explicitly by Equations (61), (62). Also, equating coefficients of $x^{\mu+1}$ and $x^{\mu+2}$ in (63) gives*

$$A_{4\mu+4}^* = -\alpha_{\mu+1} \tag{64}$$

and

$$A_{4\mu+8}^* = -\alpha_{\mu+2} + (4\mu + 46)\alpha_{\mu+1}. \tag{65}$$

Now $\alpha_{\mu+1} < 0$ and $A_{4\mu+4}^* > 0$ follow immediately from (61), giving another proof of the upper bound in Theorem 13. But the next coefficient becomes negative:

Corollary 16. (Mallows et al. [891].) $A_{4\mu+8}^* < 0$ *for all $n = 24\mu$ sufficiently large.*

Proof. After some messy algebra, (61) implies that $|\alpha_{\mu+2}/\alpha_{\mu+1}|$ is bounded for large μ, and so $A_{4\mu+8}^* < 0$ follows from (65). The details are omitted.
Q.E.D.

The proof shows that $A_{4\mu+8}^*$ first goes negative when n is about 3720. Indeed, when $n = 3720$ a computer was used to show that

$$W^*(x, y) = x^{3720} + A_{624}^* x^{3096} y^{624} + A_{628}^* x^{3092} y^{628} + \cdots$$

where $A_{624}^* = 1.163 \ldots \cdot 10^{170}$, $A_{628}^* = -5.848 \ldots \cdot 10^{170}$, and $A_i^* > 0$ for $632 \leqslant i \leqslant 3088$.

The same argument can be used to show that Corollary 16 holds also in the cases $n = 24\mu + 8$ and $n = 24\mu + 16$. An alternative method of showing that $|\alpha_{\mu+2}/\alpha_{\mu+1}|$ is bounded is given in [891]. This uses the saddle-point method to expand the expression inside the square brackets in (61).

So far we have just considered codes of type (ii) of Theorem 1. Corresponding results hold for the other types, with a similar proof.

Theorem 17. *The minimum distance of a self-dual code over* $GF(q)$ *with all weights divisible by* t *is at most* d^*, *where*:

q	t	d^*	
2	2	$2\left[\dfrac{n}{8}\right] + 2$	(66)
2	4	$4\left[\dfrac{n}{24}\right] + 4$	(67)
3	3	$3\left[\dfrac{n}{12}\right] + 3$	(68)
4	2	$2\left[\dfrac{n}{6}\right] + 2$	(69)

But for all sufficiently large n there is no code meeting these bounds.

Note that the McEliece et al. bound (Theorem 37 of Ch. 17) for binary codes of rate $\frac{1}{2}$ is $d/n \leqslant 0.178$, and the Elias bound (Theorem 34 of Ch. 17) for codes of rate $\frac{1}{2}$ is $d/n \leqslant 0.281$ (over $GF(3)$) and $d/n \leqslant 0.331$ (over $GF(4)$), as $n \to \infty$. Thus (67), (68) are stronger bounds, while (66), (69) violate the latter bounds.

The bound (66) is met by self-dual codes of lengths 2, 4, 6, 8, 12, 14, 22, 24 but no other values of n (see [893, 1058, 1062, 1063], and Ward [1389] for the proof). Formally self-dual codes meeting (66) exist in additon for $n = 10$, 16, 18, 28 and possibly other values ([893, 1389]).

The bound (68) is met for $n = 4$, 8, 12 (the Golay code), 24, 36, 48, and 60 (QR and symmetry codes), 16 and 40 (using a code with generator matrix $[I \mid H]$ where H is a Hadamard matrix – H.N. Ward, private communication), and possibly other values (see [892]). Much less is known about codes meeting (69), but see [1478].

Research Problem (19.4). What is the largest n for which codes exist meeting the bounds of Theorem 17?

The following stronger result is also proved in [891]. For any constant b there is an $n_0(b)$ such that all self-dual codes of length greater than $n_0(b)$ have minimum distance less than $d^* - b$, where d^* is given by **Theorem 17**.

§6. Good self-dual codes exist

This section counts self-dual codes in various ways and shows that some of them meet the Gilbert–Varshamov bound. For simplicity only the binary case is considered – for the general case see Mallows et al. [892], Pless [1053], [1054], Pless and Pierce [1061], and also Zhe-Xian [1457].

If $\mathscr{C} \subset \mathscr{C}^{\perp}$ then $\mathscr{C}$ is called *weakly self-dual* (w.s.d.)

Theorem 18. (MacWilliams et al. [887].) *Let $n = 2t$ and suppose $\mathscr{C}$ is an $[n, k]$ w.s.d. (binary) code, with $k \geqslant 1$. Then the number of $[n, t]$ self-dual codes containing $\mathscr{C}$ is*

$$\prod_{i=1}^{t-k} (2^i + 1). \tag{70}$$

Proof. Let $\sigma_{n,m}$, $k \leqslant m < t$, be the number of $[n, m]$ w.s.d. codes which contain $\mathscr{C}$. We establish a recursion formula for $\sigma_{n,m}$. An $[n, m]$ w.s.d. code $\mathscr{D}$ containing $\mathscr{C}$ can be extended to an $[n, m + 1]$ w.s.d. code containing $\mathscr{C}$ by adjoining any vector of $\mathscr{D}^{\perp}$ not already in $\mathscr{D}$. Write $\mathscr{D}^{\perp}$ as the union of $2^{n-m}/2^m$ cosets of $\mathscr{D}$,

$$\mathscr{D}^{\perp} = \mathscr{D} \cup (h_1 + \mathscr{D}) \cup \cdots \cup (h_l + \mathscr{D}),$$

where $l = 2^{n-2m} - 1$. There are l different extensions of $\mathscr{D}$, namely $\mathscr{D} \cup (h_j + \mathscr{D})$ for $j = 1, 2, \ldots, l$. In each of these extensions there are $2^{m+1-k} - 1$ $[n, m]$ subcodes $\mathscr{D}'$ which contain $\mathscr{C}$, since that is the number of nonzero vectors in $\mathscr{D} \cup (h_j + \mathscr{D})$ which are orthogonal to $\mathscr{C}$. Thus for $k \leqslant m < t$,

$$\sigma_{n,m+1} = \sigma_{n,m} \cdot \frac{2^{n-2m} - 1}{2^{m+1-k} - 1}. \tag{71}$$

Starting from $\sigma_{n,k} = 1$ gives (70).   Q.E.D.

Corollary 19. *The total number of binary self-dual codes of length n is*

$$\prod_{i=1}^{\frac{1}{2}n-1} (2^i + 1). \tag{72}$$

Proof. Take $\mathscr{C} = \{0, 1\}$ in Theorem 18.   Q.E.D.

Corollary 20. *Let v be a binary vector of even weight other than 0 or 1. The number of self-dual codes containing v is*

$$\prod_{i=1}^{\frac{1}{2}n-2} (2^i + 1). \tag{73}$$

Theorem 21. ([887].) *There exist long binary self-dual codes which meet the Gilbert–Varshamov bound.*

Proof. This follows from Corollaries 19, 20 in the usual way – see Theorem 31 of Ch. 17.   Q.E.D.

A corresponding argument shows that the same result holds for self-dual codes over GF(q) ([1061]). The hardest case is that of even self-dual codes, and the reader is referred to [887] for the proof of the next result.

Theorem 22. *Let n be a multiple of 8, and suppose $\mathscr{C}$ is an $[n, k]$ w.s.d. binary code with all weights divisible by 4. The number of $[n, \frac{1}{2}n]$ even self-dual codes containing $\mathscr{C}$ is*

$$2 \prod_{i=0}^{\frac{1}{2}n-k-1} (2^i + 1).\tag{74}$$

Corollary 23. *The total number of even self-dual codes of length n is*

$$2 \prod_{i=0}^{\frac{1}{2}n-2} (2^i + 1).\tag{75}$$

The number which contain a given vector v other than $\mathbf{0}$ or $\mathbf{1}$, with $\mathrm{wt}(v) \equiv 0$ (mod 4), is

$$2 \prod_{i=0}^{\frac{1}{2}n-3} (2^i + 1).\tag{76}$$

Theorem 24 (Thompson [1320a]; see also [887].) *There exist long even self-dual codes which meet the Gilbert–Varshamov bound.*

For the next six problems, let $\Phi_{n,k}$ = the class of w.s.d. binary $[n, k]$ codes, $\Phi'_{n,k}$ = the subclass of $\Phi_{n,k}$ of codes containing **1**, for $0 \leqslant k \leqslant n/2$ (see [1063]).

Problems. (14) Let n be even and $\mathscr{C} \in \Phi'_{n,s}$. Show that the number of codes in $\Phi'_{n,k}(k \geqslant s)$ which contain $\mathscr{C}$ is

$$\prod_{j=0}^{k-s-1} \frac{2^{n-2s-2j} - 1}{2^{j+1} - 1}.$$

(15) Show that the total number of codes in $\Phi'_{n,k}$ is

$$\prod_{j=1}^{k-1} \frac{2^{n-2j} - 1}{2^j - 1} \quad \text{if } n \text{ even, 0 if } n \text{ odd.}$$

(16) Let $\mathscr{C} \in \Phi_{n,s} - \Phi'_{n,s}$. Show that the number of codes in $\Phi_{n,k} - \Phi'_{n,k}$ $(k \geqslant s)$ which contain $\mathscr{C}$, is

$$2^{k-s} \prod_{j=1}^{k-s} \frac{2^{n-2s-2j} - 1}{2^j - 1} \quad (n \text{ even}),$$

$$\prod_{j=1}^{k-s} \frac{2^{n-2s-2j+1} - 1}{2^j - 1} \quad (n \text{ odd}).$$

(17) The total number of codes in $\Phi_{n,k} - \Phi'_{n,k}$ is

$$2^k \prod_{j=1}^{k} \frac{2^{n-2j} - 1}{2^j - 1} \quad (n \text{ even}),$$

$$\prod_{j=1}^{k} \frac{2^{n-2j+1} - 1}{2^j - 1} \quad (n \text{ odd}).$$

(18) Let n be even and $\mathscr{C} \in \Phi_{n,s} - \Phi'_{n,s}$. The number of codes in $\Phi_{n,k}$ $(k > s)$ which contain $\mathscr{C}$ is

$$(2^{n-k-s} - 1) \prod_{j=1}^{k-s-1} (2^{n-2s-2j} - 1) \Big/ \prod_{j=1}^{k-s} (2^j - 1).$$

(19) If n is even, the total number of codes in $\Phi_{n,k}$ is

$$(2^{n-k} - 1) \prod_{j=1}^{k-1} (2^{n-2j} - 1) \Big/ \prod_{j=1}^{k} (2^j - 1).$$

(20) Show that the sum of the weight enumerators of all self-dual codes of length n is (for n even)

$$\prod_{j=1}^{(n/2)-2} (2^j + 1) \left[2^{(n/2)-1}(x^n + y^n) + \sum_{2|i} \binom{n}{i} x^{n-i} y^i \right].$$

The same sum for even self-dual codes is (if n is divisible by 8)

$$\prod_{j=0}^{(n/2)-3} (2^j + 1) \left[2^{(n/2)-2}(x^n + y^n) + \sum_{4|i} \binom{n}{i} x^{n-i} y^i \right].$$

(21) ([1053, 1054].) Let $\mathscr{C}$ be an $[n, k]$ w.s.d. code over GF(3) which is maximal in the sense of not being contained in any longer w.s.d. code of the same length. (a) Show that

$$k = \begin{cases} \frac{1}{2}n & \text{if } n \equiv 0 \pmod{4}, \\ \frac{1}{2}(n - 1) & \text{if } n \text{ is odd}, \\ \frac{1}{2}(n - 2) & \text{if } n \equiv 2 \pmod{4}. \end{cases}$$

(b) The number of such maximal codes is

$$2 \prod_{i=1}^{(n-2)/2} (3^i + 1) \quad \text{if } n \equiv 0 \text{ (mod. 4)},$$

$$\prod_{i=1}^{(n-1)/2} (3^i + 1) \quad \text{if } n \text{ is odd},$$

$$\prod_{i=2}^{n/2} (3^i + 1) \quad \text{if } n \equiv 2 \text{ (mod. 4)}.$$

(c) Thus a self-dual code exists iff n is a multiple of 4.

(22) ([1054].) More generally, show that an $[n, \frac{1}{2}n]$ self-dual code exists over GF(q) iff one of the following holds: (i) q and n are both even, (ii) $q \equiv 1$ (mod. 4) and n is even, and (iii) $q \equiv 3$ (mod. 4) and n is a multiple of 4.

(23) ([1054].) Show that if the conditions of the previous problem are satisfied then the number of $[n, \frac{1}{2}n]$ self-dual codes over GF(q) is

$$b \prod_{i=1}^{\frac{1}{2}n-1} (q^i + 1),$$

where $b = 1$ if q is even, $b = 2$ if q is odd.

Research Problem (19.5). How many inequivalent self-dual codes of length n are there? (See [1058, 1059, 1062, 1063, 892] for small values of n.)

Notes on Chapter 19

There are many parallels between self-dual codes and certain types of lattice sphere packings. There are analogous theorems about lattices to most of the theorems of this chapter. See Berlekamp et al. [129], Broué [200], Broué and Enguehard [201], Conway [302–306], Gunning [570], Leech [804, 805, 807–809], Leech and Sloane [810], Mallows et al. [891], Milnor and Husemoller [962], Niemeier [992] and Serre [1184].

The ALTRAN system (Brown [205], Hall [579]) for rational function manipulation makes it very easy to work with weight enumerators on a computer.

This chapter is based in part on the survey [1231]; see also [1228]. Another reference dealing with self-dual codes is [141]. For more about self-dual and formally self-dual codes over GF(4) see [1478].

20

The Golay codes

§1. Introduction

In this chapter we complete the study of the Golay codes by showing that their automorphism groups are the Mathieu groups, and that these codes are unique. There are four of these important codes: [23, 12, 7] and [24, 12, 8] binary codes, and [11, 6, 5] and [12, 6, 6] ternary codes, denoted by $\mathscr{G}_{23}$, $\mathscr{G}_{24}$, $\mathscr{G}_{11}$, $\mathscr{G}_{12}$ respectively.

Properties of the binary Golay codes. We begin with $\mathscr{G}_{24}$ (rather than $\mathscr{G}_{23}$) since this has the larger automorphism group and is therefore more fundamental. The [24, 12, 8] code $\mathscr{G}_{24}$ may be defined by any of the generator matrices given in Fig. 2.13, Equations (41), (48) of Ch. 16, or Equation (6) below; or by the $|a + x|b + x|a + b + x|$ construction (Theorem 12 of Ch. 18); or by adding an overall parity check to $\mathscr{G}_{23}$. $\mathscr{G}_{24}$ is self-dual (Lemma 18 of Ch. 2, Theorem 7 of Ch. 16), has all weights divisible by 4 (Lemma 19 of Ch. 2, Theorem 8 of Ch. 16), and has weight distribution:

$$
\begin{array}{llllll}
i: & 0 \ 8 & 12 & 16 & 24 & \\
A_i: & 1 \ 759 & 2576 & 759 & 1 & \quad\quad (1)
\end{array}
$$

The codewords of weight 8 in $\mathscr{G}_{24}$ form the blocks of a Steiner system $S(5, 8, 24)$ (Corollary 23 of Ch. 2; see also Corollary 25 and Theorem 26 of Ch. 2). These codewords are called *octads*, and the same name is also used for the set of eight coordinates where the codeword is nonzero. Let $\mathcal{O} = \{x_1 x_2 \cdots x_8\}$ be an octad and denote the number of octads which contain $x_1 \cdots x_j$ but not $x_{j+1} \cdots x_i$ by λ_{ij} $(0 \le j \le i \le 8)$. These numbers are independent of the choice of $\mathcal{O}$ (Theorem 10 of Ch. 2), and are tabulated in Fig. 2.14. We will show below (Theorem 9) that the Steiner system $S(5, 8, 24)$ is unique. Since there is a generator matrix for $\mathcal{G}_{24}$ all of whose rows have weight 8, it follows that the octads of $S(5, 8, 24)$ generate $\mathcal{G}_{24}$. This is used in §6 to show that $\mathcal{G}_{24}$ is itself unique (Theorem 14). The codewords of weight 12 in $\mathcal{G}_{24}$ are called *dodecads*.

The automorphism group of $\mathcal{G}_{24}$ contains $PSL_2(23)$ (Theorem 10 of Ch. 16), but is in fact equal to the much larger Mathieu group M_{24} (Theorem 1). This is a 5-fold transitive group, of order $24 \cdot 23 \cdot 22 \cdot 21 \cdot 20 \cdot 48 = 244823040$ and is generated by $PSL_2(23)$ and an additional permutation W given in Equation (8).

The $[23, 12, 7]$ perfect code $\mathcal{G}_{23}$ may be obtained by deleting any coordinate of $\mathcal{G}_{24}$ (it doesn't matter which, by Corollary 11 of Ch. 16), or as a quadratic residue code with idempotent equal to either of the polynomials in Equation (10) of Ch. 16, and generator polynomial given in Equation (4) of Ch. 16. The weight distribution is:

$$
\begin{array}{ccccccccc}
i: & 0 & 7 & 8 & 11 & 12 & 15 & 16 & 23 \\
A_i: & 1 & 253 & 506 & 1288 & 1288 & 506 & 253 & 1
\end{array}
\tag{2}
$$

The codewords of weight 7 in $\mathcal{G}_{23}$ form the blocks of a Steiner system $S(4, 7, 23)$, and these codewords generate $\mathcal{G}_{23}$. Both $\mathcal{G}_{23}$ and $S(4, 7, 23)$ are unique (Corollary 16 and Problem 14).

The full automorphism group of $\mathcal{G}_{23}$ is the Mathieu group M_{23} (Corollary 8). This is 4-fold transitive group of order $23 \cdot 22 \cdot 21 \cdot 20 \cdot 48 = 10200960$.

Decoding methods for these codes are given in §9 of Ch. 16.

Properties of the ternary Golay codes. The $[12, 6, 6]$ code $\mathcal{G}_{12}$ may be defined by any of the generator matrices given in Equations (25) or (61) of Ch. 16, or Equation (13) below; or by adding an overall parity check to $\mathcal{G}_{11}$. $\mathcal{G}_{12}$ is self-dual (Theorem 7 of Ch. 16), so has all weights divisible by 3. The minimum distance is 6 (from Theorem 1 of Ch. 16), hence the Hamming and complete weight enumerators are as given in Equation (6) of Ch. 19. The supports (the nonzero coordinates) of codewords of weight 6 form the 132 blocks of the Steiner system $S(5, 6, 12)$ (see Fig. 16.9). Both $\mathcal{G}_{12}$ and $S(5, 6, 12)$ are unique (Theorem 20 and Problem 18).

The automorphism group $\mathrm{Aut}(\mathcal{G}_{12})$† (p. 493 of Ch. 16) contains a group isomorphic to $PSL_2(11)$ (Theorem 12 of Ch. 16), but is in fact isomorphic to the much larger Mathieu group M_{12} (Theorem 18). This is a 5-fold transitive group of order $12 \cdot 11 \cdot 10 \cdot 9 \cdot 8 = 95040$.

The $[11, 6, 5]$ perfect code $\mathcal{G}_{11}$ may be obtained by deleting any coordinate of $\mathcal{G}_{12}$ (again it doesn't matter which), or as a quadratic residue code with idempotent and generator polynomials given by Equations (16), (5) of Ch. 16. The Hamming and complete weight enumerators of $\mathcal{G}_{11}$ are given by Mallows et al. [892]. The supports of the codewords of weight 5 form the blocks of the Steiner system $S(4, 5, 11)$. $\mathcal{G}_{11}$ and $S(4, 5, 11)$ are unique (Corollary 21 and Problem 18).

§2. The Mathieu group M_{24}

In this section the Mathieu group M_{24} is defined and shown to preserve $\mathcal{G}_{24}$.

Notation. The coordinates of $\mathcal{G}_{23}$ will be labeled $\{0, 1, \ldots, 22\}$, and the coordinates of $\mathcal{G}_{24}$ by $\Omega = \{0, 1, \ldots, 22, \infty\}$, the last coordinate containing the overall parity check. Also

$$Q = \{1, 2, 3, 4, 6, 8, 9, 12, 13, 16, 18\},$$

$$N = \{5, 7, 10, 11, 14, 15, 17, 19, 20, 21, 22\} \tag{3}$$

denote the quadratic residues and nonresidues mod 23.

For concreteness we take $\mathcal{G}_{23}$ to be the cyclic code with idempotent

$$\theta(x) = \sum_{i \in N} x^i \tag{4}$$

and generator polynomial

$$(1 + x + x^{20})\theta(x) = 1 + x^2 + x^4 + x^5 + x^6 + x^{10} + x^{11}. \tag{5}$$

$\mathcal{G}_{24}$ is then obtained by adding an overall parity check to $\mathcal{G}_{23}$, and has generator matrix

$$\begin{bmatrix} & & \begin{matrix} 1 \\ 1 \\ \vdots \end{matrix} \\ \Pi & & \\ \hline 11 \cdots 1 & & 1 \end{bmatrix}, \tag{6}$$

where Π is the 23×23 circulant whose first row corresponds to $\theta(x)$. The $(i+1)^{\text{th}}$ row of (6) is $|x^i \theta(x)| 1|$, $0 \le i \le 22$.

From Theorem 10 of Ch. 16, $\mathcal{G}_{24}$ is preserved by the group $PSL_2(23)$, which has order $\frac{1}{2} \cdot 23 \cdot (23^2 - 1) = 6072$, and is generated by the following permutations of Ω (Equation (30) of Ch. 16):

$$S: \quad i \to i + 1,$$
$$V: \quad i \to 2i,$$
$$T: \quad i \to -\frac{1}{i}. \tag{7}$$

In other words,

$$S = (\infty)(0\ 1\ 2\ 3\ \cdots\ 22),$$
$$V = (\infty)(0)(1\ 2\ 4\ 8\ 16\ 9\ 18\ 13\ 3\ 6\ 12)$$
$$(5\ 10\ 20\ 17\ 11\ 22\ 21\ 19\ 15\ 7\ 14),$$
$$T = (\infty\ 0)(1\ 22)(2\ 11)(3\ 15)(4\ 17)(5\ 9)(6\ 19)$$
$$(7\ 13)(8\ 20)(10\ 16)(12\ 21)(14\ 18).$$

Definition. The *Mathieu group M_{24}* is the group generated by S, V, T and W, where

$$W \text{ sends} \begin{cases} \infty \text{ to } 0, & 0 \text{ to } \infty, \\ i \text{ to } -(\tfrac{1}{2}i)^2 & \text{if } i \in Q, \\ i \text{ to } (2i)^2 & \text{if } i \in N, \end{cases} \tag{8}$$

or equivalently

$$W = (\infty \ 0)(3 \ 15)(1 \ 17 \ 6 \ 14 \ 2 \ 22 \ 4 \ 19 \ 18 \ 11)$$
$$(5 \ 8 \ 7 \ 12 \ 10 \ 9 \ 20 \ 13 \ 21 \ 16).$$

Theorem 1. *M_{24} preserves $\mathscr{C}_{24}$.*

Proof. We have only to check that W fixes $\mathscr{C}_{24}$. It is easily verified that

$$W(|\theta(x)|1|) = |\theta(x)|1| + 1 \in \mathscr{C}_{24}.$$
$$W(|x\theta(x)|1|) = |x^2\theta(x) + x^{11}\theta(x) + x^{20}\theta(x)|1| \in \mathscr{C}_{24},$$
$$W(|x^{22}\theta(x)|1|) = |\theta(x) + x\theta(x) + x^{20}\theta(x) + x^{22}\theta(x)|0| \in \mathscr{C}_{24}.$$

We now make use of the identity

$$VW = WV^2 = (\infty \ 0)(18 \ 21)(1 \ 22 \ 16 \ 20 \ 6 \ 10 \ 13 \ 15 \ 12 \ 17)$$
$$(2 \ 19 \ 3 \ 14 \ 8 \ 5 \ 9 \ 11 \ 4 \ 7). \tag{9}$$

Since $V(|x^i\theta(x)|1|) = |x^{2i}\theta(x)|1|$, we have

$$W(|x^{2i}\theta(x)|1|) = (VW)(|x^i\theta(x)|1|) = (WV^2)(|x^i\theta(x)|1|),$$

and so W transforms every row of (6) into a codeword of $\mathscr{C}_{24}$.   Q.E.D.

§3. M_{24} is five-fold transitive

Theorem 2. *M_{24} is five-fold transitive.*

Proof. M_{24} contains the permutation $U = W^{-2}$:

$$U = (\infty)(0)(3)(15)(1 \ 18 \ 4 \ 2 \ 6)(5 \ 21 \ 20 \ 10 \ 7)(8 \ 16 \ 13 \ 9 \ 12)(11 \ 19 \ 22 \ 14 \ 17). \tag{10}$$

M_{24} is generated by S, T, U, V, since $W = TU^2$. By multiplying the generators we find permutations of the following cycle types: $1 \cdot 23$, $1^2 \cdot 11^2$, $1^3 \cdot 7^3$, $1^4 \cdot 5^4$, $1^8 \cdot 2^8$, 2^{12} and 4^6. (For example, S, V, US^2, U, $(SU)^3$, T and $(S^{13}TU^2)^3$.)

The permutations of cycle types $1 \cdot 23$ and 2^{12} show that M_{24} is transitive. By conjugating (using Problem 1), we see that the *stabilizer* of any point (the subgroup leaving that point fixed) contains a permutation of type $1 \cdot 23$, so is transitive on the remaining 23 points. Therefore M_{24} is doubly transitive. Again by conjugating, the stabilizer of two points contains the types $1^2 \cdot 11^2$ and $1^3 \cdot 7^3$, so is transitive on the remaining 22 points. Therefore M_{24} is triply transitive. Similarly the stabilizer of three points contains the types $1^3 \cdot 7^3$ and $1^4 \cdot 5^4$, so is transitive on the remaining 21 points. Therefore M_{24} is quadruply transitive. The subgroup fixing a set of 4 points as a whole contains the types $1^4 \cdot 5^4$ and 4^6, so is transitive on the remaining 20 points. Therefore M_{24} is transitive on 5-element subsets of Ω. The subgroup fixing any 5-element subset as a whole contains the types $1^4 \cdot 5^4$ and $1^8 \cdot 2^8$, which induce permutations of types 5 and $1^3 \cdot 2$ inside the 5-set (again conjugating using Problem 1). Since the latter two permutations generate the full symmetric group on 5 symbols (Problem 2), M_{24} is quintuply transitive.   Q.E.D.

Problems. (1) Let π, σ be permutations, with π written as a product of disjoint cycles. Show that the *conjugate* of π by σ, $\sigma^{-1}\pi\sigma$, is obtained from π by applying σ to the symbols in π. E.g. if $\pi = (12)(345)$, $\sigma = (1524)$, then $\sigma^{-1}\pi\sigma = (54)(312)$.

(2) Show that any permutation of $\{1, 2, \ldots, n\}$ is generated by the permutations $(12 \cdots n)$ and (12).

(3) Show $T^{-1}VT = V^{-1}$ and $U^{-1}VU = V^3$.

(4) Show $T = W^5$, so M_{24} is generated by S, V and W.

(5) Show that U sends $x \to x^3/9$ if $x = 0, \infty$ or $x \in Q$, and $x \to 9x^3$ if $x \in N$.

(6) Show that M_{24} is transitive on octads. [Hint: use Theorem 2.]

§4. The order of M_{24} is $24 \cdot 23 \cdot 22 \cdot 21 \cdot 20 \cdot 48$

First we need a lemma. Let Γ be a group of permutations of a set S, and let T be a subset of S. $\gamma \in \Gamma$ sends T into $\gamma(T) = \{\gamma(t): t \in T\}$. The set of all $\gamma(T)$ is called the *orbit* of T under Γ, and is denoted by T^{Γ}. Let Γ_T be the subgroup of Γ fixing T setwise (i.e. if $t \in T$ and $\gamma \in \Gamma_T$, $\gamma(t) \in T$).

Lemma 3.

$$|\Gamma| = |\Gamma_T| \cdot |T^{\Gamma}|.$$

The most important case is when T consists of a single point.

Proof. For $\gamma, \delta \in \Gamma$ we have $\gamma(T) = \delta(T)$ iff $\gamma\delta^{-1} \in \Gamma_T$. Thus $|T^{\Gamma}|$ is equal to the number of cosets of Γ_T in Γ, which is $|\Gamma|/|\Gamma_T|$.   Q.E.D.

Theorem 4.

$$|M_{24}| = 24 \cdot 23 \cdot 22 \cdot 21 \cdot 20 \cdot 48 = 244823040.$$

Proof. Suppose $\{a, b, c, d, e, f, g, h\}$ is an octad. Let Γ be the subgroup of M_{24} fixing this octad setwise, and let H be the subgroup of Γ which fixes in addition a ninth point i. Since M_{24} is transitive on octads (Problem 6), $|M_{24}| = 759|\Gamma|$ by Lemma 3. The calculation of $|\Gamma|$ is in four steps. (i) We show H is a subgroup of Γ of index 16, so $|\Gamma| = 16|H|$. (ii) By looking at the action of H on the remaining 15 points, we show H is isomorphic to a subgroup of $GL(4, 2)$, so $|H| \leq |GL(4, 2)| = 20160$. (iii) By looking at the action of H inside the octad, we find H contains a subgroup isomorphic to the alternating group $\mathcal{A}_8$, of order $\frac{1}{2} 8! = 20160$. (iv) Therefore $H \cong GL(4, 2) \cong \mathcal{A}_8$, $|H| = 20160$, $|\Gamma| = 16 \cdot 20160$, and $|M_{24}| = 759 \cdot 16 \cdot 20160 = 244823040$.

Step (i). A permutation π of Aut $(\mathcal{G}_{24})$ which fixes 5 points setwise must fix the octad $\mathcal{O}$ containing them (or else wt $(\mathcal{O} + \pi(\mathcal{O})) < 8$). M_{24} contains a permutation of type $1 \cdot 3 \cdot 5 \cdot 15$, e.g. US^{11}. The octad containing the 5-cycle in this permutation is fixed by it and so must be the union of the 3- and the 5-cycles. Then by conjugating US^{11}, using Problem 1, we may assume Γ contains the permutation $\lambda = (abcde)(fgh)(i)(jkl \cdots x)$. M_{24} also contains a permutation of type $1^2 \cdot 2 \cdot 4 \cdot 8^2$, e.g. US^5. The octad containing one of the fixed points and the 4-cycle must be the union of the 4-, 2-, and 1-cycles. Therefore Γ contains a permutation which fixes the octad $\{a, \ldots, h\}$ setwise and permutes the remaining 16 points $\{i, j, k, \ldots, x\}$ in two cycles of length 8. Thus Γ is transitive on $\{i, \ldots, x\}$, and so H, the subgroup fixing i, has index 16 in Γ by Lemma 3.

Step (ii). Let $\mathcal{C}$ be the code of length 15 obtained from those codewords of $\mathcal{G}_{24}$ which are zero on the coordinates $\{a, b, \ldots, i\}$. Since $\mathcal{G}_{24}$ is self-dual we know exactly the dependencies in $\mathcal{G}_{24}$ among these 9 coordinates – there is just one, corresponding to the octad $\{a, b, \ldots, h\}$. Therefore $\mathcal{C}$ has size $2^{12}/2^8 = 2^4$. Furthermore $\mathcal{C}$ only contains codewords of weights 8 and 12, and 12 is impossible or else $\mathcal{G}_{24}$ would contain a word of weight 20. Therefore $\mathcal{C}$ is a [15, 4, 8] code containing 15 codewords of weight 8. It follows (Problem 7) that $\mathcal{C}$ is equivalent to the simplex code with generator matrix

$$\begin{pmatrix} 111111110000000 \\ 111100001111000 \\ 110011001100110 \\ 101010101010101 \end{pmatrix} \tag{11}$$

and has automorphism group $GL(4, 2)$.

Each nontrivial permutation in H induces a nontrivial permutation of $\mathcal{C}$.

(For if $h, h' \in H$ induce the same permutation on $\mathscr{C}$, then $h^{-1}h'$ has 16 fixed points, and (Problem 8) only the identity permutation in M_{24} can fix 16 points.) Thus H is isomorphic to a subgroup of $GL(4, 2)$.

Step (iii). Let H_2 be the group of permutations on the octad $\{a, \ldots, h\}$ induced by H. The 5th power of the permutation λ defined in Step (i) gives $(fgh) \in H_2$, and by conjugating (fgh) we can get all 3-cycles. Therefore H_2 contains the alternating group $\mathscr{A}_8$ (which is generated by 3-cycles, by Problem 9).   Q.E.D.

This theorem has a number of important corollaries.

Corollary 5. *M_{24} is the full automorphism group of $\mathscr{G}_{24}$.*

Proof. Let $M = \text{Aut}(\mathscr{G}_{24})$. We know $M \supseteq M_{24}$. The proof of Theorem 4 goes through unchanged if M_{24} is replaced throughout by M. Hence $|M| = 244823040$ and $M = M_{24}$.   Q.E.D.

Corollary 6. *The subgroup of M_{24} fixing an octad setwise has order 8.8!*

Corollary 7. $GL(4, 2) \cong \mathscr{A}_8$ (Corollaries 6 and 7 follow directly from the proof of Theorem 4.)

Definition. The Mathieu group M_{23} consists of the permutations in M_{24} fixing a point in Ω. Thus M_{23} is a 4-fold transitive group of order $23 \cdot 22 \cdot 21 \cdot 20 \cdot 48$.

Corollary 8. *M_{23} is the full automorphism group of $\mathscr{G}_{23}$.*

Problems. (7.) Let $\mathscr{C}$ be any [15, 4, 8] code containing 15 codewords of weight 8. Show that $\mathscr{C}$ is equivalent to the simplex code with generator matrix (11). Hence show that the automorphism group of $\mathscr{C}$ is $GL(4, 2)$. [Hint: Theorem 24 of Ch. 13.]

(8) Show that a permutation π in M_{24} which fixes 16 points (individually), where the other 8 form an octad, must fix all 24 points. [Hint: Suppose π fixes $8, \ldots, \infty$ where $t = \{0, \ldots, 7\}$ is an octad. An octad s with $|s \cap t| = 4$ is transformed by π into s or $t + s$. By considering all such s, show π is the identity.]

(9) Show that the alternating group $\mathscr{A}_n$ is generated by all 3-cycles (ijk).

(10) Show that the pointwise stabilizer of an octad is an elementary abelian group of order 16, and is transitive on the remaining 16 points; and that the

pointwise stabilizer of any five points has order 48. [Hint: in the first part use the permutation $(US^5)^4$, and in the second use λ^5].

(11) Show that M_{24} is transitive on dodecads.

§5. The Steiner system $S(5, 8, 24)$ is unique

The goal of this section is to prove:

Theorem 9. *There is a unique Steiner system $S(5, 8, 24)$. More precisely, if there are two Steiner systems $S(5, 8, 24)$, $\mathcal{A}$ and $\mathcal{B}$ say, then there is a permutation of the 24 points which sends the octads of $\mathcal{A}$ onto the octads of $\mathcal{B}$. (In this section an octad means a block of an $S(5, 8, 24)$.)*

The proof makes good use of the table of λ_{ij}'s given in Fig. 2.14. In particular, the last row of this table implies that two octads meet in either 0, 2 or 4 points. We begin with some lemmas.

Lemma 10. (Todd's lemma.) *In an $S(5, 8, 24)$ if B and C are octads meeting in 4 points then $B + C$ is also an octad.*

Proof. Let $B = \{abcdefgh\}$, $C = \{abcdijkl\}$ and suppose $B + C$ is not an octad. Then the octad D containing $\{efghi\}$ must contain just one more point of C, say $D = \{efghijmn\}$. Similarly the octad containing $\{efghk\}$ is $E = \{efghklop\}$. But now it is impossible to find an octad containing $\{efgik\}$ and meeting B, C, D, E in 0, 2 or 4 points. This is a contradiction, since there must be an octad containing any five points.   Q.E.D.

Lemma 11. *If the 280 octads meeting a given octad $\mathcal{O}$ in four points are known, then all 759 octads are determined.*

Proof. From Fig. 2.14 we must find 30 octads disjoint from $\mathcal{O}$ and 448 meeting $\mathcal{O}$ in two points, i.e. 16 octads meeting $\mathcal{O}$ in any two specified points a and b.

Let $\mathcal{O} = \{xyzab \cdots\}$. There are four octads, besides $\mathcal{O}$, through $\{xyza\}$, say A_1, A_2, A_3, A_4, and four through $\{xyzb\}$, say B_1, B_2, B_3, B_4. The sums $A_i + B_j$ are octads by Lemma 10, are readily seen to be distinct, and are the 16 octads through a and b. The 6 sums $A_i + A_j$ are octads which are disjoint from $\mathcal{O}$. Clearly these are distinct and $A_i + A_j \neq B_i + B_j$. Since there are 5 choices for a these give the 30 octads disjoint from $\mathcal{O}$.   Q.E.D.

Definition of sextet. Any four points *abcd* define a partition of the 24 points into 6 sets of 4, called *tetrads*, with the property that the union of any two tetrads is an octad. This set of 6 tetrads is called a *sextet* (= *six tetrads*). (To see this, pick a fifth point *e*. There is a unique octad containing {*abcde*}, say {*abcdefgh*}. Then {*efgh*} is the second tetrad. A ninth point *i* determines the octad {*abcdijkl*} and the tetrad {*ijkl*}. By Todd's lemma {*efghijkl*} is an octad, and so on.)

Lemma 12. *An octad intersects the 6 tetrads of a sextet either* $3 \cdot 1^5$, $4^2 \cdot 0^4$ *or* $2^4 \cdot 0^2$. *(The first of these means that the octad intersects one tetrad in three points and the other five in one point.)*

Proof. Two octads meet in 0, 2 or 4 points.   Q.E.D.

Lemma 13. *The intersection matrix for the tetrads of two sextets is one of the following:*

$$\begin{bmatrix} 400000 \\ 040000 \\ 004000 \\ 000400 \\ 000040 \\ 000004 \end{bmatrix}, \begin{bmatrix} 220000 \\ 220000 \\ 002200 \\ 002200 \\ 000022 \\ 000022 \end{bmatrix}, \begin{bmatrix} 200011 \\ 020011 \\ 002011 \\ 000211 \\ 111100 \\ 111100 \end{bmatrix}, \begin{bmatrix} 310000 \\ 130000 \\ 001111 \\ 001111 \\ 001111 \\ 001111 \end{bmatrix}.$$

Proof. From Lemma 12 and the definition of a sextet.   Q.E.D.

Proof of Theorem 9. In a Steiner system $S(5, 8, 24)$ let $\mathcal{O}$ be a fixed octad. The idea of the proof is to determine uniquely all the octads meeting $\mathcal{O}$ in 4 points; the theorem then follows from Lemma 11. To find these octads we shall construct 7 sextets $S_1, \ldots, S_7$.

Let $x_1 \cdots x_6$ be six points of $\mathcal{O}$ and x_7 a point not in $\mathcal{O}$. Suppose S_1 is the sextet defined by $\{x_1 x_2 x_3 x_4\}$ and let its tetrads be the columns of the 4×6 array:

$$S_1 = \begin{bmatrix} x_1 & x_5 & x_7 & z_4 & z_8 & z_{12} \\ x_2 & x_6 & z_1 & z_5 & z_9 & z_{13} \\ x_3 & y_1 & z_2 & z_6 & z_{10} & z_{14} \\ x_4 & y_2 & z_3 & z_7 & z_{11} & z_{15} \end{bmatrix}.$$

Thus $\mathcal{O}$ consists of the two left-hand columns.

The octad containing $\{x_2 x_3 x_4 x_5 x_7\}$ must intersect S_1 $3 \cdot 1^5$, by Lemma 12, and so, after relabeling the z_i, can be taken to be $\{x_2 x_3 x_4 x_5 x_7 z_4 z_8 z_{12}\}$. The tetrad

$\{x_2x_3x_4x_5\}$ determines a sextet which by Lemma 13 can be taken to be

$$S_2 = \begin{bmatrix} 2 & 1 & 3 & 3 & 3 & 3 \\ 1 & 2 & 4 & 4 & 4 & 4 \\ 1 & 2 & 5 & 5 & 5 & 5 \\ 1 & 2 & 6 & 6 & 6 & 6 \end{bmatrix}.$$

This diagram means that the tetrads of S_2 are $\{x_2x_3x_4x_5\}$, $\{x_1x_6y_1y_2\}$, $\{x_7z_4z_8z_{12}\}$, $\{z_1z_5z_9z_{13}\}$, $\{z_2z_6z_{10}z_{14}\}$ and $\{z_3z_7z_{11}z_{15}\}$. In this notation

$$S_1 = \begin{bmatrix} 1 & 2 & 3 & 4 & 5 & 6 \\ 1 & 2 & 3 & 4 & 5 & 6 \\ 1 & 2 & 3 & 4 & 5 & 6 \\ 1 & 2 & 3 & 4 & 5 & 6 \end{bmatrix}.$$

Remark. We have now identified the 30 octads disjoint from $\mathcal{O}$. They are (i) the sums of any two rows of

$$\begin{bmatrix} x_7 & z_4 & z_8 & z_{12} \\ z_1 & z_5 & z_9 & z_{13} \\ z_2 & z_6 & z_{10} & z_{14} \\ z_3 & z_7 & z_{11} & z_{15} \end{bmatrix},$$

(ii) the sums of any two columns of this array, and (iii) the sums of (i) and (ii). These are illustrated by

The octad containing $\{x_1x_3x_4x_5x_7\}$ intersects both S_1 and S_2 $3 \cdot 1^5$, and so can be taken to be $\{x_1x_3x_4x_5x_7z_5z_{10}z_{15}\}$. The tetrad $\{x_1x_3x_4x_5\}$ defines a sextet which, using Lemma 13 and considering the intersections with the 30 octads mentioned in the above Remark, can be taken to be

$$S_3 = \begin{bmatrix} 1 & 1 & 3 & 4 & 6 & 5 \\ 2 & 2 & 4 & 3 & 5 & 6 \\ 1 & 2 & 6 & 5 & 3 & 4 \\ 1 & 2 & 5 & 6 & 4 & 3 \end{bmatrix}.$$

At this stage we observe that the sextets S_1, S_2 and S_3 are preserved by the following permutations of $\{y_1y_2z_1z_2 \cdots z_{15}\}$:

$$\pi = (z_4z_8z_{12})(z_5z_{10}z_{15})(z_1z_2z_3)(z_9z_{14}z_7)(z_6z_{11}z_{13}),$$

$$\sigma = (z_4z_8)(z_7z_{11})(z_1z_2)(z_{13}z_{14})(z_5z_{10})(z_6z_9),$$

$$\varphi = (y_1y_2).$$

The octad containing $\{x_1x_2x_3x_5x_7\}$ cannot contain any of $z_1z_2z_3z_4z_8z_{12}z_5z_{10}$ or z_{15} because of the intersections with the previous octads, and so must be either

$$
\begin{array}{|cc|c|c|c|c|}
\hline
x & x & x & & & \\
x & & & x & & \\
x & & & & x & \\
& & x & & & \\
\hline
\end{array}
\quad \text{or} \quad
\begin{array}{|cc|c|c|c|c|}
\hline
x & x & x & & & \\
x & & & & x & \\
x & & & x & & \\
& & & & & x \\
\hline
\end{array}
$$

Since these are equivalent under σ the octad can be taken to be $\{x_1x_2x_3x_5x_7z_7z_9z_{14}\}$. The tetrad $\{x_1x_2x_3x_5\}$ then determines the sextet

$$
S_4 = \begin{bmatrix}
1 & 1 & 3 & 4 & 5 & 6 \\
1 & 2 & 5 & 6 & 3 & 4 \\
1 & 2 & 6 & 5 & 4 & 3 \\
2 & 2 & 4 & 3 & 6 & 5
\end{bmatrix}.
$$

In the same way we obtain

$$
S_5 = \begin{bmatrix}
1 & 1 & 3 & 4 & 5 & 6 \\
1 & 2 & 6 & 5 & 4 & 3 \\
2 & 2 & 4 & 3 & 6 & 5 \\
1 & 2 & 5 & 6 & 3 & 4
\end{bmatrix}.
$$

Now the octad containing $\{x_1x_2x_5x_6x_7\}$ cuts S_1 $2^4 \cdot 0^2$ and so, using π, may be assumed to intersect the first four columns of S_1 2^4. So we can take this octad to be $\{x_1x_2x_5x_6x_7z_1z_4z_5\}$. This gives the sextet

$$
S_6 = \begin{bmatrix}
1 & 1 & 3 & 3 & 4 & 4 \\
1 & 1 & 3 & 3 & 4 & 4 \\
2 & 2 & 5 & 5 & 6 & 6 \\
2 & 2 & 5 & 5 & 6 & 6
\end{bmatrix}.
$$

Similarly, using φ, we can take the octad containing $\{x_1x_3x_5x_7z_2\}$ to be $\{x_1x_3x_5x_7y_1z_2z_4z_6\}$, giving the sextet

$$
S_7 = \begin{bmatrix}
1 & 1 & 3 & 3 & 5 & 5 \\
2 & 2 & 4 & 4 & 6 & 6 \\
1 & 1 & 3 & 3 & 5 & 5 \\
2 & 2 & 4 & 4 & 6 & 6
\end{bmatrix}.
$$

It remains to show that the 280 octads meeting $\mathcal{O}$ in 4 points are determined by $S_1, \ldots, S_7$.

If two sextets meet evenly (i.e. if their tetrads intersect in the third matrix of Lemma 13), then we can add suitable octads in them to get a new octad and sextet. For example, from S_6 and S_7, octad   +   octad   =   new octad

$$\begin{bmatrix} 2\ 2 & 3\ 3 & 5\ 5 \\ 1\ 1 & 4\ 4 & 6\ 6 \\ 1\ 1 & 4\ 4 & 6\ 6 \\ 2\ 2 & 3\ 3 & 5\ 5 \end{bmatrix} = \text{new sextet.}$$

It is easily checked that in this way we obtain all $\frac{1}{2}\binom{8}{4} = 35$ sextets defined by four points of $\mathcal{O}$. But these sextets give all the octads meeting $\mathcal{O}$ in 4 points.

Q.E.D.

Witt's original proof of Theorem 9 was to successively show the uniqueness of $S(2,5,21)$, $S(3,6,22)$, $S(4,7,23)$ and finally $S(5,8,24)$. The starting point is:

Problems. (12) Show that the affine planes $S(2, m, m^2)$ are unique for $m = 2, 3, 4, 5$ (see §5 of Ch. 2 and Appendix B, Theorem 11). [Hint: Pick a family of m nonintersecting blocks in $S(2, m, m^2)$ and call them "horizontal lines," and another family of nonintersecting blocks called "vertical lines." The point at the intersection of vertical line x and horizontal line y is given the coordinates (x, y). Each of the $m^2 - m$ remaining blocks consists of points $(1, a_1)(2, a_2) \cdots (m, a_m)$. Form the $(m^2 - m) \times m$ matrix $A(m)$ whose rows are the vectors $(a_1 a_2 \cdots a_m)$. Show that $A(m)$ is essentially unique for $m = 2$–5.]

(13) Show that the projective planes $S(2, m + 1, m^2 + m + 1)$ are unique for $m = 2$–5. [Hint: Since the affine planes are unique they are the planes constructed from finite fields (given in §2 of Appendix B). There is a unique way to extend such a plane to a projective plane.]

(14) Show that the Steiner systems $S(3, 6, 22)$ and $S(4, 7, 23)$ are unique. [Hint: Let P be a fixed point of an $S(3, 6, 22)$. The idea is to show that the blocks containing P belong to an $S(2, 5, 21)$, unique from Problem 13, and the other blocks form ovals (p. 330 of Ch. 11) in $S(2, 5, 21)$. See Witt [1424] or Lüneburg [866].]

(15) (a) Show that a Hadamard matrix H_8 of order 8 is unique up to equivalence (cf. p. 48 of Ch. 2). [Hint: Use the fact that $S(2, 3, 7)$ is unique.] (b) Show that H_{12} is unique.

(16) Show that the Steiner system $S(3, 4, 8)$ is unique, and hence so is the $[8, 4, 4]$ extended Hamming code.

§6. The Golay codes $\mathscr{G}_{23}$ and $\mathscr{G}_{24}$ are unique

Theorem 14. *Let $\mathscr{C}$ be any binary code of length 24 and minimum distance 8. Then (i) $|\mathscr{C}| \leqslant 2^{12}$. (ii) If $|\mathscr{C}| = 2^{12}$, $\mathscr{C}$ is equivalent to $\mathscr{G}_{24}$.*

Proof. (i) This follows from linear programming (§4 of Ch. 17, see also Ch. 17, Problem 16) or from the sphere-packing bound (Theorem 6 of Ch. 1). (ii) Suppose $|\mathscr{C}| = 2^{12}$. Then the linear programming bound shows that the distance distribution of $\mathscr{C}$ is

$$B_0 = B_{24} = 1, \qquad B_8 = B_{16} = 759, \qquad B_{12} = 2576; \tag{12}$$

i.e. is the same as that of $\mathscr{G}_{24}$. The transform of (12), Equation (23) of Ch. 17, coincides with (12). Hence it follows from Theorem 3 of Ch. 6 that the weight and distance distributions of $\mathscr{C}$ coincide, assuming $\mathscr{C}$ contains **0**.

Let a, b be two codewords of $\mathscr{C}$. From the formula (Equation (16) of Ch. 1)

$$\text{dist}\,(a, b) = \text{wt}\,(a) + \text{wt}\,(b) - 2\text{wt}\,(a * b)$$

it follows that $\text{wt}\,(a * b)$ is even (since all the other terms are divisible by 4). Hence every codeword of $\mathscr{C}$ is orthogonal to itself and to every other codeword. The following simple lemma now implies that $\mathscr{C}$ is linear.

Lemma 15.
Let A, B be subsets of F^n which are mutually orthogonal, i.e.

$$\sum_{i=1}^{n} a_i b_i = 0 \quad \text{for all } (a_1 \cdots a_n) \in A, (b_1 \cdots b_n) \in B$$

Suppose further that $|A| = 2^k$ and $|B| \geqslant 2^{n-k-1} + 1$. Then A is a linear code.

Proof. Let $\bar{A}$ and $\bar{B}$ be the linear spans of A and B. Clearly $\bar{A}, \bar{B}$ are mutually orthogonal, hence

$$\dim \bar{A} + \dim \bar{B} \leqslant n.$$

But by hypothesis $\dim \bar{A} \geqslant k$ and $\dim \bar{B} \geqslant n - k$. Thus $\dim \bar{A} = k$ and $A = \bar{A}$.
Q.E.D.

Theorem 22 of Ch. 2 proves that the codewords of weight 8 in $\mathscr{C}$ are the octads of a Steiner system $S(5, 8, 24)$. By Theorem 9 this system is unique, and as mentioned above generates a code equivalent to $\mathscr{G}_{24}$. Hence $\mathscr{C}$ is equivalent to $\mathscr{G}_{24}$.
Q.E.D.

Corollary 16. *Any $(23, 2^{12}, 7)$ binary code is equivalent to $\mathscr{G}_{23}$.*

Proof. Add an overall parity check and use Theorem 14.   Q.E.D.

§7. The automorphism groups of the ternary Golay codes

In this section we show that the automorphism group Aut $(\mathcal{G}_{12})\dagger$ is isomorphic to the Mathieu group M_{12}. Recall from Ch. 16 that if $\mathcal{C}$ is a ternary code, Aut $(\mathcal{C})$ consists of all monomial matrices A (permutations with $\pm$ signs attached) which preserve $\mathcal{C}$, and

$$\text{Aut } (\mathcal{C})\dagger = \text{Aut } (\mathcal{C})/\{\pm I\},$$

which is the quotient group obtained by identifying A and $-A$. Thus $|\text{Aut } (\mathcal{C})| = 2 |\text{Aut } (\mathcal{C})\dagger|$.

Notation. The coordinates of $\mathcal{G}_{12}$ are labeled by $\Omega = \{0, 1, \ldots, 9, X, \infty\}$, and $Q = \{1, 3, 4, 5, 9\}$, $N = \{2, 6, 7, 8, X\}$. As the generator matrix for $\mathcal{G}_{12}$ we take

$$T = \left[\begin{array}{c|c} & \begin{matrix} 1 \\ 1 \\ \vdots \\ 1 \end{matrix} \\ \Pi & \\ \hline 11 \cdots 1 & 1 \end{array} \right], \tag{13}$$

where Π is the 11×11 circulant whose first row corresponds to

$$-1 - \sum_{i \in Q} x^i + \sum_{i \in N} x^i,$$

i.e. is the vector $(--1---111-1)$, writing $-$ for -1. (Check that this matrix does generate $\mathcal{G}_{12}$ – see p. 487.)

From Theorem 12 of Ch. 16, $\mathcal{G}_{12}$ is preserved by the group, isomorphic to $\text{PSL}_2(11)$, which is generated by S, V and T', where S and V are the permutations

$$S: \quad i \to i + 1,$$
$$V: \quad i \to 3i, \tag{14}$$

and T' is the monomial transformation which sends the element in position i to position $-1/i$ after multiplying it by 1 if $i = \infty$ or $i \in Q$, or by -1 if $i = 0$ or $i \in N$.

In addition, $\mathcal{G}_{12}$ is also preserved by the permutation

$$\Delta = (\infty)(0)(1)(2X)(34)(59)(67)(8). \tag{15}$$

For it is readily checked that if $w_i(i = 0, 1, \ldots, 10)$ is the $(i + 1)^{\text{th}}$ row of (13), then

$$\Delta(w_i) = w_{\delta(i)},$$

where

$$\delta = (\infty)(0)(3)(X)(19)(26)(45)(78). \tag{16}$$

The monomial group generated by S, V, T' and Δ will be denoted by M'_{12}; we have shown that

$$\text{Aut}\,(\mathcal{G}_{12})\dagger \supseteq M'_{12}.$$

Definition. The *Mathieu group* M_{12} is the permutation group on Ω obtained by ignoring the signs in M'_{12}. Thus M_{12} is generated by S, V, T and Δ, where T sends i to $-1/i$. Clearly M_{12} is isomorphic to M'_{12}.

By imitating the proof of Theorems 2 and 4 we obtain Theorems 17 and 18.

Theorem 17. M_{12} *is 5-fold transitive, and has order* $12 \cdot 11 \cdot 10 \cdot 9 \cdot 8 = 95040$.

Theorem 18. M'_{12} *is the full automorphism group* $\text{Aut}\,(\mathcal{G}_{12})\dagger$.

Corollary 19. $\text{Aut}\,(\mathcal{G}_{11})\dagger$ *is isomorphic to the Mathieu group* M_{11} *consisting of the permutations in* M_{12} *fixing a point of* Ω. M_{11} *is a 4-fold transitive group of order* $11 \cdot 10 \cdot 9 \cdot 8$.

Problem. (17) Show that the subgroup of M_{24} fixing a dodecad setwise is isomorphic to M_{12}.

§8. The Golay codes $\mathcal{G}_{11}$ and $\mathcal{G}_{12}$ are unique

Theorem 20. *Any* $(12, 3^6, 6)$ *ternary code is equivalent to* $\mathcal{G}_{12}$.

Sketch of Proof. (1) By linear programming we find that the largest code of length 12 and distance 6 contains 3^6 codewords, and has the same distance distribution as $\mathcal{G}_{12}$.

(2) As before (though with somewhat more trouble) it can be shown that the code is orthogonal to itself and hence must be linear.

(3) The total number of ternary self-dual codes of length 12 is $N_0 = 2^9 \cdot 5 \cdot 7 \cdot 41 \cdot 61$ (Problem 21 of Ch. 19). There are two inequivalent $[12, 6, 3]$ codes, namely $\mathcal{C}_1$, the direct sum of 3 copies of the $[4, 2, 3]$ code #6 of Ch. 1,

and $\mathscr{C}_2$, a code with generator matrix

$$
\begin{bmatrix}
1 & 1 & 1 & & & & & & \\
 & & 1 & 1 & 1 & & & & \\
 & & & & 1 & 1 & 1 & & \\
 & & & & & & 1 & 1 & 1 \\
1 & - & & 1 & - & & & 1 & - \\
- & 1 & & & & 1 & - & 1 & -
\end{bmatrix}. \tag{17}
$$

Then $|\mathrm{Aut}\,(\mathscr{C}_1)| = 2^3 \cdot 3!(4!)^3$ and $|\mathrm{Aut}\,(\mathscr{C}_2)| = 2(3!)^4 4!$ From Lemma 3, there are $N_1 = 2^{12} \cdot 12!/2^3 \cdot 3!(4!)^3 = 2^9 \cdot 3 \cdot 5^2 \cdot 7 \cdot 11$ codes equivalent to $\mathscr{C}_1$, and $N_2 = 2^{12} \cdot 12!/2(3!)^4 \cdot 4! = 2^{14} \cdot 5^2 \cdot 7 \cdot 11$ codes equivalent to $\mathscr{C}_2$. Finally, there are $N_3 = 2^{12} \cdot 12!/2|M_{12}| = 2^7 \cdot 3^3 \cdot 5 \cdot 11$ codes equivalent to $\mathscr{G}_{12}$. Since $N_0 = N_1 + N_2 + N_3$, all self-dual codes of length 12 are accounted for, and $\mathscr{G}_{12}$ is unique.   Q.E.D.

Corollary 21. *Any* $(11, 3^6, 5)$ *ternary code is equivalent to* $\mathscr{G}_{11}$.

Problem. (18) Show that the Steiner systems $S(3, 4, 10)$, $S(4, 5, 11)$ and $S(5, 6, 12)$ are unique. [Hint: Witt [1424], Lüneburg [866].]

Notes on Chapter 20

The Golay codes were discovered by Golay [506]. Our sources for this chapter are as follows. For all of Sections 2, 3, 4 and 7, Conway [306]. Theorem 9 is due to Witt [1424], but the proof given here is from Curtis [322]. Another proof is given by Jónsson [699]. Lemma 10 is from Todd [1330]. The latter paper gives a list of all 759 octads. Problems 12, 13, 14, 16 are from Witt (op. cit.), but see also Lüneburg [865, 866]. Theorem 14 was first proved by Snover [1247], but the proof given here is that of Delsarte and Goethals [363] and Pless [1054]. Theorem 20 is also from [363] and [1054]. For more about the enumeration of self-dual codes over GF(3) see Mallows et al. [892]. The counting method could also be used to prove that $\mathscr{G}_{24}$ is unique, although there are many more codes to be considered – see Pless and Sloane [1063].

The Mathieu groups were discovered by Mathieu in 1861 ([925, 926]) and have since accumulated a rich literature – see for example Assmus and Mattson [35, 36], Berlekamp [120], Biggs [143], Conway [306], Curtis [322, 323], Garbe and Mennicke [466], Greenberg [559], Hall [584, 590], James [689], Jónsson [699], Leech [806], Lüneberg [865, 866], Paige [1018], Rasala [1095], Stanton [1264], Todd [1329, 1330], Ward [1388], Whitelaw [1412], and Witt [1423, 1424]. These groups are *simple*, i.e., have no normal subgroups (for a proof see for example Biggs [143]), and do not belong to any of the known infinite families of simple groups. Many properties of M_{24} can be obtained from $\mathcal{G}_{24}$, for example the maximal subgroups of M_{24} can be simply described in this way (Conway [306], Curtis [323].) Wielandt [1415] is a good general reference on permutation groups.

Finally, three very interesting topics not described in this chapter are:

(i) Curtis' Miracle Octad Generator, or MOG, [322], which miraculously finds the octad containing 5 given points in the $S(5, 8, 24)$,

(ii) Conway's diagram (Fig. 1 of [306], p. 41 of [322]) showing the action of M_{24} on all binary vectors of length 24, and

(iii) the Leech lattice (see Leech [804], Conway [302–306], Leech and Sloane [810]), which is a very dense sphere-packing in 24 dimensional space constructed from $\mathcal{G}_{24}$. The automorphism group of this lattice is very large indeed – see Conway [302, 303].

21

Association schemes

§1. Introduction

This chapter gives the basic theory of association schemes. An association scheme is a set with relations defined on it satisfying certain properties (see §2). A number of problems in coding and combinatorics, such as finding the largest code or the largest constant weight code with a given minimum distance can be naturally stated in terms of finding the largest subset of an association scheme. In many cases this leads to a linear programming bound for such problems (using Theorem 12).

Association schemes were first introduced by statisticians in connection with the design of experiments (Bose and Mesner [183], Bose and Shimamoto [186], James [688], Ogasawara [1007], Ogawa [1008], Yamamoto et al. [1444], Raghavarao [1085]), and have since proved very useful in the study of permutation groups (see Notes) and graphs (Biggs [144–147] and Damerell [327]). The applications to coding theory given in this chapter are due to Delsarte [352–358, 361 and 364].

The chapter is arranged as follows. §2 defines association schemes and gives the basic theory. §§3–6 discuss the three most important examples for coding theory, the Hamming, Johnson and symplectic form schemes. We will see that a code is a subset of a Hamming scheme, and a constant weight code is a subset of a Johnson scheme. Section 7 studies the properties of subsets of an arbitrary association scheme. Finally §§8, 9 deal with subsets of symplectic forms and with t-designs.

§2. Association schemes

This section gives the basic theory.

Definition. An *association scheme with n classes* (or relations) consists of a finite set X of v points together with $n + 1$ relations $R_0, R_1, \ldots, R_n$ defined on X which satisfy:

(i) Each R_i is symmetric: $(x, y) \in R_i \Rightarrow (y, x) \in R_i$.

(ii) For every $x, y \in X$, $(x, y) \in R_i$ for exactly one i.

(iii) $R_0 = \{(x, x): x \in X\}$ is the identity relation.

(iv) If $(x, y) \in R_k$, the number of $z \in X$ such that $(x, z) \in R_i$ and $(y, z) \in R_j$ is a constant c_{ijk} depending on i, j, k but not on the particular choice of x and y.

Two points x and y are called i^{th} *associates* if $(x, y) \in R_i$. In words, the definition states that if x and y are i^{th} associates so are y and x; every pair of points are i^{th} associates for exactly one i; each point is its own zeroth associate while distinct points are never zeroth associates; and finally if x and y are k^{th} associates then the number of points z which are both i^{th} associates of x and j^{th} associates of y is a constant c_{ijk}.

It is sometimes helpful to visualize an association scheme as a complete graph with labeled edges. The graph has v vertices, one for each point of X, and the edge joining vertices x and y is labeled i if x and y are i^{th} associates. Each edge has a unique label, and the number of triangles with a fixed base labeled k having the other edges labeled i and j is a constant c_{ijk}, depending on i, j, k but not on the choice of the base. In particular, each vertex is incident with exactly $c_{ii0} = v_i$ (say) edges labeled i; v_i is the *valency* of the relation R_i.

There are also loops labeled 0 at each vertex x, corresponding to R_0.

Examples of association schemes are given in the following sections. The most important for coding theory are the Hamming, Johnson, and symplectic form schemes.

The c_{ijk} satisfy several identities:

Theorem 1.

$$c_{ijk} = c_{jik}, \qquad c_{0jk} = \delta_{jk}, \tag{1}$$

$$v_k c_{ijk} = v_i c_{kji},$$

$$\sum_{j=0}^{n} c_{ijk} = v_i,$$

$$\sum_{m=0}^{n} c_{ijm} c_{mkl} = \sum_{h=0}^{n} c_{ihl} c_{jkh}. \tag{2}$$

Proof. The last identity follows from counting the quadrilaterals

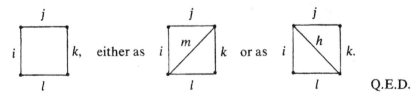

QED.

We describe the relations by their *adjacency matrices*. D_i, the adjacency matrix of R_i (for $i = 0, \ldots, n$), is the $v \times v$ matrix with rows and columns labeled by the points of X, defined by

$$(D_i)_{x,y} = \begin{cases} 1 & \text{if } (x, y) \in R_i, \\ 0 & \text{otherwise.} \end{cases}$$

The definition of an association scheme is equivalent to saying that the D_i are $v \times v$ $(0, 1)$-matrices which satisfy

(i) D_i is symmetric,

(ii) $\sum_{i=0}^{n} D_i = J$ (the all-ones matrix), $\qquad\qquad$ (3)

(iii) $D_0 = I$, $\qquad\qquad$ (4)

(iv) $D_i D_j = \sum_{k=0}^{n} c_{ijk} D_k = D_j D_i$, $\quad i, j = 0, \ldots, n$. $\qquad$ (5)

Indeed the $(x, y)^{\text{th}}$ entry of the left side of Equation (5) is the number of paths $x \circ \!\!\overset{i}{-\!\!-\!\!-}\!\! \circ \!\!\overset{j}{-\!\!-\!\!-}\!\! \circ\, y$ in the graph. Also the rows and columns of D_i contain v_i 1's:

$$D_i J = J D_i = v_i J. \qquad\qquad (6)$$

The Bose–Mesner Algebra. Let us consider the vector space $\mathcal{A}$ consisting of all matrices $\sum_{i=0}^{n} a_i D_i$, a_i real. From (i), these matrices are symmetric. From (ii), $D_0, \ldots, D_n$ are linearly independent, and the dimension of $\mathcal{A}$ is $n + 1$. From (iv), $\mathcal{A}$ is closed under multiplication, and multiplication is commutative. Multiplication is associative (matrix multiplication is always associative; alternatively, associativity follows from (2)). This associative commutative algebra $\mathcal{A}$ is called the *Bose–Mesner algebra* of the association scheme.

Since the matrices in $\mathcal{A}$ are symmetric and commute with each other, they can be simultaneously diagonalized (Marcus and Minc [915]). I.e. there is a matrix S such that to each $A \in \mathcal{A}$ there is a diagonal matrix Λ_A with

$$S^{-1} A S = \Lambda_A. \qquad\qquad (7)$$

Therefore $\mathcal{A}$ is semisimple and has a unique basis of primitive idempotents $J_0, \ldots, J_n$. These are real $n \times n$ matrices satisfying (see Burrow [212], Ogawa [1008], Wedderburn [1393])

$$J_i^2 = J_i, \quad i = 0, \ldots, n, \qquad\qquad (8)$$
$$J_i J_k = 0, \quad i \neq k, \qquad\qquad (9)$$
$$\sum_{i=0}^{n} J_i = I. \qquad\qquad (10)$$

From Equations (3), (6), $(1/v)J$ is a primitive idempotent, so we shall always choose

$$J_0 = \frac{1}{v} J.$$

Now we have two bases for $\mathcal{A}$, so let's express one in terms of the other, say

$$D_k = \sum_{i=0}^{n} p_k(i)J_i, \quad k = 0, \dots, n \tag{11}$$

for some uniquely defined *real* numbers $p_k(i)$. Equations (8), (9), (11) imply

$$D_k J_i = p_k(i)J_i. \tag{12}$$

Therefore the $p_k(i)$, $i = 0, \dots, n$, are the eigenvalues of D_k. Also the columns of the J_i span the eigenspaces of all the matrices D_k. Let rank $J_i = \mu_i$ (say) be the dimension of the ith eigenspace, i.e. the multiplicity of the eigenvalue $p_k(i)$.

Conversely, to express the J_k in terms of the D_i, let P be the real $(n+1) \times (n+1)$ matrix

$$P = \begin{bmatrix} p_0(0) & p_1(0) & \cdots & p_n(0) \\ p_0(1) & p_1(1) & \cdots & p_n(1) \\ \cdots\cdots\cdots\cdots\cdots\cdots\cdots \\ p_0(n) & p_1(n) & \cdots & p_n(n) \end{bmatrix} \tag{13}$$

and let

$$Q = vP^{-1} = \begin{bmatrix} q_0(0) & q_1(0) & \cdots & q_n(0) \\ q_0(1) & q_1(1) & \cdots & q_n(1) \\ \cdots\cdots\cdots\cdots\cdots\cdots\cdots \\ q_0(n) & q_1(n) & \cdots & q_n(n) \end{bmatrix} \quad \text{(say)}. \tag{14}$$

We call P and Q the eigenmatrices of the scheme. Then

$$J_k = \frac{1}{v} \sum_{i=0}^{n} q_k(i)D_i, \quad k = 0, \dots, n. \tag{15}$$

Theorem 2.

$$p_0(i) = q_0(i) = 1, \qquad p_k(0) = v_k, \qquad q_k(0) = \mu_k.$$

Proof. Only the last equation is not immediate. Since J_k is an idempotent the diagonal entries of $S^{-1}J_kS$ (see Equation (7)) are 0 and 1. Therefore

$$\text{trace } S^{-1}J_kS = \text{trace } J_k = \text{rank } J_k = \mu_k.$$

Since trace $D_i = v\delta_{0i}$, (15) implies $\mu_k = q_k(0)$.   Q.E.D.

Problem. (1) Show that $|p_k(i)| \le v_k$.

Theorem 3. *The eigenvalues $p_k(i)$ and $q_k(i)$ satisfy the orthogonality conditions*:

$$\sum_{i=0}^{n} \mu_i p_k(i) p_l(i) = v v_k \delta_{kl}, \tag{16}$$

$$\sum_{i=0}^{n} v_i q_k(i) q_l(i) = v \mu_k \delta_{kl}. \tag{17}$$

Also

$$\mu_i p_i(j) = v_i q_j(i), \quad i, j = 0, \dots, n. \tag{18}$$

In matrix terminology these are

$$P^T \Delta_\mu P = v \Delta_v, \tag{19}$$

$$Q^T \Delta_v Q = v \Delta_\mu, \tag{20}$$

where $\Delta_v = \text{diag}\{v_0, v_1, \dots, v_n\}$, $\Delta_\mu = \text{diag}\{\mu_0, \mu_1, \dots, \mu_n\}$.

Proof. The eigenvalues of $D_k D_l$ are $p_k(i) p_l(i)$ with multiplicities μ_i. From (5),

$$v v_k \delta_{kl} = \text{trace } D_k D_l = \sum_{i=0}^{n} \mu_i p_k(i) p_l(i),$$

which proves (16) and (19). From (19),

$$Q = v P^{-1} = \Delta_v^{-1} P^T \Delta_\mu$$

which gives (17), (18) and (20).   Q.E.D.

Corollary 4.

$$p_i(s) p_j(s) = \sum_{k=0}^{n} c_{ijk} p_k(s), \quad s = 0, \dots, n.$$

Proof. Equate eigenvalues in (5).   Q.E.D.

An isomorphic algebra of $(n+1) \times (n+1)$ matrices. We briefly mention that there is an algebra of $(n+1) \times (n+1)$ matrices which is isomorphic to $\mathcal{A}$, and is often easier to work with. Let

$$L_i = \begin{bmatrix} c_{i00} & c_{i10} & \cdots & c_{in0} \\ c_{i01} & c_{i11} & \cdots & c_{in1} \\ \cdots\cdots\cdots\cdots\cdots\cdots \\ c_{i0n} & c_{i1n} & \cdots & c_{inn} \end{bmatrix}, \quad i = 0, \dots, n.$$

Then Equation (2) implies

$$L_i L_j = \sum_{k=0}^{n} c_{ijk} L_k. \tag{21}$$

Thus the L_i multiply in the same manner as the D_i. Since $c_{ik0} = \delta_{ik}$, it follows that $L_0, \ldots, L_n$ are linearly independent. Therefore the algebra $\mathscr{B}$ consisting of all matrices $\sum_{i=0}^{n} a_i L_i$ (a_i real) is an associative commutative algebra, which is isomorphic to $\mathscr{A}$ under the mapping $D_i \to L_i$.

From Corollary 4

$$P L_k P^{-1} = \operatorname{diag}\{p_k(0), \ldots, p_k(n)\}, \tag{22}$$

which implies that the $p_k(i)$ ($i = 0, \ldots, n$) are the eigenvalues of L_k. (Alternatively since $\mathscr{A}$ and $\mathscr{B}$ are isomorphic, D_k and L_k have the same eigenvalues.) Since L_k is much smaller than D_k it is sometimes easier to find the $p_k(i)$ in this way.

Now it is time for some examples.

§3. The Hamming association scheme

The *Hamming* or *hypercubic* association scheme is the most important example for coding theory. In this scheme $X = F^n$, the set of binary vectors of length n, and two vectors $x, y \in F^n$ are i^{th} associates if they have Hamming distance i apart. Clearly conditions (i), (ii) and (iii) of the definition of an association scheme are satisfied. Problem 19 of Ch. 1 shows that (iv) holds with

$$c_{ijk} = \begin{cases} \left(\dfrac{k}{\dfrac{i-j+k}{2}} \right) \left(\dfrac{n-k}{\dfrac{i+j-k}{2}} \right) & \text{if } i+j-k \text{ is even,} \\[2em] 0 & \text{if } i+j-k \text{ is odd.} \end{cases}$$

Also $v = |X| = 2^n$ and $v_i = \binom{n}{i}$. The matrices in the Bose–Mesner algebra $\mathscr{A}$ are $2^n \times 2^n$ matrices, with rows and columns labeled by vectors $x \in F^n$. In particular the $(x, y)^{\text{th}}$ entry of D_k is 1 if and only if $\operatorname{dist}(x, y) = k$.

For example, let $n = 3$ and label the rows and columns of the matrices by $000, 001, 010, 100, 011, 101, 110, 111$. The graph which has edges labeled 1 is of course the cube (Fig. 21.1).

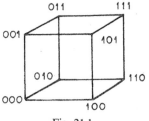

Fig. 21.1.

The diagonals of the faces of the cube are labeled 2, and the four main diagonals are labeled 3. The adjacency matrices are $D_0 = I$,

$$
D_1 = \begin{bmatrix}
0 & 1 & 1 & 1 & 0 & 0 & 0 & 0 \\
1 & 0 & 0 & 0 & 1 & 1 & 0 & 0 \\
1 & 0 & 0 & 0 & 1 & 0 & 1 & 0 \\
1 & 0 & 0 & 0 & 0 & 1 & 1 & 0 \\
0 & 1 & 1 & 0 & 0 & 0 & 0 & 1 \\
0 & 1 & 0 & 1 & 0 & 0 & 0 & 1 \\
0 & 0 & 1 & 1 & 0 & 0 & 0 & 1 \\
0 & 0 & 0 & 0 & 1 & 1 & 1 & 0
\end{bmatrix},
\quad
D_2 = \begin{bmatrix}
0 & 0 & 0 & 0 & 1 & 1 & 1 & 0 \\
0 & 0 & 1 & 1 & 0 & 0 & 0 & 1 \\
0 & 1 & 0 & 1 & 0 & 0 & 0 & 1 \\
0 & 1 & 1 & 0 & 0 & 0 & 0 & 1 \\
1 & 0 & 0 & 0 & 0 & 1 & 1 & 0 \\
1 & 0 & 0 & 0 & 1 & 0 & 1 & 0 \\
1 & 0 & 0 & 0 & 1 & 1 & 0 & 0 \\
0 & 1 & 1 & 1 & 0 & 0 & 0 & 0
\end{bmatrix},
\quad
D_3 = \begin{bmatrix}
0 & 0 & 0 & 0 & 0 & 0 & 0 & 1 \\
0 & 0 & 0 & 0 & 0 & 0 & 1 & 0 \\
0 & 0 & 0 & 0 & 0 & 1 & 0 & 0 \\
0 & 0 & 0 & 0 & 1 & 0 & 0 & 0 \\
0 & 0 & 0 & 1 & 0 & 0 & 0 & 0 \\
0 & 0 & 1 & 0 & 0 & 0 & 0 & 0 \\
0 & 1 & 0 & 0 & 0 & 0 & 0 & 0 \\
1 & 0 & 0 & 0 & 0 & 0 & 0 & 0
\end{bmatrix}.
$$

Note that

$$D_1^2 = 3I + 2D_2,$$

$$D_1 D_2 = 2D_1 + 3D_3, \tag{23}$$

$c_{110} = v_1 = 3$, $c_{111} = 0$, $c_{112} = 2$, $c_{113} = 0$, $c_{220} = v_2 = 3$, $c_{330} = v_3 = 1$, and so on.

Theorem 5. *In a Hamming scheme the primitive idempotent $J_k(k = 0, \ldots, n)$ is the matrix with $(x, y)^{\text{th}}$ entry*

$$\frac{1}{2^n} \sum_{\text{wt}(z)=k} (-1)^{(x+y)\cdot z}. \tag{24}$$

The eigenvalues are given by

$$p_k(i) = q_k(i) = P_k(i; n), \tag{25}$$

where $P_k(x; n)$ is the Krawtchouk polynomial defined in Equation (14) of Ch. 5.

Proof. Let A_k denote the matrix with $(x, y)^{\text{th}}$ entry (24). We show

$$A_k = \frac{1}{2^n} \sum_{i=0}^{n} P_k(i; n) D_i, \tag{26}$$

so $A_k \in \mathcal{A}$. We then show (8)–(10) hold, so the A_k are the primitive idempotents. Then (26) implies $q_k(i) = P_k(i; n)$, and from (18) (and Theorem 17 of Ch. 5)

$$p_k(i) = q_i(k) \binom{n}{k} \Big/ \binom{n}{i}$$

$$= P_i(k; n) \binom{n}{k} \Big/ \binom{n}{i} = P_k(i; n).$$

To prove (26), Problem 14 of Ch. 5 implies

$$(A_k)_{x,y} = \frac{1}{2^n} P_k(\text{wt}(x + y); n),$$

hence

$$A_k = \frac{1}{2^n} \sum_{i=0}^{n} P_k(i; n) D_i.$$

The $(x, y)^{\text{th}}$ entry of $A_k A_l$ is

$$\frac{1}{2^{2n}} \sum_{u \in F^n} \sum_{\text{wt}(v)=k} (-1)^{(x+u) \cdot v} \sum_{\text{wt}(w)=l} (-1)^{(u+y) \cdot w},$$

which simplifies to

$$\frac{1}{2^n} \delta_{kl} \sum_{\text{wt}(v)=k} (-1)^{(x+y) \cdot v},$$

the $(x, y)^{\text{th}}$ entry of $\delta_{kl} A_k$, and proves (8), (9). (10) follows easily.   Q.E.D.

Equation (16) now becomes the familiar orthogonality relation for Krawtchouk polynomials (Theorem 16 of Ch. 5).

For example, when $n = 3$, $J_0 = \frac{1}{8} J$,

$$J_1 = \frac{1}{8} \begin{bmatrix} 3 & 1 & 1 & 1 & -1 & -1 & -1 & -3 \\ 1 & 3 & -1 & -1 & 1 & 1 & -3 & -1 \\ 1 & -1 & 3 & -1 & 1 & -3 & 1 & -1 \\ 1 & -1 & -1 & 3 & -3 & 1 & 1 & -1 \\ -1 & 1 & 1 & -3 & 3 & -1 & -1 & 1 \\ -1 & 1 & -3 & 1 & -1 & 3 & -1 & 1 \\ -1 & -3 & 1 & 1 & -1 & -1 & 3 & 1 \\ -3 & -1 & -1 & -1 & 1 & 1 & 1 & 3 \end{bmatrix} \quad J_2 = \frac{1}{8} \begin{bmatrix} 3 & -1 & -1 & -1 & -1 & -1 & -1 & 3 \\ -1 & 3 & -1 & -1 & -1 & -1 & 3 & -1 \\ -1 & -1 & 3 & -1 & -1 & 3 & -1 & -1 \\ -1 & -1 & -1 & 3 & 3 & -1 & -1 & -1 \\ -1 & -1 & -1 & 3 & 3 & -1 & -1 & -1 \\ -1 & -1 & 3 & -1 & -1 & 3 & -1 & -1 \\ -1 & 3 & -1 & -1 & -1 & -1 & 3 & -1 \\ 3 & -1 & -1 & -1 & -1 & -1 & -1 & 3 \end{bmatrix},$$

$$J_3 = \frac{1}{8} \begin{bmatrix} 1 & -1 & -1 & -1 & 1 & 1 & 1 & -1 \\ -1 & 1 & 1 & 1 & -1 & -1 & -1 & 1 \\ -1 & 1 & 1 & 1 & -1 & -1 & -1 & 1 \\ -1 & 1 & 1 & 1 & -1 & -1 & -1 & 1 \\ 1 & -1 & -1 & -1 & 1 & 1 & 1 & -1 \\ 1 & -1 & -1 & -1 & 1 & 1 & 1 & -1 \\ 1 & -1 & -1 & -1 & 1 & 1 & 1 & -1 \\ -1 & 1 & 1 & 1 & -1 & -1 & -1 & 1 \end{bmatrix}.$$

Then

$$I = J_0 + J_1 + J_2 + J_3,$$

$$D_1 = 3J_0 + J_1 - J_2 - 3J_3,$$

$$D_2 = 3J_0 - J_1 - J_2 + 3J_3,$$

$$D_3 = J_0 - J_1 + J_2 - J_3,$$

and

$$P = Q = \begin{bmatrix} 1 & 3 & 3 & 1 \\ 1 & 1 & -1 & -1 \\ 1 & -1 & -1 & 1 \\ 1 & -3 & 3 & -1 \end{bmatrix}, \qquad L_0 = I,$$

$$L_1 = \begin{bmatrix} 0 & 1 & 0 & 0 \\ 3 & 0 & 2 & 0 \\ 0 & 2 & 0 & 3 \\ 0 & 0 & 1 & 0 \end{bmatrix}, \qquad L_2 = \begin{bmatrix} 0 & 0 & 1 & 0 \\ 0 & 2 & 0 & 3 \\ 3 & 0 & 2 & 0 \\ 0 & 1 & 0 & 0 \end{bmatrix}, \qquad L_3 = \begin{bmatrix} 0 & 0 & 0 & 1 \\ 0 & 0 & 1 & 0 \\ 0 & 1 & 0 & 0 \\ 1 & 0 & 0 & 0 \end{bmatrix}.$$

Problems. (2) Verify that the eigenvalues of L_i are the $(i+1)^{st}$ column of P. (3) Find P when $n = 4$.

§4. Metric schemes

The Hamming scheme and others given below are examples of metric schemes, which are defined by graphs.

Let Γ be a connected graph with v vertices, containing no loops or multiple edges, and let X be the set of vertices. The *distance* $\rho(x, y)$ between $x, y \in X$ is the number of edges on the shortest path joining them. The maximum distance n (say) between any two vertices is called the *diameter* of Γ.

Definition. The graph Γ is called *metrically regular* (or *perfectly-* or *distance-regular*) if, for any $x, y \in X$ with $\rho(x, y) = k$, the number of $z \in X$ such that $\rho(x, z) = i$ and $\rho(y, z) = j$ is a constant c_{ijk} independent of the choice of x and y.

Clearly we obtain an association scheme with n classes from a metrically regular graph of diameter n by calling $x, y \in X$ i^{th} associates if $\rho(x, y) = i$. Association schemes which can be obtained in this way are called *metric schemes*. To recover the graph from the scheme, define x and y to be adjacent iff $(x, y) \in R_1$.

For example the Hamming scheme is metric, and the graph is the skeleton of a unit cube in n-dimensions (see Fig. 21.1).

Metrically regular graphs of diameter two are called *strongly regular graphs*.

Problems. (4) Show that any association scheme with two classes is metric (and is obtained from a strongly regular graph).

But not all association schemes are metric.

(5) Show that a scheme is metric iff (i) $c_{1ii+1} \neq 0$ and (ii) $c_{1ij} \neq 0 \Rightarrow i - 1 \leq j \leq i + 1$. [Hint: (only if) Let Γ be the graph defined by R_1, and show $\rho(x, y) = i$ iff $(x, y) \in R_i$.]

Thus if an association scheme is metric then for $k = 1, 2, \ldots$

$$D_1 D_k = c_{1,k,k+1} D_{k+1} + c_{1kk} D_k + c_{1,k,k-1} D_{k-1}, \tag{27}$$

and hence D_k is a polynomial in D_1 of degree k. $\mathcal{A}$ is the algebra of polynomials in D_1, and all the eigenvalues are determined when the eigenvalues of D_1 (or L_1) have been found. In the above example $2L_2 = L_1^2 - 3I$ (from (23)), so if α is any eigenvalue of L_1, $(\alpha^2 - 3)/2$ is an eigenvalue of L_2. In general, from (27)

$$p_1(i)p_k(i) = c_{1kk+1}p_{k+1}(i) + c_{1kk}p_{k+1}(i) + c_{1kk-1}p_{k-1}(i), \tag{28}$$

for $i = 0, \ldots, n$.

Problem. (6) Prove that if D_k is a polynomial in D_1 of degree k then the scheme is metric.

The most interesting property of a metric scheme is that the eigenvalues $p_k(i)$ are obtained from a family of orthogonal polynomials.

Definition. An association scheme is called a *P-polynomial* scheme if there exist nonnegative real numbers $z_0 = 0$, $z_1, \ldots, z_n$ and real polynomials $\Phi_0(z), \Phi_1(z), \ldots, \Phi_k(z)$, where $\deg \Phi_k(z) = k$, such that

$$p_k(i) = \Phi_k(z_i), \quad i, k = 0, \ldots, n. \tag{29}$$

Theorem 3 implies that the $\Phi_k(z)$ are a family of orthogonal polynomials:

$$\sum_{i=0}^{n} \mu_i \Phi_k(z_i) \Phi_l(z_i) = v v_k \delta_{kl}. \tag{30}$$

A *Q-polynomial* scheme is defined similarly.

For example, a Hamming scheme is both a *P-* and *Q*-polynomial scheme (with $z_i = i$, $\Phi_k(z) = P_k(z; n)$).

Theorem 6. (Delsarte.) *An association scheme is metric iff it is a P-polynomial scheme.*

Proof. If the scheme is metric, then $p_k(i)$ is a polynomial in $p_1(i)$ of degree k. Set $z_i = v_1 - p_1(i)$. Then there exist polynomials $\Phi_k(z)$ such that $p_k(i) = \Phi_k(z_i)$

and deg $\Phi_k(z) = k$. Also $z_0 = 0$ and $z_i \geqslant 0$ by Problem 1. The z_i are distinct because $z_i = z_j$ implies $p_k(i) = p_k(j)$, $k = 0, \ldots, n$, which is impossible since P is invertible. Hence the scheme is P-polynomial.

Conversely, suppose the scheme is P-polynomial, and write $\Phi_1(z) = b - (1/a)z$ ($a > 0$). Then $p_1(i) = \Phi_1(z_i)$ and $p_1(0) = v_i$ imply $b = v_i$ and $z_i = a(v_i - p_1(i))$. Since the $\Phi_k(z)$ are orthogonal they satisfy a 3-term recurrence (Szegö [1297, p. 42]) say

$$\alpha_{k+1}\Phi_{k+1}(z) = (\beta_k - z)\Phi_k(z) - \gamma_k\Phi_{k-1}(z)$$

where $\alpha_{k+1} > 0$. (The coefficients must satisfy various conditions that do not concern us here.) Evaluating this recurrence at z_i gives

$$-z_i\Phi_k(z_i) = \alpha_{k+1}\Phi_{k+1}(z_i) - \beta_k\Phi_k(z_i) + \gamma_k\Phi_{k-1}(z_i),$$

$$p_1(i)p_k(i) = c_{1kk+1}p_{k+1}(i) + c_{1kk}p_k(i) + c_{1kk-1}p_{k-1}(i),$$

for suitably defined c_{ijk}. Problem 5 now implies the scheme is metric.   Q.E.D.

Research Problem (21.1). Give a similar characterization for Q-polynomial schemes.

*§5. Symplectic forms

In this section we construct the association scheme of symplectic forms, essential for studying subcodes of the second-order Reed–Muller code (see Ch. 15).

The set X consists of all binary symplectic forms in m variables, or equivalently all $m \times m$ symmetric binary matrices with zero diagonal (§2 of Ch. 15). Thus $|X| = v = 2^{m(m-1)/2}$. We define $(x, y) \in R_i$ iff the rank of the matrix $x + y$ is $2i$, for $i = 0, 1, \ldots, n = [m/2]$. The number of symplectic forms of rank $2i$, v_i, is given by Theorem 2 of Ch. 15. Let D_i be the adjacency matrix of R_i.

We shall prove that this *is* an association scheme by showing that (3)–(5) hold, and at the same time construct the Bose–Mesner algebra $\mathcal{A}$. The treatment is parallel to that of the Hamming scheme, but uses Gaussian binomial coefficients with $b = 4$ (see Problem 3 of Ch. 15).

For matrices $x = (x_{ij})$, $y = (y_{ij}) \in X$, define the inner product

$$\langle x, y \rangle = (-1)^{\sum_{i=1}^{m} \sum_{j=i}^{m} x_{ij}y_{ij}} \tag{31}$$

Problem. (7) Show that

$$\sum_{x \in X} \langle x, y \rangle = v\delta_{0y}.$$

Theorem 7. *If x has rank 2i then*

$$\sum_{\text{rank } y=2k} \langle x, y \rangle = \sum_{j=0}^{k} (-1)^{k-j} b^{\binom{k-j}{2}} \begin{bmatrix} n-j \\ n-k \end{bmatrix} \begin{bmatrix} n-i \\ j \end{bmatrix} c^{j}, \tag{32}$$

where $b = 4$ and $c = 2^{m}$ (m odd) or $c = 2^{m-1}$ (m even).

Sketch of Proof. We first show that

$$\sum_{\text{rank } (y)=2k} \langle x, y \rangle$$

depends only on the rank of x. If R is an invertible $m \times m$ matrix, the map $y \to RyR^{T}$ is a permutation of the set of symplectic matrices which preserves rank. Hence

$$\sum_{\text{rank }(y)=2k} \langle x, y \rangle = \sum_{\text{rank }(y)=2k} \langle x, RyR^{T} \rangle.$$

Also $\langle x, RyR^{T} \rangle = \langle RxR^{T}, y \rangle$. By Theorem 4 of Ch. 15 there is an R such that

$$RxR^{T} = \begin{bmatrix} G & & & 0 \\ & G & & & 0 \\ & & \ddots & \\ 0 & & & G & \\ \hline & & 0 & & 0 \end{bmatrix} = N_{i} \quad \text{(say)}, \tag{33}$$

where

$$G = \begin{bmatrix} 0 & 1 \\ 1 & 0 \end{bmatrix}$$

and there are $i = \text{rank}(x)$ G's down the diagonal. Then

$$\sum_{\text{rank }(y)=2k} \langle x, y \rangle = \sum_{\text{rank }(y)=2k} \langle N_{i}, y \rangle = p_{k}^{(m)}(i) \quad \text{(say)},$$

depending only on i, k and m.

The next step is to prove a recurrence for $p_{k}^{(m)}(i)$:

$$p_{0}^{(m)}(i) = 1, \qquad p_{1}^{(m)}(0) = \frac{1}{3}(2^{m} - 1)(2^{m-1} - 1), \tag{34}$$

$$p_{k}^{(m)}(i) = p_{k}^{(m)}(i-1) - 2^{2m-2i-1} p_{(k-1)}^{(m-2)}(i-1). \tag{35}$$

(34) follows from Theorem 2 of Ch. 15, and then by induction on i

$$p_{1}^{(m)}(i) = \frac{1}{3}(2^{2m-2i-1} - 3.2^{m-1} + 1). \tag{36}$$

Let $x = N_{i}$, and define M_{i} to be the matrix obtained from N_{i} by changing the first G to

$$\begin{bmatrix} 0 & 0 \\ 0 & 0 \end{bmatrix}.$$

Then

$$p_k^{(m)}(i-1) - p_k^{(m)}(i) = \sum_{\text{rank}(y)=2k} (\langle M_i, y \rangle - \langle N_i, y \rangle)$$

$$= \sum_{\text{rank}(y)=2k} \langle M_i, y \rangle (1 - (-1)^{y_{12}})$$

where y_{12} is the appropriate entry in y,

$$= 2 \sum_{\substack{\text{rank}(y)=2k \\ \text{and } y_{12}=1}} \langle M_i, y \rangle.$$

Now $\langle M_i, y \rangle = \langle A, D \rangle$ where A, D are of size $(m-2) \times (m-2)$, and

$$y = \begin{bmatrix} G & Z \\ Z^T & D \end{bmatrix}, \quad Z = \begin{bmatrix} z_{11} & \cdots & z_{1m-2} \\ z_{21} & \cdots & z_{2m-2} \end{bmatrix}, \quad z_{ij} = 0 \text{ or } 1. \tag{37}$$

If

$$L = \begin{bmatrix} I_2 & GZ \\ 0 & I_{m-2} \end{bmatrix} \quad \text{then} \quad L^T y L = \begin{bmatrix} G & 0 \\ 0 & F \end{bmatrix},$$

where $F = D + Z^T G Z$ is an $(m-2) \times (m-2)$ symplectic matrix of rank $2k-2$. Thus

$$p_k^{(m)}(i-1) - p_k^{(m)}(i) = 2 \sum_{\substack{\text{rank}(F)=2k-2, \\ \text{all } Z}} \langle A, F + Z^T G Z \rangle.$$

Now $Z^T G Z$ is either 0 or of rank 2. By Problem 8 the number of Z such that $Z^T G Z = 0$ is $3.2^{m-2} - 2$, and the number of Z such that $Z^T G Z$ is a particular matrix of rank 2 is 6. Thus

$$p_k^{(m)}(i-1) - p_k^{(m)}(i) = (3.2^{m-1} - 4) \sum_{\text{rank}(F)=2k-2} \langle A, F \rangle$$

$$+ 12 \sum_{\text{rank}(F)=2k-2} \langle A, F \rangle \sum_{\text{rank}(E)=2} \langle A, E \rangle$$

$$= (3.2^{m-1} - 4) p_{k-1}^{(m-2)}(i-1) + 12 p_{k-1}^{(m-2)}(i-1) p_1^{(m-2)}(i-1). \tag{38}$$

Combining (36) and (38) gives the recurrence (35). Finally (32) is the solution to (34) and (35); we omit the details. Q.E.D.

The RHS of (32) will be denoted by $p_k(i)$, for it will turn out that these *are* the eigenvalues of the scheme.

Define matrices J_k, $k = 0, \ldots, n$, by

$$(J_k)_{x,y} = \frac{1}{v} \sum_{\text{rank } z = 2k} \langle x + y, z \rangle. \tag{39}$$

An argument similar to that used in the proof of Theorem 5 shows that the J_k

satisfy (8)–(10) and hence are primitive idempotent matrices, and that

$$J_k = \frac{1}{v} \sum_{i=0}^{n} p_k(i) D_i \tag{40}$$

where $p_k(i)$ is given by (32).

Lemma 8.

$$v_i p_k(i) = v_k p_i(k).$$

Proof. Consider in two ways the sum

$$\sum_{\text{rank } x = 2i} \sum_{\text{rank } y = 2k} \langle x, y \rangle. \qquad \text{Q.E.D.}$$

Theorem 9. (The orthogonality relation.)

$$\sum_{i=0}^{n} v_i p_k(i) p_l(i) = v v_k \delta_{kl}.$$

Proof. Consider in two ways the sum

$$\sum_{x \in X} \sum_{\text{rank } y = 2k}^{v} \sum_{\text{rank } z = 2l} \langle x, y \rangle \langle x, z \rangle,$$

and use Problem 7. $\qquad \text{Q.E.D.}$

From Lemma 8 and Theorem 9,

$$\sum_{i=0}^{n} p_k(i) p_i(l) = v \delta_{kl},$$

hence the matrix P with $(i, k)^{\text{th}}$ entry $p_k(i)$ satisfies $P^2 = vI$, $P^{-1} = vP$. Therefore (40) implies

$$D_k = \sum_{i=0}^{n} p_k(i) J_i. \tag{41}$$

Let $\mathcal{A}$ be the commutative algebra generated by the J_k. From (10) this has dimension $n + 1$, and from (40), (41) the D_k are also a basis. Therefore (3)–(5) hold, and this is an association scheme. From (40), (41) the eigenvalues $p_k(i) = q_k(i)$ of (11), (15) are indeed given by (32), as claimed.

Problems. (8) Show that (i) the number of Z, given by Equation (37), such that $Z^T G Z = 0$ is $3.2^{m-2} - 2$; and (ii) if rank $(B) = 2$, there are 6 Z such that $Z^T G Z = B$.

(9) Show that $p_k(i)$ is a polynomial of degree k in the variable b^{-i}, and hence from Theorem 6 that this is a metric association scheme.

§6. The Johnson scheme

Our third example is the *Johnson* or *triangular* association scheme. Here X is the set of all binary vectors of length l and weight n, so $v = |X| = \binom{l}{n}$. Two vectors $x, y \in X$ are called i^{th} associates if dist $(x, y) = 2i$, for $i = 0, 1, \ldots, n$.

Problem. (10) Show that this is an association scheme with n classes, and find the c_{ijk}. Show that

$$v_i = \binom{n}{i}\binom{l-n}{i}.$$

Theorem 10. *The eigenvalues are given by*

$$p_k(i) = E_k(i), \qquad q_k(i) = \frac{\mu_k}{v_i} E_i(k), \tag{42}$$

where

$$\mu_i = \frac{l - 2i + 1}{l - i + 1}\binom{l}{i} \tag{43}$$

and $E_k(x)$ is an Eberlein polynomial defined by

$$E_k(x) = \sum_{j=0}^{k}(-1)^j\binom{x}{j}\binom{n-x}{k-j}\binom{l-n-x}{k-j}, \quad k = 0, \ldots, n. \tag{44}$$

For the proof see Yamamoto et al. [1444] and Delsarte [352].

Theorem 11. (Properties of Eberlein polynomials.)
  (i) $E_k(x)$ *is a polynomial of degree $2k$ in x.*
  (ii) $E_k(x)$ *is a polynomial of degree k in the variable $z = x(l + 1 - x)$. Hence this is a P-polynomial scheme.*
  (iii)

$$E_k(x) = \sum_{j=0}^{k}(-1)^j\binom{x}{j}\binom{n-x}{k-j}\binom{l-n-x}{k-j}.$$

  (iv) *A recurrence:*

$$E_0(x) = 1, \qquad E_1(x) = n(l - n) - x(l + 1 - x),$$

$$(k + 1)^2 E_{k+1}(x) = \{E_1(x) - k(l - 2k)\}E_k(x) - (n - k + 1)(l - n - k + 1)E_{k-1}(x).$$

For the proof see Delsarte [352, 361].

Problem. (11) Let A_i be the $\binom{l}{n} \times \binom{l}{i}$ matrix with rows labeled by the binary vectors x of weight n and columns by the binary vectors ξ of weight i, with

$$(A_i)_{x,\xi} = \begin{cases} 1 & \text{if } \xi \subset x \\ 0 & \text{if not.} \end{cases}$$

Also let $C_i = A_i A_i^T$. Show that

$$C_i = \sum_{k=i}^{n} \binom{k}{i} D_{n-k}.$$

Hence $C_0, \ldots, C_n$ is a basis for $\mathcal{A}$.

§7. Subsets of association schemes

A code is a subset of a Hamming scheme. In this section we consider a nonempty subset Y of an arbitrary association scheme X with relations $R_0, R_1, \ldots, R_n$.

Suppose $|Y| = M > 0$. The *inner distribution* of Y is the $(n+1)$-tuple of rational numbers $(B_0, B_1, \ldots, B_n)$, where

$$B_i = \frac{1}{M} |R_i \cap Y^2|$$

is the average number of $z \in Y$ which are i^{th} associates of a point $y \in Y$. In the Hamming scheme $(B_0, \ldots, B_n)$ is the distance distribution (§1 of Ch. 2) of the code Y.

Of course $B_0 = 1$,

$$B_i \geq 0, \quad i = 1, \ldots, n, \tag{45}$$

and

$$B_0 + B_1 + \cdots + B_n = M. \tag{46}$$

Delsarte has observed that certain linear combinations of the B_i are also nonnegative:

Theorem 12. (Delsarte.)

$$B_k' = \frac{1}{M} \sum_{i=0}^{n} q_k(i) B_i \geq 0 \tag{47}$$

for $k = 0, \ldots, n$, where the $q_k(i)$ are the eigenvalues defined in §2.

Proof. Let u be a vector indicating which elements of X belong to Y:

$$u_x = 1 \quad \text{if } x \in Y, \qquad u_x = 0 \quad \text{if } x \notin Y.$$

Then

$$B_i = \frac{1}{M} u D_i u^T,$$

$$B_k' = \frac{1}{M^2} u \Big(\sum_{i=0}^{n} q_k(i) D_i \Big) u^T,$$

$$= \frac{v}{M^2} \cdot u J_k u^T, \quad \text{from (15)}. \tag{48}$$

Since J_k is idempotent, its eigenvalues are 0 or 1, and hence J_k is nonnegative definite. Therefore $B_k' \geq 0$.   Q.E.D.

For example, applying Theorem 12 to codes we have another proof of Theorem 6 of Ch. 5. Applying it to the Johnson scheme we obtain the results about constant weight codes used on page 545 of Ch. 17. In general, Equation (46) makes it possible to apply linear programming to many problems which involve finding the largest subset of an association scheme subject to constraints on the B_i and B_k'. Such a problem can be stated as: Maximize $B_0 + B_1 + \cdots + B_n$ $(= M)$, subject to (45), (47) and any additional constraints. For examples see pages 538 and 546 of Ch. 17.

*§8. Subsets of symplectic forms

Let X be the set of symplectic forms in m variables (see §5), and Y a subset of X which has the property that if $y, y' \in Y$ then the form $y + y'$ has rank at least $2d$. Such a set Y is called an (m, d)-set. In this section we derive an upper bound on the size of an (m, d)-set.

Let $B_0, B_1, \ldots, B_n$, $n = [m/2]$, be the inner distribution of Y. Then $B_1 = \cdots = B_{d-1} = 0$.

Theorem 13. (Delsarte and Goethals.) *For any (m, d)-set Y, $|Y| \leq c^{n-d+1}$, where c was defined in Theorem* 7.

Proof. It was shown in §5 that for the association scheme of symplectic forms

$$q_k(i) = \sum_{j=0}^{k} (-1)^{k-j} b^{\binom{k-j}{2}} \begin{bmatrix} n-j \\ n-k \end{bmatrix} \begin{bmatrix} n-i \\ j \end{bmatrix} c^j.$$

Using parts (b), (c), (g) of Problem 3 of Ch. 15, this implies

$$\sum_{k=0}^{j} \begin{bmatrix} n-k \\ n-j \end{bmatrix} q_k(i) = \begin{bmatrix} n-i \\ j \end{bmatrix} c^j, \quad j = 0, \ldots, n. \tag{49}$$

Then

$$B'_k = \sum_{i=0}^{n} q_k(i) B_i,$$

$$\sum_{k=0}^{n-d+1} \begin{bmatrix} n-k \\ d-1 \end{bmatrix} B'_k = \sum_{i=0}^{n} B_i \sum_{k=0}^{n-d+1} \begin{bmatrix} n-k \\ d-1 \end{bmatrix} q_k(i)$$

$$= \sum_{i=0}^{n} B_i \begin{bmatrix} n-i \\ n-d+1 \end{bmatrix} c^{n-d+1}. \tag{50}$$

Now $B_1 = \cdots = B_{d-1} = 0$ and

$$\begin{bmatrix} n-i \\ n-d+1 \end{bmatrix} = 0 \quad \text{for } i \geq d,$$

so this becomes

$$\begin{bmatrix} n \\ d-1 \end{bmatrix} B'_0 + \sum_{k=1}^{n-d+1} \begin{bmatrix} n-k \\ d-1 \end{bmatrix} B'_k = \begin{bmatrix} n \\ d-1 \end{bmatrix} c^{n-d+1}.$$

Also, from Theorem 2,

$$B'_0 = \sum_{i=0}^{n} q_0(i) B_i = \sum_{i=0}^{n} B_i = |Y|.$$

Therefore

$$\sum_{k=1}^{n-d+1} \begin{bmatrix} n-k \\ d-1 \end{bmatrix} B'_k = \begin{bmatrix} n \\ d-1 \end{bmatrix} (c^{n-d+1} - |Y|). \tag{51}$$

By Theorem 12 $B'_k \geq 0$, hence $|Y| \leq c^{n-d+1}$.   Q.E.D.

This theorem shows that the sets of symplectic forms described in Ch. 15 are indeed maximal sets.

Theorem 14. *If an (m, d)-set Y is such that $|Y| = c^{n-d+1}$, then the inner distribution of Y is given by*

$$B_{n-i} = \sum_{j=i}^{n-d} (-1)^{j-i} b^{\binom{j-i}{2}} \begin{bmatrix} j \\ i \end{bmatrix} \begin{bmatrix} n \\ j \end{bmatrix} (c^{n-d+1-j} - 1) \tag{52}$$

for $i = 0, 1, \ldots, n-d$.

Proof. If Y attains the bound every term on the LHS of (51) is zero, i.e.

$$B'_0 = |Y| = c^{n-d+1}, \qquad B'_k = 0, \quad k = 1, \ldots, n-d+1.$$

Now for $j = 0, \ldots, n$

$$\sum_{k=0}^{j} \begin{bmatrix} n-k \\ n-j \end{bmatrix} B'_k = \sum_{i=0}^{n} B_i \sum_{k=0}^{j} \begin{bmatrix} n-k \\ n-j \end{bmatrix} q_k(i).$$

For $j \le n - d$ all terms on the left except the first vanish, and using (49) this becomes

$$\begin{bmatrix} n \\ j \end{bmatrix} c^{n-d+1} = c^j \sum_{i=0}^{n} \begin{bmatrix} n - i \\ j \end{bmatrix} B_i, \quad j = 0, \ldots, n - d.$$

But $B_0 = \cdots = B_{d-1} = 0$, so

$$\sum_{s=0}^{n-d} \begin{bmatrix} s \\ j \end{bmatrix} B_{n-s} = \sum_{s=0}^{n-d} \begin{bmatrix} n \\ j \end{bmatrix} (c^{n-d+1-j} - 1).$$

The stated result now follows from Problem 3g of Ch. 15.   Q.E.D.

This theorem gives the distance distributions of the codes constructed from maximal (m, d)-sets in Ch. 15.

Problem. (12) (Kasami, [727].) Use Theorem 14 to show that if $m \ge 5$ is odd then the dual of the triple-error-correcting BCH code has parameters

$$[2^m - 1, 3m, 2^{m-1} - 2^{(m+1)/2}]$$

and weight distribution

i	A_i
0	1
$2^{m-1} \pm 2^{(m+1)/2}$	$\frac{1}{3} \cdot 2^{(m-5)/2}(2^{(m-3)/2} \mp 1)(2^m - 1)(2^{m-1} - 1)$
$2^{m-1} \pm 2^{(m-1)/2}$	$\frac{1}{3} \cdot 2^{(m-3)/2}(2^{(m-1)/2} \mp 1)(2^m - 1)(5.2^{m-1} + 4)$
2^{m-1}	$(2^m - 1)(9 \cdot 2^{2m-4} + 3 \cdot 2^{m-3} + 1)$.

Remark. For even m this method doesn't work. However in this case the weight distribution has been found by Berlekamp ([113, Table 16.5], [114] and [118]). The dual of the extended triple-error-correcting BCH code has parameters

$$[N = 2^m, 3m + 1, 2^{m-1} - 2^{(m+2)/2}], \quad \text{for even } m \ge 6,$$

and weight distribution

i	A_i
$0, 2^m$	1
$2^{m-1} \pm 2^{(m+2)/2}$	$N(N - 1)(N - 4)/960$
$2^{m-1} \pm 2^{m/2}$	$7N^2(N - 1)/48$
$2^{m-1} \pm 2^{(m-2)/2}$	$2N(N - 1)(3N + 8)/15$
2^{m-1}	$(N - 1)(29N^2 - 4N + 64)/32$.

Problem. (13) Apply construction $Y1$ of §9.1 of Ch. 18 with $\mathscr{C}$ equal to a $[2^{m+1}, 2^{m+1} - 3m - 4, 8]$ extended triple-error-correcting BCH code, and obtain codes with distance 7, redundancy $3m + 2$, and length $2^m + 2^{(m+2)/2} - 1$ if m is even, or length $2^m + 2^{(m+3)/2} - 1$ if m is odd, for all $m \geqslant 3$.

§9. t-Designs and orthogonal arrays

A subset Y of X is called a *t-design* if $B_1' = \cdots = B_t' = 0$. The justification for this definition comes from:

Theorem 15. *A subset Y in the Johnson scheme is a t-design iff the vectors of Y form an ordinary t-(l, n, λ) design for some λ.*

Proof. Note from (48) that in any association scheme Y is a t-design iff

$$uJ_k u^T = 0 \quad \text{for } k = 1, \ldots, t \tag{53}$$

where u is the indicator vector of Y as defined in the proof of Theorem 12.

Suppose that the vectors of Y form a t-(l, n, λ) design. For $1 \leqslant i \leqslant t$ let A_i and C_i be as in Problem 11. Then $A_i^T u^T$ is a column vector in which the ξth entry is the number of vectors in Y which contain the vector ξ of weight i. Since the vectors of Y also form an i-(l, n, λ_i) design, with $\lambda_i = |Y|\binom{i}{i}/\binom{l}{i}$,

$$A_i^T u^T = \lambda_i \begin{pmatrix} 1 \\ 1 \\ \vdots \\ 1 \end{pmatrix} = \frac{|Y|}{\binom{l}{n}} A_i^T \begin{pmatrix} 1 \\ 1 \\ \vdots \\ 1 \end{pmatrix}.$$

Since $C_i = A_i A_i^T$,

$$C_i u^T = \frac{|Y|}{\binom{l}{n}} C_i \begin{pmatrix} 1 \\ 1 \\ \vdots \\ 1 \end{pmatrix}. \tag{54}$$

As the C_i are a basis for $\mathscr{A}$, (54) holds for all the matrices in $\mathscr{A}$, and in particular:

$$J_k u^T = \frac{|Y|}{\binom{l}{n}} J_k \begin{pmatrix} 1 \\ 1 \\ \vdots \\ 1 \end{pmatrix}.$$

But

$$J_0 = \frac{1}{\binom{l}{n}} J \quad \text{and} \quad J_k J_0 = 0 \ (k > 0),$$

hence

$$J_k u^T = 0 \quad \text{for } 1 \leq k \leq t,$$
$$u J_k u^T = 0 \quad \text{for } 1 \leq k \leq t,$$

and Y is a t-design in the association scheme by (53).

Conversely, if Y is a t-design in the association scheme, $u J_k u^T = 0$, hence $J_k u^T = 0$ since J_k is nonnegative definite. Reversing the above proof shows that Y is an ordinary t-design.   Q.E.D.

For example, let Y consist of the 14 codewords of weight 4 in the $[8, 4, 4]$ Hamming code. The inner distribution and its transform are

$$B_0 = 1, \qquad B_2 = 12, \qquad B_4 = 1,$$
$$B_1' = B_2' = B_3' = 0, \qquad B_4' = 56,$$

and indeed these codewords form a 3-$(8, 4, 1)$ design.

Theorem 16. *Y is a t-design in the Hamming scheme iff the vectors of Y form an orthogonal array of size $|Y|$, n constraints, 2 levels, strength t and index $|Y|/2^t$.*

Proof. See Theorem 8 of Ch. 5 and §8 of Ch. 11.   Q.E.D.

Notes on Chapter 21

§1. Association schemes were introduced by Bose and Shimamoto [186] and the algebra $\mathcal{A}$ by James [688] and Bose and Mesner [183]. For further properties see Yang [103, 104], Blackwelder [154], Bose [178], Yamamoto et al. [1422–1444], Wan [1458] and Wan and Yang [1459].

The group case. An important class of association schemes arise from permutation groups. Let $\mathcal{G}$ be a transitive permutation group on a set X containing v points. X is called a *homogeneous space* of $\mathcal{G}$. Let $\mathcal{G}_x$ be the subgroup of permutations fixing $x \in X$, and let $S_0 = \{x\}, S_1, \ldots, S_n$ be the orbits in X under $\mathcal{G}_x$. Then $\mathcal{G}$ is said to be of *rank* $n + 1$.

Define the action of $\mathcal{G}$ on $X \times X$ by $g(x, y) = (g(x), g(y))$, $g \in \mathcal{G}$, $x, y \in X$. Then $X \times X$ is partitioned into the orbits $R_0 = \{(x, x): x \in X\}$, $R_1, \ldots, R_n$ under $\mathcal{G}$, and there is a 1-1-correspondence between the R_i and the S_i. $(X, R_0, \ldots, R_n)$ is a homogeneous configuration in Higman's terminology. The R_i are relations on X which however need not be symmetric. If they are symmetric the configuration is called *coherent*, and forms an association scheme with n classes. The algebra $\mathcal{A}$ can be defined even if the R_i are not symmetric. $\mathcal{A}$ is the set of all complex $v \times v$ matrices which commute with all the permutation matrices representing the elements of $\mathcal{G}$, and is called the *centralizer ring* of $\mathcal{G}$. Then $\mathcal{A}$ is commutative iff the R_i are symmetric, in which case $\mathcal{A}$ coincides with the Bose–Mesner algebra defined in §2. Higman [652] has proved

Theorem 17. $\mathcal{A}$ *is commutative if* $n \leqslant 5$

For much more about this subject see for example Cameron [232], Hestenes [644], Hestenes and Higman [645], Higman [646–653], Koornwinder [776] and Wielandt [1415].

For example, the Hamming, Johnson and symplectic schemes are coherent configurations corresponding to the following groups (i) the group of the n-cube, of order $2^n n!$, (ii) the symmetric group $\mathcal{S}_t$, and (iii) the group of all $m \times m$ invertible matrices (cf. Theorem 4 of Ch. 15).

A combinatorial problem closely related to association schemes is that of finding the largest number of lines in n-dimensional Euclidean space having a given number of angles between them – see Cameron et al. [233], Delsarte et al. [367], Hale and Shult [578], Lemmens and Seidel [812] and Van Lint and Seidel [856].

§4. For metrically regular graphs see Biggs [144–147], Damerell [327], Doob [379], Higman [649] and Smith [1240–1242]. Strongly regular graphs are described by Berlekamp et al. [128], Bose [175], Bussemaker and Seidel [221], Chakravarti et al. [260, 261], Delsarte [349], Goethals and Seidel [502, 503], Higman [646], Hubaut [672] and Seidel [1175–1177].

§§5, 8 are based on Delsarte [364], which also gives the theory of symplectic forms over $GF(q)$ (in which case B is a skew-symmetric matrix). The association scheme of bilinear forms over $GF(q)$ is described by Delsarte in [355].

§6. The Eberlein polynomials are defined in [401]. They are closely related to the dual Hahn polynomials – see Hahn [574] and Karlin and McGregor [719]. The association scheme of subspaces of a vector space is closely related to the Johnson scheme – see Delsarte [357, 361].

A preliminary version of this chapter appeared in [1229].

Tables of the best
codes known

§1. Introduction

This Appendix contains three tables of the best codes known (to us). Figure 1 is a table of upper and lower bounds on $A(n, d)$, the number of codewords in the largest possible (linear *or* nonlinear) binary code of length n and minimum distance d, for $n \leqslant 23$ and $d \leqslant 9$. Figure 2 is a more extensive table of the best codes known, covering the range $n \leqslant 512$, $d \leqslant 29$. Thus Fig. 2 gives lower bounds on $A(n, d)$. Figure 3 is a table of $A(n, d, w)$, the number of codewords in the largest possible binary code of length n, distance d, and *constant weight* w, for $n \leqslant 24$ and $d \leqslant 10$.

The purpose of these tables is to serve as a reference list of very good codes, as bench-marks for judging new codes, and as illustrations of the constructions given throughout the book. We would greatly appreciate hearing of improvements (send them to N. J. A. Sloane, Math. Research Center, Bell Labs, Murray Hill, New Jersey, 07974). Many of the lower bounds in Fig. 3 are extremely weak (or nonexistent).

General comments.

(1) It is enough to give tables for odd values of d only (*or* even values of d only) in view of Theorem 10a and Theorem 1a of Ch. 17.

(2) A code given without reference is often obtained by adding zeros to a shorter code. (Thus $A(n + 1), d) \geqslant A(n, d)$ and $A(n + 1, d, w) \geqslant A(n, d, w)$.)

(3) Some other tables worth mentioning are: Berlekamp's table [113] of selected linear codes of length < 100 and their weight distributions; Chen's table [266] of all cyclic codes of length $\leqslant 65$; Delsarte et al.'s table [368] of upper bounds on $A(n, d, w)$; Helgert and Stinaff's table [636] of upper and

n	$d = 3$	$d = 5$	$d = 7$	$d = 9$
5	4	2		
6	8^d	2		
7	16	2	2	
8	20^a	4	2	
9	38^b–40	6	2	2
10	72^b–80	12	2	2
11	144^b–160	24	4	2
12	256	32^e	4	2
13	512	64	8	2
14	1024	128	16	4
15	2048^d	256^f	32	4
16	2560^b–3276	256–340	36^h–37^i	6
17	5120^b–6552	512–680	64–74	10
18	9728^b–13104	1024–1288	128–144	20
19	19456^b–26208	2048^g–2372	256–279	40
20	·	2560^g–4096	512	40–55
21	·	4096–6942	1024	48^l–90
22	·	8192–13774	2048	64–150
23	·	16384^g–24106	4096^j	128^k–280

Fig. 1. Values of $A(n, d)$.

Key to Fig. 1.

a See p. 541 of Ch. 17.
b Constructed in §9 of Ch. 2.
d A Hamming code, §7 of Ch. 1.
e See p. 538 of Ch. 17.
f The Nordstrom–Robinson code (§8 of Ch. 2).

g See §7.3 of Ch. 18.
h A conference matrix code (§4 of Ch. 2).
i From Best et al. [140].
j The Golay code $\mathcal{G}_{24}$ (§6 of Ch. 2).
k Constructed by W. O. Alltop [26].

l From a Hadamard code $\mathcal{C}_{24}$ (§3 of Ch. 2).

lower bounds on *linear* codes of length ≤ 127; Johnson's tables [695], [696] of upper bounds on $A(n, d)$ and $A(n, d, w)$; and McEliece et al.'s table [946] of upper bounds on $A(n, d)$. Also Peterson and Weldon [1040] contains a number of useful tables.

Research Problem (A1). An extended version of Berlekamp's table would be useful, giving weight distributions of a number of the best linear and distance-invariant nonlinear codes of length up to 512.

FIGURE 2
THE BEST CODES KNOWN OF LENGTH UP TO 512
AND MINIMUM DISTANCE UP TO 29.

FOR EACH LENGTH N AND MINIMUM DISTANCE D,
THE TABLE GIVES THE SMALLEST REDUNDANCY
R = LENGTH - LOG (NUMBER OF CODEWORDS)
OF ANY KNOWN BINARY CODE.

REMARKS. LAST REVISED AUGUST 12, 1976. THE CODES NEED
NOT BE LINEAR. AN EARLIER VERSION OF THIS TABLE AP-
PEARED IN [1225].

DISTANCE D = 3
(SEE SECT. 9 OF CH 2 AND
SECT. 7.3 OF CH 18)

N		R	TYPE	REF
4-	7	3	HG	[592]
	8	3.678	SW	[509]
	9	3.752	SW	[509]
10-	11	3.830	SW	[701]
12-	15	4	HG	[592]
16-	17	4.678	SW	[1239]
18-	19	4.752	SW	[1239]
20-	23	4.830	SW	[1239]
24-	31	5	HG	[592]
32-	35	5.678	SW	[1239]
36-	39	5.752	SW	[1239]
40-	47	5.830	SW	[1239]
48-	63	6	HG	[592]
64-	71	6.678	SW	[1239]
72-	79	6.752	SW	[1239]
80-	95	6.830	SW	[1239]
96-	127	7	HG	[592]
128-	143	7.678	SW	[1239]
144-	159	7.752	SW	[1239]
160-	191	7.830	SW	[1239]
192-	255	8	HG	[592]
256-	287	8.678	SW	[1239]
288-	319	8.752	SW	[1239]
320-	383	8.830	SW	[1239]
384-	511	9	HG	[592]
	512	9.678	SW	[1239]

DISTANCE D = 5
(SEE SECT. 7.3 OF CH 18)

N		R	TYPE	REF
7-	8	6	LI	[308]
9-	11	6.415	HD	[819]
12-	15	7	PR	[1002]
16-	19	8	XP	[1237]
	20	8.678	X4	[1237]
21-	23	9	LI	[1378]
24-	32	10	B	[518]
33-	63	11	PR	[1081]
64-	70	12	X4	[1237]
71-	74	13	AL	[633]
75-	128	14	B	[1237]
129-	255	15	PR	[1081]
256-	271	16	X4	[1237]
272-	278	17	GB	[286]
279-	512	18	B	[1237]

DISTANCE D = 7

N	R	TYPE	REF
10- 11	9	LI	[308]
12- 15	10	RM	
16	10.830	CM	[1239]
17- 23	11	GO	[506]
24	12		
25- 27	13	DC	[715]
28- 30	14	DC	[715]
31- 35	15	LI	[1094]
36- 63	16	IM	[495]
64	17		
65- 67	18	XC	[1237]
68- 72	19	LI	[625]
73- 87	20	ZV	[1470]
88-128	21	GP	[536]
129-255	22	IM	[495]
256-257	23- 24		
258-265	25	GP	[1291]
266-311	26	SV	[631]
312-512	27	GP	[536]

DISTANCE D = 9

N	R	TYPE	REF
13- 14	12	LI	[308]
15	13		
16	13.415	HD	[819]
17- 19	13.678	HD	[819]
20	14.678		
21	15.415	HD	[819]
22- 23	16	LI	[26]
24- 26	17	LI	[625]
27- 30	18	PT	[1047]
31- 35	18.415	Y2	[1237]
36	19.415		
37- 41	20	QR	[113]
42- 45	21	QR	
46- 47	22	LI	[625]
48- 49	22.193	Y2	[1237]
50- 52	23	SV	[631]
53- 73	24	B	[741]
74	25		
75- 76	26	XQ	[721]
77- 91	27	AL	[633]
92-128	28	B	[1237]
129-135	29	XB	[1225]
136-142	30	XQ	[721]
143-167	31	AL	[633]
168-256	32	B	[1237]
257-265	33	GP	[1291]
266-274	34	GP	[1291]
275-311	35	SV	[631]
312-512	36	B	[1237]

DISTANCE D = 11

N	R	TYPE	REF
16- 17	15	LI	[308]
18	16		
19	16.415	HD	[819]
20	17	B	[518]
21- 23	17.415	HD	[819]
24	18.300	CM	[1238]
25- 26	19	HS	[636]
27- 31	20	B	
32	21		
33- 35	22	Y1	[1237]
36- 47	23	QR	
48- 50	24- 26		
51- 63	27	B	
64- 67	28	XB	[1225]
68	29		
69- 71	30	ZV	[1470]
72	31		
73- 77	32	GP	[1291]
78- 89	33	CY	[1083A]
90- 96	34	GB	[286]
97-128	35	GP	[536]
129-135	36	XB	[1225]
136-142	37	XQ	[721]
143-149	38	GP	[1291]
150-191	39	Y1	[637]
192-256	40	GP	[536]
257-264	41	XB	[1225]
265-272	42	XQ	[721]
273-280	43	GP	[1291]
281-311	44	SV	[631]
312-512	45	GP	[536]

DISTANCE D = 13

N	R	TYPE	REF
19- 20	18	LI	[308]
21- 22	19- 20		
23	20.415	HD	[819]
24	21	BV	[102]
25- 27	21.193	HD	[819]
28	22.093	CM	[1238]
29	23	RM	
30- 31	24	GP	[625A]
32- 34	25	PT	[1047]
35- 37	26	XB	[1225]
38	27		
39- 43	28	CY	[113]
44- 45	29- 30		
46- 55	31	Y2	[1237]
56	32		
57- 63	33	B	
64- 70	34	XB	[1225]
71- 73	35	XQ	[721]
74	36		
75- 77	37	QR	[715]
78- 79	38	KS	[722]
80- 90	39	CY	[1083A]
91	40		
92- 99	41	GB	[286]
100-128	42	B	[1237]
129-135	43	XB	[1225]
136-142	44	XQ	[721]
143-149	45	GP	[1291]
150-156	46	GP	[1291]
157-191	47	Y1	[637]
192-256	48	B	[1237]
257-264	49	XB	[1225]
265-272	50	XQ	[721]
273-280	51	GP	[1291]
281-288	52	GP	[1291]
289-327	53	Y1	[637]
328-512	54	B	[1237]

DISTANCE D = 15

N	R	TYPE	REF
22- 23	21	LI	[308]
24- 25	22- 23		
26	23.415	HD	[819]
27	24	BV	[102]
28	24.678	HD	[819]
29- 31	25	RM	
32	26		
33	26.830	HD	[819]
34	27.752	CM	[1238]
35	28	Z	[1225]
36- 39	29	GP	[1291]
40- 41	30	XC	[1237]
42- 44	31	LI	[859A]
45- 47	32	AX	[1237]
48- 50	33	CY	[113]
51- 55	34	CY	[113]
56- 63	35	CY	[744]
64- 72	36	CY	[1083A]
73- 74	37- 38		
75- 79	39	QR	[715]
80- 81	40- 41		
82- 87	42	QR	[715]
88- 90	43- 45		
91- 92	46	CY	[1083A]
93- 95	46.678	KS	[722]
96- 99	47	QR	[715]
100-101	47.678	KS	[722]
102-104	48	GB	[286]
105-128	49	GP	[536]
129-135	50	XB	[1225]
136-142	51	XQ	[721]
143-149	52	GP	[1291]
150-156	53	GP	[1291]
157-163	54	GP	[1291]
164-191	55	Y1	[637]
192-256	56	GP	[536]
257-264	57	XB	[1225]
265-272	58	XQ	[721]
273-280	59	GP	[1291]
281-288	60	GP	[1291]
289-296	61	GP	[1291]
297-327	62	Y1	[637]
328-512	63	GP	[536]

DISTANCE D = 17

N	R	TYPE	REF
25- 26	24	LI	[308]
27- 28	25- 26		
29	26.415	HD	[819]
30	27.415		
31	28	BV	[102]
32	28.415	HD	[819]
33- 35	28.830	HD	[819]
36	29.752	CM	[1238]
37	30.678	HD	[819]
38	31.608	CM	[1238]
39	32.541	HD	[819]
40- 41	33	GP	[26]
42	34		
43- 44	35	GP	[1291]
45- 46	36	XQ	[721]
47- 49	37	LI	[717]
50	38		
51- 53	39	LI	[715]
54- 55	39.142	Y3	[1237]
56	40	Y1	[1237]
57- 62	41	CY	[266]
63- 66	42	XC	[1237]
67- 71	43	Y1	[1237]
72- 89	44	QR	[715]
90- 93	45- 48		
94-101	49	QR	[715]
102	50		
103-105	51	DC	[715]
106-107	52- 53		
108-125	54	B	
126	55		
127-128	56	B	[1237]
129-135	57	XB	[1225]
136-142	58	XQ	[721]
143-149	59	GP	[1291]
150-156	60	GP	[1291]
157-163	61	GP	[1291]
164-170	62	GP	[1291]
171-191	63	Y1	[637]
192-256	64	B	[1237]
257-264	65	XB	[1225]
265-272	66	XQ	[721]
273-280	67	GP	[1291]
281-288	68	GP	[1291]
289-296	69	GP	[1291]
297-304	70	GP	[1291]
305-331	71	Y1	[637]
332-512	72	B	[1237]

DISTANCE D = 19

N	R	TYPE	REF
28- 29	27	LI	[308]
30- 32	28- 30		
33	30.415	HD	[819]
34	31	CY	[266]
35	31.678	HD	[819]
36	32.415	HD	[819]
37- 39	32.678	HD	[819]
40	33.608	CM	[1238]
41	34.541	HD	[819]
42	35.541		
43- 44	36	AX	[26]
45	37		
46- 48	38	HS	[636]
49- 51	39	LI	[717]
52	40		
53- 55	41	Y2	[1237]
56- 59	42	DG	[364]
60- 61	43	B	
62- 63	44	CY	[266]
64- 72	45	CY	[1083A]
73- 76	46- 49		
77- 83	50	Y.1	[1237]
84-103	51	QR	[715]
104	52		
105-107	53	DC	[715]
108-109	54- 55		
110-127	56	B	
128-129	57- 58		
130-131	59	GP	[1291]
132-135	60	GP	[1291]
136-139	61	XC	[1237]
140-141	62- 63		
142-143	64	NL	[1291]
144-147	65	NL	[1291]
148-151	66	GP	[1291]
152-191	67	Y1	[637]
192-255	68	B	
256-260	69	XB	[1225]
261-263	70- 72		
264-271	73	GP	[1291]
272-273	74	GP	[1291]
274-280	75	GP	[1291]
281-288	76	GP	[1291]
289-296	77	GP	[1291]
297-304	78	GP	[1291]
305-312	79	GP	[1291]
313-339	80	Y1	[637]
340-512	81	GP	[536]

DISTANCE D = 21

N	R	TYPE	REF
31- 32	30	LI	[308]
33- 35	31- 33		
36	33.415	HD	[819]
37	34.415		
38	35	BV	[102]
39	35.678	HD	[819]
40	36.193	HD	[819]
41- 43	36.541	HD	[819]
44	37.541		
45	38.415	HD	[819]
46- 47	39	GP	[26]
48- 50	40- 42		
51- 53	43	ZV	[1470]
54- 57	43.415	Y2	[1237]
58- 61	44	DG	[364]
62- 63	45	B	
64- 67	46- 49		
68- 69	50	GP	[1291]
70	51		
71- 72	52	GP	[1291]
73- 77	53	XC	[1237]
78	54		
79- 86	55	B	[618]
87- 90	56- 59		
91- 95	60	ZV	[1470]
96	61		
97-101	61.678	ZV	[1470]
102-105	62	Y1	[637]
106-127	63	B	
128-135	64	XB	[1225]
136-138	65- 67		
139-142	68	GP	[1291]
143-146	69	XQ	[721]
147-148	70- 71		
149-150	72	NL	[1291]
151-154	73	NL	[1291]
155-158	74	XQ	[721]
159-191	75	Y1	[637]
192-255	76	B	
256-264	77	XB	[1225]
265-268	78	XQ	[721]
269-270	79- 80		
271-272	81	X3	CH.18
273-279	82	GP	[1291]
280-281	83	GP	[1291]
282-288	84	GP	[1291]
289-296	85	GP	[1291]
297-304	86	GP	[1291]
305-312	87	GP	[1291]
313-320	88	GP	[1291]
321-339	89	Y1	[637]
340-512	90	B	[1237]

DISTANCE D = 23

N	R	TYPE	REF
34- 35	33	LI	[308]
36- 38	34- 36		
39	36.415	HD	[819]
40	37.415		
41	38	BV	[102]
42	39		
43	39.415	HD	[819]
44	40	CY	[266]
45- 47	40.415	HD	[819]
48	41.356	CM	[1238]
49- 50	42	CY	[266]
51- 52	43- 44		
53- 55	45	ZV	[1470]
56- 63	46	DG	[364]
64	47		
65- 66	48	XB	[1225]
67- 72	49- 54		
73- 74	55	ZV	[1470]
75	56		
76- 87	57	LI	[715]
88- 91	58- 61		
92- 95	62	ZV	[1470]
96- 98	63- 65		
99-101	65.678	ZV	[1470]
102-104	66	ZV	[1470]
105-111	67	ZV	[1470]
112-113	68- 69		
114-127	70	B	
128-135	71	XB	[1225]
136-142	72	XQ	[721]
143-145	73- 75		
146-149	76	GP	[1291]
150-153	77	GP	[1291]
154-155	78- 79		
156-161	80	KS	[722]
162-167	81	KS	[722]
168-173	82	KS	[722]
174-207	83	Y1	[637]
208-255	84	B	
256-264	85	XB	[1225]
265-272	86	XQ	[721]
273-276	87	GP	[1291]
277-279	88- 90		
280-287	91	GP	[1291]
288-289	92	GP	[1291]
290-296	93	GP	[1291]
297-304	94	GP	[1291]
305-312	95	GP	[1291]
313-320	96	GP	[1291]
321-328	97	GP	[1291]
329-347	98	Y1	[637]
348-512	99	GP	[536]

DISTANCE D = 25

N	R	TYPE	REF
37- 38	36	LI	[308]
39- 42	37- 40		
43	40.415	HD	[819]
44	41.415		
45	42	BV	[102]
46	42.678	HD	[819]
47	43.415	HD	[819]
48	44	HD	[819]
49- 51	44.300	HD	[819]
52	45.245	CM	[1238]
53	46.193	HD	[819]
54	47.193		
55	48	LI	[25]
56- 61	49	DG	[758]
62- 63	50- 51		
64- 66	52	LI	[715]
67	53		
68- 71	54	DG	CH.18
72- 73	55- 56		
74- 75	57	X	[26]
76- 78	58- 60		
79- 81	61	PT	[1047]
82	62		
83- 86	63	B	[637]
87- 91	64- 68		
92- 93	69	LI	CH.10
94	70		
95- 99	71	GP	[1291]
100-101	72	KS	[722]
102	73		
103-109	74	Y1	[637]
110-125	75	B	
126-127	76- 77		
128-135	78	XB	[1225]
136-142	79	XQ	[721]
143-149	80	GP	[1291]
150-152	81- 83		
153-156	84	GP	[1291]
157-161	85	KS	[722]
162-167	86	KS	[722]
168-173	87	KS	[722]
174-179	88	KS	[722]
180-185	89	KS	[722]
186-191	90	KS	[722]
192-207	91	Y1	[637]
208-255	92	B	
256-264	93	XB	[1225]
265-272	94	XQ	[721]
273-280	95	GP	[1291]
281-284	96	GP	[1291]
285-287	97- 99		
288-295	100	GP	[1291]
296-297	101	GP	[1291]
298-304	102	GP	[1291]
305-312	103	GP	[1291]
313-320	104	GP	[1291]
321-328	105	GP	[1291]
329-336	106	GP	[1291]
337-347	107	Y1	[637]
348-512	108	B	[1237]

DISTANCE D = 27

N	R	TYPE	REF
40- 41	39	LI	[308]
42- 45	40- 43		
46	43.415	HD	[819]
47	44.415		
48	45	CY	[266]
49	46		
50	46.678	HD	[819]
51	47.193	HD	[819]
52	47.830	HD	[819]
53- 55	48.193	HD	[819]
56	49.193		
57	50.093	HD	[819]
58- 63	51	DG	[758]
64- 67	52- 55		
68- 73	56	DG	CH.18
74- 77	57- 60		
78- 79	61	NL	[1291]
80- 81	62- 63		
82- 84	64	CY	[1083A]
85- 88	65	B	[637]
89- 93	66- 70		
94- 95	71	LI	CH.10
96- 97	72- 73		
98- 99	74	ZV	[1470]
100-103	75	Y4	[637]
104-111	76	Y1	[637]
112-127	77	B	
128-129	78- 79		
130-131	80	GP	[1291]
132-135	81	GP	[1291]
136-139	82	XC	[1237]
140-141	83- 84		
142-143	85	NL	[1291]
144-147	86	NL	[1291]
148-151	87	GP	[1291]
152-156	88	GP	[1291]
157-158	89- 90		
159-167	91	KS	[722]

N	R	TYPE	REF
168-173	92	KS	[722]
174-179	93	KS	[722]
180-185	94	KS	[722]
186-191	95	KS	[722]
192-197	96	KS	[722]
198-199	97- 98		
200-207	99	Y1	[637]
208-255	100	B	
256-264	101	XB	[1225]
265-272	102	XQ	[721]
273-280	103	GP	[1291]
281-288	104	GP	[1291]
289-292	105	GP	[1291]
293-295	106-108		
296-303	109	GP	[1291]
304-305	110	GP	[1291]
306-312	111	GP	[1291]
313-320	112	GP	[1291]
321-328	113	GP	[1291]
329-336	114	GP	[1291]
337-344	115	GP	[1291]
345-352	116	GP	[1291]
353-512	117	GP	[536]

DISTANCE D = 29

N	R	TYPE	REF
43- 44	42	LI	[308]
45- 48	43- 46		
49	46.415	HD	[819]
50	47.415		
51	48.415		
52	49	BV	[102]
53	49.678	HD	[819]
54	50.415	HD	[819]
55	51.193	HD	[819]
56	51.678	HD	[819]
57- 59	52.093	HD	[819]
60	53.046	CM	[1238]
61	54	RM	
62	55		
63	55.913	HD	[819]
64	56.913		
65- 69	57	X	CH.18
70- 74	58- 62		
75- 77	62.752	ZV	[1470]
78	63.752		
79	64	NL	[1291]
80- 81	65	GP	[1291]
82- 83	66- 67		
84- 86	68	GP	[1291]
87	69		
88- 89	70	ZV	[1470]
90	71		
91- 95	72	ZV	[1470]
96- 99	73- 76		
100-101	77	LI	CH.10
102-104	78	ZV	[1470]
105	79		
106-111	80	ZV	[1470]
112	81		
113-125	82	B	[744]
126-127	83- 84		
128-135	85	XB	[1225]
136-138	86- 88		
139-142	89	GP	[1291]
143-146	90	XQ	[721]
147-148	91- 92		
149-150	93	NL	[1291]
151-154	94	NL	[1291]
155-158	95	GP	[1291]
159-163	96	GP	[1291]
164-167	97	KS	[722]
168-179	98	KS	[722]
180-185	99	KS	[722]
186-191	100	KS	[722]
192-197	101	KS	[722]
198-202	102-106		
203-207	107	Y1	[637]
208-255	108	B	
256-264	109	XB	[1225]
265-272	110	XQ	[721]
273-280	111	GP	[1291]
281-288	112	GP	[1291]
289-296	113	GP	[1291]
297-300	114	GP	[1291]
301-303	115-117		
304-311	118	GP	[1291]
312-313	119	GP	[1291]
314-320	120	GP	[1291]
321-328	121	GP	[1291]
329-336	122	GP	[1291]
337-344	123	GP	[1291]
345-352	124	GP	[1291]
353-360	125	GP	[1291]
361-512	126	B	[1237]

THE NUMBER OF CODEWORDS IN FIG. 2

IF THE REDUNDANCY $R = I + X$ WHERE I IS AN INTEGER AND X IS BETWEEN 0 AND 1, THE NUMBER OF CODEWORDS IS
$$F.2**(N-I-5)$$
WHERE F IS GIVEN BY

X:	0	.046	.039	.142	.193	.245	300	.356
F:	32	31	30	29	28	27	26	25

X:	.415	.476	.541	.608	.678	.752	.830	.913
F:	24	23	22	21	20	19	18	17

TYPES OF CODES IN FIG. 2

AL = ALTERNANT CODE (LINEAR) (CH. 12, [633])

AX = THE |A+X|B+X|A+B+X| CONSTRUCTION (LINEAR) (SECT. 7.4 OF CH. 18, [1237])

B = BCH OR SHORTENED BCH CODE (LINEAR) (CH. 3,7,9)

BV = BELOV ET AL.'S LINEAR CODES WHICH MEET THE GRIESMER BOUND (SECT. 6 OF CH. 17, [102])

CM = CONFERENCE MATRIX CODE (NONLINEAR) (SECT. 4 OF CH. 2, [1238])

CY = CYCLIC OR SHORTENED CYCLIC LINEAR CODE.

DC = DOUBLE CIRCULANT CODE (LINEAR) (SECT. 7 OF CH. 16, [715])

DG = DELSARTE-GOETHALS GENERALIZED KERDOCK CODE (NONLINEAR) (SECT. 5 OF CH. 15, [364])

GB = GENERALIZED BCH CODE (LINEAR) (SECT. 7 OF CH. 12, [286])

GO = GOLAY CODE (LINEAR) (SECT. 6 OF CH. 2, [506])

GP = GOPPA OR MODIFIED GOPPA CODE (LINEAR) (SECT. 3 OF CH. 12, [536], [1291])

HD = HADAMARD MATRIX CODE (NONLINEAR) (SECT. 3 OF CH. 2, [819])

HG = HAMMING CODE (LINEAR) (SECT. 7 OF CH. 1, [592])

HS = HELGERT AND STINAFF'S CONSTRUCTION A (SECT. 9.2 OF CH. 18, [636])

IM = GOETHALS NONLINEAR CODE I(M) (SECT. 7 OF CH. 15, [495])

KS = KASAHARA ET AL. S MODIFIED CONCATENATED CODES (LINEAR OR NONLINEAR) (SECT. 8.1 OF CH. 18, [722])

LI = LINEAR CODE.

NL = NONLINEAR CODE.

PR = PREPARATA CODE (NONLINEAR) (SECT. 6 OF CH. 15, [1081])

PT = PIRET'S CONSTRUCTION (LINEAR) (SECT. 7.5 OF CH. 18, [1047])

QR = QUADRATIC RESIDUE CODE (LINEAR) (CH. 16)

RM = REED-MULLER OR SHORTENED REED-MULLER CODE (LINEAR) (CH. 13)

SV = SRIVASTAVA CODE (LINEAR) (SECT. 6 OF CH. 12)

SW = NONLINEAR SINGLE-ERROR-CORRECTING CODE (SECT. 9 OF CH. 2, [1239])

X = CONSTRUCTION X (LINEAR OR NONLINEAR) (SECT. 7.7 OF CH. 18, [1237])

XB = APPLY CONSTRUCTION X TO A BCH CODE (LINEAR) (SECT. 7.1 OF CH. 18, [1237])

XC = APPLY CONSTRUCTION X TO A CYCLIC CODE (LINEAR) (SECT. 7.1 OF CH. 18, [1237])

XP = APPLY CONSTRUCTION X TO A PREPARATA CODE (NONLINEAR) (SECT. 7.1 OF CH. 18, [1237])

XQ = KASAHARA ET AL.'S EXTENDED BCH CODES (LINEAR) (PROBLEM 14 OF CH. 18, [721])

X3 = CONSTRUCTION X3 (LINEAR OR NONLINEAR) (PROBLEM 14 OF CH. 18)

X4 = CONSTRUCTION X4 (LINEAR OR NONLINEAR) (SECT. 7.2 OF CH. 18, [1237])

Y1 = CONSTRUCTION Y1 (LINEAR) (SECT. 9.1 OF CH. 18.)

Y2 = CONSTRUCTION Y2 (NONLINEAR) (PROBLEM 29 OF CH. 18)

Y3 = CONSTRUCTION Y3 (NONLINEAR) (PROBLEM 29 OF CH. 18)

Y4 = CONSTRUCTION Y4 (LINEAR) (PROBLEM 30 OF CH. 18, [637])

Z = THE |U|U+V| CONSTRUCTION (LINEAR OR NONLINEAR) (SECT. 9 OF CH. 2, [1239])

ZV = ZINOVIEV'S CONSTRUCTION (LINEAR OR NONLINEAR) (SECT. 8.2 OF CH. 18, [1470])

§2. Figure 1, a small table of $A(n, d)$

Figure 1 gives upper and lower bounds on $A(n, d)$. The Plotkin–Levenshtein theorem (Theorem 11 of Ch. 17) gives all codes in this figure on and above the line $n = 2d + 1$. Below the line $n = 2d + 1$ unless indicated otherwise the upper bounds are obtained by linear programming (§4 of Ch. 17) or Theorem 10b of Ch. 17. The lower bounds are continued in Fig. 2.

Research Problem (A2). Of course all the undecided entries in these figures are research problems. One particularly interesting code which might exist is a $(20, 4096, 5)$ code, corresponding to $A(20, 5) \gtrless 4096$. The weight distribution of the $(21, 4096, 6)$ extended code would be, from linear programming (§4 of Ch. 17): $B_0 = 1$, $B_6 = 314$, $B_8 = 595$, $B_{10} = 1596$, $B_{12} = 1015$, $B_{14} = 490$, $B_{16} = 84$, $B_{20} = 1$.

Distance 4: $A(n, 4, w)$

n/w	2	3[a]	4[a]	5	6	7	8	9	10	11	12
4	2	1	1								
5	2	2	1	1							
6	3	4	3	1	1						
7	3	7	7	3	1	1					
8	4	8	14	8	4	1	1				
9	4	12	18	18	12	4	1	1			
10	5	13	30	36	30	13	5	1	1		
11	5	17	35	66	66	35	17	5	1	1	
12	6	20	51	[b]73–84	[c]132	73–84	51	20	6	1	1
13	6	26	65	[b]99–132	[d]143–182	143–182	99–132	65	26	6	1
14	7	28	91	[d]143–182	[d]210–308	[d]216–364	210–308	143–182	91	28	7
15	7	35	105	[d]213–272[f]	321–455	[d]435–660	435–660	321–455	213–273	105	35
16	8	37	140	[d]305–336	513–725	?–1040	?–1320	?–1040	513–725	305–336	140
17	8	44	154–157	[d]424–476	792–952	?–1760	?–2210	?–2210	?–1760	792–952	424–476

Fig. 3. A table of $A(n, d, w)$. For the latest versions of the tables of $A(n, d, w)$ see "Lower bounds for constant weight codes" by R. L. Graham and N. J. A. Sloane, IEEE Trans. Info. Theory, vol. IT-26 (January 1980) 37–43.

Distance 4: $A(n, 4, w)$

18	9	48	[a]198	[e]480–565	[d]1188–1428	?–2448	?–3960	?–4420	?–3960	?–2448	1188–1428
19	9	57	228	612–752	1428–1789	?–3876	?–5814	?–8360	?–8360	?–5814	?–3876
20	10	60	285	816–912	[c]2040–2506	?–5111	?–9690	?–12920	?–16720	?–12920	?–9690
21	10	70	315	[c]1071–1197	[c]2856–3192	?–7518	?–13416	?–22610	?–27132	?–27132	?–22610
22	11	73	385	1386	3927–4389	?–10032	?–20674	?–32794	?–49742	?–54264	?–49742
23	11	83	415–419	1771	5313	?–14421	?–28842	?–52833	?–75426	?–104006	?–104006
24	12	88	498	[d]1859–2011	[e]7084	?–18216	?–43263	?–76912	?–126799	?–164565	?–208012

(*Fig. 3. contd.*)

Distance 6: $A(n, 6, w)$

n/w	2	3	4	5	6	7	8	9	10	11	12
6	1	2	1	1	1						
7	1	2	2	1	1	1					
8	1	2	2	2	1	1	1				
9	1	3	3	3	3	1	1	1			
10	1	3	5	6	5	3	1	1	1		
11	1	3	6	11	11	6	3	1	1	1	
12	1	4	9	12^q	22	12	9	4	1	1	1
13	1	4	$^g13$	$^h18^q$	$^g26$	26	18	13	4	1	1
14	1	4	14	$^h28$	$^h42$	$^d42$–51^r	42	28	14	4	1
15	1	5	15	$^h42$	$^h70$	$^i60$–88^L	60–88	70	42	15	5
16	1	5	20	48	$^h112$	$^i90$–156	$^i120$–150^L	90–156	112	48	20
17	1	5	20^d	$^e68$	112–136	112–244^L	125–283	125–283	112–244	112–136	68
18	1	6	$^b22$	68–72	$^d144$–203^f	$^i160$–349	$^i232$–428^L	249–425^L	232–428^L	160–349	144–203
19	1	6	$^c25^s$	$^i72$–83	$^i172$–228	$^i228$–520^L	$^i332$–739	$^d472$–789^L	472–789	332–739	228–520
20	1	6	$^a30$	$^i81$–100	$^i232$–276	310–651	$^i492$–1199^L	$^d672$–1363^L	$^d944$–1421^L	672–1363	492–1199

(*Fig. 3. contd.*)

686

Distance 8: $A(n, 8, w)$

n/w	2	3	4	5	6	7	8	9	10	11	12
8	1	1	2	1	1	1	1				
9	1	1	2	2	1	1	1	1			
10	1	1	2	2	2	1	1	1	1		
11	1	1	2	2	2	2	1	1	1	1	
12	1	1	3	3	4	3	3	1	1	1	1
13	1	1	3	3	4	4	3	3	1	1	1
14	1	1	3	4	7	8	7	4	3	1	1
15	1	1	3	6	[k]10	15	15	10	6	3	1
16	1	1	4	6	[k]16	16–22	30	16–22	16	6	4
17	1	1	4	7	[b]17	[i]21–31[L]	[m]34–35	34–35	21–31	17	7
21	1	7	[a]31	[d]102–126	[i]253–350	465–828	[d]668–1708	[d]1068–2364[L]	[d]1286–2702[L]	1286–2702	1068–2364
22	1	7	[a]37	[d]132–136	294–462	675–1100	[d]708–2277	[d]1288–3775[L]	[d]1450–4416[L]	[d]1574–5064[L]	1450–4416
23	1	7	[a]40	[d]147–170	399–521	969–1518	?–3162	?–5819	?–7521[L]	?–7953[L]	?–7953
24	1	8	[e]42	[e]168–192	[e]532–680	[e]1368–1786	?–4554	?–8432	?–12418[L]	?–14682	?–15906[L]

(Fig. 3. contd.)

Distance 8: $A(n, 8, w)$

n											
18	1	1	4	[i]9	[i]20–21	[i]33–41[L]	[i]46–63	[i]48–70	46–63	33–41	20–21
19	1	1	4	[i]12	[i]28	[i]52–57	[i]78–97	[i]88–122[L]	88–122	78–97	52–57
20	1	1	5	[i]16	[i]40	[i]80	[i]130–142	[i]160–215	[i]176–244[L]	160–215	130–142
21	1	1	5	[i]21	[i]56	[i]120	[i]210	[i]280–331	[i]336–399[L]	336–399	280–331
22	1	1	5	21[b]	[i]77	[i]176	[i]330	280–497[L]	[i]616–728	[i]672–798	616–728
23	1	1	5	[g]23	77–80	[i]253	[i]506	[i]400–816	616–1111[L]	[i]1288–1417[L]	1288–1417
24	1	1	6	°24	77–92	253–274	[i]759	[i]640–1160[L]	[i]960–1639[L]	1288–2305[L]	[i]2576

(Fig. 3. contd.)

Distance 10: $A(n, 10, w)$

n/w	2	3	4	5	6	7	8	9	10	11	12
10	1	1	1	2	1	1	1	1	1		
11	1	1	1	2	2	1	1	1	1	1	
12	1	1	1	2	2	2	1	1	1	1	1
13	1	1	1	2	2	2	2	1	1	1	1
14	1	1	1	2	2	2	2	2	1	1	1
15	1	1	1	3	3	3	3	3	3	1	1
16	1	1	1	3	3	3	4	3	3	3	1
17	1	1	1	3	3	5	6	6	5	3	3
18	1	1	1	3	4	6	9	10	9	6	4
19	1	1	1	3	4	8	[b]12	19	19	12	8
20	1	1	1	4	5	[n]10	[b]17–18	[g]20–24	38	20–24	17–18
21	1	1	1	4	7	13[b]	[g]21–26	[g]21–41[L]	38–49[L]	38–49	21–41
22	1	1	1	4	7	15–19	[g]22–35	[g]22–57[L]	?–74[L]	?–82[L]	?–74
23	1	1	1	4	8	[b]16–23	[g]23–50[L]	[g]23–87[L]	?–117[L]	?–135[L]	?–135
24	1	1	1	4	9	[g]24–27	[g]24–69	?–119[L]	?–171[L]	?–223[L]	?–247[L]

(Fig. 3. contd.)

689

Key to Fig. 3

a Theorems 6, 7 of Ch. 17.
b Shen Lin [831].
c H. R. Phinney [1043].
d A. E. Brouwer [201a].
e From Problem 6 of Ch. 17 and the Steiner systems $S(5, 6, 12)$, $S(3, 5, 17)$, $S(3, 6, 26)$, $S(5, 6, 24)$, $S(5, 7, 28)$ (Chen [277], Denniston [372], Doyen and Rosa [386]).
f From the nonexistence of Steiner systems $S(4, 5, 15)$, $S(4, 6, 18)$ ([386]).
g A cyclic code.
h From the t-design 3-(16, 6, 4) (from the Nordstrom–Robinson code).
i From translates of the Nordstrom–Robinson code (see Fig. 6.2 and [201a]).
j From the [24, 12, 8] Golay code.
k See Problem 1.
L Linear programming bound (§4 of Ch. 17).
m From a conference matrix.
n A quasi-cyclic code.
q Problems 7 and 8 of Ch. 17.
r W. G. Valiant, personal communication.
s D. Stinson, personal communication.
t See [140].

§3. Figure 2, an extended table of the best codes known

Let M be the size of the largest known (linear *or* nonlinear) binary code of length $N \leqslant 512$ and minimum distance $D \leqslant 29$. Figure 2 tabulates the redundancy

$$R = N - \log_2 M$$

of this code.

Since the code may be nonlinear, R need not be an integer. A small table below Figure 2 makes it easy to find M given R. If $R = I + X$, where I is an integer and $0 \leqslant X < 1$, the number of codewords is

$$M = F \cdot 2^{N-I-5}.$$

where F is given in the table. For example, consider the entry

$$N = 8, \qquad R = 3.678, \qquad \text{Type} = SW, \quad \text{Ref. [509]}$$

near the beginning of the figure. Here $I = 3$, $X = 0.678$, $F = 20$ and the number of codewords is

$$M = 20 \cdot 2^0 = 20.$$

This is an $(8, 20, 3)$ code, of type SW (see the list of types below the figure), found by Golay [509] – see §7 of Chapter 2.

To save space, several entries have been compressed into one. For example, under Distance $D = 11$, the entry

$$48\text{–}50 \qquad 24\text{–}26$$

is a contraction of

48	24
49	25
50	26.

Shortened codes have been given the same name as the parent code. For example CY denotes a cyclic code or a shortened cyclic code.

§4. Figure 3, a table of $A(n, d, w)$

This table is important because it leads to bounds on $A(n, d)$ (§3 of Ch. 17), and in its own right for providing constant weight codes (§2 of Ch. 17) and as the solution to a packing problem. The packing problem is sometimes stated as follows. Find $D(t, k, n)$, the maximum number of k-sets from an n-set S such that every t-set from S is contained in at most one k-set (see for example Gardner [467], Kalbfleisch et al. [708–711], Mills [957–959], Stanton [1265], Stanton et al. [1267–1271] and Swift [1295]. See also the football pool problem mentioned in Ch. 7. Since $D(t, k, n) = A(n, 2k - 2t + 2, k)$, Fig. 3 is also a table of $D(t, k, n)$.

Unless indicated otherwise the upper bounds in Fig. 3 are from the Johnson bounds Theorems 1–4 and Problem 2 of Ch. 17, or from Problem 4 of Ch. 17. Unmarked lower bounds are also from Theorem 4 and Problem 2 of Ch. 17, or can be found by easy constructions. The small letters are explained in the key on p. 690. Letters on the left refer to lower bounds, on the right to upper bounds. There may be several ways to construct one of these codes, but only the simplest is mentioned. A slightly later version of this table will appear in Best et al. [140].

Problem. (1)

Use the last row of Fig. 5 of Ch. 14 to show that $A(16, 8, 6) = 16$ and $A(15, 8, 6) = 10$.

Research Problem (A3). A 2-(17, 6, 15) design certainly exists (Hanani [600]). But is there a 2-$(17, 6, 15)$ where the blocks form a code proving that $A(17, 6, 6) = 136$?

Research Problem (A4). Some more constant weight codes which might exist, together with their distance distributions (found by linear programming)

$$A(21, 4, 5) \stackrel{?}{=} 1197 \ (B_0 = 1, B_4 = 80, B_6 = 320, B_8 = 540, B_{10} = 256),$$

$$A(16, 6, 8) \stackrel{?}{=} 150 \ (B_0 = 1, B_6 = 64, B_8 = 20, B_{10} = 64, B_{16} = 1).$$

Research Problem (A5). Is it true that if $w_1 < w_2 \leqslant \frac{1}{2} n$ then $A(n, d, w_1) \leqslant A(n, d, w_2)$?

Finite geometries

§1. Introduction

Finite geometries are large combinational objects just as codes are, and therefore it is not surprising that they have turned up in many chapters of this book (see especially Ch. 13). In this Appendix we sketch the basic theory of these geometries, beginning in §2 with the definitions of projective and affine geometries. The most important examples (for us) are the projective geometry $PG(m, q)$ and the affine or Euclidean geometry $EG(m, q)$ of dimension m constructed from a finite field $GF(q)$. For dimension $m \geqslant 3$ there are no other geometries (Theorem 1).

In §3 we study some of the properties of $PG(m, q)$ and $EG(m, q)$, especially their collineation groups (Theorem 7 and Corollary 9) and the number of subspaces of each dimension (Theorems 4–6).

In dimension 2 things are more complicated. A projective *plane* is equivalent to a Steiner system $S(2, n + 1, n^2 + n + 1)$, for some $n \geqslant 2$, and an affine plane to an $S(2, n, n^2)$ for some $n \geqslant 2$ (Theorem 10). But now other kinds of planes exist besides $PG(2, q)$ and $EG(2, q)$ – see §4.

§2. Finite geometries, $PG(m, q)$ and $EG(m, q)$

Definition. A finite *projective geometry* consists of a finite set Ω of *points* $p, q, \ldots$ together with a collection of subsets $L, M, \ldots$ of Ω called *lines*, which satisfies axioms (i)–(iv). (If $p \in L$ we say that p lies on L or L passes through p.)

(i) There is a unique line (denoted by (pq)) passing through any two distinct points p and q.

(ii) Every line contains at least 3 points.

(iii) If distinct lines L, M have a common point p, and if q, r are points of L not equal to p, and s, t are points of M not equal to p, then the lines (qt) and (rs) also have a common point (see Fig. 1).

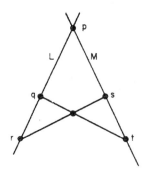

Fig. 1. Axiom (iii).

(iv) For any point p there are at least two lines not containing p, and for any line L there are at least two points not on L.

A *subspace* of the projective geometry is a subset S of Ω such that

(v) If p, q are distinct points of S then S contains all points of the line (pq).

Examples of subspaces are the points and lines of Ω and Ω itself. A *hyperplane H* is a maximal proper subspace, so that Ω is the only subspace which properly contains H.

Definition. An *affine* or *Euclidean geometry* is obtained by deleting the points of a fixed hyperplane H (called the hyperplane at infinity) from the subspaces of a projective geometry. The resulting sets are called the subspaces of the affine geometry.

A set T of points in a projective or affine geometry is called *independent* if, for every $x \in T$, x does not belong to the smallest subspace which contains $T - \{x\}$. For example, any three points not on a line are independent. The *dimension* of a subspace S is $r - 1$, where r is the size of the largest set of independent points in S. In particular, if $S = \Omega$ this defines the dimension of the projective geometry.

The projective geometry $\mathrm{PG}(m, q)$. The most important examples of projective and affine geometries are those obtained from finite fields.

Let $\mathrm{GF}(q)$ be a finite field (see Chs. 3, 4) and suppose $m \geq 2$. The points of Ω are taken to be the nonzero $(m + 1)$-tuples

$$(a_0, a_1, \ldots, a_m), \quad a_i \in \mathrm{GF}(q),$$

with the rule that

$$(a_0, \ldots, a_m) \quad \text{and} \quad (\lambda a_0, \ldots, \lambda a_m)$$

are the same point, where λ is any nonzero element of GF(q). These are called *homogeneous coordinates* for the points. There are $q^{m+1} - 1$ nonzero $(m+1)$-tuples, and each point appears $q - 1$ times, so the number of points in Ω is $(q^{m+1} - 1)/(q - 1)$.

The line through two distinct points $(a_0, \ldots, a_m)$ and $(b_0, \ldots, b_m)$ consists of the points

$$(\lambda a_0 + \mu b_0, \ldots, \lambda a_m + b\mu_m), \tag{1}$$

where $\lambda, \mu \in$ GF(q) are not both zero. A line contains $q + 1$ points since there are $q^2 - 1$ choices for λ, μ and each point appears $q - 1$ times in (1).

Axioms (i), (ii) are clearly satisfied.

Problem. (1) Check that (iii) and (iv) hold.

The projective geometry defined in this way is denoted by PG(m, q).

Problem. (2) Show that PG(m, q) has dimension m.

Examples. (1) If $m = q = 2$, the projective plane PG(2, 2) contains 7 points labeled (001), (010), (100), (011), (101), (110), (111), and 7 lines, as shown in Fig. 2 (cf. Fig. 2.12).

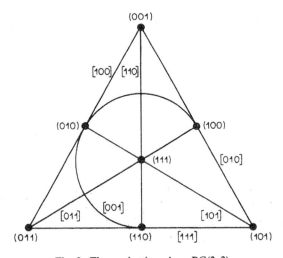

Fig. 2. The projective plane PG(2, 2).

(2) If $m = 2$, $q = 3$ we obtain the projective plane PG(2, 3), containing $3^2 + 3 + 1 = 13$ points

(001)   (010)   (011)   (012)
(100)   (101)   (102)   (110)
(111)   (112)   (120)   (121)
(122),

and 13 lines, nine of which are shown in Fig. 3.

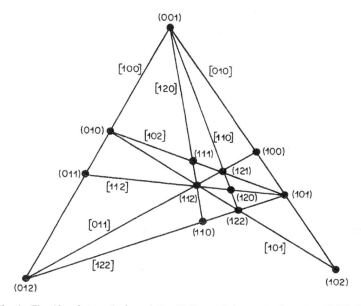

Fig. 3. The 13 points and nine of the 13 lines of the projective plane PG(2, 3).

It is convenient to extend the definition of PG(m, q) to include the values $m = -1$, 0 and 1, even though these degenerate geometries do not satisfy axiom (iv). Thus PG(-1, q) is the empty set, PG(0, q) is a point, and PG(1, q) is a line.

A *hyperplane* or subspace of dimension $m - 1$ in PG(m, q) consists of those points $(a_0, \ldots, a_m)$ which satisfy a homogeneous linear equation

$$\lambda_0 a_0 + \lambda_1 a_1 + \cdots + \lambda_m a_m = 0, \quad \lambda_i \in \mathrm{GF}(q).$$

Such a hyperplane is in fact a PG($m - 1$, q), and will be denoted by $[\lambda_0, \ldots, \lambda_m]$. Note that $[\lambda_0, \ldots, \lambda_m]$ and $[\mu\lambda_0, \ldots, \mu\lambda_m]$, $\mu \neq 0$, represent the same hyperplane. The lines (i.e. hyperplanes) in Figs. 2, 3 have been labeled in this way. Clearly a point $(a_0, \ldots, a_m)$ is on the hyperplane $[\lambda_0, \ldots, \lambda_m]$ iff $\sum_{i=0}^{m} \lambda_i a_i = 0$.

Problems. (3) Show that the points of $PG(m, q)$ can be uniquely labeled by making the left-most nonzero coordinate equal to 1 (as in Example 2).

(4) Show that $PG(m, q)$ is constructed from a vector space V of dimension $m + 1$ over $GF(q)$ by taking the 1-dimensional subspaces of V to be the points of $PG(m, q)$ and the 2-dimensional subspaces to be the lines.

(5) Find the four missing lines in Fig. 3.

(6) Construct $PG(3, 2)$.

The affine or Euclidean projective geometry $EG(m, q)$. This is obtained from $PG(m, q)$ by deleting the points of a hyperplane H (it doesn't matter which one, by Corollary 8). For example, deleting the line [100] from Fig. 2 gives the $EG(2, 2)$

In general if we choose H to be the hyperplane $[1 \ 0 \cdots 0]$ consisting of all points with $a_0 = 0$, we are left with the points whose coordinates can be taken to be $(1, a_1, \ldots, a_m)$. In this way the q^m points of $EG(m, q)$ can be labeled by the m-tuples $(a_1, \ldots, a_m)$, $a_i \in GF(q)$.

Again we make the convention that $EG(-1, q)$ is empty, $EG(0, q)$ is the point 0, and $EG(1, q)$ is a line.

Problem. (7) Show that the dimension (as defined above) of $EG(m, q)$ is equal to m. Show that $EG(m, q)$ is also a vector space of dimension m over $GF(q)$.

Remark. The nonzero elements of $GF(q^{m+1})$ represent the points of $PG(m, q)$, but there are $q - 1$ elements sitting on each point. For example, take $GF(4) = \{0, 1, \omega, \omega^2\}$. The elements $100, \omega00, \omega^2 00$ of $GF(4^3)$ all represent the point (100) of $PG(2, 4)$. The line through (100) and (010) contains the five points

$$
\begin{array}{ll}
100 & (\text{or } \omega00 \text{ or } \omega^2 00), \\
010 & (\text{or } 0\omega0 \text{ or } 0\omega^2 0), \\
110 & (\text{or } \omega\omega0 \text{ or } \omega^2 \omega^2 0), \\
1\omega0 & (\text{or } \omega\omega^2 0 \text{ or } \omega^2 10), \\
1\omega^2 0 & (\text{or } \omega10 \text{ or } \omega^2 \omega0).
\end{array}
$$

The points of the affine geometry are all elements of $GF(q^m)$.

Desarguesian geometries. If the dimension exceeds 2 all projective and affine geometries come from finite fields. But in dimension 2 things can be more complicated.

Theorem 1. *If* $m \geqslant 3$ *then a finite projective geometry of dimension* m *is a* $PG(m, q)$ *for some* q, *and an affine geometry of dimension* m *is an* $EG(m, q)$ *for some* q.

$PG(m, q)$ is called a *Desarguesian* geometry since Desargues' theorem holds there.

The proof of Theorem 1 is in three steps. (i) A projective geometry of dimension $m > 2$ is one in which Desargues' theorem holds; (ii) the points of a Desarguesian geometry can be given coordinates from a possibly noncommutative field S; (iii) if S is finite it is commutative, hence $S = GF(q)$ for some q. For the details see Artin [29, Ch. 2], Baer [56, Ch. 7], Dembowski [370, Ch. 1], Herstein [642, p. 70], Veblen and Young [1368, Vol. 1, Ch. 2].

§3. Properties of PG(m, q) and EG(m, q)

Subspaces of PG(m, q)

Problem. (8) Show that if S is a subspace of $PG(m, q)$ then S is a $PG(r, q)$ for some r, $0 \leqslant r \leqslant m$, and that S may be defined as the set of points satisfying $m - r$ independent homogeneous linear equations.

This implies that the intersection of two distinct $PG(m - 1, q)$'s in a $PG(m, q)$ is a $PG(m - 2, q)$ (since the points satisfy two linear equations). The intersection of a $PG(m - 1, q)$ and a $PG(m - 2, q)$ is either the $PG(m - 2, q)$ or a $PG(m - 3, q)$, and so on. The intersection of a $PG(m - 1, q)$ and a line is either the line or a point.

In general, the intersection of a $PG(r, q)$ (defined by $m - r$ equations) and a $PG(s, q)$ ($m - s$ equations) has dimension $r, r - 1, \ldots$, or $r - m + s$, supposing $s \geqslant r$. If $r - m + s < 0$, the subspaces may be disjoint.

Principle of duality. Since points and hyperplanes in a $PG(m, q)$ are both represented by $(m + 1)$-tuples, there is a natural $1 - 1$-correspondence between them, with the point p corresponding to the *dual* hyperplane $[p]$. Similarly there is a $1 - 1$-correspondence between lines and subspaces $PG(m - 2, q)$, with the line (pq) corresponding to the dual subspace $[p] \cap [q]$. This correspondence has the property that if p is on the line (qr) then the dual hyperplane $[p]$ contains the dual subspace $[q] \cap [r]$.

Similarly there is a $1 - 1$-correspondence (the technical term is a *correlation*) between subspaces of dimension r and subspaces of dimension $m - r - 1$, which preserves incidence. For example if two $PG(r, q)$'s meet in a point then the dual $PG(m - r - 1, q)$'s span a hyperplane.

This correspondence justifies the *principle of duality*, which says that any statement about $PG(m, q)$ remains true if we interchange "point" and "hyperplane," "$PG(r, q)$" and "$PG(m - r - 1, q)$," "intersect" and "span," and "contained in" and "contained by."

An important application of this principle is:

Theorem 2. *If $s \geqslant r$, the number of subspaces $PG(s, q)$ in $PG(m, q)$ which contain a given $PG(r, q)$ is equal to the number of $PG(m - s - 1, q)$ contained in a given $PG(m - r - 1, q)$.*

Problem. (9) Prove directly the special case of Theorem 2 which says that the number of lines through a point is equal to the number of points on a hyperplane.

The number of subspaces

Theorem 3. *The number of $PG(r, q)$ contained in a $PG(m, q)$ is*

$$\frac{(q^{m+1} - 1)(q^{m+1} - q) \cdots (q^{m+1} - q^r)}{(q^{r+1} - 1)(q^{r+1} - q) \cdots (q^{r+1} - q^r)} = \begin{bmatrix} m + 1 \\ r + 1 \end{bmatrix}, \tag{2}$$

where

$$\begin{bmatrix} m + 1 \\ r + 1 \end{bmatrix}$$

is a Gaussian binomial coefficient defined in Problem 3 of Ch. 15.

Proof. The numerator of (2) is the number of ways of picking $r + 1$ independent points in $PG(m, q)$, to define a $PG(r, q)$. However, many of these sets of points determine the same $PG(r, q)$, so we must divide by the denominator of (2), which is the number of ways of picking $r + 1$ independent points in a $PG(r, q)$. Q.E.D.

A similar argument proves:

Theorem 4. *In $PG(m, q)$ let $R = PG(r, q) \subset S = PG(s, q)$. The number of subspaces T of dimension t with $R \subset T \subset S$ is*

$$\begin{bmatrix} s - r \\ t - r \end{bmatrix}.$$

Problem. (10) Use Theorem 3 to show that the number of $PG(r, q)$ contained in $PG(m, q)$ is equal to the number of $PG(m - r - 1, q)$ contained in $PG(m, q)$.

Subspaces and flats of EG(m, q). A subspace S of EG(m, q) is called a *flat*.

Problem. (11) Show that if a flat contains the origin then it is a linear subspace of EG(m, q) regarded as a vector space; and that a flat not containing the origin is a coset of a linear subspace.

Thus a flat of dimension r in EG(m, q) is a coset of an EG(r, q), and will be referred to as an EG(r, q) or an *r-flat*. A subspace PG(r, q) of PG(m, q) is also called an r-flat.

Theorem 5. *The number of* EG(r, q) *in an* EG(m, q) *is*

$$q^{m-r}\begin{bmatrix} m \\ r \end{bmatrix}.$$

Proof. Let EG(m, q) be obtained from PG(m, q) by deleting the hyperplane H. A PG(r, q) either meets H in a PG$(r - 1, q)$ or is contained in H. Thus the desired number is the difference between the number of PG(r, q) in PG(m, q) and the number of PG(r, q) in H. By Theorem 3 this is

$$\begin{bmatrix} m + 1 \\ r + 1 \end{bmatrix} - \begin{bmatrix} m \\ r + 1 \end{bmatrix} = q^{m-r}\begin{bmatrix} m \\ r \end{bmatrix},$$

by Problems 3(b), 3(e) of Ch. 15. Q.E.D.

Theorem 6. *In* EG(m, q) *let* $R = $ EG$(r, q) \subset S = $ EG(s, q), *where* $r \geqslant 1$. *The number of flats* T *of dimension* t *with* $R \subset T \subset S$ *is*

$$\begin{bmatrix} s - r \\ t - r \end{bmatrix}.$$

Proof. Follows from Theorem 4. Q.E.D.

Note that in a projective geometry two hyperplanes always meet in a subspace of dimension $m - 2$, whereas in an affine geometry two hyperplanes may meet in a subspace of dimension $m - 2$ or not at all. Disjoint hyperplanes are called *parallel*.

Problem. (12) Show that EG(m, q) can be decomposed into q mutually parallel hyperplanes.

The collineation group of PG(m, q).

Definition. A *collineation* of a projective or affine geometry is a permutation of its points which maps lines onto lines. It follows that every subspace is mapped onto a subspace of the same dimension.

For example, the permutation ((011), (010), (001)) ((110), (111), (100)) is a collineation of PG(2, 2) – see Fig. 2.

The set of all collineations of PG(m, q) forms its *collineation group*. Suppose $q = p^s$ where p is a prime.

Recall from Theorem 12 of Ch. 4 that the automorphism group of the field GF(p^s) is a cyclic group of order s generated by

$$\sigma_p : \beta \to \beta^p, \qquad \beta \in \mathrm{GF}(p^s).$$

Clearly σ_p is a collineation of PG(m, p^s).

Let C be an invertible $(m + 1) \times (m + 1)$ matrix over GF(p^s). Then the permutation of the points of PG(m, p^s) given by

$$(a_0, \ldots, a_m) \to (a_0, \ldots, a_m)C$$

is also a collineation. Clearly C and λC, $\lambda \neq 0$, are the same collineation.

Together σ_p and the matrices C generate a group consisting of the permutations

$$(a_0, \ldots, a_m) \to (a_0^{p^i}, \ldots, a_m^{p^i})C, \quad 0 \le i < s. \tag{3}$$

There are $s \, \Pi_{i=0}^{m} (q^{m+1} - q^i)$ such permutations, but only

$$\frac{s}{q - 1} \prod_{i=0}^{m} (q^{m+1} - q^i) \tag{4}$$

distinct collineations. This group of collineations is denoted by $P\Gamma L_{m+1}(q)$, $q = p^s$.

Theorem 7. (The fundamental theorem of projective geometry.) $P\Gamma L_{m+1}(q)$ *is the full collineation group of* PG(m, q).

For the proof see for example Artin [29, p. 88], Baer [56, Ch. 3] or Carmichael [250, p. 360].

Since $P\Gamma L_{m+1}(q)$ is doubly transitive we have:

Corollary 8. *There is essentially only one way of obtaining* EG(m, q) *from* PG(m, q).

Corollary 9. *The full collineation group of* EG(m, q) *is the subgroup of* $P\Gamma L_{m+1}(q)$ *which fixes the hyperplane at infinity (setwise), and has order*

$$s \prod_{i=0}^{m-1} (q^m - q^i). \tag{5}$$

(See for example Carmichael [250, p. 374]).

Problem. (13) Given EG(m, q) show that there is essentially only one way to add a hyperplane and obtain PG(m, q).

§4. Projective and affine planes

A projective geometry of dimension 2 is a projective *plane*. Unlike the situation in higher dimensions, a projective plane need not be a PG(2, q) for any q.

In §5 of Ch. 2 we defined a projective plane to be a Steiner system $S(2, n + 1, n^2 + n + 1)$ for some $n \geq 2$, or in other words (Definition 2): a collection of $n^2 + n + 1$ points and $n^2 + n + 1$ lines, with $n + 1$ points on each line and a unique line containing any two points.

However, the best definition of a projective plane is this. Definition 3. A projective plane is a collection of points and lines satisfying (i) there is a unique line containing any two points, (ii) any two distinct lines meet at a unique point, and (iii) there exist four points no three of which lie on a line.

Theorem 10. *The three definitions of a projective plane are equivalent.*

Sketch of Proof. Definition 1 (§2) $\Rightarrow$ Definition 3. It is only necessary to show that any two lines meet. This follows because otherwise the two lines would contain four independent points and the dimension would not be 2.

Definition 3 $\Rightarrow$ Definition 2. Take two points p, q and a line L not containing them. Then the number of lines through p (or through q) is equal to the number of points on L. Call this number $n + 1$. Then the total number of points (or lines) is $n(n + 1) + 1 = n^2 + n + 1$.

Definition 2 $\Rightarrow$ Definition 1. To prove (iii) we show that any two lines meet. This follows from evaluating in two ways the sum of $\chi(p, L, M)$ over all points p and distinct lines L, M, where $\chi(p, L, M) = 1$ if $p = L \cap M$, $= 0$ otherwise. The dimension is 2, for if p, q, r, s are independent points then the lines (pq) and (rs) do not meet.   Q.E.D.

A Steiner system $S(2, n + 1, n^2 + n + 1)$ is called a projective plane of *order* n. Thus Figs. 2, 3 show projective planes of orders 2 and 3. In general a PG(2, q) is a projective plane of order q. From Theorem 7 of Ch. 4, this gives Desarguesian projective planes of all prime power orders.

However, not all projective planes are Desarguesian. In fact non-Desarguesian planes are known of all orders $n = p^e > 8$ where p is a prime and $e > 1$. For $n \leqslant 8$ we have

Theorem 11. *The projective planes of orders* $n = 2, 3, 4, 5, 7, 8$ *are unique* (*and are the Desarguesian planes* $\mathrm{PG}(2, n)$).

For the proof see Problem 13 of Ch. 20, and the references on p. 144 of Dembowski [370].

We know from Problem 11 of Ch. 19 that there is no projective plane of order 6. This is a special case of

Theorem 12. (Bruck and Ryser.) *If* $n \equiv 1$ *or* $2 \pmod 4$ *and if* n *is not the sum of two squares then there is no projective plane of order* n.

For the proof see Hall [587, p. 175] or Hughes and Piper [674, p. 87].

Thus planes are known of orders $2, 3, 4, 5, 7, 8, 9, 11, 13, 16, 17, 19, \dots$, orders $6, 14, 21, \dots$ do not exist by Theorem 12, and orders $10, 12, 15, 18, 20, \dots$ are undecided. For the connection between codes and orders $n \equiv 2 \pmod 4$ see Problem 11 of Ch. 19.

Affine or euclidean planes. An affine geometry of order 2 is an *affine* plane, and is obtained by deleting the points of a fixed line from a projective plane. A second definition was given in §5 of Ch. 2: an affine plane is an $S(2, n, n^2)$, $n \geqslant 2$. A third definition is this. An affine plane is a collection of points and lines satisfying (i) there is a unique line containing any two points, (ii) given any line L and any point $p \notin L$ there is a unique line through p which does not meet L, and (iii) there exist three points not on a line. Again the three definitions agree, and we call an $S(2, n, n^2)$ an affine plane of *order n*. Then the results given above about the possible orders of projective planes apply also to affine planes.

Notes on Appendix B

Projective geometries are discussed by Artin [29], Baer [56], Biggs [143], Birkhoff [152, Ch. 8], Carmichael [250], Dembowski [370], Hall [583, Ch. 12], MacNeish [869], Segre [1173], and Veblen and Young [1368]. References on projective planes are Albert and Sandler [20], Hall [582, Ch. 20 and 587, Ch. 12], Segre [1173] and especially Dembowski [370] and Hughes and Piper [674]. For the numbers of subspaces see for example Carmichael [250] or Goldman and Rota [519]. See also the series of papers by Dai, Feng, Wan and Yang [103, 104, 1441, 1457–1460, 1474].

Bibliography

At the end of each entry is a number in square brackets giving the chapter (or appendix) which refers to this entry. Although we give references to the English journals whenever possible, note that Problems of Information Transmission, Soviet Mathematics Doklady, Electronics and Communications in Japan, and others are translations of journals originally published in another language.

A

[1] N. M. Abramson, A class of systematic codes for non-independent errors, *IEEE Trans. Info. Theory*, 5 (1959) 150–157 [7].

[2] N. Abramson, Error correcting codes from linear sequential networks, in: C. Cherry, ed., *Fourth London Symposium on Info. Theory*, (Butterworths, London, 1961) [7].

[3] N. Abramson, *Information Theory and Coding*, (McGraw-Hill, New York, 1963) [1].

[4] N. Abramson, Encoding and decoding cyclic code groups, *Problems of Information Transmission*, 6 (2) (1970) 148–154 [18].

[5] L. E. Adelson, R. Alter and T. B. Curtz, Long snakes and a characterization of maximal snakes on the d-cube, *Proc. 4th S-E Conf. Combinatorics, Graph Theory, & Computing*, (Utilitas Math., Winnipeg, 1973) pp. 111–124 [Intro.].

[6] L. E. Adelson, R. Alter and T. B. Curtz, Computation of d-dimensional snakes, *Proc. 4th S-E Conf. Combinatorics, Graph Theory, & Computing* (Utilitas Math., Winnipeg, 1973) pp. 135–139 [Intro.].

[7] V. K. Agrawala and J. G. Belinfante, An algorithm for computing SU(n) invariants, *BIT* 11 (1971) 1–15 [19].

[8] S. V. Ahamed, The design and embodiment of magnetic domain encoders and single-error correcting decoders for cyclic block codes, *Bell Syst. Tech. J.*, 51 (1972) 461–485 [16].

[9] S. V. Ahamed, Applications of multidimensional polynomial algebra to bubble circuits, *Bell Syst. Tech. J.*, 51 (1972) 1559–1580 [16].

[10] S. V. Ahamed, Extension of multidimensional polynomial algebra to domain circuits with multiple propagation velocities, *Bell Syst. Tech. J.*, 51 (1972) 1919–1922 [16].

[11] S. V. Ahamed, Serial coding for cyclic block codes, *Bell Syst. Tech. J.*, 59 (1980) 269–276 [16].

[12] N. Ahmed, R. M. Bates, and K. R. Rao, Multidimensional BIFORE Transform, *Electronics Letters*, 6 (1970) 237–238 [2].

[13] N. Ahmed and K. R. Rao, Discrete Fourier and Hadamard transforms, *Electronics Letters*, **6** (1970) 221–224 [2].

[14] N. Ahmed and K. R. Rao, Complex BIFORE transform, *Electronics Letters*, **6** (1970) 256–258 and 387 [2].

[15] N. Ahmed and K. R. Rao, Additional properties of complex BIFORE transform, *IEEE Trans. Audio Electroacoustics*, **19** (1971) 252–253 [2].

[16] A. V. Aho, J. E. Hopcroft and J. D. Ullman, *The Design and Analysis of Computer Algorithms* (Addison-Wesley, Reading, MA., 1974) [12].

[17] J. D. Alanen and D. E. Knuth, Tables of finite fields, *Sankhyā, Series A*, **26** (1964) 305–328 [4].

[18] A. A. Albert, Symmetric and alternating matrices in an arbitrary field, *Trans. Amer. Math. Soc.*, **43** (1938) 386–436 [15].

[19] A. A. Albert, *Fundamental Concepts of Higher Algebra*, (University of Chicago Press, Chicago, 1956) [3, 4].

[20] A. A. Albert and R. Sandler, *An Introduction to Finite Projective Planes* (Holt, New York, 1968) [B].

[21] W. O. Alltop, An infinite class of 4-designs, *J. Comb. Theory*, **6** (1969) 320–322 [2, 17].

[22] W. O. Alltop, Some 3-designs and a 4-design, *J. Comb. Theory* **11** A (1971) 190–195 [2, 17].

[23] W. O. Alltop, 5-designs in affine space, *Pacific J. Math.*, **39** (1971) 547–551 [2, 17].

[24] W. O. Alltop, An infinite class of 5-designs, *J. Comb. Theory*, **12** A (1972) 390–395 [2, 17].

[25] W. O. Alltop, Binary codes with improved minimum weights, *IEEE Trans. Info. Theory*, **22** (1976) 241–243 [17, 18, A].

[26] W. O. Alltop, private communication [A].

[27] H. L. Althaus and R. J. Leake, Inverse of a finite-field Vandermonde matrix, *IEEE Trans Info. Theory*, **15** (1969) 173 [4].

[28] D. R. Anderson, A new class of cyclic codes, *SIAM J. Appl. Math.*, **16** (1968) 181–197 [9].

[29] E. Artin, *Geometric Algebra* (Wiley, New York, 1957) [B].

[30] E. Artin, *Galois Theory* (Notre Dame, IN, 2nd edition, 1959) [4].

[31] T. Asabe, H. Nakanishi and Y. Tezuka, On cyclic multilevel codes, *Electron. Commun. Japan*, **55-A** (2) (1972) 8–15 [1].

[32] R. Ash, *Information Theory* (Wiley, New York, 1965) [10].

[33] E. F. Assmus, Jr., J.-M. Goethals, and H. F. Mattson, Jr., Generalized *t*-designs and majority logic decoding of linear codes, *Info. and Control*, **32** (1976) 43–60 [16].

[34] E. F. Assmus, Jr., and H. F. Mattson, Jr., Error-correcting codes: An axiomatic approach, *Info. and Control*, **6** (1963) 315–330 [16].

[35] E. F. Assmus, Jr., and H. F. Mattson, Jr., Disjoint Steiner systems associated with the Mathieu groups, *Bull. Am. Math. Soc.* **72** (1966) 843–845 [20].

[36] E. F. Assmus, Jr., and H. F. Mattson, Jr., Perfect codes and the Mathieu groups, *Archiv der Math.*, **17** (1966) 121–135 [20].

[37] E. F. Assmus, Jr., and H. F. Mattson, Jr., On tactical configurations and error-correcting codes, *J. Comb. Theory*, **2** (1967) 243–257 [6, 16].

[38] E. F. Assmus, Jr., and H. F. Mattson, Jr., Research to Develop the Algebraic Theory of Codes, Report AFCRL-68-0478, Air Force Cambridge Res. Labs., Bedford, Mass., September, 1968 [16].

[39] E. F. Assmus, Jr., and H. F. Mattson, Jr., On the automorphism groups of Paley-Hadamard matrices, in: R. C. Bose and T. A. Dowling, eds. *Combinatorial Mathematics and Its Applications*, (Univ. North Carolina Press, Chapel Hill, NC 1969) [16].

[40] E. F. Assmus Jr., and H. F. Mattson, Jr., Some $(3p, p)$ codes, in: Information Processing 68, (North-Holland, Amsterdam, 1969) pp. 205–209 [18].

[41] E. F. Assmus, Jr., and H. F. Mattson, Jr., New 5-designs, *J. Comb. Theory*, **6** (1969) 122–151 [6, 11, 16].

[42] E. F. Assmus, Jr., and H. F. Mattson, Jr., Algebraic Theory of Codes II, Report AFCRL-69-0461, Air Force Cambridge Res. Labs., Bedford, Mass., 15 October, 1969 [16].

[43] E. F. Assmus, Jr., and H. F. Mattson, Jr., Algebraic Theory of Codes II, Report AFCRL-71-0013, Air Force Cambridge Res. Labs., Bedford, Mass., 15 October, 1970 [16, 19].

[44] E. F. Assmus, Jr., and H. F. Mattson, Jr., On weights in quadratic-residue codes, *Discrete Math.*, **3** (1972) 1–20 [16].

[45] E. F. Assmus, Jr., and H. F. Mattson, Jr., Contractions of self-orthogonal codes, *Discrete Math.*, 3 (1972) 21–32 [16].

[46] E. F. Assmus, Jr., and H. F. Mattson, Jr., Error-Correcting Codes, Report AFCRL-72-0504, Air Force Cambridge Res. Labs., Bedford, Mass., August 31, 1972 [16].

[47] E. F. Assmus, Jr., and H. F. Mattson, Jr., Coding and combinatorics, *SIAM Review*, 16 (1974) 349–388 [5, 6, 16, 19].

[48] E. F. Assmus, Jr., H. F. Mattson, Jr., and M. Garcia, Self-orthogonal Steiner systems and projective planes, *Math. Z.*, 138 (1974) 89–96 [19].

[49] E. F. Assmus, Jr., H. F. Mattson, Jr., and H. E. Sachar, A new form of the square-root bound, *SIAM J. Appl. Math.*, 30 (1976) 352–354 [16].

[50] E. F. Assmus, Jr., H. F. Mattson, Jr., and R. J. Turyn, Cyclic Codes, Air Force Cambridge Research Labs., Report AFCRL-65-332, Bedford Mass., April 28, 1965 [11, 16].

[51] E. F. Assmus, Jr., H. F. Mattson, Jr., and R. J. Turyn, Cyclic Codes, Report AFCRL-66-348, Air Force Cambridge Res. Labs., Bedford, Mass., April 28, 1966 [16].

[52] E. F. Assmus, Jr., H. F. Mattson, Jr., and R. J. Turyn, Low Weights in Quadratic Residue Codes, Inst. of Stat. Mimeo Series 484.3, Dept. of Stat., Univ. of N. Carolina, Chapel Hill, N. C., 1966 [16].

[53] E. F. Assmus, Jr., H. F. Mattson, Jr., and R. J. Turyn, Research to Develop the Algebraic Theory of Codes, Report AFCRL-67-0365, Air Force Cambridge Res. Labs., Bedford, Mass., June 1967 [16, 19].

[54] J. Astola, On the nonexistence of certain perfect Lee-error-correcting codes, *Ann. Univ. Turku.*, Ser. A, (1975) 167 [6].

[55] S. Azumi and T. Kasami, On the optimal modified Hamming codes, *Trans. Inst. Elect. and Commun. Engineers*, (Japan) 58 (6) (1958) 325–330 [7].

B

[56] R. Baer, *Linear Algebra and Projective Geometry*, (Academic Press, New York, 1952) [B].

[57] L. R. Bahl, and R. T. Chien, On Gilbert burst-error-correcting codes, *IEEE Trans. Info. Theory* 15 (1969) 431–433 [10].

[58] L. R. Bahl and R. T. Chien, Single – and multiple – burst-correcting properties of a class of cyclic product codes, *IEEE Trans. Info. Theory*, 17, (1971) 594–600 [18].

[59] B. G. Bajoga and W. J. Walbesser, Decoder complexity for BCH codes, *Proc. IEE* 120 (1973) 429–432 [Intro., 9].

[60] A. H. Ball and L. J. Cummings, The comma-free codes with words of length two, *Bull Austral. Math. Soc.*, 14 (1976) 249–258 [Intro.].

[61] A. H. Ball and L. J. Cummings, Extremal digraphs and comma-free codes, *Ars Combinat.*, 1 (1976) 239–251 [Intro.].

[62] J. R. Ball, A. H. Spittle and H. T. Liu, High-speed m-sequence generation: a further note, *Electronics Letters*, 11 (1975) 107–108 [14].

[63] C. Balza, A. Fromageot, and M. Maniere, Four-level pseudo-random sequences, *Electronics Letters*, 3 (1967) 313–315 [14].

[64] R. P. Bambah, D. D. Joshi and I. S. Luthar, Some lower bounds on the number of code points in a minimum distance binary code, *Information and Control*, 4 (1961) 313–323 [17, 18].

[65] K. S. Banerjee, An introduction to the weighing problem, *Chemica Scripta* 6 (1974) 158–162 [2].

[66] K. S. Banerjee, *Weighing Designs for Chemistry, Medicine, Economics, Operations Research, Statistics* (Marcel Dekker, New York, 1975) [2].

[67] R. B. Banerji, A decoding procedure for double-error correcting Bose–Ray–Chaudhuri codes, *Proc. IEEE*, 49 (1961) 1585 [9].

[68] R. H. Barker, Group synchronizing of binary digital sequences, in: *Communication Theory* (Butterworth, London, 1953) pp. 273–287 [14].

[69] A. Barlotti, Some topics in finite geometrical structures, Institute of Statistics Mimeo Series No. 439, Univ. of N. Carolina, Chapel Hill, N. Carolina, August 1965 [11].

[70] A. Barlotti, Some classical and modern topics in finite geometrical structures, in: J. N. Srivastava et al., eds., *A Survey of Combinatorial Theory*, (North-Holland, Amsterdam, 1973) [11].

[71] T. C. Bartee and D. I. Schneider, An electronic decoder for Bose–Chaudhuri–Hocquenghem codes, *IEEE Trans. Info. Theory*, **8** (1962) 17–24 [9].

[72] T. C. Bartee and D. I. Schneider, Computation with finite fields, *Info. Control* **6** (1963) 79–98 [3].

[73] D. E. Barton and C. L. Mallows, Some aspects of the random sequence, *Annals Math. Stat.* **36** (1965) 236–260 [14].

[74] J. E. Bartow, A reduced upper bound on the error correction ability of codes, *IEEE Trans. Info. Theory*, **9** (1963) 46 [17].

[75] J. E. Bartow, An upper bound on the error correction ability of codes, *IEEE Trans. Info. Theory*, **9** (1963) 290 [17].

[76] L. A. Bassalygo, New upper bounds for error correcting codes, *Problems of Information Transmission*, **1** (4) (1965) 32–35 [17].

[76a] L. A. Bassalygo, A generalization of Lloyd's Theorem to the case of any alphabet, *Problems of Control and Information Theory*, **2** (2) (1973) 25–28 [6].

[77] L. A. Bassalygo, A necessary condition for the existence of perfect codes in the Lee metric, *Math. Notes*, **15** (1974) 178–181 [6].

[78] L. A. Bassalygo, G. V. Zaitsev and V. A. Zinov'ev, Uniformly packed codes, *Problems of Info. Transmission*, **10** (1) (1974) 6–9 [15].

[79] L. A. Bassalygo and V. A. Zinov'ev, Remarks on uniformly packed codes, *Problemy Peredachi Informatsii*, **13** (No. 3, 1977) 22–25 [15].

[80] S. A. Butman and R. J. McEliece, The ultimate limits of binary coding for a wideband Gaussian channel, *JPL Progress Report* 42–22 (1975) 78–79 [1].

[81] L. D. Baumert, Cyclic Hadamard matrices, *JPL Space Programs Summary*, 37–43–IV (1967) 311–314 and 338 [2].

[82] L. D. Baumert, *Cyclic Difference Sets*, Lecture Notes in Math. No. 182 (Springer, Berlin, New York, 1971) [13].

[83] L. D. Baumert and J. Mykkeltveit, Weight Distributions of Some Irreducible Cyclic Codes, *JPL Technical Report*, 32–1526 (1973) 128–131 [8].

[84] L. D. Baumert et al., A combinatorial packing problem, in: G. Birkhoff and M. Hall, Jr., eds., *Computers in Algebra and Number Theory*, Am. Math. Soc. Providence, R. I. (1971) pp. 97–108 [6].

[85] L. D. Baumert and D. Cantor, Evolution and coding: inverting the genetic code, *JPL Space Programs Summary*, 37–65–III (Oct. 31, 1970) 41–44 [Intro.].

[86] L. D. Baumert and R. J. McEliece, Weights of irreducible cyclic codes, *Inf. Control*, **20** (1972) 158–175 [8].

[87] L. D. Baumert and R. J. McEliece, A note on the Griesmer bound, *IEEE Trans. Info. Theory*, **19** (1973) 134–135 [17].

[88] L. D. Baumert and R. J. McEliece, A Golay – Viterbi Concatenated Coding Scheme for MJS'77, *JPL Technical Report*, 32–1526 (1973) 76–83 [1].

[89] L. D. Baumert and R. J. McEliece, Golay – Viterbi Decoding: Results of The MVM'73 X-Band Telemetry Experiment, *JPL Progress Report*, 42–25 (1974) 108–110 [1].

[90] L. D. Baumert and H. C. Rumsey, Jr., The index of comma freedom for the Mariner Mars 1969 high data rate telemetry code, *JPL Space Programs Summary*, 37–46–IV (1967) 221–226 and 249–250 [14].

[91] L. D. Baumert and H. C. Rumsey, Jr., The maximum indices of comma freedom for the high-data-rate telemetry codes, *JPL Space Programs Summary* 37–51–III (June 1968), 215–217 [14].

[92] J. T. B. Beard, Jr., Computing in GF(q), *Math. Comp.*, **28**(1974) 1159–1168 [3].

[93] J. T. B. Beard, Jr., and K. I. West, Some primitive polynomials of the third kind, *Math. Comp.*, **28**(1974) 1166–1167 [4].

[94] E. F. Beckenbach and R. Bellman, *Inequalities* (Springer, Berlin, 1961) [6].

[95] V. Belevitch, Theory of 2n-terminal networks with applications to conference telephony, *Electronics Comm.*, **27** (1950) 231–244 [2].

[96] V. Belevitch, Synthesis of four-wire conference networks and related problems, in: *Proc. Symposium Modern Network Synthesis*, Polytech. Inst. Brooklyn, (April 1956), pp. 175–195 [2].

[97] V. Belevitch, Conference networks and Hadamard matrices, *Ann. Soc. Scientifique Bruxelles*, **82** (I) (1968) 13–32 [2].

[98] D. A. Bell, Class of binary codes, *Proc. IEE*, **122** (1975) 47 [18].

[99] D. A. Bell and R. Laxton, Some BCH codes are optimum, *Electronics Letters*, **11** (1975) 296–297 [13].

[100] R. Bellman, *Introduction to Matrix Analysis* (McGraw-Hill, New York, 1960) [2, 14, 16].

[101] R. Bellman, On various versions of the defective coin problem, *Info. and Control*, **4** (1961) 118–131 [7].

[102] B. I. Belov, V. N. Logachev, and V. P. Sandimirov, Construction of a class of linear binary codes achieving the Varshamov-Griesmer bound, *Problems of Info. Transmission*, **10** (3) (1974) 211–217 [17, A].

[103] B.-F. Yang, Studies in finite geometries and the construction of incomplete block designs VII, *Acta Math. Sinica*, **15** (1965) 812–825 [21, B].

[104] B.-F. Yang, Studies in finite geometries and the construction of incomplete block designs VIII, *Acta Math. Sinica*, **15** (1965) 826–841 [21, B].

[105] E. R. Berger, Some additional upper bounds for fixed-weight codes of specified minimum distance, *IEEE Trans. Info. Theory*, **13** (1967) 307–308 [17].

[106] T. Berger and J. A. van der Horst, see Ref. 663.

[107] T. R. Berger and I. Reiner, A proof of the normal basis theorem, *Amer. Math. Monthly*, **82** (1975) 915–918 [4].

[108] G. D. Bergland, A guided tour of the fast Fourier transform, *IEEE Spectrum*, **6** (July, 1969), 41–52 [14].

[109] E. R. Berlekamp, Distribution of cyclic matrices in a finite field, *Duke Math. J.* **33** (1965) 45–48 [16].

[110] E. R. Berlekamp, On decoding binary Bose–Chaudhuri–Hocquenghem codes, *IEEE Trans. Info. Theory*, **11** (1965) 577–579 [9].

[111] E. R. Berlekamp, Factoring polynomials over finite fields, *Bell Syst. Tech. J.*, **46** (1967) 1853–1859 [9].

[112] E. R. Berlekamp, The enumeration of information symbols in BCH codes, *Bell Syst. Tech. J.*, **46** (1967) 1861–1880 [9].

[113] E. R. Berlekamp, *Algebraic coding theory* (McGraw-Hill, New York, 1968) [many references].

[114] E. R. Berlekamp, Weight enumeration theorems, *Proc. Sixth Allerton Conf. Circuit and Systems Theory*, (Univ. of Illinois Press, Urbana, IL, 1968) pp. 161–170 [8, 21].

[115] E. R. Berlekamp, Negacyclic codes for the Lee metric, in: R. C. Bose and T. A. Dowling, eds., *Combinatorial Mathematics*, (Univ. of N. Carolina Press, 1969) pp. 298–316 [5].

[116] E. R. Berlekamp, *A Survey of Algebraic Coding Theory*, (International Centre for Mechanical Sciences, Udine, Italy, 1970) [1, 9].

[117] E. R. Berlekamp, Factoring polynomials over large finite fields, *Math. Comp.*, **24** (1970) 713–735 [9].

[118] E. R. Berlekamp, The weight enumerators for certain subcodes of the second order Reed–Muller codes, *Info. and Control*, **17** (1970) 485–500 [8, 15, 21].

[119] E. R. Berlekamp, Some mathematical properties of a scheme for reducing the bandwidth of motion pictures by Hadamard smearing, *Bell Syst. Tech. J.*, **49** (1970) 969–986 [2, 14].

[120] E. R. Berlekamp, Coding theory and the Mathieu groups, *Info. Control*, **18** (1971) [2, 15, 20].

[121] E. R. Berlekamp, Factoring polynomials, in: *Proc. 3rd S-E Conf. Combinatorics, Graph Theory and Computing*, (Utilitas Math., Winnipeg, 1972) pp. 1–7 [9].

[122] E. R. Berlekamp, Long primitive binary BCH codes have distance $d \sim 2n \ln R^{-1}/\log n \ldots$, *IEEE Trans. Info. Theory*, **18** (1972) 415–426 [9].

[123] E. R. Berlekamp, Decoding the Golay code, JPL Technical Report 32-1526, Vol. IX, pp. 81–85, Jet Propulsion Laboratory, Pasadena, Calif., 1972 [16].

[124] E. R. Berlekamp, Block coding for the binary symmetric channel with noiseless, delayless feedback, pp. 61–85 in: H. B. Mann ed., *Error Correcting Codes*, (Wiley, New York, 1973) pp. 61–85 [Intro.].

[125] E. R. Berlekamp, Goppa codes, *IEEE Trans. Info. Theory* **19** (1973) 590–592 [12].

[126] E. R. Berlekamp, editor, *Key Papers in the Development of Coding Theory* (IEEE Press, New York, 1974) [Intro.].

[127] E. R. Berlekamp and J. Justesen, Some long cyclic linear binary codes are not so bad, *IEEE Trans. Info. Theory*, **20** (1974) 351–356 [7, 18].

[128] E. R. Berlekamp, J. H. van Lint, and J. J. Seidel, A strongly regular graph derived from the perfect ternary Golay code, in: J. N. Srivastava et al., eds., A Survey of Combinatorial Theory (North-Holland, Amsterdam, 1973) Ch. 3 [21].

[129] E. R. Berlekamp, F. J. MacWilliams, and N. J. A. Sloane, Gleason's theorem on self-dual codes, IEEE Trans. Info. Theory, 18 (1972) 409–414 [19].

[130] E. R. Berlekamp and O. Moreno, Extended double-error-correcting binary Goppa codes are cyclic, IEEE Trans. Info. Theory, 19 (1973) 817–818 [12].

[131] E. R. Berlekamp, H. Rumsey, and G. Solomon, On the solutions of algebraic equations over finite fields, Inform. Control, 10 (1967) 553–564 [4, 9].

[132] E. R. Berlekamp and N. J. A. Sloane, Restrictions on weight distribution of Reed–Muller codes, Info. Control, 14 (1969) 442–456 [15].

[133] E. R. Berlekamp and L. R. Welch, Weight distributions of the cosets of the (32, 6) Reed–Muller code, IEEE Trans. Info. Theory, 18 (1972) 203–207 [1, 14, 19].

[134] G. Berman and K. D. Fryer, Introduction to Combinatorics (Academic Press, New York, 1972) [15].

[135] S. D. Berman, On the theory of group codes, Cybernetics 3 (1) (1967) 25–31 [18].

[136] S. D. Berman, Semisimple cyclic and abelian codes II, Cybernetics 3 (3) (1967) 17–23 [18].

[137] S. D. Berman and A. B. Yudanina, Codes with generalized majority decoding and convolutional codes, Problems Info. Trans., 6 (1) (1970) 1–12 [13].

[138] S. D. Berman and A. B. Yudanina, Irreducible configurations of minimum-weight elements in second order Reed–Muller codes, in: B. N. Petrov and F. Csáki, eds., 2nd Internat. Symp. Info. Theory, (Tsahkadsor, 1971) (Akadémiai Kiadó, Budapest, 1973) pp. 151–168 [15].

[139] M. R. Best and A. E. Brouwer, The triply shortened binary Hamming code is optimal, Discrete Math., 17 (1977) 235–245.

[140] M. R. Best, A. E. Brouwer, F. J. MacWilliams, A. M. Odlyzko and N. J. A. Sloane, Bounds for binary codes of length less than 25, IEEE Trans. Info. Theory, 24 (1978) 81 – 93 [17, A].

[141] V. K. Bhargava and J. M. Stein, (v, k, λ) configurations and self-dual codes, Info. Control, 28 (1975) 352–355 [16, 19].

[142] V. K. Bhargava, S. E. Tavares, and S. G. S. Shiva, Difference sets of the Hadamard type and quasi-cyclic codes, Info. and Control, 26 (1974) 341–350 [16].

[143] N. Biggs, Finite Groups of Automorphisms, London Math. Soc. Lecture Note Series, No. 6, (Cambridge Univ. Press, London 1971) [20, B].

[144] N. L. Biggs, Perfect codes in graphs, J. Comb. Theory. 15B (1973) 289–296 [6, 21].

[145] N. L. Biggs, Algebraic Graph Theory (Cambridge Univ. Press, London, 1974) [6, 21].

[146] N. L. Biggs, Perfect Codes and Distance-transitive Graphs, in: Combinatorics, T. P. McDonough and V. C. Mavron eds., London Math. Soc., Lecture Notes No. 13 (Cambridge Univ. Press, London 1974) pp. 1–8 [6, 21].

[147] N. Biggs, Designs, factors and codes in graphs, Quart. J. Math. Oxford (2) 26 (1975) 113–119 [6, 21].

[148] E. Biglieri and M. Elia, On the existence of group codes for the Gaussian channel, IEEE Trans. on Info. Theory, 18 (1972) 399–402 [Intro.].

[149] E. Biglieri and M. Elia, Cyclic-group codes for the Gaussian channel, IEEE Trans. Info. Theory. 22 (1976) 624–629 [Intro.].

[150] E. Biglieri and M. Elia, Optimum permutation modulation codes, and their asymptotic performance, IEEE. Trans. Info. Theory, 22 (1976) 750–753 [Intro.].

[151] E. Biglieri and M. Elia, Some results on symmetric-group codes for the Gaussian channel, preprint [Intro.].

[152] G. Birkhoff, Lattice Theory, 2nd Edition, Colloq. Publications vol. 25, (Amer. Math Soc., Providence, RI, 1961) [B].

[153] G. Birkoff and T. C. Bartee, Modern Applied Algebra, (McGraw-Hill, New York, 1970) [4].

[154] W. C. Blackwelder, On constructing balanced incomplete block designs from association matrices with special reference to association schemes of two and three classes, J. Combinat. Theory, 7 (1969) 15–36 [21].

[155] I. F. Blake, The Leech lattice as a code for the Gaussian Channel, Info. Control, 19 (1971) 66–74 [Intro.].

[156] I. F. Blake, Distance properties of group codes for the Gaussian channel, SIAM J. Appl. Math. 23 (1972) 312–324 [Intro.].

[157] I. F. Blake, editor, *Algebraic Coding Theory: History and Development* (Dowden, Stroudsburg, P.A., 1973) [Intro.].

[158] I. F. Blake, Configuration matrices of group codes, *IEEE Trans. Info. Theory*, **20** (1974) 95–100 [Intro.].

[159] I. F. Blake, Permutation codes for discrete channels, *IEEE Trans. Info. Theory*, **20** (1974) 138–140 [18].

[160] I. F. Blake, Properties of generalized Pless codes, in: *Proc. 12th Allerton Conference on Circuit and System Theory*, Univ. Illinois, Urbana, Ill. (1974) pp. 787–789 [16].

[161] I. F. Blake, On a generalization of the Pless symmetry codes, *Info. and Control*, **27** (1975) 369–373 [16].

[162] I. F. Blake and R. C. Mullin, *The Mathematical Theory of Coding*, (Academic Press, New York 1975) [Intro.].

[163] H. F. Blichfeldt, The minimum value of quadratic forms, and the closest packing of spheres, *Math. Ann.*, **101** (1929) 605–608 [17].

[164] E. L. Blokh, On a decoding method for 3-error-correcting Bose-Chaudhuri codes, *Izv. Akad. Sci. USSR (Tekh. Kibern.)*, No. 3, Sect. IV, (1964) 30–37 (in Russian) [9].

[165] E. L. Blokh and V. V. Zyablov, Existence of linear concatenated binary codes with optimal error-correcting properties, *Problems of Info. Trans.*, **9** (4) (1973) 271–276 [10].

[166] E. L. Blokh and V. V. Zyablov, Coding of generalized concatenated codes, *Problems of Info. Trans.*, **10** (3) (1974) 218–222 [10].

[167] J. H. Blythe and K. Edgcombe, Net coding gain of error-correcting codes, *Proc. IEE*, **122** (1975) 609–614 [1].

[168] L. S. Borrow, Decoding augmented cutset codes, *IEEE Trans. Info. Theory*, **17** (1971) 218–220 [18].

[169] L. S. Bobrow and B. M. Franaszczuk, On cyclic codes generated by graphs, *Info. Control* **22** (1973) 296–301 [18].

[170] L. S. Bobrow and S. L. Hakimi, Graph theoretic *Q*-ary codes, *IEEE Trans. Info. Theory* **17** (1971) 215–218 [18].

[171] D. Borel, Application of the theory of cyclic codes to data transmission (in French), *Onde Elec.*, **48** (1968) 55–59 [1].

[172] L. F. Borodin, Some problems in the theory of group codes, *Radio Engineering and Electronic Physics*, **7** (8) (1962) 1199–1207 [11].

[173] R. C. Bose, On the construction of balanced incomplete block designs, *Ann. Eugenics*, **9** (1939) 353–399 [2].

[174] R. C. Bose, On some connections between the design of experiments and information theory, *Bull. Intern. Stat. Inst.*, **38** (1961) 257–271 [6].

[175] R. C. Bose, Strongly regular graphs, partial geometries and partially balanced designs, *Pac. J. Math.*, **13** (1963) 389–419 [21].

[176] R..C. Bose, Error correcting, error detecting and error locating codes, in: *Essays in Probability and Statistics*, Univ. of North Carolina Press, Chapel Hill, N.C., (1970) pp. 147–178 [1].

[177] R. C. Bose, Graphs and designs, in: *Finite Geometric Structures and their Applications* (Edizioni Cremonese, Rome, 1973) pp. 1–104 [2].

[178] R. C. Bose, Characterization problems of combinatorial graph theory, in: J. N. Srivastava et al., eds., *A Survey of Combinatorial Theory* (North-Holland, Amsterdam, 1973) [21].

[179] R. C. Bose, C. T. Abraham and S. P. Ghosh, File organization of records with multiple-valued attributes for multi-attribute queries, in: R. C. Bose and T. A. Dowling, eds., *Combinatorial Mathematics and its Applications*, (U. of N. Carolina Press, 1969) pp. 277–297 [9].

[180] R. C. Bose and R. C. Burton, A characterization of flat spaces in a finite geometry, and the uniqueness of the Hamming and MacDonald codes, *J. Combinatorial Theory*, **1** (1966) 96–104 [7].

[181] R. C. Bose and K. A. Bush, Orthogonal arrays of strength two and three, *Ann. Math. Stat.*, **23** (1952) 508–524 [5].

[182] R. C. Bose and J. G. Caldwell, Synchronizable error-correcting codes, *Info. Control*, **10** (1967) 616–630 [1].

[183] R. C. Bose and D. M. Mesner, On linear associative algebras corresponding to association schemes of partially balanced designs, *Ann. Math. Stat.*, **30** (1959) 21–38 [21].

[184] R. C. Bose and D. K. Ray – Chaudhuri, On a class of error correcting binary group codes, *Info. and Control*, **3** (1960) 68–79 [7].

[185] R. C. Bose and D. K. Ray – Chaudhuri, Further results on error correcting binary group codes, *Info. and Control*, **3** (1960) 279–290 [7].

[186] R. C. Bose and T. Shimamoto, Classification and analysis of partially balanced incomplete block designs with two associate classes, *J. Amer. Stat. Assoc.*, **47** (1952) 151–184 [21].

[187] R. C. Bose and S. S. Shrikhande, A note on a result in the theory of code construction, *Info. and Control*, **2** (1959) 183–194 [2].

[188] D. C. Bossen, b-adjacent error correction, *IBM J. Res. Devel.*, **14** (1970) 402–408 [10].

[189] D. C. Bossen and S. S. Yau, Redundant residue polynomial codes, *Info. and Control*, **13** (1968) 597–618 [10].

[190] N. Bourbaki, *Groups et Algèbres de Lie*, (Hermann, Paris, 1968) Ch. 4, 5 and 6, [19].

[191] K. Brayer, Error control techniques using binary symbol burst codes, *IEEE Trans. Commun.*, **16** (1968) 199–214 [1].

[192] K. Brayer, Improvement of digital HF communication through coding, *IEEE Trans. Commun.*, **16** (1968) 771–786 [1].

[193] K. Brayer, Error correction code performance on HF, troposcatter, and satellite channels, *IEEE Trans. Commun.*, **19** (1971) 781–789 [1].

[194] K. Brayer and O. Cardinale, Evaluation of error correction block encoding for high-speed HF data, *IEEE Trans. Commun.*, **15** (1967) 371–382 [1].

[195] J. G. Bredeson and S. L. Hakimi, Decoding of graph theoretic codes, *IEEE Trans. Info. Theory*, **13** (1967) 348–349 [18].

[196] J. Brenner and L. Cummings, The Hadamard maximum determinant problem, *Amer. Math. Mnthly*, **79** (1972) 626–630 [2].

[197] J. D. Bridewell and J. K. Wolf, Burst distance and multiple burst correction, *Bell Syst. Tech. J.*, **49** (1970) 889–909 [10].

[198] P. A. N. Briggs and K. R. Godfrey, Autocorrelation function of a 4-level m-sequence, *Electronics Letters*, **4** (1963) 232–233 [14].

[199] E. O. Brigham, *The Fast Fourier Transform* (Prentice-Hall, Englewood Cliffs, 1974) [14].

[200] M. Broué, Codes correcteurs d'erreurs auto-orthogonaux sur le corps à deux éléments et formes quadratiques entières définies positives à discriminant +1, in: *Comptes Rendus des Journées Mathématiques de la Société Math. de France* (Univ. Sci. Tech. Languedoc, Montpellier 1974) pp. 71–108 [19]. See also *Discrete Math.*, **17** (1977) 247–269.

[201] M. Broué and M. Enguehard, Polynômes des poids de certains codes et fonctions thêta de certains réseaux, *Ann. Sciènt. Ec. Norm. Sup.*, **5** (1972) 157–181 [19].

[201a] A. E. Brouwer, personal communication [17, A].

[202] D. A. H. Brown, Some error-correcting codes for certain transposition and transcription errors in decimal integers, *Computer Journal*, **17** (1974) 9–12 [1].

[203] D. T. Brown and F. F. Sellers, Jr., Error correction for IBM 800-bit-per-inch magnetic tape. *IBM J. Res. Dev.*, **14** (1970) 384–389 [1].

[204] T. A. Brown and J. H. Spencer, Minimization of ±1 matrices under line shifts, *Colloq. Math.*, **23** (1971) 165–171 [14].

[205] W. S. Brown, *ALTRAN User's Manual*, 4th ed. (Bell Laboratories, N.J., 1977) [19].

[206] N. G. de Bruijn, A combinatorial problem, *Nederl. Akad. Wetensch. Proc. Ser. A.* **49** (1946) 758–764 (= Indag. Math. **8** (1946) 461–467) [14].

[207] K. A. Brusilovskii, Hardware simulation of pseudorandom binary sequences, *Automation and Remote Control* **36** (5) (1975) 836–845 [14].

[208] M. M. Buchner, Jr., Computing the spectrum of a binary group code, *Bell Syst. Tech. J.*, **45** (1966) 441–449 [8].

[209] M. M. Buchner, Jr., Coding for numerical data transmission, *Bell Syst. Tech. J.*, **46** (1967) 1025–1041 [1].

[210] M. M. Buchner, Jr., The equivalence of certain Harper codes, *Bell Syst. Tech. J.*, **48** (1969) 3113–3130 [Intro.].

[211] W. Burnside, *Theory of Groups of Finite Order*, 2nd ed. (Dover, New York, 1955) [19].

[212] M. Burrow, *Representation Theory of Finite Groups* (Academic Press, New York, 1965) [8, 21].

[213] N. E. Burrowes, The timing of remote events using periodic telemetry samples, *Proc. I.R.E.E. Australia*, **27** (1966) 244–253 [14].

[214] H. O. Burton, A survey of error correcting techniques for data on telephone facilities, in: *Proc. Intern. Commun. Conf.*, San Francisco, Calif., 1970, pp. 16–25 to 16–32 [1].

[215] H. O. Burton, Some asymptotically optimal burst-correcting codes and their relation to single-error-correcting Reed-Solomon codes, *IEEE Trans. Info. Theory,* 17 (1971) 92–95 [10].

[216] H. O. Burton, Inversionless decoding of binary BCH codes, *IEEE Trans. Info. Theory,* 17 (1971) 464–466 [9].

[217] H. O. Burton and D. D. Sullivan, Errors and error control, *Proc. IEEE* 60 (1970) 1293–1301 [1].

[218] H. O. Burton, D. D. Sullivan, and S. Y. Tong, Generalized burst-trapping codes, *IEEE Trans. Info. Theory,* 17 (1971) 736–742 [10].

[219] H. O. Burton and E. J. Weldon, Jr., Cyclic product codes, *IEEE Trans. Info. Theory,* 11 (1965) 433–439 [18].

[220] K. A. Bush, Orthogonal arrays of index unity, *Ann. Math. Stat.,* 23 (1952) 426–434 [5, 10, 11].

[221] F. C. Bussemaker and J. J. Seidel, Symmetric Hadamard matrices of order 36, *Ann. New York Acad. Sci.,* 175 (Article 1) (1970) 66–79 [2, 21].

[222] W. H. Bussey, Galois field tables for $p^n \le 169$, *Bull. Amer. Math. Soc.,* 12 (1905) 22–38 [4].

[223] **W. H. Bussey, Tables of Galois fields of order less than 1,000, *Bull. Amer. Math. Soc.,* 16 (1910) 188–206 [4].**

C

[224] J. B. Cain and R. Simpson, Average digit error probability in decoding linear block codes, *Proc. IEEE* 56 (1968) 1731–1732 [1].

[225] T. W. Cairns, On the fast Fourier transform on finite abelian groups, *IEEE Trans. Computers,* 20 (1971) 569–571 [14].

[226] L. Calabi and W. E. Hartnett, A family of codes for the correction of substitution and synchronization errors, *IEEE Trans. Info. Theory,* 15 (1969) 102–106 [1].

[227] L. Calabi and W. E. Hartnett, Some general results of coding theory with applications to the study of codes for the correction of synchronization errors, *Info. Control,* 15 (1969) 235–249 [1].

[228] L. Calabi and E. Myrvaagnes, On the minimal weight of binary group codes, *IEEE Trans. Info. Theory,* 10 (1964) 385–387 [2, A].

[229] D. Calabro and J. K. Wolf, On the synthesis of two-dimensional arrays with desirable correlation properties, *Info. and Control,* 11 (1968) 537–560 [18].

[230] P. Calingaert, Two-dimensional parity checking, *J. ACM,* 8 (1961) 186–200 [18].

[231] P. J. Cameron, Biplanes, *Math. Zeit.,* 131 (1973) 85–101 [14].

[232] P. J. Cameron, Suborbits in transitive permutation groups, in: M. Hall, Jr., and J. H. Van Lint, eds., *Combinatorics,* (Reidel, Dordrecht, 1975) pp. 419–450 [21].

[233] P. J. Cameron, J.-M. Goethals, J. J. Seidel, and E. E. Shult, Line graphs, root systems and elliptic geometry, *J. Algebra,* 43 (1976) 305–327 [21].

[234] P. J. Cameron and J. H. Van Lint, *Graph Theory, Coding Theory and Block Designs,* London Math. Soc. Lecture Note Series, No. 19 (Cambridge Univ. Press, London, 1975) [Intro.].

[235] P. J. Cameron and J. J. Seidel, Quadratic forms over GF(2), *Koninkl. Nederl. Akad. Wetensch., Proc., Ser. A,* 76 (1973) 1–8 [15].

[236] P. J. Cameron, J. A. Thas and S. E. Payne, Polarities of generalized hexagons and perfect codes, *Geometriae Dedicata,* 5 (1976) 525–528 [6].

[237] P. Camion, A proof of some properties of Reed-Muller codes by means of the normal basis theorem, in: R. C. Bose and T. A. Dowling, eds., *Combinatorial Mathematics and its Applicants,* (Univ. North Carolina Press, Chapel Hill, N.C. 1969) Ch. 22 [9, 13].

[238] P. Camion, Abelian codes, Math. Res. Center, Univ. of Wisconsin, Rept. 1059 (1970) [18].

[239] P. Camion, Linear codes with given automorphism groups, *Discrete Math.,* 3 (1972) 33–45 [8].

[240] C. N. Campopiano, Construction of relatively maximal, systematic codes of specified minimum distance from linear recurring sequences of maximal period, *IEEE Trans. Info. Theory,* 6 (1960) 523–528 [18].

[241] J. R. Caprio, Strictly complex impulse-equivalent codes and subsets with very uniform amplitude distributions, *IEEE Trans. Info. Theory,* 15 (1969) 695–706 [14].

[242] L. Carlitz, Some problems involving primitive roots in a finite field, *Proc. Nat. Ac. Sci.*, **38** (1952) 314–318 [4].

[243] L. Carlitz, Primitive roots in a finite field, *Trans. Amer. Math. Soc.*, **73** (1952) 373–382 [4].

[244] L. Carlitz, Distribution of primitive roots in a finite field, *Quart. J. Math. Oxford*, (2) **4** (1953) 4–10 [4].

[245] L. Carlitz, The distribution of irreducible polynomials in several indeterminates, *Ill. J. Math.*, **7** (1963) 371–375 [4].

[246] L. Carlitz, The distribution of irreducible polynomials in several indeterminates II, *Canad. J. Math.*, **17** (1965) 261–266 [4].

[247] L. Carlitz and J. H. Hodges, Distribution of matrices in a finite field, *Pacific J. Math.* **6** (1956) 225–230 [16].

[248] L. Carlitz and J. H. Hodges, Distribution of bordered symmetric, skew and hermitian matrices in a finite field, *J. Reine Angew. Math.* **195** (1956) 192–201 [16].

[249] L. Carlitz and S. Uchiyama, Bounds for exponential sums, *Duke Math. J.*, **24** (1957) 37–41 [9].

[250] R. Carmichael, *Introduction to the Theory of Groups of Finite Order* (Dover, New York, 1956) [2, B].

[251] D. E. Carter, On the generation of pseudo-noise codes, *IEEE Trans. Aerospace and Electronic Systems*, **10** (1974) 898–899 [14].

[252] L. R. A. Casse, A solution to Beniamino Segre's "Problem $I_{r,q}$" for q even, *Atti Accad. Naz. Lincei, Rend. Cl. Sc. Fis. Mat. Natur.*, **46** (1969) 13–20 [11].

[253] S. R. Cavior, An upper bound associated with errors in Gray code, *IEEE Trans. Info. Theory*, **21** (1975) 596–599 [Intro.].

[254] C. Cazacu and D. Simovici, A new approach to some problems concerning polynomials over finite fields, *Info. and Control*, **22** (1973) 503–511 [4].

[255] A. G. Cerveira, On a class of wide-sense binary BCH codes whose minimum distances exceed the BCH bound, *IEEE Trans. Info. Theory*, **14** (1968) 784–785 [9].

[256] H. D. Chadwick and L. Kurz, Rank permutation group codes based on Kendall's correlation statistic, *IEEE Trans. Info. Theory*, **15** (1969) 306–315 [Intro.].

[257] H. D. Chadwick and I. S. Reed, The equivalence of rank permutation codes to a new class of binary codes, *IEEE Trans. Info. Theory*, **16** (1970) 640–641 [Intro.].

[257a] G. J. Chaitin, On the length of programs for computing finite binary sequences: statistical considerations, *J. ACM*, **16** (1969) 145–159 [Intro.].

[258] G. J. Chaitin, On the difficulty of computations, *IEEE Trans. Info. Theory*, **16** (1970) 5–9 [Intro.].

[258a] G. J. Chaitin, Information-theoretic computational complexity, *IEEE Trans. Info. Theory*, **20** (1974) 10–15 [Intro.].

[259] I. M. Chakravarti, Bounds on error correcting codes (non-random), in: R. C. Bose, ed., *Essays in Probability and Statistics*, (Univ. of North Carolina, 1970) Ch. 9 [17].

[260] I. M. Chakravarti, Some properties and applications of Hermitian varieties in a finite projective space $PG(N, q^2)$ in the construction of strongly regular graphs (Two-class association schemes) and block designs, *J. Combin. Theory*, **11B** (1971) 268–283 [21].

[261] I. M. Chakravarti and S. Ikeda, Construction of association schemes and designs from finite groups, *J. Combin. Theory*, **13A** (1972) 207–219 [21].

[262] J. A. Chang, Generation of 5-level maximal-length sequences, *Electronics Letters*, **2** (1966) 258 [14].

[263] R. W. Chang and E. Y. Ho, On fast start-up data communication systems using pseudo-random training sequences, *Bell Syst. Tech. J.*, **51** (1972) 2013–2027 [14].

[264] D. Chase, A class of algorithms for decoding block codes with channel measurement information, *IEEE Trans. Info. Theory*, **18** (1970) 170–182 [1, 16].

[265] D. Chase, A combined coding and modulation approach for communication over dispersive channels, *IEEE Trans. Commun.*, **21** (1973) 159–174 [1].

[266] C.-L. Chen, Computer results on the minimum distance of some binary cyclic codes, *IEEE Trans. Info. Theory*, **16** (1970) 359–360, and Appendix D of Peterson and Weldon, Ref. [1040]; [7, 9, 18, A].

[267] C.-L. Chen, On decoding Euclidean geometry codes, *Proc. 4th Hawaii Int. Conf. Systems Sciences*, (1971) pp. 111–113 [13].

[268] C.-L. Chen, On majority-logic decoding of finite geometry codes, *IEEE Trans. Info. Theory*, **17** (1971) 332–336 [13].

[269] C.-L. Chen, Note on majority-logic decoding of finite geometry codes, *IEEE Trans. Info. Theory*, **18** (1972) 539–541 [13].

[270] C.-L. Chen, On shortened finite geometry codes, *Info. and Control*, **20** (1972) 216–221 [13].

[271] C.-L. Chen, Some efficient majority-logic decodable codes, *NTC'73 Conference Record*, 1973 [13].

[273] C.-L. Chen and R. T. Chien, Recent developments in majority-logic decoding, *Proc. 26th Annual National Electronics Conf.*, (1970) pp. 458–461 [13].

[274] C.-L. Chen and S. Lin, Further results on polynomial codes, *Info. and Control*, **15** (1969) 38–60 [9, 13].

[275] C.-L. Chen, W. W. Peterson, and E. J. Weldon, Jr., Some results on quasi-cyclic codes, *Info. and Control*, **15** (1969) 407–423 [16].

[276] C.-L. Chen and W. T. Warren, A note on one-step majority-logic decodable codes, *IEEE Trans. Info. Theory*, **19** (1973) 135–137 [13].

[277] Y. Chen, The Steiner system S(3, 6, 26), *J. of Geometry*, **2** (1) (1972) 7–28 [17, A].

[278] R. T. Chien, Orthogonal matrices, error-correcting codes and load-sharing matrix switches, *IEEE Trans. Computers*, **8** (1959) 400 [1].

[279] R. T. Chien, Cyclic decoding procedure for the Bose–Chaudhuri–Hocquenghem codes, *IEEE Trans. Info. Theory*, **10** (1964) 357–363 [9].

[280] R. T. Chien, Burst-correcting codes with high-speed decoding, *IEEE Trans. Info. Theory*, **15** (1969) 109–113 [10].

[281] R. T. Chien, Coding for error control in a computer-manufacturers environment: A foreword, *IBM J. Res. Devel.*, **14** (1970) 342 [1].

[282] R. T. Chien, Block-coding techniques for reliable data transmission, *IEEE Trans. Comm.*, **19** (1971) 743–751 [1, 9].

[283] R. T. Chien, A new proof of the BCH bound, *IEEE Trans. Info. Theory*, **18** (1972) 541 [8].

[284] R. T. Chien, Memory error control: beyond parity, *Spectrum*, **10** (7) (1973) 18–23 [1].

[285] R. T. Chien, L. R. Bahl and D. T. Tang, Correction of two erasure bursts, *IEEE Trans. Info. Theory*, **15** (1969) 186–187 [10].

[286] R. T. Chien and D. M. Choy, Algebraic generalization of BCH–Goppa–Helgert codes, *IEEE Trans. Info. Theory*, **21** (1975) 70–79 [8, 12, 18, A].

[287] R. T. Chien, V. E. Clayton, P. E. Boudreau and R. R. Locke, Error correction in a radio-based data communications system, *IEEE Trans. Commun.*, **23** (1975) 458–462 [1].

[288] R. T. Chien, B. D. Cunningham, and I. B. Oldham, Hybrid methods for finding roots of a polynomial – with application to BCH decoding, *IEEE Trans. Info. Theory*, **15** (1969) 329–335 [9].

[289] R. T. Chien and W. D. Frazer, An application of coding theory to document retrieval, *IEEE Trans. Info. Theory*, **12** (1966) 92–96 [9].

[290] R. T. Chien, C. V. Freiman and D. T. Tang, Error correction and circuits on the n-cube, *Proc. 2nd Allerton Conf. Circuit Syst. Theory*, Sept. 1964, Univ. of Ill., Monticello, Ill., pp. 899–912 [Intro.].

[291] R. T. Chien and V. Lum, On Golay's perfect codes and step-by-step decoding, *IEEE Trans. Info. Theory*, **12** (1966) 403–404 [16].

[292] R. T. Chien and S. W. Ng, Dual product codes for correction of multiple low-density burst errors, *IEEE Trans. Info. Theory*, **19** (1973) 672–677 [10, 18].

[293] R. T. Chien and K. K. Tzeng. Iterative decoding of BCH codes, Coordinated Science Lab, Univ. of Illinois, Urbana, Ill., Tech. Report, Feb. 1968 [9].

[294] D. K. Chow, On threshold decoding of cyclic codes, *Info. and Control*, **13**, (1968) 471–483 [13].

[295] G. C. Clark, Jr., and R. C. Davis, A decoding algorithm for group codes and convolution codes based on the fast Fourier–Hadamard transform, unpublished paper presented at IEEE 1969 Intl. Symp. on Info. Theory, Ellenville, NY [14].

[296] J. Cocke, Lossless symbol coding with nonprimes, *IEEE Trans. Info. Theory*, **5** (1959) 33–34 [7].

[297] G. Cohen, P. Godlewski, and S. Perrine, Sur les idempotents des codes, *C. R. Acad. Sci. Paris, Sér. A–B*, **284** (1977) A509–A512.

[298] J. H. E. Cohn, On the value of determinants, *Proc. Amer. Math. Soc.*, **14** (1963) 581–588 [2].

[299] M. Cohn, Affine m-ary Gray codes, *Info. and Control*, **6** (1963) 70–78 [Intro.].

[300] R. J. Collens, A listing of balanced incomplete block designs, in: *Proc. 4th S-E Conf. Combinatorics, Graph Theory and Computing*, (Utilitas Math., Winnipeg, 1973) pp. 187–231 [2, 17].

[301] J. H. Conway, A tabulation of some information concerning finite fields, in: R. F. Churchhouse and J.-C. Herz eds., *Computers in Mathematical Research*, (North-Holland, Amsterdam, 1968) pp. 37–50 [3, 4].

[302] J. H. Conway, A perfect group of order 8,315,553,613,086,720,000 and the sporadic simple groups, *Proc. Nat. Acad. Sci.*, **61** (1968) 398–400 [19, 20].

[303] J. H. Conway, A group of order 8,315,553,613,086,720,000, *Bull. London Math. Soc.*, **1** (1969) 79–88 [19, 20].

[304] J. H. Conway, A characterization of Leech's lattice, *Invent. Math.*, **7** (1969) 137–142 [19, 20].

[305] J. H. Conway, Groups, lattices, and quadratic forms, in: *Computers in Algebra and Number Theory*, SIAM-AMS Proc. IV, (Am. Math. Soc., Providence, R.I. 1971) pp. 135–139 [19, 20].

[306] J. H. Conway, Three lectures on exceptional groups, in: *Finite Simple Groups*, ed. M. B. Powell and G. Higman (Academic Press, New York, 1971) pp. 215–247 [16, 19, 20].

[307] A. B. Cooper and W. C. Gore, A recent result concerning the dual of polynomial codes, *IEEE Trans. on Info. Theory*, **16** (1970) 638–640 [13].

[308] J. T. Cordaro and T. J. Wagner, Optimum $(N, 2)$ codes for small values of channel error probability, *IEEE Trans. Info. Theory*, **13** (1967) 349–350 [A].

[309] F. P. Corr, Statistical evaluation of error detection cyclic codes for data transmission, *IEEE Trans. Commun. Systems*, **12** (1964) 211–216 [1].

[310] T. M. Cover, Enumerative source encoding, *IEEE Trans. Info. Theory*, **19** (1973) 73–77 [17].

[311] W. R. Cowell, The use of group codes in error detection and message retransmission, *IEEE Trans. Info. Theory*, **7** (1961) 168–171 [1].

[312] J. W. Cowles and G. I. Davida, Decoding of triple-error-correcting BCH codes, *Electronics Letters*, **8** (1972) 584 [9].

[313] H. S. M. Coxeter, *Regular Polytopes*, 2nd Edition, (MacMillan, New York, 1963) [14].

[314] H. S. M. Coxeter and W. O. J. Moser, *Generators and Relations for Discrete Groups*, 2nd Edition, (Springer, Berlin, 1965) [9, 16].

[315] H. H. Crapo and G.-C. Rota, *On the Foundations of Combinatorial Theory: Combinatorial Geometries*, (MIT Press, Cambridge, Mass., 1970) (Preliminary version) [17].

[316] H. C. Crick, J. S. Griffiths and L. E. Orgel, Codes without commas, *Proc. Nat. Acad. Sci.*, **43** (1957) 416–421 [Intro.].

[317] T. R. Crimmins, Minimization of mean-square error for data transmitted via group codes, *IEEE Trans. Info. Theory*, **15** (1969) 72–78 [1].

[318] T. R. Crimmins and H. M. Horwitz, Mean-square-error optimum coset leaders for group codes, *IEEE Trans. Info. Theory*, **16** (1970) 429–432 [1].

[319] T. R. Crimmins, H. M. Horwitz, C. J. Palermo and R. V. Palermo, Minimization of mean-square error for data transmission via group codes, *IEEE Trans. Info. Theory*, **13** (1967) 72–78 [1].

[320] I. G. Cumming, Autocorrelation function and spectrum of a filtered, pseudo-random binary sequence, *Proc. IEE*, **114** (1967) 1360–1362 [14].

[321] C. W. Curtis and I. Reiner, *Representation Theory of Finite Groups and Associative Algebras*, (Wiley, New York, 1962) [8].

[322] R. T. Curtis, A new combinatorial approach to M_{24}, *Math. Proc. Camb. Phil. Soc.*, **79** (1976) 25–41 [18, 20].

[323] R. T. Curtis, The maximal subgroups of M_{24}, *Math Proc. Cambridge Phil. Soc.*, **81** (1977) 185–192.

D

[324] E. C. Dade, Answer to a question of R. Brauer, *J. Algebra*, **1** (1964) 1–4 [19].

[325] E. C. Dade, private communication [19].

[326] G. Dagnino, On a new class of binary group codes, *Calcolo*, **5** (1968) 277–294 [18].

[327] R. M. Damerell, On Moore graphs, *Proc. Comb. Phil. Soc.*, **74** (1973) 227–236 [21].

[328] L. Danzer and V. Klee, Length of snakes in boxes, *J. Combin. Theory*, **2** (1967) 258–265 [Intro.].

[329] H. Davenport, On primitive roots in finite fields, *Quart. J. Math. Oxford*, (1) **8** (1937) 308–312 [4].

[330] H. Davenport, Bases for finite fields, *J. London Math. Soc.*, **43** (1968) 21–39 [4]. See also Addendum, **44** (1969) p. 378.

[331] F. N. David and D. E. Barton, *Combinatorial Chance*, (Hafner, New York, 1962) [8].

[332] F. N. David, M. G. Kendall and D. E. Barton, *Symmetric Function and Allied Tables*, (Cambridge Univ. Press, London, 1966) [8].

[333] G. I. Davida, Decoding of BCH codes, *Electronics Letters*, **7** (1971) 664 [9].

[334] G. I. Davida, A new error-locating polynomial for decoding of BCH codes, *IEEE Trans. Info. Theory*, **21** (1975) 235–236 [9].

[335] D. W. Davies, Longest "separated" paths and loops in an N-cube, *IEEE Trans. Computers*, **14** (1965) 261 [Intro.].

[336] W. D. T. Davies, Using the binary maximum length sequence for the identification of system dynamics, *Proc. IEE*, **114** (1967) 1582–1584 [14].

[337] A. A. Davydov, Higher data-storage reliability with the aid of programmable error-correcting codes, *Automation and Remote Control*, **34** (1) (1973) 137–143 [1].

[338] A. A. Davydov and G. M. Tenegol'ts, Codes correcting errors in the exchange of information between computers, *Eng. Cyber.*, **9** (1971) 700–706 [1].

[339] D. E. Daykin, Distribution of bordered persymmetric matrices in a finite field, *J. Reine Angew. Math.*, **203** (1960) 47–54 [16].

[340] D. E. Daykin, Generation of irreducible polynomials over a finite field, *Amer. Math. Mnthly*, **72** (1965) 646–648 [4].

[341] J. A. Decker, Jr., Hadamard-transform spectrometry: A new analytical technique, *Analytical Chemistry*, **44** (1972) 127A–134A [2].

[342] P. Deligne, La conjecture de Weil I, Institut des Hautes Etudes Scientifiques, *Publ. Math.* No. 43, pp. 273–307 [9].

[343] P. Delsarte, A geometrical approach to a class of cyclic codes, *J. Combin. Theory*, **6** (1969) 340–358 [13].

[344] P. Delsarte, On cyclic codes that are invariant under the general linear group, *IEEE Trans. Info. Theory*, **16** (1970) 760–769 [8, 13].

[345] P. Delsarte, Automorphisms of abelian codes, *Philips Res. Rept.*, **25** (1970) 389–402 [18].

[346] P. Delsarte, BCH bounds for a class of cyclic codes, *SIAM Jnl. Applied Math.*, **19** (1970) 420–429 [8].

[347] P. Delsarte, Weights of p-ary Abelian codes, *Philips Res. Repts.*, **26** (1971) 145–153 [15].

[348] P. Delsarte, Majority-logic decodable codes derived from finite inversive planes, *Info. and Control*, **18** (1971) 319–325 [13].

[349] P. Delsarte, Weights of linear codes and strongly regular normed spaces, *Discrete Math.*, **3** (1972) 47–64 [21].

[350] P. Delsarte, Bounds for unrestricted codes, by linear programming, *Philips Res. Reports*, **27** (1972) 272–289 [5, 6, 17].

[351] P. Delsarte, Four fundamental parameters of a code and their combinatorial significance, *Info. and Control*, **23** (1973) 407–438 [5, 6, 17].

[352] P. Delsarte, An algebraic approach to the association schemes of coding theory, *Philips Research Reports Supplements*, No. 10 (1973) [5, 6, 17, 21].

[353] P. Delsarte, Association schemes in certain lattices, Report R 241, MBLE Res. Lab., Brussels, 1974 [21].

[354] P. Delsarte, Association schemes and t-designs in regular semi-lattices, *J. Combin. Theory*, **20** A (1976) 230–243 [21].

[355] P. Delsarte, Bilinear forms over a finite field, with applications to coding theory, *J. Combin. Theory*, **25A** (1978) 226–241 [21].

[356] P. Delsarte, The association schemes of coding theory, in: M. Hall, Jr. and J. H. van Lint, *Combinatorics*, (Reidel, Dordrecht, 1975) pp. 143–161 [21].

[357] P. Delsarte, Hahn polynomials, discrete harmonics and t-designs, *SIAM J. Appl. Math.*, **34** (1978) 157–166.

[358] P. Delsarte, Regular schemes over an abelian group, Report R302, MBLE Res. Lab., Brussels, 1975 [21].

[359] P. Delsarte, On subfield subcodes of Reed–Solomon codes, *IEEE Trans. Info. Theory*, **21** (1975) 575–576 [7, 10, 12].

[360] P. Delsarte, Distance distribution of functions over Hamming spaces, *Philips Res. Reports*, **30** (1975) 1–8 [9].

[361] P. Delsarte, Properties and applications of the recurrence $F(i+1, k+1, n+1) = q^{k+1}F(i, k+1, n) - q^k F(i, k, n)$, *SIAM J. Appl. Math.*, **31** (1976) 262–270 [21].

[362] P. Delsarte and J.-M. Goethals, Irreducible binary cyclic codes of even dimension, in: *Combinatorial Mathematics and its Applications*, Proc. Second Chapel Hill Conference, May 1970 (Univ. of N. Carolina, Chapel Hill, N.C., 1970) pp. 100–113 [8, 15].

[363] P. Delsarte and J.-M. Goethals, Unrestricted codes with the Golay parameters are unique, *Discrete Math.*, **12** (1975) 211–224 [20].

[364] P. Delsarte and J.-M. Goethals, Alternating bilinear forms over GF(q), *J. Combin. Theory*, **19A** (1975) 26–50 [15, 21, A].

[365] P. Delsarte, J.-M. Goethals and F. J. MacWilliams, On generalized Reed–Muller codes and their relatives, *Info. and Control*, **16** (1974) 403–442 [13].

[366] P. Delsarte, J.-M. Goethals, and J. J. Seidel, Orthogonal matrices with zero diagonal II., *Canad. J. Math.*, **23** (1971) 816–832 [2].

[367] P. Delsarte, J.-M. Goethals, and J. J. Seidel, Bounds for systems of lines and Jacobi polynomials, *Philips Research Reports*, **30** (1975) 91*–105* [21].

[368] P. Delsarte, W. Haemers and C. Weug, Unpublished tables, 1974 [A].

[369] P. Delsarte and R. J. McEliece, Zeroes of functions in finite abelian group algebras, *Amer. J. Math.*, **98** (1976) 197–224 [15].

[370] P. Dembowski, *Finite Geometries* (Springer, Berlin, 1968) [2, 17, B].

[371] J. Dênes and A. D. Keedwell, *Latin Squares and their Applications*, (Academic Press, New York, 1974) [11, 18].

[372] **R. H. F. Denniston, Some new 5-designs, *Bull. London Math. Soc.*, 8 (1976) 263–267 [17, A].**

[373] M. Deza, Une propriété extrémale des plans projectifs finis dans une classe de code équidistants, *Discrete Math.*, **6** (1973) 343–352 [1].

[374] L. E. Dickson, *Linear groups with an exposition of the Galois field theory* (Dover, New York, 1958) [15].

[375] J. Dieudonnè and J. B. Carrell, *Invariant Theory, Old and New*, (Academic Press, New York, 1971) [19].

[376] J. F. Dillon, Elementary Hadamard difference sets, Ph.D. Thesis, Univ. of Maryland, 1974 [14].

[377] J. F. Dillon, Elementary Hadamard difference sets, in: *Proc. 6th S-E Conf. Combinatorics, Graph Theory and Computing*, (Utilitas Math., Winnepeg, 1975) pp. 237–249 [14].

[377a] *Discrete Mathematics*, Special triple issue on coding theory, 3 (Numbers 1, 2, 3, 1972) [Intro.].

[378] R. L. Dobrushin, Survey of Soviet research in information theory, *IEEE Trans. Info. Theory*, **18** (1972) 703–724 [1, 13].

[379] M. Doob, On graph products and association schemes, *Utilitas Math.*, **1** (1972) 291–302 [21].

[380] B. G. Dorsch, A decoding algorithm for binary block codes and j-ary output channels, *IEEE Trans. Info. Theory*, **20** (1974) 391–394 [1].

[381] P. Doubilet, G.-C. Rota, and J. Stein, On the foundations of combinatorial theory: IX Combinatorial methods in invariant theory, *Studies in Appl. Math.*, **53** (1974) 185–216 [19].

[382] R. J. Douglas, Some results on the maximum length of circuits of spread k in the d-cube, *J. Combin. Theory*, **6** (1969) 323–339 [Intro.].

[383] R. J. Douglas, Upper bounds on the length of circuits of even spread in the d-cube, *J. Combin. Theory*, **7** (1969) 206–214 [Intro.].

[384] T. A. Dowling, A class of tri-weight cyclic codes, *Instit. Statistic Mimeo series No. 600.3*, (Univ. North Carolina, Chapel Hill, N.C., Jan. 1969) [15].

[385] T. A. Dowling, Codes, packings and the critical problem, *Atti Del Convegno Di Geometria Combinatoria e Sue Applicazioni*, Università Degli Studi di Perugia, Istituto di Matematica (1971) 209–224 [11, 17].

[386] J. Doyen and A. Rosa, A bibliography and survey of Steiner systems, *Bollettino Unione Matematica Italiana*, **7** (1973) 392–419 [2, 17, A].

[387] N. Q. Duc, Pseudostep orthogonalization: a new threshold decoding algorithm, *IEEE Trans. Info. Theory*, 17 (1971) 766-768 [13].

[388] N. Q. Duc, On a necessary condition for L-step orthogonalization of linear codes and its applications, *Info. and Control*, 22 (1973) 123-131 [13].

[389] N. Q. Duc, On the Lin-Weldon majority-logic decoding algorithm for product codes, *IEEE Trans. Info. Theory*, 19 (1973) 581-583 [18].

[390] N. Q. Duc and L. V. Skattebol, Algorithms for majority-logic check equations of maximal-length codes, *Electron. Letters*, 5 (1969) 577-579 [13].

[391] N. Q. Duc and L. V. Skattebol, Further results on majority-logic decoding of product codes, *IEEE Trans. Info. Theory*, 18 (1972) 308-310 [18].

[392] J. O. Duffy, Detailed design of a Reed-Muller (32, 6) block encoder, *JPL Space Programs Summary*, Vol. 37-47-III, (1967) 263-267 [14].

[393] C. F. Dunkl, A Krawtchouk polynomial addition theorem and wreath products of symmetric groups, *Indiana Univ. Math. J.*, 25 (1976) 335-358 [5].

[394] C. F. Dunkl and D. E. Ramirez, Krawchouk polynomials and the symmetrization of hypergroups, *SIAM J. Math. Anal.*, 5 (1974) 351-366 [5].

[395] P. F. Duvall and R. E. Kibler, On the parity of the frequency of cycle lengths of shift register sequences, *J. Combin. Theory*, 18A (1975) 357-361 [14].

[396] H. Dym and H. P. McKean, *Fourier Series and Integrals*, (Academic Press, New York, 1972) [5, 8].

[397] V. I. Dyn'kin and G. M. Tenengol'ts, A class of cyclic codes with a majority decoding scheme, *Problems Info. Trans.*, 5 (No. 1, 1969) 1-11 [13].

E

[398] G. K. Eagleson, A characterization theorem for positive definite sequences on the Krawtchouk polynomials, *Australian J. Stat.*, 11 (1969) 29-38 [5].

[399] W. L. Eastman, On the construction of comma-free codes, *IEEE Trans. Info. Theory*, 11 (1965) 263-267 [Intro.].

[400] W. L. Eastman and S. Even, On synchronizable and PSK-synchronizable block codes, *IEEE Trans. Info. Theory*, 10 (1964) 351-356 [1].

[401] P. J. Eberlein, A two parameter test matrix, *Math. Comp.*, 18 (1964) 296-298 [21].

[402] P. M. Ebert and S. Y. Tong, Convolutional Reed-Solomon codes, *Bell Syst. Tech. J.*, 48 (1969) 729-742 [10].

[403] H. Ehlich, Determinantenabschätzung für binäre Matrizen mit $n \equiv 3$ mod 4, *Math. Zeit.*, 84 (1964) 438-447 [2].

[404] H. Ehlich and K. Zeller, Binäre Matrizen, *Zeit. Angew. Math. Mech.*, 42 (1962) 20-21 [2].

[405] R. W. Ehrich and S. S. Yau, A class of high-speed decoders for linear cyclic binary codes, *IEEE Trans. Info. Theory*, 15 (1969) 113-117 [16].

[406] P. Elias, Error-free coding, *IEEE Trans. Info. Theory*, 4 (1954) 29-37 [18].

[407] E. O. Elliott, Estimates of error rates for codes on burst-noise channels, *Bell Syst. Tech. J.*, 42 (1963) 1977-1997 [10].

[408] E. O. Elliott, On the use and performance of error-voiding and error-marking codes, *Bell Syst. Tech. J.*, 45 (1966) 1273-1283 [1].

[409] B. Elspas, A conjecture on binary nongroup codes, *IEEE Trans. Info. Theory*, 11 (1965) 599-600 [17].

[410] A. Erdélyi et al., *Higher Transcendental Functions*, Vols. 1-3, (McGraw-Hill, New York, 1953) [5].

[411] P. Erdös and H. Hanani, On a limit theorem in combinatorical analysis, *Publ. Math. Debrecen*, 10 (1963) 10-13 [17].

[412] J. V. Evans and T. Hagfors, eds., *Radar Astronomy*, (McGraw-Hill, New York, 1968) [14].

[413] S. Even, Snake-in-the box codes, *IEEE Trans. Computers*, 12 (1963) 18 [Intro.].

F

[414] D. D. Falconer, A hybrid coding scheme for discrete memoryless channels, *Bell Syst. Tech. J.*, 48 (1969) 691-728 [1].

[415] R. M. Fano, *Transmission of Information*, (MIT Press, Cambridge, MA, 1961) Appendix B [10].

[416] S. Farber, On signal selection for the incoherent channel, Abstract 4.5, *International Symposuim on Info. Theory, Program and Abstracts of Papers*, (IEEE New York, 1969) [1].

[417] E. H. Farr, unpublished [9].

[418] P. G. Farrell, Linear binary anticodes, *Electronics Letters*, 6 (1970) 419–421 [17].

[419] P. G. Farrell and Z. Al-Bandar, Multilevel single-error-correcting codes, *Electronics Letters*, 10 (1974) 347–348 [7].

[420] P. G. Farrell and A. A. M. Farrag, Further properties of linear binary anticodes, *Electronics Letters*, 10 (1974) 340–341 [17].

[421] P. G. Farrell and A. A. M. Farrag, New error-control codes derived from anti-codes, in preparation [17].

[422] H. Feistel, Cryptography and computer privacy, *Scientific American*, 228 (No. 5, 1973) 15–23 [1].

[423] H. Feistel, W. A. Notz and J. L. Smith, Some cryptographic techniques for machine-to-machine data communications, *Proc. IEEE*, 63 (1975) 1545–1554 [1, 14].

[424] W. Feit, Some remarks on weight functions of spaces over GF(2), unpublished (1972) [19].

[425] W. Feit, On weight functions of self orthogonal spaces over GF(3), unpublished (1972) [19].

[426] W. Feit, A self-dual even (96, 48, 16) code, *IEEE Trans. Info. Theory*, 20 (1974) 136–138 [19].

[427] W. Feller, *An Introduction to Probability Theory and its Applications*, Vol. 2 (Wiley, New York, 1966) [9].

[428] F. A. Ficken, *The Simplex Method of Linear Programming*, (Holt, Rinehart and Winston, New York, 1961) [17].

[429] J. P. Fillmore and M. L. Marx, Linear recursive sequences, *SIAM Review*, 10 (1968) 342–353 [14].

[430] M. A. Fischler, Minimax combinational coding, *IEEE Trans. Info. Theory*, 14 (1968) 139–150 [1].

[431] C. S. Fisher, The death of a mathematical theory: A study in the sociology of knowledge, *Arch. Hist. Exact Sci.*, 3 (1967) 136–159 [19].

[432] I. Flores, Reflected number systems, *IEEE Trans. Computers*, 5 (1956) 79–82 [Intro.].

[433] A. B. Fontaine and W. W. Peterson, On coding for the binary symmetric channel, *Trans. AIEE*, 77 Part I (1958) 638–646 [18].

[434] A. B. Fontaine and W. W. Peterson, Group code equivalence and optimum codes, *IEEE Trans. Info. Theory*, 5 (Special Supplement) (1959) 60–70 [2, 18, 19].

[435] G. D. Forney, Jr., On decoding BCH codes, *IEEE Trans. Info. Theory*, 11 (1965) 549–557 [8, 9].

[436] G. D. Forney, Jr., *Concatenated Codes*, (M.I.T. Press, Cambridge, MA., 1966) [10, 11, 18].

[437] G. D. Forney, Jr., Generalized minimum distance decoding, *IEEE Trans. Info. Theory*, 12 (1966) 125–131 [1].

[438] G. D. Forney, Jr., Coding and its application in space communications, *IEEE Spectrum*, 7 (June 1970) 47–58 [1].

[439] G. D. Forney, Jr., Convolutional codes I: algebraic structure, *IEEE Trans. Info. Theory*, 16 (1970) 720–738 and 17 (1971) 360 [1].

[440] G. D. Forney, Jr., Burst-correcting codes for the classic bursty channel, *IEEE Trans. Commun.*, 19 (1971) 772–781 [10].

[441] G. D. Forney, Jr., Maximum likelihood estimation of digital sequences in the presence of intersymbol interference, *IEEE Trans. Info. Theory*, 18 (1972) 363–378 [1].

[442] G. D. Forney, Jr., The Viterbi algorithm, *Proc. IEEE*, 61 (1973) 268–278 [1].

[443] G. D. Forney, Jr., Convolution codes, II: maximum likelihood decoding, *Info. and Control*, 25 (1974) 222–226 [1].

[444] G. D. Forney, Jr., Convolutional codes III: sequential decoding, *Info. and Control*, 25 (1974) 267–297 [1].

[445] G. D. Forney, Jr., and E. K. Bower, A high-speed sequential decoder: prototype design and test, *IEEE Trans. on Commun.*, 19 (1971) 821–835 [1].

[446] M. K. Fort, Jr., and G. A. Hedlund, Minimal coverings of pairs by triples, *Pacific J. Math.*, 8 (1958) 709–719 [17].

[447] H. O. Foulkes, Theorems of Kakeya and Pólya on power-sums, *Math. Zeit.*, **65** (1956) 345–352 [8].

[448] P. A. Franaszek, Sequence-state coding for digital transmission, *Bell Syst. Tech. J.*, **47** (1968) 143–157 [1].

[449] A. G. Franco and L. J. Saporta, Performance of random error correcting codes on the switched telephone network, *IEEE Trans. Commun.*, **15** (1967) 861–864 [1].

[450] H. Fredricksen, Error Correction for Deep Space Network Teletype Circuits, Report 32-1275, Jet Propulsion Lab., Pasadena, CA., June 1968 [1].

[451] H. Fredricksen, The lexicographically least de Bruijn cycle, *J. Combin. Theory*, **9** (1970) 1–5 [14].

[452] H. Fredricksen, Generation of the Ford sequence of length 2^n, n large, *J. Combin. Theory*, **12A** (1972) 153–154 [14].

[453] H. Fredricksen, A class of nonlinear de Bruijn cycles, *J. Combin. Theory*, **19A** (1975) 192–199 [14].

[454] H. Fredricksen and I. Kessler, Lexicographic compositions and de Bruijn sequences, in: *Proc. 6th S-E Conf. Combinatorics, Graph Theory and Computing*, (Utilitas Math., Winnipeg, 1975) pp. 315–339 [14].

[455] S. Fredricsson, Pseudo-randomness properties of binary shift register sequences, *IEEE Trans. Info. Theory*, **21** (1975) 115–120 [14].

[456] C. V. Freiman, Optimal error detection codes for completely asymmetric binary channels, *Info. and Control*, **5** (1962) 64–71 [1, 17].

[457] C. V. Freiman, Upper bounds for fixed-weight codes of specified minimum distance, *IEEE Trans. Info. Theory*, **10** (1964) 246–248 [17].

[458] C. V. Freiman, and J. P. Robinson, A comparison of block and recurrent codes for the correction of independent errors, *IEEE Trans. Info. Theory*, **11** (1965) 445–449 [1].

[459] C. V. Freiman and A. D. Wyner, Optimum block codes for noiseless input restricted channels, *Info. and Control*, **7** (1964) 398–415 [1].

[460] A. H. Frey and R. E. Kavanaugh, Every data bit counts in transmission cleanup, *Electronics*, (Jan. 22 1968) 77–83 [1].

[461] G. Frobenius, Über die Charaktere der mehrfach transitiven Gruppen, *Sitzungberichte Ak. Berlin*, (1904) 558–571 [20].

[462] C. Fujiwara, M. Kasahara, Y. Tezuka and Y. Kasahara, On codes for burst-error correction, *Electron. Commun. Japan*, **53-A** (7) (1970) 1–7 [10].

G

[463] H. F. Gaines, *Cryptanalysis*, (Dover, New York, 1956) [1].

[464] R. G. Gallagher, Information theory and reliable communication (Wiley, New York, 1968) [Intro., 1, 9, 17].

[465] R. G. Gallager, The random coding bound is tight for the average code, *IEEE Trans. Info. Theory*, **19** (1973) 244–246 [17].

[466] D. Garbe and J. L. Mennicke, Some remarks on the Mathieu groups, *Canad. Math. Bull.*, **7** (1964) 201–212 [20].

[467] B. Gardner, Results on coverings of pairs with special reference to coverings by quintuples, in: *Proc. Manitoba Conf. Num. Math.*, (Dept. of Computer Sci., Univ. of Manitoba, Winnipeg, 1971) pp. 169–178 [A].

[468] M. Gardner, The curious properties of the Gray code..., *Scientific American*, **227** (August) (1972) 106–109 [Intro.].

[469] P. R. Geffe, An open letter to communication engineers, *Proc. IEEE*, **55** (1967) 2173 [1, 14].

[470] P. R. Geffe, How to protect data with ciphers that are really hard to break, *Electronics*, **46** (Jan. 4) (1973) 99–101 [1, 14].

[471] S. I. Gelfand, R. L. Dobrushin and M. S. Pinsker, On the complexity of coding, in: B. N. Petrov and F. Csáki, eds., *2nd Internat. Symp. Info. Theory*, Akad. Kiadó, Budapest, 1973 pp. 177–184 [Intro.].

[472] W. M. Gentleman, Matrix multiplication and fast Fourier transforms, *Bell Syst. Tech. J.*, **47** (1968) 1099–1103 [14].

[473] A. V. Geramita, J. M. Geramita and J. S. Wallis, Orthogonal designs, *Linear and Multilinear Algebra*, **3** (1976) 281–306 [2].

[474] A. V. Geramita, N. J. Pullman and J. S. Wallis, Families of weighing matrices, *Bull. Aust. Math. Soc.*, **10** (1974) 119–122 [2].

[474a] A. V. Geramita and J. H. Verner, Orthogonal designs with zero diagonal, ·*Canad. J. Math.*, **28** (1976) 215–224 [2].

[475] A. V. Geramita and J. S. Wallis, Orthogonal designs II, *Aequat. Math.*, **13** (1975) 299–313 [2].

[476] A. V. Geramita and J. S. Wallis, Orthogonal designs III; weighing matrices, *Utilitas Mathematics*, **6** (1974) 209–236 [2].

[476a] A. V. Geramita and J. S. Wallis, A survey of orthogonal designs, in: *Proc. 4th Manitoba Conf. Num. Math.*, (1974) pp. 121–168 [2].

[477] A. V. Geramita and J. S. Wallis, Orthogonal designs IV; existence questions, *J. Combin. Theory*, **19A** (1975) 66–83 [2].

[478] J. E. Gibbs and H. A. Gebbie, Application of Walsh functions to transform spectroscopy, *Nature*, **224** (1969) 1012–1013 [2].

[479] E. N. Gilbert, A comparison of signalling alphabets, *Bell Syst. Tech. Jnl.*, **31** (1952) 504–522 [1].

[480] E. N. Gilbert, Unpublished notes, 1953 [14].

[481] E. N. Gilbert, Gray codes and paths on the *n*-cube, *Bell Syst. Tech. J.*, **37** (1958) 815–826 [Intro.].

[482] E. N. Gilbert, Synchronization of binary messages, *IEEE Trans. Info. Theory*, **6** (1960) 470–477 [1].

[483] E. N. Gilbert, Cyclically permutable error-correcting codes, *IEEE Trans. Info. Theory*, **9** (1963) 175–182 [17].

[484] E. N. Gilbert, F. J. MacWilliams and N. J. A. Sloane, Codes which detect deception, *Bell Syst. Tech. J.*, **53** (1974) 405–424 [1].

[485] A. Gill, *Linear Sequential circuits: Analysis, Synthesis, and Applications* (McGraw-Hill, New York, 1966) [3].

[486] A. M. Gleason, Weight polynomials of self-dual codes and the MacWilliams identities, in: *Actes Congrès Internl. de Mathematique*, **3** 1970 (Gauthier–Villars, Paris, 1971) 211–215 [19].

[487] K. R. Godfrey, Three-level *m*-sequences, *Electronics Letters*, **2** (1966) 241–243 [14].

[488] J.-M. Goethals, Analysis of weight distribution in binary cyclic codes, *IEEE Trans. Info. Theory*, **12** (1966) 401–402 [8].

[489] J.-M. Goethals, Cyclic error-locating codes, *Info. and Control*, **10** (1967) 378–385 [18].

[490] J.-M. Goethals, Factorization of cyclic codes, *IEEE Trans. Info. Theory*, **13** (1967) 242–246 [18].

[491] J.-M. Goethals, A polynomial approach to linear codes, *Philips Res. Reports*, **24** (1969) 145–159 [11, 13].

[492] J.-M. Goethals, On *t*-designs and threshold decoding, Institute of Statistics Mimeo Series, No. 600.29, (Univ. of North Carolina, Chapel Hill, N.C., June, 1970) [16].

[493] J.-M. Goethals, On the Golay perfect binary code, *J. Combin. Theory*, **11** (1971) 178–186 [2, 15, 16].

[494] J.-M. Goethals, Some combinatorial aspects of coding theory, Ch. 17 in: J. N. Srivastava et al., eds., *A Survey of Combinatorial Theory*, (North-Holland, Amsterdam, 1973) [1, 13].

[495] J.-M. Goethals, Two Dual Families of Nonlinear Binary Codes, Electronics Letters, **10** (1974) 471–472 [15, A].

[496] J.-M. Goethals, Nonlinear Codes Defined by Quadratic Forms Over GF(2), *Information and Control*, **31** (1976) 43–74 [15, A].

[497] J.-M. Goethals, The extended Nadler code is unique, *IEEE Trans. Info. Theory*, **23** (1977) 132–135 [2].

[498] J.-M. Goethals, personal communication [15].

[499] J.-M. Goethals and P. Delsarte, On a class of majority-logic decodable cyclic codes, *IEEE Trans. Info. Theory*, **14** (1968) 182–188 [13].

[500] J.-M. Goethals and J. J. Seidel, Orthogonal matrices with zero diagonal, *Canad. J. Math.*, **19** (1967) 1001–1010 [2].

[501] J.-M. Goethals and J. J. Seidel, A skew Hadamard matrix of order 36, *J. Aust. Math. Soc.*, **11** (1970) 343–344 [2].

[502] J.-M. Goethals and J. J. Seidel, Strongly regular graphs derived from combinatorial designs, *Canad. J. Math.*, **22** (1970) 597–614 [21].

[503] J.-M. Goethals and J. J. Seidel, The regular two-graph on 276 vertices, *Discrete Math.*, **12** (1975) 143–158 [21].

[504] J.-M. Goethals and S. L. Snover, Nearly perfect binary codes, *Discrete Math.*, **3** (1972) 65–88 [15, 17].

[505] J.-M. Goethals and H. C. A. van Tilborg, Uniformly packed codes, *Philips Research Reports*, **30** (1975) 9–36 [15].

[506] M. J. E. Golay, Notes on digital coding, *Proc. IEEE*, **37** (1949) 657 [1, 2, 20, A].

[507] M. J. E. Golay, Multi-slit spectrometry, *J. Optical Soc. America*, **39** (1949) 437–444 [2].

[508] M. J. E. Golay, Static multislit spectrometry, and its application to the panoramic display of infrared spectra, *J. Optical Soc. America*, **41** (1951) 468–472 [2].

[509] M. J. E. Golay, Binary coding, *IEEE Trans. Info. Theory*, **4** (1954) 23–28 [2, A].

[510] M. J. E. Golay, Notes on the penny-weighing problem, lossless symbol coding with nonprimes, etc., *IEEE Trans. Info. Theory*, **4** (1958) 103–109 [7].

[511] M. J. E. Golay, Complementary series, *IEEE Trans. Info. Theory*, **7** (1961) 82–87 [14].

[512] M. J. E. Golay, Note on lossless coding with nonprimes, *IEEE Trans. Info. Theory*, **14** (1968) 605 [7].

[513] M. J. E. Golay, A class of finite binary sequences with alternate autocorrelation values equal to zero, *IEEE Trans. Info. Theory*, **18** (1972) 449–450 [14].

[514] M. J. E. Golay, Hybrid low autocorrelation sequences, *IEEE Trans. Info. Theory*, **21** (1975) 460–462 [14].

[514a] M. J. E. Golay, Anent codes, priorities, patents, etc., *Proc. IEEE*, **64** (1976) 572 [1].

[515] R. Gold, Characteristic linear sequences and their coset functions, *J. SIAM Appl. Math.*, **14** (1966) 980–985 [14].

[516] R. Gold, Maximal recursive sequences with 3-valued recursive cross-correlation functions, *IEEE Trans. Info. Theory*, **14** (1968) 154–156 [14].

[517] M. Goldberg, Augmentation techniques for a class of product codes, *IEEE Trans. Info. Theory*, **19** (1973) 666–672 [18].

[518] H. D. Goldman, M. Kliman, and H. Smola, The weight structure of some Bose–Chaudhuri codes, *IEEE Trans. Info. Theory*, **14** (1968) 167–169 [9, A].

[519] J. Goldman and G.-C. Rota, The number of subspaces of a vector space, in: W. T. Tutte, ed., *Recent Progress in Combinatorics*, (Academic Press, New York, 1969) pp. 75–83 [15, B].

[520] A. J. Goldstein, A residue operator in formal power series, preprint [19].

[521] R. M. Goldstein and N. Zierler, On trinomial recurrences, *IEEE Trans. Info. Theory*, **14** (1968) 150–151 [14].

[522] S. W. Golomb, ed., *Digital Communications with Space Applications*, (Prentice-Hall, Englewood Cliffs, NJ, 1964) [1, 2, 14].

[523] S. W. Golomb, *Shift Register Sequences*, (Holden-Day. San Francisco, 1967) [Intro., 3, 9, 14].

[524] S. W. Golomb, A general formulation of error metrics, *IEEE Trans. Info. Theory*, **15** (1969) 425–426 [5, 6].

[524a] S. W. Golomb, Irreducible polynomials, synchronization codes, and the cyclotomic algebra, in: R. C. Bose and T. A. Dowling, eds., *Combinatorial Mathematics and Its Applications*, (Univ. North Carolina Press, Chapel Hill, NC, 1969) pp. 358–370 [4].

[525] S. W. Golomb, Sphere packing, coding metrics, and chess puzzles, in: *Proc. Second Chapel Hill Conf. on Combin. Math. Applic.*, (Univ. of North Carolina, Chapel Hill, NC, 1970) pp. 176–189 [5, 6].

[526] S. W. Golomb et al., Synchronization, *IEEE Trans. Commun.*, **11** (1963) 481–491 [1].

[527] S. W. Golomb and B. Gordon, Codes with bounded synchronization delay, *Info. and Control*, **8** (1965) 355–372 [1].

[528] S. W. Golomb, B. Gordon and L. R. Welch, Comma-free codes, *Canad. J. Math.*, **10** (1958) 202–209 [Intro.].

[529] S. W. Golomb and E. C. Posner, Rook domains, Latin squares, affine planes, and error-distribution codes, *IEEE Trans. Info. Theory*, **10** (1964) 196–208 [5, 6].

[530] S. W. Golomb and R. A. Scholtz, Generalized Barker sequences, *IEEE Trans. Info. Theory*, **11** (1965) 533–537 [14].

[531] S. W. Golomb and L. R. Welch, Algebraic coding and the Lee metric, in: H. B. Mann, ed., *Error Correcting Codes*, (Wiley, New York, 1968) pp. 175–194 [5, 6].

[532] S. W. Golomb and L. R. Welch, Perfect codes in the Lee metric and the packing of polyominoes, *SIAM J. Appl. Math.*, **18** (1970) 302–317 [5, 6].

[533] I. J. Good, Generalizations to several variables of Lagranges's expansion, with applications to stochastic processes, *Proc. Camb. Phil. Soc.*, **56** (1960) 367–380 [19].

[534] R. M. F. Goodman, Binary codes with disjoint codebooks and mutual Hamming distance, *Electronic Letters*, **10** (1974) 390–391 [8].

[535] R. M. F. Goodman and P. G. Farrell, Data transmission with variable-redundancy error control over a high-frequency channel, *Proc. IEE*, **122** (1975) 113–118 [1].

[536] V. D. Goppa, A new class of linear error-correcting codes, *Problems of Info. Transmission*, **6** (3) (1970) 207–212 [12, A].

[537] V. D. Goppa, Rational representation of codes and (L, g) – codes, *Problems of Info. Transmission*, **7** (3) (1971) 223–229 [12].

[538] V. D. Goppa, Codes constructed on the basis of (L, g) – codes, *Problems of Info. Transmission*, **8** (2) (1972) 165–166 [12, 18].

[538a] V. D. Goppa, Binary symmetric channel capacity is attained with irreducible codes, *Problems of Info. Transmission*, **10** (1) (1974) 89–90 [12].

[539] V. D. Goppa, Correction of arbitrary error patterns by irreducible codes, *Problems of Info. Transmission*, **10** (3) (1974) 277–278 [12].

[540] V. D. Goppa, On generating idempotents of cyclic codes, preprint [8].

[541] B. Gordon, On the existence of perfect maps, *IEEE Trans. Info. Theory*, **12** (1966) 486–487 [18].

[542] Y. Gordon and H. S. Witsenhausen, On extensions of the Gale–Berlekamp switching problem and constants of l_p-spaces, *Israel J. Mathematics*, **11** (1972) 216–229 [14].

[543] W. C. Gore, Generalized threshold decoding and the Reed–Solomon codes, *IEEE Trans. Info. Theory*, **15** (1969) 78–81 [9, 10].

[544] W. C. Gore, The equivalence of L-step orthogonalization and a Reed decoding procedure, *IEEE Trans. Info. Theory*, **15** (1969) 184–186 [13].

[545] W. C. Gore, Generalized threshold decoding of linear codes, *IEEE Trans. Info. Theory*, **15** (1969) 590–592 [13].

[546] W. C. Gore, Further results on product codes, *IEEE Trans. Info. Theory*, **16** (1970) 446–451 [18].

[547] W. C. Gore, Transmitting binary symbols with Reed–Solomon codes, *Proc. 7th Annual Princeton Conf. Information Sciences and Systems*, (Dept. of Electrical Engineering, Princeton University, Princeton, NJ, 1973) pp. 495–499 [9, 10].

[548] W. C. Gore and A. B. Cooper, Comments on polynomial codes, *IEEE Trans. Info. Theory*, **16** (1971) 635–638 [13].

[549] W. C. Gore, and C. C. Kilgus, Cyclic codes with unequal error protection, *IEEE Trans. Info. Theory*, **17** (1971) 214–215 [Intro.].

[550] D. C. Gorenstein, W. W. Peterson, and N. Zierler, Two-error correcting Bose–Chaudhuri codes are quasi-perfect, *Info. and Control*, **3** (1960) 291–294 [9].

[551] D. C. Gorenstein and E. Weiss, An acquirable code, *Info. and Control*, **7** (1964) 315–319 [14].

[552] D. C. Gorenstein and N. Zierler, A class of error-correcting codes in p^m symbols, *J. Soc. Indus. Applied Math.*, **9** (1961) 207–214 [7].

[553] R. L. Graham and F. J. MacWilliams, On the number of information symbols in difference-set cyclic codes, *Bell Syst. Tech. J.*, **45** (1966) 1057–1070 [13].

[554] D. H. Green and I. S. Taylor, Irreducible polynomials over composite Galois fields and their applications in coding techniques, *Proc. IEE*, **121** (1974) 935–939 [4].

[555] J. H. Green, Jr., and R. L. San Souci, An error-correcting encoder and decoder of high efficiency, *Proc. IEEE*, **46** (1958) 1741–1744 [13].

[556] M. W. Green, Two heuristic techniques for block-code construction, *IEEE Trans. Info. Theory*, **12** (1966) 273 [2].

[557] R. R. Green, A serial orthogonal decoder, *JPL Space Programs Summary*, Vol. 37–39–IV (1966) 247–253 [14].

[558] R. R. Green, Analysis of a serial orthogonal decoder, *JPL Space Programs Summary*, Vol. 37–53–III (1968) 185–187 [14].

[559] P. Greenberg, *Mathieu Groups*, (Courant Institute of Math. Sci., NY, 1973) [20].

[560] C. Greene, Weight enumeration and the geometry of linear codes, *Studies in Applied Math.*, **55** (1976) 119–128 [5].

[561] L. D. Grey, Some bounds for error-correcting codes, *IEEE Trans. Info. Theory*, **8** (1962) 200–202 and 355 [17].

[562] J. H. Griesmer, A bound for error-correcting codes, *IBM J. Res. Develop.*, **4** (1960) 532–542 [17].

[563] A. J. Gross, Some augmentations of Bose–Chaudhuri error correcting codes, in: J. N. Srivastava et al., eds., *A Survey of Combinatorial Theory*, (North-Holland, Amsterdam, 1973) Ch. 18 [10].

[564] E. J. Groth, Generation of binary sequences with controllable complexity, *IEEE Trans. Info. Theory*, **17** (1971) 288–296 [Intro.].

[565] B. R. Gulati, On maximal (k, t)-sets, *Annals Inst. Statistical Math.*, **23** (1971) 279–292 and 527–529 [11].

[566] B. R. Gulati, More about maximal (N, R)-sets, *Info. and Control*, **20** (1972) 188–191 [11].

[567] B. R. Gulati, B. M. Johnson and U. Koehn, On maximal t-linearly independent sets, *J. Comb. Theory*, **15A** (1973) 45–53 [11].

[568] B. R. Gulati and E. G. Kounias, On bounds useful in the theory of symmetrical fractional designs, *J. Roy. Statist. Soc., Ser. B*, **32** (1970) 123–133 [11].

[569] B. R. Gulati and E. G. Kounias, Maximal sets of points in finite projective space, no t-linearly dependent, *J. Comb. Theory*, **15A** (1973) 54–65 [11].

[570] R. C. Gunning, *Lectures on Modular Forms*, (Princeton Univ. Press, Princeton, NJ, 1962) [19].

[571] R. K. Guy, Twenty odd questions in combinatorics, in: R. C. Bose and T. A. Dowling, eds., *Combinatorial Mathematics and its Applications*, (Univ. N. Carolina Press, 1970) [Intro.].

H

[572] J. Hadamard, Résolution d'une question relative aux déterminants, *Bull. Sci. Math.*, (2) **17** (1893) 240–248 [2].

[573] D. Hagelbarger, Recurrent codes; easily mechanized burst-correcting binary codes, *Bell Syst. Tech. J.*, **38** (1959) 969–984 [10].

[574] W. Hahn, Über Orthogonalpolynome, die q-Differenzengleichungen genügen, *Math. Nachr.*, **2** (1949) 4–34 [21].

[575] S. L. Hakimi and J. G. Bredeson, Graph theoretic error-correcting codes, *IEEE Trans. Info. Theory*, **14** (1968) 584–591 [18].

[576] S. L. Hakimi and J. G. Bredeson, Ternary graph theoretic error-correcting codes, *IEEE Trans. Info. Theory*, **15** (1969) 435–437 [18].

[577] S. L. Hakimi and H. Frank, Cut-set matrices and linear codes, *IEEE Trans. Info. Theory*, **11** (1965) 457–458 [18].

[578] M. P. Hale, Jr. and E. E. Shult, Equiangular lines, the graph extension theorem, and transfer in triply transitive groups, *Math. Zeit.*, **135** (1974) 111–123 [21].

[579] A. D. Hall, Jr., The ALTRAN system for rational function manipulation – A survey, *Comm. Assoc. Comput. Mach.*, **14** (1971) 517–521 [19].

[580] J. I. Hall, Bounds for equidistant codes and partial projective planes, *Discrete Math.*, **17** (1977) 85–94 [1].

[581] J. I. Hall, A. J. E. M. Janssen, A. W. J. Kolen and J. H. van Lint, Equidistant codes with distance 12, *Discrete Math.*, **17** (1977) 71–83 [1].

[582] M. Hall, Jr., *The Theory of Groups* (Macmillan, New York, 1959) [B].

[583] M. Hall, Jr., Hadamard matrices of order 16, *Research Summary*, No. 36-10 (Jet Propulsion Lab., Pasadena, CA, Sept. 1, 1961) Vol. 1, pp. 21–26 [2].

[584] M. Hall, Jr., Note on the Mathieu group M_{12}, *Arch. Math.*, **13** (1962) 334–340 [20].

[585] M. Hall Jr., Block designs, in: E. F. Bechenbach, ed., *Applied Combinatorial Mathematics*, (Wiley, New York, 1964) pp. 369–405 [2].

[586] M. Hall, Jr., Hadamard matrices of order 20, *Technical report 32-761*, (Jet Propulsion Lab., Pasadena, CA, Nov. 1, 1965) [2].

[587] M. Hall, Jr., *Combinatorial Theory* (Blaisdell, Waltham, MA, 1967) [Intro., 2, 5, A, B].

[588] M. Hall, Jr., Automorphisms of Hadamard matrices, *SIAM J. Appl. Math.*, **17** (1969) 1094–1101 [2, 16].

[589] M. Hall, Jr., Difference sets, in: M. Hall, Jr., and J. H. van Lint, eds., *Combinatorics*, (Reidel, Dordrecht, Holland 1975) pp. 321–346 [13].

[590] M. Hall, Jr., Semi-automorphisms of Hadamard matrices, *Math. Proc. Camb. Phil. Soc.*, **77** (1975) 459–473 [2, 16, 20].

[591] N. Hamada, On the *p*-rank of the incidence matrix of a balanced or partially balanced incomplete block design and its applications to error correcting codes, *Hiroshima Math. J.*, **3** (1973) 153–226 [13].

[592] R. W. Hamming, Error detecting and error correcting codes, *Bell Syst. Tech. J.*, **29** (1950) 147–160 [1, A].

[593] P. Hammond, Perfect codes in the graphs O_k, *J. Comb. Theory*, **19B** (1975) 239–255 [6].

[594] P. Hammond, Nearly perfect codes in distance-regular graphs, *Discrete Math.*, **14** (1976) 41–56 [6].

[595] H. Hanani, On quadruple systems, *Canad. J. Math.*, **12** (1960) 145–157 [2, 17].

[596] M. Hanani, The existence and construction of balanced incomplete block designs, *Annals Math. Stat.*, **32** (1961) 361–386 [2, 17].

[597] H. Hanani, On some tactical configurations, *Canad. J. Math.*, **15** (1963) 702–722 [2, 17].

[598] H. Hanani, A balanced incomplete block design, *Annals Math. Stat.*, **36** (1965) 711 [2, 17].

[599] H. Hanani, On balanced incomplete block designs with blocks having five elements, *J. Comb. Theory*, **12A** (1972) 184–201 [2, 17].

[600] H. Hanani, Balanced incomplete block designs and related designs, *Discrete Math.*, **11** (1975) 255–369 [2, 17, A].

[601] G. H. Hardy, J. E. Littlewood and G. Pólya, *Inequalities*, (Cambridge Univ. Press, Cambridge, 1959) [2].

[602] G. H. Hardy and E. M. Wright, *An Introduction to the Theory of Numbers*, (Oxford University Press, Oxford, 3rd edition, 1954) [4].

[603] H. F. Harmuth, Applications of Walsh functions in communications, *IEEE Spectrum*, **6** (November, 1969) 81–91 [2].

[604] H. F. Harmuth, *Transmission of Information by Orthogonal Functions*, (Springer, New York, 1970) [2].

[605] L. H. Harper, Combinatorial coding theory, in: R. C. Bose and T. A. Dowling, eds., *Combinatorial Mathematics and its Applications*, (Univ. of North Carolina Press, Chapel Hill, NC, 1970) pp. 252–260 [Intro.].

[606] M. A. Harrison, *Introduction to Switching and Automata Theory*, (McGraw-Hill, New York, 1965) [13].

[607] C. R. P. Hartmann, A note on the decoding of double-error-correcting binary BCH codes of primitive length, *IEEE Trans. Info. Theory*, **17** (1971) 765–766 [9].

[608] C. R. P. Hartmann, A note on the minimum distance structure of cyclic codes, *IEEE Trans. Info. Theory*, **18** (1972) 439–440 [7].

[609] C. R. P. Hartmann, Decoding beyond the BCH bound, *IEEE Trans. Info. Theory*, **18** (1972) 441–444 [9].

[610] C. R. P. Hartmann, Theorems on the minimum distance structure of binary cyclic codes, in: B. N. Petrov and F. Csáki, eds., *2nd Internat. Sympos. on Info. Theory*, (Akad. Kiadó, Budapest, 1973) pp. 185–190 [7].

[611] C. R. P. Hartmann, J. B. Ducey and L. D. Rudolph, On the structure of generalized finite-geometry codes, *IEEE Trans. Info. Theory*, **20** (1974) 240–252 [13].

[612] C. R. P. Hartmann, L. Kerschberg and M. A. Diamond, Some results on single error correcting binary BCH codes, Tech. Rep. 6–74, Syracuse Univ., 1974 [9].

[613] C. R. P. Hartmann, J. R. Riek, Jr. and R. J. Longobardi, Weight distributions of some classes of binary cyclic codes, *IEEE Trans. Info. Theory*, **21** (1975) 345–350 [8].

[614] C. R. P. Hartmann and L. D. Rudolph, An optimum symbol-by-symbol decoding rule for linear codes, *IEEE Trans. Info. Theory* **22** (1976) 514–517 [16].

[615] C. R. P. Hartmann and K. K. Tzeng, Generalizations of the BCH bound, *Info. and Control*, **20** (1972) 489–498 [7].

[616] C. R. P. Hartmann and K. K. Tzeng, A bound for cyclic codes of composite length, *IEEE Trans. Info. Theory*, **18** (1972) 307–308 [18].

[617] C. R. P. Hartmann, K. K. Tzeng and R. T. Chien, Some results on the minimum distance structure of cyclic codes, *IEEE Trans. Info. Theory*, 18 (1972) 402–409 [18].

[618] C. R. P. Hartmann and K. K. Tzeng, On some classes of cyclic codes of composite length, *IEEE Trans. Info. Theory*, 19 (1973) 820–822 [18, A].

[619] C. R. P. Hartmann and K. K. Tzeng, Decoding beyond the BCH bound using multiple sets of syndrome sequences, *IEEE Trans. Info. Theory*, 20 (1974) 292–295 [9].

[620] W. E. Hartnett, ed., *Foundations of Coding Theory*, (Reidel, Holland, 1974) [Intro.].

[621] J. T. Harvey, High-speed *m*-sequence generation, *Electronics Letters*, 10 (1974) 480–481 [14].

[622] M. Harwit, P. G. Phillips, T. Fine and N. J. A. Sloane, Doubly multiplexed dispersive spectrometers, *Applied Optics*, 9 (1970) 1149–1154 [2].

[623] A. A. Hashim, New families of error-correcting codes generated by modification of other linear binary block codes, *Proc. IEE*, 121 (1974) 1480–1485 [18].

[624] A. A. Hashim and A. G. Constantinides, Class of linear binary codes, *Proc. IEE*, 121 (1974) 555–558 [18].

[625] A. A. Hashim and A. G. Constantinides, Some new results on binary linear block codes, *Electronics Letters*, 10 (1974) 31–33 [18, A].

[625a] A. A. Hashim and V. S. Pozdniakov, Computerized search for linear binary codes, *Electronics Letters*, 12 (1976) 350–351 [A].

[626] T. R. Hatcher, On minimal distance, shortest length and greatest number of elements for binary group codes, Report TM-6-3826, Parke Math. Labs., Inc., Carlisle, Mass., Sept. 1964 [17].

[627] T. R. Hatcher, On a family of error-correcting and synchronizable codes, *IEEE Trans. Info. Theory*, 15 (1969) 620–624 [1].

[628] O. Heden, Perfect codes in antipodal distance-transitive graphs, *Math. Scand.*, 35 (1974) 29–37 [6].

[629] O. Heden, A generalized Lloyd theorem and mixed perfect codes, *Math. Scand.*, 37 (1975) 13–26 [6].

[630] O. Heden, A new construction of group and nongroup perfect codes, *Info. Control*, 34 (1977) 314–323.

[631] H. J. Helgert, Srivastava codes, *IEEE Trans. Info. Theory*, 18 (1972) 292–297 [12, A].

[632] H. J. Helgert, Noncyclic generalizations of BCH and Srivastava codes, *Info. and Control*, 21 (1972) 280–290 [12].

[633] H. J. Helgert, Alternant codes, *Info. and Control*, 26 (1974) 369–380 [12, 18, A].

[634] H. J. Helgert, Binary primitive alternant codes, *Info. and Control*, 27 (1975) 101–108 [12].

[635] H. J. Helgert, Alternant Codes, Talk given at Information Theory Workshop, Lenox, Mass., June 1975 [12].

[636] H. J. Helgert and R. D. Stinaff, Minimum-distance bounds for binary linear codes, *IEEE Trans. Info. Theory*, 19 (1973) 344–356 [2, 17, 18, A].

[637] H. J. Helgert and R. D. Stinaff, Shortened BCH codes, *IEEE Trans. Info. Theory*, 19 (1973) 818–820 [18, A].

[638] J. A. Heller and I. M. Jacobs, Viterbi decoding for satellite and space communications, *IEEE Trans. Commun. Theory*, 14 (1971) 835–848 [1].

[639] M. E. Hellman, On using natural redundancy for error detection, *IEEE Trans. on Computers*, 22 ·(1974) 1690–1693 [1].

[640] M. E. Hellman, Error detection in the presence of synchronization loss, *IEEE Trans. Commun.*, 23 (1975) 538–539 [1].

[640a] M. E. Hellman and W. Diffie, New directions in cryptography, *IEEE Trans. Info. Theory*, 22 (1976), 644–654 [1].

[641] U. Henriksson, On a scrambling property of feedback shift registers, *IEEE Trans. Commun.*, 20 (1972) 998–1001 [14].

[642] I. N. Herstein, *Noncommutative rings*, Carus Monog. 15, (Math. Assoc. America and Wiley, N.Y., 1968) [B].

[643] M. Herzog and J. Schönheim, Linear and nonlinear single-error-correcting perfect mixed codes, *Info. Control*, 18 (1971) 364–368 [6].

[644] M. D. Hestenes, On the use of graphs in group theory, in: F. Harary, ed., *New Directions in the Theory of Graphs*, (Academic Press, New York, 1973) pp. 97–128 [21].

[645] M. D. Hestenes and D. G. Higman, Rank 3 groups and strongly regular graphs, in:

SIAM-AMS Proc. Vol. IV, Computers in Algebra and Number Theory, (Amer. Math. Soc., Providence, 1971) pp. 141–159 [21].

[646] D. G. Higman, Finite permutation groups of rank 3, *Math. Zeit.*, **86** (1964) 145–156 [21].

[647] D. G. Higman, Intersection matrices for finite permutation groups, *J. Algebra* **4** (1967) 22–42 [21].

[648] D. G. Higman, Characterization of families of rank 3 permutation groups by the subdegrees, I and II, *Archiv. Math.* **21** (1970) 151–156 and 353–361 [21].

[649] D. G. Higman, *Combinatorial Considerations about Permutation Groups*, Lecture notes, Math. Inst., Oxford, (1972) [21].

[650] D. G. Higman, Coherent configurations and generalized polygons, pp. 1–5 of *Combinatorial Mathematics*, Lecture Notes in Math. 403, Springer-Verlag N.Y. 1974 [21].

[651] D. G. Higman, Invariant relations, coherent configurations, and generalized polygons, in: M. Hall, Jr., and J. H. van Lint, eds., *Combinatorics*, (Reidel, Dordrecht, 1975) pp. 347–363 [21].

[652] D. G. Higman, Coherent configurations, part I: ordinary representation theory, Geometriae Dedicata, **4** (1975) 1–32 [21].

[653] D. G. Higman, Coherent configurations, part II: weights, *Geometriae Dedicata*, **5** (1976) 413–424 [21].

[654] D. Hilbert and S. Cohn-Vossen, *Geometry and the Imagination*, (Chelsea, N.Y. 1952) [17].

[655] J. W. P. Hirschfeld, Rational curves on quadrics over finite fields of characteristic two, *Rendiconti di Matematica*, **3** (1971) 772–795 [11].

[656] C. F. Hobbs, Approximating the performance of a binary group code, *IEEE Trans. Info. Theory*, **11** (1965) 142–144 [1, 14].

[657] M. Hochster and J. A. Eagon, Cohen-Macaulay rings, invariant theory, and the generic perfection of determinantal loci, *Amer. J. Math.*, **93** (1971) 1020–1058 [19].

[658] A. Hocquenghem, Codes correcteurs d'erreurs, *Chiffres* (Paris), **2** (1959) 147–156 [7].

[659] C. W. Hoffner and S. M. Reddy, Circulant bases for cyclic codes, *IEEE Trans. Info. Theory*, **16** (1970) 511–512 [16].

[660] J. K. Holmes, A note on some efficient estimates of the noise variance for first order Reed-Muller codes, *IEEE Trans. Info. Theory*, **17** (1971) 628–630 [14].

[661] S. J. Hong and D. C. Bossen, On some properties of self-reciprocal polynomials, *IEEE Trans. Info. Theory*, **21** (1975) 462–464 [7].

[662] S. J. Hong and A. M. Patel, A general class of maximal codes for computer applications, *IEEE Trans. Comput.*, **21** (1972) 1322–1331 [1].

[663] J. A. van der Horst and T. Berger, Complete decoding of triple-error-correcting binary BCH codes, *IEEE Trans. Info. Theory*, **22** (1976) 138–147 [9].

[664] M. Horstein, Sequential transmission using noiseless feedback, *IEEE Trans. Info. Theory*, **9** (1963) 136–143 [Intro.].

[665] H. Hotelling, Some improvements in weighing and other experimental techniques, *Annals Math. Stat.*, **15** (1944) 297–306 [2].

[666] M. Y. Hsiao, On calling station codes, *IEEE Trans. Info. Theory*, **15** (1969) 736–737 [17].

[667] M. Y. Hsiao, A class of optimal minimum odd-weight-column SEC-DED codes, *IBM J. Res. Dev.*, **14** (1970) 395–401 [7].

[668] M. Y. Hsiao, Incomplete block design codes for ultra high speed computer applications, *Discrete Math.*, **3** (1972) 89–108 [1, 18].

[669] M. Y. Hsiao and D. C. Bossen, Orthogonal Latin square configuration for LSI memory yield and reliability enhancement, *IEEE Trans. on Computers*, **24** (1975) 512–516 [1, 18].

[670] M. Y. Hsiao, D. C. Bossen and R. T. Chien, Orthogonal Latin square codes, *IBM J. Res. Devel.*, **14** (1970) 390–394 [1, 18].

[671] H. T. Hsu and T. Kasami, Error-correcting codes for a compound channel, *IEEE Trans. Info. Theory*, **14** (1968) 135–139 [1].

[672] X. L. Hubaut, Strongly regular graphs, *Discrete Math.*, **13** (1975) 357–381 [21].

[673] D. R. Hughes, *t*-designs and groups, *Amer. J. Math.*, **87** (1965) 761–778 [2].

[674] D. R. Hughes and F. C. Piper, *Projective Planes*, (Springer, New York, 1973) [B].

[675] D. R. Hughes and F. C. Piper, *Block designs*, in preparation [2].

[676] B. Huppert, *Endliche Gruppen I*, (Springer, New York, 1967) [12, 16].

[677] W. J. Hurd, Efficient generation of statistically good pseudonoise by linearly interconnected shift registers, Report 32-1526, vol. XI, Jet Propulsion Labs, Pasadena, 1973 [14].

I

[678] *IBM Journal of Research and Development*, Special issue on Coding for Error Control, **14** (July 1970) [Intro., 1].

[679] *IEEE Trans. Commun. Technology*, Special Issue on Error Correcting Codes, **19** (October 1971, Part II) [Intro., 1].

[680] *IEEE Transactions on Electromagnetic Compatibility*, Special Issue on Applications of Walsh functions, **13** (August 1971) [2].

[681] T. Ikai and Y. Kojima, Two-dimensional cyclic codes, *Electron. Commun. Japan*, **57-A** (No. 4, 1974) 27–35 [18].

[682] H. Imai, Two-Dimensional Fire Codes, *IEEE Trans. Info. Theory*, **19** (1973) 796–806 [18].

[683] I. Ingemarsson, Commutative group codes for the Gaussian channel, *IEEE Trans. Info. Theory*, **19** (1973) 215–219 [Intro.].

[684] Y. Iwadare, A unified treatment of burst-correcting codes, *Electron. Commun. Japan*, **52-A** (No. 8, 1969) 12–19 [10].

[685] Y. Iwadare, A class of high-speed decodable burst-correcting codes, *IEEE Trans. Info. Theory*, **18** (1972) 817–821 [10].

J

[686] I. M. Jacobs, Practical applications of coding, *IEEE Trans. Info. Theory*, **20** (1974) 305–310 [1].

[687] N. Jacobson, *Lectures in Abstract Algebra*, (Van Nostrand, Princeton, New Jersey; Vol. 1, 1951; Vol. 2, 1953; Vol. 3, 1964) [4, 19].

[688] A. T. James, The relationship algebra of an experimental design, *Ann. Math. Stat.*, **28** (1957) 993–1002 [21].

[689] G. D. James, The modular characters of the Mathieu groups, *J. Algebra*, **27** (1973) 57–111 [20].

[690] F. Jelinek, *Probabilistic Information Theory*, (McGraw-Hill New York, 1968) [1, 10].

[691] F. Jelinek, Fast sequential decoding algorithm using a stack, *IBM J. Res. Devel.*, **13** (1969) 675–685 [1].

[692] B. H. Jiggs, Recent results in comma-free codes, *Canad. J. Math.*, **15** (1963) 178–187 [Intro.].

[693] S. M. Johnson, A new upper bound for error-correcting codes, *IEEE Trans. Info. Theory*, **8** (1962) 203–207 [17].

[694] S. M. Johnson, Improved asymptotic bounds for error-correcting codes, *IEEE Trans. Info. Theory*, **9** (1963) 198–205 [17].

[695] S. M. Johnson, Unpublished tables, 1970 [17, A].

[696] S. M. Johnson, On upper bounds for unrestricted binary error-correcting codes, *IEEE Trans. Info. Theory*, **17** (1971) 466–478 [17, A].

[697] S. M. Johnson, Upper bounds for constant weight error correcting codes, *Discrete Math.*, **3** (1972) 109–124 [17].

[698] S. M. Johnson, A new lower bound for coverings by rook domains, *Utilitas Math.*, **1** (1972) 121–140 [6].

[699] W. Jónsson, On the Mathieu groups M_{22}, M_{23}, M_{24} and the uniqueness of the associated Steiner systems, *Math. Z.*, **125** (1972) 193–214 [20].

[700] D. D. Joshi, A note on upper bounds for minimum distance bounds, *Info. Control*, **1** (1958) 289–295 [1, 17].

[701] D. Julin, Two improved block codes, *IEEE Trans. Info. Theory*, **11** (1965) 459 [2, A].

[702] H. W. E. Jung, Über die kleinste Kugel, die eine raumliche Figur einschliesst, *J. reine angew. Math.*, **123** (1901) 241–257 [6].

[703] R. R. Jurick, An algorithm for determining the largest maximally independent set of vectors from an r-dimensional vector space over a Galois field of n elements, Tech. Rep. ASD-TR-68-40, Air Force Systems Command, Wright-Patterson Air Force Base, Ohio, September 1968 [11].

[704] J. Justesen, A class of constructive asymptotically good algebraic codes, *IEEE Trans. Info. Theory*, **18** (1972) 652–656 [10].

[705] J. Justesen, New convolutional code construction and a class of asymptotically good time varying codes, *IEEE Trans. Info. Theory,* **19** (1973) 220–225 [10].

[706] J. Justesen, On the complexity of decoding Reed-Solomon codes, *IEEE Trans. Info. Theory,* **22** (1976) 237–238 [Intro., 12].

K

[707] D. Kahn, *The Codebreakers,* (Macmillan, New York, 1967) [1].

[708] J. G. Kalbfleisch and R. G. Stanton, Maximal and minimal coverings of $(k-1)$-tuples by k-tuples, *Pacific J. Math.,* **26** (1968) 131–140 [17, A].

[709] J. G. Kalbfleisch and R. G. Stanton, A combinatorial problem in matching, *J. London Math. Soc.,* **44** (1969) 60–64 [A].

[710] J. G. Kalbfleisch, R. G. Stanton and J. D. Horton, On covering sets and error-correcting codes, *J. Combinat. Theory,* **11A** (1971) 233–250 [A].

[711] J. G. Kalbfleisch and P. H. Weiland, Some new results for the covering problem, in: W. T. Tutte, ed., *Recent Progress in Combinatorics,* (Academic Press, New York, 1969) pp. 37–45 [A].

[712] H. J. L. Kamps and J. H. van Lint, The football pool problem for 5 matches, *J. Comb. Theory,* **3** (1967) 315–325 [7].

[713] H. J. L. Kamps and J. H. van Lint, A covering problem, Colloquia Math. Soc. János Bolyai, Vol. **4**: *Combinatorial Theory and its Applications,* (Balatonfured, Hungary, 1969) pp. 679–685 [7].

[714] W. M. Kantor, Automorphism groups of Hadamard matrices, *J. Comb. Theory,* **6** (1969) 279–281 [16].

[715] M. Karlin, New binary coding results by circulants, *IEEE Trans. Info. Theory,* **15** (1969) 81–92 [16, A].

[716] M. Karlin, Decoding of circulant codes, *IEEE Trans. Info. Theory,* **16** (1970) 797–802 [16].

[717] M. Karlin, personal communication [16, A].

[718] M. Karlin and F. J. MacWilliams, On finding low weight vectors in quadratic residue codes for $p = 8m - 1$, *SIAM J. Appl. Math.,* **25** (1973) 95–104 [16].

[719] S. Karlin and J. L. McGregor, The Hahn polynomials, formulas and an application, *Scripta Math.,* **26** (1961) 33–46 [21].

[720] M. Kasahara, H. Nakonishi, Y. Tezuka and Y. Kasahara, Error correction of data using computers, *Electron. Commun. Japan,* **50** (1967) 637–642 [1].

[721] M. Kasahara, Y. Sugiyama, S. Hirasawa and T. Namekawa, A new class of binary codes constructed on the basis of BCH codes, *IEEE Trans. Info. Theory,* **21** (1975) 582–585 [18, A].

[722] M. Kasahara, Y. Sugiyama, S. Hirasawa and T. Namekawa, New Classes of Binary Codes constructed on the basis of concatenated codes and product codes, *IEEE Trans. Info. Theory,* **22** (1976), 462–468 [18, A].

[723] T. Kasami, Optimum shortened cyclic codes for burst-error correction, *IEEE Trans. Info. Theory,* **9** (1963) 105–109 [10].

[724] T. Kasami, Error-correcting and detecting codes, *Electron. Commun. Japan,* **46** (9) (1963) 90–102 [1].

[725] T. Kasami, A decoding procedure for multiple-error-correcting cyclic codes, *IEEE Trans. Info. Theory,* **10** (1964) 134–138 [9, 16].

[726] T. Kasami, Some lower bounds on the minimum weight of cyclic codes of composite length, *IEEE Trans. Info. Theory,* **14** (1968) 814–818 [18].

[727] T. Kasami, Weight distributions of Bose-Chaudhuri-Hocquenghem Codes, in: R. C. Bose and T. A. Dowling, eds., *Combinatorial Math. and its Applications,* (Univ. of North Carolina Press, Chapel Hill, NC, 1969) Ch. 20 [8, 9, 15, 21].

[728] T. Kasami, An upper bound on k/n for affine invariant codes with fixed d/n, *IEEE Trans. Info. Theory,* **15** (1969) 174–176 [7, 9].

[729] T. Kasami, The weight enumerators for several classes of subcodes of the 2nd order binary Reed-Muller codes, *Info. and Control,* **18** (1971) 369–394 [8, 15].

[730] T. Kasami, A Gilbert-Varshamov bound for quasi-cyclic codes of rate 1/2, *IEEE Trans. Info. Theory,* **20** (1974) 679 [16].

[731] T. Kasami, Construction and decomposition of cyclic codes of composite length, *IEEE Trans. Info. Theory*, **20** (1974) 680–683 [18].

[732] T. Kasami and S. Lin, On majority-logic decoding for duals of primitive polynomial codes, *IEEE Trans. Info. Theory*, **17** (1971) 322–331 [13].

[733] T. Kasami and S. Lin, The construction of a class of majority-logic decodable codes, *IEEE Trans. Info. Theory*, **17** (1971) 600–610 [13, 18].

[734] T. Kasami and S. Lin, Some results on the minimum weight of BCH codes, *IEEE Trans. Info. Theory*, **18** (1972) 824–825 [9].

[735] T. Kasami and S. Lin, Coding for a multiple-access channel, *IEEE Trans. Info. Theory*, **22** (1976) 129–137 [1].

[736] T. Kasami, S. Lin and W. W. Peterson, Some results on weight distributions of BCH codes, *IEEE Trans. Info. Theory*, **12** (1966) 274 [11].

[737] T. Kasami, S. Lin and W. W. Peterson, Linear codes which are invariant under the affine group and some results on minimum weights in BCH codes, *Electron. Commun. Japan*, **50** (No. 9, 1967), 100–106 [9].

[738] T. Kasami, S. Lin and W. W. Peterson, Some results on cyclic codes which are invariant under the affine group and their applications, *Info. and Control*, **11** (1968) 475–496 [8, 13].

[739] T. Kasami, S. Lin and W. W. Peterson, Generalized Reed-Muller codes, *Electron. Commun. Japan*, **51**-C(3) (1968) 96–104 [13].

[740] T. Kasami, S. Lin and W. W. Peterson, New generalizations of the Reed-Muller codes, Part I: Primitive codes, *IEEE Trans. Info. Theory*, **14** (1968) 189–199 [13].

[741] T. Kasami, S. Lin and W. W. Peterson, Polynomial codes, *IEEE Trans. Info. Theory*, **14** (1968) 807–814 and **16** (1970) 635 [13, A].

[742] T. Kasami, S. Lin and S. Yamamura, Further results on coding for a multiple-access channel, *Trans. of Colloquium on Info. Theory*, (Keszthely, August 25–29, 1975) [1].

[743] T. Kasami and S. Matora, Some efficient shortened cyclic codes for burst-error correction, *IEEE Trans. Info. Theory*, **10** (1964) 252 [10].

[744] T. Kasami and N. Tokura, Some remarks on BCH bounds and minimum weights of binary primitive BCH codes, *IEEE Trans. Info. Theory*, **15** (1969) 408–413 [9, A].

[745] T. Kasami and N. Tokura, On the weight structure of Reed-Muller codes, *IEEE Trans. Info. Theory*, **16** (1970) 752–759 [15].

[746] T. Kasami, N. Tokura and S. Azumi, *On the Weight Enumeration of Weights Less than 2.5d of Reed-Muller codes*, (Faculty of Engineering Science, Osaka University, Osaka, Japan, June 1974) [15].

[747] T. Katayama, Enumeration of linear dependence structures in the *n*-cube, *Electron. Commun. Japan*, **55**-D(2) (1972) pp. 121–129 [5].

[748] G. O. H. Katona, Combinatorial search problems, in: J. N. Srivastava et al., eds., *A Survey of Combinatorial Theory*, (North-Holland, Amsterdam, 1973) pp. 285–308 [7].

[749] W. H. Kautz, Codes and coding circuitry for automatic error correction within digital systems, in: R. H. Wilcox and W. C. Mann, eds., *Redundancy Techniques for Computing Systems*, (Spartan Books, Washington, 1962) pp. 152–195 [16].

[750] W. H. Kautz, ed., *Linear sequential switching circuits: Selected papers*, (Holden-Day, San Francisco, 1965) [3, 14].

[751] W. H. Kautz, Fibonacci codes for synchronization control, *IEEE Trans. Info. Theory*, **11** (1965) 284–292 [1].

[752] W. H. Kautz and B. Elspas, Single-error-correcting codes for constant-weight data words, *IEEE Trans. Info. Theory*, **11** (1965) 132–141 [17].

[753] W. H. Kautz and K. N. Levitt, A Survey or progress in coding theory in the Soviet Union, *IEEE Trans. Info. Theory*, **15** (1969) 197–244 [1].

[754] D. Kelly and G.-C. Rota, Some problems in combinatorial geometry, Ch. 24 in: J. N. Srivastava et al., eds., *A Survey of Combinatorial Theory*, (North-Holland, 1973) [17].

[755] J. G. Kemeny, H. Mirkil, J. L. Snell and G. L. Thompson, *Finite Mathematical Structures*, (Prentice-Hall, Englewood Cliffs, N.J., 1959) [13].

[756] M. G. Kendall and A. Stuart, *The Advanced Theory of Statistics*, Vol. 1, (Hafner, New York, 1969) [8, 19].

[757] B. L. N. Kennett, A note on the finite Walsh transform, *IEEE Trans. Info. Theory*, **16** (1970) 489–491 [2].

[758] A. M. Kerdock, A class of low-rate nonlinear codes, *Info. and Control*, **20** (1972) 182–187 [15, A].

[759] A. M. Kerdock, F. J. MacWilliams and A. M. Odlyzko, A new theorem about the Mattson-Solomon polynomial and some applications, *IEEE Trans. Info. Theory*, **20** (1974) 85–89 [8].

[760] C. C. Kilgus, Pseudonoise code acquisition using majority logic decoding, *IEEE Trans. Commun.*, **21** (1973) 772–774 [14].

[761] C. C. Kilgus and W. C. Gore A class of cyclic unequal-error-protection codes, *IEEE Trans. Info. Theory*, **18** (1972) 687–690 [Intro.].

[762] Z. Kiyasu, Information Theory, *Electron. Commun. Japan*, **50**(10) (1967) 109–117 [1].

[763] G. K. Kladov, Majority decoding of linear codes, *Problems of Info. Trans.*, **8**(3) (1972) 194–198 [13].

[764] V. Klee, Long paths and circuits on polytopes, in: B. Grünbaum et al., eds., *Complex Polytopes*, (Wiley New York, 1967) Ch. 17 [Intro.].

[765] V. Klee, A method for constructing circuit codes, *J. Assoc. Comput. Math.*, **14** (1967) 520–528 [Intro.].

[766] V. Klee, What is the maximum length of a *d*-dimensional snake? *Amer. Math. Monthly*, **77** (1970) 63–65 [Intro.].

[767] V. Klee, The use of circuit codes in analog-to-digital conversion, in: B. Harris, ed., *Graph Theory and its Applications*, (Academic Press, New York, 1970) pp. 121–131 [Intro.].

[768] F. Klein, *Lectures on the Icosahedron and the Solution of Equations of the Fifth Degree*, 2nd rev. ed., 1913. New York: Dover, 1956, 1st German ed., 1884 [19].

[769] T. J. Klein and J. K. Wolf, On the use of channel introduced redundancy for error correction, *IEEE Trans. on Commun. Tech.*, **19** (1971) 396–402 [1].

[770] D. Knee and H. D. Goldman, Quasi-self-reciprocal polynomials and potentially large minimum distance BCH Codes, *IEEE Trans. Info. Theory*, **15** (1969) 118–121 [9].

[771] K. Knowlton, private communication [18].

[772] D. E. Knuth, *The Art of Computer Programming*, Vol. 1, *Fundamental Algorithms*, (Addison-Wesley, Reading, MA, 1969) [4, 11, 14].

[773] V. D. Kolesnik, Probabilistic decoding of majority codes, *Problems of Info. Trans.*, **7**(3) (1971) 193–200 [13].

[774] V. D. Kolesnik and E. T. Mironchikov, Cyclic Reed-Muller codes and their decoding, *Problems of Info. Trans.*, **4** (1968) 15–19 [13].

[774a] A. N. Kolmogorov, Three approaches to the quantitative definition of information, *Problems of Info. Trans.*, **1**(1) (1965) 1–7 [Intro.].

[775] Y. Komamiya, Application of logical mathematics to information theory, *Proc. 3rd Japan. Natl. Cong. Applied Math.*, (1953) 437 [1].

[776] T. H. Koornwinder, Homogeneous spaces, spherical functions and association schemes, unpublished notes, 1974 [21].

[777] V. I. Korzhik, The correlation between the properties of a binary group code and those of its null space, *Problems of Info. Trans.*, **2**(1) (1966) 70–74 [5].

[778] V. I. Korzhik, An estimate of d_{min} for cyclic codes, *Problems of Info. Trans.*, **2**(2) (1966) 78 [7].

[779] V. N. Koshelev, On some properties of random group codes of great length, *Problems of Info. Trans.*, **1**(4) (1965) 35–38 [17].

[780] Y. I. Kotov, Correlation function of composite sequences constructed from two M-sequences, *Radio Eng. Elect. Phys.*, **19** (1974) 128–130 [14].

[781] M. V. Kozlov, The correcting capacities of linear codes, *Soviet Physics-Doklady*, **14** (1969) 413–415 [17].

[782] M. Krawtchouk, Sur une généralisation des polynomes d'Hermite, *Comptes Rendus*, **189** (1929) 620–622 [5].

[783] M. Krawtchouk, Sur la distribution des racines des polynomes orthogonaux, *Comptes Rendus*, **196** (1933) 739–741 [5].

[784] R. E. Krichevskii, On the number of errors which can be corrected by the Reed–Muller code, *Soviet Physics-Doklady*, **15**(3) (1970) 220–222 [13].

[785] J. B. Kruskal, Golay's complementary series, *IEEE Trans. Info. Theory*, **7** (1961) 273–276 [14].

[786] V. S. Kugurakov, Redundancy of linear codes with orthogonal and λ-connected checks, *Problems of Info. Trans.*, **8**(1) (1972) 25–32 [13].

[787] H. Kuroda, Note on error-control systems using product codes, *Electron. Commun. Japan*, **57**-A(2) (1974) 22–28 [18].

[788] R. P. Kurshan, What is a cyclotomic polynomial? preprint [7, 8].

[789] R. P. Kurshan and N. J. A. Sloane, Coset analysis of Reed-Muller codes via translates of finite vector spaces, *Info. and Control*, **20** (1972) 410–414 [14].

L

[790] T. Y. Lam, *The Algebraic Theory of Quadratic Forms*, (Benjamin, Reading, MA, 1973) [1].

[791] H. J. Landau, How does a porcupine separate its quills?, *IEEE Trans. Info. Theory*, **17** (1971) 157–161 [Intro.].

[792] H. J. Landau and D. Slepian, On the optimality of the regular simplex code, *Bell Syst. Tech. J.*, **45** (1966) 1247–1272 [1].

[793] S. Lang and A. Weil, Number of points of varieties in finite fields, *Amer. J. Math.*, **76** (1954) 819–827 [9].

[794] V. S. Lapin, The problem of grouped errors on magnetic tape, *Problems of Info. Trans.*, **4**(1) (1968) 28–34 [1].

[795] B. A. Laws, Jr., A parallel BCH decoder, (Technical Report, Bozeman Electronics Research Lab, Montana State Univ., June 1970) [9].

[796] B. A. Laws, Jr. and C. K. Rushforth, A cellular-array multiplier for $GF(2^m)$, *IEEE Trans. Computers*, **C-20** (1971) 1573–1578 [3].

[797] R. J. Lechner, Affine Equivalence of Switching Functions, Ph.D. Thesis, Applied Mathematics, Harvard University, January 1963 = Report BL-33, "Theory of Switching", Computation Laboratory of Harvard University, December 1963 [1, 14].

[798] R. J. Lechner, A correspondence between equivalence classes of switching functions and group codes, *IEEE Trans. Computers*, **16** (1967) 621–624 [1, 14].

[799] R. J. Lechner, Harmonic analysis of switching functions, in: A. Mukhopadhyay, ed., *Recent Developments in Switching Theory*, (Academic Press, New York, 1971) pp. 121–228 [1, 14].

[800] P. Lecointe, Généralisation de la notion de réseau géométriques et application à la construction de codes correcteurs, *Comptes Rendus*, (A) **265** (1967) 196–199 [18].

[801] C. Y. Lee, Some properties of non-binary error-correcting codes, *IEEE Trans. Info. Theory*, **4** (1958) 77–82 [5].

[802] Y. L. Lee and M. C. Cheng, Cyclic mappings of product codes, *IEEE Trans. Info. Theory*, **21** (1975) 233–235 [18].

[803] J. Leech, Some sphere packings in higher space, *Canad. J. Math.*, **16** (1964) 657–682 [Intro., 2, 16].

[804] J. Leech, Notes on sphere packings, *Canad. J. Math.*, **19** (1967) 251–267 [Intro., 19, 20].

[805] J. Leech, Five-dimensional nonlattice sphere packings, *Canad. Math. Bull.*, **10** (1967) 387–393 [Intro., 19].

[806] J. Leech, A presentation of the Mathieu group M_{12}, *Canad. Math. Bull.*, **12** (1969) 41–43 [20].

[807] J. Leech, Six and seven dimensional nonlattice sphere packings, *Canad. Math. Bull.*, **12** (1969) 151–155 [Intro., 19].

[808] J. Leech and N. J. A. Sloane, New sphere packings in dimensions 9–15, *Bull. Amer. Math. Soc.*, **76** (1970) 1006–1010 [Intro., 19].

[809] J. Leech and N. J. A. Sloane, New sphere packings in more than thirty-two dimensions, in: *Proc. Second Chapel Hill Conference on Comb. Math. and Applic.*, (Chapel Hill, NC, 1970) pp. 345–355 [Intro., 19].

[810] J. Leech and N. J. A. Sloane, Sphere packings and error-correcting-codes, *Canad. J. Math.*, **23** (1971) 718–745 [Intro., 2, 19, 20].

[811] E. Lehmer, On the magnitude of the coefficients of the cyclotomic polynomials, *Bull. Am. Math. Soc.*, **42** (1936) 389–392 [7].

[812] P. W. H. Lemmens and J. J. Seidel, Equiangular lines, *J. Algebra*, **24** (1973) 494–513 [21].

[813] A. Lempel, Analysis and synthesis of polynomials and sequences over GF(2), *IEEE Trans. Info. Theory*, **17** (1971) 297–303 [8].

[814] A. Lempel, Matrix factorization over GF(2) and trace-orthogonal bases of GF(2^n), *SIAM J. Comput.*, **4** (1975) 175–186 [4].

[815] H. W. Lenstra, Jr., Two theorems on perfect codes, *Discrete Math.*, **3** (1972) 125–132 [6].

[816] V. K. Leont'ev, A conjecture about Bose-Chaudhuri codes, *Problems of Infor. Trans.*, **4**(1) (1968) 83–85 [9].

[817] V. K. Leont'ev, Error-detecting encoding, *Problems of Info. Trans.*, **8**(2) (1972) 86–92 [1].

[818] T. Lerner, Analysis of digital communications system using binary error correcting codes, *IEEE Trans. Commun.*, **15** (1967) 17–22 [1].

[819] V. I. Levenshtein, The application of Hadamard matrices to a problem in coding, *Problemy Kibernetiki*, **5** (1961) 123–136. English translation in *Problems of Cybernetics*, **5** (1964) 166–184 [2, A].

[820] V. I. Levenshtein, Binary codes capable of correcting spurious insertions and deletions of ones, *Problems of Info. Trans.*, **1**(1) (1965) 8–17 [1].

[821] V. I. Levenshtein, Binary codes capable of correcting deletions, insertions, and reversals, *Soviet Physics-Doklady*, **10**(8) (1966) 707–710 [1].

[822] V. I. Levenshtein, A method of constructing quasilinear codes providing synchronization in the presence of errors, *Problems of Info. Trans.*, **7**(3) (1971) 215–222 [1].

[823] V. I. Levenshtein, Upper-bound estimates for fixed-weight codes, *Problems of Info. Trans.*, **7**(4) (1971) 281–287 [17].

[824] V. I. Levenshtein, Minimum redundancy of binary error-correcting codes, *Problems of Info. Trans.*, **10**(2) (1974) 110–123; and *Info. and Control*, **28** (1975) 268–291 [17].

[825] W. J. LeVeque, *Topics in Number Theory*, 2 vols., (Addison-Wesley, Reading, MA., 1956) [2, 16].

[826] N. Levinson, Coding theory: a counterexample to G. H. Hardy's conception of applied mathematics, *Amer. Math. Monthly*, **77** (1970) 249–258 [7].

[827] K. N. Levitt and W. H. Kautz, Cellular arrays for the parallel implementation of binary error-correcting codes, *IEEE Trans. Info. Theory*, **15** (1969) 597–607 [3].

[828] K. N. Levitt and J. K. Wolf, A class of nonlinear error-correcting codes based upon interleaved two-level sequences, *IEEE Trans. Info. Theory*, **13** (1967) 335–336 [18].

[829] J. E. Levy, Self synchronizing codes derived from binary cyclic codes, *IEEE Trans. Info. Theory*, **12** (1966) 286–291 [1].

[830] J. E. Levy, A weight-distribution bound for linear codes, *IEEE Trans. Info. Theory*, **14** (1968) 487–490 [17].

[831] Shen Lin, personal communication [A].

[832] S. Lin, Some codes which are invariant under a transitive permutation group, and their connection with balanced incomplete block designs, in: R. C. Bose and T. A. Dowling, eds., *Combinatorial Mathematics and its Applications*, (Univ. of North Carolina Press, Chapel Hill, NC, 1969) pp. 388–401 [8].

[833] S. Lin, On a class of cyclic codes, in: H. B. Mann, ed., *Error correcting codes*, (Wiley, New York, 1969) pp. 131–148 [13].

[834] S. Lin, *An Introduction to Error-Correcting Codes*, (Prentice-Hall, Englewood Cliffs, N.J. 1970) [Intro.].

[835] S. Lin, Shortened finite geometry codes, *IEEE Trans. Info. Theory*, **18** (1972) 692–696 [13].

[836] S. Lin, On the number of information symbols in polynomial codes, *IEEE Trans. Info. Theory*, **18** (1972) 785–794 [13].

[837] S. Lin, Multifold euclidean geometry codes, *IEEE Trans. Info. Theory*, **19** (1973) 537–548 [13].

[838] S. Lin and E. J. Weldon, Jr., Long BCH codes are bad, *Info. and Control*, **11** (1967) 445–451 [9].

[839] S. Lin and E. J. Weldon, Jr., Further results on cyclic product codes, *IEEE Trans. Info. Theory*, **16** (1970) 452–495 [9, 18].

[840] S. Lin and K.-P. Yiu, An improvement to multifold euclidean geometry codes, *Info. and Control*, **28** (1975) 221–265 [13].

[841] J. Lindner, Binary sequences up to length 40 with best possible autocorrelation function, *Electronics Letters*, **11** (1975) 507 [14].

[842] B. Lindström, On group and nongroup perfect codes in q symbols, *Math. Scand.*, **25** (1969) 149–158 [6].

[843] B. Linström, Group partition and mixed perfect codes, *Canad. Math. Bull.*, **18** (1975) 57–60 [6].

[844] K. Lindström, The nonexistence of unknown nearly perfect binary codes, *Ann. Univ. Turku.*, *Ser. A, No. 169*, (1975), 3–28 [17].

[844a] K. Lindström, All nearly perfect codes are known, *Info. and Control*, **35** (1977) 40–47 [15].

[844b] K. Lindström and M. Aaltonen, The nonexistence of nearly perfect nonbinary codes for $1 \leq e \leq 10$, *Ann. Univ. Turku, Ser. A1, No. 172* (1976), 19 pages [15].

[845] J. H. van Lint, 1967–1969 Report of the Discrete Mathematics Group, Report 69-WSK-04 of the Technological University, Eindhoven, Netherlands (1969) [6].

[846] J. H. van Lint, On the nonexistence of perfect 2- and 3-Hamming-error-correcting codes over GF(q), *Info. and Control*, **16** (1970) 396–401 [6].

[847] J. H. van Lint, On the Nonexistence of Perfect 5-, 6-, and 7-Hamming-Error-Correcting Codes over GF(q), Report 70-WSK-06 of the Technological University, Eindhoven, Netherlands, 1970 [6].

[848] J. H. van Lint, *Coding Theory*, (Springer, New York, 1971) [many references].

[849] J. H. van Lint, On the nonexistence of certain perfect codes, in: A. O. L. Atkin and B. J. Birch, eds., *Computers in Number Theory*, (Academic Press, New York, 1971) pp. 227–282 [6].

[850] J. H. van Lint, Nonexistence theorems for perfect error-correcting-codes, in: *Computers in Algebra and Number Theory*, Vol. IV, (SIAM-AMS Proceedings, 1971) [6].

[851] J. H. van Lint, A new description of the Nadler code, *IEEE Trans. Info. Theory*, **18** (1972) 825–826 [2].

[852] J. H. van Lint, A theorem on equidistant codes, *Discrete Math.*, **6** (1973) 353–358 [1].

[853] J. H. van Lint, *Combinatorial Theory Seminar Eindhoven University of Technology*, Lecture Notes in Mathematics 382, (Springer, Berlin, 1974) [2].

[854] J. H. van Lint, Recent results on perfect codes and related topics, in: M. Hall, Jr. and J. H. van Lint, eds., *Combinatorics*, (Reidel, Dordrecht, Holland, 1975) [6].

[855] J. H. van Lint, A survey of perfect codes, *Rocky Mountain J. of Mathematics*, **5** (1975) 199–224 [6, 7].

[856] J. H. van Lint and J. J. Seidel, Equilateral point sets in elliptic geometry, *Kon. Ned. Akad. Wetensch. Proc.* A **69** (= Indag. Math. **28**) (1966) 335–348 [2, 21].

[857] D. E. Littlewood, *A University Algebra*, 2nd ed, (Dover, New York, 1970) [19].

[858] C. L. Liu, B. G. Ong and G. R. Ruth, A construction scheme for linear and nonlinear codes, *Discrete Math.*, **4** (1973) 171–184 [2].

[859] S. P. Lloyd, Binary block coding, *Bell Syst. Tech. J.*, **36** (1957) 517–535 [6].

[859a] S. J. Lomonaco, Jr., private communication [A].

[860] R. J. Longobardi, L. D. Rudolph and C. R. P. Hartmann, On a basic problem in majority decoding, Abstracts of papers presented at IEEE International Symposium on Information Theory, October 1974, IEEE Press, N.Y., 1974 p. 15 [13].

[862] M. E. Lucas, Sur les congruences des nombres Euleriennes, et des coefficients différentials des functions trigonométriques, suivant un-module premier, *Bull. Soc. Math. France*, **6** (1878) 49–54 [13].

[863] V. Lum, Comments on "The weight structure of some Bose-Chaudhuri codes", *IEEE Trans. Info. Theory*, **15** (1969) 618–619 [7].

[864] V. Lum and R. T. Chien, On the minimum distance of Bose-Chaudhuri-Hocquenghem codes, *SIAM J. Appl. Math.*, **16** (1968) 1325–1337 [9].

[865] H. Lüneburg, Über die Gruppen von Mathieu, *J. Algebra*, **10** (1968) 194–210 [20].

[866] H. Lüneburg, *Transitive Erweiterungen endlicher Permutationsgruppen*, Lecture Notes in Math. 84, Springer-Verlag, N.Y. 1969 [20].

[867] F. E. Lytle, Hamming type codes applied to learning machine determinations of molecular formulas, *Analytical Chemistry*, **44** (1972) 1867–1869 [7].

M

[868] J. E. MacDonald, Design methods for maximum minimum-distance error-correcting codes, *IBM J. Res. Devel.*, **4** (1960) 43–57 [17].

[869] H. F. MacNeish, Four finite geometries, *Amer. Math Monthly*, **49** (1942) 15–23 [B].

[870] F. J. MacWilliams, Error correcting codes for multiple level transmission, *Bell Syst. Tech. J.*, **40** (1961) 281–308 [8].

[871] F. J. MacWilliams, Combinatorial problems of elementary group theory, Ph.D. Thesis, Department of Math., Harvard University., May 1962 [5, 6, 8].

[872] F. J. MacWilliams, A theorem on the distribution of weights in a systematic code, *Bell Syst. Tech. J.*, **42** (1963) 79–94 [5, 6].

[873] F. J. MacWilliams, Permutation decoding of systematic codes, *Bell Syst. Tech. J.*, **43** (1964) 485–505 [16].

[874] F. J. MacWilliams, The structure and properties of binary cyclic alphabets, *Bell Syst. Tech. J.*, **44** (1965) 303–332 [8].

[875] F. J. MacWilliams, An example of two cyclically orthogonal sequences with maximum period, *IEEE Trans. Info. Theory*, **13** (1967) 338–339 [4].

[876] F. J. MacWilliams, Error-correcting codes – An Historical Survey, in: H. B. Mann, ed., *Error-Correcting Codes*, (Wiley New York, 1969) pp. 3–13 [1, 2].

[877] F. J. MacWilliams, Codes and ideals in group algebras, in: R. C. Bose and T. A. Dowling, eds., *Combinatorial Mathematics and its Applications*, (Univ. North Carolina Press, Chapel Hill, 1969) Ch. 18 [18].

[878] F. J. MacWilliams, Orthogonal matrices over finite fields, *Amer. Math. Monthly*, **76** (1969) 152–164 [15, 16].

[879] F. J. MacWilliams, On binary cyclic codes which are also cyclic codes over $GF(2^s)$, *SIAM J. Applied Math.*, **19** (1970) 75–95 [10].

[880] F. J. MacWilliams, Binary codes which are ideals in the group algebra of an Abelian group, *Bell Syst. Tech. J.*, **49** (1970) 987–1011 [18].

[881] F. J. MacWilliams, Orthogonal circulant matrices over finite fields, and how to find them, *J. Comb. Theory*, **10** (1971) 1–17 [16].

[882] F. J. MacWilliams, Cyclotomic numbers, coding theory, and orthogonal polynomials, *Discrete Math.*, **3** (1972) 133–151 [8].

[883] F. J. MacWilliams, C. L. Mallows and N. J. A. Sloane, Generalization of Gleason's theorem on weight enumerators of self-dual codes, *IEEE Trans. Info. Theory*, **18** (1972) 794–805 [5, 19].

[884] F. J. MacWilliams and H. B. Mann, On the p-rank of the design matrix of a difference set, *Info. and Control*, **12** (1968) 474–488 [13].

[885] F. J. MacWilliams and N. J. A. Sloane, Pseudo-random sequences and arrays, *Proc. IEEE* **64** (1976) 1715–1729 [18].

[886] F. J. MacWilliams, N. J. A. Sloane and J.-M. Goethals, The MacWilliams identities for nonlinear codes, *Bell Syst. Tech. J.*, **51** (1972) 803–819 [5].

[887] F. J. MacWilliams, N. J. A. Sloane and J. G. Thompson, Good self-dual codes exist, *Discrete Math.*, **3** (1972) 153–162 [19].

[888] F. J. MacWilliams, N. J. A. Sloane and J. G. Thompson, On the existence of a projective plane of order 10, *J. Comb. Theory*, **14A** (1973) 66–78 [19].

[889] G. K. Maki and J. H. Tracey, Maximum-distance linear codes, *IEEE Trans. Info. Theory*, **17** (1971) 632 [17].

[890] V. K. Malhotra and R. D. Fisher, A double error-correction scheme for peripheral systems, *IEEE Trans. Info. Theory*, **25** (1976) 105–114 [1].

[891] C. L. Mallows, A. M. Odlyzko and N. J. A. Sloane, Upper bounds for modular forms, lattices, and codes, *J. Algebra*, **36** (1975) 68–76 [19].

[892] C. L. Mallows, V. Pless and N. J. A. Sloane, Self-dual codes over GF(3), *SIAM J. Applied Math.*, **31** (1976) 649–666 [19, 20].

[893] C. L. Mallows and N. J. A. Sloane, An upper bound for self-dual codes, *Info. and Control*, **22** (1973) 188–200 [19].

[894] C. L. Mallows and N. J. A. Sloane, On the invariants of a linear group of order 336, *Proc. Camb. Phil. Soc.*, **74** (1973) 435–440 [19].

[895] C. L. Mallows and N. J. A. Sloane, Weight enumerators of self-orthogonal codes, *Discrete Math.*, **9** (1974) 391–400 [5, 19].

[896] D. Mandelbaum, A method of coding for multiple errors, *IEEE Trans. Info. Theory*, **14** (1968) 518–521 [9, 10].

[897] D. Mandelbaum, Note on Tong's burst-trapping technique, *IEEE Trans. Info. Theory*, **17** (1971) 358–360 [10].

[898] D. Mandelbaum, On decoding of Reed–Solomon codes, *IEEE Trans. Info. Theory*, **17** (1971) 707–712 [10].

[899] D. Mandelbaum, Some results in decoding of certain maximal-distance and BCH codes, *Info. and Control*, **20** (1972) 232–243 [9, 10].

[900] D. M. Mandelbaum, Synchronization of codes by means of Kautz's Fibonacci encoding, *IEEE Trans. Info. Theory*, **18** (1972) 281–285 [1].

[901] D. M. Mandelbaum, Unequal error-protection derived from difference sets, *IEEE Trans. Info. Theory*, **18** (1972) 686–687 [Intro.].

[902] D. M. Mandelbaum, On the derivation of Goppa codes, *IEEE Trans. Info. Theory*, **21** (1975) 110–111 [12].

[903] D. M. Mandelbaum, A method for decoding of generalized Goppa codes, *IEEE Trans. Info. Theory*, **23** (1977) 137–140 [12].

[904] C. Maneri and R. Silverman, A vector-space packing problem, *J. Algebra*, 4 (1966) 321–330 [11].

[905] C. Maneri and R. Silverman, A combinatorial problem with applications to geometry, *J. Comb. Theory*, **11A** (1971) 118–121 [11].

[906] H. B. Mann, On the number of information symbols in Bose-Chaudhuri codes, *Info. and Control*, **5** (1962) 153–162 [9].

[907] H. B. Mann, *Addition Theorems*, (Wiley, New York, 1965) [5, 7].

[908] H. B. Mann, Recent advances in difference sets, *Amer. Math. Monthly*, **74** (1967) 229–235 [13].

[909] H. B. Mann, ed., *Error Correcting Codes*, (Wiley, New York, 1969) [Intro.].

[910] H. B. Mann, On canonical bases for subgroups of an Abelian group, in: R. C. Bose and T. H. Dowling, eds., *Combinatorial Mathematics and its Applications*, (Univ. of North Carolina Press, Chapel Hill, NC, 1969) pp. 38–54 [4].

[911] D. E. Maracle and C. T. Wolverton, Generating cyclically permutable codes, *IEEE Trans. Info. Theory*, **20** (1974) 554–555 [17].

[912] A. S. Marchukov, Summation of the products of codes, *Problems of Info. Trans.*, **4**(2) (1968) 8–15 [18].

[913] A. B. Marcovitz, Sequential generation and decoding of the P-nary Hamming code, *IEEE Trans. Info. Theory*, 7 (1961) 53–54 [7].

[914] M. Marcus, *Basic Theorems in Matrix Theory*, National Bureau of Standards Applied Math. Ser. No. 57, U.S. Department of Commerce, Washington, D.C., 1960 [2, 14].

[915] M. Marcus and H. Minc, *A Survey of Matrix Theory and Matrix Inequalitities*, (Allyn and Bacon, Boston) 1964 [21].

[916] M. A. Marguinaud, Application des algorithms rapides au decodage de codes polynomiaux, *Journée d'Etudes Sur les Codes Correcteurs d'Erreurs*, Paris, 10th March, 1971 [Intro.].

[917] R. W. Marsh, Table of Irreducible Polynomials over GF(2) Through Degree 19, Office of Technical Services, Dept. of Commerce, Washington D.C., October 24, 1957 [4].

[917a] P. Martin-Löf, The definition of random sequences, *Info. and Control*, 9 (1966) 602–619 [Intro.].

[918] J. L. Massey, *Threshold decoding*, (MIT Press, Cambridge, MA, 1963) [13].

[919] J. L. Massey, Reversible codes, *Info. and Control*, 7 (1964) 369–380 [7].

[920] J. L. Massey, Step-by-step decoding of the Bose–Chaudhuri–Hocquenghem codes, *IEEE Trans. Info. Theory*, **11** (1965) 580–585 [9].

[921] J. L. Massey, Advances in threshold decoding, in: A. V. Balakrishnan, ed., *Advances in Communication Systems*, vol. 2, (Academic Press, N.Y., 1968) pp. 91–115 [13].

[922] J. L. Massey, Shift-register synthesis and BCH decoding, *IEEE Trans. Info. Theory*, **15** (1969) 122–127 [9, 12].

[922a] J. L. Massey, *Notes on Coding Theory*, (Waltham Research Center, General Telephone and Electronics, Inc., Waltham, Mass., 1969) [Intro., 9].

[923] J. L. Massey, On the fractional weight of distinct binary n-tuples, *IEEE Trans. Info. Theory*, **20** (1974) 131 [10].

[924] J. L. Massey, D. J. Costello and J. Justesen, Polynomial weights and code constructions, *IEEE Trans. Info. Theory*, **19** (1973) 101–110 [13, 18].

[925] E. Mathieu, Mémoire sur l'étude des fonctions de plusiers quantités, *J. Math. p. et a.*, **6** (1861) 241–323 [20].

[926] E. Mathieu, Sur la fonction cinq fois transitive de 24 quantités, *J. Math. p. et a.*, **18** (1873) 25–46 [20].

[927] H. F. Mattson, Jr. and E. F. Assmus, Jr., Research program to extend the theory of weight distribution and related problems for cyclic error-correcting codes, Report AFCRL-64-605, Air Force Cambridge Res. Labs., Bedford, Mass., July 1964 [16].

[928] H. F. Mattson, Jr. and G. Solomon, A new treatment of Bose–Chaudhuri codes, *J. Soc. Indust. Appl. Math.*, **9** (1961) 654–669 [8].

[929] M. V. Matveeva, On the 3-error-correcting Bose–Chaudhuri codes over the field GF(3), *Problems of Info. Trans.*, **4**(1) (1968) 20–27 [9].

[930] M. V. Matveeva, On a solution of a cubic equation in a field of characteristic 3, *Problems of Info. Trans.*, **4**(4) (1968) 76–78 [9].

[931] E. A. Mayo, Efficient computer decoding of pseudorandom radar signal codes, *IEEE Trans. Info. Theory*, **18** (1972) 680–681 [14].

[932] L. E. Mazur, A class of polynomial codes, *Problems of Info. Trans.*, **8**(4) (1972) 351–533 [9].

[933] L. E. Mazur, Minimum code distance of a particular subcode class of Reed–Solomon codes, *Problems of Info. Trans.*, **9**(2) (1973) 169–171 [9, 11].

[934] L. E. Mazur, Codes correcting errors of large weight in the Lee metric, *Problems of Info. Trans.*, **9**(4) (1973) 277–281 [5].

[935] E. J. McCluskey, *Introduction to the Theory of Switching Circuits*, (McGraw-Hill, New York, 1965) [13].

[936] R. J. McEliece, Quadratic forms over finite fields and second-order Reed–Muller codes, *JPL Space Programs Summary*, **37-58-III** (1969) 28–33 [15].

[937] R. J. McEliece, Factorization of polynomials over finite fields, *Math. Comp.*, **23** (1969) 861–867 [9].

[938] R. J. McEliece, On the symmetry of good nonlinear codes, *IEEE Trans. Info. Theory*, **16** (1970) 609–611 [9].

[939] R. J. McEliece, On periodic sequences from GF(q), *J. Comb. Theory*, **10A** (1971) 80–91 [15].

[940] R. J. McEliece, Weights modulo 8 in binary cyclic codes, *JPL Technical Report 32-1526*, **XI**, (Jet Propulsion Lab., Pasadena, Calif. Oct. 1972) pp. 86–88 [8].

[941] R. J. McEliece, Weight congruences for p-ary cyclic codes, *Discrete Math.*, **3** (1972) 177–192 [8, 15].

[942] R. J. McEliece, A nonlinear, nonfield version of the MacWilliams identities, unpublished notes, 1972 [5].

[943] R. J. McEliece, Comment on "A class of codes for asymmetric channels and a problem from the additive theory of numbers", *IEEE Trans. Info. Theory*, **19** (1973) 137 [1].

[944] R. J. McEliece, An application of linear programming to a problem in coding theory, unpublished notes, 1973 [17].

[945] R. J. McEliece, Irreducible cyclic codes and Gauss sums, in: M. Hall, Jr. and J. H. van Lint, eds., *Combinatorics*, (Reidel, Dordrecht, 1975) pp. 185–202 [8].

[946] R. J. McEliece, E. R. Rodemich, H. C. Rumsey, Jr. and L. R. Welch, unpublished tables, 1972, [A].

[947] R. J. McEliece, E. R. Rodemich, H. C. Rumsey, Jr. and L. R. Welch, New upper bounds on the rate of a code via the Delsarte-MacWilliams inequalities, *IEEE Trans. Info. Theory*, **23** (1977) 157–166 [17].

[948] R. J. McEliece and H. C. Rumsey, Jr., Sphere-Packing in the Hamming Metric, *Bull. Amer. Math. Soc.*, **75** (1969) 32–34 [17].

[949] R. J. McEliece and H. C. Rumsey, Jr., Euler products, cyclotomy, and coding, *J. Number Theory*, **4** (1972) 302–311 [8].

[950] R. L. McFarland, A family of difference sets in non-cyclic groups, *J. Comb. Theory*, **15A** (1973) 1–10 [14].

[951] P. Mecklenburg, W. K. Pehlert, Jr. and D. D. Sullivan, Correction of errors in multilevel Gray-coded data, *IEEE Trans. Info. Theory*, **19** (1973) 336–340 [Intro.].

[952] J. E. Meggitt, Error correcting codes for correcting bursts of errors, *IBM J. Res. Devel.*, **4** (1960) 329–334 [13].

[953] C. M. Melas, A cyclic code for double error correction, *IBM J. Res. Devel.*, **4** (1960) 364–366 [7].

[954] A. M. Michelson, Computer implementation of decoders for several BCH codes, in: *Proc. of Symp. on Computer Processing in Communications*, (Polytech. Inst. Brooklyn, New York, 1969) pp. 401–413 [9].

[955] G. A. Miller, H. F. Blichfeldt and L. E. Dickson, *Theory and Applications of Finite Groups*, (Dover, New York, 1961) [19].

[956] W. H. Mills, Some complete cycles on the n-cube, *Proc. Am. Math. Soc.*, **14** (1963) 640–643 [Intro.].

[957] W. H. Mills, On the covering of pairs by quadruples, *J. Comb. Theory*, **13A** (1972) 55–78 and **15A** (1973) 138–166 [A].

[958] W. H. Mills, Covering problems, in: *Proc. 4ᵗʰ S-E Conf. Combinatorics, Graph Theory and Computing*, (Utilitas Math., Winnipeg 1973) pp. 23–52 [A].

[959] W. H. Mills, On the covering of triples by quadruples, in: *Proc. 5ᵗʰ S-E Conf. on Combinatorics, Graph Theory and Computing*, (Utilitas Math. Publ., Winnipeg, 1974) pp. 563–581 [A].

[960] W. H. Mills, Continued fractions and linear recurrences, *Math. Comp.*, **29** (1975) 173–180 [12].

[961] W. H. Mills and N. Zierler, On a conjecture of Golomb, *Pacific J. Math.*, **28** (1969) 635–640 [9].

[962] J. Milnor and D. Husemoller, *Symmetric Bilinear Forms*, (Springer, Berlin, 1973) [19].

[963] N. Mitani, On the transmission of numbers in a sequential computer, delivered at the National Convention of the Inst. of Elect. Engineers of Japan, (November, 1951) [13].

[964] M. E. Mitchell, Simple decoders and correlators for cyclic error-correcting codes, *IEEE Trans. Info. Theory*, **8** (1962) 284–292 [16].

[965] H. Miyakawa, H. Imai and I. Nakajima, Modified Preparata codes-optimum systematic nonlinear double-error-correcting codes, *Electron. Commun. in Japan*, **53A**,(10) (1970) 25–32 [18].

[966] H. Miyakawa and T. Kaneko, Decoding algorithms for error-correcting codes by use of analog weights, *Electron. Commun. Japan*, **58A**(1) (1975) 18–27 [16].

[967] H. Miyakawa and T. Moriya, Methods of construction of comma-free codes with fixed block sync patterns, *Electron. Commun. Japan*, **52A**(5) (1969) 10–18 [Intro.].

[968] P. S. Moharir, Ternary Barker codes, *Electronics Letters*, **10** (1974) 460–461 [14].

[969] P. S. Moharir and A. Selvarajan, Optical Barker codes, *Electronics Letters*, **10** (1974) 154–155 [14].

[970] P. S. Moharir and A. Selvarajan, Systematic search for optical Barker codes with minimum length, *Electronics Letters*, **10** (1974) 245–246 [14].

[971] T. Molien, Über die Invarianten der linear Substitutions-gruppe, *Sitzungsber. König. Preuss. Akad. Wiss.*, (1897) 1152–1156 [19].

[972] A. M. Mood, On Hotelling's weighing problem, *Ann. Math. Stat.*, **17** (1946) 432–446 [2].

[973] S. Mossige, Table of irreducible polynomials over GF(2) of degrees 10 through 20, *Math. Comp.*, **26** (1972) 1007–1009 [4].

[974] T. Muir, *A Treatise on the Theory of Determinants*, (revised by W. H. Metzler, Dover, New York, 1960) [7, 12, 16].

[975] D. E. Muller, Application of Boolean algebra to switching circuit design and to error detection, *IEEE Trans. Computers*, **3** (1954) 6–12 [13].

[976] D. Mumford, *Geometric Invariant Theory*, (Springer, N.Y. 1965) [19].

[977] B. R. N. Murthy, An efficient parity checking scheme for random and burst errors, *IEEE Trans. Commun.*, **24** (1976) 249–254 [1].

[978] J. Mykkeltveit, A note on Kerdock codes, Technical Report 32–1526, Vol IX., June 15, 1972, pp. 82–83, Jet Propulsion Lab, Pasadena, Calif. [15].

[979] J. Mykkeltveit, C. Lam and R. J. McEliece, On the weight enumerators of quadratic residue codes, JPL Technical Report 32–1526, Vol. XII, pp. 161–166, Jet Propulsion Laboratory, Pasadena, Calif., 1974 [16].

[980] E. Myravaagnes, On maximum-weight codes, *IEEE Trans. Info. Theory*, **9** (1963) 289–290 [17].

[981] E. Myrvaagnes, The weight distribution of two cyclic codes of length 73, *IEEE Trans. Info. Theory*, **11** (1965) 316 [9].

N

[982] M. Nadler, A 32-point $n = 12$, $d = 5$ code, *IEEE Trans. Info. Theory*, **8** (1962) 58 [2].

[983] G. Nakamura, N. Ikeno and K. Naemura, Optimality of constant-weight codes, *Electron. Commun. Japan*, **55**-A(7) (1972) 32–37 [17].

[984] K. Nakamura and Y. Iwadare, Data scramblers for multilevel pulse sequences, *NEC Research and Development*, (26) (July 1972) pp. 53–63 [14, 18].

[985] M. Nesenbergs, Comparison of the 3-out-of-7 ARQ with Bose–Chaudhuri–Hocquenghem coding systems, *IEEE Trans. Comm.*, **11** (1963) 202–212 [9].

[986] P. G. Neumann, A note on cyclic permutation error-correcting codes, *Info. and Control*, **5** (1962) 72–86 [17].

[987] P. G. Neumann, On a class of cyclically permutable error-correcting codes, *IEEE Trans. Info. Theory*, **10** (1964) 75–78 [17].

[988] P. M. Neumann, Transitive permutation groups of prime degree, in: *Proc. Second. Internat. Conf. on Theory of Groups*, (Springer-Verlag, N.Y. 1975) pp. 520–535 [16, 19].

[989] P. M. Neumann, Transitive permutation groups of prime degree (IV), *Proc. London Math. Soc.*, **32** (1976) 52–62 [16, 19].

[990] S. W. Ng, On Rudolph's majority-logic decoding algorithm, *IEEE Trans. Info. Theory*, **16** (1970) 651–652 [13].

[991] P. J. Nicholson, Algebraic theory of finite Fourier transforms, *J. Computer and Syst. Sci.*, **5** (1971) 524–547 [14].

[992] H.-V. Niemeier, Definite Quadratische Formen der Dimension 24 und Diskriminante 1, *J. Number Theory*, **5** (1973) 142–178 [19].

[993] Y. Niho, On maximal comma-free codes, *IEEE Trans. Info. Theory*, **19** (1973) 580–581 [Intro.].

[994] H. Nili, Matrixschaltungen zur Codierung und Decodierung von Gruppen-Codes, *Archiv für Elekt. Übertragung*, **18** (1964) 555–564 [16].

[995] I. Niven and H. S. Zuckerman, *An Introduction to the Theory of Numbers*, 2nd ed., (Wiley, New York, 1966) [10, 12].

[996] S. Niven, On the number of k-tuples in maximal systems $m(k, l, n,)$, in: R. K. Guy et al., eds., *Combinatorial Structures and their Applications*, (Gordon and Breach, New York, 1970) pp. 303–306 [17].

[997] E. Noether, Der Endlichkeitssatz der Invarianten endlicher Gruppen, *Math. Ann.*, **77** (1916) 89–92 [19].

[998] T. Nomura and A. Fukuda, Linear recurring planes and two-dimensional cyclic codes, *Electron. Commun. Japan*, **54**A(3) (1971) 23–30 [18].

[999] T. Nomura, H. Miyakawa, H. Imai and A. Fukuda, A method of construction and some properties of planes having maximum area matrix, *Electron. Commun. Japan*, **54**A(5) (1971) 18–25 [18].

[1000] T. Nomura, H. Miyakawa, H. Imai and A. Fukuda, Some properties of the $\gamma\beta$-plane and its extension to three-dimensional space, *Electron. Commun. Japan*, **54**A(8) (1971) 27–34 [18].

[1001] T. Nomura, H. Miyakawa, H. Imai and A. Fukuda, A theory of two-dimensional linear recurring arrays, *IEEE Trans. Info. Theory*, **18** (1972) 775–785 [18].

[1002] A. W. Nordstrom and J. P. Robinson, An optimum nonlinear code, *Info. and Control*, **11** (1967) 613–616 [2, A].

O

[1003] S. Sh. Oganesyan, V. I. Tairyan and V. G. Yagdzyan, Decomposition of cyclic codes into equal-weight classes, *Problems of Control and Info. Theory*, 3(2) (1974) 13–21 [8].

[1004] S. Sh. Oganesyan and V. G. Yagdzhyan, Weight spectra of certain classes of cyclic correcting codes, *Problems of Info. Trans.*, **6**(3) (1970) 213–218 [8].

[1005] S. Sh. Oganesyan and V. G. Yagdzhyan, Class of optimum cyclic codes with base p, *Problems of Info. Tran.*, **8**(2) (1972) 167–169 [8].

[1006] S. Sh. Oganesyan, V. G. Yagdzhyan and V. I. Tairyan, On a class of optimal cyclic codes, in: B. N. Petrov and F. Csáki, eds., *2nd International Symposium on Information Theory*, (Akadémiai Kiadó, Budapest, 1973) pp. 219–224 [8].

[1007] M. Ogasawara, A necessary condition for the existence of regular and symmetrical PBIB designs of T_m type, *Inst. of Stat. Mimeo. Ser. No.* 418 (Univ. of North Carolina, Chapel Hill, NC, Feb. 1965) [21].

[1008] J. Ogawa, The theory of the association algebra and the relationship algebra of a partially balanced incomplete block design, *Institute of Stat. Mimeo. Ser. No.* 224 (Univ. of North Carolina, Chapel Hill, NC, April 1959) [21].

[1009] K. Ohno, H. Nakanishi and Y. Tezuka, A method for combining several codes and its application to data transmission, Technical Report of the Institute of Electronics and Communication Engineers of Japan, AL 73–56, Nov. 27, 1973 [18, A].

[1010] G. B. Olderogge, On some special correcting codes of matrix type, *Radiotechniques*, **18**(7) (1963) 14–19 (See Dênes and Keedwell [371, p. 356]) [18].

[1011] I. B. Oldham, R. T. Chien and D. T. Tang, Error detection and correction in a photo-digital storage system, *IBM J. Res. Devel.,* **12** (1968) 422–430 [1].

[1012] J. K. Omura, A probabilistic decoding algorithm for binary group codes (Abstract), *IEEE Trans. Info. Theory*, **16** (1970) p. 123 [16].

[1013] J. K. Omura, Iterative decoding of linear codes by a modulo-2 linear program, *Discrete Math.*, **3** (1972) 193–208 [16].

[1014] A. Ong, A note on the nonbinary BCH code, *IEEE Trans. Info. Theory*, **15** (1969) 735–736 [9].

[1015] O. Ore, On a special class of polynomials, *Trans. Amer. Math. Soc.*, **35** (1933) 559–584 **36** (1934) 275 [4].

[1016] O. Ore, Contributions to the theory of finite fields, *Trans. Amer. Math. Soc.*, **36** (1934) 243–274 [4].

[1017] R. Ottoson, Group codes for phase- and amplitude-modulated signals on a Gaussian channel, *IEEE Trans. Info. Theory*, **17** (1971) 315–321 [Intro.].

P

[1018] L. F. Paige, A note on the Mathieu groups, *Canad. J. Math.*, **9** (1956) 15–18 [6, 20].

[1019] R. E. A. C. Paley, On orthogonal matrices, *J. Math. and Phys.*, **12** (1933) 311–320 [2].

[1020] J. W. di Paola, J. S. Wallis and W. D. Wallis, A list of (v, b, r, k, λ) designs for $r \leq 30$, in: *Proc. 4th S-E Conf. Combinatorics, Graph Theory and Computing*, (Utilitas Math., Winnipeg 1973) pp. 249–258 [2, 17].

[1021] J. H. Park, Jr., Inductive proof of an important inequality, *IEEE Trans. Info. Theory*, **15** (1969) 618 [10].

[1022] Parke Mathematical Laboratories Staff, Annotated Bibliography on Error Correcting Codes, Report AFCRL-62-101, Parke Math. Labs., Carlisle, Mass., Dec. 1961 [Intro.].

[1023] E. T. Parker and P. J. Nikolai, A search for analogues of the Mathieu groups, *Math. Comp.*, **12** (1968) 38–43 [16].

[1024] M. B. Pursley and D. V. Sarwate, Bounds on aperiodic crosscorrelation for binary sequences, *Electronics Letters*, **12** (1976) 304–305 [14].

[1025] R. H. Paschburg, Software implementation of error-correcting codes, Coordinated Science Lab., Report R-659, University of Illinois, Urbana, Ill., 1974 [13].

[1026] R. H. Paschburg, C. L. Chen and D. V. Sarwate, Comparison of various software coding systems, Abstracts of papers presented at IEEE Internat. Symposium on Info. Theory, October 1974, IEEE Press, N.Y. 1974 p. 80 [13].

[1027] A. M. Patel, Maximal group codes with specified minimum distance, *IBM J. Res. Devel.*, **14** (1970) 434–443 [17].

[1028] A. M. Patel, Maximal q-nary linear codes with large minimum distance, *IEEE Trans. Info. Theory*, **21** (1975) 106–110 [17].

[1029] A. M. Patel and S. J. Hong, Optimal rectangular code for high density magnetic tapes, *IBM J. Res. Devel.*, **18** (1974) 579–588 [1, 18].

[1030] N. J. Patterson, The algebraic decoding of Goppa codes, *IEEE Trans. Info. Theory*, **21** (1975) 203–207 [12].

[1031] N. J. Patterson, A 4-dimensional Kerdock set over GF(3), *J. Comb. Theory*, **20A** (1976) 365–366 [15].

[1032] S. E. Payne, On maximizing det $(A^T A)$, *Discrete Math.*, **10** (1974) 145–158 [2].

[1033] W. K. Pehlert, Jr., Design and evaluation of a generalized burst-trapping error control system, *IEEE Trans. on Comm. Theory*, **19** (1971) 863–868 [10].

[1034] R. L. Pele, Some remarks on the vector subspaces of a finite field, *Acta Math. Acad. Sci. Hungar.*, **20** (1969) 237–240 [4].

[1035] O. Perron, Bemerkungen über die Verteilung der quadratischen Reste, *Math. Zeit.*, **56** (1952) 122–130 [2, 16].

[1036] W. W. Peterson, Encoding and error-correction procedures for the Bose–Chaudhuri codes, *IEEE Trans. Info. Theory*, **6** (1960) 459–470 [9].

[1036a] W. W. Peterson, *Error-Correcting Codes*, (MIT Press, Cambridge, MA, 1961) [Intro.].

[1037] W. W. Peterson, Bounds for error-correcting codes, *IEEE Trans. Info. Theory*, **8** (1962) 60 [17].

[1038] W. W. Peterson, On the weight structure and symmetry of BCH codes, *J. Inst. Elec. Commun. Engrs. Japan*, **50** (1967) 1183–1190 (in Japanese). For English version see Scientific Report AFCRL-65-515, Air Force Cambridge Research Labs., Bedford, Mass., July 1965 [8, 9].

[1039] W. W. Peterson, Some new results on finite fields with applications to BCH codes, in: R. C. Bose and T. A. Dowling, eds., *Combinatorial Mathematics and its Applications*, (Univ. North Carolina Press, Chapel Hill, NC, 1969) Ch. 19 [9].

[1040] W. W. Peterson and E. J. Weldon, Jr., *Error-Correcting Codes*, 2nd ed., (MIT Press, 1972) [many references].

[1041] G. H. Pettengill, Radar astronomy, in: M. I. Skolnik, ed., *Radar Handbook*, (McGraw-Hill, New York, 1970) [14].

[1042] P. G. Phillips, M. Harwit and N. J, A. Sloane, A new multiplexing spectrometer with large throughput, in: *Proc. Aspen Internat. Conf. Fourier Spectroscopy*, (Air Force Systems Command, 1970) pp. 441–444 [2].

[1043] H. R. Phinney, personal communication [A].

[1044] J. N. Pierce, Limit distribution of the minimum distance of random linear codes, *IEEE Trans. Info. Theory*, **13** (1967) 595–599 [17].

[1045] J. R. Pierce, Synchronizing digital networks, *Bell Syst. Tech. J.*, **48** (1969) 615–636 [1].

[1046] W. H. Pierce, Linear-real codes and coders, *Bell Syst. Tech. J.*, **47** (1968) 1065–1097 [16].

[1046a] M. S. Pinsker, On the complexity of decoding, *Problems of Info. Trans.*, **1**(1) (1965) 84–86 [Intro.].

[1047] P. Piret, Good block codes derived from cyclic codes, *Electronics Letters*, **10** (1974) 391–392 [18, A].

[1048] P. Piret, On a class of alternating cyclic convolutional codes, *IEEE Trans. Info. Theory*, **21** (1975) 64–69 [18].

[1049] P. Piret, Structure and constructions of cyclic convolutional codes, *IEEE Trans. Info. Theory*, **22** (1976) 147–155 [18].

[1050] P. Piret, Some optimal AMC codes, *IEEE Trans. Info. Theory*, **22** (1976) 247–248 [18].

[1051] V. Pless, Power moment identities on weight distributions in error correcting codes, *Info. and Control*, **6** (1963) 147–152 [5].

[1052] V. Pless, Weight Distribution of the Quadratic Residue (71; 35) Code, Air Force Cambridge Research Lab. Report 64–697 (1964) [16].

[1053] V. Pless, The number of isotropic subspaces in a finite geometry, *Rend. Cl. Scienze fisiche, matematiche e naturali, Accad. Naz. Lincei*, **39** (1965) 418–421 [19].

[1054] V. Pless, On the uniqueness of the Golay codes, *J. Comb. Theory*, **5** (1968) 215–228 [19, 20].

[1055] V. Pless, On a new family of symmetry codes and related new five-designs, *Bull. Am. Math. Soc.*, **75** (1969) 1339–1342 [16].

[1056] V. Pless, The weight of the symmetry code for $p = 29$ and the 5-designs contained therein, *Annals N.Y. Acad. Sci.*, **175** (Article 1) (1970) 310–313 [16].

[1057] V. Pless, Symmetry codes over GF(3) and new five-designs, *J. Comb. Theory*, **12** (1972) 119–142 [16].

[1058] V. Pless, A classification of self-orthogonal codes over GF(2), *Discrete Math.*, **3** (1972) 209–246 [19].

[1059] V. Pless, Formulas and enumeration theorems in coding: origins and uses, *Proc. 4th Manitoba Conf. Numerical Math.*, 1974 pp. 29–51 [19].

[1060] V. Pless, Symmetry codes and their invariant subcodes, *J. Comb. Theory*, **18** (1975) 116–125 [16].

[1061] V. Pless and J. N. Pierce, Self-dual codes over GF(q) satisfy a modified Varshamov bound, *Info. and Control*, **23** (1973) 35–40 [19].

[1062] V. Pless and N. J. A. Sloane, Binary self-dual codes of length 24, *Bull. Amer. Math. Soc.*, **80** (1974) 1173–117f [19].

[1063] V. Pless and N. J. A. Sloane, On the classification and enumeration of self-dual codes, *J. Combin. Theory*, **18A** (1975) 313–335 [19, 20].

[1064] M. Plotkin, Binary codes with specified minimum distances, *IEEE Trans. Info. Theory*, **6** (1960) 445–450. (This paper had already appeared as a research report in 1951.) [2].

[1065] F. Polkinghorn, Jr., Decoding of double and triple error-correcting Bose–Chaudhuri codes, *IEEE Trans. Info. Theory*, **12** (1966) 480–481 [9].

[1066] G. Pólya and G. L. Alexanderson, Gaussian binomial coefficients, *Elemente der Mathematik*, **26** (1971) 102–109 [15].

[1067] G. Pólya and G. Szegö, *Aufgaben und Lehrsätze aus der Analysis*, 2 vols., 2nd ed., (Springer, Berlin, 1964) [11].

[1068] E. C. Posner, Nonbinary codes and projective planes, *JPL Calif. Inst. Tech., Space Programs Summary*, **37-16-IV** (Pasadena, California, August 1962) pp. 42–45 [11].

[1069] E. C. Posner, Simultaneous error-correction and burst-error detection using binary linear cyclic codes, *SIAM J. Appl. Math.*, **13** (1965) 1087–1095 [10].

[1070] E. C. Posner, Properties of error-correcting codes at low signal-to-noise ratios, *SIAM J. Appl. Math.*, **15** (1967) 775–798 [1].

[1071] E. C. Posner, Combinatorial structures in planetary reconnaissance, in: H. B. Mann, ed., *Error Correcting Codes*, (Wiley, New York, 1969) pp. 15–46 [1, 14].

[1072] K. A. Post, Nonexistence theorems on perfect Lee codes over large alphabets, *Info. and Control*, **29** (1975) 369–380 [6].

[1073] A. Potton, Implementing an error locating code, *Electronics Letters*, **5**(6) (1969) 121–122 [1].

[1074] E. Prange, The following technical notes issued by Air Force Cambridge Research Labs, Bedford, Mass: Cyclic Error-Correcting Codes in Two Symbols, *TN-57-103*, (September, 1957) [7].

[1075] Some cyclic error-correcting codes with simple decoding algorithms, *TN-58-156*, (April 1958) [7].

[1076] The use of coset equivalence in the analysis and decoding of group codes, *TN-59-164*, (1959) [7].

[1077] An algorism for factoring $x^n - 1$ over a finite field, *TN-59-175*, (October 1959) [7, 8].

[1078] E. Prange, The use of information sets in decoding cyclic codes, *IEEE Trans. Info. Theory*, **8**(5) (1962) pp. S5–S9 [16].

[1079] W. K. Pratt, J. Kane, and H. C. Andrews, Hadamard transform image coding, *Proc. IEEE*, **57** (1969) 57–68 [2].

[1080] F. P. Preparata, Weight and distance structure of Nordstrom–Robinson quadratic code, *Info. and Control*, **12** (1968) 466–473; **13** (1968) 172 [15].

[1081] F. P. Preparata, A class of optimum nonlinear double-error correcting codes, *Info. and Control*, **13** (1968) 378–400 [15, A].

[1082] F. P. Preparata, A new look at the Golay (23, 12) code, *IEEE Trans. Info. Theory*, **16** (1970) 510–511 [15].

[1083] F. P. Preparata and J. Nievergelt, Difference-preserving codes, *IEEE Trans. Info. Theory*, **20** (1974) 643–649 [Intro.].

[1083a] G. Promhouse and S. E. Tavares, private communication [A].

R

[1084] A. O. H. Racsmány, Perfect single-Lee-error-correcting code, *Studia Scient. Math. Hung.*, **9** (1974) 73–75 [6].

[1085] D. Raghavarao, *Constructions and Combinatorial Problems in Design of Experiments*, (Wiley, New York, 1971) [2, 5, 11, 13, 21].

[1086] M. Rahman and I. F. Blake, Majority logic decoding using combinatorial designs, *IEEE Trans. Info. Theory*, **21** (1975) 585–587 [16].

[1087] S. J. Rallis, New and old results in invariant theory with applications to arithmetic groups, in: W. M. Boothby and G. L. Weiss, eds., *Symmetric Spaces*, (Dekker, New York, 1972) pp. 443–458 [19].

[1088] J. D. Ralphs, Limitations of error-detection coding at high error rates, *Proc. IEE*, **118** (1971) 409–416 [1].

[1089] R. A. Rankin, On the closest packing of spheres in n dimensions, *Annals of Math.*, **48** (1947) 1062–1081 [17].

[1090] R. A. Rankin, The closest packing of spherical caps in n-dimensions, *Proc. Glasgow Math. Assoc.*, **2** (1955) 139–144 [17].

[1091] R. A. Rankin, On the minimal points of positive definite quadratic forms, *Mathematika*, **3** (1956) 15–24 [17].

[1092] C. R. Rao, Cyclical generation of linear subspaces in finite geometries, in: R. C. Bose and T. A. Dowling, eds., *Combinatorial Mathematics and Its Applications*, (Chapel Hill, NC, 1969) [13].

[1093] T. R. N. Rao, *Error Coding for Arithmetic Processors*, (Academic Press, New York, 1974) [1].

[1094] V. V. Rao and S. M. Reddy, A (48, 31, 8) linear code, *IEEE Trans. Info. Theory*, **19** (1973) 709–711 [18, A].

[1095] R. Rasala, Split codes and the Mathieu groups, *J. Algebra*, **42** (1976) 422–471 [20].

[1096] S. M. Reddy, A note on decoding of block codes, *IEEE Trans. Info. Theory*, **15** (1969) 627–628 [9].

[1097] S. M. Reddy, On decoding iterated codes, *IEEE Trans. Info. Theory*, **16** (1970) 624–627 [18].

[1098] S. M. Reddy, On block codes with specified maximum distance, *IEEE Trans. Info. Theory*, **18** (1972) 823–824 [17].

[1099] S. M. Reddy, Further results on decoders for q-ary output channels, *IEEE Trans. Info. Theory*, **20** (1974) 552–554 [18].

[1100] S. M. Reddy and J. P. Robinson, Random error and burst correction by iterated codes, *IEEE Trans. Info. Theory*, **18** (1972) 182–185 [18].

[1101] G. R. Redinbo, Generalized bandpass filters for decoding block codes, *IEEE Trans. Info. Theory*, **21** (1975) 417–422 [13].

[1102] G. R. Redinbo and J. R. Wolcott, Systematic construction of cyclically permutable code words, *IEEE Trans. Commun.*, **23** (1975) 786–789 [Intro.].

[1103] G. R. Redinbo and G. A. Wolf, On minimum mean-square error linear block codes when the data have q-adic weighting, *Info. and Control*, **26** (1974) 154–177 [1].

[1104] I. S. Reed, A class of multiple-error-correcting codes and the decoding scheme, *IEEE Trans. Info. Theory*, **4** (1954) 38–49 [13].

[1105] I. S. Reed, k^{th} order near-orthogonal codes, *IEEE Trans. Info. Theory*, **17** (1971) 116–117 [10].

[1106] I. S. Reed and G. Solomon, Polynomial codes over certain finite fields, *J. SIAM*, **8** (1960) 300–304 [10].

[1106a] I. S. Reed and R. M. Stewart, Note on the existence of perfect maps, *IEEE Trans. Info. Theory*, **8** (1962) 10–12 [18].

[1107] I. S. Reed and C. T. Wolverton, The systematic selection of cyclically equivalent codes, *IEEE Trans. Info. Theory*, **18** (1972) 304–307 [10].

[1108] C. Reid, *Hilbert*, (Springer, New York, 1970) [19].

[1109] C. T. Retter, Decoding Goppa codes with a BCH decoder, *IEEE Trans. Info. Theory*, **21** (1975) 112 [12].

[1110] C. T. Retter, Correcting burst and random errors with Goppa codes, *IEEE Trans. Info. Theory*, **22** (1976) 84 [12].

[1111] P. Ribenboim, *Algebraic Numbers*, (Wiley, New York, 1972) [2, 16].

[1112] J. R. Riek, Jr., C. R. P. Hartmann and L. D. Rudolph, Majority decoding of some classes of binary cyclic codes, *IEEE Trans. Info. Theory*, **20** (1974) 637–643 [13].

[1113] J. Riordan, *An Introduction to Combinatorial Analysis*, (Wiley, New York, 1958) [5].

[1114] J. Riordan, *Combinatorial Identities*, (Wiley, New York, 1968) [1].

[1115] W. E. Robbins, Weighted-majority-logic decoding using nonorthogonal parity checks – an application, in: *Proc. 12th Allerton Conference on Circuit and System Theory*, (Dept. of Elect. Engin., Univ. Illinois, Urbana, IL, 1974) pp. 567–573 [13].

[1116] P. D. Roberts and R. H. Davis, Statistical properties of smoothed maximal-length linear binary sequences, *Proc. IEE*, 113 (1966) 190–196 [14].

[1117] P. Robillard, Some results on the weight distribution of linear codes, *IEEE Trans. Info. Theory*, 15 (1969) 706–709 [8].

[1118] P. Robillard, Optimal codes, *IEEE Trans. Info. Theory*, 15 (1969) 734–735 [11].

[1119] J. P. Robinson, Analysis of Nordstrom's optimum quadratic code, *Proc. Hawaii Intern. Conf. System Sciences*, (1968) pp. 157–161 [2].

[1120] E. Y. Rocher and R. L. Pickholtz, An analysis of the effectiveness of hybrid transmission schemes, *IBM J. Res. Devel.*, 14 (1970) 426–433 [1].

[1121] W. F. Rogers, A practical class of polynomial codes, *IBM J. Res. Devel.*, 10 (1966) 158–161 [1].

[1122] B. Rokowska, Some new constructions of 4-tuple systems, *Colloq. Math.*, 17 (1967) 111–121 [2, 17].

[1123] J. E. Roos, An algebraic study of group and nongroup error-correcting codes, *Info. and Control*, 8 (1965) 195–214 [6].

[1124] J. Rosenstark, On a sequence of subcodes of the quadratic residue code for $p = 8m + 1$, *SIAM J. Appl. Math.*, 28 (1975) 252–264 [16].

[1125] G.-C. Rota, Combinatorial theory and invariant theory, Lecture notes, Bowdoin College, Maine, Summer 1971 [19].

[1126] H. H. Roth, Linear binary shift register circuits utilizing a minimum number of mod-2 adders, *IEEE Trans. Info. Theory*, 11 (1965) 215–220 [4].

[1127] O. S. Rothaus, On "bent" functions, *J. Comb. Theory*, 20A (1976) 300–305 [14].

[1128] B. L. Rothschild and J. H. van Lint, Characterizing finite subspaces, *J. Comb. Theory*, 16A (1974) 97–110 [13].

[1129] B. Rudner, Construction of minimum-redundancy codes with an optimum synchronizing property, *IEEE Trans. Info. Theory*, 17 (1971) 478–487 [1].

[1130] L. D. Rudolph, A class of majority-logic decodable codes, *IEEE Trans. Info. Theory*, 13 (1967) 305–307 [13].

[1131] L. D. Rudolph, Threshold decoding of cyclic codes, *IEEE Trans. Info. Theory*, 15 (1969) 414–418 [13].

[1132] L. D. Rudolph and C. R. P. Hartmann, Decoding by sequential code reduction, *IEEE Trans. Info. Theory*, 19 (1973) 549–555 [13].

[1133] L. D. Rudolph and M. E. Mitchell, Implementation of decoders for cyclic codes, *IEEE Trans. Info. Theory*, 10 (1964) 259–260 [16].

[1134] L. D. Rudolph and W. E. Robbins, One-step weighted-majority decoding, *IEEE Trans. Info. Theory*, 18 (1972) 446–448 [13].

[1135] C. K. Rushforth, Fast Fourier–Hadamard decoding of orthogonal codes, *Info. and Control*, 15 (1969) 33–37 [14].

[1136] H. J. Ryser, *Combinatorial Mathematics*, Carus Monograph 14, Math. Assoc. America, 1963 [2].

S

[1137] R. A. Sack, Interpretation of Lagrange's expansion and its generalization to several variables as integration formulas, *J. SIAM*, 13 (1965) 47–59 [19].

[1138] R. A. Sack, Generalization of Lagrange's expansion for functions of several implicitly defined variables, *J. SIAM*, 13 (1965) 913–926 [19].

[1139] R. A. Sack, Factorization of Lagrange's expansion by means of exponential generating functions, *J. SIAM*, 14 (1966) 1–15 [19].

[1140] G. E. Sacks, Multiple error correction by means of parity checks, *IEEE Trans. Info. Theory*, 4 (1958) 145–147 [1, 17].

[1141] M. K. Sain, Minimal torsion spaces and the practical input/output problem, *Info. and Control*, 29 (1975) 103–124 [12].

[1142] C. Saltzer, Topological codes, in: H. B. Mann, ed., *Error-Correcting Codes*, (Wiley, New York, 1968) pp. 111–129 [18].

[1143] P. V. Sankar and E. V. Krishnamurthy, Error correction in nonbinary words, *Proc. IEEE*, 61 (1973) 507–508 [7].

[1144] D. V. Sarwate, Weight Enumeration of Reed–Muller Codes and Cosets, Ph.D. Thesis, Dept. of Elec. Engin., Princeton, N.J, August 1973 [1, 14, 15].

[1145] D. V. Sarwate, On the complexity of decoding Goppa codes, *IEEE Trans. Info. Theory*, **23** (1977) 515–516 **[Intro., 12]**.

[1147] N. Sauer, On the existence of regular *n*-graphs with given girth, *J. Comb. Theory*, **9** (1970) 144–147 [18].

[1148] J. E. Savage, Some simple self-synchronizing digital data scramblers, *Bell Syst. Tech. J.*, **46** (1967) 449–487 [14].

[1149] J. E. Savage, The complexity of decoders: I – Classes of decoding rules, *IEEE Trans. Info. Theory*, **15** (1969) 689–695 [Intro.].

[1150] J. E. Savage, Three measures of decoder complexity, *IBM J. Res. Devel.*, **14** (1970) 417–425 [Intro.].

[1151] J. E. Savage, A note on the performance of concatenated codes, *IEEE Trans. Info. Theory*, **16** (1970) 512–513 [10].

[1151a] J. E. Savage, Coding efficiency and decoder complexity, *JPL Space Programs Summary*, **37-62-II** (1970) 73–75 [Intro.].

[1151b] J. E. Savage, The asymptotic complexity of the Green decoding procedure, *JPL Space Programs Summary*, **37-64-II** (1970) 29–32 [Intro.].

[1152] J. E. Savage, The complexity of decoders: II – Computational work and decoding time, *IEEE Trans. Info. Theory*, **17** (1971) 77–85 [Intro.].

[1152a] J. A. Savage, Computational work and time on finite machines, *J. ACM*, **19** (1972) 660–674 [Intro.].

[1153] J. P. M. Schalkwijk, A class of simple and optimal strategies for block coding on the binary symmetric channel with noiseless feedback, *IEEE Trans. Info. Theory*, **17** (1971) 283–287 [Intro.].

[1154] J. P. M. Schalkwijk and T. Kailath, A coding scheme for additive noise channels with feedback, *IEEE Trans. Info. Theory*, **12** (1966) 172–189 [Intro.].

[1155] J. P. M. Schalkwijk and K. A. Post, On the error probability for a class of binary recursive feedback strategies, *IEEE Trans. Info. Theory*, **19** (1973) 498–511 and **20** (1974) 284 [Intro.].

[1156] F. D. Schmandt, Coding need: A system's concept, in: J. N. Srivastava et al., eds., *A Survey of Combinatorial Theory*, (North-Holland, Amsterdam, 1973) Ch. 32 [1].

[1157] K. S. Schneider and R. S. Orr, Aperiodic correlation constraints on large binary sequence sets, *IEEE Trans. Info. Theory*, **21** (1975) 79–84 [14].

[1158] J. Schönheim, On maximal systems of *k*-tuples, *Stud. Sci. Math. Hungar.*, **1** (1966) 363–368 [17].

[1159] J. Schönheim, On linear and nonlinear single-error-correcting *q*-nary perfect codes, *Info. and Control*, **12** (1968) 23–26 [6].

[1160] J. Schönheim, Semilinear codes and some combinatorial applications of them, *Info. and Control*, **15** (1969) 61–66 [6].

[1161] J. Schönheim, On the number of mutually disjoint triples in Steiner systems and related maximal packing and minimal covering systems, in: W. T. Tutte, ed., *Recent Progress in Combinatorics*, (Academic Press, New York, 1969) pp. 311–318 [17].

[1162] J. Schönheim, A new perfect single error-correcting group code. Mixed code, in: R. Guy et al., eds., *Combinatorial Structures and their Applications*, (Gordon and Breach, New York, 1970) p. 385 [6].

[1163] R. A. Scholtz, Codes with synchronization capability, *IEEE Trans. Info. Theory*, **12** (1966) 135–142 [1].

[1164] R. A. Scholtz, Maximal and variable word-length comma-free codes, *IEEE Trans. Info. Theory*, **15** (1969) 300–306 [1].

[1165] R. A. Scholtz and R. M. Storwick, Block codes for statistical synchronization, *IEEE Trans. Info. Theory*, **16** (1970) 432–438 [1].

[1166] R. A. Scholtz and L. R. Welch, Mechanization of codes with bounded synchronization delays, *IEEE Trans. Info. Theory*, **16** (1970) 438–446 [1].

[1167] P. H. R. Scholefield, Shift registers generating maximum – length sequences, *Electronic Technology*, **37** (1960) 389–394 [14].

[1168] M. R. Schroeder, Sound diffusion by maximum – length sequences, *J. Acoustical Soc. Amer.*, **57** (1975) 149–150 [18].

[1169] J. W. Schwartz, A note on asynchronous multiplexing, *IEEE Trans. Info. Theory*, **12** (1966) 396–397 [17].

[1170] B. Segre, Curve razionali normali e k-archi negli spazi finiti, *Ann. Mat. Pura Appl.*, **39** (1955) 357–379 [11].

[1171] B. Segre, Ovals in a finite projective plane, *Canad. J. Math.*, **7** (1955) 414–416 [11].

[1172] B. Segre, Le geometrie di Galois, *Annali di Mat.*, **48** (1959) 1–97 [11].

[1173] B. Segre, *Lectures on Modern Geometry*, (Edizioni Cremonese, Rome, 1961) [11, B].

[1174] G. Seguin, On the weight distribution of cyclic codes, *IEEE Trans. Info. Theory*, **16** (1970) 358 [8].

[1175] J. J. Seidel, Strongly regular graphs, in: W. T. Tutte, ed., *Recent Progress in Combinatorics*, (Academic Press, New York, 1969) pp. 185–198 [2, 21].

[1176] J. J. Seidel, A Survey of two-graphs, in: *Colloq. Internaz. Teorie Combinatorie*, Atti dei Convegno Lincei **17**, Tomo 1, Roma 1976, pp. 481–511 [2, 21].

[1177] J. J. Seidel, Graphs and two-graphs, in: *Proc. 5ᵗʰ S-E Conf. Combinatorics, Graph Theory and Computing*, (Utilitas Math. Winnipeg, 1974) pp. 125–143 [2, 21].

[1178] E. S. Selmer, *Linear recurrence relations over finite fields*, (Dept. of Math., Univ. of Bergen, Norway, 1966) [14].

[1179] N. V. Semakov and V. A. Zinov'ev, Equidistant maximal q-ary codes and resolvable balanced incomplete block designs, *Problems of Info. Trans.*, **4**(2) (1968) 1–7 [2].

[1180] N. V. Semakov and V. A. Zinov'ev, Complete and quasi-complete balanced codes, *Problems of Info. Trans.*, **5**(2) (1969) 11–13 [2, 6].

[1181] N. V. Semakov and V. A. Zinov'ev, Balanced codes and tactical configurations, *Problems of Info. Trans.*, **5**(3) (1969) 22–28 [2, 6, 15].

[1182] N. V. Semakov, V. A. Zinov'ev and G. V. Zaitsev, A class of maximum equidistant codes, *Problems of Info. Trans.*, **5**(2) (1969) 65–68 [2].

[1183] N. V. Semakov, V. A. Zinov'ev and G. V. Zaitsev, Uniformly packed codes, *Problems of Info. Trans.*, **7**(1) (1971) 30–39 [15].

[1184] J. P. Serre, *Cours d'arithmétique*, (Presses Univ. de France, Paris, 1970, English translation published by Springer, Berlin, 1973) [19].

[1185] J. P. Serre, *Représentations Linéaires des Groupes Finis*, 2nd ed., (Hermann, Paris, 1971) [19].

[1186] D. Shanks, *Solved and Unsolved Problems in Number Theory*, Vol. I, (Spartan Books, Washington, 1962) [16].

[1187] J. L. Shanks, Computation of the fast Walsh–Fourier transform, *IEEE Trans. Computers*, **18** (1969) 457–459 [2, 14].

[1188] C. E. Shannon, A mathematical theory of communication, *Bell Syst. Tech. J.*, **27** (1948) pp. 379–423 and 623–656. Reprinted in: C. E. Shannon and W. Weaver, eds., *A Mathematical Theory of Communication*, (Univ. of Illinois Press, Urbana, Illinois, 1963) [1].

[1189] C. E. Shannon, Communication in the presence of noise, *Proc. IEEE*, **37** (1949) 10–21 [1].

[1190] C. E. Shannon, Communication theory of secrecy systems, *Bell Syst. Tech. J.*, **28** (1949) 656–715 [1].

[1191] C. E. Shannon, Probability of error for optimal codes in a Gaussian channel, *Bell Syst. Tech. J.*, **38** (1959) 611–656 [Intro.].

[1192] C. E. Shannon, R. G. Gallager and E. R. Berlekamp, Lower bounds to error probability for coding on discrete memoryless channels, *Info. and Control*, **10** (1967) 65–103 and 522–552 [17].

[1193] E. P. Shaughnessy, Codes with simple automorphism groups, *Archiv der Math.*, **22** (1971) 459–466 [16].

[1194] E. P. Shaughnessy, Automorphism groups of the $(l+1, (l+1)/2)$ extended QR codes for l and $(l-1)/2$ prime, $5 < l < 4079$. Abstract 722-A5, *Notices Am. Math. Soc.*, **22** (1975) A–348 [16].

[1195] E. P. Shaughnessy, Disjoint Steiner triple systems, Preprint [6].

[1196] G. C. Shephard and J. A. Todd, Finite unitary reflection groups, *Canad. J. Math.*, **6** (1954) 274–304 [19].

[1197] J. Shimo, H. Nakanishi and Y. Tezuka, Word synchronization for data transmission using cyclic codes, *Electron. Commun. Japan*, **55-A**(1) (1972) 17–24 [1].

[1198] S. G. S. Shiva, Certain group codes, *Proc. IEEE*, **55** (1967) 2162–2163 [18].

[1199] S. G. S. Shiva, Some results on binary codes with equidistant words, *IEEE Trans. Info. Theory*, **15** (1969) 328–329 [1].

[1200] S. G. S. Shiva and P. E. Allard, A few useful details about a known technique for factoring $1 + X^{2q-1}$, *IEEE Trans. Info. Theory*, **16** (1970) 234–235 [7].

[1201] S. G. S. Shiva and K. C. Fung, Permutation decoding of certain triple-error-correcting binary codes, *IEEE Trans. Info. Theory*, **18** (1972) 444–446 [16].

[1202] S. G. S. Shiva, K. C. Fung and H. S. Y. Tan, On permutation decoding of binary cyclic double-error-correcting codes of certain lengths, *IEEE Trans. Info. Theory*, **16** (1970) 641–643 [16].

[1203] G. S. S. Shiva and G. Seguin, Synchronizable error-correcting binary codes, *IEEE Trans. Info. Theory*, **16** (1970) 241–242 [1].

[1204] S. G. S. Shiva and C. L. Sheng, Multiple solid burst-error-correcting binary codes, *IEEE Trans. Info. Theory*, **15** (1969) 188–189 [10].

[1205] S. G. S. Shiva and S. E. Tavares, On binary majority-logic decodable codes, *IEEE Trans. Info. Theory*, **20** (1974) 131–133 [13].

[1206] S. S. Shrikhande, A note on minimum distance binary codes, *Calcutta Statistical Assoc. Bull.*, (1962) 94–97 [18].

[1207] V. M. Sidel'nikov, Some k-valued pseudo-random sequences and nearly equidistant codes, *Problems of Info. Trans.*, **5**(1) (1969) 12–16 [14].

[1208] V. M. Sidel'nikov, Weight spectrum of binary Bose–Chaudhuri–Hocquenghem codes, *Problems of Info. Trans.*, **7**(1) (1971) 11–17 [9].

[1209] V. M. Sidel'nikov, On the densest packing of balls on the surface of an n-dimensional Euclidean sphere and the number of binary code vectors with a given code distance, *Soviet Math. Doklady*, **14** (1973) 1851–1855 [17].

[1210] V. M. Sidel'nikov, Upper bounds for the number of points of a binary code with a specified code distance, *Problems of Info. Trans.*, **10**(2) (1974) 124–131; and *Info. Control*, **28** (1975) 292–303 [17].

[1211] M. Simonnard, *Linear Programming*, (Prentice-Hall, Englewood Cliffs, N.J., 1966) [17].

[1212] G. J. Simmons, A constructive analysis of the aperiodic binary correlation function, *IEEE Trans. on Info. Theory*, **15** (1969) 340–345 [14].

[1213] J. Singer, A theorem in finite projective geometry, and some applications to number theory, *Trans. Amer. Math. Soc.*, **43** (1938) 377–385 [13].

[1214] R. C. Singleton, Maximum distance q-nary codes, *IEEE Trans. Info. Theory*, **10** (1964) 116–118 [1, 11].

[1215] R. C. Singleton, Generalized snake-in-the-box codes, *IEEE Trans. Computers*, **15** (1966) 596–602 [Intro.].

[1216] D. Singmaster, Notes on binomial congruences: I – A generalization of Lucas' congruence, *J. London Math. Soc.*, **8** (1974) 545–548 [13].

[1217] D. Slepian, A class of binary signaling alphabets, *Bell Syst. Tech. J.*, **35** (1956) 203–234 [1].

[1218] D. Slepian, A note on two binary signaling alphabets, *IEEE Trans. Info. Theory*, **2** (1956) 84–86 [1].

[1219] D. Slepian, Some further theory of group codes, *Bell Syst. Tech. J.*, **39** (1960) 1219–1252 [1].

[1220] D. Slepian, Bounds on communication, *Bell Syst. Tech. J.*, **42** (1963) 681–707 [1].

[1221] D. Slepian, Permutation modulation, *Proc. IEEE*, **53** (1965) 228–236 [Intro.].

[1222] D. Slepian, Group codes for the Gaussian channel, *Bell Syst. Tech. J.*, **47** (1968) 575–602 [Intro.].

[1223] D. Slepian, On neighbor distances and symmetry in group codes, *IEEE Trans. Info. Theory*, **17** (1971) 630–632 [Intro.].

[1224] D. Slepian, ed., *Key Papers in The Development of Information Theory*, (IEEE Press, New York, 1974) [Intro., 1].

[1225] N. J. A. Sloane, A survey of constructive coding theory, and a table of binary codes of highest known rate, *Discrete Math.*, **3** (1972) 265–294 [1, A].

[1226] N. J. A. Sloane, Sphere packings constructed from BCH and Justesen codes, *Mathematika*, **19** (1972) 183–190 [Intro.].

[1227] N. J. A. Sloane, Is there a (72, 36) $d = 16$ self-dual code? *IEEE Trans. Info. Theory*, **19** (1973) p. 251 [19].

[1227a] N. J. A. Sloane, *A short course on error-correcting codes* (Lectures given at International Centre for Mechanical Sciences, Udine, 1973), Springer-Verlag, N.Y. 1975 [Intro.].

[1228] N. J. A. Sloane, Weight enumerators of codes, in: M. Hall, Jr. and J. H. van Lint, eds., *Combinatorics*, (Reidel Publishing Co., Dordrecht, Holland, 1975) pp. 115–142 [19].

[1229] N. J. A. Sloane, An introduction to association schemes and coding theory, in: R. A. Askey, ed., *Theory and Application of Special Functions*, (Academic Press, New York, 1975) pp. 225–260 [21].

[1230] N. J. A. Sloane, A simple description of an error-correcting code for high density magnetic tape, *Bell Syst. Tech. J.*, **55** (1976) 157–165 [1, 18].

[1231] N. J. A. Sloane, Error-correcting codes and invariant theory: new applications of a nineteenth-century technique, *Amer. Math. Monthly*, **84** (1977) 82–107 [19].

[1232] N. J. A. Sloane and E. R. Berlekamp, Weight enumerator for second-order Reed–Muller codes, *IEEE Trans. Info. Theory*, **16** (1970) 745–751 [15].

[1233] N. J. A. Sloane and R. J. Dick, On the enumeration of cosets of first order Reed–Muller codes, *IEEE Intern. Conf. on Commun.*, **7** (Montreal 1971) 36-2 to 36-6 [1, 14].

[1234] N. J. A. Sloane, T. Fine and P. G. Phillips, New methods for grating spectrometers, *Optical Spectra*, **4** (1970) 50–53 [2, 18].

[1235] N. J. A. Sloane, T. Fine, P. G. Phillips and M. Harwit, Codes for multislit spectrometry, *Applied Optics*, **8** (1969) 2103–2106 [2, 18].

[1236] N. J. A. Sloane and M. Harwit, Masks for Hadamard transform optics, and weighing designs, *Applied Optics*, **15** (1976) 107–114 [2, 14, 18].

[1237] N. J. A. Sloane, S. M. Reddy and C. L. Chen, New binary codes, *IEEE Trans. Info. Theory*, **18** (1972) 503–510 [18, A].

[1238] N. J. A. Sloane and J. J. Seidel, A new family of nonlinear codes obtained from conference matrices, *Annals N.Y. Acad. Sci.*, **175** (Article 1, 1970) 363–365 [2, A].

[1239] N. J. A. Sloane and D. S. Whitehead, A new family of single-error-correcting codes, *IEEE Trans. Info. Theory*, **16** (1970) 717–719 [2, A].

[1240] D. H. Smith, Bounding the diameter of a distance-transitive graph, *J. Comb. Theory*, **16B** (1974) 139–144 [21].

[1241] D. H. Smith, Distance-transitive graphs, in *Combinatorics*, edited by T. P. McDonough and V. C. Mavron, London Math. Soc., Lecture Notes No. 13, Cambridge Univ. Press, Cambridge, 1974, pp. 145–153 [21].

[1242] D. H. Smith, Distance-transitive graphs of valency four, *J. London Math. Soc.*, **8**(2) (1974) 377–384 [21].

[1243] K. J. C. Smith, On the rank of incidence matrices in finite geometries, Univ. of North Carolina, Institute of Statistics Mimeo Series No. 555, Chapel Hill N.C., 1967 [13].

[1244] K. J. C. Smith, Majority decodable codes derived from finite geometries, Univ. North Carolina, Institute of Statistics Mimeo Series No. 561, Chapel Hill, N.C., 1967 [13].

[1245] K. J. C. Smith, An application of incomplete block designs to the construction of error-correcting codes, Univ. of North Carolina, Institute of Statistics Mimeo Series No. 587, Chapel Hill, N.C., 1968 [13].

[1246] K. J. C. Smith, On the p-rank of the incidence matrix of points and hyperplanes in a finite projective geometry, *J. Comb. Theory*, **7** (1969) 122–129 [13].

[1247] S. L. Snover, The Uniqueness of the Nordstrom-Robinson and the Golay Binary Codes, Ph.D. Thesis, Dept. of Mathematics, Michigan State Univ., 1973 [2, 15, 20].

[1248] G. Solomon, A note on a new class of codes, *Info. Control*, **4** (1961) 364–370 [16].

[1249] G. Solomon, A weight formula for group codes, *IEEE Trans. Info. Theory*, **8** (1962) S1–S4 [8].

[1250] G. Solomon, Self-synchronizing Reed–Solomon codes, *IEEE Trans. Info. Theory*, **14** (1968) 608–609 [10].

[1251] G. Solomon, Algebraic coding theory, in: A. V. Balakrishnan, ed., *Communication Theory*, (McGraw-Hill, New York, 1968) Ch. 6 [Intro.].

[1252] G. Solomon, The (7,5) R-S code over $GF(2^3)$ is a (21, 15) BCH code, *IEEE Trans. Info. Theory*, **15** (1969) 619–620 [10].

[1253] G. Solomon, Coding with character, in: R. C. Bose and T. H. Dowling, eds., *Combinatorial Mathematics and its Applications*, (Univ. of North Carolina Press, Chapel Hill, NC, 1970) pp. 405–415 [2].

[1254] G. Solomon, A note on alphabet codes and fields of computation, *Info. and Control*, **25** (1974) 395–398 [10].

[1255] G. Solomon, Unified theory of block and convolutional codes, Preprint [16].

[1256] G. Solomon and R. J. McEliece, Weights of cyclic codes, *J. Comb. Theory*, 1 (1966) 459–475 [8, 15].

[1257] G. Solomon and J. J. Stiffler, Algebraically punctured cyclic codes, *Info. and Control*, 8 (1965) 170–179 [17].

[1258] R. Spann, A two-dimensional correlation property of pseudorandom maximal-length sequences, *Proc. IEEE*, 53 (1965) 2137 [18].

[1259] J. Spencer, Maximal consistent families of triples, *J. Comb. Theory*, 5 (1968) 1–8 [17].

[1260] W. Stahnke, Primitive binary polynomials, *Math. Comp.*, 27 (1973) 977–980 [4].

[1261] R. Stanley, Moments of weight distributions, *JPL Space Programs Summary*, 37-40-IV (Pasadena, California, 1966) pp. 214–216 [5].

[1262] R. P. Stanley, Private communication [19].

[1263] R. P. Stanley and M. F. Yoder, A study of Varshamov codes for asymmetric channels, Report 32-1526, Vol. 14, pp. 117–123, Jet Propulsion Labs., Pasadena, Calif., 1973 [1].

[1264] R. G. Stanton, The Mathieu groups, *Canad. J. Math.*, 3 (1951) 164–174 [20].

[1265] R. G. Stanton, Covering theorems in groups (or: how to win at football pools), in: W. T. Tutte, ed., *Recent Progress in Combinatorics*, (Academic Press, New York, 1969) pp. 21–36 [7, A].

[1266] R. G. Stanton and R. J. Collens, A computer system for research on the family classification of BIBDs, in: *Colloq. Internaz. Teorie Combinatorie*, Atti del Convegni Lincei 17, Tomo 1, Roma 1976, pp. 133–169 [2, 17].

[1267] R. G. Stanton, J. D. Horton and J. G. Kalbfleisch, Covering theorems for vectors with special reference to the case of four or five components, *J. London Math. Soc.*, 2 (1969) 493–499 [7, A].

[1268] R. G. Stanton and J. G. Kalbfleisch, Covering problems for dichotomized matchings, *Aequationes mathematicae*, 1 (1968) 94–103 [A].

[1269] R. G. Stanton and J. G. Kalbfleisch, Intersection inequalities for the covering problem, *SIAM J. Appl. Math.*, 17 (1969) 1311–1316 [A].

[1270] R. G. Stanton and J. G. Kalbfleisch, Coverings of pairs by k-sets, *Annals N.Y. Acad. Sci.*, 175 (1970) 366–369 [17, A].

[1271] R. G. Stanton, J. G. Kalbfleisch and R. C. Mullin, Covering and packing designs, in: R. C. Bose et al., eds., *Proc. Second Chapel Hill Conference on Comb. Math. and Applic.*, (Chapel Hill, NC, 1970) pp. 428–450 [A].

[1272] H. M. Stark, *An Introduction to Number Theory*, (Markham, Chicago, 1970) [12].

[1273] J. M. Stein and V. K. Bhargava, Equivalent rate 1/2 quasi-cyclic codes, *IEEE Trans. Info. Theory*, 21 (1975) 588–589 [16].

[1274] J. M. Stein and V. K. Bhargava, Two quaternary quadratic residue codes, *Proc. IEEE*, 63 (1975) 202 [16].

[1275] J. M. Stein, V. K. Bhargava and S. E. Tavares, Weight distribution of some "best" (3m, 2m) binary quasi-cyclic codes, *IEEE Trans. Info. Theory*, 21 (1975) 708–711 [8].

[1276] J. P. Stenbit, Table of generators for Bose–Chaudhuri codes, *IEEE Trans. Info. Theory*, 10 (1964) 390–391 [9].

[1277] R. F. Stevens and W. G. Bouricius, The heuristic generation of large error-correcting codes, *IBM Res. Memo.*, (Yorktown Heights, New York, August 1, 1959) [2].

[1278] J. J. Stiffler, Comma-free error-correcting codes, *IEEE Trans. Info. Theory*, 11 (1965) 107–112 [1].

[1279] J. J. Stiffler, Rapid acquistion sequences, *IEEE Trans. Info. Theory*, 14 (1968) 221–225 [1].

[1280] J. J. Stiffler, *Theory of Synchronous Communication*, (Prentice-Hall, Englewood Cliffs, N.J., 1971) [1, 14].

[1281] L. R. Stirzaker and C. K. Yuen, A note on error-correction in nonbinary words, *Proc. IEEE*, 62 (1974) 1399–1400 [7].

[1282] J. J. Stone, Multiple-burst error-correction with the Chinese remainder theorem, *J. SIAM*, 11 (1963) 74–81 [10].

[1283] P. T. Strait, A lower bound for error-detecting and error-correcting codes, *IEEE Trans. Info. Theory*, 7 (1961) 114–118 [17].

[1284] M. Sugino, Y. Ienaga, M. Tokura and T. Kasami, Weight distribution of (128, 64) Reed–Muller code, *IEEE Trans. Info. Theory*, 17 (1971) 627–628 [15].

[1285] Y. Sugiyama and M. Kasahara, A class of asymptotically good algebraic codes using second-order concatentation, *Electron. Commun. in Japan*, 57-A (2) (1974) 47–53 [10].

[1286] Y. Sugiyama and M. Kasahara, Personal communication [10].

[1287] Y. Sugiyama, M. Kasahara, S. Hirasawa and T. Namekawa, A modification of the constructive asymptotically good codes of Justesen for low rates, *Info. and Control,* **25** (1974) 341–350 [10].

[1288] Y. Sugiyama, M. Kasahara, S. Hirawawa and T. Namekawa, A method for solving key equation for decoding Goppa codes, *Info. and Control,* **27** (1975) 87–99 [12].

[1289] Y. Sugiyama, M. Kasahara, S. Hirasawa and T. Namekawa, Some efficient binary codes constructed using Srivastava codes, *IEEE Trans. Info. Theory,* **21** (1975) 581–582 [12, A].

[1290] Y. Sugiyama, M. Kasahara, S. Hirasawa and T. Namekawa, An erasures-and-errors decoding algorithm for Goppa codes, *IEEE Trans. Info. Theory,* **22** (1976) 238–241 [12].

[1291] Y. Sugiyama, M. Kasahara, S. Hirawawa and T. Namekawa, Further results on Goppa codes and their applications to constructing efficient binary codes, *IEEE Trans. Info. Theory,* **22** (1976) 518–526 [12, 18, A].

[1292] D. D. Sullivan, A fundamental inequality between the probabilities of binary subgroups and cosets, *IEEE Trans. Info. Theory,* **13** (1967) 91–95 [1].

[1293] D. D. Sullivan, A branching control circuit for Berlekamp's BCH decoding algorithm, *IEEE Trans. Info. Theory,* **18** (1972) 690–692 [9].

[1294] J. D. Swift, Quasi-Steiner systems, Accad. Nazionale dei Lincei, *Rend. Cl. Sci. fis. mat. e nat.,* (8) **44** (1968) 41–44 [17].

[1295] J. D. Swift, On (*k, l*)-coverings and disjoint systems, pp. 223–228 of *Proc. Symp. Pure Math.,* **19** *Combinatorics,* (Amer. Math. Soc., Providence 1971) [A].

[1296] J. J. Sylvester, Thoughts on inverse orthogonal matrices, simultaneous sign successions, and tesselated pavements in two or more colors, with applications to Newton's rule, ornamental tile-work, and the theory of numbers, *Phil. Mag.,* **34** (1867) 461–475 [2].

[1297] G. Szegö, *Orthogonal Polynomials,* Colloquium Publications, Vol. 23, New York: Amer. Math. Soc., revised edition 1959, pp. 35–37 [5, 17, 21].

[1298] Z. Szwaja, On step-by-step decoding of the BCH binary codes, *IEEE Trans. Info. Theory,* **13** (1967) 350–351 [9].

T

[1298a] M. H. Tai, M. Harwit and N. J. A. Sloane, Errors in Hadamard spectroscopy or imaging caused by imperfect masks, *Applied Optics,* **14** (1975) 2678–2686 [14].

[1299] H. Tanaka and S. Kaneku, A fuzzy decoding procedure for error-correcting codes, *Electron. Commun. Japan,* **57-A** (7) (1974) 26–31 [16].

[1300] H. Tanaka, M. Kasahara, Y. Tezuka and Y. Kasahara, Computation over Galois fields using shift registers, *Info. and Control,* **13** (1968) 75–84 [3].

[1301] H. Tanaka, M. Kasahara, Y. Tezuka and Y. Kasahara, An error-correction procedure for the binary Bose–Chaudhuri–Hocquenghem codes, *Electron. Commun. Japan,* **51-C** (1968) 102–109 [9].

[1302] H. Tanaka and T. Kasai, Synchronization and substitution error-correcting codes for the Levenshtein metric, *IEEE Trans. Info. Theory,* **22** (1976) 156–162 [1].

[1303] H. Tanaka and F. Nishida, A construction of a polynomial code, *Electron. Commun. Japan,* **53-A** (8) (1970) 24–31 [10].

[1304] H. Tanaka and F. Nishida, A decoding procedure for polynomial codes, *Systems, Computers, Controls,* **1**(3) (1970) 58–59 [9].

[1305] H. Tanaka and F. Nishida, Burst-trapping codes constructed by connecting error-correcting block codes, *Electron. Commun. Japan,* **55-A** (6) (1972) 17–23 [10].

[1306] D. T. Tang and R. T. Chien, Cyclic product codes and their implementation, *Info. and Control,* **9** (1966) 196–209 [18].

[1307] D. T. Tang and C. N. Liu, Distance-2 cyclic chaining of constant-weight codes, *IEEE Trans. Computers,* **22** (1973) 176–180 [Intro., 17].

[1308] R. M. Tanner, Contributions to the simplex code conjecture, Report 6151-8, Center for Systems Research, Stanford Univ., Dec. 1970 [1].

[1309] R. M. Tanner, Some content maximizing properties of the regular simplex, *Pac. J. Math.,* **52** (1974) 611–616 [1].

[1310] G. L. Tauglikh, A polynomial weighted code correcting multiple group errors, *Problems of Info. Trans.,* **8**(2) (1972) 93–97 [10].

[1311] S. E. Tavares, V. K. Bhargava and S. G. S. Shiva, Some rate-$p/(p+1)$ quasi-cyclic codes, *IEEE Trans. Info. Theory,* **20** (1974) 133–135 [16].

[1312] S. E. Tavares and M. Fukada, Matrix approach to synchronization recovery for binary cyclic code, *IEEE Trans. Info. Theory*, 15 (1969) 93–101 [1].

[1313] S. E. Tavares and M. Fukada, Further results on the synchronization of binary cyclic codes, *IEEE Trans. Info. Theory*, 16 (1970) 238–241 [1].

[1314] S. E. Tavares and S. G. S. Shiva, Detecting and correcting multiple bursts for binary cyclic codes, *IEEE Trans. Info. Theory*, 16 (1970) 643–644 [10].

[1315] Y. Tezuka and H. Nakanishi, Personal communication [18].

[1316] J. A. Thas, Normal rational curves and k-arcs in Galois spaces, *Rendiconti di Matematica*, 1 (1968) 331–334 [11].

[1317] J. A. Thas, Normal rational curves and $(q+2)$−arcs in a Galois space $S_{q-2,q}$ $(q = 2^h)$, *Atti Accad. Naz. Lincei, Rend. Cl. Sc. Fis. Mat. Natur.*, (8) 47 (1969) 115–118 [11].

[1318] J. A. Thas, Connection between the Grassmannian $G_{k-1;n}$ and the set of the k-arcs of the Galois space $S_{n,q}$, *Rendiconti di Matematica*, 2 (1969) 121–134 [11].

[1319] J. A. Thas, Two infinite classes of perfect codes in metrically regular graphs, *J. Combin. Theory*, 23B (1977) 236–238 [6].

[1320] R. Thoene and S. W. Golomb, Search for cyclic Hadamard matrices, *JPL Space Programs Summary*, 37-40-IV (1966) 207–208 [2].

[1320a] J. G. Thompson, Weighted averages associated to some codes, *Scripta Math.*, 29 (1973) 449–452 [19].

[1321] A. Tietäväinen, On the nonexistence of perfect 4-Hamming-error-correcting codes, *Ann. Ac. Sci. Fennicae, Ser.*, A. I, No. 485, 1970 [6].

[1322] A. Tietäväinen, On the nonexistence of perfect codes over finite fields, *SIAM J. Appl. Math.*, 24 (1973) 88–96 [6].

[1323] A. Tietäväinen, A short proof for the nonexistence of unknown perfect codes over GF(q), $q > 2$, *Annales Acad. Scient. Fennicae*, Ser. A, (I. Mathematica), No. 580, (Helsinki, 1974) pp. 1–6 [6].

[1324] A. Tietäväinen and A. Perko, There are no unknown perfect binary codes, *Ann. Univ. Turku*, Ser. A, I 148 (1971) 3–10 [6].

[1325] H. C. A. van Tilborg, On weight in codes, Report 71-WSK-03, Department of Mathematics, Technological University of Eindhoven, Netherlands, December 1971 [15, 16].

[1326] H. C. A. van Tilborg, Binary uniformly packed codes, Thesis, Tech. Univ. Eindhoven, 1976 [15].

[1327] H. C. A. van Tilborg, All binary, (n, e, r)-uniformly packed codes are known, Preprint [15].

[1328] J. A. Todd, A combinatorial problem, *J. Math. and Phys.*, 12 (1933) 321–333 [2].

[1329] J. A. Todd, On representations of the Mathieu groups as collineation groups, *J. London Math. Soc.*, 34 (1959) 406–416 [20].

[1330] J. A. Todd, A representation of the Mathieu group as a collineation group, *Ann. Mat. Pura Appl.*, Ser. 4, 71 (1966) 199–238 [20].

[1331] N. Tokura, K. Taniguchi and T. Kasami, A search procedure for finding optimum group codes for the binary symmetric channel, *IEEE Trans. Info. Theory*, 13 (1967) 587–594 [18].

[1332] N. Tokura and T. Kasami, Shortened cyclic codes for multiple burst-error correction, *Electron. Commun. Japan*, 48(1) (1965) 77–84 [7].

[1333] S. Y. Tong, Synchronization recovery techniques for binary cyclic codes, *Bell Syst. Tech. J.*, 45 (1966) 561–596 [1].

[1334] S. Y. Tong, Correction of synchronization errors with burst-error-correcting cyclic codes, *IEEE Trans. Info. Theory*, 15 (1969) 106–109 [1, 10].

[1335] S. Y. Tong, Burst-trapping techniques for a compound channel, *IEEE Trans. Info. Theory*, 15 (1969) 710–715 [1, 10].

[1336] S. Y. Tong, Performance of burst-trapping codes, *Bell Syst. Tech. J.*, 49 (1970) 477–519 [1, 10].

[1337] R. L. Townsend and R. N. Watts, Effectiveness of error control in data communication over the switched telephone network, *Bell Syst. Tech. J.*, 43 (1964) 2611–2638 [1].

[1338] R. L. Townsend and E. J. Weldon, Jr., Self-orthogonal quasi-cyclic codes, *IEEE Trans. Info. Theory*, 13 (1967) 183–195 [16].

[1339] S. A. Tretter, Properties of PN^2 sequences, *IEEE Trans. Info. Theory*, 20 (1974) 295–297 [14].

[1340] K. Tröndle, Die Verbesserung der Geräuschwelle pulscodemodulierter Signale durch Anwendung fehlerkorrigierender Codes, *Arch. Elektron Ubertr.*, 26 (1972) 207–212 [1].

[1341] S. H. Tsao, Generation of delayed replicas of maximal-length linear binary sequences, *Proc. IEE*, 111 (1964) 1803–1806 [14].

[1341] H. W. Turnbull, *Theory of Equations*, (Oliver and Boyd, Edinburgh, 1947) [8].

[1343] R. J. Turyn, The correlation function of a sequence of roots of 1, *IEEE Trans. Info. Theory*, **13** (1967) 524–525 [14].

[1344] R. J. Turyn, Sequences with small correlation, in: H. B. Mann, ed., *Error Correcting Codes*, (Wiley, New York, 1969) pp. 195–228 [14].

[1345] R. J. Turyn, Complex Hadamard matrices, in: R. Guy et al., eds., *Combinatorial Structures and Their Applications*, (Gordon and Breach, New York, 1970) pp. 435–437 [2].

[1346] R. J. Turyn, On *C*-matrices of arbitrary powers, *Canad. J. Math.*, **23** (1971) 531–535 [2].

[1347] R. J. Turyn, Hadamard matrices, Baumert-Hall units, four-symbol sequences, pulse compression, and surface wave encodings, *J. Comb. Theory*, **16A** (1974) 313–333 [2].

[1348] R. Turyn and J. Storer, On binary sequences, *Proc. Amer. Math. Soc.*, **12** (1961) 394–399 [14].

[1349] C. C. Tseng and C. L. Liu, Complementary sets of sequences, *IEEE Trans. Info. Theory*, **18** (1972) 649–652 [14].

[1350] K. K. Tzeng and C. R. P. Hartmann, On the minimum distance of certain reversible cyclic codes, *IEEE Trans. Info. Theory*, **16** (1970) 644–646 [7].

[1351] K. K. Tzeng and C. R. P. Hartmann, Generalized BCH decoding, Preprint [9].

[1352] K. K. Tzeng, C. R. P. Hartmann and R. T. Chien, Some notes on iterative decoding, *Proc. Ninth Allerton Conf. on Circuit and Systems Theory*, October, 1971, pp. 689–695 [9].

[1353] K. K. Tzeng and C. Y. Yu, Characterization theorems for extending Goppa codes to cyclic codes, Preprint [12].

[1354] K. K. Tzeng and K. Zimmermann, On full power decoding of cyclic codes, *Proc. Sixth Annual Princeton Conference on Info. Sci. and Syst.*, (Princeton, 1972) pp. 404–407 [16].

[1355] K. K. Tzeng and K. P. Zimmermann, On extending Goppa codes to cyclic codes, *IEEE Trans. Info. Theory*, **21** (1975) 712–716 [12].

[1356] K. K. Tzeng and K. P. Zimmermann, Lagrange's interpolation formula and generalized Goppa codes, Preprint [12].

U

[1357] J. D. Ullman, Decoding of cyclic codes using position invariant functions, *IBM J. Res. Devel.*, **9** (1965) 233–240 [9].

[1358] J. D. Ullman, Near optimal, single synchronization error-correcting codes, *IEEE Trans. Info. Theory*, **12** (1966) 418–425 [1].

[1359] J. V. Uspensky and M. A. Heaslet, *Elementary Number Theory*, (McGraw-Hill, New York, 1939) [2, 10, 12].

V

[1360] S. Vajda, *Patterns and Configurations in Finite Spaces*, (Hafner, New York, 1967) [2].

[1361] S. Vajda, *The Mathematics of Experimental Design*, (Hafner, New York, 1967) [2].

[1362] R. R. Varshamov, Estimate of the number of signals in error correcting codes, *Dokl. Akad. Nauk SSSR*, **117** (1957) 739–741 [1].

[1363] R. R. Varshamov, Some features of linear codes that correct asymmetric errors, *Soviet Physics-Doklady*, **9**(7) (1965) 538–540 [1].

[1364] R. R. Varshamov, On the theory of asymmetric codes, *Soviet Physics-Doklady*, **10**(10) (1966) 901–903 [1].

[1365] R. R. Varshamov, A class of codes for asymmetric channels and a problem from the additive theory of numbers, *IEEE Trans. Info. Theory*, **19** (1973) 92–95 [1].

[1366] J. L. Vasil'ev, On nongroup close-packed codes (in Russian), *Probl. Kibernet.*, **8** (1962) 337–339, translated in *Probleme der Kybernetik*, **8** (1965) 375–378 [2].

[1367] J. L. Vasil'ev, On the length of a cycle in an *n*-dimensional unit cube, *Soviet Math. Doklady*, **4** (1963) 160–163 [Intro.].

[1368] O. Veblen and J. W. Young, *Projective Geometry*, 2 vols., (Ginn and Co., Boston, 1916–1917) [B].

[1369] G. Venturini, Construction of maximal linear binary codes with large distances, (in Russian) *Probl. Kibern.*, **16** (1966) 231–238 [17].

[1370] D. Vere-Jones, Finite bivariate distributions and semi-groups of nonnegative matrices, *Q. J. Math. Oxford*, (2) **22** (1971) 247–270 [5].

[1371] I. M. Vinogradov, *Elements of Number Theory*, (Dover, New York, 1954) [16].

[1372] A. J. Viterbi, Error bounds for convolutional codes and an asymptotically optimum decoding algorithm, *IEEE Trans. Info. Theory*, **13** (1967) 260–269 [1].

[1373] A. J. Viterbi, Convolutional codes and their performance in communication systems, *IEEE Trans. Commun. Theory*, **19** (1971) 751–772 [1].

[1374] A. J. Viterbi, Information theory in the Sixties, *IEEE Trans. Info. Theory*, **19** (1973) 257–262 [1].

W

[1375] D. G. Wadbrook and D. J. Wollons, Implementation of 2-dimensional Walsh transforms for pattern recognition, *Electronics Letters*, **8** (1972) 134–136 [2].

[1376] B. L. van der Waerden, *Modern Algebra*, 3rd ed., 2 vols., (Ungar, New York, 1950) [3, 4, 8].

[1377] T. J. Wagner, A remark concerning the minimum distance of binary group codes, *IEEE Trans. Info. Theory*, **11** (1965) 458 [18].

[1378] T. J. Wagner, A search technique for quasi-perfect codes, *Info. and Control*, **9** (1966) 94–99 [18, A].

[1379] T. J. Wagner, A remark concerning the existence of binary quasi-perfect codes, *IEEE Trans. Info. Theory*, **12** (1966) 401 [18].

[1380] T. J. Wagner, Some additional quasi-perfect codes, *Info. and Control*, **10** (1967) 334 [18].

[1381] S. Wainberg, Error-erasure decoding of product codes, *IEEE Trans. Info. Theory*, **18** (1972) 821–823 [18].

[1382] S. Wainberg and J. K. Wolf, Burst decoding of binary block codes on Q-ary output channels, *IEEE Trans. Info. Theory*, **18** (1972) 684–686 [10].

[1384] S. Wainberg and J. K. Wolf, Algebraic decoding of block codes over a q-ary input, Q-ary output channel, $Q > q$, *Info. and Control*, **22** (1973) 232–247 [10].

[1385] J. S. Wallis, Families of codes from orthogonal $(0, 1, -1)$-matrices. *Proc. of the Third Manitoba Conf. on Numerical Math.*, (Winnipeg, Man., 1973), pp. 419–426. (Utilitas Math., Winnipeg, Man., 1974) [18].

[1386] W. D. Wallis, A. P. Street and J. S. Wallis, *Combinatorics: Room Squares, Sum-Free Sets, Hadamard Matrices*, Lecture Notes in Mathematics 292, Springer, Berlin, 1972 [2].

[1387] H. N. Ward, Quadratic residue codes and sympletic groups, *J. Algebra*, **29** (1974) 150–171 [16].

[1388] H. N. Ward, A form for M_{11}, *J. Algebra*, **37** (1975) 340–361 [20].

[1389] H. N. Ward, A restriction on the weight enumerator of a self-dual code, *J. Comb. Theory*, **21A** (1976) 253–255 [19].

[1390] W. T. Warren and C. L. Chen, On efficient majority logic decodable codes, *IEEE Trans. Info. Theory*, **22** (1976) 737–745 [13].

[1391] N. Wax, On upper bounds for error detecting and error correcting codes of finite length, *IEEE Trans. Info. Theory*, **5** (1959) 168–174 [17].

[1392] G. D. Weathers, E. R. Graf and G. R. Wallace, The subsequence weight distribution of summed maximum length digital sequences, *IEEE Trans. Commun.*, **22** (1974) 997–1004 [14].

[1393] J. H. M. Wedderburn, *Lectures on Matrices*, (Dover, New York, 1964) [21].

[1394] A. Weil, On some exponential sums, *Proc. Nat. Acad. Sci. USA*, **34** (1948) 204–207 [9].

[1395] S. B. Weinstein, In Galois fields, *IEEE Trans. Info. Theory*, **17** (1971) 220 [4].

[1396] E. Weiss, Generalized Reed–Muller codes, *Info. and Control*, **5** (1962) 213–222 [13].

[1397] E. Weiss, Linear codes of constant weight, *J. SIAM Applied Math.*, **14** (1966) 106–111; **15** (1967) 229–230 [8].

[1398] L. R. Welch, Computation of finite Fourier series, *JPL Space Programs Summary*, **37-37-IV** (1966) 295–297 [14].

[1399] L. R. Welch, R. J. McEliece and H. C. Rumsey, Jr., A low-rate improvement on the Elias bound, *IEEE Trans. Info. Theory*, **20** (1974) 676–678 [17].

[1400] E. J. Weldon, Jr., Difference set cyclic codes, *Bell Syst. Tech. J.*, **45** (1966) 1045–1055 [13].

[1401] E. J. Weldon, Jr., New generalizations of the Reed–Muller codes, Part II: Nonprimitive codes, *IEEE Trans. Info. Theory*, **14** (1968) 199–205 and 521 [13].

[1402] E. J. Weldon, Jr., Euclidean geometry cyclic codes, in: R. C. Bose and T. A. Dowling, eds., *Combinatorial Mathematics and Its Applications*, (Univ. of North Carolina, Chapel Hill, NC, 1969) Ch. 23 [13].

[1403] E. J. Weldon, Jr., Some results on majority-logic decoding, in: H. B. Mann, ed., *Error-Correcting Codes*, (Wiley, New York, 1969) pp. 149–162 [13].

[1404] E. J. Weldon, Jr., Decoding binary block codes on q-ary output channels, *IEEE Trans. Info. Theory*, **17** (1971) 713–718 [16, 18].

[1405] E. J. Weldon, Jr., Justesen's construction – the low-rate case, *IEEE Trans. Info. Theory*, **19** (1973) 711–713 [10].

[1406] E. J. Weldon, Jr., Some results on the problem of constructing asymptotically good error-correcting codes, *IEEE Trans. Info. Theory*, **21** (1975) 412–417 [10].

[1407] L.-J. Weng, Decomposition of M-sequences and its applications, *IEEE Trans. Info. Theory*, **17** (1971) 457–463 [14, 18].

[1408] L.-J. Weng and G. H. Sollman, Variable redundancy product codes, *IEEE Trans. Commun.*, **15** (1967) 835–838 [18].

[1409] H. Weyl, Invariants, *Duke Math. J.*, **5** (1939) 489–502 [19].

[1410] H. Weyl, *The Classical Groups*, (Princeton University Press, Princeton, NJ, 1946) [19].

[1411] N. White, The critical problem and coding theory, *JPL Space Programs Summary*, **37-66-III** (December 31, 1970), 36–42 [17].

[1412] T. A. Whitelaw, On the Mathieu group of degree twelve, *Proc. Camb. Phil. Soc.*, **62** (1966) 351–364 [20].

[1413] A. L. Whiteman, Skew Hadamard matrices of Goethals–Seidel type, *Discrete Math.*, **2** (1972) 397–405 [2].

[1414] E. T. Whittaker and G. N. Watson, *A Course of Modern Analysis*, 4th ed., (Cambridge Univ. Press, 1963) [19].

[1415] H. Wielandt, *Finite Permutation Groups*, (Academic Press, New York, 1964) [20, 21].

[1416] M. C. Willett, The minimum polynomial for a given solution of a linear recursion, *Duke Math. J.*, **39** (1972) 101–104 [14].

[1417] M. C. Willett, The index of an m-sequence, *SIAM J. Appl. Math.*, **25** (1973) 24–27 [14].

[1418] M. C. Willett, Cycle representatives for minimal cyclic codes, *IEEE Trans. Info. Theory*, **21** (1975) 716–718 [8].

[1419] K. S. Williams, Exponential sums over $GF(2^n)$, *Pac. J. Math.*, **40** (1972) 511–519 [10].

[1420] R. M. Wilson, An existence theory for pairwise balanced designs, *J. Comb. Theory*, **13A** (1972) 220–273 [2, 17].

[1421] R. M. Wilson, The necessary conditions for t-designs are sufficient for something, *Utilitas Math.*, **4** (1973) 207–215 [2, 17].

[1422] R. M. Wilson, An existence theory for pairwise balanced designs: III – Proof of the existence conjectures, *J. Comb. Theory*, **18A** (1975) 71–79 [2, 17].

[1423] E. Witt, Die 5-fach transitiven Gruppen von Mathieu, *Abh. Math. Sem. Hamb.*, **12** (1938) 256–265 [2, 17, 20].

[1424] E. Witt, Über Steinersche Systeme, *Abh. Math. Sem. Hamb.*, **12** (1938) 265–275 [2, 17, 20].

[1425] J. K. Wolf, On codes derivable from the tensor product of check matrices, *IEEE Trans. Info. Theory*, **11** (1965) 281–284 [18].

[1426] J. K. Wolf, Decoding of Bose–Chaurhuri–Hocquenghem codes and Prony's method for curve fitting, *IEEE Trans. Info. Theory*, **13** (1967) 608 [9].

[1427] J. K. Wolf, Adding two information symbols to certain nonbinary BCH codes and some applications, *Bell Syst. Tech. J.*, **48** (1969) 2405–2424 [10].

[1428] J. K. Wolf, Nonbinary random error-correcting codes, *IEEE Trans. Info. Theory*, **16** (1970) 236–237 [9, 18].

[1429] J. K. Wolf, A survey of coding theory 1967–1972, *IEEE Trans. Info. Theory*, **19** (1973) 381–389 [1].

[1430] J. K. Wolf and B. Elspas, Error-locating codes – a new concept in error control, *IEEE Trans. Info. Theory*, **9** (1963) 113–117 [18].

[1431] M. J. Wolfmann, Formes quadratiques et codes à deux poids, *Comptes Rendus*, (A) **281** (1975) 533–535 [15].

[1432] C. T. Wolverton, Probability of decoding error for the (7, 2) Reed–Solomon code, *Proc. IEEE*, **60** (1972) 918–919 [10].

[1433] J. M. Wozencraft and I. M. Jacobs, *Principles of Communication Engineering*, (Wiley, New York, 1965) [1].

[1434] E. P. G. Wright, Error correction: Relative merits of different methods, *Electrical Commun.*, **48** (1973) 134–145 [1].

[1435] W. W. Wu, New convolutional codes–Part I, *IEEE Trans. Commun.*, **23** (1975) 942–956 [1].

[1436] W. W. Wu, New convolutional codes–Part II, *IEEE Trans. Commun.*, **24** (1976) 19–33 [1].

[1437] A. D. Wyner, Capabilities of bounded discrepancy decoding, *Bell Syst. Tech. J.*, **54** (1965) 1061–1122 [1].

[1438] A. D. Wyner, Random packings and coverings of the unit *n*-sphere, *Bell Syst. Tech. J.*, **46** (1967) 2111–2118 [1].

[1439] A. D. Wyner, On coding and information theory, *SIAM Review*, **11** (1969) 317–346 [1].

[1440] A. D. Wyner, Note on circuits and chains of spread *k* in the *n*-cube, *IEEE Trans. Computers*, **20** (1971) 474 [Intro.].

X

[1441] X.-N. Feng and Z.-D. Dai, Studies in finite geometries and the construction of incomplete block designs V, *Acta Math. Sinica*, **15** (1965) 664–652 [B].

Y

[1442] S. Yamamoto, Some aspects of the composition of relationship algebras of experimental designs, *J. Sci. Hiroshima Univ.*, Ser. A-I, **28** (1964) 167–197 [21].

[1443] S. Yamamoto and Y. Fuji, Analysis of partially balanced incomplete block designs, *J. Sci. Hiroshima Univ.*, Ser. A-I, **27** (1963) 119–135 [21].

[1444] S. Yamamoto, Y. Fujii and N. Hamada, Composition of some series of association algebras, *J. Sci. Hiroshima Univ.*, Ser. A-I, **29** (1965) 181–215 [21].

[1445] C. H. Yang, A construction for maximal (+1, −1)-matrix of order 54, *Bull. Amer. Math. Soc.*, **72** (1966) 293 [2].

[1446] S. S. Yau and Y.-C. Liu, On decoding of maximum-distance separable linear codes, *IEEE Trans. Info. Theory*, **17** (1971) 487–491 [10].

[1447] P.-W. Yip, S. G. S. Shiva and E. L. Cohen, Permutation-decodable binary cyclic codes, *Electronics Letters*, **10** (1974) 467–468 [16].

[1448] C. K. Yuen, The separability of Gray codes, *IEEE Trans. Info. Theory*, **20** (1974) 668 [Intro.].

[1449] C. K. Yuen, Comments on "Correction of errors in multilevel Gray coded data", *IEEE Trans. Info. Theory*, **20** (1974) 283–284 [Intro.].

Z

[1450] L. A. Zadeh, ed., Report on progress in information theory in the U.S.A. 1960–1963, *IEEE Trans. Info. Theory*, **9** (1963) 221–264 [1].

[1451] G. V. Zaitsev, V. A. Zinov'ev and N. V. Semakov, Interrelation of Preparata and Hamming codes and extension of Hamming codes to new double-error-correcting codes, in: B. N. Petrov and F. Csáki, eds., *2nd International Symposium on Information Theory*, (Akadémiai Kiadó, Budapest, 1973) pp. 257–263 [15, 18].

[1452] S. K. Zaremba, A covering theorem for abelian groups, *J. London Math. Soc.*, **26** (1950) 70–71 [7].

[1453] S. K. Zaremba, Covering problems concerning abelian groups, *J. London Math. Soc.*, **27** (1952) 242–246 [7].

[1454] O. Zariski and P. Samuel, *Commutative Algebra*, 2 Vols. (van Nostrand, Princeton, New Jersey, 1960) [4].

[1455] L. H. Zetterberg, Cyclic codes from irreducible polynomials for correction of multiple errors, *IEEE Trans. Info. Theory*, **8** (1962) 13–20 [7].

[1456] L. H. Zetterberg, A class of codes for polyphase signals on a bandlimited gaussian channel, *IEEE Trans. Info. Theory*, 11 (1965) 385–395 [Intro.].

[1457] Z.-X. Wan, Studies in finite geometries and the construction of incomplete block designs I, *Acta math. Sinica*, 15 (1965) 354–361 [19, B].

[1458] Z.-X. Wan, Studies in finite geometries and construction of incomplete block designs, II, *Acta Math. Sinica*, 15 (1965) 362–371 [21, B].

[1459] Z.-X. Wan and B.-F. Yang, Studies in finite geometry and the construction of incomplete block designs III, *Acta Math. Sinica*, 15 (1965) 533–544 [21, B].

[1460] Z.-X. Wan, Studies in finite geometries and the construction of incomplete block designs VI, (In Chinese), *Progress in Math.*, 8 (1965) 293–302 [B].

[1461] N. Zierler, Linear recurring sequences, *J. Soc. Indust. Appl. Math.*, 7 (1959) 31–48 [14].

[1462] N. Zierler, On decoding linear error-correcting codes-I, *IEEE Trans. Info. Theory*, 6 (1960) 450–459 [16].

[1463] N. Zierler, On decoding linear error-correcting codes-II, in: C. Cherry, ed., *Fourth London Symp. Info. Theory*, (Butterworths, London, 1961) pp. 61–67 [16].

[1464] N. Zierler, A note on the mean square weight for group codes, *Info. and Control*, 5 (1962) 87–89 [5].

[1465] N. Zierler, Linear recurring sequences and error-correcting codes, in: H. B. Mann, ed., *Error Correcting Codes*, (Wiley, New York, 1969) pp. 47–59 [14].

[1466] N. Zierler, Primitive trinomials whose degree is a Mersenne exponent, *Info. and Control*, 15 (1969) 67–69 [4].

[1467] N. Zierler, On $x^n + x + 1$ over GF(2), *Info. and Control*, 16 (1970) 502–505 [4].

[1468] N. Zierler, On the MacWilliams identity, *J. Comb. Theory*, 15A (1973) 333–337 [5].

[1469] N. Zierler and J. Brillhart, On primitive trinomials (mod 2), *Info. and Control*, 13 (1968) 541–554; 14 (1969) 566–569 [4].

[1470] V. A. Zinov'ev, Generalized concatenated codes, *Problemy Peredachi Informatsii*, 12(1) (1976) 5–15 [18, A].

[1471] V. A. Zinov'ev and V. K. Leont'ev, On perfect codes, *Problems of Info. Trans.*, 8(1) (1972) 17–24 [6].

[1472] V. A. Zinov'ev and V. K. Leont'ev, The nonexistence of perfect codes over Galois fields, *Problems of Control and Info. Theory*, 2(2) (1973) 123–132 [6].

[1473] Yu. S. Zlotnikov and I. Z. Kaiser, Computer simulation of decoding algorithms for error-correcting cyclical codes, *Telecommun. Radio Eng.*, 26 (1972) 54–57 [16].

[1474] Z.-D. Dai and X.-N. Feng, Studies in finite geometries and the construction of incomplete block design IV, *Acta Math. Sinica*, 15 (1965) 545–558 [B].

[1475] V. V. Zyablov, An estimate of the complexity of constructing binary linear cascade codes, *Problems of Info. Trans.*, 7(1) (1971) 3–10 [10].

[1476] V. V. Zyablov, Optimization of concatenated decoding algorithms, *Problems of Info. Trans.*, 9(1) (1973) 19–24. [10].

[1477] R. Mathon, Symmetric conference matrices of order $pq^2 + 1$, preprint [2].

[1478] F. J. MacWilliams, A. M. Odlyzko, N. J. A. Sloane, and H. N. Ward, Self-dual codes over GF(4), *J. Comb. Theory*, 25A (1978) 288–318 [19].

Index

Note that the bibliography on pages 703–755 serves as a comprehensive index of author's names, since each entry in the bibliography is followed by a list of the chapters where it is cited.

Printed and bound by CPI Group (UK) Ltd, Croydon, CR0 4YY

03/10/2024

01040428-0007